ROUTLEDGE HANDBOOK OF STRATEGIC CULTURE

This handbook offers a collection of cutting-edge essays on all aspects of strategic culture by a mix of international scholars, consultants, military officers, and policymakers.

The volume explicitly addresses the analytical conundrums faced by scholars who wish to employ or generate strategic cultural insights, with substantive commentary on defining and scoping strategic culture, analytic frameworks and approaches, levels of analysis, sources of strategic culture, and modalities of change in strategic culture. The chapters engage strategic culture at the civilizational, regional, supra-national, national, non-state actor, and organizational levels. The volume is divided into five thematic parts, which will appeal to both students who are new to the subject and scholars who wish to incorporate strategic culture into their toolbox of analytical techniques. Part I assesses the evolving theoretical strengths and weaknesses of the field. Part II lays out elements of the theoretical and methodological foundations of the field, including sources and components of strategic culture. Part III presents a number of national strategic cultural profiles, representing the state of contemporary strategic culture scholarship. Part IV addresses the utility of strategic culture for practitioners and scholars. Part V summarizes the key theoretical and practical insights offered by the volume's contributors.

This handbook will be of much interest to students of strategic studies, defense studies, security studies, and international relations in general, as well as to professional practitioners.

Kerry M. Kartchner teaches strategic culture courses for Missouri State University's Graduate Department of Defense and Strategic Studies and for Johns Hopkins University. He is the editor/author of several books, including *Crossing Nuclear Thresholds* (2019), *On Limited Nuclear War in the 21st Century* (2014), and *Strategic Culture and Weapons of Mass Destruction* (2008).

Briana D. Bowen is the Associate Director and Cofounder of the Center for Anticipatory Intelligence at Utah State University.

Jeannie L. Johnson is the Director of Utah State University's Center for Anticipatory Intelligence and an Associate Professor within the Political Science Department. She is also the author of *The Marines, Counterinsurgency, and Strategic Culture* (2018).

ROUTLEDGE HANDBOOK OF STRATEGIC CULTURE

Edited by Kerry M. Kartchner, Briana D. Bowen, and Jeannie L. Johnson

LONDON AND NEW YORK

Cover image: Bell tower stairwell in Old Main, Utah State University campus, Logan, Utah. Photo by Kerry M. Kartchner.

First published 2024
by Routledge
4 Park Square, Milton Park, Abingdon, Oxon OX14 4RN

and by Routledge
605 Third Avenue, New York, NY 10158

Routledge is an imprint of the Taylor & Francis Group, an informa business

British Library Cataloguing-in-Publication Data
A catalogue record for this book is available from the British Library

Library of Congress Cataloguing-in-Publication Data
Names: Kartchner, Kerry M., 1956- editor. | Bowen, Briana D., editor. | Johnson, Jeannie L., editor.
Title: Routledge handbook of strategic culture / edited by Kerry M. Kartchner, Briana D. Bowen, and Jeannie L. Johnson.
Description: Abingdon, Oxon ; New York : Routledge, 2023. | Includes bibliographical references and index.
Identifiers: LCCN 2023015759 (print) | LCCN 2023015760 (ebook) | ISBN 9780367445485 (hardback) | ISBN 9781032565149 (paperback) | ISBN 9781003010302 (ebook)
Subjects: LCSH: Strategic culture. | Military policy. | Strategy.
Classification: LCC U21.2 .R67 2023 (print) | LCC U21.2 (ebook) | DDC 355.02--dc23/eng/20230622
LC record available at https://lccn.loc.gov/2023015759
LC ebook record available at https://lccn.loc.gov/2023015760

ISBN: 978-0-367-44548-5 (hbk)
ISBN: 978-1-032-56514-9 (pbk)
ISBN: 978-1-003-01030-2 (ebk)

DOI: 10.4324/9781003010302

Typeset in Sabon
by MPS Limited, Dehradun

Kerry dedicates this volume to:
Britt, always, and (in order of accession)
Brayden, Max, and Seth
James and Eden
Penelope
Maren and Ezra
Clarke

Briana dedicates this volume to her parents, Chuck and Jill,
who are an exceptional force for good in the world
and even more so in her life.

Jeannie dedicates this volume to Colin S. Gray,
an exceptional mentor, colleague, intellectual giant, and dear friend.

IN MEMORIAM

Colin S. Gray
1943–2020

The scholarship of Colin S. Gray is foundational to the entire discipline of strategic culture. His passing in 2020 was felt with profound force by family, friends, colleagues, and students whose intellectual lives have been shaped by the privilege of having been mentored by him. Gray was a strategic genius. His insights remain profound and prescient, even decades after they were written. We are grateful for the wealth of scholarship he left to us. We are better as scholars, as practitioners, and as nations.

Edward D. Last
1983–2021

Edward D. Last sadly passed away in 2021, having recently completed work on the chapter contained in this book, and as such, this will be his final contribution to the literature on this subject which had absorbed and interested him throughout his adult life. His family is grateful to the publishers and editors for continuing with its publication as originally intended. He had a thirst for knowledge to the end and plans for further research, continuing to read books and articles, and study Arabic until the week he died. We hope his legacy will inspire others to overcome their own adversities.

CONTENTS

ILLUSTRATIONS

Table

Figures

Box

CONTRIBUTORS

Arshad Ali is the Director of the India Study Centre at the Institute of Strategic Studies, Islamabad. Dr. Ali holds a PhD in international relations from the University of Otago, Dunedin, New Zealand. He is the author of *Pakistan's National Security Approach and Post-Cold War Security: Uneasy Co-existence* (London and New York: Routledge, 2021).

Michael Barron is a Clinical Professor of Anesthesiology within the Department of Anesthesiology at the Miller School of Medicine of the University of Miami. He has been appointed to several medical and administrative leadership roles within the University of Miami healthcare system and its affiliates, including serving as Chief Medical Officer of UHealth, as well as its faculty physician practice. Dr. Barron held the office of president of the Medical Staff at one of the nation's largest hospital systems, The Jackson Health System, where he contributed to numerous innovative clinical and educational programs. His wide-ranging research experience centers around organizational learning, culture, and strategy, particularly the interdisciplinary application of theory and practices. Dr. Barron's contribution to this volume reflects his interest in how cultural contexts shape conditions and motivate actor behaviors, as well as drive key strategic decisions. He holds a BS and MD from Wayne State University, a Certificate in Policy Strategy from the Brookings Institution, and an MA in Global Security Studies from Johns Hopkins University.

Thomas Biggs is an analyst at ManTech and a graduate student in Johns Hopkins University's Master of Science in Intelligence Analysis Program. He graduated from George Washington University in May 2019 with a BA in International Affairs, concentrating on Security Policy, and with minors in History and German Studies. In the spring of 2018, he studied German language and culture at the Humboldt University of Berlin. Mr. Biggs has also conducted research as an intern for the Naval History and Heritage Command and the German Historical Institute, Washington. In addition to his studies, he serves in the Pennsylvania Army National Guard.

George E. Bogden is a Strategy & Policy Fellow funded by the Smith Richardson Foundation, a Senior Visiting Researcher at Bard College, an Olin Fellow at Columbia

Law School, and the Helmut Schmidt Fellow at the German Marshall Fund of the United States. He previously served as an inaugural Senior Fellow at the Hungary Foundation, in residence in Budapest, as well as the first Associate Director of the Center for the Future of Liberal Society at the Hudson Institute. Dr. Bogden's commentary has appeared in the *Wall Street Journal, The Atlantic, War on the Rocks,* GMFUS's *Paper Series, The American Interest, The Wavell Room, Presidential Studies Quarterly,* and the *Marine Corps University Press.* Before defending his dissertation, he undertook a Fulbright Fellowship in Kosovo. Dr. Bogden's research has been recognized by the Brussels Forum Young Writers Award and the Trench Gascoigne Essay Prize. After earning his BA from Yale, he served as the university's Joseph C. Fox Fellow in Istanbul.

Briana D. Bowen is the Associate Director of the Center for Anticipatory Intelligence at Utah State University, an interdisciplinary nexus focused on emergent security issues that she cofounded with colleagues Jeannie Johnson and Matt Berrett. Her teaching and research concentrate on Russian security affairs, weapons of mass destruction, strategic culture, and anticipatory intelligence. Ms. Bowen served as a co-Principal Investigator on a 2022 Minerva Research Initiative grant focusing on human governance implications of the proliferation of artificial intelligence-enabled surveillance technologies. Her previous work has included running the Oxford University Strategic Studies Group and supporting research on three federally funded grants applying sociocultural methodologies to the study of weapons of mass destruction. Ms. Bowen is a Truman Scholar and holds a BA in Political Science from Utah State University and an MPhil in Russian and East European Studies from the University of Oxford.

Michael Eisenstadt is the Kahn Fellow and director of the Military and Security Studies Program at The Washington Institute for Near East Policy. He is a specialist in the Persian Gulf and Arab-Israeli security affairs. He has published widely on both irregular and conventional warfare, as well as nuclear weapons proliferation, in the Middle East. A former US government military analyst, he served for 26 years as an officer in the US Army Reserve with active-duty stints in Iraq, Israel, the West Bank, Jordan, and Turkey. He has also served in a civilian capacity on the Multinational Force–Iraq/US Embassy Baghdad Joint Campaign Plan Assessment Team and as an advisor to the Congressionally mandated Iraq Study Group, the State Department's Future of Iraq defense policy working group, and the Gulf War Air Power Survey. He has an MA in Arab Studies from Georgetown University and a BA in political science from the State University of New York at Binghamton.

Theo Farrell is a Professor and Deputy Vice-Chancellor (Academic and Student Life) at the University of Wollongong, Australia. Professor Farrell was previously Head of the Department of War Studies at King's College London, where he remains a visiting professor. He is a Fellow of the Academy of Social Sciences (UK) and a Senior Associate Fellow of the Royal United Services Institute (UK). Professor Farrell's research focuses on transformative change in the character of armed conflict. His latest monograph, *Unwinnable: Britain's War in Afghanistan* (Penguin, Random House, 2018) was shortlisted for three national book awards and selected book of the year by *The Sunday Times*.

Nathan Fedorchak is a recent graduate of Missouri State University's Master's Program in Defense and Strategic Studies. He received a Bachelor's degree from Michigan State University in Political Theory & Constitutional Democracy in 2013. He works in Washington, DC and, reflecting his contribution to this volume, remains deeply interested in Singapore and international relations in Southeast Asia more generally.

Alastair Finlan is a Professor of War Studies at the Swedish Defence University. He is the author of numerous books on military culture, Special Forces, and modern warfare, including *The Collapse of Yugoslavia 1991–99, 2nd Edition* (Bloomsbury, 2022), *Contemporary Military Strategy and the Global War on Terror: US and UK Armed Forces in Afghanistan and Iraq 2001–2012* (Bloomsbury, 2014), *Contemporary Military Culture and Strategic Studies* (Routledge, 2013), and *Special Forces, Strategy and the War on Terror: Warfare by Other Means* (Routledge, 2008), among others.

Roger Z. George spent 30 years as a Central Intelligence Agency analyst and subsequently has taught intelligence and national security at Georgetown University, the National War College, and Occidental College. During his government career, he served as the National Intelligence Officer for Europe, as well as a policy planner at the State and Defense Departments. After his retirement from government, he has also been a consultant to the RAND Corporation on European security topics. Dr. George received his BA in political science from Occidental College and his MA and PhD from the Fletcher School of Law and Diplomacy at Tufts University. He has published numerous articles on intelligence and national security and is the coeditor (with Harvey Rishikof) of *The National Security Enterprise: Navigating the Labyrinth, 2nd Edition* (Georgetown University Press, 2016). His latest book is *Intelligence in the National Security Enterprise: An Introduction* (Georgetown University Press, 2019).

Colin S. Gray died on 27 February 2020. He was a British-American writer on geopolitics and professor of International Relations and Strategic Studies at the University of Reading, where he was the director of the Centre for Strategic Studies. He cofounded the National Institute for Public Policy in Washington, DC, and was also Fellow at the Center for Technology and Strategy at the Air University, Maxwell AFB. For many years, he advised the American and British governments, dividing his time between the two countries. He was the author of 30 books on military history and strategic studies, as well as numerous articles. Among his books are a trilogy on strategy with Oxford University Press: *The Strategy Bridge: Theory for Practice* (2010); *Perspectives on Strategy* (2013); and *Strategy and Defence Planning: Meeting the Challenge of Uncertainty* (2014). His last book, also with Oxford University Press, is *Theory of Strategy* (2018).

Maria Hellman is an Associate Professor at the Department of Economic History and International Relations, Stockholm University. Her research interests include global and political communication, security studies, and international relations. Hellman has written about communication during wars and crises, identity and strategic narratives, and French foreign policy orientation and political culture. She has published her work in journals such as *International Political Sociology*, *International Journal of Cultural Studies*, and

New Media and Society. In Britz (ed.) *European Participation in International Operations: The Role of Strategic Culture* (2016) she contributed with the chapter "Assuming Great Power Responsibility: French Strategic Culture and International Operations."

Beatrice Heuser holds the Chair in International Relations at the University of Glasgow. She previously taught at the Department of War Studies, King's College London; then at the University of Reading. She has also taught at French and German universities (most recently the universities of Sorbonne and Paris-Panthéon-Assas, and at Sciences Po' Paris). From 1997 to 1998, she worked at NATO HQ in Brussels. She holds degrees from the Universities of London (BA, MA), Oxford (DPhil), and Marburg (Habilitation). Her publications include *The Evolution of Strategy* (2010), *Reading Clausewitz* (2002), *Strategy before Clausewitz* (2017), with Eitan Shamir (eds.) *Insurgencies and Counterinsurgencies: National Styles and Strategic Cultures* (2017), and, most recently, *WAR: A Genealogy of Western Ideas and Practices* (2022). She has also worked for the Bundeswehr, most recently lecturing at its General Staff College (*Führungsakademie*) in Hamburg, is a Senior Associate Fellow at the Royal United Services Institution, and a Visiting Fellow at the Royal Navy Strategic Studies Centre.

Brigitte E. Hugh holds an MS in Political Science with an emphasis in Anticipatory Intelligence and a BA in Political Science from Utah State University. During her Master's degree, she was part of the inaugural cohort of the Center for Anticipatory Intelligence at Utah State University, which emphasizes cross-disciplinary skills, methods, and cooperation for anticipating and building resilience against emergent and future security risks. Hugh currently works to address the complex range of threats posed by a changing climate at the Center for Climate and Security, an institute of the Council on Strategic Risks.

Jeannie L. Johnson is the Director of Utah State University's Center for Anticipatory Intelligence and an Associate Professor in the Political Science Department. Dr. Johnson cofounded the Center for Anticipatory Intelligence in 2018 with the vision of creating an interdisciplinary nexus that fuses expertise in national security and geopolitics with leading-edge instruction in cyber threats, data analytics, and emergent technology. Prior to her academic career, Dr. Johnson worked within the Central Intelligence Agency's Directorate of Intelligence. She has conducted in-depth research on US national and military service cultures, including critical blind spots in foreign and security policy. She has also focused on the application of strategic culture analysis to nuclear weapons issues and has coedited two books on that topic. Dr. Johnson received her PhD in strategic studies from the University of Reading.

Judith Meister Johnston is an independent researcher and President of Johnston Analytics, Inc. Her research interests include intelligence analysis methods and means of establishing analytic rigor, problem-solving in complex environments, and creative methods for knowledge management. Dr. Johnston has written about the complexity of the intelligence analysis and production cycle and its impact on performance, as well as the challenges of interagency research in the US Intelligence Community. She has published her work in *Analytic Culture in the US Intelligence Community: An Ethnographic Study*

(Center for the Study of Intelligence) in *National Intelligence University's Role in Interagency Research: Recommendations from the Intelligence Community* (RAND.org), and various government-sponsored technical reports.

Rob Johnston is the Chief Scientist of Johnston Analytics, Inc., a quantitative and qualitative research service and data science consultancy. His research interests include computational social science, ethnographic research, and knowledge management in complex organizations. Dr. Johnston has written about the challenges of high-risk and high-reward professions, ranging from surgeons to special forces to intelligence community officers, and the impact complexity has on human performance. He is the author of *Analytic Culture in the US Intelligence Community: An Ethnographic Study* and the producer of *Extraordinary Fidelity*, an award-winning documentary.

Kerry M. Kartchner served for 32 years in the US Departments of State and Defense, where he was primarily responsible for nuclear weapons policy and nonproliferation issues. He currently teaches courses on strategic culture and the ethics of war and peace for Missouri State University's Graduate Department of Defense and Strategic Studies, and Johns Hopkins University's Krieger School of Arts and Sciences. Dr. Kartchner is the coeditor and contributor to several books on US arms control and nuclear weapons policy, including *On Limited Nuclear War in the 21st Century* (with Jeffrey A. Larsen, 2014). He is also the coeditor of two previous books on strategic culture, *Strategic Culture and Weapons of Mass Destruction: Culturally Based Insights into Comparative National Security Policymaking* (with Jeannie L. Johnson and Jeffrey A. Larsen, 2009), and *Crossing Nuclear Thresholds: Leveraging Sociocultural Insights into Nuclear Decisionmaking* (with Jeannie L. Johnson and Marilyn J. Maines, 2018). His MA and PhD in international relations are from the University of Southern California, and his BA in international relations is from Brigham Young University.

Jeffrey S. Lantis is a Professor of Political Science at the College of Wooster. His research specializations include foreign policy analysis, emerging technologies, international norm contestation theory, and strategic culture. A former Fulbright Senior Scholar in Australia, Dr. Lantis is the author of several recent books, including *United States Foreign Policy in Action, 2nd Edition*, with Patrick Homan (Routledge, 2022), *Congressional Foreign Policy Advocacy and Entrepreneurship* (University of Michigan Press, 2019), and *Arms and Influence: U.S. Technology Innovations and the Evolution of International Security Norms* (Stanford University Press, 2016). In addition, Lantis has published numerous academic journal articles and book chapters. He is coeditor of the journal *International Studies Perspectives*, and in 2020, he received the Distinguished Teacher-Scholar Award from the Active Learning in International Affairs Section of the International Studies Association. Lantis holds a PhD in political science from The Ohio State University.

Jeffrey A. Larsen is a Research Professor in the Department of National Security Affairs at the Naval Postgraduate School (Monterey, CA), and President of Larsen Consulting Group in Colorado Springs, CO. He was director of the Research Division at the NATO Defense College in Rome, Italy, from 2013–2018. Prior to that he was a senior policy analyst focusing on strategic issues for Science Applications International Corporation.

A retired Air Force Lt Colonel, he served as a command pilot, associate professor at the Air Force Academy, and founding director of the Air Force Institute for National Security Studies. Dr Larsen holds a PhD in politics and MA in international relations from Princeton University, an MA in European area studies from the Naval Postgraduate School, and a Bachelor's in Soviet area studies from the Air Force Academy. He is a graduate of the Defense Language Institute (German), was a NATO Manfred Wörner Fellow and a Fulbright research fellow, and is the author or editor of more than 150 publications.

Edward D. Last completed his PhD in politics and international relations at the University of Southampton, UK in 2018. In 2020, his book *Strategic Culture and Violent Non-State Actors: A Comparative Study of Salafi-Jihadist Groups* (Routledge) was published, and he was awarded a Visiting Fellowship at the University of Southampton. Dr. Last sadly died in 2021, having recently completed work on the chapter contained in this book, and as such, this will be his final contribution to the literature on this subject which had absorbed and interested him throughout his adult life. His family is grateful to the publishers and editors for continuing with its publication as originally intended. He had a thirst for knowledge to the end and plans for further research, continuing to read books and articles, and study Arabic until the week he died. We hope his legacy will inspire others to overcome their own adversities.

Thomas G. Mahnken is currently the President and Chief Executive Officer of the Center for Strategic and Budgetary Assessments and a Senior Research Professor at the Philip Merrill Center for Strategic Studies at Johns Hopkins University's Paul H. Nitze School of Advanced International Studies (SAIS). He previously taught for 20 years at the US Naval War College, where he served as the Jerome E. Levy Chair of Economic Geography and National Security. Dr. Mahnken served as the Deputy Assistant Secretary of Defense for Policy Planning from 2006 to 2009. He earned his Master's degree and doctorate in international affairs from SAIS and was a National Security Fellow at the John M. Olin Institute for Strategic Studies at Harvard University. He was a summa cum laude graduate of the University of Southern California with Bachelor's degrees in history and international relations (with highest honors) and a certificate in defense and strategic studies.

Harrison Menke is a Strategic Analyst at the Defense Threat Reduction Agency (DTRA). His work focuses on nuclear forces, deterrence, and regional conflict and escalation. Prior to joining DTRA, Mr. Menke was an Assistant Research Fellow at the Center for the Study of Weapons of Mass Destruction and the Science Applications International Corporation. Mr. Menke is currently a doctoral candidate at Missouri State University's Department of Defense and Strategic Studies. The views expressed by Mr. Menke are those of the author and do not reflect the official policy or position of DTRA, the Department of Defense, or the US government.

Christoph O. Meyer is a Full Professor of European & International Politics at King's College London. He studied political science and sociology in Hamburg before completing an MPhil (1997) and a PhD in international relations at the University of Cambridge (2001). Christoph worked as a visiting doctoral researcher at the Max Planck Institute for the Study of Societies, a Research Associate at the University of Cologne, a Marie Curie Postdoctoral Research Fellow at the Centre of European Policy Studies in Brussels, and a Lecturer at the Birkbeck College London before joining King's in 2007. He was elected as a Fellow to the UK Academy of Social Sciences in 2020. Professor Meyer's book *Warning about War* (2020) won the 2021 best book prize of the International Studies Association (ISA) and the 2021 best book prize in international communication of the ISA section. His latest coedited book with Edinburgh University Press (2022) is titled *Estimative Intelligence in European Foreign Policymaking: Learning Lessons from an Era of Surprise.*

Megan F. Moore is a recent Master's graduate of Johns Hopkins University's School for Advanced Governmental Studies. Her academic focus is global security and intelligence. Ms. Moore is in federal service and began her career as a Peace Corps volunteer. She has served in various national security and intelligence roles in the US Department of Homeland Security, Department of the Treasury, and Department of Defense.

Neil Munro holds a BA (combined honors) in Chinese and Russian (Queensland, 1990) and a PhD in public policy from the University of Strathclyde (2004). His research focuses on governance in post-Communist and developing societies. Dr. Munro has published on a wide range of themes ranging from acceptance of bureaucratic norms through national identity, public participation, regime legitimacy, trust, and social cohesion. He has held research positions at the Universities of Strathclyde and Aberdeen and is currently senior lecturer in Chinese Politics at the University of Glasgow.

Ali Parchami is a Senior Lecturer in Defence and International Affairs at the Royal Military Academy Sandhurst. Having received his MA, MPhil, and DPhil from the University of Oxford, he was a Lecturer in Politics at Exeter College, Oxford, and had a stint as a Visiting Lecturer in International Relations at the University of Reading. His research is interdisciplinary, focusing on the history of ideas, strategic culture, and conflict and stability in the Middle East. His publications range across Classical Antiquity, Modern History, Regional Studies, and International Relations. His most recent works include a book chapter on the "Crisis of the Liberal International Order" and an article on "The Dynamics of Sino-Iranian Relations."

Kenneth Payne is a Professor of Strategy in the School of Security Studies at King's College, London. Professor Payne is the author of four books on strategy, most recently *I, Warbot: The Dawn of Artificially Intelligent Conflict.* A political psychologist, he is interested in the connection between mind and conflict, including artificial minds. His next book, with the working title *Know the Enemy, Know Yourself,* reflects on the role of empathy in strategy.

Muhammad Shoaib Pervez is the Chair of the Department of Political Science and International Relations at the University of Management and Technology, Lahore, Pakistan. He is the author of the award-winning book *Security Community in South Asia* (Routledge, 2013). His most recent book is titled *Radicalization in Pakistan: A Critical Perspective* (Routledge, 2021). His peer-reviewed articles on security communities, strategic culture, and nuclear proliferation have been published in journals including

International Politics and South East Asian Research. Dr. Pervez holds a PhD from Leiden University and completed his postdoc at Columbia University.

Itai Shapira (Col, res.) is a former Israeli Defense Intelligence Officer, with more than 25 years of experience in analytical and management roles on the strategic, operational, and tactical levels. Among his assignments are the Deputy Head of the Israeli Defense Forces Research and Analysis Division and the Head of the "Red Team" Department. He is a PhD candidate at the University of Leicester, studying Israeli national intelligence culture. He holds a BA and an MBA from Tel Aviv University and is also a graduate of the Israeli National Defense College. He has published works on intelligence, strategy, and Middle Eastern affairs in *Intelligence and National Security*, *Defense One*, *War on the Rocks*, *RealClear Defense*, *RUSI Commentary*, *19FortyFive*, *Small Wars Journal*, *The National Interest*, and *Strategic Assessment*.

Steve S. Sin is the Director of the Unconventional Weapons and Technology Division of the National Consortium for the Study of Terrorism and Responses to Terrorism, headquartered at the University of Maryland. Dr. Sin develops, leads, and manages interdisciplinary research projects spanning across a broad range of national and homeland security challenges. His expertise includes countering weapons of mass destruction; operations in the information environment; adversary decision modeling; and Northeast Asia regional security. His expertise in Northeast Asia regional security is focused on North Korea, including its nuclear program; cyber capabilities; intelligence apparatus; and regime survival. Dr. Sin's extensive experience also includes a career as a US Army Officer. He holds a PhD in political science from the University at Albany (SUNY) and is multilingual in English, Korean, Mandarin Chinese, and Japanese.

Mette Skak is an Associate Professor of Political Science at Aarhus University profiled on Russian foreign and security policy as her key research interest and teaches International Relations and related topics. She holds an MA in Russian Language and History and a PhD from Copenhagen University. She is the author of *From Empire to Anarchy: Postcommunist Foreign Policy and International Relations* (1996), and the editor and coauthor of *Putin's Rusland* (2022, in Danish). She has authored numerous other book chapters and articles on Russian affairs, including "Russia's New 'Monroe Doctrine'" in *Russian Foreign Policy in the 21st Century* (2011); "Russian Strategic Culture: The Role of Today's Chekisty" in the journal *Contemporary Politics*; and "Russian Strategic Culture: The Generational Approach and the Counter-Intelligence State Thesis" in *Routledge Handbook of Russian Security* (2019).

Briana Marie Stephan is the Director of Continuity Programs, Office of Emergency Management, National Nuclear Security Administration within the US Department of Energy. Prior to this position, she served as a Senior Advisor to the Assistant Secretary for Preparedness and Response at the Administration for Strategic Preparedness and Response, within the US Department of Health and Human Services. During that time, she was also a Graduate Fellow with the National Defense University/Missouri State University where she earned her second Master's in Countering Weapons of Mass Destruction. She began federal service in 2010 and holds other degrees including a Master's in Forensic Psychology (emphasis in law enforcement and responder psychology), Emergency Medical Technicians Certification, and a Bachelor's in Psychology.

Nicholas Taylor is a Senior Principal Analyst in the Exploration Division of the UK Ministry of Defence Science and Technology Laboratory (Dstl). In this role, he leads research and analysis on deterrence and other strategic effects. He has pioneered a new approach to the development and implementation of cross-government deterrence strategies, and now heads its application in the UK Government and advises senior decisionmakers on deterrence issues. Mr. Taylor conducts regular collaboration with the US Department of Defense and other NATO allies. He also maintains a community of interest of over 20 UK universities, with the aim of stimulating research and teaching on deterrence and related topics. Mr. Taylor continues to lead studies at Dstl on conceptual and theoretical aspects of deterrence, with a current emphasis on how the effective application of strategic culture can improve deterrence planning.

Dodeye Uduak Williams is an Associate Professor in the Department of Political Science, University of Calabar, Nigeria. She holds a Bachelor's degree in Political Science (Nigeria), a Master's of Science in International Relations (Nigeria), a Master's of Arts in Contemporary Global Security (Sheffield, United Kingdom), and a Doctorate degree in International Relations (Nigeria). She is an academic with over 20 years of work experience. Dr. Williams is a former Commonwealth Scholar (2008–2009 Edinburgh, UK). She teaches strategic studies at the undergraduate and postgraduate levels. She has published several academic articles in reputable journals and contributed several book chapters. Her current research interests tend to interrogate the complexities of the phenomena of politics, terrorism, violent extremism, counterterrorism, and counter-extremism, particularly religious terrorism and extremism in Africa. She is also currently a Research Fellow at the Institute for Gender Studies, College of Human Sciences, University of South Africa, UNISA.

James J. Wirtz has served as a Professor, Department Chairman, and Dean at the Naval Postgraduate School (NPS), Monterey, California. He played a pivotal role in creating the Center for Homeland Defense and Security at NPS, which educates state, local, federal, and tribal officials to better respond to natural and deliberate threats. Professor Wirtz's written work includes scores of books and articles on national security, intelligence affairs, and foreign matters. His textbook published by Oxford University Press, *Strategy in the Contemporary World*, is in its seventh edition and has introduced thousands of students to the study of strategy and international security. During his career, he has served as a consultant to US agencies and allied governments and has led mobile education teams in locations as varied as Tegucigalpa, Honduras, and Nur-Sultan, Kazakhstan. A native of New Jersey, Dr. Wirtz graduated from Columbia University and the University of Delaware and was a recipient of an Olin Fellowship at Harvard University. In 2016, the International Studies Association honored him as a Distinguished Scholar.

ACKNOWLEDGMENTS

The collective efforts represented in this volume have been nearly 20 years in the making. Our intellectual journey began in 2004 when one of our coeditors advocated for and initiated a US Department of Defense-funded project to develop a comparative strategic culture training and education curriculum with an emphasis on exploring the applicability and viability of strategic culture as an analytical tool to better understand critical desiderata among foreign leaders related to weapons of mass destruction, nonproliferation, and counterproliferation. Crucial funding for this early initiative was provided by the now-defunct Advanced Systems and Concepts Office at the Defense Threat Reduction Agency, with the support of Dick Gullickson, Dave Hamon, Jennifer Perry, and Mike Urena. This project culminated in a conference held in Monterey, California, in September 2005, managed and hosted by the Center for Contemporary Conflict at the Naval Postgraduate School and its staff, including Anne Clunan, Peter Lavoy, Elizabeth Stone, and Christopher Twomey, with the participation of key strategic culture scholars, including William D. Casebeer, Steve Coll, Michael C. Desch, Kevin Farrell, Theo Farrell, Robert Hickson, Darryl Howlett, Feroz Hassan Khan, Vali Nasr, and Andrew Scobell.

A follow-on second phase was implemented in 2006 with contract support provided by Lew Dunn, Paul Bernstein, and Thomas Skypek from Science Applications International Corporation. Workshops in connection with this phase were held in McLean, Virginia in February 2006, and Park City, Utah, in May 2006, hosted by Utah State University. The US Air Force Academy's Institute for National Security Studies (INSS) has long been a key institutional advocate and sponsor for strategic culture scholarship, and hosted a two-day workshop on strategic futures and Al-Qaeda's strategic culture in June 2006 in McLean, Virginia, managed by INSS Director James Smith. Colin Gray hosted the final conference of this project, at Reading University in the UK in August 2006. This conference engaged an international audience, including Simon Anglim, Jeremy Black, Stewart Brewer, J.H. Choi, Tony Coates, Christopher Coker, Theo Farrell, Babak Ganji, Andrew Garner, Bastien Giegerich, Sebastian L.v. Gorka, Darryl Howlett, Keith Payne, Glen Segell, Geoffrey Sloan, Mark Smith, Jeremy Stocker, and Rashed Uz Zaman. Huiyan Feng, Jerry Mark Long, and Bradley Thayer also contributed to this phase of our strategic culture journey. These conferences, colloquia, and workshops ultimately culminated in a book titled *Strategic Culture and Weapons of Mass Destruction: Culturally Based Insights into Comparative National Security Policymaking*, published by Palgrave Macmillan in 2009.

The third phase of our collective journey was again supported through the US Department of Defense and a grant administered by the US Naval Postgraduate School for a project titled "Scoping Future Nuclear Proliferation Risks: Leveraging Emerging Trends in Socio-Cultural Modeling and Analysis," in 2014. Many of our previous collaborators also participated in a "Workshop on Consolidating Recent Developments in Socio-Cultural Modeling and Analysis for Intelligence and Threat Assessment," as well as a workshop on assessing risks of future nuclear proliferation on the basis of contemporary sociocultural modeling and analytical methodologies. This latter group effort resulted in an April 2014 report that provided a strategic cultural assessment of the nuclear proliferation risks posed by Saudi Arabia, Iran, Israel, and North Korea.

A fourth phase involved an important collaboration with the Center for the Advanced Study of Language (CASL) at the University of Maryland – since renamed the Applied Research Laboratory for Intelligence and Security (ARLIS) – and included important contributions and leadership from Marilyn Maines, Joe Danks, John Walker, Kathy Faleris, Anne Wright, Steve Fetter, and Mike Bunting. This project was generously sponsored by a grant from the US Intelligence Community's National Center for Counterproliferation (NCPC). This collaboration involved not only further development of our analytical tools and methodologies but also included deep research into native-language materials on the strategic cultures of South Korea, Saudi Arabia, and Egypt. This year-long collaboration culminated in the publishing of a November 2017 report titled "A Cultural Approach to Assessment of Nuclear Intentions of Over-The-Horizon Countries." J.E. Petersen, Çan Kasapoğlu, Hannah Benninger, Scott Weiner, Adam Weinstein, Salma Bouziani, Steve Sin, and Sophia Sin, provided substantive regional and language research in support of this effort. Much of the scholarship produced by this project with the University of Maryland was eventually published in a book titled *Crossing Nuclear Thresholds: Socio-Cultural Insights into Nuclear Decisionmaking*, published by Palgrave Macmillan in 2018, edited by Jeannie Johnson, Kerry Kartchner, and Marilyn Maines.

Other friends and colleagues whose support and contributions along the course of our strategic culture journey should also be gratefully acknowledged include the following: Gregory F. Giles, Rich Delewski, Rich Love, Fritz Ermarth, David G. Haglund, Rodney Jones, Joseph S. Bermudez, Jr., Jeffrey Larsen, Willis Stanley, Nima Gerami, Dimitry (Dima) Adamsky, Ekaterina Svyatets, Shane Smith, and Polina Sinovets. Nik Taylor graciously represented the UK's Ministry of Defence at a workshop held with this volume's editors at a lodge in the deep snow of the Uintah Mountains south of Park City, Utah in February 2022.

We offer special thanks to Hannah Stevens and Kai Phillips, two wonderful Utah State University students in their respective doctoral and undergraduate programs during 2022 who provided much-appreciated technical support in preparing this lengthy manuscript. We wish you both all the best in your future pursuits. We also express appreciation to Joe Ward, Dean of the College of Humanities and Social Sciences at Utah State University (USU), for his leadership and support in launching USU's Center for Anticipatory Intelligence, enabling it to serve as a crucial incubator for many of the thoughts and themes in this volume. We warmly note that those students who have taken the center's strategic culture course have in many ways also contributed to refining our approach to the discipline and its methodologies across the years.

Our special gratitude is also extended to Angela Anderson for her permission to republish several articles from the 2022 special issue of the *Journal of Advanced Military Studies*, published by the Marine Corps University Press, which can be found here: https://muse.jhu.edu/issue/47612.

The editors express sincere appreciation for the unflagging support of Andrew Humphrys, Senior Editor for Military, Strategic and Security Studies at Routledge Publishers, and his amiable, effective, and responsive editorial assistant Devon Harvey, for their encouragement, long-suffering patience, and editorial service, advice, and guidance.

It is essential to note that the views expressed in this volume are those of the editors and contributors alone, and should not be construed as representing the views or policies of Science Applications International Corporation (SAIC), the University of Maryland, Utah State University, the Defense Threat Reduction Agency, the Naval Postgraduate School, the US Air Force Institute for National Security Studies, the US Department of State, or any other academic or US government agency.

Kerry M. Kartchner
Briana D. Bowen
Jeannie L. Johnston

FOREWORD

Thomas G. Mahnken

Center for Strategic and Budgetary Assessments, Johns Hopkins University SAIS

The notion that different polities develop distinct strategic cultures that flow from their geography, demography, and traditions dates back millennia. Indeed, the chapters that comprise Part I of this handbook ably discuss the evolution of the strategic culture paradigm. On the eve of the Peloponnesian War, for example, the Athenian *strategos* Pericles and the Spartan leader Archidamus sought to assess the balance between Athens and Sparta in an effort to determine whether or not it made sense to go to war with their rival. Their appraisals of the balance between the two city-states included not only economic and military factors but also contained an appreciation of the social and cultural influences that would shape Athenian and Spartan behavior in times of war.[1]

The study of strategic culture is, however, a neglected field in the United States. Indeed, one could argue that American strategic culture is profoundly anti-strategic culture.[2] In part, this is because Americans all too often fall prey to ethnocentrism and mirror-imaging. Related to these widespread cognitive processes is a general lack of knowledge of other regions, cultures, and ways of life. Many Americans view the world like the famous *New Yorker* cover portrays the United States: Manhattan looms large in the foreground bounded by the mighty Hudson River, with the rest of the country off in the middle distance, and with the rest of the world (marked "Russia," "Japan," "China") off in the far distance.[3] Americans tend to assume that other regions are unimportant (a manifestation of ethnocentrism) or that others are "just like us" (a function of mirror-imaging).

Beyond these common cognitive processes, however, belief in the universality of scientific rationalism is deeply ingrained in American strategic culture. That is, Americans believe that the world is governed by universal truths that apply to all people at all times. Moreover, American strategic culture tends to be ahistorical. That is, Americans tend to discount the role of history in shaping the present. As the essays that comprise Part IV of this handbook argue persuasively, an appreciation of strategic culture is nonetheless an important resource for both scholars and practitioners and certainly not limited to those in the United States. It offers a powerful lens through which we can view the behavior of political and military decisionmakers. To the extent that strategic culture conditions action, it will manifest itself in a consistent set of predilections and proclivities over time. Indeed, an appreciation of strategic culture may help us understand behavior that on the face of it seems arbitrary or even anomalous. The chapters that form Part III of this handbook discuss this with respect to a broad range of states.

Strategic cultures may, of course, share features. For example, just because a belief in the efficacy of deception is a feature of Chinese strategic culture, it does not follow that such a belief is *uniquely* Chinese. Other cultures, such as that of Persia/Iran, also put great weight on stratagem.[4] Similarly, just because culture is influential, it is not necessarily determinative. As Alexander L. George wrote in the late 1960s, culture "influences, but does not unilaterally determine, decision-making; it is an important, but not the only, variable that shapes decision-making behavior."[5] Or, as the Chinese military authors Peng Guangqian and Yao Youzhi put it, "Culture is not tantamount to behavior, nor is tradition to realities. However, cultural tradition, no doubt, may influence on choice of behavior in realities."[6]

Another reason to study strategic culture is that, unlike the United States, many cultures take it very seriously. For example, Chinese officers in the People's Liberation Army (PLA) see strategic culture as an important determinant of both their own behavior and that of others, including the United States. Indeed, the PLA published a volume on the subject, entitled *Analysis of China's Strategic Culture*, which explores Chinese strategic culture in depth and contrasts it with "Western" strategic culture.[7] In the words of Lieutenant General Li Jijun, the former vice president of the Chinese Academy of Military Sciences, "Culture is the root and foundation of strategy. Strategic thinking, in the process of its evolutionary history, flows into the mainstream of a country or a nation's culture. Each country or nation's strategic culture cannot but bear the imprint of cultural traditions, which in a subconscious and complex way, prescribes and defines strategy making."[8] Similarly, the authors of the Chinese handbook *The Science of Military Strategy* argue that, "Strategic thought is always formed on the basis of certain historical and national cultural tradition, and formulation and performance of strategy by strategists are always controlled and driven by certain cultural ideology and historical cultural complex."[9]

Most Chinese military leaders believe that ancient Chinese values and warfighting principles remain relevant today. Classical Chinese philosophical texts, military treatises such as *The Art of War* and the *Seven Military Classics*, and literary classics such as *Journey to the West* and *The Water Margin* are central to the identity of the PLA. The PLA teaches that its members are heirs to an ancient Chinese legacy and the Chinese Communist Party's revolutionary struggles. PLA military handbooks routinely refer to battles fought 4,000 years ago as object lessons, and PLA leaders seek guidance from 2,500-year-old writings for modern operations.[10] Indeed, even today, Chinese officers freely distribute translations of the Chinese military classics to their hosts.[11] Chinese strategists also view others in terms of their culture. For example, they contrast an expansionist, bellicose "Western" way of war with what they see as their own pacifist tradition.[12]

Culture must, of course, be applied carefully. Not to do so "risks replacing strategy with stereotypes."[13] One of the central challenges that the student of any state's strategic culture faces is determining which institutions serve as the keepers and transmitters of strategic culture. Is it the state? The military as a whole? Or some subset of the military? Another lies in identifying the content of strategic culture: the most salient beliefs and attitudes that comprise culture. Last but not least is the problem of determining the extent to which strategic culture, rather than power considerations, actually determines attitudes and behavior.[14]

If employed properly, however, the cultural lens is a powerful one. As I have argued elsewhere, strategic culture can usefully be considered on (at least) three levels: those of the nation, the military, and the military service.[15] At the national level, strategic culture reflects a society's values regarding the use of force. At the military level, strategic culture (or a nation's "way of war") is an expression of how the nation's military *wants* to fight wars. Although practice does not have to conform to this desire, success in waging wars that run counter to national ways of war requires painful adaptation. Finally, strategic culture at the service level represents the organizational culture of the particular service – those values, missions, and technologies that the institution holds dear.[16] The essays that comprise Part II of this handbook, which focus on dimensions and levels of culture, contribute greatly to this aspect of the field.

Each of these levels can yield insights relevant to scholars and practitioners. A compelling case is the clash of national strategic cultures the world is witnessing between the United States, and the West more broadly, on the one hand, and China on the other. There is a stark contrast between Western views of war, peace, and strategy, and those embraced by the Chinese leadership.

First, the West and China have divergent concepts of war and peace. Central to Western strategic thought is the notion that war and peace are distinct. Either one is at peace, or one is at war. Indeed, the Western liberal tradition sees peace as the natural order of things and war as a temporary, regrettable deviation from that norm. This notion informs Western legal systems as well as the authorities granted to Western government institutions.[17] By contrast, as elaborated below, Chinese strategy admits no clear dichotomy between war and peace. Rather, it envisions a spectrum that runs from cooperation through competition to confrontation.

Second, the United States and other Western states conceive of the various instruments of national strategy, and the institutions dedicated to them, as separate and distinct. For example, the widely used American acronym "DIME" conceives of the "Diplomatic, Informational, Military, and Economic" instruments of strategy as discrete. By contrast, the Chinese Communist Party pursues an integrated strategy, one that deploys the full spectrum of elements of national power in various combinations to achieve its ends. The Chinese Communist Party follows a "combined arms" approach to achieving its aims, albeit one that is much broader than Western concepts.

This broad approach has deep historical and cultural roots in China. Most elements of Beijing's recent hybrid warfare operations have echoes in the guerilla and insurgent campaigns that the Chinese Communists conducted in the first half of the twentieth century and even as far back as Sun Tzu in 500 BC. This is a type of conflict with which Beijing is very familiar. China's approach also reflects a Marxist-Leninist worldview that is markedly different from the strategic cultures prevailing in the West. China's security planners and operational commanders think, and often act, differently from those in the West. In particular, China's leaders are well practiced in designing and implementing campaigns and sequencing various political, nonmilitary, and information instruments into integrated campaigns.

Furthermore, current events demonstrate the importance of military strategic culture or national ways of war. The performance of the Russian Federation's armed forces in Ukraine following Russia's 2022 invasion demonstrates strong continuity not only with that of previous campaigns but also with the performance of the Soviet Red Army or, indeed, the Imperial Russian Army that preceded it. The Russian way of war, with its emphasis on defending the motherland against aggressors real and imagined, its substitution of quantity of men and materiel for quality, and its emphasis on firepower and attrition, appears to be alive and well. An understanding of organizational culture also offers important insight into these unfolding developments, as the deep imprint of organizational culture appears to have stymied efforts to reform the Russian military in the wake of the 2008 Georgia War. As one analysis concluded, "The basic problem was that the military culture of the Soviet Union endured, despite the lack of men and means to sustain it."[18]

Similarly, the challenges experienced by the United States in Iraq and Afghanistan across the earlier years of the twenty-first century betrayed a fundamental mismatch between counterinsurgency strategy and an American way of war based upon wars for unlimited aims, an emphasis on technology, and the lavish use of firepower.[19] At times, America's performance in Iraq and Afghanistan bore more than a passing resemblance to that in Vietnam.[20] As the American armed services evolve and innovate amid a pivot toward renewed great power competition, the cultures of the individual services have shaped and will continue to shape the missions and technologies they are interested in, and those they are disinterested in; even, perhaps, when these present less than perfect matches to emergent challenges.[21]

Strategic culture offers a powerful lens for better understanding the behavior of key decisionmakers in an era of great consequence for global security. And although it does not provide a formulaic prediction of all future action, as the authors in this volume readily admit, its contributions can refine both the accuracy and exquisiteness of insights for scholars and practitioners around the world as they seek to navigate the complexities of the twenty-first-century security landscape. This volume thus offers an indispensable antidote to the twin challenges of ethnocentrism and mirror-imaging, pathologies that could otherwise thwart these insights, and a comprehensive guide to using strategic culture to navigate these complexities.

Notes

1 Robert B. Strassler, ed., *The Landmark Thucydides* (New York: The Free Press, 1996), pp. 1.80–81, 1.141–42.
2 I am grateful to Kerry Kartchner for this formulation.
3 *The New Yorker*, 29 March 1976, cover.
4 See, for example, Andrew Campbell, "Iran and Deception Modalities: The Reach of *Taqiyya*, *Kitman*, *Khod'eh* and *Taarof*," *National Observer* 70 (Spring 2006), 25–48.
5 Alexander L. George, "The 'Operational Code': A Neglected Approach to the Study of Political Leaders and Decision-Making," *International Studies Quarterly* 13, no. 2 (June 1969), 191.
6 Peng Guangqian and Yao Youzhi, eds., *The Science of Military Strategy* (Beijing: Military Science Publishing House, 2005), p. 74.
7 Gong Yuzhen, *Analysis of China's Strategic Culture* (Beijing: Military Science Press, 2002).
8 Quoted in Andrew Scobell, *China and Strategic Culture* (Carlisle: SSI, 2002), p. 1.
9 Peng Guangqian and Yao Youzhi, eds., *The Science of Military Strategy* (Beijing: Military Science Publishing House, 2005), p. 31. See also pp. 72–77.
10 *The Culture of the Chinese People's Liberation Army* (Quantico, VA: Marine Corps Intelligence Activity, 2009), p. 5. See also Guangqian and Youzhi, eds., *The Science of Military Strategy*, p. 87.
11 During an October 2009 visit to Washington, DC, for example, General Xu Caihou, vice chairman of the People's Liberation Army Central Military Commission, gave out hundreds of copies of "The Wisdom of Sun Tzu" to his hosts, including the author.
12 Scobell, *China and Strategic Culture*, 17.

13 Patrick Porter, "Good Anthropology, Bad History: The Cultural Turn in Studying War," *Parameters* 37, no. 2 (Summer 2007), 46.
14 Alan Macmillan, Ken Booth, and Russell Trood, "Strategic Culture" in Ken Booth and Russell Trood, eds., *Strategic Cultures in the Asia-Pacific Region* (New York: St. Martin's, 1999), pp. 8–12.
15 Thomas G. Mahnken, "U.S. Strategic and Organizational Subcultures" in Jeannie L. Johnson, Kerry M. Kartchner, and Jeffrey A. Larsen, eds., *Strategic Culture and Weapons of Mass Destruction: Culturally Based Insights into Comparative National Security Policymaking* (New York: Palgrave Macmillan, 2009), pp. 69–84.
16 As Edgar Schein puts it, organizational culture is "The pattern of basic assumptions that a given group has invented, discovered, or developed in learning to cope with its problems of external adaptation and internal integration, and that have worked well enough to be considered valid, and, therefore, to be taught to new members as the correct way to perceive, think, and feel in relation to those problems." Edgar H. Schein, "Coming to a New Awareness of Organizational Culture," *Sloan Management Review* 25, no. 2 (Winter 1984), 3.
17 See, for example, Thomas G. Mahnken, "U.S. Strategic and Organizational Subcultures" in Jeannie L. Johnson, Kerry M. Kartchner and Jeffrey A. Larsen, eds., *Strategic Culture and Weapons of Mass Destruction: Culturally Based Insights into Comparative National Security Policymaking* (New York: Palgrave Macmillan, 2008), pp. 69–84.
18 Neil MacFarquhar, "Russia Sought to Create a Leaner, Meaner Military. But Entrenched, Soviet-Era Practices Endure," *New York Times*, 16 May 2022.
19 See, for example, Todd Greentree, "The Accidental Counterinsurgents: U.S. Performance in Afghanistan" in Thomas G. Mahnken, ed., *Learning the Lessons of Modern War* (Stanford, CA: Stanford University Press, 2020), pp. 157–179.
20 Todd Greentree, "Bureaucracy Does Its Thing: U.S. Performance and the Institutional Dimension of Strategy in Afghanistan," *The Journal of Strategic Studies* 36, no. 3 (June 2013), 325–356.
21 Thomas G. Mahnken, *Technology and the American Way of War since 1945* (New York: Columbia University Press, 2008).

PART I

Evolution of the Strategic Culture Paradigm

1

DEFINING AND SCOPING STRATEGIC CULTURE

Promises, Challenges, and Conundrums

Kerry M. Kartchner

Culture as Culprit

When events go catastrophically wrong, culture is often the first thing to be blamed. If nations or their governments are involved, commissions will be formed to investigate and report, and invariably the culture of the liable organizations will be found at fault. If the catastrophe involves private enterprises, then stockholders and boards of directors will demand that outside firms or consultants audit, analyze, and assess, again often concluding that the business's culture resulted in failure to understand the risks involved, or to adapt to changing market conditions, or to navigate cross-cultural work environments, or to modernize technology and procedures, or to hold risk-prone employees and management accountable.[1]

This tendency to blame culture is true not only at the organizational level, but at the level of military and defense organizations, executive and legislative branches, governments, and administrations. If we assume that "strategic" culture is that cultural and ideational arena where national security stakes are involved, it is not a stretch to say that, in certain cases, failures of strategic culture occur.[2] The business management literature is replete with case studies of business culture failures, lessons learned, and schemes, classes, and publications on how to correspondingly fix corporate culture. But examples can also be drawn from serious national security events, and from national responses to large-scale natural catastrophes that fully qualify as "strategic" in level and scope. Three examples may suffice to illustrate.

In the aftermath of the terrorist attacks of 9/11 – which resulted in the single greatest loss of life from a terrorist attack in history – a commission was formed by US Congressional mandate. In trying to determine how the events leading to this tragedy had gone unanticipated and unprevented, the commission concluded that "[a]cross the government, there were failures of imagination, policy, capabilities, and management." But among these, "[t]he most important failure was one of imagination."[3] It was therefore proposed by the commission that the US Intelligence Community "find a way of routinizing, even bureaucratizing, the exercise of imagination."[4] That is, the commission

DOI: 10.4324/9781003010302-2

called for changes to the analytical culture of the US Intelligence Community to overcome the boundaries of its own culturally bounded ethnocentric threat assessments, calling specifically for "the development of an institutional culture imbued with a deep expertise in intelligence and national security."[5]

The word "imagination" is not to be found in the canon of traditional international relations realism, which concerns itself only with objective and measurable constructs, nor is there a curriculum within Western political science or international relations programs that explicitly routinizes and bureaucratizes the training and exercise of imagination. It is not hard to explain these absences. Imagination defies traditional Western empirical examination because it is a function of culture, which exists only in the collective minds of a society.[6] In this context, the commission reviewing 9/11 concluded that the US Intelligence Community had not fostered a culture where "outside the box" thinking could have imagined turning commercial airliners into suicidal attack missiles, and therefore not only failed to plan for and to prepare for such an eventuality but also failed to even conceive of it. Culture (and by extension, strategic culture) can *bind* a community, but it can also *blind* that community, because cultural communities draw on action palettes that include preferentially emphasizing those courses of action that they themselves would consider acceptable and legitimate for that community (or that are "the way we've always done it").[7]

For the second example, on 11 March 2011, a magnitude 9.0 earthquake off the coast of Japan automatically triggered the shutdown of four power-generating fission reactors at the Fukushima Daiichi Nuclear Power Plant in Ōkuma, Fukushima, Japan. According to the US Geological Survey, it was the most powerful earthquake ever recorded in Japan and the fourth most powerful earthquake in the world since recordkeeping began in 1900.[8] Some estimates suggest that the earthquake and its effects were responsible for nearly 20,000 deaths.[9] The Fukushima plant's safety measures functioned as designed in reacting to the earthquake. The reactors were shut down, and backup generators began circulating cool water to siphon off the residual fission heat. However, the earthquake was followed shortly by a powerful tsunami measuring 46 feet high that swept over the plant's seawall and flooded the lower parts of reactors one through four, causing the failure of the emergency generators and the complete loss of power to the circulating pumps. The subsequent loss of reactor core cooling resulted in three nuclear meltdowns, three hydrogen explosions, and the release of radioactive contamination for days afterward. Large swaths of territory around the plant were contaminated, and eventually 154,000 people had to be evacuated. The cleanup program was estimated to take 30–40 years to complete.[10]

It is fair to ask: what does this disaster and Japan's mitigation response have to do with strategic culture? Subsequent investigations into the Fukushima disaster concluded that the consequences of the catastrophe had their roots in a "network of corruption, collusion, and nepotism" that contributed to poor safety design and oversight, and weak disaster mitigation capabilities.[11] Kiyoshi Kurokawa, Chairman of the National Diet of Japan Fukushima Nuclear Accident Independent Investigation Commission, wrote in his foreword to the official report that "this was a disaster 'Made in Japan.' Its fundamental causes are to be found in the ingrained conventions of Japanese culture: our reflexive obedience; our reluctance to question authority; our devotion to 'sticking with the program'; our groupism; and our insularity."[12] Noted Japan scholar Brad Glosserman suggested that "the roots of the breakdown that day originated in the same beliefs and the same institutions that had propelled Japan to its greatness."[13] Thus, while nature was at fault for the earthquake and tsunami, the response of the Japanese government, the power company, and local governing authorities was hamstrung by cultural values, norms, and ways of thinking.

Third, failures at the level of strategic culture have also been associated with the Chernobyl disaster. Russian physicist Sergei Kapitza has argued that broader social and psychological factors greatly exacerbated the catastrophe and that cultural issues may even have been a fundamental cause of the Chernobyl accident. Among these, he lists disregard for human life, excessive secrecy, isolation from the broader international nuclear safety community, a tendency to promote experts "chosen not so much for their scientific or technical excellence but for their political loyalty," and finally the suppression of ingenuity and initiative in a compartmentalized and centrally controlled atmosphere where "it is better to fail according to the rules than succeed by breaking them."[14] Other authors have also highlighted how more fundamental origins of the Chernobyl disaster lie in the nature of the Soviet system, institutional inadequacies in the Soviet nuclear industry, control of scientific information, emphasis on economic production over humanitarian and environmental concerns, and the authoritarian character of communist party rule.[15] Students of strategic culture will recognize elements of strong but ultimately counterproductive cultural values and norms at play here.

Numerous other examples could be cited. But what do these failures reflect about strategic culture? If problematic patterns and habits are endemic to certain national and organizational cultures, the same must be correspondingly true for strategic culture. Some initial conclusions might serve. First, these failures were the result of mistakes made at the strategic level, in terms of the national security issues involved, the group mindsets of the given societies and institutions, and the regional, global, and society-wide effects of these disasters. Second, these failures could have been prevented or at least foreseen and anticipated (in the case of the 9/11 attacks), or their effects mitigated (in the case of the Fukushima disaster), by better understanding the strategic culture of one's prospective adversaries, or by knowing the limitations of one's own strategic cultural proclivities and habits of mind.

If there are aspects of strategic culture that contribute to catastrophic failures, it is incumbent on students and scholars of strategic culture to make deliberate efforts to systematically study strategic culture, to more fully develop analytical frameworks for assessing and shaping strategic culture, and to advance the dissemination of theoretical and practical applications of strategic cultural analysis.

Conundrums in Defining Strategic Culture

This volume raises and aspires in some modest way to contribute to resolving the common criticism that strategic culture lacks a consensual definition. We note that the author of Chapter 3 herein, Brigitte E. Hugh, has taken on the task of discussing some of these definitional challenges, suggesting that if strategic culture is considered more of a tool or method rather than a theory or paradigm, this may relieve some definitional pressures. There has been a tendency within the literature for authors who are new to the subject to take the initiative to offer their own definition of strategic culture, sometimes in ways that further muddy the waters rather than consolidate a definitional consensus. Unless those authors can explain why the dozens of previous definitions are inadequate to their specific analytical project, it is singularly unhelpful to proliferate yet more definitions of strategic culture. Notwithstanding these definitional conundrums, this volume seeks to be mindful of David Haglund's admonition that "our failure to come to agreement over the term's meaning and applicability should be neither surprising nor particularly discouraging; it simply illustrates we are dealing with an interesting concept that, as with all such concepts, can be used to advantage if used carefully."[16] To underscore the challenge of defining strategic culture, some of the more often cited definitions are cataloged in Table 1.1 for quick comparison.[17]

Table 1.1 Selected Definitions of Strategic Culture

Author	*Definition*
– *Jack Snyder*	"[T] he sum total of ideas, conditioned emotional responses, and patterns of habitual behavior that members of a national strategic community have acquired through instruction or imitation and share with each other with regard to nuclear strategy ... a set of general beliefs, attitudes, and behavioral patterns with regard to nuclear strategy [that] has achieved a state of semipermanence that places them on the level of 'culture' rather than mere 'policy.'"[18]
– *Ken Booth*	"The concept of strategic culture refers to a nation's traditions, values, attitudes, patterns of behavior, habits, symbols, achievements and particular ways of adapting to the environment and solving problems with respect to the threat or use of force."[19]
– *Colin Gray*	"[T]he persisting (though not eternal) socially transmitted ideas, attitudes, traditions, habits of mind, and preferred methods of operation that are more or less specific to a particular geographically based security community that has had a necessarily unique historical experience."[20]
– *Alastair Iain Johnston*	"[A]n ideational milieu which limits behavioral choices ... [from which] one ought to be able to derive specific predictions about strategic choice.... An integrated 'system of symbols (e.g., argumentation structures, languages, analogies, metaphors) which acts to establish pervasive and long-lasting strategic preferences by formulating concepts of the role and efficacy of military force in interstate political affairs ...'"[21]
– *Robert Cassidy*	"[M]ilitary-strategic culture is a set of beliefs, attitudes, and values within the military establishment that shape collective (shared) preferences of how and when military means should be used to accomplish strategic aims. It is derived or developed as a result of historical experience, geography, and political culture."[22]
– *Stephen Peter Rosen*	"[B]eliefs and assumptions that frame ... decisions [to] go to war, preferences for offensive, expansionist or defensive modes of warfare, and levels of wartime casualties that would be acceptable."[23]
– *John Duffield*	"The overall effect of national security culture is to predispose societies in general and political elites in particular toward certain actions and policies over others. Some options will simply not be imagined ... some are more likely to be rejected as inappropriate or ineffective than others."[24]
– *Jeannie Johnson, Kerry Kartchner, and Jeffrey Larsen*	"[T]hat set of shared beliefs, assumptions, and modes of behavior, derived from common experiences and accepted narratives (both oral and written), that shape collective identity and relationships to other groups, and which determine appropriate ends and means for achieving security objectives."[25]

Several common themes arise from this list. First, strategic culture is associated with groups at various levels, and not with a given individual's attributes (which would fall into the category of psychology). Second, strategic culture exists in the realm of ideas, values, beliefs, and perceptions. Members of a group, whether at the family, tribal, organizational, regional, national, or civilizational level, are enculturated with that group's sense of identity, values, norms, and even its way of thinking and perceiving the world around them.[26] Individuals are not born with these attributes; they acquire them along the way. Third, as a distinct subset of culture, strategic culture concerns itself with matters of national security and ways of war, including those values, norms, intentions, and objectives that bear on how a culture relates its own conceptualized security ends to its own uniquely manifested security means. Fourth, strategic culture is recognized as comprising habits of thinking, ways of organizing, and patterns and rituals of behavior that although they evolve as circumstances dictate, they collectively resist change. This may explain the oft-repeated refrain from the business management community that cultural change is one of the most daunting challenges to corporate improvement.

Reasons to Study Strategic Culture

Strategic culture can be a powerful tool for providing insights into the context and collective cognition, which suggests meaningful motivations for studying strategic culture. As suggested at the outset, the study of strategic culture and the implementation of its analytical outputs may potentially help prevent or mitigate catastrophic events or their consequences. Drawing on a rich body of strategic culture research and publication, we propose in this volume the following reasons for studying strategic culture and taking strategic culture into account when shaping policymaking or assessing adversary intentions: [27]

1 Realism is a powerful explanatory paradigm for international relations, but it cannot shed light on nuances in the motivations and intentions of leaders and communities, nor provide insights into the role that identity, values, norms, and cognitive processes play in shaping behaviors, policies, and outcomes. Strategic culture can supplement realism by providing such context and awareness.[28]

2 Strategic culture can offset or ameliorate the impact of ethnocentrism as a factor contributing to misperception, misunderstanding, and misconstruing messages in diplomacy and strategy.

3 Strategic culture can provide insights into the motivations, intentions, and beliefs of both allies and potential adversaries, including helping to identify the foundational and iterative narratives that shape their thinking.

4 Strategic culture can sensitize the analyst to dimensions and sources of strategic behavior that are absent from realist theories, such as the role of history, geography, and religion as sources of a culture's identity, beliefs, and behaviors.

5 Strategic culture highlights the role of domestic factors in foreign and security policies in ways that realism cannot. As Ken Booth has noted, strategic culture "[breaks] down the artificial boundary between the domestic environment within which policies are made and the external security environment. It reminds us that decisionmaking structures, military establishments and decisionmaking processes all operate in peculiar political cultures. It therefore draws attention to the differences between nations when the 'science' in political science is tempted to play them down."[29]

6 Strategic culture reminds us that "rationality" may be construed differently across cultures and societies.[30] For example, nuclear deterrence has been widely conceived as a calculus in an adversary's mind between costs and ends. But how does a prospective opponent construe "costs" and "ends"? How does it calculate their relative value? Only strategic culture can offer descriptive explanations of how these might differ from one's own calculus.

Of course, many other reasons could be added to this list, including justifications specific to certain vocations such as diplomats needing to resolve conflicts or negotiate agreements across cultural divides, or intelligence analysts needing to parse intentions when assessing the implications of dual-use technologies that could have either legitimate peaceful purposes or that could be diverted or converted to military purposes. In these cases, strategic culture can provide additional context or hypotheses for intent.[31] The utility of strategic culture has broadened over the course of its conceptual evolution, and further applications may yet be explored.[32]

A Note on the Origins of the Term "Strategic Culture"

Definitions of and the potential benefits of studying strategic culture have been contested since the earliest use of the term. A strong consensus within the field of strategic studies recognizes Jack Snyder's 1977 RAND report on Soviet "strategic culture" as the origin of the term and as the source of its most often cited definition.[33] However, the context of Snyder's invention of this term is either forgotten or rarely explained in the subsequent literature on strategic culture.[34] Yet it is important to understand this context to appreciate not only his own subsequent renunciation of how the term came to be applied but also the ways in which Snyder contributed to creating a foundation for the so-called first generation of strategic culture studies. Charting a future for strategic culture studies should also be grounded in and informed by a nuanced understanding of the term and the field's origins.

Significantly, the term "strategic culture" was born out of a necessity to explain substantive and persistent differences in US and Soviet strategic nuclear doctrine and practices during the Cold War. By the late 1970s, it was becoming difficult to avoid acknowledging that the long-promised "cold dawn" of a convergence[35] between US and Soviet nuclear strategies around the American notion of "Mutually Assured Destruction" had not in fact occurred. Furthermore, US and Soviet strategic nuclear forces had achieved a condition of parity in numbers of nuclear warheads by the early 1970s, and although Western analysts had long predicted that a stable condition of parity would then prevail thereafter, Soviet strategic nuclear force modernization defied these predictions of doctrinal convergence and perpetual parity as the Soviet Union achieved numerical superiority over the United States by the late 1970s and then continued to increase its nuclear forces even further.[36] This Soviet nuclear force buildup was consistently underestimated by the Central Intelligence Agency,[37] and an outside group of experts convened in 1976 called "Team B" found that pervasive "mirror imaging" and lack of accounting for cultural differences between US and Soviet nuclear doctrine and intentions had consistently resulted in mischaracterizing the objectives of Soviet strategic nuclear force modernization programs.[38] In a quirk of history, the head of "Team B," Richard Pipes, published an essay based on the team's analysis in *Commentary* magazine in July 1977, the same month the Snyder report was published by

RAND, in which he argued that ignoring and misunderstanding Soviet culture had resulted in mistaken ethnocentric assumptions within the US Intelligence Community.[39] So, in fact, while Snyder may have been the first to use the term, he was not the only scholar to attribute rigid Soviet nuclear doctrine to cultural factors.

Nevertheless, in his seminal report, Snyder was explicitly tasked to explore whether the Soviet Union was likely to reciprocate developments from the early 1970s in US strategic nuclear doctrine ("the Schlesinger Shift") that came to place greater emphasis on flexible and limited nuclear targeting options to enhance the credibility of deterrence and strengthen options for limiting nuclear war should it occur. Snyder concluded that the Soviets were organizationally and institutionally culture-bound by distinct ways of thinking about nuclear strategy, and that they were unlikely to adopt an American-style approach to either developing more flexible nuclear attack options or adopting the distinctly American approach to nuclear damage limitation.

On the basis of his study of Soviet strategic nuclear doctrine, Snyder found that individual members of the Soviet nuclear policymaking elite "are socialized into a distinctively Soviet mode of strategic thinking. As a result of this socialization process, a set of general beliefs, attitudes, and behavioral patterns with regard to nuclear strategy has achieved a state of semipermanence that places them on the level of 'culture' rather than mere 'policy.'" He then defined strategic culture as "the sum total of ideas, conditioned emotional responses, and patterns of habitual behavior that members of a national strategic community have acquired through instruction or imitation and share with each other with regard to nuclear strategy."[40]

Snyder made three other less often noted contributions to the concept of strategic culture that would form the basis of the field of strategic culture studies and later be expanded upon and applied in subsequent literature by diverse authors. First, he acknowledged a problem that has bedeviled strategic culturalists ever since, namely the question of whether strategic cultures are susceptible to change over time. Snyder responded in the affirmative. He noted that "[p]roblems of continuity, change, and the adaptation of doctrine to new technologies take on new light when considered in a cultural context. We assume that strategic cultures, like cultures in general, change as objective conditions change. But we also assume a large residual degree of continuity … that [evolves] only marginally over time."[41] A key theme of Snyder's analysis is that Soviet nuclear doctrine had become mired in a sort of bureaucratic inertia. Later strategic culture and organization process scholars would amplify this theme of cultural or organizational inertia to explain the lack of change or the opposition and resistance to change. In fact, in some cases, those scholars who were introduced to the concept of strategic culture later in their professional careers, or who were discovering strategic culture for the first time, have remarked that the most important factor explained by strategic culture was not what behaviors and options strategic culture was responsible for but what it took off the table, or the courses of action that were excluded by strategic culture. For example, according to Michael Desch, "Cultural variables may explain the lag between structural change and alterations in state behavior … [and] may account for why some states behave irrationally and suffer the consequences of failing to adapt to the constraints of the international system."[42]

Other scholars have suggested additional modalities of change in strategic culture.[43] Military historians Theo Farrell and Terry Terriff, in their coedited volume *The Sources of Military Change: Culture, Politics, Technology*, explicitly addressed mechanisms of strategic cultural change in military organizations and doctrine, arguing that there are two major pathways of cultural change: planned or deliberate efforts (sometimes driven by new political, technological, or geopolitical developments), and changes as a result of an external shock.[44] Other sources of change identified in the literature have included change through accretion by association with other cultures, by adoption or conglomeration when two or more cultures intermingle (through immigration, for example, or through engagement in prolonged conflict), and by emulation (as when Arab rulers of the eighteenth century set out to adopt Western ways of military strategy and tactics).

Second, Snyder introduced the concept of "subcultures," and defined a subculture as "a subsection of the broader strategic community with reasonably distinct beliefs and attitudes on strategic issues, with a distinct and historically traceable analytical tradition."[45] This theme has become a hallmark of fourth-generation strategic cultural research, with its greater focus on organizational or institutional subcultures such as the cultural differences among services within the US military.[46]

Third, Snyder may be the first scholar to have recognized that strategic culture endows a community with a lens through which members of that community view the world, and through which they process information, including how they confront new challenges that might arise. As Snyder put it, "attitudes may change as a result of changes in technology and the international environment. However, new problems are not assessed objectively. Rather, they are seen through the perceptual lens provided by the strategic culture."[47] It is significant that this dimension of strategic culture incorporates a cognitive function (how members of a community think about and process what is perceived around them) that manifests itself not only in how matters of national security are framed for internal consumption, but how perceptions of threat and alliance, friends and foes, are constituted.[48]

Snyder later renounced the way subsequent authors had employed and expanded on the term "strategic culture," stating he meant only to apply the term narrowly to explain apparent organizational and institutional inertia in Soviet nuclear policy, asserting that the term is too vague, and that "culture is an explanation of last resort ... to be used only when all else fails."[49] In a sense then, Snyder's retraction of his own contributions to the field of strategic culture, and his subsequent criticisms, served as prelude to common themes that later arose among critics of the field.[50]

In a rejoinder to Snyder's renunciation, British scholar Ken Booth, whose own writings helped establish strategic culture as a viable field of study, argued that "whatever Snyder himself originally meant when he used the term 'strategic culture' his accidental coinage gave focus to a set of ideas which were not new but whose intellectual time had finally come."[51] Booth dismissed Snyder's renunciation as a "distinctive political science approach which is not comfortable with 'cultural' explanations that are not always easy to identify or relate to other considerations." Finally, Booth claimed that Snyder uses the term "culture" in an idiosyncratic way, and does not provide the cultural context necessary to make his analysis truly "cultural."[52]

The origins of the field of strategic culture were thus situated in an effort to explain apparent discrepancies between US and Soviet approaches to strategic nuclear doctrine and deterrence. The earliest writings to use this term (the so-called first generation of strategic culture) were therefore understandably focused on respective national styles in strategic nuclear thinking and doctrine.[53] However, later generations of strategic culture scholars have built on the foundation laid by Jack Snyder – with or without his blessing – especially with respect to the modalities of change or immutability in strategic culture, the relevance of identifying and explicating subcultures or sub-identities within any given strategic culture, and the cognitive dimensions of culturally endowed perceptual lenses.

Objectives and Overview of This Handbook

The editors and authors of this Handbook have been motivated by several factors. First, interest in strategic culture as an analytical construct has been growing. Books and articles continue to be published in ever-greater numbers that incorporate or develop a strategic cultural theme. Practical frameworks using strategic culture are being applied to a growing number of academic and policy issues, including nuclear nonproliferation, tailored deterrence strategies, negotiating style, intelligence analysis, the diffusion of technology, and anticipating strategic surprise. The US military and foreign service communities, among others, have increasingly sought to incorporate strategic cultural analysis into their respective curriculums and professional development programs. Furthermore, the number of countries and groups that have been the subject of published strategic culture profiles is expanding, including concerted efforts to develop case studies of South American countries,[54] countries in the Asia-Pacific region,[55] and the nations of Europe.[56]

Second, despite the proliferation of various case studies, books, articles, and monographs applying a strategic cultural framework, little consensus has developed on the definition of "strategic culture," the appropriate level of analysis, the applied value of strategic cultural insights to policymaking, or the methodological framework for strategic cultural research. In other words, the field of strategic culture currently lacks the requisite cumulativeness to develop into a more rigorous discipline with more robust theory and policy relevance.[57] Moreover, several recent innovative developments in strategic cultural analysis could benefit substantially from greater consensual validation in a *Handbook of Strategic Culture* that bridges the disparate communities advancing the study of strategic culture. Part of the challenge is that strategic culture implicitly requires a multidisciplinary approach that is disavowed in most academic political science or international relations programs.

Third, a number of distinct academic communities have arisen that have developed an emphasis on incorporating strategic cultural analysis and expanding the scope of national strategic cultural profiles, but remain unconnected to, and possibly even unaware of, each other. One community is a European group of scholars pursuing the study of EU strategic culture, with its own distinct approach to defining and conceptualizing "strategic culture."[58] This community's focus has been on the degree to which the national strategic cultures of disparate European states are converging enough to contribute positively to developing a common European security and defense policy. Another community has emerged centered in the UK, primarily originating at the University of Reading and at the University of Southampton, where there has been an emphasis on creating strategic cultural profiles of various states and non-state actors. A third locus of strategic cultural analysis later emerged in the United States, with some impetus provided by a 2006 research program sponsored by the US Defense Threat Reduction Agency focused on weapons of mass destruction and strategic culture from a deterrence and nonproliferation perspective. The respective bodies of literature of these three groups only rarely cross-reference the literature and analytical frameworks of the others. This circumstance feeds into the claims by critics of strategic culture that the field lacks consensus on definitions, modalities of change, and robustness of theoretical claims and hypotheses. This *Handbook of Strategic Culture* brings together representatives of these various communities with the aim to help resolve this academic "stove-piping" and respond to these criticisms.

Fourth, and finally, few general textbooks exist for the student of strategic culture, and those that do exist (for example, Katzenstein, *The Culture of National Security: Norms and Identity in World Politics*, Columbia University Press, 1996) are somewhat outdated, narrowly focused on a specific region or a select number of countries, and do not contribute to building a consensus on how strategic culture should be defined or applied.

The editors of this volume have deliberately structured this Handbook to appeal to both students new to the subject and to scholars who may wish to incorporate strategic culture in their toolbox of analytical techniques. We have made an explicit effort to draw in a range of contributing authors to this volume, leveraging a rich international academic community in this field, and including both experienced scholars and junior contributors in the early stages of their careers. In Part I, essays are presented on the concept and definition of strategic culture, and its evolution as a "research programme" (to use the Lakatosian term versus the Kuhnian "paradigm shift"). In this Part, the chapter authors confront the theoretical strengths and weaknesses of the field, including acknowledging those problems identified by critics of the concept, such as insufficient characterization of the functions of strategic culture, inadequately clarifying the modalities and dynamics of cultural change, and the sometimes overly deterministic quality of some predictions and assertions within the early strategic culture literature. These essays address questions such as the contestation of identity narratives within complex strategic cultures, the nature and degree of change in strategic cultural norms and values, and the impact of external shocks versus internal domestic imperatives as sources of cultural change.

Part II continues laying out elements of the theoretical and methodological foundations of the field by examining the key dimensions of strategic culture, such as sources and components of strategic culture. Essays in this section address the relationships among strategic culture, political culture, and military culture. Other essays develop the major sources of strategic culture, including the often ignored or discounted role of religion as a source of strategic cultural identity, values, norms, and perceptual lens. We address the cognitive dimension and function of strategic culture in this section, as well as the issues of how, when, and to what degree strategic cultures change. One of the enduring conundrums in the field concerns how to engage the appropriate level of analysis for strategic culture research and applications.[59] A major theme of this Handbook is the importance of applying a multi-level approach to any given policy or theoretical issue. Part II thus also includes essays that look at the presence of subcultures within larger polities, the strategic cultures of non-state actors, the value of taking a "supranational" perspective on strategic culture (as in the European Union [EU], or within the NATO alliance), and the functions of strategic culture at the organizational level.

Much of the existing literature in the field is given over to case studies of different countries, and the approach of "profiling" nations often seems to be the main objective of strategic cultural analysis. We do not see a compelling need to replicate this well-developed body of literature and instead have devoted Part III to a series of short essays that constitute reviews of the state of the scholarship on several key countries and entities. These essays go beyond simple literature reviews, however, by highlighting key conceptual and substantive themes found in the existing literature related to each of the selected countries. The purpose of this approach is to provide a solid set of resources for students and academics who might want to expand the scope of the given body of knowledge on these countries by incorporating newly developed methodologies, as presented, for example, in Parts I and II of this volume. In this section, we have included review essays on the strategic culture literature on Russia, China, France, India, Pakistan, Nigeria, North and South Korea, and others. This section of the *Handbook of Strategic Culture* captures, rather than duplicates, the literature of existing case studies and synthesizes it for the student of strategic culture.

Part IV is devoted to showing how strategic culture insights have been and can be applied in relevant professional domains. This section opens with essays on the utility of strategic culture for scholars that explore where strategic culture fits within the constructivist paradigm and its use as a supplement to other schools of thought in international relations, including realism. For policymakers and practitioners, we have included essays contextualizing debates on tailored deterrence and intelligence analysis, among others. Several chapters on how strategic culture can contribute insights to military and intelligence operators round out this part of the volume.

Part V provides a chapter on conclusions, a summary of key principles and insights from contributions across the volume, and some contemplation on the future of the field of strategic culture, with recommendations for areas and approaches that warrant further study.

Notes

1 The author thanks Jeannie Johnson and Briana Bowen for their comments and suggestions on an early draft of this chapter.
2 "Strategic culture" is further defined in Chapter 3 of this volume by Brigitte E. Hugh.
3 Executive Summary, *Final Report of the National Commission on Terrorist Attacks Upon the United States* (9/11 Commission Report), p. 9. https://www.govinfo.gov/app/details/GPO-911REPORT/context
4 *Final Report of the National Commission on Terrorist Attacks Upon the United States* (9/11 Commission Report), p. 344.
5 *Final Report of the National Commission on Terrorist Attacks Upon the United States* (9/11 Commission Report), p. 426. Jim Wirtz explores comparisons and contrasts between the commissions charged with assessing the Japanese attack on Pearl Harbor and the 9/11 commission in "Responding to Surprise," *Annual Review of Political Science*, 9 (2006): 45–65.
6 This analogy draws on Geert Hofstede, Gert Jan Hofstede, and Michael Minkov, *Cultures and Organizations: Software of the Mind: Intercultural Cooperation and Its Importance for Survival* (New York: McGraw-Hill, 2010).
7 This is why the West in general, and the United States in particular, has such a difficult time understanding the phenomenon of suicide terrorism, and often completely misconstrues the origins and devotional intensity of such terrorism, as in the case of the 9/11 terrorists, by attributing it to socioeconomic causes and dismissing the role of religion. See, for example, Robert A. Pape, *Dying to Win: The Strategic Logic of Suicide Terrorism* (New York: Random House, 2005).
8 Associated Press, "New USGS Number Puts Japan Quake at 4th Largest," *CBS News*, 14 March 2011.
9 John P. Rafferty, "Japan Earthquake and Tsunami of 2011," *Britannica online*. https://www.britannica.com/event/Japan-earthquake-and-tsunami-of-2011, updated 4 March 2022.
10 Justin McCurry, "Fukushima Operator May Have to Dump Contaminated Water into Pacific," *The Guardian*, 10 March 2014.
11 Norimitsu Onishi and Ken Belson, "Culture of Complicity tied to Stricken Nuclear Plant," *New York Times*, 26 April 2011, https://www.nytimes.com/2011/04/27/world/asia/27collusion.html?

pagewanted=2&_r=4&src=me; and, Bloomberg, "Fukushima Nuclear Accident 'Man-Made', Not Natural Disaster," *The Sydney Morning Herald*, 5 July 2012. https://www.smh.com.au/world/fukushima-nuclear-accident–manmade-not-natural–disaster-20120705-21jrl.html

12 Kiyoshi Kurokawa, Chairman of *The National Diet of Japan Fukushima Nuclear Accident Independent Investigation Commission*, 2012, p. 9. http://large.stanford.edu/courses/2013/ph241/mori1/docs/NAIIC_report_hi_res10.pdf

13 Brad Glosserman, *Peak Japan: The End of Great Ambitions* (Washington, DC: Georgetown University Press, 2019), "Higashi Nihon Daishinsai, Or The 'Great East Japan Earthquake'," pp. 136.

14 Sergei P. Kapitza, "Lessons of Chernobyl: The Cultural Causes of the Meltdown *Foreign Affairs*, 72, no. 3 (Summer 1993): 7–11. https://www.jstor.org/stable/20045618

15 See, for example, Serhii Plokhy, *Chernobyl: The History of a Nuclear Catastrophe* (New York: Basic Books, 2018).

16 David G. Haglund, "What Good Is Strategic Culture: A Modest Defense of an Immodest Concept," in *Strategic Culture and Weapons of Mass Destruction: Culturally Based Insights into Comparative National Security Policymaking*, Johnson, Jeannie L., Kerry M. Kartchner, and Jeffrey A. Larsen, eds. (London: New York: Palgrave Macmillan, 2009), p. 15.

17 For other worthy efforts at grappling with the definitional challenges of the field, please note: Arjun Sen, "Defining Strategic Culture," *Air Power Journal*, 5, no. 4 (Winter 2010): 31–54; and, Melanie Graham, *Redefining Strategic Culture*, MA thesis, The University of Northern British Columbia, June 2014. https://www.academia.edu/38225968/REDEFINING_STRATEGIC_CULTURE_Graham_MA_Thesis_June_2014.pdf

18 Jack L. Snyder, *The Soviet Strategic Culture: Implications for Limited Nuclear Operations* (Santa Monica, CA: RAND Corporation, September 1977). https://www.rand.org/content/dam/rand/pubs/reports/2005/R2154.pdf

19 Ken Booth, "The Concept of Strategic Culture Affirmed," in *Strategic Power, USA/USSR*, Carl G. Jacobsen, ed. (New York: St. Martin's Press, 1990), p. 121.

20 Colin S. Gray, "Strategic Culture as Context," in Colin S. Gray, *Modern Strategy* (Oxford University Press, 1999), p. 131.

21 Alastair Iain Johnston, "Thinking about Strategic Culture," *International Security*, 19, no. 4 (Spring 1995): 46.

22 Robert Cassidy, *Russia in Afghanistan and Chechnya: Military Strategic Culture and the Paradoxes of Asymmetric Conflict* (Carlisle, PA:Army War College, 2003).

23 Stephen Peter Rosen, "Military Effectiveness: Why Society Matters," *International Security* 19, no. 4 (Spring 1995): 5–31.

24 John S. Duffield, *World Power Forsaken: Political Culture, International Institutions, and German Security Policy After Unification* (Stanford, CA: Stanford University Press, 1999), p. 2.

25 *Comparative Strategic Cultures Curriculum Project: Assessing Strategic Culture as a Methodological Approach to Understanding WMD Decision-Making by States and Non-State Actors*, Defense Threat Reduction Agency, Advanced Systems and Concepts Office, October 2006. https://apps.dtic.mil/sti/citations/ADA521640

26 An excellent and exhaustive survey of contrasts in cognition and reasoning across cultures and civilizations can be found in Julian Baggini, *How the World Thinks: A Global History of Philosophy* (London: Granta Publications, 2018). See also Wayne H. Brekhus, *Culture and Cognition: Patterns in the Social Cognition of Reality* (Cambridge: Policy Press, 2015).

27 Ken Booth, "Strategic Culture: Validity and Validation," *Oxford Journal on Good Governance*, 2, 1 (2006): 25–28. https://www.files.ethz.ch/isn/110308/OJGG_Vol_2_No_1.pdf. See also Ken Booth, "The Concept of Strategic Culture Affirmed," in *Strategic Power USA/USSR*, C. G. Jacobsen, ed. (London: Macmillan, 1990), pp. 121–128.

28 For a more complete discussion of this issue, see John Glenn, Darryl Howlett, and Stuart Poore, eds., *Neorealism versus Strategic Culture* (London: Ashgate, 2004).

29 Ken Booth, "Strategic Culture: Validity and Validation," *Oxford Journal on Good Governance*, 2, no. 1 (2006): 25–28. https://www.files.ethz.ch/isn/110308/OJGG_Vol_2_No_1.pdf. See also Ken Booth, "The Concept of Strategic Culture Affirmed," in *Strategic Power USA/USSR*, C.G. Jacobsen, ed. (London: Macmillan, 1990), pp. 121–128.

30 David R. Olson, "Narrative, Cognition, and Rationality," in *Routledge Handbook of Discourse Analysis*, James Paul Gee and Michael Handford, eds. (Abingdon, UK: Routledge), 2012, pp. 604–615.

31 On ethnocentric distortions in assessing intent, see James H. Lebovic, "Perception and Politics in Intelligence Assessment: U.S. Estimates of the Soviet and 'Rogue-State' Nuclear Threats," *International Studies Perspectives* 10, no. 4 (2009): 394–412. https://www.jstor.org/stable/44218612
32 The editors explore further possible professional applications of strategic culture in Chapter 35 of this volume.
33 Snyder, *The Soviet Strategic Culture*; see also Brigitte E. Hugh's Chapter 3 in this volume.
34 Mette Skak's Chapter 12 in this volume is an exception, and provides additional significant context for the origins of the term "strategic culture."
35 For an early articulation of this "convergence theory," see John Newhouse, *Cold Dawn: The Story of SALT* (New York: Holt, Rinehart and Winston, 1973), pp. 3–4.
36 This history is summarized in Michaela Dodge, "History Shows U.S. Nuclear Restraint Is A One-Way Street," *War on the Rocks*, 18 November 2020. https://warontherocks.com/2020/11/history-shows-u-s-nuclear-restraint-is-a-one-way-street/.
37 See "Annex: Soviet Strategic Nuclear Forces as Perceived by the NIEs, 1962–1975." https://nsarchive2.gwu.edu/NSAEBB/NSAEBB139/nitze10.pdf
38 "Intelligence Report of Team B," in *Foreign Relations of the United States, 1969–1976, Volume XXXV, National Security Policy, 1973–1976* (Washington, DC: U.S. Department of State, Office of the Historian). https://history.state.gov/historicaldocuments/frus1969-76v35/d171. For a photocopy of the original report, see National Security Archives, "Soviet Strategic Objectives: An Alternative View, Report of Team B," https://nsarchive2.gwu.edu/NSAEBB/NSAEBB139/nitze10.pdf.
39 See Richard Pipes, "Why the Soviet Union Thinks It Could Fight and Win a Nuclear War," *Commentary*, 64, no. 1 (July 1977): 21–34.
40 Snyder, *The Soviet Strategic Culture*, pp. v, 8–9.
41 Snyder, *The Soviet Strategic Culture*, p. 9.
42 Michael Desch, "Culture versus Structure in Post-9/11 Security Studies," *Strategic Insights*, 4, no. 10 (October 2005). http://citeseerx.ist.psu.edu/viewdoc/download?doi=10.1.1.904.8204&rep=rep1&type=pdf
43 See the discussion of how strategic culture changes in Brigitte E. Hugh's Chapter 3 in this volume.
44 Theo Farrell and Terry Terriff, "The Sources of Military Change," in *The Sources of Military Change: Culture, Politics, Technology*, Theo Farrell and Terry Terriff, eds. (Boulder: Lynne Rienner, 2002), pp. 3–10.
45 Snyder, *The Soviet Strategic Culture*, p. 10.
46 See, for example, Thomas G. Mahnken, "U.S. Strategic and Organizational Subcultures," in *Strategic Culture and Weapons of Mass Destruction: Culturally Based Insights into Comparative National Security Policymaking*, Jeannie L. Johnson, Kerry M. Kartchner, and Jeffrey A. Larsen, eds. (London: New York: Palgrave Macmillan, 2009), pp. 69–84; and, Jeannie L. Johnson, *The Marines, Counterinsurgency, and Strategic Culture* (Washington, DC: Georgetown University Press, 2018).
47 Snyder, *The Soviet Strategic Culture*, p. v.
48 Dima Adamsky presents an example of featuring this cognitive dimension of strategic culture in *The Culture of Military Innovation: The Impact of Cultural Factors on the Revolution in Military Affairs in Russia, the US, and Israel* (Stanford: Stanford University Press, 2010), pp. 15–23. On the impact of culture on perceptions of security or insecurity, see also Jutta Weldes, Mark Laffey, Hugh Gusterson, Raymond Duvall, eds., *Cultures of Insecurity: States, Communities, and the Production of Danger* (Minneapolis: University of Minnesota Press, 1999); and, Wolfgang Schivelbusch, *The Culture of Defeat: On National Trauma, Mourning, and Recovery* (New York: Picador, 2003).
49 Jack Snyder, "The Concept of Strategic Culture: Caveat Emptor," in *Strategic Power, USA/USSR*, Carl G. Jacobsen, ed. (New York: St. Martin's Press, 1990), p. 4.
50 Two important critiques of strategic culture are Christopher P. Twomey, "Lacunae in the Study of Culture in International Security," *Contemporary Security Policy*, 29, 2 (August 2008): 338–357; and Antulio J. Echevarria II, *Reconsidering the American Way of War: US Military Practice from the Revolution to Afghanistan* (Washington, DC: Georgetown University Press, 2014), which

explicitly criticizes strategic culture as assuming that nations have only one distinctive way of war, when in fact they have several.

51 Ken Booth, "The Concept of Strategic Culture Affirmed," in *Strategic Power, USA/USSR*, Carl G. Jacobsen, ed. (New York: St. Martin's Press, 1990), pp. 122–124.

52 Booth, "The Concept of Strategic Culture Affirmed," p. 122.

53 See notably, Colin S. Gray, "National Style in Strategy: The American Example," *International Security*, 6, no. 2 (Fall 1981): 21–47; and his subsequent book comparing both US and Soviet strategic cultures, *Nuclear Strategy and National Style* (New York: Madison Books 1986).

54 See, for example, Marcos Rosas Degaut Pontes, *Ideas, Beliefs, Strategic Culture, and Foreign Policy: Understanding Brazil's Geopolitical Though*t, University of Central Florida, Doctoral Dissertation, 2016, https://stars.library.ucf.edu/cgi/viewcontent.cgi?referer=https://www.google.com/&httpsredir=1&article=6105&context=etd

55 Jeffrey S., Lantis, ed. *Strategic Cultures and Security Policies in the Asia-Pacific* (London: Routledge, 2015). See also Ashley J. Tellis, Alison Szalwinski, and Michael Wills, eds., *Strategic Asia 2016–2017: Understanding Strategic Cultures in the Asia-Pacific* (Washington, DC: The National Bureau of Asian Research, 2016).

56 Heiko Biehl, Bastian Giegerich, Alexandra Jonas, eds., *Strategic Cultures in Europe: Security and Defence Policies across the Continent* (Potsdam, Germany: Springer VS, 2013).

57 It is important for those of us writing in this field to be mindful, in this regard, of the critics of strategic culture who have highlighted these shortcomings. See especially, Christopher P. Twomey, "Lacunae in the Study of Culture in International Security," *Contemporary Security Policy*, 29, no. 2 (2008), 338–357. See also Chapter 2 in Antulio J. Echavarria II, *Reconsidering the American Way of War: US Military Practice from the Revolution to Afghanistan* (Washington, DC: Georgetown University Press, 2014), pp. 32–46; and Chapter 7 in Colin S. Gray, *Strategy and Politics* (Routledge, 2017), pp. 95–110.

58 An early example of this subfield focus on EU strategic culture is Christoph O. Meyer, *Theorising European Strategic Culture: Between Convergence and the Persistence of National Diversity*, CEPS Working Document No. 204, June 2004, available at: http://aei.pitt.edu/6634/1/1126_204.pdf. For a review of this literature, see Paul Cornish and Geoffrey Edwards, "The Strategic Culture of the European Union: a Progress Report," *International Affairs*, 81, 4, *Special Issue on Britain and Europe: Continuity and Change* (July 2005): 801–820.

59 For an early discussion of this issue, see Michael C. Desch, "Culture Clash: Assessing the Importance of Ideas in Security Studies," *International Security*, 23, no. 1 (Summer 1998): 141–170. See also John Glenn, Darryl Howlett, and Stuart Poore, eds., *Neorealism versus Strategic Culture* (London: Ashgate, 2004), pp. 226–227.

2
THE CONCEPTUAL HERITAGE OF STRATEGIC CULTURE AND COLLECTIVE MENTALITY

Beatrice Heuser

This chapter gives a survey of two hugely important conceptual aids to understanding an adversary, but also our allies, and ourselves: "strategic culture" and "collective mentality." This discussion needs to be prefaced with a methodological point. Neither of these concepts pretends to be an "independent variable." Much to the contrary, both are highly dependent on many, many other variables, which themselves are in part interdependent. There is no way any of these concepts could be packaged into a neat physics or chemistry formula spelling out that "whenever A, then B." Such neat rules governing the natural sciences can rarely, if ever, predict collective human actions or interactions. Conceptual aids do not work the way Newton's theorems do.

Nevertheless, there is some predictive quality to these concepts: they identify patterns and potentials. Most importantly, they can allow us to understand the background and subtexts of policymaking, especially where these are not subconsciously accessible to us, which is usually the case for our own culture and collective mentality. This greater understanding is very important, often crucial, or even vital, to international relations.

"Strategic Culture"

"Culture" is a term that has for a long time been used to describe variations among societies, although it was not originally used in a comparative sense.[1] Beloved by anthropologists, the study of cultures brought forward the subdiscipline of comparative cultural studies, also known as comparative ethnology or *Vergleichende Kulturwissenschaften,* in the German-speaking countries (with "ethnology" thus being rendered in German as "academic knowledge about cultures"[2]).

The American political scientist Jack Snyder seems to have been the first, in a 1977 paper for the US Air Force, to have coined the term "strategic culture," which he defined as

> The sum total of ideas, conditioned emotional responses and patterns of habitual behavior that members of a *national strategic community* have acquired through instruction or imitation with each other with regard to nuclear strategy. [emphasis added][3]

DOI: 10.4324/9781003010302-3

Snyder explained whom he meant by "the strategic community:" the "strategy-making elite and those writing about strategy." In the United States, the former would include decisionmakers within government (both military and civilians), and single or collective authors writing on strategy within and outside the government; the latter referred to elsewhere as "academic strategists."[4] As Snyder was writing specifically about the Soviet Union's nuclear strategy, he was dealing with a tiny number of military officers (even Soviet defense ministers were always military officers) plus "certain members of the political leadership and the research institutes of the Academy of Sciences."[5] The Soviet "national strategic community" was thus a small elite. Snyder explicitly acknowledged this in a section of his report in which he commented on "The [Soviet] Military's Monopoly on Doctrinal Elaboration and Technical Information."[6] Even if applied to the United States, "strategic community" thus designated a small elite segment of society. It did not even extend to the entire military, which is more appropriately covered by research into "military culture." Subsequent works that used the term "strategic culture" were also focused especially on nuclear strategy, but they tended to draw a wider circle around members of this culture. They tend to talk about "national style" in warfare or war-related activities on the one hand,[7] and on the other, about "national public culture" or "national strategic culture," as the larger setting in which decisionmakers act.[8] Indeed, the term has made its way into wider use, as when a French defense official commented that American and British public references to information obtained by intelligence were very far from French "culture stratégique."[9] A common "culture stratégique" is also what the European Union, especially under the French presidency of January-June 2022, seeks in conjunction with the publication of a European "Strategic Compass."[10]

Strategic culture overlaps with another concept, introduced between the world wars by the British strategist Captain Basil Liddell Hart: a particular nation's "way of warfare."[11] Applied initially to Britain, that term was then picked up in an analogous work on American warfare by the US military historian Russell Weigley,[12] and later by numerous authors either responding to Weigley or attaching the same "way of war" label to many other nations and their warfare.

But the question remained of how to capture a larger culture, beliefs, values, and opinions held across an entire civic nation, on behalf of which governments make policy. In democracies in particular, governments are not only supposed to represent these nations' beliefs and values; they are also obliged to take them into account in policy- (and strategy-) making, even if it is not known what – if anything – the average citizen thinks about matters of defense. To cover this, a term suggested was that of "national style." In 1979, another influential American strategist, Edward Luttwak, wrote a short article focusing on the United States itself, entitled "The American Style of Warfare and the Military Balance" between the United States and the USSR.[13] Two years later, his friend and colleague, the late Colin S. Gray, published an article entitled "National Styles in Strategy: The American Example."[14] The terms "national way of war" and "national style" in warfare came to be used interchangeably, and the number of those who underscored the importance of such "national" variations grew.[15] This subsumed "strategic culture," and "military culture," but of course applied mainly to the particularities of conducting wars.

Luttwak and Gray's concept of "national style" specifically countered the underlying assumption of the prevailing theory of international relations, that of so-called, supposed "realism" (which unrealistically eclipses cultural variations, or its more recent clone, neorealism), namely (in the words of a critic) "that states are functionally undifferentiated units that seek to optimize their ... power."[16] Realism is blind to culture-specific beliefs, interpretations of historical experiences, traditions, influences of ideologies including religions, assumptions specific to a particular group/tribe/ethnicity/nation, and a host of other factors that make for cultural differences between them. Realism, in sum, is blind to different mindsets or collective mentalities.[17]

National Public Culture or Mentalité/Mentality

An alternative term to what Jeannie Johnson calls "national public culture" is "collective mentality," a term owed to the French. As Snyder, Luttwak, and Gray developed their ideas of strategic culture and national style, the Annales School of historiography had long flourished in France. This school included the French medievalists Marc Bloch, a hero of the French World War II *Résistance*, Jacques Le Goff, Georges Duby, and Emmanuel Le Roy Ladurie; specialists on the French Revolution like Lucien Febvre and Michel Vovelle; and historians exploring the *longue durée* (a long time span) of phenomena, such as Philippe Arriès and Fernand Braudel.[18] The *Annalistes* had identified the need to explain phenomena such as the crusades or the French Revolution against the background of how entire societies thought at the time, rather than reconstructing only the ideas of key individual actors (which of course would be influenced by more general collectively held views). Such clusters of views or collective mentalities, they argued, conditioned the judgments and decisions of individuals. Even if individuals argued and acted against *mentalités collectives*, these would be the framework of thinking within which or against which the individuals had to operate. These collective mindsets were the landscape – the topography of thought – in which the individuals lived and moved.

In the early 1970s, Jean-Baptiste Duroselle, together with Pierre Renouvin, one of the parents of modern French diplomatic history, engaged with the historiography of the Annales School to use it for an analysis of foreign policymaking. He produced a remarkable work of classification.[19] He defined *mentalité collective* as relating to relatively stable attitudes of mind or mindsets as well as images and stereotypes shared by a group containing value judgments or echoing the value judgments of others. He distinguished it from doctrine (that which is taught by an authoritative institution like a church or the military), from a coherent philosophy, and from an ideology or religion, even though all of these might leave their imprint on such a mentality. It could include a firm belief in the legitimacy and utility of an institution (e.g., a monarchy or parliament) or the desirability of a policy or economic order (a colonial empire or anti-colonialism, pacifism or bellicosity), be favorable to reform or the preservation of the status quo or even the return to an earlier order, and include views on the role of institutions (especially those of the state) in certain contexts.[20]

In modern times, collective mentalities or mindsets are created and passed on to new generations by views heard or expressed in the family or by peers, taught at school, in government or other propaganda, and in the media. Such a mentality might exist over a *longue durée*; it could also change, adapt gradually, or be catalyzed suddenly by an event, be "fabricated" – coaxed into being by a government through incessant propaganda or brainwashing – or a mixture of these.[21]

Applying all of this to the analysis of foreign policy, Duroselle emphasized that politicians at home and abroad had to take collective mentalities into account – in their own countries, in allied countries, and in countries under adversarial regimes. At the same time, he demonstrated how politicians could contribute to shaping them by using them or condemning them.[22] Historical examples illustrate that collective mentality may or may not be a decisive factor, and it may be the collective views of a minority that push a decisionmaker one way or the other, or that another may have a hold on office sufficiently safe and personal convictions sufficiently strong to ignore majority opinion. Duroselle allowed for convictions by groups within a society to be held with varying degrees of fervor. When it comes to going to war, a society with universal conscription cannot but develop strong feelings, either for or against it.[23]

Duroselle listed the sources an historian would draw upon to reconstruct the collective mentality or mentalities of societies in more recent history, noting that opinion polls had only been invented in the late 1930s. Other sources include parliamentary debates, the press, and other publications of all sorts, and more recently, debates on radio or television as indicators of collective views.[24] For contemporary or very recent subjects of research it is important to add social media and interviews. The same sources would nowadays be used by political scientists, who would call them "data," if they sought to understand a society and its subcultures, including attitudes toward any subject. The methodology would be much the same; where the "data" are digitalized, they might lend themselves to quantitative evaluation on the condition that the "data" always use the same words for what is analyzed. Note, of course, that governments, individual politicians, and embassies abroad have used this sort of evaluation of public opinion for a long time, accompanied by the gifted politician's intuitive feelings about what their electorate would or would not go along with.

Redefinitions of Strategic Culture

Despite attempts to build on and import the Annales School terminology of *mentalités* into the Anglosphere,[25] this approach never made it across the great linguistic and subject divide. Yet both the Annales School and many sociologists took inspiration from cultural anthropologists – the most frequently quoted in the Anglophone world is Clifford Geertz, though his work mainly postdates that of the Annales School – but never read one another's work. Thus, Anglophone literature came up with a parallel solution: after giving a nod to Snyder's narrow definition of strategic culture, it simply redefined and thus reappropriated the term. The notable definitions that offer helpful insights into its nature and effects include Alastair Iain Johnston's, which noted that "strategic culture" includes "basic assumptions about ... the strategic environment" (Is it hostile or benign?) and "the role of war in human affairs" (Whether it is inevitable or an aberration?), "about the nature of the adversary and the threat it poses," whether relations are such that if one side wins the other loses (a "zero-sum" outcome), or whether one can imagine outcomes in which both sides stand to lose or gain something ("variable-sum"), and "about the efficacy of the use of force."[26]

Eleven years later, Jeannie Johnson defined strategic culture as "those inherited conceptions and shared beliefs that shape a nation's collective identity, the values that color how a country evaluates its interests, and the norms that influence a state's understanding of the means by which it can best realize its destiny in a competitive international system."[27] Also regarding its effects, Johnston noted that culture-specific assumptions will condition not only the way the world, its workings, an adversary, and the use of force are seen, but it will also condition preferences within the decisionmaking process.[28]

Ashley Tellis, building on Ann Swindler, pointed out that "strategic culture shapes outcomes both by 'providing the ultimate values toward which action is orientated' and by 'shaping a repertoire or tool kit of habits, skills, and styles from which people construct strategies of action.'"[29] He also, very importantly, stressed that understanding a society's culture where it pertains to things military does not have a predictive capability to the extent that the laws of physics have, but that anything to do with the interaction of polities is subject to the interplay of multiple variables – something that already Clausewitz stressed (and which drew him into a fruitless debate with his colleagues at the Prussian War College from which he withdrew in frustration).[30]

Subcultures or Sub-(Mind)sets

This chapter's use of the term "subculture" requires further explanation. The *Annalistes* and indeed Duroselle noted that a collective mentality does not necessarily map perfectly onto one nation. Aspects of collective mentality could straddle class or group boundaries within one nation and could divide individual parties.[31] Jack Goldstone, Professor of Public Policy at the University of Wisconsin – Madison, was thus rediscovering the wheel 30 years later when he noted, using the term "culture" rather than mentality, that

> Any society may be thought of as having not a single culture, but a variety of cultural "themes," various interpretations and heterodoxies of the core culture, plus whatever borrowings and incursions have grown up around it.[32]

"Cultural themes," however, seem less clear as a term than "subcultures" or "sub-(mind)sets," subsets of views that overlap but mainly exist alongside each other, on a political spectrum on which values and ideas form clusters. These clusters are the most appropriate designators of the different subcultures. In the political discourse, sets of ideas are often knitted together into narratives, which again are often more closely related to a subculture than to a nationwide culture or collective mentality; generally, it is the government that provides a dominant narrative, but its opposition parties and others can provide alternative narratives. As Jeannie Johnson noted,

> The strategic culture of any nation state comprises competing narratives. Strategic culture, then, is best understood as a bundle of narratives, some held at the national level and some cultivated within particular organizations. Which of these is dominant at any point in time and for any particular issue is driven by a number of impacting factors including the nation's most recent formative events, the sway of particularly charismatic political figures, popular perceptions of the nation's future security, and a national sense of status and role on the world stage.[33]

There is *prima facie* no limit to the number of subcultures, but it is unhelpful for foreign- or military-policy analytical purposes to multiply them infinitely. As noted, they do not always map entirely onto political parties, but until recent years, political parties (and in the case of big popular parties, their main two or three wings) used to be good indicators as to where the clusters can be found.

Many diplomatic historians – or historians concerned with the history of war and peace – whose expertise like Duroselle's lay in the twentieth century, used the same methodology.

Key works like Volker Berghahn's or Donald Cameron Watt's that explained the origins of the First and Second World Wars, respectively, thus went to great lengths to explore the collective mentalities of nations and subgroups within these, as well as among the ruling elites they were studying.[34] Indeed, one might say that integrating this dimension among others has come to be entirely standard in the historiography of interstate relations.

Using the same approach of bringing in collective mentalities as an explanation, this chapter's author explained why with the same exposure to potential Soviet nuclear attacks by aircraft or missiles, the United Kingdom, France, and West Germany had such significantly different preferences for nuclear strategies during the Cold War.[35] Indeed, along with geography (which of course is a crucial factor shaping the historical experience that in turn forms collective mentalities), differences in collective mentalities provided the most convincing answers. That work specifically did not focus on any military culture, the culture of merely the tiny elite involved in drawing up the strategies, nor the decisionmaking on the procurement and deployment of particular types of nuclear weapons over others. It embedded these decisions in the larger context of what, in each country, were widely held or at least acceptable views, and showed which policies and strategies clashed with those of vocal subcultures opposing them. Looking exclusively at published materials from books and articles to parliamentary debates and the press, this author's work identified divergent views among populations held about their own nation and its role in the international system, about adversaries, about the mechanisms of hostile or alliance relations between states, about the war itself, about the price of peace, and the price of freedom, all of which came together to create the cultural contexts in which decisionmakers, regardless of their own beliefs, had to operate. Not only did this approach go a long way to explain the different preferences for nuclear strategy but it also served to construct a picture of the collective mentalities in each country beyond the mere subject of defense, with a great amount of convergence in France, and much less convergence within the respective bodies politic of the United Kingdom and West Germany.[36]

Collective Memory and Its Myths and Narratives

One subarea of research into collective mentality, or strategic culture as per the definitions of Johnson, should be stressed, as it has brought forth a particularly rich harvest of publications in its overlap with the burgeoning field of research on nationalism and identity politics. The focus of this research is less on how collective mentality has confined, influenced, or even determined policymaking in the particular area of interest – military policy. Instead, its focus is on how collective mentalities have been created, especially through the construction of "collective memory" of significant past events, and within that, "myths" in the sense of a "primitive explanation of the world" (Friedrich von Schelling) or a "narrative" (Ernst Cassirer) – often tied to actual past events, but distorting them considerably – that seeks to make sense of the world and at the same time give guidance for future behavior.[37] Spread by schools, politicians, and the media, such guiding myths or narratives and the actionable message-prescription they carry may be common to several nations, or particular to just one.

One example is the myth of uniqueness, of difference, or of separateness in some form. Paradoxically, this is found in several countries. One is Sweden with its devotion to neutrality that until Russia's invasion of Ukraine in 2022 made it impossible for Sweden to join NATO despite strong interest in this option among the Swedish strategic community and a strategic culture that would strongly favor it, and despite past close but secret defense cooperation with the United States.[38] Similarly, there is in Norway a guiding myth of independence, translating itself into staunch public opposition to EU membership, again despite support for such membership among the Norwegian strategic community, and despite the de facto arrangements deeply integrating Norway into EU norms and regulations.[39]

A third is Britain with its myth of separateness from Europe that has some roots in balance-of-power politics and anti-Catholic sentiments dating from the sixteenth to the eighteenth centuries, and which crystallized properly in the imperial Victorian Age. That was an exceptional period in over 2000 years of British history, as more than ever before and after, the United Kingdom was politically and militarily disengaged from continental Europe in the period of Queen Victoria's reign (1837–1901).[40] It lived on among those opposed to British integration into the "Catholic" European Communities in the 1950s and then again in the 1970s. Then in reaction to austerity following the 2008 financial crisis, the allure of returning to a glorious past – when Britain turned its back on Europe to face the world with commerce (and an Empire encompassing a large part of the globe) – was deftly revived by the "Brexiteers," who succeeded in turning a slight majority of British voters to choose to exit the EU in the referendum of 2016. This myth of "global Britain" is complemented by that of the special British-US relationship, one not largely shared in America outside of the strategic community (and even there, not by all), but constituting a firm part of the British collective self-perception since its creation in the Second World War. Its notion of the special relationship grossly overestimates British influence on policymaking in Washington and overlooks many other countries' claims – from Ireland to Israel, from Canada to Germany – to have an equally special relationship with America.[41]

Another myth paradoxically found in several countries is that of being the ultimate, self-sacrificing defender of Christendom against the onslaught of Islam. Nations with this self-perception include the Portuguese, the Spanish, the French, and the Italians (against the Arabs), the Hungarians, Croatians, Serbs, Bulgarians, Austrians, and even Poles, and of course the Greeks (against the Turks). In each case, there was some historical battle that can be pointed to, but the carefully fostered memory of it is hardly helpful for the integration of Muslim immigrants in the late twentieth and twenty-first centuries and exacerbated the refugee crises of 2015–2016.

There are also many nations each seeing themselves as successor to the Ancient Israelites as God's Chosen People (including the French and the Americans), or as successors to Ancient Rome and its Empire (the Byzantines/Greeks, Germans, and Austrians with their claim to be the heirs of the Holy Roman Empire, the Russians with Moscow as Third Rome, and indeed the Ottomans themselves, despite having decapitated Eastern Christianity with their conquest of Constantinople in 1453). All of these self-perceptions supposedly bestowed upon the respective nations a leadership role, in Europe, in the larger Euro-Mediterranean area, or even the wider world. Similar myths of exceptional collective identity exist in other parts of the world, of course – from Iran to China and beyond.

Some myths are guiding stars for the security policymaking not just of one, but of several countries. Examples of this include the myth – this time "myth" in the sense of untruth – fabricated by Charles de Gaulle that the United States and the USSR "divided" the world into spheres of influence at their meeting in Yalta in early 1945, held up as a caveat against superpower collusion at the expense of the smaller powers.[42] "Yalta" is evoked whenever somebody cautions against letting America and Russia between them decide the fate of Europe, at the expense of the Europeans. The evocation of the shame of Munich in 1938 serves similar political aims: but as a *guiding* myth, it warns against making deals with dictators, but also against selling smaller countries down the river.[43] Another example of a guiding myth, in the sense of guidance for future behavior, concerns the crucial importance of Franco-German cooperation if war is to be banished from Europe, an interpretation of history widely shared throughout Europe and not just anchored, since the early 1950s, in the collective mentalities on both sides of the Rhine.[44] For the time being, no government in either country would risk a serious breakdown of this relationship, and no third country seriously opposes it even if there are occasional grumbles about its exclusivity: the strategic communities of EU Member States understand that peace in Europe depends on this.

Such guiding myths have also included particular interpretations of past events that had a major impact on a country's history, including a selection of battles that have reached epic importance in collective memory, much massaged by political leaders. Such battles may have been won by the side commemorating them such as the Battle of Kahlenberg in 1683 that ended the last Turkish siege of Vienna, or even lost, such as the Battles of Kosovo Polje in 1389 or Mohacs in 1526 against the relentlessly expansionist Turks, commemorated by the Serbs and the Hungarians, respectively; the latter two were subsequently portrayed as metaphysical, moral triumphs.[45] These mythical battles may have been decisive or perhaps not, as the Battle of Leipzig of 1813, which did not prevent the comeback of Napoleon two years later, but would be commemorated by the narrative of German nationalism as the event where German national consciousness was finally born.

That interest in such battle myths is not just academic, as can be seen in a few recent examples. The commemorations by Britain of the hundredth anniversary of the Battle of the Somme and by France and Germany of the Battle of Verdun were remarkably different. The former practically airbrushed the French (who also fought in large numbers on the Somme) and even the German enemy out of the picture and seemed to be a celebration of British stoicism and forbearance, coupled of course with mourning for the dead.[46] The latter continued the tradition established in the 1960s of casting Verdun as the negative founding myth of the need for Franco-German reconciliation.[47] While the British commemoration of the Battle of the Somme would not have given a precise prediction of the outcome of the Brexit referendum in the same year, it certainly showed that in Britain, the lesson learned from the First World War was not that membership of a European inter- or supra-state (con)federation that replaced war with negotiations was vital to the avoidance of war. The glaring absence of the guiding foundation myth of the EU – that it alone guarantees the peaceful coexistence of its members – characterizes British perception of European integration, which has since the 1960s always been about commercial interests alone.

For a final example of the enduring importance of how mythogenic battles have been and are being commemorated, a vivid illustration is offered by the propaganda surrounding the 950th anniversary of the Battle of Manzikert, won in 1071 by Turks against the combined forces of the Byzantine Empire and the Armenians, after which the first Turkish State was established in Eastern Anatolia. It has been taken up by the Turkish movie industry (with *Fetih 1453* – English: *The Conquest 1453*, directed by Faruk Aksoy and released in 2012) but also specifically by the Turkish government, fitting well into the neo-imperialism with pan-Islamic leadership overtones of President Recep Tayyip Erdoğan.[48]

The Turkish government spokesman posted a neat little propaganda film on Twitter with the comment:

> For us, the Golden Apple is a big and powerful Turkey. It is the blessed march of our nation who wrote epics from Malazgirt (Mazikert) to the 15th of July [what is presumably meant is the purging of the Turkish army in 2016]. The Golden Apple is the great plane tree under which many oppressed people find cool shade. It is what all humanity, from Gibraltar to the Hejaz, from the Balkans to Asia is longing for.[49]

In other words, the instrumentalization of battles as foundation myths for collective mentalities and national identity, and as directives for policies in the present and future, is still alive and kicking.

Critiques

There have been critiques of "strategic culture" and of analyses of wider political cultures and collective mentalities. Most are unreasonably founded on the entirely misguided assumption that the complexities of social interrelations can be reduced to monocausal "testable" explanations as though one is dealing with machines. One such comprehensive attempt to press mentality explanations through the sausage machine of monocausal theory was made by Alastair Iain Johnston.[50] The criticism he levels against an analytical approach, namely that it does not lend itself to any *ceteris paribus* explanations, is absurd. Take this analogy. In a family with several children, the youngest, a teenager, comes home later than the curfew of 10:00 p.m. that is allowed by the family rules. He remains unpunished although the parents grumble, while ten years earlier, a major family fracas erupted around a similarly late return of the oldest daughter, resulting in her running away from home for a week. Why did the parents react so differently in the latter case? Was it (a) because social mores have evolved and 10:00 p.m. is no longer regarded as a reasonable curfew for teenagers, or (b) because double standards are applied by the parents to their daughters and sons, or (c) because the parents had "learnt" from the earlier event that the effects of their attitude had been worse than the offense, so they did not want to repeat their mistake, or (d) they had grown older and simply did not have the energy for a big fight,[51] or (e) it was a particular one-off situation where the younger son relied on somebody else for safe transport, with no alternative way of getting home earlier? The most likely answer is probably: all or most of the above.

Similarly, it is pointless to try to test the validity of the claim that collective "culture," "mindset," or "mentality" has an influence on defense policy or strategy-making by attempting to construct a case where no other influences (threat perceptions, reports on the military build-up of adversaries, pressure from the defense industry, pressure from members of parliament with defense industries in their constituencies, (un)employment figures, alarming reports in the media, and so forth) can be found. Reality is complex, and theories of international relations are at best useful pointers to the many interrelating influences, the many variables that influence decisionmaking and in a wider sense, interstate relations themselves, many of which in turn are interdependent. Theories of international relations are misleading at best, and dangerous at worst when they pretend to offer monocausal, fully falsifiable explanations.

Quite a different criticism of Snyder's original employment of "strategic culture" (as of later ways of describing *mentalité* or national culture/style) is very valid, however. Snyder contrasted "generic rational man" with the *homo sovieticus* to make the point that the logic followed by Soviet nuclear strategy deviated from that of US strategists. The flaw lies in assuming that American strategists were "prudent 'generic men,'" while only Soviet leaders were "not culture-free, preconception-free." Snyder was right in admonishing game theorists not to assume that all nuclear strategists the world over would think through a problem of strategy in the same way.[52] This does not, however, mean that given different fundamental assumptions to start with, views held by other strategic communities are irrational or illogical. The notion that "we" are "rational" while others have "culture" was rightly slammed down by Gray when he pointed out that every nation and every strategic community has its own preconceptions, none is simply "irrational."[53]

Conclusion

Let us conclude, then, with some proposals for the instrumentalization of an analysis of strategic culture and collective mentality for the purposes of military policy and defense planning. To recapitulate, "strategic culture" in Snyder's original conception applies to strategic communities (leading military officers and authors, civil servants, the key political decisionmakers, and to a greater – in the United States – and lesser – in the United Kingdom – extent defense academics), and "collective mentality" applies to entire nations or subcultures within them (e.g., pacifist movements, independence movements, groups pushing for casting off trade restrictions, etc.). While many scholars tend to blend the two under the strategic culture label, collective mentality is the larger whole encompassing strategic communities with their particular "strategic cultures," but also many other subcultures. Each of these concepts has its own analytical uses.

Tracking both is important both for gauging what one can expect from one's own nation (in view of its collective mentality), one's allies, or one's adversaries and their respective governments. They are malleable by politicians and by government propaganda, but only to some extent, and only over time. They can be transformed or catalyzed by exogenous shocks (as Britain's and France's appeasement gave way to determination to resist Hitler's expansionism at the cost of a world war, by the war scare of early 1939), or they can change slowly through (re-)education. The bellicose Swedes of the seventeenth and eighteenth centuries were transformed in the nineteenth century into a nation with pacifism writ on its banner; the Germans and the Japanese were equally transformed by intense reeducation from the 1950s onward. But the world also saw the war-weary Germans of the 1920s instilled with revanchism and bellicosity under the influence of national socialism that made them ready to fight another devastating war by 1939. The commemorations of battles in that period would have given observers a pretty good measure of the rise of vindictive nationalism,[54] and taking such observations more seriously might have given more influence to the critics of Appeasement in Whitehall and in the Quai d'Orsay.

As such analysis and monitoring of collective mentalities is based entirely on open sources – press and other media – and as even studies of strategic communities can to a large extent rely on openly available material, this is an area where defense practitioners can and should compare notes with experts on the region (ethnographers, regional specialists, historians) at regular intervals. When the two German states were about to be united into one at the end of the Cold War, Prime Minister Margaret Thatcher invited a number of historians and specialists on Germany to Chequers for a conference to seek their opinion on whether such a united Germany might become a threat to Britain and the rest of Europe. This conference, when it was leaked to the press, was much ridiculed or even described as scandalous at the time. Wrongly so: the majority of academics assembled told her that such a development was unlikely, but that it would be crucial to embed such a united Germany firmly not only in NATO but also in the European institutions.[55] Thirty years later, there is no sign that they were mistaken. Regular conferences of this sort, mainly at a working level and with less press attention, would make the best use of the enormous, varied, and rich potential of scholars in all parts of academia and would hedge against intragovernmental groupthink.

Notes

1 Since Immanuel Kant and the late eighteenth century, the term "culture" was mostly used to designate that which is made by man as opposed to nature.
2 For the confusion of the meanings of the words "art" and "science" and the mistranslation of "Wissenschaft," see Beatrice Heuser, "Theory and Practice, Art of Science of Warfare: An Etymological Note," in Daniel Marston and Tamara Leahy, eds., *War, History and Strategy; Essays in Honour of Professor Robert O'Neill* (Canberra: ANU Press, 2016), pp. 179–196.
3 Jack L. Snyder, *The Soviet Strategic Culture: Implications for Limited Nuclear Options* RAND Research Paper R-2154-AF (September 1977), p. 8.
4 Snyder, *The Soviet Strategic Culture,* p. 9.
5 Snyder, *The Soviet Strategic Culture*, p. 10.
6 Snyder, *The Soviet Strategic Culture*, p. 30f.
7 Colin S. Gray, *Nuclear Strategy and National Style* (New York: The Hudson Institute, 1981).
8 Jeannie L. Johnson, Kerry M. Kartchner, and Jeffrey A. Larsen, eds., *Strategic Culture and Weapons of Mass Destruction: Culturally Based Insights into Comparative National Security Policymaking* (London: Palgrave Macmillan, 2009); Jeannie Johnson and Marilyn Maines, "The Cultural Topography Analytic Framework," in Jeannie Johnson, Kerry Kartchner and Marilyn Maines, eds., *Crossing Nuclear Thresholds: Leveraging Sociocultural Insights into Nuclear Decisionmaking* (London: Palgrave Macmillan, 2018), p. 32; Jeannie Johnson, "Fit for Future Conflict? American Strategic Culture in the Context of Great Power Competition," *Journal of Advanced Military Studies* Vol. 11 No. 1 (2020), p. 186.
9 Cécile Decourtieux et al., "Guerre en Ukraine: un déni européen," *Le Monde* (20–21 March 2022).
10 Une boussole stratégique pour renforcer la sécurité et la défense de l'UE au cours de la prochaine décennie - Présidence française du Conseil de l'Union européenne 2022 (europa.eu).
11 B.H. Liddell Hart, *The British Way of Warfare* (London: Faber and Faber, 1932).
12 Russell F. Weigley, *The American Way of War: A History of United States Military Strategy and Policy* (London: Macmillan and Bloomington, IN: Indiana University Press, 1973).
13 Edward Luttwak, "The American Style of Warfare and the Military Balance," *Survival* Vol. 21 No. 1 (1979), pp. 57–60.
14 Colin Gray, "National Styles in Strategy: The American Example," *International Security* Vol. 6 No. 2 (Fall 1981), pp. 21–47, followed by Colin Gray, *Nuclear Strategy and National Style* (Lanham, MD: Hamilton Press, 1986).
15 See, for example, Joseph Nye and Sean M. Lynn-Jones, "International Security Studies: A Report of a Conference on the State of the Field," *International Security* Vol. 12 No. 4 (1988), pp. 5–27.
16 Alastair Iain Johnston, "Thinking about Strategic Culture," *International Security* Vol. 19 No. 4 (1995), p. 35.
17 David J. Elkins, "A Cause in Search of Its Effect, or What Does Political Culture Explain?" *Comparative Politics* Vol. 11 No. 2 (1979), p. 128.

18 For a short introduction, see Michael Gismondi, "'The Gift of Theory': A Critique of the *histoire des mentalités*," in *Social History* Vol. 10 No. 2 (May 1985), pp. 211–230.
19 Jean-Baptiste Duroselle, "Opinion, attitude, mentalité, mythe, idéologie: essai de clarification," *Relations Internationales* No. 2 (1974), pp. 3–23.
20 Duroselle, "Opinion, attitude, mentalité."
21 Duroselle, "Opinion, attitude, mentalité," pp. 4–6, 8.
22 Duroselle, "Opinion, attitude, mentalité," pp. 15, 18–20.
23 Duroselle, "Opinion, attitude, mentalité," p. 16f.
24 Duroselle, "Opinion, attitude, mentalité," p. 16.
25 Beatrice Heuser, *Nuclear Mentalities? Strategies and Beliefs in Britain, France and the FRG* (London: Macmillan, and New York: St Martin's Press, 1998).
26 Alastair Iain Johnston, "Thinking about Strategic Culture," *International Security* Vol. 19 No. 4 (1995), p. 46.
27 Jeannie L. Johnson, "Conclusion: Toward a Standard Methodological Approach," in Jeannie L. Johnson, Kerry M. Kartchner, and Jeffrey A. Larsen, eds., *Strategic Culture and Weapons of Mass Destruction: Culturally Based Insights into Comparative National Security Policymaking* (New York: Palgrave Macmillan, 2009), pp. 243–257.
28 Johnston, "Thinking about Strategic Culture," p. 34.
29 Ashley Tellis, "Overview," in *Strategic Asia 2016–17* (The National Bureau of Asia Research, 2016), p. 7.
30 A sentiment that many serious students of strategy can relate to. On Clausewitz, see Beatrice Heuser, "What Clausewitz Read," in *Strategy before Clausewitz: Linking Warfare and Statecraft, 1400–1830* (Abingdon: Routledge, 2017), pp. 193–196.
31 Duroselle, "Opinion, attitude, mentalité," p. 7f. The criticism that the concept of *mentalités collectives* does not allow for a subset of view is thus entirely unjustified. It is nevertheless made repeatedly, see Gismondi, "'The gift of theory'"; and see Florence Hulak, "En avons-nuous fini avc l'histoire des mentalités?," *Philosorbonne* No. 2 (2007–2008), pp. 89–108.
32 Jack Goldstone, quoted in Emily Goldman, "Cultural Foundations of Military Diffusion," *Review of International Studies* Vol. 32 No. 1 (Jan. 2006), p. 75.
33 Johnson and Maines, "Cultural Topography Analytic Framework," p. 32.
34 Volker Berghahn, *Imperial Germany, 1871–1918: Economy, Society, Culture, and Politics* (2nd ed. New York: Berghahn Books, 2005); Donald Cameron Watt, *How War Came* (London: Heinemann, 1989).
35 Beatrice Heuser, *NATO, Britain, France and the FRG: Nuclear Strategies and Forces for Europe, 1949–2000* (London: Macmillan, and New York: St Martin's Press, 1997).
36 Heuser, *Nuclear Mentalities?*
37 Cyril Buffet and Beatrice Heuser, "Introduction," in Cyril Buffet and Beatrice Heuser, eds., *Haunted by History* (Oxford: Berghahn, 1998), pp. vii–x.
38 Ann-Sofie Dahl, "The Myth of Swedish Neutrality," in Buffet and Heuser, eds., *Haunted by History,* pp. 28–40; Robert Dalsjö, *Life-Line Lost: The Rise and Fall of 'Neutral' Sweden's Secret Reserve Option of Wartime Help from the West* (Stockholm: Sanders Press, 2006).
39 Kjell A. Eliassen and Nick Sitter, "Ever Closer Cooperation? The Limits of the 'Norwegian Method'of European Integration," *Scandinavian Political Studies* Vol. 26 No. 2 (2003), pp. 125–144; Håkon Lunde Saxi, *Norwegian and Danish Defence Policy. A Comparative Study of the Post-Cold War Era*, in the series *Norwegian Security Studies* No 1 (Oslo: Norwegian Institute for Defence Studies, 2010); Håkon Lunde Saxi, "So Similar, Yet So Different: Explaining Divergence in Nordic Defence Policies," in Robin Allers, Carlo Marsala and Rolf Tamnes, eds., *Common or Divided Security: German and Norwegian Perspectives on Euro-Atlantic Security* (Frankfurt am Main: Peter Lang, 2014), pp. 257–277.
40 Antonio Varsori, "Is Britain Part of Europe? The Myth of British 'Differentness,'" in Buffet and Heuser, eds., *Haunted by History,* pp. 135–156; Beatrice Heuser, *Brexit in History: Sovereignty vs. European Union* (London: Hurst, 2019).
41 John Baylis, "The 'Special Relationship': A Diverting British Myth," in Buffet and Heuser, eds., *Haunted by History,* pp. 117–134.
42 Reiner Marcowitz, "Yalta, the Division of the World," in Buffet and Heuser, eds., *Haunted by History,* pp. 65–79.

43 David Chuter, "Munich, or the Blood of Others," in Buffet and Heuser, eds., *Haunted by History,* pp. 65–87.
44 Cyril Buffet and Beatrice Heuser, "Marianne and Michel: The Franco-German Couple," in Buffet and Heuser, eds., *Haunted by History,* pp. 175–208.
45 For these battles and their impact on collective mentalities, see Beatrice Heuser and Athena Leoussi, eds., *Famous Battles and How They Shaped the Modern World,* 2 vols (Barnsley, Yorkshire: Pen and Sword, 2018).
46 Mungo Melvin, "Busting the Myths of the Somme, 1916," in Heuser and Leoussi, *Famous Battles* Vol. 2, pp. 115–132.
47 Isabelle Davion, "Verdun," in Isabelle Davion and Beatrice Heuser, eds., *Batailles: Une histoire des grands mythes nationaux* (Paris: Belin, 2020), pp. 265–282.
48 See https://www.dailysabah.com/turkey/battle-of-manzikert-a-celebration-for-the-landmark-victory-in-turkey/news (25 Aug. 2020), accessed on 28 Feb. 2021.
49 "Bizim için #Kızılelma büyük ve güçlü Türkiye'dir. Malazgirt'ten 15 Temmuz'a destanlar yazan milletimizin kutlu yürüyüşüdür. Kızılelma, gölgesinde nice mazlumun serinlediği ulu çınardır. Cebeli Tarık'tan Hicaz'a Balkanlardan Asya'ya tüm insanlığın hasretle beklediğidir." https://twitter.com/fahrettinaltun/status/1297971223591358465. There are variations on this little film on the Turkish Armed Forces' website, for example: https://www.youtube.com/watch?v=8-YoA4ZalVQ
50 Johnston, "Thinking about Strategic Culture," pp. 32–64.
51 Incidentally an observable pattern in government policymaking: governments *may* be reluctant to give in to opposition against new policies they impose, but if that opposition results in major upheavals, strikes, and demonstrations, they will feel less inclined to resort to similar measures in the future. Note the *may.*
52 Snyder, *The Soviet Strategic Culture,* p. 4.
53 Gray, "National Styles in Strategy, The American Example."
54 Sandra Petermann, *Rituale machen Räume: Zum kollektiven Gedenken der Schlacht von Verdun und der Landung in der Normandie* (Bielefeld: transcript Verlag, 2015.
55 For the transcript of this conference, see https://c59574e9047e61130f13-3f71d0fe2b653c4f00f32175760e96e7.ssl.cf1.rackcdn.com/9D2F1EFFB60C4F10879687C160B4B529.pdf accessed on 28 Feb. 2021.

Selected Bibliography

Berghahn, Volker, *Imperial Germany, 1871–1918: Economy, Society, Culture, and Politics* (2nd Edition), New York: Berghahn Books, 2005.
Buffet, Cyril and Beatrice Heuser, eds. *Haunted by History*, Oxford: Berghahn Books, 1998.
Dalsjö, Robert, *Life-Line Lost: The Rise and Fall of 'Neutral' Sweden's Secret Reserve Option of Wartime Help from the West*, Stockholm: Sanders Press, 2006.
Davion, Isabelle, and Beatrice Heuser, eds. *Batailles: Une Histoire des Grands Mythes Nationaux*, Paris: Belin, 2020, 265–282.
Duroselle, Jean-Baptiste, "Opinion, Attitude, Mentalité, Mythe, Idéologie: Essai de Clarification," *Relations Internationales* No. 2 (1974), pp. 3–23.
Elkins, David J., "A Cause in Search of Its Effect, or What Does Political Culture Explain?" *Comparative Politics* Vol. 11 No. 2 (1979), pp. 127–145.
Gismondi, Michael, "'The Gift of Theory': A Critique of the *Histoire des Mentalités*," in *Social History* Vol. 10 No. 2 (May 1985), pp. 211–230.
Goldman, Emily, "Cultural Foundations of Military Diffusion," *Review of International Studies* Vol. 32 No. 1 (Jan. 2006), pp. 69–91.
Gray, Colin S., *Nuclear Strategy and National Style*, New York: The Hudson Institute, 1981.
Gray, Colin S., "National Styles in Strategy: The American Example," *International Security* Vol. 6 No. 2 (Fall 1981), pp. 21–47.
Heuser, Beatrice and Athena Leoussi, eds., *Famous Battles and How They Shaped the Modern World*, 2 vols, Barnsley, Yorkshire: Pen & Sword, 2018.

Heuser, Beatrice, "Theory and Practice, Art of Science of Warfare: An Etymological Note," in Daniel Marston and Tamara Leahy, eds. *War, History and Strategy; Essays in Honour of Professor Robert O'Neill*, Canberra: ANU Press, 2016, pp. 179–196.
Heuser, Beatrice, *Brexit in History: Sovereignty vs. European Union*, London: Hurst, 2019.
Heuser, Beatrice, *NATO, Britain, France and the FRG: Nuclear Strategies and Forces for Europe, 1949–2000*, London: Macmillan, and New York: St Martin's Press, 1997.
Heuser, Beatrice, *Nuclear Mentalities? Strategies and Beliefs in Britain, France and the FRG*, London: Macmillan, and New York: St Martin's Press, 1998.
Hulak, Florence, "En Avons-nuous Fini avc l'Histoire des Mentalités?", *Philosorbonne* No. 2 (2007–2008), pp. 89–108.
Johnson, Jeannie L., and Marilyn J. Maines, "The Cultural Topography Analytic Framework," in Jeannie L. Johnson, Kerry M. Kartchner and Marilyn J. Maines, eds. *Crossing Nuclear Thresholds: Leveraging Sociocultural Insights into Nuclear Decisionmaking*, London: Palgrave Macmillan, 2018.
Johnson, Jeannie L., Kerry M. Kartchner, and Jeffrey A. Larsen, eds. *Strategic Culture and Weapons of Mass Destruction: Culturally Based Insights into Comparative National Security Policymaking*, London: Palgrave Macmillan, 2009.
Johnson, Jeannie L., "Fit for Future Conflict? American Strategic Culture in the Context of Great Power Competition," *Journal of Advanced Military Studies* Vol. 11 No. 1 (2020), pp. 185–208.
Johnson, Jeannie L., "Strategic Culture: Refining the Theoretical Construct," Sciences Applications International Corporation Report for the Defense Threat Reductions Agency (31 Oct. 2006).
Johnston, Alastair Iain, "Thinking About Strategic Culture," *International Security* Vol. 19 No.4 (1995), pp. 32–64.
Liddell Hart, B.H., *The British Way of Warfare*, London: Faber & Faber, 1932.
Luttwak, Edward, "The American Style of Warfare and the Military Balance," *Survival* Vol. 21 No. 1 (1979).
Nye, Joseph and Sean M. Lynn-Jones, "International Security Studies: A Report of a Conference on the State of the Field," *International Security* Vol. 12 No. 4 (1988), pp. 5–27.
Petermann, Sandra, *Rituale machen Räume: Zum kollektiven Gedenken der Schlacht von Verdun und der Landung in der Normandie*, Bielefeld: transcript Verlag, 2015.
Snyder, Jack L., *The Soviet Strategic Culture: Implications for Limited Nuclear Options* RAND Research Paper R-2154-AF (September 1977).
Tamnes, Rolf, ed. *Common or Divided Security: German and Norwegian Perspectives on Euro-Atlantic Security*, Frankfurt am Main: Peter Lang, 2014.
Tellis, Ashley, "Overview," in *Strategic Asia 2016–17*, The National Bureau of Asia Research, 2016.
Watt, Donald Cameron, *How War Came*, London: Heinemann, 1989.
Weigley, Russell F., *The American Way of War: A History of United States Military Strategy and Policy*, London: Macmillan and Boomington, IN: Indiana Univeristy Press, 1973.

3

GROWTH AND REFINEMENT ACROSS THE FIELD OF STRATEGIC CULTURE

From First Generation to Fourth

Brigitte E. Hugh

Introduction

While the influence of culture on the actions of warring groups has been discussed by strategists throughout history,[1] the concept of strategic culture – the impact of ideas, symbols, myths, and historical experience on foreign policy or the military decisions made by influential individuals in a group's deciding body – is relatively new to the field of international relations. This modern incarnation of strategic culture appeared in 1977 when Snyder referred to "the Soviet strategic culture" in conducting an analysis of the USSR's limited nuclear strategy.[2] The concept evolved over the next four decades, with the early scholarship generally classified into three generations as characterized by Johnston in 1995.[3] The continued evolution of the discipline places current strategic culture scholarship in the fourth generation, which is the focus of this literature review.

The principal criticism of strategic culture as an international relations paradigm is that the concept is hard to define and even more difficult to operationalize. As anthropologists struggle to agree on the definition of culture itself, the derivative field of strategic culture continues to search for a comprehensive and manageable definition acceptable to the scholarly community.[4]

Strategic culture, as originally defined by Snyder, is:

> The sum total of ideas, conditioned emotional responses, and patterns of habitual behavior that members of a national strategic community have acquired through instruction or imitation and share with each other with regard to [national security] strategy.[5]

This definition is sufficient to provide a basic understanding of the concept before examining its history and ideological evolutions. As current scholars tend to individually tailor the definition of strategic culture to their specific methodological approach, this literature review does not attempt to accumulate an exhaustive list of definitions.

DOI: 10.4324/9781003010302-4

Beyond the difficulty of settling on a universal definition, strategic culture poses a challenge for international relations scholars trying to classify the concept as a theory, when according to Johnson, it should be considered a method.[6] Because strategic culture may lend support to any of several major international relations theories (including realism, neorealism, constructivism, and liberalism),[7] it is more usefully regarded as a paradigm that can be used to understand the motives and bounded rationality of the actor under study in partnership with the primary theoretical constructs of the field. The lens of strategic culture helps refine contextual reasoning for decisions and provides further insight into foreign policy actions or previously inexplicable breaks from international norms by nation-states.

In the policy realm, strategic culture offers decisionmakers a potent analytic tool regardless of worldview. Understanding an adversary's – or ally's – strategic culture will not necessarily change the end goal of neorealist decisionmakers who seek power in the international arena for the sake of their state's security, but it may help those decisionmakers identify historically weak points in an opponent's strategic culture that may be exploited. In another vein, policymakers subscribing to a constructivist worldview may employ the concept to identify how an adversarial state sees the world and illuminate how reciprocal action has created an international system reflective of both parties' understanding of themselves and their opponents. Strategic culture is both a paradigm and method employed to help scholars and practitioners understand interlocutors on the international stage more completely and better inform a spectrum of effective policy options.

Three Previous Generations of Strategic Culture

In 1995, Johnston organized the strategic culture scholarship to date into three distinct generations. The *first generation* viewed strategic culture as a macro-level concept principally informed by variables such as history and geography. This generation believed that strategic culture was not only found among the elite populations of the country but also among the general population. During this period, strategic culture remained loosely defined and most studies focused on the United States and the Soviet Union with particular regard to the superpowers' nuclear strategies.[8] The first generation had its faults: Stone points out the tendency of first-generation scholars to assume that states had one monolithic strategic culture and state actions were based on a single critical cultural factor[9] – a criticism still levied on contemporary strategic culture literature as current scholars work to correct the tendency toward oversimplification.

The *second generation* of strategic culture scholarship attempted to steer away from broad causational theories and instead focused on how strategic culture could be used as an instrument of strategic decisionmaking. This generation of scholarship viewed strategic culture as a mechanism used to fulfill states' needs based on their interests.[10] However, it struggled to define the exact influence of strategic culture on policy decisionmaking – failing to link cultural causes to manifest behaviors of states.[11]

The *third generation* of strategic culture scholars used a more scientific approach in assigning independent and dependent variables where possible, thereby focusing on causal mechanisms and becoming more deterministic in their construction of strategic culture paradigms. Due to this narrower approach, the third generation tended to base analyses on specific influential policy decisions.[12] One of the weaknesses of this approach was its inability to identify when strategic culture has little or no impact on state behavior.[13]

Literature Review

There is significant divergence across the approaches of contemporary strategic culture scholars, leaving ample room for debate and further scholarship – including some debate over the discipline's evolution into a fourth generation of strategic culture.[14] However, there has been a decided shift in the tenor of the scholarship which has been noted briefly by other scholars.[15] The fourth, and current, generation of strategic culture scholarship tends to focus on a few key areas of debate and research.

The first area focuses on the conception of strategic culture as either largely monolithic or as multiple subcultures within a state from which one strategic culture narrative emerges to dominate decisionmaking. Second, within this focus, a smaller debate is trained on whether strategic culture should be issue-oriented or focused on broad, state-based studies. Third, some scholars deliberate the existence of strategic culture outside of the generally accepted state construct, such as in the context of a supra- or sub-state actor. Fourth, scholars disagree over strategic culture as constant or continually changing. Fifth, methods for assessing cultural data to provide an effective map of strategic culture were few and far between in early scholarship, but each fourth-generation strategic culture method uses clearly defined criteria to organize and analyze data. A few criteria appear across multiple methods and seem to provide the most useful data; these are presented along with the key methodologies of the fourth generation.

Many Strategic Cultures

One of the most frequent critiques of the first three generations of strategic culture scholarship is that it is reductionist and masks important internal contradictions in a search for continuity over time. Many of these criticisms are appropriate. Viewed empirically, a monolithic conception of strategic culture does not bear out. People and localized cultures are constantly changing in small and incremental ways and these shifts are reflected in the often multiple and competing discourses within a single strategic culture. In the field of international relations this kind of absolutism, in which a single causal variable is sought as an explanation, has become somewhat common – a tendency Tellis calls a "commitment to reductionist explanations."[16] This preference for a simplified explanation from concrete data makes assessing state strategic culture – a highly elastic and flexible construct[17] – challenging to conduct in a way acceptable to the social science community.

Early generations of the strategic culture paradigm struggled to conceptualize multiple causal variables and competing discourses. Instead, scholars sought to profile strategic cultures according to their most dominant and consistent variables. The change was assumed to be either slow or nonexistent.[18] Many later scholars have pushed back against the oversimplifications that emerged from this process and have endeavored to prove that strategic culture is neither monolithic nor constant. These scholars point out that strategic culture, much like community culture, is not simple or uniform. Just as an individual is part of several cultures (religion, race, community, education, social class, etc.), a nation is made up of many cultures and strategic culture narratives.

Longhurst and Dalgaard-Nielsen suggest one theory for how these multiple narratives function: they view culture as a series of concentric circles, which they compare to the spiraling, nestled leaves of a cabbage. Subcultures represent the "leaves" which slowly spiral into the center, with the most influential factors existing near the core. As different subcultures approach the "core" or mainstream of a society, once-radical narratives can become accepted modes of behavior that influence the adoption of previously unacceptable policies.[19]

Bloomfield, Lantis, Johnson, and Berrett all conceptualize a nation having multiple strategic subcultures whose advocates compete for dominance.[20] Lantis draws on Mead's analysis of the cultural strands which have impacted American grand strategy to explain the concept of multiple strategic subcultures. Mead identifies four fundamentally different strands of American culture, each of which has dominated US foreign policy at different points in history, ranging from isolationism to active interventionism.[21] Pirani and Bloomfield both emphasize that the element of competition and opposition between strategic subcultures has been left out of earlier strategic culture scholarship.[22]

Johnson and Berrett go one step further and provide a technique for dealing with these competing strategic subcultures in their research methodology, Cultural Topography (or CTops). In acknowledging that there are multiple competitive narratives within a group or state, Johnson and Berrett posit that the dominance of a certain strategic culture narrative will depend on the substantive nature of the situation under study. The CTops method is structured such that the analyst or scholar is not only tasked with identifying features of the shared strategic culture of the group they are studying but they must also identify which single narrative of the several existent narratives will dominate on the particular topic in question.[23]

The takeaway, as pointed out by both Bloomfield and Johnson, is that strategic culture is a toolbox to the very states and organizations "experiencing" it. When faced with a situation, a state may be able to reach into the toolbox and retrieve the tool (subcultural narrative) that will be most acceptable and effective in dealing with *that* particular circumstance.[24] A nuanced understanding of strategic culture helps the scholar or analyst to determine which tools will be acceptable for a group to use at a given time and in a given situation.[25] In order for the idea of multiple subcultural narratives to be accurate, strategic culture logically cannot be an impenetrable monolith; it must be malleable and changeable so it is able to adapt to new circumstances and challenges that propel different subcultural narratives to the fore.

Expanded vs. Issue-Based Strategic Culture

As noted previously, at the genesis of strategic culture scholarship in the 1970s, scholars attempted to identify the strategic culture of entire countries. Usually, these scholars were already regional experts and had a fairly comprehensive understanding of the country and many facets of its culture. Current scholarship has pivoted away from the monolithic understanding of strategic culture and is based on more nuanced understandings of multiple subcultures. This may be, in part, because current analysts and scholars are often not approaching these problem sets with the same level of state or regional expertise as their first-generation (often Sovietologist) forebears. As a result, attempting to identify an entire state's strategic culture is an expansive and virtually unconquerable task from a research standpoint. These points contribute to an ongoing debate – how can one conduct strategic culture research in a way that is both useful and not insurmountably overwhelming?

While the *expanded strategic culture* approach – wherein one attempts to discern broad strategic culture elements belonging to an entire state – has some applications, it is less likely to provide results that would be actionable for those wishing to define solid insights into state behavior or make policy decisions. Instead, some more recently developed methodologies choose to assess focused aspects of strategic culture by directing research with a specific issue in mind. This type of analysis is also referred to as *issue-based strategic culture.*

Bloomfield discusses the importance of both issue-based and expanded strategic culture approaches. This author explains that *strategic behavior* is likely more reflective of short-term decisionmaking (i.e., first reactions when an individual or state is faced with an imminent threat), while *strategic policy* is more indicative of the long-term context in which the decision is being made, and thus likely more revealing of the influence of strategic culture. The critical nature of this bifurcation is that strategic behavior and strategic policy *may at times conflict*: in order to understand the significance of these conflicts, it is important to understand the function of both behavior and policy.[26]

Frank posits that issue-based strategic culture case studies provide a more nuanced depiction of strategic culture in action, while the expanded strategic culture profile increases understanding of the group's strategic culture across time.[27] Both types of studies are important and will reveal different results, especially if short-term actions founded in instinct are contradictory to the group's overall strategic culture.

While several fourth-generation scholars have focused on the expanded strategic culture track, at least two major studies have endeavored to build a methodology oriented toward issue-based strategic culture. First, Johnson, Kartchner, and Larsen developed a framework to conduct an issue-based analysis of countries and their strategic culture with respect to weapons of mass destruction acquisition, proliferation, use, and international norm compliance.[28] Second, Johnson and Berrett developed the Cultural Topography framework (CTops, as noted above) for the US Intelligence Community.[29] The CTops methodology requires the researcher to first choose an *issue of interest*, whether for intelligence or scholarship. The researcher then selects the *actor group* to study through the lens of that issue.[30] The actor might be the most obvious key group playing in the culture's decisionmaking structure, the "keeper" of strategic culture who will make the final decision on the matter at hand, or even an understudied supporting actor.[31] The CTops method is designed to generate intelligence insights and can prove particularly useful in understanding the specific issue-based strategic cultures of different groups. This kind of analysis is also applicable to state-level actors such as governments, especially if iterated for several constituent components of the collective actor (e.g., assessing how the parliamentary majority, the military, and a key ethnic faction each might culturally respond to a specific shared issue).

Levels of Strategic Culture

While some scholars resist the possibility that strategic culture exists outside of the generally accepted state construct, methods like Johnson, Kartchner, and Larsen's and Johnson and Berrett's are premised on the concept that groups other than nation-states can have a strategic culture. Recent research seeks to identify other groups or levels of society that may have a strategic culture: for instance, Biava, Drent, and Herd,[32] as well as Becker and Malesky,[33] show that supranational organizations – such as the European Union and NATO – can also develop their own strategic cultures, although they are even more prone to having competitive state subcultures. These subcultures, in turn, have an impact on the supranational strategic culture.

Johnson, Berrett,[34] and Lantis[35] argue that sub-state actors (such as large businesses, political parties, tribes, religious organizations, ethnic groups, etc.) can all develop their own strategic cultures which influence critical decisionmaking. Depending upon the national context and historical experiences of these groups, some can play a significant role in a state's strategic actions and therefore influence a state's strategic culture. Thus, for some issue sets, understanding relevant sub-state actors' strategic culture can be imperative.

Interestingly, Frank shows that relatively new countries, which lack a long history to draw upon (one of the commonly accepted critical components of strategic culture), can also manifest distinctive strategic cultures.[36] Their strategic cultures tend to be mostly based upon traditions of the dominant tribes or demographic groups and the past experience of leaders.[37]

Continuity vs. Change

In addition to the debate over strategic culture as one monolithic entity or several competing subcultures, current strategic culture scholars wrestle with a second major oversimplification found in previous generations: that strategic culture is unchanging and unadaptable. History itself offers challenges to this claim, as major cultural shifts, some at the national and some at the organizational level, have taken place in relatively short periods of time. The two most notable are Japan and Germany's dramatic shifts in strategic culture after World War II. A less studied but equally compelling example is the US Marine Corps' shift from frontal assault to maneuver warfare during the 1990s.[38]

Though early strategic culture scholarship did acknowledge the ability of strategic culture to change, it often emphasized a major external shock as a necessary catalyst. These external shocks are defined by Lantis as occurrences that "fundamentally challenge existing beliefs and undermine past historical narratives."[39] However, Lantis also notes an increasing acceptance of change as a feature of strategic culture, especially in scholarship rooted in the ideological base of constructivism that assumes culture as a constantly evolving construct in which participants codevelop new understandings and shared meanings.[40]

Frank points out that if strategic culture is largely based upon the norms – the accepted and expected modes of behavior[41] – of a given society, then it must be understood that over time these norms will change, and with them, the overarching strategic culture. In addition, groups' histories are likely to be reinterpreted through new normative lenses as these shifts occur. Given the importance of history to strategic culture, any reinterpretation through new normative lenses directly impacts strategic culture.[42]

Major shocks as the catalyst for rapid change in strategic culture narratives are much easier to track and understand, which is likely why some scholars gravitate to the analysis of that particular source of change. Incremental adjustments and alterations to strategic narratives which take place over a long stretch of time are more difficult to identify and analyze.

Frank develops the concept of "path dependency," in which an action is taken over and over again in response to similar problem sets. Over time, that action becomes a historical precedent and a habitual response.[43] Hudson argues that "culture provides its members a repertoire or palette of adaptive responses from which members build off-the-shelf strategies of action."[44] While path dependency and action templates could perhaps be viewed as arguments *against* possible change in strategic culture – as they develop established, consistent responses to certain foreign policy issues – the development of these paths or action templates is a long, drawn-out process in which small tweaks and changes are made to the overall policy process according to contemporary interpretations of strategic culture. Over time, these templates can provide critical information on the nature and time frame of strategic culture shifts.

Hines suggests that strategic culture and human action form a feedback loop in which strategic culture informs and shapes human action, which then informs and changes strategic culture.[45] Both Hines and Pirani believe that the political elites they label "keepers of strategic culture" have a significant capacity to change the strategic culture. The concept of "keepers of strategic culture" is present in other scholarship, and it is important to note that Pirani, Hines, Lantis, and Johnson all believe to some extent that keepers of strategic culture have significant impact on whether or not the strategic culture of a group evolves or maintains the status quo.[46]

In Das's analysis of India's nuclear insecurities, the author relies on Campbell's "representations of danger" – the idea that a state's sense of insecurity is socially and culturally produced and is a catalyst for cultural change over time. Das's larger claim is that the state constructs its own insecurity, which then reinforces its identity as well as the perception (and perhaps reality) that the state is threatened and in need of a security policy. This author then conceptualizes strategic culture as based on a state's internally generated insecurities and how they translate into interactions with other states.[47] Das highlights the fact that insecurities themselves are not stable, nor are the relationships between states, and if insecurities are constantly changing, so will strategic culture at large.

Das's scholarship parallels some features in Bloomfield's work, which asserts that every strategic culture is based on a simple, culturally informed classification of other nations as enemy or ally, or the "friend/foe calculus." Bloomfield argues that groups tend to place cultures similar to their own into the friend category, while cultures that are seen as different are placed in the foe category. This calculus can be subject to change due to fluctuating perceptions and relationships, but Bloomfield argues that in the end, all strategic decisions will be made based on the friend/foe calculus.[48]

An innovative fourth-generation contribution comes from Choi, who reorganizes the strategic culture paradigm into two dimensions: the horizontal and the vertical. Choi's "vertical" dimension focuses on the keepers of strategic culture, which the author subdivides into two levels: the higher – members of the political and strategic elite, the strategic decisionmakers; and the lower – members of the defense organizations, who actually practice the strategy.[49] Choi argues that most strategic culture research tends to focus on the elites (notably, the opposite of what Pirani argues just seven years later[50]), and consequently misses the critical perspective of the people carrying out strategic decisions.

In Choi's bifurcation of "higher" and "lower" levels of keepers, the author argues that understanding the vertical dimension of strategic culture is important because it allows observers to track the changes in a strategic culture and better define its character. Observing the "higher" and "lower" groups separately enables simpler identification of the changes in strategic decisions, and from which level the change originates (or if they are equally involved).[51] This also provides easier identification of which group outside actors should attempt to influence in order to bring about change.[52]

Choi also argues that state elites should not be the sole guardians of strategic culture and that general society may be a more valid keeper in the long run, a point also made by Macmillan.[53] Adding to the debate, Lantis argues that elites might just be "users" of strategic culture, and not necessarily the ones who maintain the narratives and pass them down from one generation to another.[54] Lantis suggests that there are other actors who could serve as the keepers of strategic culture, including families, business groups, parties and domestic coalitions, military institutions, or the political elites.[55] All these arguments suggest that analysis should be expanded beyond elites and practitioners to include an examination of significant subgroups within the general populace as keepers and users of strategic culture.

Choi's second aspect, the "horizontal" dimension, consists of two forms of culture. *Hard strategic culture* comprises the "material, non-ideational" elements of the culture, which may include arts, crafts, styles, technology, and even language. *Soft strategic culture* is composed of ideational elements, such as the identity of the group, their worldview, and their preferred methods of responding to threats.[56] Choi argues that the study of soft strategic culture has dominated the field, and that hard strategic culture needs to be considered in tandem. The author claims that analyzing both elements together helps to identify whether one is dominating in decisionmaking, if the dominant type of culture is changing over time, and what outcomes might come from a situation in which soft and hard strategic cultures suggest opposing action templates.[57]

Fourth-generation scholarship has made clear that the discipline has come to accept that strategic cultures are not stagnant and unchanging. However, when it comes to identifying ways to capture that change, the literature will likely continue to develop. Choi's work provides a useful final note on this subject, offering a typology of change consisting of eight questions meant to better understand how a change in strategic culture is occurring and why. These include: "subject of change; cause of change; speed of change; width and depth of change; momentum of change; manner of change; attitude of change; and result of change."[58] Choi's delineation of each of these questions can help scholars and analysts more methodically capture, document, and analyze important cultural changes.

Sources of Strategic Culture

Most scholars agree that there are certain "qualities" that constitute a strategic culture. However, the number of these qualities, the methods of capturing the pertinent data, and how to understand their impact on strategic culture are debated frequently in the literature. Of the core qualities associated with strategic culture, *history* and *geography* are the most broadly accepted.[59] With the exception of Frank, each strategic culture scholar surveyed for this review places history as a prominent source of strategic culture, and Frank dissents only in arguing that strategic culture can also exist in nations that do not have a shared historical narrative. Frank's work points out that strategic culture does not rest as a concept upon history alone and that other qualities play an enormous role.[60] An in-exhaustive list of other agreed-upon sources of strategic culture includes: *relevant resources, myths and symbols, political structure and defense organizations, climate, habits and traditions*, and *rituals and ceremonies*.[61] The more controversial qualities are those such as worldview and values, which are agreed upon as important but debated in terms of definition and effective methods of capture.

As traditionally understood, values are those moral standards to which a group or individual holds themselves accountable. The immense subjectivity and variability of values across populations can make them difficult to identify. Choi specifies that hard strategic culture can be identified through "published lists of values," although that means relying on the overt values portrayed by the group which may conflict with the underlying unspoken values (soft strategic culture) and provide useful gaps for analysis.[62]

Johnson and Berrett's CTops methodology focuses on four categories: identity, values, norms, and perceptual lens.[63] The traditional definition of values as acceptable or moral action is captured in their norms category, which focuses on behavior. Values within the CTops methodology are defined somewhat unconventionally. For Johnson and Berrett, values are "material or ideational goods that are honored or that confer increased status to members."[64] Their definition serves the purposes of policymakers who seek salient (valued) incentives, including material inducements or correct modes of praise when dealing with a culturally distinct group.

Across the current literature, one area that is gaining increased attention is a group's *worldview*. Bloomfield's theorizing of the friend/foe calculus is a component of worldview or perceptual lens since it impacts who a group will classify as enemies or allies and influences consequent behavior.[65] This author's analysis points out:

> A schema's cultural orientation identifies a state's friends and enemies and may also encompass a wider range of social relations, such as who can be trusted, what/who "we want to be like," what "role the state sees itself and others 'playing' in international politics" and so on.[66]

Tremblay and Bentley refer to worldview by the German word "*weltanschauung*" and posit that worldview is made up of culture, religion, and ideology,[67] though the absence of a compelling definition of "culture" makes their concept somewhat tautological. Johnson and Berrett use the alternate label of "perceptual lens," or "the filter through which this group determines 'facts' about others" to capture much the same idea. Their definition is broken down further into multiple levels of analysis: first, how the group sees the world at large; second, how it perceives its close friends and neighbors, as well as more distant or adversarial states; and finally how it sees itself with respect to the rest of the world and the global role it sees itself filling.[68]

Given that the goal of strategic culture is to identify how actors will interact with the world based on their specific cultures, worldview has proved useful in identifying how those interactions will play out. The addition of worldview as an analytical category in strategic culture scholarship has been an effort to make the implicit become explicit and therefore make the field's contributions more relevant overall.

Fourth-Generation Methodologies

One of the concerted efforts of recent strategic culture scholarship has been to develop methodologies that are effective in gathering and organizing pertinent cultural data in the process of assessing a group's strategic culture. Most of the scholarship comes to the same basic conclusion: establishing at least a few categories for organizing this cultural data is necessary, but these categories differ across methodologies and studies.

Choi's research focuses around a set of three fundamental sources of strategic culture: geography and resources; history and experience; and political structure and defense organization.[69] As noted above, upon gathering materials in these three categories Choi interprets the cultural data through the "horizontal dimension," assessing the sources and impacts of strategic culture; and the "vertical dimension," assessing the influential groups.[70] When integrated, Choi's dimensions aim to provide a holistic vision of a country's strategic culture and an organizational method for understanding key cultural factors.[71]

Biava, Drent, and Herd's examination of the European Union's strategic culture argues that one can determine strategic culture based on the magnitude, frequency, and purpose of military missions using six categories of analysis.[72] First, one must assess the strategic guidelines, or framework for action, which usually set the parameters for the strategic documents produced by the group. Second, examine strategic threats or the security threats against which the group would need to act. Third, account for military and civilian tool-building or the physical capability to act – including the building and strengthening of both diplomatic and military tools. Fourth, analyze relevant institutional machinery in order to determine the capacity to act or to be able to decide to act. Fifth, identify and characterize norms, which are defined as the "legitimacy for acting" and which are the rationalizations for undertaking action. Sixth, track key military and civilian operations, or the actual action that is carried out.[73]

Tremblay and Bentley choose to organize their research into four main categories. The first two categories, geography and history, are both consistent with discipline consensus. The third category – a mega-cluster of culture, religion, ideology, and governance – is posited as impacting the worldview of a country. However, this category is "both elemental and vast," and the authors do not provide direction for how to effectively assess it, which may pose some challenges to researchers attempting to duplicate their method. Tremblay and Bentley's fourth category is governance, or the structure of the military institutions, the system of governance, and the way that civil-military interactions play out.[74]

In Rosyidin's scholarship on the strategic culture of the South China Sea dispute, the author does not develop a novel methodology but instead uses Lantis' three main categories – geography, politics, and social experience – and adds that history, ideology, and leader beliefs play an important role in the development of strategic culture.[75] Though these categories appear to some extent in other literature, they are notably clear in Rosyidin's work and represent an effort to build on and refine previous methods.

Frank's assessment of sub-Saharan strategic culture focuses broadly on history, experience, and politics, but then further clarifies the author's methodology using Booth's 1990 definition of strategic culture:[76]

> The concept of strategic culture refers to a nation's traditions, values, attributes, patterns of behavior, habits, symbols, achievements and particular ways of adapting to the environment and solving problems with respect to the threat and use of force.[77]

Frank uses each of these attributes as categorical repositories for the information gathered.[78] This author also notes that some strategic cultures will be more "mature" than others – defining maturity as the extent to which a strategic culture influences decisionmaking and how absorbed it is in the national security apparatus.[79]

Biehl, Giergerich, and Jonas's volume examines European countries with the goal of providing an index of European strategic cultures that can be easily compared and contrasted. The methodology developed asked four questions of the contributing scholars: the state's level of ambition in international security policy, the scope of action for the state's executive in decisionmaking, the state's foreign policy orientation, and its willingness to use military force.[80] Biehl, Giergerich, and Jonas acknowledge that some authors were forced to drop aspects of a nation's strategic culture in order to fit the methodology, but that in doing so they were able to more effectively create comparable strategic culture summaries.[81]

Lantis' volume on Asia-Pacific strategic cultures developed a methodology assessing the "actors, factors, norms and ideas that impact security policy development."[82] This assessment is done by exploring the historical background of the state with respect to geography and resources, history, and conflict experiences. The methodology then turns to the identification of the keepers of strategic culture, especially those who guide the development of strategic culture, the institutions, strategic subcultures that are present, and the geostrategic changes in the region through the past decade.[83] Lastly, researchers were asked to examine the factors which Lantis calls the "sources of strategic culture," including geopolitical considerations and structural pressures, geography, climate, resources, narratives of historical experiences or "roles," the nature of the country's political structure and defense organizations, interstate loyalties (such as tribes, religions, or ethnicities), myths and symbols, and key texts.[84]

Finally, Johnson and Berrett's CTops methodology identifies four research categories that impact strategic culture. *Identity* is "the way that the group sees itself and the reputation it would like to pursue, as well as the individual roles it assigns to its members." *Norms* are defined as "the accepted and expected modes of behavior, as well as taboo behaviors." *Values,* as noted in the previous section, are those "material or ideational goods that are honored or that give increased status to members," and *perceptual lens* is "the filter through which this group determines facts about itself and others."[85]

The CTops method aims to whittle down the mass of available cultural information to a few "critical cultural factors" by asking researchers to assess the "three Rs" of each variable: *relevance* (whether or not this particular cultural attribute is relevant to this group, on this issue, at this time), *robustness* (how prevalent this cultural factor is within the group), and the likelihood to provoke a *response* (the impact an engagement on this particular issue would have in the group and what kind of reaction, conflictual or cooperative, it would incite).[86] As a package, the CTops approach offers a rigorous method for locating germane cultural data and making sense of it for the academic or intelligence questions at hand.

Conclusion

Recent contributions to the strategic culture paradigm have moved the discipline forward in innovative and useful directions. The current fourth generation of strategic culture scholarship is characterized by a commitment to the messy state of strategic culture and is unafraid to distance itself from the "reductionist explanations" that characterized some of the earlier generations of strategic culture scholarship. This generation accepts, on the whole, that strategic culture is not one constant cultural narrative but is composed of many subcultural narratives that are continually changing and adapting, meaning that the strategic culture narrative impacting contemporary action is often situationally determined.

This generation is also differentiated by a desire to develop methodologies that are replicable and more effectively enable scholars and analysts to capture strategic culture and assess how it influences policy decisionmaking. In some ways, the fourth generation is shadowing its forebears in the second generation with an intent to identify how an understanding of strategic culture can be operationalized and leveraged for the benefit of the group that is conducting the study. Very recently, two scholars applied a machine learning approach designed to track historical changes in national strategic culture and demonstrate their impact on military engagements.[87] With the development of new methodologies and more careful evaluation of the sources and keepers of strategic culture, fourth-generation scholars are attempting to mitigate the "messiness" of strategic culture research by providing clear frameworks for organized and methodical pursuits of cultural data.

The importance and value of understanding strategic culture is increasingly evident as states and other significant groups make strategic decisions that are otherwise inexplicable according to rational choice models. Ongoing developments in the strategic culture paradigm demonstrate a commitment to scholarship that is academically sound and pragmatically useful.

Notes

1 Beatrice Heuser and Jeannie Johnson, "Introduction: National Styles and Strategic Culture," in Beatrice Heuser and Eitan Shamir eds. *Insurgencies and Counterinsurgencies: National Styles and Strategic Cultures* (Cambridge: Cambridge University Press, 2016), 1–22.
2 Jack L. Snyder, *The Soviet Strategic Culture: Implications for Limited Nuclear Operations,* (Santa Monica, CA: RAND Corporation, September 1977), https://www.rand.org/pubs/reports/R2154.html.
3 Alastair Iain Johnston, *Cultural Realism: Strategic Culture and Grand Strategy in Chinese History* (Princeton, New Jersey: Princeton University Press, 1995), 5–22.
4 Jeannie L. Johnson, "Bounding Strategic Culture: Method, Not Theory," October 2009, 2.
5 Snyder, *The Soviet Strategic Culture,* 8.
6 Johnson, "Bounding Strategic Culture: Method, Not Theory."
7 Jeannie L. Johnson, "Strategic Culture: Refining the Theoretical Construct" (Defense Threat Reduction Agency Advance Systems and Concepts Office, 31 October 2006), https://fas.org/irp/agency/dod/dtra/strat-culture.pdf.
8 Johnston, *Cultural Realism: Strategic Culture and Grand Strategy in Chinese History*, 5–14; Robert Lincoln Hines, "Change and Continuity in Chinese Strategic Culture – Chinese Decision-Making in the Taiwan Strait" (MA, United States – District of Columbia, American University, 2012), 16–17, https://search-proquest-com.dist.lib.usu.edu/pqdtglobal/docview/1418463621/abstract/DAF247364277410CPQ/1.
9 Elizabeth Stone, "Comparative Strategic Cultures Literature Review (Part 1)" (Defense Threat Reduction Agency Advance Systems and Concepts Office, 31 October 2006), 1, https://fas.org/irp/agency/dod/dtra/stratcult-comp.pdf.
10 Stone, 1; Johnston, *Cultural Realism: Strategic Culture and Grand Strategy in Chinese History*, 15–18.
11 Johnston, *Cultural Realism: Strategic Culture and Grand Strategy in Chinese History*, 15.
12 Stone, "Comparative Strategic Cultures Literature Review (Part 1)," 1; Johnston, *Cultural Realism: Strategic Culture and Grand Strategy in Chinese History*, 18–22.
13 Stone, "Comparative Strategic Cultures Literature Review (Part 1)," 1.

14 Tamir Libel, "Rethinking Strategic Culture: A Computational (Social Science) Discursive-Institutionalist Approach," *Journal of Strategic Studies* 43, no. 5 (28 July 2020): 686–709, https://doi.org/10.1080/01402390.2018.1545645.
15 David G. Haglund, "What Can Strategic Culture Contribute to Our Understanding of Security Policies in the Asia-Pacific Region?," in *Strategic Cultures and Security Policies in the Asia-Pacific*, 1st ed. (Routledge, 2014), 152–157; Fredrik Doeser and Joakim Eidenfalk, "Using Strategic Culture to Understand Participation in Expeditionary Operations: Australia, Poland, and the Coalition against the Islamic State," *Contemporary Security Policy* 40, no. 1 (2 January 2019): 6, https://doi.org/10.1080/13523260.2018.1469709; Andrew Williams, "Strategic Culture and Cyber Warfare: A Methodology for Comparative Analysis," *Journal of Information Warfare* 19, no. 1 (March 2020): 115.
16 Ashley J. Tellis, "Overview," in *Understanding Strategic Cultures in the Asia-Pacific* (Seattle, WA: The National Bureau of Asian Research, 2016), 6.
17 Jeffrey S. Lantis, *Strategic Cultures and Security Policies in the Asia-Pacific*, 1st ed. (Routledge, 2014), 2–3; Tellis, "Overview," 6.
18 Pietro Pirani, "Elites in Action: Change and Continuity in Strategic Culture," *Political Studies Review* 14, no. 4 (November 2016): 512–520, https://doi.org/10.1111/1478-9302.12058.
19 Pirani, "Elites in Action," 513.
20 Alan Bloomfield, "Australia's Strategic Culture: An Investigation of the Concept of Strategic Culture and Its Application to the Australian Case" (Ph.D., Canada, Queen's University (Canada), 2011), https://search-proquest-com.dist.lib.usu.edu/pqdtglobal/docview/1886372340/abstract/26A4C50665EB43D8PQ/1; Lantis, *Strategic Cultures and Security Policies in the Asia-Pacific*; Jeannie L. Johnson and Matthew T. Berrett, "Cultural Topography: A New Research Tool for Intelligence Analysis" *Studies in Intelligence,* Vol. 55, No. 2 (Extracts, June 2011), accessed 19 February 2018, https://www.cia.gov/static/811a5292007dbd5e9a73844c113f257b/Cultural-Topography.pdf.
21 Lantis, *Strategic Cultures and Security Policies in the Asia-Pacific*, 13.
22 Pirani, "Elites in Action"; Bloomfield, "Australia's Strategic Culture."
23 Johnson and Berrett, "Cultural Topography: A New Research Tool for Intelligence Analysis."
24 Johnson, "Bounding Strategic Culture: Method, Not Theory"; Bloomfield, "Australia's Strategic Culture," 78–79; Kevin Keasbey Frank, "Strategic Culture in Sub-Saharan Africa: The Divergent Paths of Uganda and Tanzania" (Ph.D., United States – Mississippi, The University of Southern Mississippi, 2017), 55, https://search-proquest-com.dist.lib.usu.edu/pqdtglobal/docview/1976153243/abstract/D73455332BB74E70PQ/1.
25 Frank, "Strategic Culture in Sub-Saharan Africa," 55.
26 Bloomfield, "Australia's Strategic Culture," 20.
27 Frank, "Strategic Culture in Sub-Saharan Africa," 18, 26–28.
28 Kerry M. Kartchner, "Strategic Culture and WMD Decision Making," in *Strategic Culture and Weapons of Mass Destruction: Culturally Based Insights into Comparative National Security Policymaking*, 1st ed., Initiatives in Strategic Studies: Issues and Policies (New York: Palgrave Macmillan, 2009), 55–67; Jeannie L. Johnson, Kerry M. Kartchner, and Jeffrey A. Larsen, "Introduction," in *Strategic Culture and Weapons of Mass Destruction: Culturally Based Insights into Comparative National Security Policymaking*, 1st ed., Initiatives in Strategic Studies: Issues and Policies (New York: Palgrave Macmillan, 2009), 3–14.
29 Johnson and Berrett, "Cultural Topography: A New Research Tool for Intelligence Analysis," 3–5.
30 This method cannot be applied to an individual as individual idiosyncrasies do not qualify as "culture."
31 Johnson and Berrett, "Cultural Topography: A New Research Tool for Intelligence Analysis," 6.
32 Alessia Biava, Margriet Drent, and Graeme P. Herd, "Characterizing the European Union's Strategic Culture: An Analytical Framework," *Journal of Common Market Studies* 49, no. 6 (2011): 1227–1248.
33 Jordan Becker and Edmund Malesky, "The Continent or the 'Grand Large'? Strategic Culture and Operational Burden-Sharing in NATO," *International Studies Quarterly* 61 (2017): 163–180.

34 Johnson and Berrett, "Cultural Topography: A New Research Tool for Intelligence Analysis"; Johnson, "Bounding Strategic Culture: Method, Not Theory," 14.
35 Lantis, *Strategic Cultures and Security Policies in the Asia-Pacific.*
36 Frank, "Strategic Culture in Sub-Saharan Africa," 5–6.
37 Frank, 5–6 & 45.
38 Terry Terriff, "'Innovate or Die': Organizational Culture and the Origins of Maneuver Warfare in the United States Marine Corps," *Journal of Strategic Studies* 29, no. 3 (June 2006): 475–503.
39 Pirani, "Elites in Action," 513.
40 Lantis, *Strategic Cultures and Security Policies in the Asia-Pacific*, 10–11.
41 Johnson, "Strategic Culture: Refining the Theoretical Construct."
42 Frank, "Strategic Culture in Sub-Saharan Africa," 56.
43 Frank, 45, 54.
44 Johnson, "Bounding Strategic Culture: Method, Not Theory," 4.
45 Hines, "Change and Continuity in Chinese Strategic Culture – Chinese Decision-Making in the Taiwan Strait."
46 Hines, 21; Johnson and Berrett, "Cultural Topography: A New Research Tool for Intelligence Analysis"; Lantis, *Strategic Cultures and Security Policies in the Asia-Pacific*, 9; Pirani, "Elites in Action," 515.
47 Runa Das, "Strategic Culture, Representations of Nuclear (In)Securities, and the Government of India: A Critical Constructivist Perspective," *Asian Journal of Political Science* 17, no. 2 (August 2009): 128–129, https://doi.org/10.1080/02185370903077402.
48 Bloomfield, "Australia's Strategic Culture," 63–64, 86–88.
49 J. Choi, "The Evolution of Strategic Cultures of Divided Countries: A Case Study on the Continuities and Changes of Korean Strategic Culture and Strategic Relations on the Peninsula since 1948" (Ph.D., England, The University of Reading (United Kingdom), 2009), 43–44, https://search-proquest-com.dist.lib.usu.edu/pqdtglobal/docview/898756968/F6C552D765E4E55PQ/11.
50 Pirani, "Elites in Action," 515.
51 Choi, 46.
52 Choi, 45.
53 Choi, 44.
54 Lantis, *Strategic Cultures and Security Policies in the Asia-Pacific*, 5.
55 Lantis, 6.
56 Choi, 46–47.
57 Choi, 47.
58 Choi, 48.
59 Lantis, *Strategic Cultures and Security Policies in the Asia-Pacific*, 5; Choi, "The Evolution of Strategic Cultures of Divided Countries," 24; Éric Tremblay and Bill Bentley, "Canada's Strategic Culture: Grand Strategy and the Utility of Force," *Canadian Military Journal* 15, no. 3 (2015): 2; Mohamad Rosyidin, "China's Strategic Culture and the Challenge of Security Management in the South China Sea Dispute," *East Asia* 34 (11 May 2017): 137.
60 Frank, "Strategic Culture in Sub-Saharan Africa," 5–6.
61 Choi, "The Evolution of Strategic Cultures of Divided Countries," 24, 46–48; Lantis, *Strategic Cultures and Security Policies in the Asia-Pacific*, 5–8; Johnson and Berrett, "Cultural Topography: A New Research Tool for Intelligence Analysis"; Frank, "Strategic Culture in Sub-Saharan Africa," 66–71.
62 Choi, 46–47.
63 Johnson and Berrett, "Cultural Topography: A New Research Tool for Intelligence Analysis," 2–4.
64 Johnson and Berrett, 6.
65 Bloomfield, "Australia's Strategic Culture," 63, 86.
66 Bloomfield, 83.
67 Tremblay and Bentley, "Canada's Strategic Culture: Grand Strategy and the Utility of Force," 2.
68 Johnson and Berrett, "Cultural Topography: A New Research Tool for Intelligence Analysis," 2–5.
69 Choi, 24.
70 Choi, 43–48.

71 Choi, 46–48.
72 Biava, Drent, and Herd, "Characterizing the European Union's Strategic Culture: An Analytical Framework," 1240.
73 Biava, Drent, and Herd, 1241–1243.
74 Tremblay and Bentley, "Canada's Strategic Culture: Grand Strategy and the Utility of Force," 6.
75 Rosyidin, "China's Strategic Culture and the Challenge of Security Management in the South China Sea Dispute," 137.
76 Frank, 2, 11.
77 Frank, 11.
78 Frank, 66–71.
79 Frank, 81.
80 Heiko Biehl, Bastian Giegerich, and Alexandra Jonas, "Introduction," in *Strategic Cultures in Europe: Security and Defence Policies Across the Continent* (Potsdam, Germany: Springer Fachmedien Wiesbaden, 2013), 7–17.
81 Heiko Biehl, Bastian Giegerich, and Alexandra Jonas, "Conclusion," in *Strategic Cultures in Europe: Security and Defence Policies Across the Continent* (Potsdam, Germany: Springer Fachmedien Wiesbaden, 2013), 387–393.
82 Lantis, *Strategic Cultures and Security Policies in the Asia-Pacific*, 5.
83 Lantis, 5–6.
84 Lantis, 7–8.
85 Johnson and Berrett, "Cultural Topography: A New Research Tool for Intelligence Analysis," 6–7.
86 Johnson and Berrett, 13.
87 Jonathan Tappe and Fredrik Doeser, "A Machine Learning Approach to the Study of German Strategic Culture," *Contemporary Security Policy* 42, no. 4 (2021): 450–474.

Selected Bibliography

Becker, Jordan, and Edmund Malesky. "The Continent or the 'Grand Large'? Strategic Culture and Operational Burden-Sharing in NATO." *International Studies Quarterly* 61 (2017): 163–180.

Biava, Alessia, Margriet Drent, and Graeme P. Herd. "Characterizing the European Union's Strategic Culture: An Analytical Framework." *Journal of Common Market Studies* 49, no. 6 (2011): 1227–1248.

Biehl, Heiko, Bastian Giegerich, and Alexandra Jonas. *Strategic Cultures in Europe: Security and Defence Policies Across the Continent*. Potsdam, Germany: Springer Fachmedien Wiesbaden, 2013.

Bloomfield, Alan. "Australia's Strategic Culture: An Investigation of the Concept of Strategic Culture and Its Application to the Australian Case." Ph.D., Queen's University (Canada), 2011. https://search-proquest-com.dist.lib.usu.edu/pqdtglobal/docview/1886372340/abstract/26A4C50665EB43D8PQ/1.

Choi, J. "The Evolution of Strategic Cultures of Divided Countries: A Case Study on the Continuities and Changes of Korean Strategic Culture and Strategic Relations on the Peninsula since 1948." Ph.D., The University of Reading (United Kingdom), 2009. https://search-proquest-com.dist.lib.usu.edu/pqdtglobal/docview/898756968/F6C552D765E4E55PQ/11.

Das, Runa. "Strategic Culture, Representations of Nuclear (In)Securities, and the Government of India: A Critical Constructivist Perspective." *Asian Journal of Political Science* 17, no. 2 (August 2009): 123–144. 10.1080/02185370903077402.

Doeser, Fredrik, and Joakim Eidenfalk. "Using Strategic Culture to Understand Participation in Expeditionary Operations: Australia, Poland, and the Coalition against the Islamic State." *Contemporary Security Policy* 40, no. 1 (2 January 2019): 4–29. 10.1080/13523260.2018.1469709.

Frank, Kevin Keasbey. "Strategic Culture in Sub-Saharan Africa: The Divergent Paths of Uganda and Tanzania." Ph.D., The University of Southern Mississippi, 2017. https://search-proquest-com.dist.lib.usu.edu/pqdtglobal/docview/1976153243/abstract/D73455332BB74E70PQ/1.

Haglund, David G. "What Can Strategic Culture Contribute to Our Understanding of Security Policies in the Asia-Pacific Region?" In *Strategic Cultures and Security Policies in the Asia-Pacific*, 1st ed. (New York: Routledge, 2014), 145–160.

Heuser, Beatrice, and Jeannie Johnson, "Introduction: National Styles and Strategic Culture," in Beatrice Heuser and Eitan Shamir eds. *Insurgencies and Counterinsurgencies: National Styles and Strategic Cultures* (Cambridge: Cambridge University Press, 2016), 1–22.
Hines, Robert Lincoln. "Change and Continuity in Chinese Strategic Culture – Chinese Decision-Making in the Taiwan Strait." M.A., American University, 2012. https://search-proquest-com.dist.lib.usu.edu/pqdtglobal/docview/1418463621/abstract/DAF247364277410CPQ/1.
Johnson, Jeannie L. "Bounding Strategic Culture: Method, Not Theory," October 2009.
Johnson, Jeannie L. "Strategic Culture: Refining the Theoretical Construct." Defense Threat Reduction Agency Advance Systems and Concepts Office, 31 October 2006. https://fas.org/irp/agency/dod/dtra/strat-culture.pdf.
Johnson, Jeannie L., and Berrett, Matthew T. "Cultural Topography: A New Research Tool for Intelligence Analysis." *Studies in Intelligence,* Vol. 55, No. 2 (Extracts, June 2011). Accessed 19 February 2018. https://www.cia.gov/static/811a5292007dbd5e9a73844c113f257b/Cultural-Topography.pdf.
Johnson, Jeannie L., Kerry M. Kartchner, and Jeffrey A. Larsen. *Strategic Culture and Weapons of Mass Destruction: Culturally Based Insights into Comparative National Security Policymaking*, 1st ed. Initiatives in Strategic Studies: Issues and Policies. New York : Palgrave Macmillan, 2009.
Johnston, Alastair Iain. *Cultural Realism: Strategic Culture and Grand Strategy in Chinese History.* Princeton, NJ: Princeton University Press, 1995.
Kartchner, Kerry M. "Strategic Culture and WMD Decision Making." In *Strategic Culture and Weapons of Mass Destruction: Culturally Based Insights into Comparative National Security Policymaking*, 1st ed., 3–14. Initiatives in Strategic Studies: Issues and Policies. New York: Palgrave Macmillan, 2009.
Lantis, Jeffrey S. *Strategic Cultures and Security Policies in the Asia-Pacific.* 1st ed. New York: Routledge, 2014.
Pirani, Pietro. "Elites in Action: Change and Continuity in Strategic Culture." *Political Studies Review* 14, no. 4 (November 2016): 512–520. 10.1111/1478-9302.12058.
Rosyidin, Mohamad. "China's Strategic Culture and the Challenge of Security Management in the South China Sea Dispute." *East Asia* 34 (11 May 2017): 133–145.
Snyder, Jack L. *The Soviet Strategic Culture: Implications for Limited Nuclear Operations.* Santa Monica, CA: RAND Corporation, September 1977. https://www.rand.org/pubs/reports/R2154.html.
Stone, Elizabeth. "Comparative Strategic Cultures Literature Review (Part 1)." Defense Threat Reduction Agency Advance Systems and Concepts Office, 31 October 2006. https://fas.org/irp/agency/dod/dtra/stratcult-comp.pdf.
Tellis, Ashley J. "Overview." In *Understanding Strategic Cultures in the Asia-Pacific.* Seattle, WA: The National Bureau of Asian Research, 2016.
Tremblay, Éric, and Bill Bentley. "Canada's Strategic Culture: Grand Strategy and the Utility of Force." *Canadian Military Journal* 15, no. 3 (2015): 5–17.

4

THE NATURE AND UTILITY OF STRATEGIC CULTURE SCHOLARSHIP[1]

Colin S. Gray

Introduction and Foundational Propositions

Strategic culture has become more popular than once it was. Until early this century, strategic culture was a concept known and employed only by a few academics. What strategic culture means, what one can and cannot do with it, and how it plays in national and organizational strategic thought, planning, and actual behavior are the issues on which this chapter seeks to shed some light.

Since I have been studying and writing about strategic culture for more than a quarter-century, it would be strange were I not to bring to this endeavor some reasonably definite convictions.[2] However, in recognition of the controversial nature of the subject of strategic culture in the minds of some scholars, and in honor of proper scholarly practice, my working beliefs are presented here in the form of propositions for debate and consideration, rather than claimed as facts. The purpose simply is to explain the basis for the propositions, not to justify them evidentially. A clear delineation of working assumptions about the nature and utility of strategic culture scholarship is a necessary foundation for the discussion of criticisms and cautions that will follow.

1 *Strategic culture is a valid and important concept which, when employed with care and some reservations, can enhance understanding very usefully.* To some scholars, including this one, this proposition is close to the status of a self-evident truth. It is the foundation belief of what has come to be known as the "culturalist" persuasion in Political Science, International Relations, and Strategic Studies.[3] However, the proposition is contested and, one has to concede, there are substantial grounds on which to base resistance to the new culturalism. For example, culturalist explanations of strategic behavior are likely to compete, or appear to compete, with orthodox realist explanations. Some scholars insist that attempts to deploy culture as an explanatory tool have yet to succeed in providing understanding superior to that achievable by other methods.[4] Also, as this chapter explains in the next section, there are persisting difficulties in identifying the working of culture on policy and strategy. In common with other approaches to strategic studies and defense analysis, if culturalism is applied uncritically it is worse than useless, it is actually

DOI: 10.4324/9781003010302-5

dangerous. Scholars have been known to forget that their strategic studies are really about war, peace, and security. Policymakers armed with a fallacious theory, or perhaps with an initially sound theory that loses validity when it is applied inappropriately, are capable of error that could threaten our very existence. Lest I give the impression that culturalism is uniquely perilous, I would cite the fact that modern strategic studies, a largely American-authored project, was very much an exercise in the elegant logical application of rational choice methodology. Stable deterrence, limited war, arms control, escalation, and crisis management, to cite the leading topics, were tackled and allegedly conquered by theories that their scientifically minded authors believed to have universal validity.[5] They were wrong.

2 *Everyone is encultured; there is no escape from the stamp of culture.* This theorist continues to be puzzled by those scholars who claim that culturalism can, and should, be compared in its explanatory power with some other approach, usually a variant of realism. Culture is not an optional extra for a policymaker or a soldier, neither can it be for those who seek to understand political and strategic behavior. Even the most materially minded defense analyst, a person who explicitly discounts the significance of factors typically identified as cultural, must be encultured. There are no unencultured people. I advance the point as a proposition, but really, this chapter regards the claim as a fact beyond dispute.

3 *To the scholar, nothing is simple.* We make our living, and enhance our reputations, by discovering complexity, or at least by constructing complexity. We will have good reasons to do so. We are trained to employ sophisticated methodologies, to look for multiple causes and influences, and, if we aspire to be social scientists, to build theory. Historians are apt to lose the plot of strategic history because of their undue enchantment by the rich contextuality and contingency of events. Social scientists, in some contrast, are prone to capture by big ideas that enable them to exercise their love of methodological rigor. It has long been noticed that the sophistication of scholarly method and sophistication of research findings are rarely closely correlated. In other words, huge scholarly effort, misdirected in a professionally mandated quest for testable theory, frequently has a disappointing outcome. Work on the causes of war is a prime example of this sad phenomenon. This third proposition amounts to a claim that scholarship on strategic culture, whatever its attitude toward the subject, is ever likely to stray from the productive path when it seeks to construct a general theory.[6] In International Relations a little theory goes a long way, and frequently it goes too far.

The concept of strategic culture is not a difficult idea to grasp and explain. Indeed, it is really so simple that it warrants ascription as being little more than common sense. Every society, or security community, has a more or less distinctive strategic culture which is the product of its historical experience as that experience is perceived. Moreover, every society makes war, and conducts warfare, according to its distinctive character, subject always to the discipline of circumstance. These are deliberate simplifications, but not oversimplifications. Strategic culture scholars are not claiming that strategic culture must drive a way of war, or that strategic behavior must obey some cultural command. The plot, if we may express it thus, is to the effect that societies differ in their beliefs and practices vis-à-vis war, peace, and strategy, and those differences are always likely to manifest themselves in policy, grand strategy, and military behavior, *ceteris paribus*. The scholar is entirely capable of misunderstanding the pervasiveness and importance of strategic culture, because the concept and the phenomena to which it points are resistant to rigorous scientific enquiry.

4 *Culture is a team player: there is much more to statecraft and warfare than culture, but there can be no statecraft and warfare that is innocent of culture.* All behavior is cultural in the obvious sense that it is conducted by encultured people. But all that a culturalist should claim is that culture is always present as a possible contributor to choices. Culture tends to be unusually influential in situations that allow for little analysis and deliberation when people behave all but instinctively. Culture can be important, but is not invariably so. Other members of the policymaking and executing team such as personality, material circumstances, the enemy and its culturally shaped choices, friction, and chance, for example, will also be on the strategic field of play. To recognize the existential reality of culture is not to discover the philosopher's stone for statecraft and strategy. Cultural awareness on our part can only be of assistance, it is unlikely to be the magic ingredient that enables us to comprehend all of what otherwise would be mysterious. Clausewitz tells us that "to impose our will on the enemy" is the object of war.[7] The most reliable method to effect that desirable state is by rendering "the enemy powerless," but in wars of limited purpose that should not be necessary. The master notes that "the strength of his [the enemy's] will is much less easy to determine [than are his means] and can only be gauged approximately by the strength of the motive animating it."[8] The strength of the enemy's will, and indeed the strength of ours, cannot help but be a fit subject for cultural enquiry. Recall Thucydides and his famous triptych of "fear, honor, and interest" as the dominant motives in statecraft and war.[9] Culture contributes to all three, but it rules in the realm of honor. In 1941, Imperial Japan was not deterrable because what the United States demanded was understood in Tokyo to be dishonorable. The most important battlespace in conflicts of all kinds is that in the minds of people. And human minds, all human minds, are encultured.

5 *Nations have identifiably distinct strategic cultures. While they share some characteristics, the differences are huge and they matter for cooperation in policy and strategic behavior, as well as for mutual comprehension.* Some scholars, both historians understandably impressed by the detailed variety of national strategic experience, and social scientists wedded to the authority of a universality of supposedly rational choice, contest this fifth item. Their attitude, to be brutally reductionist, is to the effect that policymakers do what they have to do in the contexts in which they find themselves. Necessity is king. Policy and strategy may be wise or foolish, successful or not, but to understand it one needs look no deeper than to the circumstances of their creation and execution. Political and strategic behavior is governed by rational calculation of costs and benefits, and that calculation is transcultural if not literally universal. Strategic culture scholars reject that view of policy and strategy, at least if the view is presented in as restricted a form as just offered. A sophisticated culturalist perspective does not claim that culture is a wholly independent variable. Culture does not figure in strategic history after the capricious manner of Homer's gods in the *Iliad*. Rather is culture ever present in the "thoughtways" of the inalienably encultured people who are making their rational, if not necessarily reasonable, calculations for policy and strategy.[10]

6 *A society does not choose its strategic culture.* The fit between a strategic culture and its social and political agents will be close and comfortable, even if that culture is less than wholly functional as a contributor to the strategic performance of its bearers. We may change our opinions, but to change our culture would be a task of far greater magnitude. The attractive claim that the truth can set us free has definite limitations. Even if we could triumph over our ethnocentrism and see ourselves more as we are, and less as we

like to believe we are, there is probably going to be little that we can do to alter our culture by an act of will.[11] We are what we are. Strategic and other dimensions of culture are not acquired lightly and casually, and neither can they be casually discarded or changed by a policy decision.

Strategic Culture and Companion Concepts

To understand a country's strategic culture it is not sufficient to examine only its wars. Strategic culture may be most obvious in evidence in wartime, but it grows, develops, and changes out of history as a whole. There are six companion concepts that cultural enquiry can find useful. Some of these are alternatives, while others are complementary.

1 *Strategic culture:* refers to the persisting socially transmitted ideas, attitudes, traditions, habits of mind, and preferred methods that are more or less specific to a particular geographically based security community that has had a unique historical experience.[12] Alternatively, but with the same meaning, strategic culture comprises "[t]he sum total of ideas, conditioned emotional responses, and patterns of habitual behavior that members of a national strategic community have acquired through instruction or imitation and share with each other with regard to ... strategy."[13]
2 *Public culture:* draws attention to the values, attitudes, beliefs, and patterns of habitual behavior common to a particular community. Strategic culture draws inspiration from public culture and should be regarded as constituting a specialized subset of items from it.
3 *Military culture:* insists that strategy is actually done, performed, by institutions that have their own cultures. In the case of military strategy, it is necessary to remember that the military agents of national military strategy will have been encultured as soldiers, sailors, marines, airmen, and space and cyber soldiers.[14] Furthermore, it is the case that within military organizations there will be attitudes, beliefs, and behaviors fairly specific to functional specialisms, which warrant identification as cultural. For example, armor, infantry, artillery, and special operations forces will have professional worldviews keyed to their missions and their demands, as well as to the institutional interests of their parent military organizations. Plainly, we need the discipline of Occam's Razor, lest we discover that bakers and drivers have military cultures all their own. Nonetheless, it is necessary to recognize that strategic culture is akin to a holding company for several not entirely harmonious military cultures.[15] And this condition is true in all societies.
4 *Way of war:* this is a concept of more ancient provenance than strategic culture, which sometimes today is deployed as a synonym for it. It is not a synonym, at least not quite. Thus far, there is next to no literature exploring the connections between strategic culture and way of war. Obviously, a national way of war should reflect, indeed express, a strategic culture, at least insofar as circumstances permit. Strategic culture is the superior concept. It is all but inconceivable that a country would choose to wage war in a manner flatly contradictory to its strategic culture. However, the two concepts are distinctive. The idea of a way of war is much closer to the action, to strategic and military behavior, than is strategic culture. It may prove useful to think of strategic culture as being reasonably permissive, within a range, of military "ways."[16]

5 *National style:* this concept suggests a country's strategic culture will find expression in a pattern of thought and especially behavior that can be conveyed meaningfully as a national style. In other words, national style is a strategic culture in action. This is a broader notion than is "way of war," since national style may encompass statecraft in general and not just warfare.[17]

6 *Strategic personality:* despite the perils of possible reification, the idea of strategic personality fits quite comfortably in the toolkit of the culturalist. If a security community has public, strategic, and military cultures, a preferred way of war and a national style, it is no great intellectual or evidential stretch to postulate a strategic personality. Some very modest amount of scholarly work has been conducted employing this concept, but thus far it has gained few adherents.[18] It is the opinion of this author that the concept of strategic personality has considerable potential and deserves a greater airing than it has secured to date.

As can be well imagined, a concept as opaque and inherently contestable as culture and its adjuncts has attracted both the worst and the best in scholarship. The grounds for scholarly doubt and quibbling, both the insightful and the merely pedantic, are legion. The next section briefly presents and examines some of these. As mentioned before, scholars hoist with professional righteousness are apt to lose the plot. I am fond of quoting science fiction writer Poul Anderson, who wrote: "I have yet to see any problem, however complicated, which, when you looked at it the right way, did not become still more complicated."[19] So it is proving with strategic culture. Anthropologists and sociologists have delighted us with in excess of 300 definitions of culture, so it is not hard to grasp why scholars can have a professional field day with the concept of, and ever-arguable evidence for, strategic culture.

There is merit in the long-familiar claim that "geography is destiny," though there are people today who would amend the claim to read, "culture is destiny."[20] In truth, it is more accurate to state that "history is destiny." But, the history of a community is dominated by the influence of its geography. It is plausible to argue that the strategic, inter alia, culture of a security community is the product of that community's responses to the challenges, including opportunities, posed by its geography.[21] As in real estate, the most vital factor of all is location. It is location that determines one's neighbors, or even whether or not one has neighbors. The location provides the terrain, latitude, and climate, and hence the economic, social, and political relations of one's society.

Culturalists are not, at least should not be, determinists. They do not claim that policy and strategy are predetermined by impersonal forces. But they do argue that a unique geography, fueling a unique history, gives rise to a distinctive way of looking at the world, both in general and in particular. Every society has founding myths and a dominant explanation of its historical experience, which will contain some truth. Patriotic history, wherein the adjective is more potent than the noun, is a universal phenomenon. Societies naturally favor the home team, right or wrong, drunk or sober. Since the primary battlespace in international relations, as in war itself, is the mind(s) of the adversary, the values, attitudes, and preferences of that mind have to be of cardinal importance. Those qualities can be conflated into the compound concept of culture. There are good reasons why culture alone will rarely suffice to explain intentions and behavior, but it is apt to be so significant a policy contributor that to ignore it is always a mistake, sometimes a fatal one.

What does culture do? Its function is to tell us who we are and what we are, in our own historically founded and geographically grounded perspective. Our strategic culture will tell us what we tend to do well, and poorly. As a consequence, culture imparts preferences for behavior in which we believe we excel. Given the ubiquity of self-deception, as suggested by the passing reference to patriotic history, it is entirely possible for us to inherit and cherish a thoroughly dysfunctional strategic culture. However, there are sharp limits to constructivism. Dysfunctional doctrine, theory, behavior, and culture are certain to be run over by unwelcome historical realities. German strategic culture, with its undue regard for the sovereignty of the operational concept, was demonstrated conclusively to be dysfunctional.[22] In short, the German way of warfare did not work well enough, notwithstanding the fighting power of German soldiers and units. Israel has a strategic culture driven both by its geography and by the recent history of the Jewish people. That culture led to repeated success in warfare, but to failure in war. That is to say, it failed to wage warfare of a character that would promote a lasting peace.

Our culture tells us who we are because of whence we came, and how we should behave. It locates us. Although culture is socially acquired, its effect upon us is closer to being the product of genetic programming than it is to the impact of merely powerful opinions. Where should we look for strategic culture? Following sociologist Raymond Williams, this chapter advises that culture is to be found (1) in a society's ideals, (2) in its documents and some other material artifacts and icons, and (3) in behavior. With respect to behavior, Williams claims that culture "is a description of a particular way of life which finds expression in institutions and ordinary behavior."[23] In other words, culture is ideals, it is the evidence of ideas, and it is behavior.

Just as strategic culture should not be assessed apart from the public culture of which it is an integral part, so that public culture can be understood only in relation to the society's history, considered broadly. But no matter how clear is one's definition of strategic culture, a host of problems beset efforts to employ the concept as an essential tool in the small arsenal of potent ideas for national and international security. We turn now to a brief review of the leading criticisms of the idea of strategic culture and its applications.

Problems with Cultural Explanation

Before we rush to our anthropology and history books, it is well to take note of the charges that have been, indeed are still, leveled at the cultural approach to the understanding of policy and strategy. By way of self-defense, although this chapter favors a cultural approach, it does not advocate such an approach exclusively, and neither is it blind to the limitations and difficulties of culturalism. To register a point of argument preemptively, it is the view of this author that to exclude cultural consideration either consciously or unconsciously is vastly more damaging to understanding than employing the admittedly troubled concept of culture. So, why are many scholars distrustful of the culturalist project?

First, it is claimed that noncultural explanations of strategic history offer as good, if not better, understanding of decisions and events than do those that lean on supposedly cultural evidence.[24] In many cases this is true. But it is also true that a cultural dimension to the explanation of policy and strategy is generally superior, and more plausible, than is straight *realpolitik*. In addition, to repeat a vital point made earlier, even the most materially minded British or American practitioner of *realpolitik* cannot help but be an encultured agent of his or her society. To butcher the old saying, you may try to take the policymaker out of culture, but you cannot remove the culture from the policymaker.

Strategic history requires explanation that includes, and sometimes should be dominated by, a cultural dimension. This is true for all communities.

Second, culture is hard to find. The broader and more inclusive the understanding of culture, the more difficult it is to identify and distinguish from noncultural phenomena. If I wish to argue that American public, strategic, and military cultures play significant roles in policy choice and military behavior, where do I look for evidence in support of the claim? Some social scientists insist, correctly of course, that if the subject for evaluation is not distinguishable from its environment, let alone from scholars themselves, careful study is impossible.[25] If strategic culture is manifest in strategic behavior, how can one assess its contribution to that behavior? This is an excellent methodological point. Fortunately, though, it is a problem only for theorists. The fact that strategic culture does not lend itself to theory building does not mean that it is lacking in truth and in value as a contributor to explanation. Since culture provides the assumptions that underpin our thoughts and behavior,[26] I sympathize with scholars who strive to find and isolate it as an explanatory variable. Plainly, theirs is mission impossible. If one asks the wrong question, one is unlikely to find a useful answer, except by accident or through fortunate serendipity. We should not be dismayed by the failure of social science to design a methodology for the identification, dissection, and measurement of the influence of strategic culture.

Third, a more troubling aspect of the culturalist project is that it is, of course, led by university-based scholars who typically have little if any practical experience of the political realities of policy and strategy making. As a consequence, they are apt to forget, or perhaps just to understate, the degree to which government runs by negotiation and compromise. Not only will decisions reflect *realpolitikal* assessments, albeit ones conducted by necessarily encultured people, assuredly they will also compromise negotiated outcomes among stakeholder institutions and among individuals with varying personal and institutional authority. The result of the normal workings of government must be influenced, however lightly, by culture, but in practice decisions on policy, grand strategy, and military strategy are likely to be heavily negotiated. The outcome will express more accurately the contemporary balance of power among key policy players than anything readily identifiable as, say, a clear manifestation of American strategic culture. This is more of a caveat than a criticism of culturalism. After all, the way in which we decide upon policy and strategy is itself a vital part of our political or public culture. Even negotiated policy outcomes will be in debt to cultural influences. Also, it is plausible to argue that public, and perhaps strategic, culture will frame the policy and strategy debate, and limit the scope of acceptable negotiable outcomes.

Fourth, as suggested already, the concept of strategic culture promises to explain too much, which has to mean that it explains too little. If its influence is pervasive, around us and within us, everything that we think, say, and do, has to be, in some measure, an expression of our culture.[27] We cannot function uncultured. This is true but heroically unhelpful. Are we the prisoners of our culture? Always, or only sometimes? And how permissive is our strategic culture? Culture begins to resemble human nature as a master narrative that bears a necessary truth that is almost wholly unenlightening. Scholars should not be daunted by this criticism, but alert to the danger indicated.

Fifth, because war is a duel on a larger scale, to quote Clausewitz again, it is a realm of necessity as well as of preference.[28] Lord Kitchener commented at a bleak moment in 1915 that "one makes war, not as one would like to, but as one must."[29] National strategic cultures and the ways of war that they favor are always hostages to events that one belligerent alone cannot control. This means that warfare and its strategic direction has to be a process of adaptation, even improvisation, as events unfold unexpectedly.[30] While belligerents may be able to adapt in a manner consistent with their cultural paradigm of warfare, it is possible that they may not. A country could be obliged by unwelcome circumstances to wage war in a manner substantially at odds with its strategic culture. One of the plainest historical examples is the British commitment to continental warfare on the largest of scales from 1915 to 1918.[31] It is true that the maritime blockade of the Central Powers eventually had an important harmful effect on public morale and health. But it is also untrue to claim that the Royal Navy won the Great War. Liddell Hart made that assertion, but he made it as an item of faith, without evidence.[32] The war could only be won on the battlefield against the main body of the German army. And that main body was on the western front. In 1918, the British Expeditionary Force was the most potent among the Allied armies, so it was compelled by circumstances in that year to bear the heaviest load in defeating the prime army of the period, the German.

A sensible culturalist does not contest the facts of adaptation and flexibility in the duel that is warfare. But he or she does insist that belligerents adapt and behave flexibly, albeit in contexts not entirely of their choice, in ways as compatible with their culture as circumstances allow. For example, even though Britain was obliged to wage continental mass warfare from 1915 to 1918, its choices in style of combat were markedly different from those of the Germans. Material explanations carry a lot of the weight for the German decision not to mass-produce tanks, for example, but so does German military culture.[33] The culturalist does not advance culture as the sole explanation for decisions and styles of behavior. Culture is only one factor, albeit a pervasive and typically unrecognized element.

Sixth, the concept of strategic culture attracts the compound charge that it encourages oversimplification, an unscholarly reductionism, and – worst of all – determinism. If seriously abused, the idea of strategic culture can indeed be misemployed to commit all those sins. But most good ideas become bad ideas if too much is asked of them. Those of us who find strategic culture to be an enlightening idea do not claim that it is the golden key that must unlock the door to a full understanding of an adversary's thoughts, plans, and behavior. We do not hold that the complexity of motives that move a community to choose to fight, and then to fight in a particular way, can be usefully reduced for our easy comprehension to cultural judgment. And least of all do we adhere to the determinists' creed which, in this context, insists, to repeat, that culture is destiny. We know from history as well as from Clausewitz that war is the realm of chance. We know also that humans are an adaptable and flexible, as well as a murderous, species. However, although these grave complaints against culturalism have little merit, it is important that they should not be denied with undue enthusiasm. Culture is a powerful influence on thought and behavior, and it is all the more so for being a silent and inalienable presence.

Seventh, some critics of strategic culture allege that the concept usually is applied in such a way that it appears to be static, unchanging. The claim is that strategic culture is in fact dynamic, constantly in motion. It can even alter radically as a result of profound systemic shock. German strategic culture was not changed by defeat in 1918, but it was transformed by the Armageddon of 1945.[34] As argued earlier in this chapter, a noteworthy degree of continuity, of stability, is required for the adjective *cultural* to be applicable. One must not claim that the strategic culture of a community is fixed for all eternity. But one must insist that because culture typically has deep roots in history and geography, it cannot be lightly amended, let alone discarded and replaced wholesale by some alternative body of assumptions, modes of thought, and behavioral preferences. Undoubtedly, however, a cultural perspective may bias an observer to privilege continuities over signals of change.

Eighth, strategic cultures are not wholly exclusive. The national ways of war of different countries will share many features. Communities in regular touch with each other, politically, strategically, economically, and technically, will be bound to adopt some, at least, of the same military ideas on contemporary best practices. Also, they will invest in proven technologies that are generally accessible and affordable. Having said that, research does demonstrate that even when societies in more or less active competition face the need to decide among common military technologies, they will tend to decide in ways and for purposes that have an unmistakable cultural dimension. National choices in the 1930s and early 1940s with respect to armored forces, mechanization, motorization, and airpower were especially revealing, for example. American armor emphasized mobility, while German stressed fighting power.[35] Both countries built armored death traps. American tanks were under-gunned and prone to go up in flames too easily because of their petrol engines. German tanks were truly lethal as combat vehicles but were notoriously mechanically unreliable. Because warfare is by definition a competitive pursuit, cultural eccentricity is apt to be disciplined by the enemy, actual or anticipated. Strategy is an inherently pragmatic activity. Any community which indulges its strategic cultural preferences as it were autarkically is likely to find that it will not trouble historians for very long.

Much of strategic lore and practice is genuinely transcultural wisdom. For example, consider the near universality of Napoleonic military organization for regular armies. Also, it is worth noting that the leading military state of an era will always attract imitation, as well as efforts to find asymmetrical compensating offsets. The accumulating enthusiasm for the concept of strategic culture and its possible practical utility had its origins in the mid-1970s in the context of some American doubts over the generally presumed commonality of assumptions between the superpowers on the subject of nuclear strategy. If the Soviet Union did not share America's official view of nuclear war, and especially its view of how such a war might be conducted in a controlled and flexibly limited fashion for early termination, America was building its nuclear war plans on the basis of false assumptions. It takes two, or three, four, and five, if we include Britain, France, and China, to control and terminate a global nuclear war at a point well short of total catastrophe. A great deal of revealed truth and practice regarding nuclear weapons is probably all but universal. But culture can intrude even in that quintessentially technological realm, where there is no room for doubt about the scale of the potential damage, and hence the totality of the stakes.

Conclusion

We conclude by noting a paradox. On the one hand, the concept of strategic culture is controversial. As was demonstrated in a previous section, scholars are troubled by the imprecision of the idea and by the difficulties attendant upon employing it to help explain strategic behavior. On the other hand, perhaps strangely, there is relatively little controversy about the content claimed for some strategic cultures. The paradox is quite startling. The scholarly disagreements at least have attracted attempts at methodological and definitional rigor, whereas most of the commentaries and histories that refer to strategic culture and ways of war are anything but rigorous. The explanation may just be that on the one hand, social scientists have tended to focus on theory building, albeit without much success to date. On the other hand, historians are apt to be hostile to grand concepts, or just to big ideas, and are uncomfortable with theory without much narrative support.[36] I must admit that I share that discomfort, even though I count myself in the ranks of the theorists.

It is important not to forget the boundary of our ambition. This chapter claims only that strategic culture is important. It is not suggested here that it is all important, or that its importance is fixed and stable from issue to issue over time. When one itemizes and explains the principal features of a strategic culture, both author and reader may forget that decisions and behavior typically are shaped and propelled by several, even many, motives. Also, the honest scholar must admit that victory is always probable in the hunt for telling illustration. When one knows what to look for, one is likely to find it.

For all of the difficulties of translating the idea of strategic culture into productive practice, it is worth the effort required. Public, strategic, and military cultural differences between nations are important for their potential to breed misunderstanding and lack of cooperation. All too often, British or American behaviors are criticized by people who fail to appreciate the cultural contexts for those behaviors. To understand is not necessarily to excuse, with respect to those occasions that demand a judgment, but at least it is to appreciate better just why particular actions were taken or declined. In addition to attempting to understand an enemy, it is scarcely less important to understand a friend and ally and, for the most difficult task of all, oneself.

Notes

1 This chapter is excerpted from an unpublished paper prepared by Colin Gray for the Jamestown Symposium, "Democracies in Partnership: 400 Years of Transatlantic Engagement," 18–19 April 2007. We are grateful to Valerie Gray for permission to publish it in memory of our dear late friend and colleague.

2 For example, Colin S. Gray: "National Style in Strategy: The American Example," *International Security*, Vol. 6, No. 2 (Fall 1981), pp. 21–47; *Modern Strategy* (Oxford: Oxford University Press, 1999), Ch.5, "Strategic Culture as Context"; and "Out of the Wilderness: Prime Time for Strategic Culture," *Comparative Strategy*, Vol. 26, No. 1 (January–March 2007), pp. 1–20.

3 International Relations in upper case refers to the scholarly field, while international relations in lower case refers to the real world.

4 See Michael C. Desch: "Assessing the Importance of Ideas in Security Studies," *International Security*, Vol. 23, No. 1 (Summer 1998), pp. 141–70. An as yet unpublished paper by Patrick Porter at Britain's JCSS, Watchfield, also is well worth reading: "Good Anthropology, Bad History: The Cultural Turn in Studying War," 2006. Beatrice Heuser provides a friendly, but characteristically hard-hitting critique of some contemporary culturalist enthusiasm in her insightful review essay, "The Cultural Revolution in Counter-Insurgency," *The Journal of Strategic Studies*, Vol. 30, No. 1 (February 2007), pp. 153–71.

5 Deterrence theory is roughly handled in Keith B. Payne, *The Fallacies of Cold War Deterrence and a New Direction* (Lexington, KY: University Press of Kentucky, 2001).

6 On the quest for testable theory, see Alastair Iain Johnston: "Thinking about Strategic Culture," *International Security*, Vol. 19, No. 4 (Spring 1995), pp. 32–64; *Cultural Realism: Strategic Culture and Grand Strategy in Chinese History* (Princeton, NJ: Princeton University Press, 1995); and "Strategic Cultures Revisited: Reply to Colin Gray," *Review of International Studies*, Vol. 25, No. 3 (July 1999), pp. 519–523.

7 Carl von Clausewitz, *On War*, ed. and trans. by Michael Howard and Peter Paret (Princeton, NJ: Princeton University Press, 1976), p. 75.

8 Ibid, p. 77.

9 Thucydides, *The Landmark Thucydides: A Comprehensive Guide to "The Peloponnesian War,"* ed. by Robert B. Strassler (New York: The Free Press, 1996), p. 43.
10 Ken Booth, *Strategy and Ethnocentrism* (London: Croom Helm, 1979), p. 14.
11 Ken Booth, *Strategy and Ethnocentrism*, has yet to be surpassed as a treatment of its topic.
12 The most convenient comparative presentation of definitions of strategic culture is offered in Lawrence Sondhaus, *Strategic Culture and Ways of War* (London: Routledge, 2006), pp. 124–25, Table 5.1.
13 Jack L. Snyder, *The Soviet Strategic Culture: Implications for Limited Nuclear Operations*, RAND R-2154-AF (Santa Monica, California: The Rand Corporation, 1977), p. 8.
14 See Williamson Murray: "Does Military Culture Matter?" *Orbis*, Vol. 43, No. 1 (Winter 1999), pp. 27–42; and "Military Culture Does Matter," *Strategic Review*, Vol. XXVII, No. 2 (Spring 1999), pp. 32–40.
15 On the contrasting world views of geographically specialized armed forces, see the brilliant short book by J.C. Wylie, *Military Strategy: A General Theory of Power Control* (Annapolis, MD: Naval Institute Press, 1989).
16 Good examples of the "way of war/warfare" literature include: Russell F. Weigley, *The American Way of War: A History of United States Military Strategy and Policy* (New York: Macmillan, 1973); Victor Davis Hanson, *The Western Way of War: Infantry Battle in Classical Greece* (London: Hodder and Stoughton, 1989); David French, *The British Way in Warfare, 1688–2000* (London: Unwin Hyman, 1990); Richard W. Harrison, *The Russian Way of War: Operational Art, 1904–1940* (Lawrence, KS: University Press of Kansas, 2001); and Robert M. Citino, *The German Way of War: From the Thirty Years' War to the Third Reich* (Lawrence, KS: University Press of Kansas, 2005).
17 For my foray into this particular terrain, see Colin S. Gray, *Nuclear Strategy and National Style* (Lanham, MD: Hamilton Press, 1986).
18 For a methodologically brave employment of the concept of strategic personality, see the pioneering study by Caroline F. Ziemke, *Strategic Personality and the Effectiveness of Nuclear Deterrence: Deterring Iraq and Iran*, IDA Paper P - 3658 (Alexandria, VA: Institute for Defense Analyses, September 2001).
19 Poul Anderson, quoted in Arthur Koestler, *The Ghost in the Machine* (London: Pan Books, 1970), p. 77.
20 John Keegan comes close to making this argument in *A History of Warfare* (London: Hutchinson, 1993).
21 Hew Strachan, for a distinguished example, claims persuasively that "[t]he key determinant of Britain's policy has been geography." "The British Way in Warfare," in David Chandler, ed., *The Oxford Illustrated History of the British Army* (Oxford: Oxford University Press, 1994), p. 425.
22 See the excellent narrative and analysis in Geoffrey P. Megargee, *War of Annihilation: Combat and Genocide on the Eastern Front, 1941* (Lanham, MD: Rowman & Littlefield, 2006), p. 26.
23 Raymond Williams, "The Analysis of Culture," in John Storey, ed., *Cultural Theory and Popular Culture: A Reader* (Hemel Hempstead, UK: Harvester-Wheatsheaf, 1994), p. 56.
24 Desch, "Assessing the Importance of Ideas in Security Studies."
25 Johnston, "Thinking About Strategic Culture."
26 A point made admirably by Jeremy Black, *Rethinking Military History* (London: Routledge, 2004), pp. 13–22.
27 This is my argument in *Modern Strategy*, Ch. 5.
28 Clausewitz, *On War*, p. 75.
29 Kitchener quoted in Michael Howard, *The Continental Commitment: The Dilemma of British Defence Policy in the Era of the Two World Wars* (London: Temple Smith, 1972), p. 126. Howard shows how Kitchener had decided by 19 August 1915 that Britain must make an open-ended, total, commitment to the western front, if France and Russia were to be saved from collapse. pp. 56–7.
30 As Clausewitz insists, "[w]ar is the realm of chance". p. 101.
31 In 1914, the British army maintained 161 regular infantry battalions. During the course of the subsequent four years, that number expanded to a total of 1,750 battalions. Robert M. Cassidy, *Counterinsurgency and the Global War on Terror: Military Culture and Irregular War* (Westport, CT: Praeger Security International, 2006), p. 96.

32 B.H. Liddell Hart, *History of the First World War* (London: Pan Books, 1972), pp. 460–464.
33 On British and German styles in land combat in World War I, see Bruce I. Gudmundsson, *Stormtroop Tactics: Innovation in the German Army, 1914–1918* (New York: Praeger Publishers, 1989); and Tim Travers, *How the War Was Won: Command and Technology in the British Army on the Western Front, 1917–1918* (London: Routledge, 1992).
34 Sondhaus, p. 39.
35 See Allan R. Millett, "Patterns of Military Innovation," in Williamson Murray and Millett, eds., *Military Innovation in the Interwar Period* (Cambridge: Cambridge University Press, 1996), p. 345.
36 Sondhaus, p. 130. Also note the hostility to grand or "meta-narratives" in Jeremy Black's many writings. Black, *Rethinking Military History,* pp. 1, 134.

PART II

Dimensions and Levels of Strategic Culture

5
RELIGION AND STRATEGIC CULTURE

Kerry M. Kartchner

Introduction

Human beings are not born with culture. They are enculturated through association with family, community, tribes, and organizations. Likewise, societies are not created from whole cloth with fully developed notions of self and others, or of values, expectations of appropriate or inappropriate behavior, action templates, or lenses for perception and reasoning. They acquire these from many sources, including history, geography, engagement with other cultures, drought, flooding, other environmental conditions, and even experiences with disease and plagues.[1] This chapter argues that one of the main sources of strategic culture is often overlooked even in the strategic culture literature, and that is religion. Religion can often be a culminating source of identity, values, norms, cognitive perceptions, and reasoning at the collective level in most if not all cultures, and its influence is heavily manifested in a culture's approach to war and peace – the ideational purview of strategic culture.

This chapter does not assume religion's ultimate veracity, nor does it seek to engage, much less settle, the political science methodological conundrum of how to parse and operationally empiricize religion and religiosity. These questions are thoroughly addressed elsewhere.[2] This chapter limits its focus to religion as a meaning-making institution and process. For the purposes of this chapter, then, religion may be considered "a social-cultural system of designated behaviors and practices, morals, beliefs, worldviews, texts, sanctified places, prophecies, ethics, or organizations, that relates humanity to supernatural, transcendental, and spiritual elements."[3] Religions have sacred histories and narratives, which may be preserved in sacred scriptures, symbols, and holy places, that aim mostly to give meaning to life. And with explicit application to the ideas explored further in this chapter, two of the most important defining features of virtually all religions are that they to some extent require faith in divine influence and they seek to organize and influence the thoughts and actions of their adherents around this faith.[4]

Religion permeates virtually all aspects of life for most societies and individuals, whether or not they think of themselves as religious. Annemarie Malefijt provides an apt and poignant observation on the pervasiveness and impact of religion, as well as its

DOI: 10.4324/9781003010302-7

essentially ideational nature, in an early book on the relationship between religion and anthropology:

> Religion ... finds expression in material culture, in human behavior, and in value systems, morals, and ethics. It interacts with systems of family organization, marriage, economics, law, and politics; it enters into the realms of medicine, science, and technology; and it has inspired rebellions and wars as well as sublime works of art. No other cultural institution presents so vast a range of expression and implication.[5]

Demographers have shown that religion can be a powerful indicator of identity. Currently, more than eight in ten people globally identify as religious, either in terms of formally affiliating with a religion, or holding religious beliefs.[6] And that proportion is expected to increase, not decrease.[7]

This chapter proceeds in the following order. First, the decline of religion in the West is explained by introducing and describing the so-called secularization thesis which presumes that influence of religion has and will decline as any given society becomes more prosperous. Second, this chapter summarizes the resurgence of religion in global affairs, and the concomitant revival of religion in social science scholarship, particularly the field of international relations theory. Third, an argument for the synergistic and syncretic relationship between religion and strategic culture is introduced, highlighting how closely intertwined that relationship is in terms of both practice and philosophy. Fourth, this chapter expounds on the specific impact of religion on four specific dimensions of strategic culture, explaining the influence of religion on the formation of identity, on the creation and shaping of values, on the delineation and propagation of norms, and on molding the prism of perceptions and cognition. This chapter concludes with a summary of the case for incorporating religion into strategic culture scholarship, and for strategic culture scholars to be mindful of secular biases in their analyses by considering the critical role of religion as a constitutive source of strategic culture.

The Secularization Thesis

Somewhere along the path to becoming a civilization of thriving modern and secular democracies, the West discarded and marginalized religion, divorcing it from the body politic, and relegating its mores and rituals to private life. With the singular exception of the United States, which has defied this trend, the richer and more prosperous a country becomes, the less religious it is.[8] Thus, a pervasive assumption arose that becoming a modern nation necessarily meant abandoning religion and supplanting it with secular sources of knowledge, values, and legitimacy.

The so-called secularization thesis asserts that as societies modernize and become more affluent, the importance and relevance of religion will decline concomitantly. As originally formulated, scholars proposed three variations on the secularization thesis: (1) that religion would decline as technology, technical standards, and scientific empiricism replaced received wisdom as the primary source of knowledge; (2) that modern society and institutions would come to substitute for the sociological functions of religion; and (3) that various explicitly anti-religious ideologies would supplant religion as "the opiate of the masses" (in Marx's iconic phrase).[9]

This so-called secularization thesis came to permeate not only the governance of societies, but the theoretical thinking about global affairs. In fact, international relations theory is a Western invention developed to describe how (Western) states do and should interact with each other (while assuming the universality of these rules).[10] It is an implicit product of the secularization thesis, having been conceived and promulgated by adamantly secular Western scholars. Consequently, the Western international relations curriculum has, until recently, virtually ignored or dismissed religion. Likewise, non-Western perspectives on international relations were also nonexistent or discounted. In some circles Western international relations scholarship has become stridently anti-religion.[11] Religious devotion has come to be regarded as "irrational" almost by definition. In addition, skepticism toward religion arose from the collective sense that religion was not a reliable source of knowledge. Some critiques of religion went further, holding that religion had brought out the worst in mankind, and had been responsible for millennia of war and bloodshed.[12] In the West – but not so elsewhere – the sources of governmental legitimation gradually devolved to secular rather than divine sources. And religion was quietly dropped from formal international relations theorizing, polite public discourse, and foreign and defense policymaking and reserved to the domain of the personal and private.

It must be noted that the advent of secularization was not (at least initially) a universal phenomenon. The Western experience of religion and secularization contrasts sharply with that of Eastern cultures and societies. Non-Western cultures did not experience these crises between secularism and religion and between modernization and stasis. Nor did they experience an equivalent to the Protestant Reformation leading to a brutal and bloody schism between entrenched and mutually hostile factions of the same faith that brought the entire notion of organized religion into widespread disrepute and delegitimization.[13] Nor did they undergo an Age of Enlightenment (in the secular Western sense) or an Industrial Revolution. The lack of equivalent civilizational upheavals, then, is part of the essential and foundational differences between the strategic cultures of the West and those of the East. Those of the West have inherited a cynical distrust of religion that has yielded wholly secularized ideals of democratic governance, sovereignty, scientific rationalism, and strict separation of church and state, while those of the East have risen from a more deferential stance toward received (religious) wisdom, an emphasis on harmony over democracy, an abiding role for religion and religious values in political governance, and a hierarchical conception of interstate relations, all based in part or in whole on contrasting East/West religious experiences.

The Global Resurgence of Religion

Notwithstanding these historically ingrained biases against taking religion seriously, a reaffirmation of religion in global affairs and a concurrent resurgence of thinking about the role and influence of religion in international relations theory have begun correcting some aspects of these Western-centric predispositions.[14] In many ways the world is experiencing a resurgence of religious zeal and influence, and international relations theory is newly discovering a vital nexus between religion and global politics.[15] Eric Patterson argues that "[t]he impact of religion on world affairs may be the most important feature of international life since the end of the Cold War."[16] This author notes five key trends:

> (1) individual religiosity is rising the world over; (2) public expression of religion by individuals and groups worldwide matters more in political discourse; (3) states are no longer the sole legitimate centers of authority and authenticity, nor are they always the most reliable providers of vital services; (4) religious actors, identities, and ideas are vigorously transnational; and (5) whether at the individual or collective level, religious impulses can transcend what scholars typically define as "rational" or material interests.[17]

To this list might be added an increase in wars, conflict, and terrorism justified by, legitimized by, and perpetuated by religion. According to a report from the Chicago Council on Global Affairs Task Force on Religion and the Making of US Foreign Policy, "Religion has been a major force in the daily lives of individuals and communities for millennia. Yet recent data show that the salience of religion is on the rise the world over." At the same time, "[w]ell-organized and well-funded extremist groups also use religion to deepen existing cultural and political fault lines and justify militancy and terrorism. Just as globalization and communication technologies have supported positive religious developments, they have also facilitated the growth of extremist religious views and the development of dangerous terrorist networks."[18] Rather than leading to further secularization, the social upheaval and economic dislocation associated with modernization has led, according to some scholars, to "a renewal of religiosity and political religious activism." Some social theorists and academics see this rise or resurgence in religiosity as a defensive reaction or backlash to globalization and modernization. They note that survey data show that "increasing numbers of people who have become disenchanted with the seemingly destabilizing effects of modernity," including "the fragmentation of society, the emptiness of consumerism and materialism, the breakdown of families, widespread socio-economic grievances, erosion of traditional morality and values, social dislocation, and culture shock," have returned to embracing religion and religious values as remedies for these social ills.[19]

Religion and the Dimensions of Strategic Culture

There is a significant confluence between the concepts of religion and strategic culture. Both originate and exist in the ideational realm of the mind. Both are collective features of groups, change slowly, and are propagated by many of the same modalities, such as through common scriptural or textual canon, oral or written narratives, teaching, training, proselyting, and personal and group development. Both are conveyed and transmitted through symbols and rituals. Both are hard to define quantitatively. Both are subject to the same criticisms by representatives of the realist, positivist, or empiricist schools of theory and pedagogy: scoping and defining them is problematic, they cannot be empirically measured, and their relevance is ignored or dismissed. Nevertheless, both are manifest in any given group's identity, values, norms, and perceptual lens. For much of the world, religious affiliation (whether through formal membership or through shared faith) is a critical component of a culture's self-identity. This shared collective identity is reinforced by values and norms held by the group that have in many cases originated in religious teachings and philosophies. Religion teaches a group what is of value, and what is to be sought after or aspired to. Religion provides a blueprint for what it means to live righteously and benevolently. Religion also is one of the most important sources of norms, rules, commandments, articles of faith, criteria for discriminating good from evil, and proscribing that which is forbidden. Religion can shape threat perceptions by demarcating the boundaries of the in-group and designating members of out-groups. Religion has left a deep imprint on international norms, especially those related to matters of war and peace. And finally, religion provides the lens through which much of the world population perceives, processes, evaluates, and acts on the social realities of their lives. In sum, religion has contributed significantly to humanity's cognitive tools for reasoning, discerning, and exercising agency.

Theologians and anthropologists have long considered religion and culture to be closely entwined, even symbiotic in their relationship.[20] In his writings, Lutheran theologian Paul Tillich went so far as to argue that religion was the soul of culture and culture was the framework of religion: "religion as ultimate concern is the meaning-giving substance of culture, and culture is the totality of forms in which the basic concern of religion expresses itself. In abbreviation: religion is the substance of culture, culture is the form of religion .... Every religious act, not only in organized religion, but also in the most intimate movement of the soul, is culturally formed."[21]

The confluence between religion and culture also manifests at the strategic cultural level. All major religions have developed some form of "just war" philosophy. However, much of the contemporary international law of armed conflict has evolved from specifically Christian teachings and doctrines for deciding whether a war is justified, and how a just war is to be conducted. Religion can dictate why wars are fought and how wars are fought. Religion can thus be a source of a society's "way of war." Religion has a complex relationship with violence – religion can be a source of conflict, a justification for conflict, or a factor exacerbating the scale, scope, and level of violence in conflict. Religion can be the rationale for believing that "the ends justify the means," or for placing constraints on what is permissible. How cultures calculate and corollate ends and means, and how they formulate and weigh costs and benefits, are often conditioned by religious factors.

Strategic culture can be dimensionalized or manifested in four categories: identity, values, norms, and perceptual lens (or cognitive reasoning).[22] This chapter argues that religion is one of the most important constitutive sources of each of those categories. What follows is a discussion of how religion can be viewed in relation to these four dimensions.

Religion and Identity

Identity at its most basic can be defined as the set of qualities and beliefs that make one person or group different from others. In the strategic culture field, identity has been defined as "the character traits a group assigns to itself, the reputation it pursues, and individual roles and statuses it designates to members."[23]

Religious identity is different from other forms of identity, such as ethnicity or nationalism, because religion grounds identity in enduring "eternal" and "sacred" "truths" that resist secular reasoning, refutation, or contestation by other potentially competing layers of identity. Its veracity cannot be proven or disproven.[24] As social psychologist Renate Ysseldyk and colleagues point out: "the unique characteristics of religion, including compelling affective experiences and a moral authority that cannot be empirically disputed, may lend this particular social identity a personal significance exceeding that of membership in other groups."[25]

In addition, religious identity provides its adherents with a distinctive worldview derived from "a belief system that offers epistemological and ontological certainty" that is unmatched in its ability to provide privileged knowledge, reassurance of meaning and purpose, a sense of self-importance, individual self-conceptualization, improved self-esteem, emotional well-being, and self-exceptionalism (being better or more pious than others).[26] Religious identity also embeds individuals and societies within a well-established support system that reinforces the bonds of association and unity, where self-sacrifice and mutual care are not only often sacred values, but are acts of belonging that reward membership, among other social functions. It is thought that religion, more so than other identities, can promote a sense of control and religious commitment.[27]

For most cultures, identity cannot be fully appreciated without reference to the origins and continued saliency of its religious aspects. Saudi Arabia's identity as the "Keeper of the Holy Sites" is probably familiar to most astute observers of international affairs, but Moscow's claim to be the "Third Rome" or a "light unto the nations" is lesser known, although it is an essential messianic element of contemporary Russian identity.[28] This narrative emphasizes the preservation of "the spiritual-cultural-civilization code of the Rus" and Moscow's destiny as the "Third Rome," which according to Dima Adamsky "positions Russia as the spiritual center responsible for the salvation of the Christian world and of all humanity."[29]

Despite their profound secularization, Western societies often have a layer of identity based on deeply held religious beliefs, values, and norms. According to Willfried Spohn, "even the most secularized Western European types of national identity are generally based on Christian foundations and shaped by the specific nationally predominating form(s) of Christianity."[30] The highly secularized nations of Western Europe also retain symbols and elements of identity (such as place names and holidays) that indicate residual religious origins. Moreover, religious identity can be a strong predictor of when religion will promote or facilitate war, according to author J. William Frost, particularly when "[a] group defines itself as a holy or chosen people with special obligations and privileges, particularly involving a right to a land."[31] He notes that such chosen people are more likely to resort to war if: "socio-political divisions are justified and enforced on religious lines; the religious group feels persecuted in the present or past and unable to obtain justice; the religious group is cohesive enough to unify politically and sees the possibility of gaining power to achieve autonomy or dominion; [or] the land itself is sacralized and contains sites of special holiness."[32]

Identity has become an important concept in global affairs and foreign policymaking, to which the literature of strategic culture has made significant contributions.[33] However, that literature has not fully leveraged the role of religion in constituting group identity, especially with regard to the latency of religious identity within supposedly secularized societies. Religion can be one of the most significant components in all aspects of identity.

Religion and Values

Values are deeply held beliefs about what is desirable, proper, and good that serve as broad guidelines for social interaction. The key noun is "beliefs" (as in an ideational construct). As one sociologist has noted, "not all beliefs are values, but all values are beliefs."[34]

Values include "material or ideational goods that are honored or confer increased status to individual actors or members of groups."[35] Values are acquired from many sources, including culture, upbringing, political beliefs, education, profession, and life experiences lived either individually or collectively. However, for many cultures around the world, religion is often the foundational source of societal and leadership values.

Religious values come from sacred texts or narratives and authoritative interpretations or discussions of these texts, from the examples of sacred role models, including collected chronicles of deeds and sayings of such role models, from philosophical reasoning, and from direct revelation. In fact, one of the most important distinctions between monotheistic and polytheistic religions is whether a religion's values and beliefs are putatively based on direct revelation, or on the reasoning of great sages passed down through the ages.[36] Others have ascribed a more secular and functional thesis to the origins of religious philosophies, identities, and values whereby religion arose because it fostered greater group cohesion and survival, and allowed societies to scale up.[37]

Religion may specify what values are to be upheld, protected, defended, or sacrificed for, in both a material sense (the material integrity of the church, its property, its resources, its people, its leadership) and a spiritual sense (the purity of the faith). Vestigial religious values may be present even in highly secularized societies.[38] Even in contemporary China, where the official ideology is atheist Marxism, Confucian ideals of benevolence and harmony are core values, as they have been for 2,000 years.[39] And this is true for other Asian nations with cultures based on Confucianism, Hinduism, and Buddhism. For much of the rest of the world, separation of church and state is categorically not a core value. Often, religion is inseparable from the state. Even largely secular countries in Western Europe subsidize specified religious groups out of state funds (as in Belgium), or require parishioners to pay a "church tax" that is collected by the government – practices that would be anathema if a strict American-style separation of church and state were enforced.

Social psychologists have found that certain resources, ideas, places, personages, and dates can transform through a process of sacralization into values that persist over time, are strongly linked to identity, transcend material trade-offs, and elicit increased anger and resentment when challenged.[40] Because these values are substantially linked with religion, they are frequently called "sacred values," although some social scientists have adopted the more secular term "protected values." A "sacred value" has been defined as "any value that a moral community implicitly or explicitly treats as possessing infinite or transcendental significance that precludes comparisons, trade-offs, or indeed any other mingling with bounded or secular values."[41]

Social and cognitive psychology research suggests that: (1) sacred values may be critically involved in sustaining seemingly intractable cultural and political conflicts; (2) sacred values appear to be intimately bound up with sentiments of personal and collective identity; (3) sacred values resist material tradeoffs; (4) sacred values are more likely to be associated with moral obligations to act rather than moral prohibitions against action; and (5) moral agents motivated by sacred values are more focused on the duty to act than on the consequences of acting.[42]

Religious experiences and the belief in religious meanings may also transform physical spaces into sacred spaces. Lily Kond notes that certain culturally derived perceptions influence the use and spiritual meaning of such sacred spaces.[43] Sacred values are a cultural phenomenon. They are created within and shared by a group, change slowly, and are not something members of a group are born with. They can have a profound and enduring impact on identity, as well as norms, values, and perceptions.

Religion and Norms of War and Peace

Norms are accepted, expected, or customary behaviors within a society or group. Norms are the mechanisms a culture uses to achieve its values. Norms are in essence a set of rules for human conduct. They are derived from constitutions, laws, rulings, directives, customs, codes of conduct, rituals, and traditions. As with other dimensions of strategic culture, religion is a major source of norms, including those governing approaches to matters of war and peace.

The literature on strategic culture has engaged the concept of norms and norm-making, especially with respect to nuclear nonproliferation and other areas of national security.[44] Military historians Theo Farrell and Terry Terriff argue that "norms regulate action by defining what is appropriate (given social rules, moral codes, etc.) and what is effective (given the laws of science)."[45] From a strategic culture perspective, norms are critical because they can shape uniquely national military styles and doctrines, including the purpose and possibilities of military change, and offer guidance on the use of force.[46]

As with values, religion is a significant source of norms even in modern largely secular and pluralistic societies.[47] Religious norms find articulation in scriptural canon, interpretive exegesis, spiritual role model examples, doctrine, tenets, articles of faith, catechisms, beatitudes, commandments, exhortations, and so forth. Again, as with values, beliefs are at the core of religious norms. Religious norms often represent a desirable way of life that achieves or points to the achievement of religious values as described above. Religion is by its nature normative and prescriptive, and membership is based on submission and sustained adherence to the rules. Religious ethics are generally deontological, meaning ethical decisions are based on applicable rules and not so much on teleological consequences or utilitarian calculations (although religious belief systems will exhibit norms informed by the full spectrum of ethical frameworks). Some sociologists and ethicists have observed that individuals and institutions in predominantly religious societies will tend toward decisionmaking based on normative guidelines, while more secular societies will make decisions based on the consequences or utility of the outcome.[48]

Among the most important normative contributions of religion at the level of strategic culture are norms of war and peace – those which determine the value of life and peace, and those which determine when it is necessary to defend life and peace by force of arms. Virtually all religions oppose war and violence while promoting peace as the ideal norm, but most make exceptions in certain circumstances, primarily for self-defense. These religious norms have shaped modern laws of war and armed conflict as these laws have undergone a process of secularization, and are now enshrined in international statutes. Furthermore, understanding religious norms of war and peace is critical to a strategic cultural analysis because religious wars differ from nonreligious wars in significant ways.

Religions down through the ages have adopted one or more of three major categories of norms on war and violence, depending on the circumstances prevailing at the time, including (1) various forms of pacifism (all violence and killing is wrong); (2) belief in "just wars" (some wars, at least, are justifiable because they are perceived to be in the interests of justice, and should therefore be fought according to just rules), or (3) "holy war" (wherein God is perceived as commanding that a religion's followers make war on the "others").[49]

Religious values and norms have contributed to shaping the modern theory of just war and the modern Law of Armed Conflict. Just war theory was first developed within early Christian writings based in part on interpretations of Greek philosophy.[50] As this concept was developed over time, it came to incorporate two sets of principles: *jus ad bellum* and *jus in bello*. *Jus ad bellum* tenets govern the decision to go to war and are based on ideas of just and legitimate authority to wage war and just causes for going to war. *Jus in bello* incorporates strict limits on the conduct of war. Jewish, Islamic, and Hindu traditions all include similar precepts meant to guide deliberations on whether to go to war, and strict rules for circumscribing the violence of war, who can engage in war, and who should be exempted or protected from war.

The modern Law of Armed Conflict is based on four principles codified in the Geneva Conventions of 1864 and 1949 and their various Protocols. They are (1) distinction (or discrimination) between the civilian population and combatants and between civilian objects and military objectives; (2) proportionality in the use of military force relative to the military objectives to be gained; (3) military necessity in selecting targets; and (4) minimizing harm to nonmilitary property and facilities (such as churches, hospitals, and schools).[51] These norms arose from religious doctrines and teachings that are fundamentally common to all major faith traditions, and which predate by centuries or even millennia the secularization and codification of the Law of Armed Conflict, as well as International Humanitarian Law.[52] The requirement to discriminate between combatants and noncombatants is well established in ancient and historical Christian, Jewish, and Islamic scripture and law. Hinduism and other Asian faith traditions are based on the central and sacred principle of *ahimsa*, or non-harm to all living things. Likewise, excessive use of force has always been prohibited or discouraged in these religious traditions, and explicit rules for circumscribing violence and destruction (such as sparing fruit-bearing trees and the environment) have been spelled out in the Qur'an and in Hindu Vedas.[53] The influence of religion on a culture's concept of "just war" is not limited to Western societies or international laws of war. Vladimir Tikhonov, Professor of Korean and East Asian Studies at the University of Oslo, asserts that "the justifications used for the sake of legitimizing warfare in the [East Asia] region today are implicitly based on Confucian assumptions."[54]

Religious wars are different from purely secular wars. Religious wars, and especially "holy wars," represent a special category with distinct features that distinguish them from other more secular forms of conflict. Holy wars last longer, tend to be bloodier, can escalate to ever-greater uses of force, and are more difficult to terminate. These features are driven primarily through the intervening action of sacred values, which cannot be compromised, and which can provoke increased anger if challenged, including in matters of war. Although it is important to know that *jihad* can have many meanings in Islamic scripture, certain forms of offensive Islamic *jihad* can also be included in the category of holy wars and can exhibit similar characteristics.[55]

Religious norms can be powerful motivations for peace, reconciliation, tolerance, humaneness, and constraint in war. They can also work to amplify intolerance, inhumanity, and hatred. Some authors term this extreme dichotomy "the ambivalence of the sacred."[56] This also means that religion can at times favor peace and diplomacy, while at other times it can enflame conflict. In any strategic cultural analysis that addresses a culture's way of war, it will be critical to parse this duality, or ambiguity, and to accurately portray the negative and positive influences of religion, to examine when religious influences facilitate or exacerbate tendencies toward war, and when they militate for peace.

Religion and Perceptual Lens (or Cognition)

The fourth dimension of the strategic cultural framework utilized here is the perceptual lens. This includes both the filter through which groups determine "facts" about others, and the cognitive processes by which groups make sense of the world around them (sometimes also referred to as cognitive style). Different cultures think and reason differently, and define and practice alternative forms of "rationalities."[57] The typical cost-benefit trade-off of Western calculus can sometimes gloss over the fact that cultures define "costs" and "benefits" differently.

Sociologists and anthropologists have long recognized that religion can affect perceptions, and have demonstrated this by showing that different religious groups may even have contrasting interpretations of simple optical illusions.[58] One group of social psychologists found that "it seems possible that religious beliefs may indeed lead to different and sometimes discrepant and incompatible interpretations of the same incident." Their study contrasted views of religious and nonreligious groups from the same regional setting:

> Dutch Calvinists and atheists, brought up in the same country and culture and controlled for race, intelligence, sex, and age, differ with respect to the way they attend to and process the global and local features of complex visual stimuli: Calvinists attend less to global aspects of perceived events, which fits with the idea that people's attentional processing style reflects possible biases rewarded by their religious belief system.[59]

Such perceptual differences may be caused, for example, by an emphasis on *individual responsibility* in Calvinism and *social solidarity* in Orthodox Judaism and Catholicism.[60]

Religion has a profound impact on how certain groups of people perceive, conceptualize, and think about reality. For these groups, "religion claims priority over reason."[61] When seen through a religious perspective, the perceptual lens can shape a culture's view of its role and situation in the global landscape. It may see itself as a "vanguard of the faith," or "safekeeper of the faith," meaning it assigns itself the role of a temporal home base where resources are allocated and prioritized for expanding and establishing a religious community at the forefront of proselyting to the world. It may see itself as a "defender of the faith," consecrated with responsibility for maintaining the purity and devotion of the community of the faithful. And sometimes it may assume the role of "avenger of the faith," going to war against infidels, apostates, and the unclean in the name of their god, and for the sake of punishing those who have offended the community of the faithful or its beliefs.

Summary and Conclusion

This chapter began by proposing that religion is one of the most important sources of strategic culture. For some cultures, religion is the principal source of identity, values, norms, and perceptual lens. Even for truly secularized Western nations, there are residual elements of religion in their belief systems, daily practices, and the holidays they celebrate.

This chapter noted the decline of religion in the West and introduced the so-called secularization thesis, which presumes the influence of religion will decline as any given society becomes more prosperous. As a consequence, religion is often missing from the

Western international relations curriculum.[62] Deeply embedded secularization has rendered most Western policymakers and scholars consistently blind to the impact of religion on their own and others' cultures. This religious blind spot has also impaired the ability of Western scholars and intelligence analysts to understand non-Western cultures.[63] The dismissal, or under-privileging, of religion has contributed to geopolitical surprise, underestimating the disruptive power of religious movements, and problematic assessments of political and military intentions of religiously inspired social uprisings.[64]

This chapter also briefly touched on the resurgence of religion in global affairs, and the concomitant revival of religion in social science scholarship, particularly in the field of international relations theory. However, such a revival is not generally reflected in strategic culture scholarship, with the exception of Dima Adamsky's recent book on religion, politics, and strategy in the Russian nuclear weapons enterprise.[65] In particular, religion is largely missing as a source of culture in the literature on strategic culture. Even eminent strategic culture scholars fail to mention religion when listing and exploring more common sources of strategic culture such as geography, history, political structure, and language. On the one hand, this is somewhat surprising, given they are closely intertwined in terms of both practices (as generators of normative ideational principles) and philosophy (as meaning-making mechanisms). On the other hand, given the workings of the secularization thesis even among strategic culture scholars, it should not be that surprising. This chapter then explored the specific impact of religion on four dimensions of strategic culture, demonstrating the influence of religion on the formation of identity, on the creation and shaping of values, on the delineation of norms (especially with respect to issues of war and peace, the nominal purview of strategic culture), and on molding the prism of perception and cognition, with special emphasis on how those dimensions are manifested at the strategic level.

In conclusion, this chapter is a plea for strategic culture scholars to examine their own religious, secular, and ethnocentric biases and to incorporate an examination of the role of religion in their own scholarship, where appropriate, by considering the critical role of religion as a constitutive source of strategic culture. As Scott Thomas wrote: "a struggle over the soul of the new world order is taking place, and taking cultural and religious pluralism seriously is now one of the most important foreign policy challenges of the twenty-first century."[66]

Notes

1 One anthropologist has noted that "some cultural phenomena can be explained in terms of the historical prevalence of communicable diseases within a particular society," Michael Minkov, *Cross-Cultural Analysis: The Science and Art of Comparing the World's Modern Societies and Their Cultures* (Los Angeles, CA: SAGE Publications, 2012), 174. Minkov cites several studies that link experience with various diseases to differences between cultures affected or unaffected by such diseases.

2 For a good survey of debates over whether religious beliefs constitute "truth," see Andrzej Bronk, "Truth and Religion Reconsidered: An Analytical Approach," The Catholic University of Lublin, accessed 11 June 2022. https://www.bu.edu/wcp/Papers/Reli/ReliBron.htm.

3 John Morreall and Tamara Sonn, "Myth 1: All Societies Have Religions," *50 Great Myths of Religion* (Wiley-Blackwell, 2017), 12–17.

4 Linda Woodhead presents a broad range of ways to conceptualize religion, including as a source of identity, values, and norms, in "Five Concepts of Religion," *International Review of Sociology* 21, no. 1 (2011): 121–143.

5 Annemarie De Waal Malefijt, *Religion and Culture: An Introduction to Anthropology of Religion* (New York: Macmillan Company, 1968), 1.
6 "Religion by Country 2023," *World Population Review*, https://worldpopulationreview.com/country-rankings/religion-by-country.
7 Jordan Wootten, "What's the Future of the Global Religious Landscape? 3 Takeaways from the Pew Center Projections," *Ethics & Religious Liberty Commission*, 21 June 2021. https://erlc.com/resource-library/articles/whats-the-future-of-the-global-religious-landscape/ See also Pew Research Center, *Religious Composition by Country*, 2010–2050, April 2015. Archived at https://web.archive.org/web/20200615053333/ https://www.pewforum.org/2015/04/02/religious-projection-table/2010/number/all/.
8 Gregory Chase, "Freedom of Religion, Religiosity, and GDP per Capital," *Journal of International Business Research and Marketing* 1, no. 5 (July 2016): 7–11; Also note the position of the United States on the chart of religion versus gross domestic product in Jeff Diamant, "With High Levels of Prayer, U.S. is an Outlier Among Wealthy Nations," *Pew Research Center*, 1 May 2019. https://www.pewresearch.org/fact-tank/2019/05/01/with-high-levels-of-prayer-u-s-is-an-outlier-among-wealthy-nations.
9 See Linda Woodhead, "The Secularisation Thesis" podcast, 12 April 2016, https://www.religiousstudiesproject.com/podcast/podcast-linda-woodhead-on-the-secularisation-thesis/. See also Bryan S. Turner, "The Secularisation Theory," Chapter 7 in *Religion and Modern Society Citizenship, Secularisation and the State* (Cambridge University Press, 2011), 127–150; and Pippa Norris and Ronald, Inglehart eds., "The Secularization Debate," in *Sacred and Secular: Religion and Politics Worldwide* (Cambridge: Cambridge University Press, 2004), 3–32.
10 Only recently have efforts been made to bring Eastern ways of thinking about international relations into the field. See Amitav Acharya and Barry Buzan, "Why Is There No Non-Western International Relations Theory? Ten Years On," *International Relations of the Asia-Pacific* 17, no. 3 (2017): 341–370. https://academic.oup.com/irap/article/17/3/341/3933493. The authors state that if international reations theory had been born somewhere other than the West, "it would look a lot different."
11 The author was present when a distinguished American dean of modern international relations theory told an audience gathered for an annual International Studies Association meeting in Washington, DC, that "the main difference between us as teachers of international relations and our students, is that our students still believe in God." This statement shockingly implies that becoming a teacher of international relations necessarily means trading religion for atheism as the secularization thesis would predetermine.
12 This is the theme of Richard Dawkins, *The God Delusion* (Boston: Houghton Mifflin, 2006).
13 Splintering into factions, of course, also occurred outside the Christian world. Most Asian religions are in some way offshoots of Hinduism, but this process of fractionation was by and large peaceful and evolutionary, rather than brutal and revolutionary. Nevertheless, in some sense, Islamic civilization may now be confronting some of these centrifugal and schismatic challenges, several centuries after Christianity, as it confronts modernity. See Bernard Lewis, *What Went Wrong? Western Impact and Middle Eastern Response* (New York: Oxford University Press, 2002).
14 For example, see Gilles Kepel, *The Revenge of God: The Resurgence of Islam, Christianity and Judaism in the Modern World* (University Park, PA: University of Pennsylvania Press, 1994); Jonathan Fox and Shmuel Sandler, *Bringing Religion into International Relations* (New York: Palgrave Macmillan, 2004). See also Scott M. Thomas, *The Global Resurgence of Religion and the Transformation of International Relations: The Struggle for the Soul of the Twenty-First Century* (New York: Palgrave Macmillan, 2005).
15 See also Monica Duffy Toft, Daniel Philpott, and Timothy Samuel Shah, *God's Century: Resurgent Religion and Global Politics* (New York: W. W. Norton, 2011); Fabio Petito and Pavlos Hatzopoulos, eds., *Religion in International Relations: The Return from Exile* (Basingstoke: Palgrave Macmillan, 2003); Vendulka Kubálková, "A 'Turn to Religion' in International Relations?" *Perspectives* 17, no. 2 (2009): 13–41; Michael C. Desch, "The Coming Reformation of Religion in International Affairs? The Demise of the Secularization Thesis and the Rise of New Thinking About Religion," in *Religion and International Relations: A Primer for Research*, edited by Michael C. Desch and Daniel Philpott (University of Notre Dame, Notre Dame, IN, 2013), 14–55.

16 Eric Patterson, *Politics in a Religious World: Building a Religiously Literate U.S. Foreign Policy* (New York: Continuum International, 2011), 5.
17 Patterson, *Politics in a Religious World*, 5–6.
18 *Engaging Religious Communities Abroad: A New Imperative of U.S. Foreign Policy*, A Report of the Task Force on Religion and the Making of U.S. Foreign Policy, R. Scott Appleby and Richard Cizik, Cochairs (Chicago Council on Global Affairs), 17. https://keough.nd.edu/wp-content/uploads/2015/12/engaging_religious_communities_abroad.pdf.
19 Chor Pharn, "Vanguard: The Global Religious Resurgence," 25 May 2009. http://www.futuresgroupsg.com/content/vanguard-globalreligiousresurgence/. See also Richard Falk, "A Worldwide Religious Resurgence in an Era of Globalization and Apocalyptic Terrorism," in *Religion in International Relations: The Return from Exile,* edited by Pavlos Hatzopoulos and Fabio Petito (New York: Palgrave Macmillan, 2003), 181–208.
20 Taylor & Francis even publishes a journal titled "Culture and Religion" to serve as a vehicle for promoting scholarship regarding this relationship. https://www.tandfonline.com/toc/rcar20/current.
21 Paul Tillich, *Theology of Culture*, Robert C. Kimball, ed. (Oxford University Press, 1959), 42.
22 For an introduction to these four dimensions, see Jeannie L. Johnson and Marilyn J. Maines, "The Cultural Topography Analytic Framework," in *Crossing Nuclear Thresholds: Leveraging Sociocultural Insights into Nuclear Decisionmaking*, edited by Jeannie L. Johnson, Kerry M. Kartchner, and Marilyn J. Maines (New York: Palgrave Macmillan, 2018), 29–60.
23 See, inter alia, Jeannie L. Johnson, *The Marines, Counterinsurgency, and Strategic Culture* (Washington, DC: Georgetown University Press, 2018), 24.
24 Renate Ysseldyk, Kimberly Matheson, and Hymie Anisman, "Religiosity as Identity: Toward an Understanding of Religion from a Social Identity Perspective," *Personality and Social Psychology Review* 14, no. 1 (February 2010): 61. https://doi.org/10.1177/1088868309349693.
25 Ysseldyk, "Religiosity as Identity," 61.
26 Ysseldyk, "Religiosity as Identity," 60.
27 Steward Harrison Oppong, "Religion and Identity," American International Journal of Contemporary Research Vol. 3 No. 6 (June 2013): 10–16. https://aijcrnet.com/journals/Vol_3_No_6_June_2013/2.pd
28 Nick Mayhew, "Moscow: The Third Rome," *Oxford Research Encyclopedias*, 26 May 2021. https://oxfordre.com/literature/view/10.1093/acrefore/9780190201098.001.0001/acrefore-9780190201098-e-1243; see also Adamsky, *Russian Nuclear Orthodoxy*, 183.
29 Dimitri (Dima) Adamsky, *Russian Nuclear Orthodoxy: Religion, Politics, and Strategy* (Stanford, CA: Stanford University Press, 2019), 188. See also Simon Franklin, "Identity and Religion," in *National Identity in Russian Culture*, edited by Simon Franklin and Emma Widdis (Cambridge, MA: Cambridge University Press, 2004), 98.
30 Willfried Spohn, "Multiple Modernity, Nationalism and Religion: A Global Perspective," *Current Sociology* 51, nos. 3–4 (May 2003): 265–286. https://doi.org/10.1177/0011392103051003007
31 J. William Frost, "Why Religions Facilitate War and How Religions Facilitate Peace," Paper prepared for the Friends Association for Higher Education Conference at Haverford College, 16–19 June 2005. https://www.swarthmore.edu/friends-historical-library/why-religions-facilitate-war-and-how-religions-facilitate-peace
32 Frost, "Why Religions Facilitate War and How Religions Facilitate Peace."
33 See, for example, Peter J. Katzenstein, ed., *The Culture of National Security: Norms and Identity in World Politics* (New York: Columbia University Press, 1996).
34 Susan Fowler, "Where Values Come From," *Huffpost*, 6 December 2017. https://www.huffpost.com/entry/where-value-come-from_b_10934032
35 Johnson and Maines, "The Cultural Topography Analytic Framework," 41.
36 It is undisputed that even polytheistic religions have accounts of deities bestowing revelatory wisdom, such as the account of Arjuna in the Hindu Vedas, but consider that the canon of monotheistic religions is composed in large part of language that is direct revelation from a unitary source.
37 Lewis Davis, "On the Origin of Religious Values," *SSRN Electronic Journal*, January 2017. https://www.researchgate.net/publication/318388968_On_the_Origin_of_Religious_Values. See, for example, the discussion of religion as a social functional creation in Joseph Henrich, *The WEIRDest People in the World: How the West Became Psychologically Peculiar and Particularly Prosperous* (New York City: Farrar, Straus and Giroux, 2020), especially 123–154.

38 For example, "Today, religion still substantially affects daily life through rituals and festivals and by informing the thought and fundamental values of the Japanese people." Soho Machida, "The Religious Traditions of Japan," in *Religion, War, and Ethics: A Sourcebook of Textual Traditions*, edited by Gregory M., Reichberg, and Henrik Sys, with Nicole M. Hartwell (New York: Cambridge University Press, 2014), 631. See also Janet Epp Buckingham, "The Relationship between Religions and a Secular Society," *Special Issue of Diversity Magazine* 9, no. 3 (Summer 2012): 1–5.

39 Zhang Lihua, "Values in Chinese Foreign Policy Culture, Leadership and Diplomacy," in *Values in Foreign Policy: Investigating Ideals and Interests*, edited by Krishnan Srinivasan, James Mayall, and Sanjay Pulipaka (New York: Rowman & Littlefield Publishers, 2019), 193–207.

40 A more complete discussion of sacred values in a tailored deterrence context can be found in Nicholas Taylor's Chapter 30, "Strategic Culture and Tailored Deterrence," in the current volume.

41 Philip E. Tetlock, "Thinking the Unthinkable: Sacred Values and Taboo Cognitions," *Trends in Cognitive Sciences* 7, no. 7 (July 2003): 320–324.

42 Daniel M. Bartels and Douglas L. Medin, "Are Morally-Motivated Decision Makers Insensitive to the Consequences of Their Choices?" Northwestern University, *Psychological Science* 18, no. 1 (2007): 24–28. http://home.uchicago.edu/bartels/papers/Bartels-Medin-2007-PsychSci.pdf.

43 Lily Kong, "Geography and Religion: Trends and Prospects," *Progress in Human Geography* 14, no. 3 (1990): 355–371.

44 See, for example, Katzenstein, *The Culture of National Security: Norms and Identity in World Politics* and, Maria Rost Rublee, *Nonproliferation Norms: Why States Choose Nuclear Restraint* (Athens: University of Georgia Press, 2009).

45 Theo Farrell and Terry Terriff, "The Sources of Military Change," in *The Sources of Military Change: Culture, Politics, Technology*, edited by Theo Farrell and Terry Terriff (Boulder: Lynne Rienner, 2002), 7.

46 Theo Farrell, *The Norms of War: Cultural Beliefs and Modern Conflict* (Boulder: Lynne Rienner, 2005).

47 William Charlton, "Religion, Society and Secular Values," *Philosophy* 91, no. 3 (2016): 321–324.

48 For a treatment of cross-cultural distinctions between deontological and teleological ethics, see Vladimir A. Lefebvre, *Algebra of Conscience: A Comparative Analysis of Western and Soviet Ethical Systems* (Dordrecht, Holland: D. Reidel, 1982).

49 An excellent and comprehensive survey of comparative religious views on war can be found in Vesselin Popovski, Gregory M. Reichberg and Nicholas Turner, eds., *World Religions and Norms of War: Conflict, Culture & Religions* (Tokyo: United Nations University Press, 2009).

50 For a good overview, see Richard Sorabji, "Just War from Ancient Origins to the Conquistadors Debate and Its Modern Relevance," in *The Ethics of War: Shared Problems in Different Traditions*, edited by Richard Sorabji and David Rodin (Burlington, VT: Ashgate Publishing Company, 2007), 13–29.

51 A concise source is International Committee of the Red Cross, *Lesson 1: The Law of Armed Conflict: Basic Knowledge*, June 2002. https://www.icrc.org/en/doc/assets/files/other/law1_final.pdf.

52 According to Islamic jurisprudence scholar Karima Bennoune, "More than a millennium before the codification of the Geneva Conventions, most of the fundamental categories of protection which the Conventions offer could be found, in a basic form, in Islamic teachings." Karima Bennoune, "As-Salāmu `Alaykum? Humanitarian Law in Islamic Jurisprudence," *Michigan Journal of International Law* 15, no. 2 (1994): 605–643, 623.

53 See chapters on the norms of war for each major faith tradition in Popovski, *World Religions and Norms of War*; and in Reichberg, *Religion, War, and Ethics*. For excellent cross-civilizational perspectives, see the chapters in Howard M. Hensel, ed., *The Prism of Just War: Asian and Western Perspectives on the Legitimate Uses of Military Force* (Burlington, VT: Ashgate Publishing, 2010); and, Paul Robinson, ed., *Just War in Comparative Perspective* (Hampshire, England: Ashgate Publishing, 2003).

54 Vladimir Tikhonov, "Chinese and Korean Religious Traditions," in *Religion, War, and Ethics: A Sourcebook of Textual Traditions*, edited by Gregory M., Reichberg, and Henrik Sys, with Nicole M. Hartwell New York: Cambridge University Press, 2014), 625.

55 John Kelsay and James Turner Johnson, eds., *Just War and Jihad: Historical and Theoretical Perspectives on War and Peace in Western and Islamic Traditions* (New York: Greenwood Press, 1991); and Joseph R. Hoffmann, ed., *The Just War and Jihad: Violence in Judaism, Christianity and Islam* (Amherst, NJ: Prometheus Books, 2006).
56 R. Scott Appleby, *The Ambivalence of the Sacred: Religion, Violence, and Reconciliation* (Lanham, MA: Rowman & Littlefield Publishers, 2000).
57 Kishore Mahbubani, *Can Asians Think? Understanding the Divide Between East and West* (Southroyalton, VT: Steerforth Press, 2002); Richard E. Nisbett, *The Geography of Thought: How Asians and Westerners Think Differently ... and Why* (New York: The Free Press, 2003).
58 The inverted arrow illusion is discussed in Jonathan Morgan, "Does Religion Influence How You See the World?" *Exploring My Religion Blog*, Research News, 18 June 2014. https://exploringmyreligion.org/blog/2014/06/18/does-religion-change-how-you-see-the-world/
59 Lorenza S. Colzato, Wery M. van den Wildenberg, and Bernhard Hommel "Losing the Big Picture: How Religion May Control Visual Attention," *PLoS One* 3, no. 11 (2008): e3679.
60 Jonathan Morgan, "Does Religion Influence How You See the World," https://exploringmyreligion.org/blog/2014/06/18/does-religion-change-how-you-see-the-world/
61 Julian Baggini, *How the World Thinks: A Global History of Philosophy* (London, UK: Granta Publications, 2018), 41.
62 Jonathan Fox, "Religion as an Overlooked Element in International Relations," *International Studies Review* 3, no. 3 (2001): 53–74. https://www.jstor.org/stable/3186242. In informal surveys of the author's strategic culture classes over the last 15 years, very few political science and international relations students report having ever been exposed to religion in their studies in the course of their college educations.
63 Anecdotally, race, religion, and ethnicity are said to be taboo subjects within the community of intelligence analysts. If true, this would substantially impair analysts' ability to assess political and military plans and intentions driven by such factors.
64 Examples of geopolitical surprises that were instigated or motivated by religion include the 1979 Iranian Revolution, the Polish Revolution and the end of the Cold War, and the terrorist attacks of 11 September 2001. As discussed in Thomas, *The Global Resurgence of Religion*, 1–10.
65 Adamsky, *Russian Nuclear Orthodoxy*.
66 Thomas, *The Global Resurgence of Religion*, 16.

6

POLITICAL AND STRATEGIC CULTURES IN THE *LONGUE DURÉE*

Insights from the Middle East

Michael Eisenstadt

If war is the continuation of policy with other means,[1] a country's political culture should shape its strategic culture; few studies, however, have systematically explored the potential nexus between the two. This chapter attempts to do so, surveying the modern polities influenced by the Middle East's ancient Arab-Islamic, Persian-Iranian, and Hebrew-Judaic civilizations. It also seeks to identify elements of continuity and change over the *longue durée* in these political and strategic cultures, and to consider how geopolitics and cross-cultural interactions have affected them.

National cultures are the beliefs, values, and practices of human communities sharing a common language, religion, history, fate, or sense of destiny. They are expressed through art, architecture, music, literature, food, and even particular national approaches to politics, diplomacy, and war. They constitute the matrix in which political and strategic cultures are formed, which shape the conduct of politics at home, as well as diplomacy and war abroad. Political cultures, then, are the beliefs, values, and practices pertaining to collective human action, the wielding of influence and power, and the allocation of resources in society. By contrast, strategic cultures are the beliefs, values, and practices regarding the use of force and diplomacy to achieve national security objectives; diplomatic styles and "ways of war" are their practical embodiment.[2]

Cultures are neither monolithic nor deterministic or immutable. Within national cultures, subcultures and countercultures coexist and contend with one another. Foreign influences, moreover, may shape a country's political and strategic cultures. And the conduct of diplomacy and war can also be influenced by the personality of leaders, organizational structures and processes, human and material resource endowments, physical geography, and the operational environment – although some of these factors may, in turn, be shaped by culture.

Culture defines an actor's perceived horizon of the possible, making certain choices, behaviors, and outcomes more likely than others. The influence of culture is expressed

 DOI: 10.4324/9781003010302-8

through tendencies and predispositions, as well as tensions and contradictions that often remain unresolved. And while cultures may change, old and new elements often coexist and comingle. Indeed, cultural attributes can persist for decades, centuries, or even millennia, and can live on long after the formative events that created them.[3] The Middle East's three major civilizations – Arab, Iranian, and Jewish – are characterized by demographic, geographic, religious, and linguistic continuities that go back thousands of years.

More than in most other parts of the world, tribal and religious values influence the political and strategic cultures of the Middle East. Nearly all human societies were once organized tribally. Europe and China created strong states that largely supplanted kinship-based structures, but in the Middle East, tribal values – and often tribal structures – continue to coexist with national and religious identities.[4] They have come to shape if not dominate Arab politics,[5] and continue to exert a vestigial but significant influence over Iranian politics – a legacy of 2,500 years of rule by tribal dynasties.[6] And while ancient Israelite society lost its tribal character long ago as a result of exile and dispersion, the modern state of Israel was one of the first places where political tribalism took root before becoming a global phenomenon.[7] Religion, and values derived from religion, are likewise central to contemporary Arab, Iranian, and Israeli politics.

How does one explain the persistence of cultural influences over centuries and millennia? Values, beliefs, and practices often long outlive the circumstances that gave rise to them: tribal values may outlast tribal structures, religious concepts may live on in a secularized form, and both may gain a degree of "immortality" by becoming embedded in national, political, and strategic cultures. Likewise, relatively fixed ecological and geographic factors may combine with persistent cultural elements such as language and religion to create enduring attitudes toward power, authority, and the use of force. A people's self-conception, as embodied in its collective beliefs and national myths, may also cause it to discern echoes of the past in the present and to see in its history a guide to the future, producing continuities in perception and conduct.

Because the Middle East is located at the crossroads of three continents and of vital land and sea lines of communication, foreign cultures and civilizations often left an imprint on the political and strategic cultures of the region. Indeed, major developments in the history of the contemporary Middle East – the emergence of modern nationalism, the "return of Islam," and the rise and fall of sectarianism – were not only responses to internal developments and crises but also to increased contact with the outside world. Today, due to modern communications, direct physical contact is no longer necessary for cultural interactions and diffusion to occur.

This chapter, then, will describe linkages between the political and strategic cultures of the modern Arab states, Iran, and Israel, highlighting elements of continuity and change while seeking to avoid the pitfalls of essentialism, overgeneralization, and cultural stereotyping. Needless to say, this brief survey will necessarily be more illustrative than comprehensive, and will focus primarily on state actors, though future research will hopefully devote more attention to non-state actors.[8]

The Arab States

With the rise and spread of Islam in the seventh century CE, the Arabs created one of the great world empires, forever altering the face of the Middle East and Southwest Asia.

After a period of efflorescence, however, the Arab peoples experienced a long decline characterized by political fragmentation, foreign conquest, and colonization. In the modern era, Arabs have aspired to restore their former prominence in world affairs.[9]

Patrimonial Power Structures and "Tribal" Politics

The Arab world – a region of more than 20 countries and nearly 300 million people – is characterized by tremendous human diversity. It is therefore impossible to speak of a unitary "Arab" political or strategic culture. But it is possible to speak of a common cultural template that shapes approaches to politics and national security.[10] Thus, the overwhelming majority of Arab governments – whether so-called republics, monarchies, or consociational democracies – are rooted in patrimonial power structures, often based on strong leaders who gain and maintain their grip on power through the mobilization and manipulation of intense *'asabiyyahs* (family, clan, tribal, ethno-sectarian, or regional solidarities).[11] The most prominent exception that proves the rule is Egypt, a relatively homogenous society in which the armed forces, possessing a strong corporate identity, is the most important institution. It is no coincidence that it was an Egyptian diplomat, Tahseen Bashir, who derisively referred to the other Arab states as "tribes with flags."[12]

The Bedouin proverb, "me against my brothers, me and my brothers against our cousins, and all of us against the stranger" captures the dynamic that animates the domestic and external politics of many Arab societies.[13] The tendency to view members of out-groups as potential rivals or adversaries is a universal human trait, but it is particularly pronounced in societies where tribal values prevail. This spirit of rivalry and conflict guides the conduct of political factions, ethno-sectarian groups, and nation-states – hindering cooperation between individuals and groups, hampering the development of inclusive political processes, and engendering conspiratorial thinking and a tendency to externalize blame.

The result is a zero-sum, winner-takes-all approach to politics, producing autocratic, *mukhabarat* (secret police) states in which domestic opponents that cannot be co-opted or marginalized are repressed or crushed (in accordance with "Hama rules").[14] Regimes whose leaders possess the ambition and means to pursue assertive foreign policies (such as the "revolutionary republics" – Gamal Abdel Nasser's Egypt, Syria under Hafiz and Bashar al-Assad, Saddam Hussein's Iraq, and Muammar al-Qaddafi's Libya) often evolved strategic cultures marked by a tendency to (1) use force to solve conflicts, engage in brinkmanship, or attempt *faits accompli*; (2) negotiate to buy time, deflect pressure, or undermine opponents; (3) conclude short-lived alliances for temporary advantage; and (4) meddle in the domestic affairs of other Arab states.

Force, Brinkmanship, and Faits Accompli

Arab autocrats tend to rely on force because it obviates the need for unpalatable compromises with enemies that could alienate members of their power base, and thus undermine the *'asabiyyahs* underpinning their rule. They are generally not bound by conventional restraints when dealing with weaker domestic and foreign enemies, and will use all available means to vanquish them. It was, however, not always so; the informal rules of bedouin raiding and warfare in early modern times traditionally sought to minimize casualties and avoid harm to noncombatants,[15] while some Arab-Muslim military leaders in earlier eras are reputed to have treated vanquished enemies with magnanimity – at least by the standards of the day.[16] One factor contributing to the brutalization of modern Arab politics and warfare was the importation from Europe of twentieth-century fascist and leftist ideologies that rationalized violence as a redemptive and regenerative force – an ethos internalized by many Arab nationalists and Islamists.[17]

Thus, Gamal Abdel Nasser's penchant for brinkmanship (president of Egypt: 1955–1970) precipitated two wars with Israel (in 1956 and 1967), while he repeatedly sought to subvert "reactionary" Arab monarchies in Iraq, Jordan, and Saudi Arabia, and dispatched Egyptian troops to fight in the Yemen Civil War.

Hafiz and Bashar al-Assad of Syria (presidents from 1971 to 2000, and 2000 to present, respectively) likewise often relied on force to achieve their goals. Thus, in the 1973 war, Syria sought simply to retake the Golan Heights from Israel. Syria's 1976 military intervention in Lebanon's civil war sought to impose a political solution that cemented its control over the country. Damascus used brute force to crush a Muslim Brotherhood uprising (1979–1982) and 30 years later to suppress, with the help of Russia, Iran, and the latter's Shiite foreign legion, a popular rebellion (2011–present). When resorting to diplomacy, it did so mainly to buy time and facilitate the use of force. However, in its interactions with more capable adversaries like Israel, Syria has generally been careful to test and probe limits before committing to a military course of action, to avoid inadvertent escalation or war.

Saddam Hussein of Iraq (president: 1979–2003), like Nasser, had a penchant for risky geopolitical gambits. He invaded Iran in 1980 to seize its oil and overthrow its nascent revolutionary government, but instead found himself mired in a costly eight-year war. He invaded Kuwait in 1990 to seize its oil wealth and thereby resolve the economic problems created by the Iran-Iraq War, but instead spurred the United States to organize a 35-nation military coalition that expelled Iraqi forces from the emirate.

Because conflicts are often perceived as existential, zero-sum struggles, every possible advantage is exploited; options are rarely foreclosed. Those Arab regimes that had established close ties to terrorist groups (e.g., Libya, Syria, and Iraq) were generally reluctant to fully sever these ties when pressed to do so, while those that had developed weapons of mass destruction and then agreed to give them up often avoided doing so. For instance, after Iraq (under Saddam), Syria, and Libya claimed to have destroyed their chemical weapons production facilities and stockpiles in response to foreign pressure (the latter two after joining the Chemical Weapons Convention), all three secretly retained elements of these programs, while Syria continued to use chemical agents against civilians and rebel forces.

Negotiate to Buy Time, Deflect Pressure, or Undermine Opponents

When military solutions are deemed too risky or prove unattainable, belligerents may negotiate to buy time, deflect pressure, seek respite through ceasefires (usually temporary), or gain through talks what could not be obtained by force. To paraphrase the former commander of US forces in the Middle East, General John Abizaid, Americans tend to prefer to fight *then* talk, while Arabs prefer to fight *and* talk.[18] As conflict is often seen as a fact of life in the Middle East, conflict resolution is generally not the goal of negotiations.

The main exceptions are talks to end bloody stalemates (e.g., the Iran-Iraq War and the Lebanese Civil War), or that involve a third-party mediator whose ongoing support is a vital interest (e.g., the Israel-Egypt and Israel-Jordan peace treaties of 1979 and 1994, respectively).

Thus, Syria entered into peace talks with Israel at the 1991 Madrid Conference in part to ensure that it would not be attacked during a period of heightened vulnerability, following the collapse of its Soviet great power patron. And during the Syrian Civil War (2011–present), Damascus talked with members of the opposition in part to sow dissension in rebel ranks and discredit those negotiating with it, while pursuing a "surrender or starve" strategy vis-à-vis rebels on the battlefield.

Short-Lived Alliances for Temporary Advantage

The zero-sum approach that makes domestic politics in so many Arab states so fraught also complicates regional politics by engendering distrust, competition, and conflict. This hinders cooperation to achieve shared goals; produces shifting regional alliances; and results in a tendency to rely on neutral third parties and outside mediators to help negotiate and guarantee agreements.[19] Moreover, the high degree of great power penetration of the region due to its strategic location and resource endowments has often led Arab states to rely on external powers as security providers or guarantors.[20]

Thus, the Arab states repeatedly failed to realize their collective military potential in their wars with Israel (1948, 1967, 1973) due to mutual distrust and their inability to subordinate their own interests to the needs of the coalition. For much the same reasons, the states of the Gulf Cooperation Council have failed to create an effective joint military force or regional missile defense architecture, and the hundreds of rebel factions fighting the Assad regime after 2011 were unable to forge themselves into a cohesive military force, even though their survival depended on it.

Arab actors often take a flexible approach to coalition formation, forging alliances of convenience with former enemies, reconciling (temporarily) and falling out as circumstances require. Coalitions are much more fluid in the Middle East than elsewhere. While the United States has sometimes sought to align itself with erstwhile adversaries – the Soviet Union during World War II and China (against the Soviets) at the height of the Cold War – such efforts are the exception rather than the rule for Washington.

Thus, Egypt and Jordan put aside their bitter, long-standing rivalry on the eve of the 1967 war to confront Israel; Syria first intervened in the Lebanese Civil War in 1976 on the side of Christian forces, but turned against them in 1978; and Iraqi insurgents joined with their erstwhile American enemy during the 2007 US surge to fight al-Qaeda in Iraq when the latter threatened to overwhelm them. While outsiders may see this as unprincipled or duplicitous behavior, regional actors see this as the kind of pragmatism needed to survive in a tough neighborhood.

Meddling and Intervention

Arab political culture – influenced by the complex interplay of tribalism, Islam, nationalism, and foreign ideologies – has also influenced the dynamics of the region's "non-Westphalian" state system.[21] The existence of a shared Arab and/or Islamic identity that transcends the "artificial" boundaries created by former colonial powers has often been used as a pretext by ambitious politicians – Arab and Iranian – to justify their meddling in the affairs of other regional states. Such interventions were a common feature of the Arab cold wars of the 1960s and 1970s and the decade that followed the Arab Spring uprisings of 2010–2011.

Moreover, since the 1950s, several Arab regimes created or sponsored Palestinian *fedayeen* organizations to advance their foreign policy objectives. In this way, they sought to control the Palestinian national movement and subordinate it to their own interests. Yet, except for Syria, which made extensive use of radical Palestinian groups against Israel, the use of proxies was never central to the "way of war" of these Arab states in the way it has been for the Islamic Republic of Iran.

Outliers: Small States, Big Personalities

The leaders of smaller or militarily less capable Arab states have generally leaned more heavily on soft power and the indirect use of force to advance their interests. Saudi Arabia has often relied on religious outreach or *dawa* (e.g., funding religious schools worldwide), "riyal politik"/checkbook diplomacy (e.g., funding the Afghan *mujahidin* in the 1980s), and its ownership of prominent media outlets (e.g., newspapers like Asharq Al-Awsat and Al-Hayat as well as satellite TV networks like Al-Arabiya and MBC) to project influence – though under the leadership of Crown Prince Mohammed bin Salman, it has also attempted to use force to quash the Houthi insurgency in Yemen (2015–present). Smaller Gulf States like Qatar have relied on religious outreach (e.g., support for the Muslim Brotherhood), checkbook diplomacy (e.g., financial support for rebel groups in Syria and Hamas in Gaza), and its ownership of media networks (e.g., Al Jazeera TV) to allow it to punch significantly above its weight. All of these states host US military bases to cultivate influence in Washington. And in recent years, several Gulf states have shown an increased propensity to project power abroad – sending special forces to work "by, with, and through" local partners in Libya (UAE and Qatar) and Yemen (UAE and Saudi Arabia).

While the Gulf Arab states have traditionally acted within security frameworks created by foreign powers like the United States, as the latter tried to disengage from the region after withdrawing from Iraq in 2011, Saudi Arabia, the UAE, and Qatar, in particular, have become more assertive and self-reliant. Indeed, it can be said that this aspect of their strategic cultures has changed, although the failed Saudi-led intervention in Yemen and unsuccessful attempts since 2017 by the UAE and Saudi Arabia to isolate Qatar have caused them to reassess these policies. They have also pursued closer ties with Israel (with the UAE and Bahrain signing normalization agreements in 2020) – more signs of geo-politically driven cultural change in the region among small states with independent-minded leaders.

Anwar Sadat of Egypt (president: 1970–1981) provides perhaps the best example of the ability of a strong personality to swim against powerful cultural currents. He rejected a zero-sum approach to the conflict with Israel (even while pursuing such an approach with domestic enemies) and sought to disengage his country from the conflict – albeit only after Egypt effaced the shame of the June 1967 defeat with its victorious crossing of the Suez Canal six years later. In so doing, he pursued a subtle and sophisticated political-military strategy during the 1973 war, which sought a limited military victory in order to catalyze diplomacy, regain lost territory, and make peace with Israel.

The Islamic Republic of Iran

Whereas most Arab states were created in the last century, Iran has existed for millennia as a distinct political entity with a diverse population and a shared identity – rooted in Persian culture and, more recently, Shiite Islam.[22] The distinctive politics of the Islamic Republic of Iran (IRI) thus combines ancient and modern elements, and has been shaped by tensions between the regime's authoritarian impulses versus the people's democratic aspirations; the regime's revolutionary religious ideology versus popular nationalism – reflecting Iran's dual identity as a "cause" and a "country" (as Henry Kissinger famously put it)[23]; and the ideological policies of unelected "deep state" actors versus the generally more pragmatic preferences of elected officials.

The IRI's leadership sees the world as a profoundly unjust and hostile place, a worldview deriving from Iran's loss of empire, its conquest by successive waves of invaders – Macedonian, Arab, Turkic, Mongol, Ottoman, Russian, and British – and to embrace of Shiite Islam in the sixteenth century CE. As Sandra Mackey has noted:

> ... the martyrdom of Ali and Hussein [the prophet Muhammad's cousin and son-in-law, and grandson, respectively – whom Shiites believe were the prophet's rightful heirs] held particular meaning for Iranians (who) saw in the Shia martyrs shadows of themselves. For they too were a defeated and humiliated people .... But even more, Ali provided the Iranians with their archetypal model for political order – the just ruler. In a parallel with Persian kingship, the Iranians accepted the cause of Ali and the house of the Prophet because it coincided with their pre-Islamic traditions of legitimacy. And in Shiism, the Iranians found the stoicism of Manichaeism and the Zoroastrian hope for the return of the savior who would restore justice on earth.[24]

Authoritarian and Factional Politics in a Hybrid System

For more than 2,500 years, Persia was ruled by absolutist kings, a legacy reflected in modern Iran's authoritarian politics, the use of ambiguity and dissimulation by Iranians as a survival strategy, and a propensity for popular uprisings and revolutions.[25] The IRI has, moreover, created a hybrid authoritarian system in which quasi-democratic processes are used to legitimize a militarized theocracy, consisting of opaque revolutionary institutions (the Supreme Leader, Guardian Council, and Islamic Revolutionary Guard Corps or IRGC) that counterbalance traditional state institutions (the presidency, parliament, and regular military).[26] This hybrid structure lends resilience to the regime, and reflects its fears of popular unrest and foreign-inspired "soft warfare."[27]

Because the regime's doctrine of "the rule of the jurisprudent" (*velayat-e faqih*) is the only ideology permitted, political differences are generally expressed through factional competition among personality- and policy-based informal networks. This also reflects the vestigial influence of tribal values, which tend to foster interpersonal and intergroup rivalries and a zero-sum approach to politics.[28] Changes in the domestic balance of power between these factions often produce policy zigzags at home and abroad.

Ayatollah Ali Khamenei (Supreme Leader: 1989–present) and before him Ayatollah Ruhollah Khomeini (Supreme Leader: 1979–1989) traditionally kept factionalism in check by making space in "the system" for all major pro-regime currents, balancing the regime's various power centers, relying on consensus decisionmaking, and promoting revolutionary

Islam as a unifying ideology. In recent years, however, Khamenei and the IRGC have marginalized reformers and pragmatic conservatives – narrowing the political base of the regime, deepening divisions between regime "insiders" and "outsiders," and further alienating the majority of Iranians from the system.

The tension between the regime's revolutionary and traditional state institutions has often worked itself out on the ideological level through the tension between Iran's identity as a cause with revolutionary commitments – embodied by the concept of resistance (*moqavamat*), and a state with pragmatic interests – embodied by the concept of "the expediency of the regime" (*maslehat-e nezam*). The IRI has often relied on proxies to carry out its revolutionary agenda through subversion and terrorism abroad.

Proxies, Indirection, and Gray Zone Activities

Reliance on proxies – a defining feature of the IRI's domestic and foreign policies – has its roots in prerevolutionary Iran. The use of street mobs and violent pressure groups goes back at least a century to Iran's Qajar dynasty, foreshadowing the Islamic Republic's use of shadowy regime-sponsored vigilante groups to repress domestic unrest and its use of foreign proxies abroad.[29] Mohammad Reza Pahlavi (Shah: 1941–1979) likewise supported Iraq's Kurds, strengthened ties with Shiite communities in Iraq and Lebanon, and supported a variety of anti-communist movements throughout the region in the 1960s and 1970s, prefiguring the IRI's ambitious regional policies.[30]

The IRI, however, has put its own stamp on these efforts. Its hybrid political system has served as a template for its project to remake the region in its own image. It has done so by creating parallel civil society and state institutions in Lebanon, Iraq, and Syria through which it can project influence; mobilizing foreign Shiite clerical networks and communities to serve as external bases of support for its policies; and creating a transnational Shiite "foreign legion" to project power and export its culture of jihad, martyrdom, and resistance. These activities have compounded the governance challenges faced by the weak and failing Arab states that Tehran hopes to influence most: Lebanon, Syria, and Iraq.

Since its inception, the IRI has relied extensively on "gray zone" activities to advance its anti-status quo agenda while managing risk, preventing escalation, and avoiding war. These include unilateral and proxy activities, covert and overt, in both the physical and cyber domains. While Iran is not unique in this regard (China and Russia have long engaged in gray zone activities – as did the United States during the Cold War) – it has in the past four decades developed its own unique asymmetric gray zone toolkit which includes: hostage taking; embassy invasions; terrorism; rocket, drone, and missile strikes; and attacks on maritime traffic. The IRGC has played a central role in developing and implementing Iran's gray zone strategy, and in bottom-up fashion has thus influenced the Islamic Republic's strategic culture.[31]

Tehran's aversion to conventional war and its preference for covert, unacknowledged, and proxy operations is not grounded in a transitory calculation of the regime's interests. Rather, it is a deeply rooted feature of the IRI's strategic culture and one of the enduring legacies of the long, traumatic, and bloody Iran-Iraq War, which reinforced a two-centuries-old historical lesson: that conventional wars have often ended badly for Iran.[32] The end of the Iran-Iraq War also marked the end of a decade of revolutionary radicalism in Iranian foreign policy. Since then, Iran has been much more cautious in its dealings with foreign powers – though risk averse does not mean risk avoidant.

Thus, Tehran typically tests adversaries to see what it can get away with. It engages in covert or proxy activities to preserve deniability and avoid becoming decisively engaged. It relies on incremental action and indirection to create ambiguity regarding its intentions and to make its enemies uncertain about how to respond. And it arranges its activities in time and space – pacing them temporally and spacing them geographically – so that adversaries do not feel pressed to respond.

Syria's reliance on elements of this approach vis-à-vis Israel (see above) begs the question: does this modus operandi reflect a prudential culture typical of minoritarian regimes (Syria's leaders are Alawites, Iran's are largely non-Arab Shiites) whose leaders have often had to live by their wits, and who are generally averse to the kind of impetuous risk-taking that has characterized some Sunni Arab leaders? Or does it simply reflect the caution of strategically lonely actors that understand their limits?

Calibrated Use of Force and Diplomatic Gambits

The IRI generally uses force in a calibrated manner. Thus, in response to the protests that followed the contested June 2009 elections, the security forces prevailed by wearing down and demoralizing the opposition over time, rather than by using brute force. It has, however, been less restrained in responding to more recent bouts of unrest – perhaps a sign of greater confidence and risk acceptance by the regime.[33]

Conversely, while Iran has enabled past mass casualty terrorist spectaculars such as the Marine Barracks bombing in Beirut (1983) that killed 241 Americans, and attacks on coalition troops in Iraq (2003–2011) that killed more than 600 Americans, in recent years it has generally been more restrained. Thus, in countering the US "maximum pressure" policy (2018–2021), it tended to rely on nonlethal means[34] – though in response to the January 2020 killing of IRGC Qods Force commander Qassem Soleimani it launched a missile strike against US troops in Iraq that could have killed scores.

Iran's leaders – like many Arab leaders – tend to see domestic and foreign policy in zero-sum terms (although Foreign Ministry officials – perhaps with an eye to foreign audiences – often speak about "win-win" outcomes).[35] They too try to avoid foreclosing options and to exploit every possible advantage, and for that reason will not abandon proxies, give up their missiles and drones, or permanently shutter their nuclear program.

Finally, the IRI's diplomats regularly attempt to drive wedges in hostile coalitions and to create countervailing power blocs. Thus during nuclear negotiations, Iran has offered lucrative oil/gas deals to some of its interlocutors to induce them to convince others to soften their terms, and used or threatened force to gain leverage.[36] And it has repeatedly threatened the United States and Europe that it would "turn East" and consolidate ties with China and Russia if demands for sanctions relief were not met. But, as former Foreign Minister Mohammad Javad Zarif has lamented, in the IRI, diplomacy ultimately serves the interests of the IRGC and the regime – not the nation.[37]

Israel

Israel is a modern state with ancient roots whose contemporary politics fuses old and new. The traumas of Jewish history combined with a fractious political culture has led to policy inertia and spawned a strategic culture that relies on daring, improvisation, tactical virtuosity, and technology to manage seemingly intractable conflicts. Yet, Israel's strategic culture also embodies several paradoxes: military risk acceptance combined with geopolitical risk aversion; a belief (paraphrasing Israel's first Prime Minister David Ben-Gurion) that to be a realist, one needs to believe in miracles; and profound pessimism regarding the zero-sum conflict with the Palestinians, combined with dark foreboding about Iran's nuclear ambitions.

Historical Traumas and a Preoccupation with Survival

Ancient Israel was a small state often riven by fratricidal tribal, sectarian, and dynastic struggles that split into northern and southern kingdoms early in its history. It rarely enjoyed full sovereignty and was often a vassal to one of the region's great empires; the threat of conquest and exile and the temptation of rebellion – driven by undercurrents of messianic expectation – were often present.[38] Repeatedly exiled from their land, the last time following an ill-fated revolt against Rome in 135 CE, tribal identities faded early on in Jewish history.[39]

Jews were not to become a majority and regain full independence in their ancient homeland once again for another 2,000 years. This long exile was marked by further expulsions from the lands of their dispersion, forced conversions, and frequent bouts of persecution – culminating in the Holocaust – which contributed to the rise of political Zionism and the birth of the state of Israel in 1948. The ordeal of exile, brought about by infighting, military weakness, and the geopolitical vulnerability of ancient Israel, has contributed to a preoccupation with survival and to negating the effects of 2,000 years of powerlessness. This has affected nearly every aspect of Israel's approach to politics and national security.[40]

Ancient Roots of a Fractious Democracy

The immigrants who built the modern state of Israel came largely from nondemocratic states in Eastern Europe and the Middle East. It is therefore understandable that the Zionist movement favored democracy, since Jews had frequently been persecuted under authoritarian regimes. But Israel's success in creating a viable democracy needs explaining, and culture provides much of the answer. First, Judaism has a strong egalitarian ethos – a vestige of the worldview of the pastoralists who were its first adherents – which eventually found expression in the ideal of equality before the law.[41] Second, the Hebrew Bible strictly limited the authority of kings (a commandment often transgressed in practice) and established a separation of powers between kings, priests, and prophets – who held kings and priests to account.[42] Third, during their long exile, Jews created representative self-governing communal institutions.[43] Thus, in building a modern state, Jews could draw on values conducive to democracy and a tradition of self-government.

On the other hand, a deep distrust of authority born of centuries of dealing with hostile non-Jewish governments has sometimes engendered an expedient attitude toward rules, the law, and the powers that be.[44] This poses challenges to Israeli democracy and the dominant Israeli ideology of *mamlachtiut* (statism) – a belief in the need for a strong, well-ordered state to ensure the survival of the Jewish people.

Moreover, in Judaism, a vigorous self-critical tradition holds the Jewish people responsible for their communal and national misfortunes, which are interpreted as divine punishment for their sins.[45] This tendency can be seen today in secularized form in the biting commentary that abounds in the Israeli press, and the commissions of inquiry often established after unsuccessful wars. This tradition ensures a degree of governmental accountability, but coexists uneasily with the tendency, embodied by contemporary political tribalism, to blame misfortunes on traitors within and enemies without.

Polarization, Tribalism, and Gridlock

The defining features of contemporary Israeli politics are fragmentation, polarization, and political gridlock. Fragmentation has been abetted by the decline of the country's formerly dominant Jewish parties and a system of proportional parliamentary representation that enables small parties to wield disproportionate power in coalition governments.[46] Polarization – between left and right, secular and religious, Jew and Arab – has in turn been fed by populist politics and political tribalism, fanned by both Jewish and Arab politicians, aggravated by debates over hot-button social and political issues, and exacerbated by periodic bouts of Israeli-Palestinian violence.[47] The result – political gridlock – has often prevented the Israeli government from addressing pressing national security issues, formulating a long-term national security strategy, or even providing clear policy guidance to the military.[48] Accordingly, the Israel Defense Forces (IDF) has played an outsized role in shaping the country's strategic culture and defining its national security policy, often in a policy vacuum.[49]

Gridlock has further entrenched the culture of ad hoc improvisation that has long prevailed in Israeli society and in the national security arena as a result of the uncertainty of life in exile and in Israel. In the diaspora, the fortunes of communities often changed overnight with the death of a benevolent king or the fall of a benign government. Survival often depended on the ability to adjust quickly to change. In Israel too, crises and wars often occurred suddenly, while the apparent intractability of conflicts with the Palestinians, some of the Arab states, and Iran has discouraged long-term planning, putting a premium on improvisation.[50]

Furthermore, an ethos of assertive nationalism born of two millennia of powerlessness, stoked by the hostility of its neighbors, has often exacerbated tensions with its Arab citizens and neighbors, and spawned a growing current of extreme nationalism. Yet, as in ancient times, divisions among Jews may pose the greatest long-term threat to Israel's survival.

Self-Reliance, Pragmatism, and Active Defense

A small, beleaguered state, Israel has sought to be self-reliant, building a domestic arms industry and a significant nuclear arsenal to ensure its survival. It has also sought to cultivate a great power patron – France through the late 1960s, the United States since then – as a source of arms and political support. And as in ancient times, diaspora Jews have sometimes played important roles as advocates for the Jewish state, and have served as emissaries to countries lacking diplomatic ties with Israel.

As the state of a people that has often survived by their wits, Israel has generally pursued a pragmatic, flexible foreign policy. Thus since the 1950s, it has pursued alliances with friendly non-Arab states on the periphery of the region (Turkey, Iran, and Ethiopia), concluded peace treaties with Arab states such as Egypt (1979) and Jordan (1994), conducted peace talks with the Palestinians and Syria (1990s–2000s), and signed normalization accords with more distant Arab states such as the UAE, Bahrain, Morocco, and Sudan (2020). Throughout this period, it also conducted secret diplomacy with Arab, Muslim, and other states that refused to deal with it openly.[51]

On the other hand, Israelis and Palestinians have developed a deep-seated conflict ethos characterized by a zero-sum approach, a powerful sense of victimhood, and a rejection of the legitimacy of the other that has complicated the resolution of their conflict.[52] For Israel, this ethos has often been expressed through a perceived need to control the Palestinians. The failure of the 1990s Oslo peace process discredited efforts to resolve the conflict and helped push Israel's politics further to the right.[53] And since 1967, Israeli policy toward the West Bank (the heart of ancient Israel) has been influenced by a streak of messianism that has both driven Israeli policy and constrained it.

Knowing that the country's lack of strategic depth meant that defeat could result in its demise, Israel's leaders have generally favored prevention or preemption and offensive action, to take the fight to the enemy, keep wars short, minimize casualties, and achieve its goals before great power diplomacy or intervention brought a halt to the fighting.[54] Israel waged preventive war in 1956 and preemptive war in 1967, and conducted sabotage campaigns and/or preventive strikes against nuclear programs in Iraq (1981), Syria (2007), and Iran (2007–present) in order to disrupt, delay, or halt these efforts. The apparent intractability of its conflicts with the Palestinians and Iran has led the IDF to adopt a policy of "mowing the grass" – periodic military operations to manage conflicts that it cannot solve.[55]

Israel likewise takes a conservative approach to defense innovation. Paradoxically, though its leaders tend to eschew technological optimism, they often pursue technological responses to operational challenges.[56] A similar paradox is evident in the realm of strategy. Because Israel cannot afford to lose a major war, its leaders will often accept significant risks in military and covert operations. Yet Israel is generally averse to geopolitical risk – an inclination reinforced by the failure of the Oslo peace process with the Palestinians – with existential fears, nationalist claims, and messianic aspirations ensuring continued control over the Palestinians in the West Bank and Gaza. This has enabled Israel to effectively manage this threat, albeit at the price of a further fraying of its domestic political fabric and its standing abroad.

Conclusion

Political and strategic cultures are linked in often complex ways, shaping the conduct of diplomacy, the waging of war, and regional conflict dynamics. While further research may provide a deeper understanding of the connections identified here, it is possible to suggest a number of tentative conclusions:

Top-down, bottom-up, and collateral influences. It is not surprising that a country's political culture may shape its strategic culture in a top-down fashion. Yet when a strong military organization dominates national security decisionmaking (e.g., the IRGC in Iran and the IDF in Israel), strategic cultures may emerge from below. Foreign technologies and concepts, military lessons learned, historical experience, and other factors may also shape the formation and evolution of strategic cultures.

Culture matters – but it is not the only thing that matters. Long after tribal structures have disappeared and the hold of religion has waned, tribal and religious values may live on in a country's political, and, by extension, its strategic culture. Yet, strong-willed leaders, organizational structures and processes, human and material resource endowments, physical geography, the operational environment, and foreign influences may, inter alia, shape the practice of politics, diplomacy, and war. Thus, in the Arab world, a shared cultural template has engendered as many political and strategic cultures as there are states. While in some the connection between political and strategic cultures is clear and direct, in others it is somewhat tenuous – and other factors may play an important role in shaping the strategic cultures of these states.

Cultural legacies, adaptive responses, and ways of war. Differences in national "ways of war" are often ascribed to efforts by states to adapt to, counter, or "design around" an adversary's strategy or way of war. However, ways of war are based, first and foremost, on cultural templates that incorporate assumptions about the character of war, onto which these adaptive responses are grafted or overlaid.[57] To view a country's way of war as primarily an adaptive response to potential threats without acknowledging deeper cultural roots risks overlooking the very factors that make an adversary tick.

In sum, a more rigorous, structured understanding of how political and strategic cultures interact and shape the conduct of states could aid intelligence analysts and policymakers, and enable more effective statecraft. Hopefully, the emerging field of cultural change research, which may explain *how* and *why* cultural evolution occurs, will facilitate such efforts while opening new vistas of inquiry.[58]

Notes

1 Carl von Clausewitz, *On War* (Princeton, NJ: Princeton University Press, 1976), p. 87.
2 Jeannie L. Johnson and Jeffrey A. Larsen, "Comparative Strategic Cultures Syllabus," prepared by SAIC for the Defense Threat Reduction Agency, 20 November 2006.
3 Stelios Michalopoulos and Elia Papaioannou, *The Long Economic and Political Shadow of History* (London: Center for Economic Policy Research Press, 2017).
4 Francis Fukuyama, *The Origins of Political Order: From Prehuman Times to the French Revolution* (New York: Farrar, Straus and Giroux, 2011).
5 See, e.g., Amatzia Baram, "Neo-Tribalism in Iraq: Saddam Hussein's Tribal Policies 1991–96," *International Journal of Middle East Studies*, Vol. 29, no. 1 (February 1997), pp. 1–31.
6 Mahmood Sariolghalam, "The Evolution of the State in Iran: A Political Culture Perspective" (Series of Studies in Social Thought No. 3, Center for Strategic and Future Studies, Kuwait University, March 2010).
7 Although the origins of modern political tribalism are different from that of traditional forms of tribalism rooted in social structures, the consequences are similar: social fragmentation and conflict. Joseph P. Forgas, "The Psychology of Populism: Tribal Challenges to Liberal Democracy," *The Centre for Independent Studies*, Occasional Paper 183, August 2021.
8 Portions of this chapter were published previously by The Washington Institute for Near East Policy and are reprinted by permission.
9 Bernard Lewis, *The Arabs in History* (New York: Harper & Row, 1967).
10 Adam Garfinkle, "An Observation on Arab Culture and Deterrence: Metaphors and Misgivings," in Efraim Inbar (ed.), *Regional Security Regimes: Israel and Its Neighbors* (Albany, New York: State University of New York Press, 1995), pp. 201–230.
11 Ibn Khaldun, *The Muqaddimah: An Introduction to History* (Princeton, NJ: Princeton University Press, 1969); Hisham Sharabi, *Neopatriarchy: A Theory of Distorted Change in Arab Society* (New York: Oxford University Press, 1988); Fuad I. Khuri, *Tents and Pyramids: Games and Ideology in Arab Culture from Backgammon to Autocratic Rule* (London: Saqi Books, 1990).
12 Tony Horwitz, *Baghdad Without a Map – And Other Misadventures in Arabia* (New York: Penguin Books, 1991), p. 51.

13 Philip Carl Salzman, "The Middle East's Tribal DNA," *Middle East Quarterly*, Vol. 15, No. 1 (Winter 2008), pp. 23–33.
14 "Hama rules" refer to the ruthless tactics used by the regime of Hafiz al-Assad to crush a Muslim Brotherhood uprising in the Syrian city by that name in 1982. Thomas L. Friedman, *From Beirut to Jerusalem* (New York: Farrar, Straus and Giroux, 1989), pp. 76–105. See also Florence Gaub, "Why Arab States Are Bad at Counterinsurgency," Lawfare (blog), 21 June 2015, https://www.lawfareblog.com/why-arab-states-are-bad-counterinsurgency.
15 However, changes in military technology and regional politics contributed to the brutalization of tribal warfare in the decades before it was eliminated with the emergence of strong centralizing states in the 1930s. William Lancaster, *The Rwala Bedouin Today* (London: Cambridge University Press, 1981), pp. 137–138, 140–144.
16 See, for instance Hugh Kennedy, *The Great Arab Conquests: How the Spread of Islam Changed the World We Live In* (Philadelphia, PA: De Capo Press, 2007), pp. 372–376.
17 Hussein Aboubakr Mansour, "The Liberation of the Arabs from the Global Left," *Tablet*, 11 July 2022.
18 Major Morgan Mann, "The Power Equation: Using Tribal Politics in Counterinsurgency," *Military Review*, Vol. 87, No. 3 (May–June 2007), p. 106.
19 Perhaps in much the same way that neutral outsiders are often called on to mediate local tribal conflicts in Arab societies.
20 L. Carl Brown, *International Politics and the Middle East: Old Rules, Dangerous Game* (London: Princeton University Press, 1984), pp. 14–18.
21 The peace of Westphalia (1648) created a norm against intervention that has become a fundamental principle of international law.
22 W.B. Fisher, "The Personality of Iran," in W.B. Fisher (ed.), *The Cambridge History of Iran* (Cambridge: Cambridge University Press, 1968), pp. 717–740.
23 David Ignatius, "Talk Boldly with Iran," *Washington Post*, 23 June 2006.
24 Sandra Mackey, *The Iranians: Persia, Islam, and the Soul of a Nation* (New York: Plume Book, 1998), p. 85.
25 Nikki R. Keddie, *Iran and the Muslim World: Resistance and Revolution* (London: Macmillan Press Ltd., 1995), especially Chapters 4 and 5, "Why Has Iran Been Revolutionary?" pp. 60–94.
26 Wilfried Buchta, *Who Rules Iran? The Structure of Power in the Islamic Republic* (Washington, DC: Washington Institute for Near East Policy and the Konrad Adenauer Stiftung, 2000).
27 Nima Adelkhah, "Iran Integrates the Concept of the 'Soft War' Into Its Strategic Planning," *Terrorism Monitor*, Vol. 8, No. 23 (12 June 2010), pp. 7-9.
28 David E. Thaler et al., *Mullahs, Guards, and Bonyads: An Exploration of Iranian Leadership Dynamics* (RAND, 2010), pp. 37–74; Sariolghalam, "The Evolution of the State in Iran: A Political Culture Perspective."
29 Michael Rubin, *Into the Shadows: Radical Vigilantes in Khatami's Iran* (Washington, DC: The Washington Institute for Near East Policy, 2001).
30 Ariane M. Tabatabai, *No Conquest, No Defeat: Iran's National Security Strategy* (New York: Oxford University Press, 2020), pp. 120–123.
31 Michael Eisenstadt, "Operating in the Gray Zone: Countering Iran's Asymmetric Way of War," Policy Focus 162 (Washington, DC: Washington Institute, 2020).
32 Tabatabai, *No Conquest, No Defeat*, pp. 297–303; Steven R. Ward, *Immortal: A Military History of Iran and Its Armed Forces* (Washington, DC: Georgetown University Press, 2009), pp. 325–326.
33 "Special Report: Iran's leader ordered crackdown on unrest – 'Do whatever it takes to end it,'" *Reuters*, 23 December 2019.
34 Khamenei reportedly approved a September 2019 drone and cruise missile attack on Saudi oil facilities on condition that no civilians or Americans be killed, presumably to avoid escalation. Michael Georgy, "'Time to Take Out Our Swords': Inside Iran's Plot to Attack Saudi Arabia," *Reuters*, 25 November 2019.
35 "Iran Nuclear Deal 'Win-Win' Says Zarif," ANSA News, 14 July 2015.
36 Julian Borger, "Iran Offered to End Attacks on British Troops in Iraq, Claims Diplomat," *Guardian*, 20 February 2009.

37 Farnaz Fassihi, "Iran's Foreign Minister, in Leaked Tape, Says Revolutionary Guards Set Policies," *New York Times*, 25 April 2021.
38 Dov S. Zakheim, "The Geopolitics of Scripture," *The American Interest*, Vol. 7, No. 6 (July/August 2012), pp. 5-16.
39 Persecution, however, has bred a strong, almost tribal sense of solidarity – hence the expression "member of the tribe" used in jest by Jews to refer to coreligionists. The sole remaining vestige of these tribal origins is the identification of Jews as Cohens, Levites, or Israelites – markers whose significance is limited to the realm of religious ritual. Cohens claim descent from Aaron, brother of the Biblical prophet Moses, of the tribe of Levi – one of the 12 original tribes of Israel. Jews who are not Cohens or Levites are Israelites.
40 Shmuel Sandler, *The Jewish Origins of Israeli Foreign Policy* (London: Routledge, 2018), pp. 165–167.
41 Joshua Berman, *Created Equal: How the Bible Broke with Ancient Political Thought* (New York: Oxford University Press, 2008).
42 Thus, kings were enjoined to not accumulate too many horses, wives, or too much wealth, and were subject to the yoke of divine law, like every other Israelite (Deuteronomy 17: 14–20).
43 Shlomo Avineri, "Democracy in Israel: Past, Present, Future" talk given at the Y. & S. Nazarian Center for Israel Studies, University of California at Los Angeles, transcript prepared by the Center for Israel Education, 12 January 2021.
44 Often expressed through such activities as queue jumping, ignoring traffic laws, and tax evasion. Alan Dowty, "Jewish Political Culture and Zionist Foreign Policy," in Abraham Ben-Zvi and Aharon Klieman (eds.), *Global Politics: Essays in Honor of David Vital* (London: Routledge, 2001), pp. 310–326.
45 According to Jewish tradition, the first and second Temples in Jerusalem were destroyed and the Jewish people were exiled as a result of the sins, respectively, of idolatry, murder, and sexual immorality, and that of groundless hatred.
46 Charles D. Freilich, *Zion's Dilemmas: How Israel Makes National Security Policy* (Ithaca: Cornell University Press, 2012).
47 According to former Israeli President Reuven Rivlin, Israeli society is made up of "four tribes": secular, national religious, and ultra-orthodox Jews, as well as the country's Arab community.
48 See, e.g., Ze'ev Schiff and Ehud Yaari, *Israel's Lebanon War* (New York: Simon & Schuster, 1984).
49 Joshua Krasna, "The Israeli National Security Constellation and Its Effects on Policymaking," Foreign Policy Research Institute Philadelphia Paper No. 17, February 2018.
50 Ira Sharkansky and Yair Zalmanovitch, "Improvisation in Public Administration and Policy Making in Israel," *Public Administration Review*, Vol. 60, No. 4 (July/August 2000), pp. 322–323.
51 Aharon Klieman, *Statecraft in the Dark: Israel's Practice of Quiet Diplomacy* (Boulder, CO: Westview Press, 1988).
52 Neta Oren, Daniel Bar-Tal, and Ohad David, "Conflict, Identity, and Ethos: The Israeli-Palestinian Case," in Yueh-Ting Lee, Clark McCauley, Fathali Moghaddam, and Stephen Worchel (eds.), *The Psychology of Ethnic and Cultural Conflict* (Westport, CT: Praeger, 2004), pp. 133–154.
53 Jonathan Rynhold, "Cultural Shift and Foreign Policy Change: Israel and the Making of the Oslo Accords," *Cooperation and Conflict*, Vol. 42, No. 4 (December 2007), pp. 419–440.
54 Eliot A. Cohen, Michael J. Eisenstadt, and Andrew J. Bacevich, *Knives, Tanks, and Missiles: Israel's Security Revolution* (Washington, DC: The Washington Institute for Near East Policy, 1998).
55 Efraim Inbar and Eitan Shamir, "'Mowing the Grass': Israel's Strategy for Protracted Intractable Conflict," *Journal of Strategic Studies*, Vol. 37, No. 1 (2014), pp. 65–90.
56 Dima Adamsky, "The Israeli Approach to Defense Innovation," Study of Innovation and Technology in China Research Brief, May 2018.
57 See, e.g., David Kilcullen, *The Dragons and the Snakes: How the Rest Learned to Fight the West* (London: Oxford University Press, 2020).
58 David Sloan Wilson et al., "Evolving the Future: Toward a Science of Intentional Change," *Behavioral and Brain Sciences*, Vol. 37, No. 4 (August 2014), 395–416.

7

THE UNITED STATES AND NATO

Is There a Western Strategic Culture?

Jeffrey A. Larsen

The North Atlantic Treaty Organization (NATO) is an institutional reflection of Western strategic culture. Born in 1949 as one element of the far-ranging post-World War II reinvention of Western institutions, the Alliance has been the premier focus of a common set of perspectives, interests, desires, and beliefs by the free states of the world. Membership has grown dramatically since NATO's founding, which shows the draw of those worldviews for new member states.

There were numerous international institutions created by the victorious West following World War II, but only NATO held the central responsibility of providing a political-military alliance to protect the freedoms cherished by the victors. Many other organizations were founded in these years and thereafter, which have played vital roles in their respective areas of responsibility. All these institutions reflect a common purpose: to bind the free world together in an interlocking web of mutually agreed commitments that will, through their comprehensive nature, prevent the outbreak of a third world war. This was the dream of the founding fathers of the postwar generation, and it has proven to be an effective vision over the past 75 years. Those men who were "present at the creation" of the postwar international system – men like Roosevelt, Truman, Acheson, Marshall, and Churchill – were determined to lead the free world in a fight against communism and tyranny. Based on the events of the 1930s and World War II, they were unwilling to accept the concept of appeasement or a balance of power system. They were, in fact, committing themselves and their countries to a grand vision of the future based on a common strategic culture, and these men went out of their way to make sure that America's wartime allies and other European states joined in that mission.[1] Their efforts resulted in a series of new international institutions that are, for the most part, still managing the world today. Each of these institutions has developed its own set of rules, procedures, and modes of behavior, which may imply a type of political or institutional culture. But not all embody a Western strategic culture. Those that do, such as NATO, are few but extremely influential.

The glue behind this effort was the set of underlying principles on which the West stood: a firm belief that the future of a secure and successful world was based on the fundamental commitment to democracy, capitalism, individual liberty, the rule of law, and a rules-based international system. As the Alliance itself has recently stated,

DOI: 10.4324/9781003010302-9

> NATO guarantees the security of our territory and our one billion citizens, our freedom, and the values we share, including individual liberty, human rights, democracy, and the rule of law. We are bound together by our common values, enshrined in the Washington Treaty, the bedrock of our unity, solidarity, and cohesion.[2]

This mission statement captures the core of Western strategic culture. These principles reflect the development of liberal thought across more than two millennia of the Western experience. They represent centuries of trial and error, learning and recognition, and form the canon of the best aspects of Western civilization. The idea of "the West" is based on an inclusionary vision of agreed values and political beliefs that remains open to all democracies, regardless of their geographic location. As John McCain put it, the West is a universalist project "open to any person or any nation that honors and upholds these values."[3] Particularly since the end of World War II, the concept has grown to include states in Asia, well outside the zone one would typically consider the inheritors of Judeo-Christian, Greco-Roman traditions. This broader definition differs from perspectives that see the concept as an ethnonationalist, religious, or civilizational approach. At the risk of overstating the importance of the development of a Western political culture and strategic culture and its positive impact on humanity's living standards, "[t]his advance in human well-being arguably rates as the greatest achievement in the history of mankind."[4] Indeed, "Western Europe and the English-speaking offshoots elsewhere were unique in developing free markets, security of property, a predictable system of law, and parliamentary limitation on the exercise of arbitrary power."[5] Another analyst has written that "[t]he idea of the West flows from its core principles – political freedom, limited government, religious tolerance, scientific reason, and individual rights – rather than from its geography."[6]

The definition of strategic culture employed across this chapter is the one set out by Jeannie Johnson, Kerry Kartchner, and Jeffrey Larsen in 2009:

> that set of shared beliefs, assumptions, and modes of behavior, derived from common experiences and accepted narratives (both oral and written), that shape collective identity and relationships to other groups, and which determine appropriate ends and means for achieving security objectives.[7]

All cultures are based on the traditions of their constituent units, allowing the elements that subscribe to their principles the freedom to change their minds when national interests are at stake. Yet a culture is not an agreement, or a treaty, or any other legal or political promise. It is deeper than any of those, and it may take some systemic shock to cause the members of that culture to reappraise whether they want to remain a part of that culture or allow it to disintegrate.

NATO has endured two major periods of adaptation to a changing world since the early 1990s, also known as "critical junctures" in the lexicon of historical institutionalism.[8] First, the Alliance had to adapt to the disappearance of the Soviet Union, the Warsaw Pact, and the bipolar threat under which the world lived for so long. For some 20 years thereafter, the Alliance put Russia out of its mind, focusing instead on new missions in faraway lands that were outside of the NATO region according to its founding treaty. But since 2014, the need for a second great adaptation became apparent as Russia returned once again as a spoiler and great power adversary along Europe's eastern frontier. The Russian invasion of Ukraine in 2022 was a continuation of this second period. This chapter assesses how well Western strategic culture has functioned to facilitate these transitions.

First Adaptation: The End of the Cold War

The first and most critical adaptation of NATO was in response to the end of the Cold War. Between 1989 and 1992 the states of the Warsaw Pact and the former Soviet Union rose up in generally peaceful revolutions against their Soviet masters, spurred by the decision of Soviet General Secretary Mikhail Gorbachev to relax Moscow's grip and introduce more political and economic freedom. Gorbachev (and thereafter Russian President Boris Yeltsin) allowed the Soviet republics and nations in the USSR's near abroad the opportunity to take as much democracy as they could handle. In short order the borders to the West were opened, the Berlin Wall fell, the Warsaw Pact disintegrated, the Soviet Union collapsed, and communism was discredited as a political concept. The aftereffects of this revolution included the appearance of 28 new nation-states with little experience in democracy, the rule of law, capitalism, or many of the values that underlie Western strategic culture. These states needed a home, however, and within a decade NATO and the European Union (EU) had determined that the best course of action was to "extend stability" to their eastern neighbors and invite them to join these organizations. Certain minimal criteria had to be met, of course, but this was the solution to prevent a power vacuum and instability along Europe's eastern frontier.[9]

When the end of the Cold War came, NATO, like the rest of the world, was initially shocked and unprepared. The sweeping and sudden end of the bipolar military confrontation with the USSR led some observers to declare that the Alliance had served its purpose: it had protected the people of Western Europe and North America from communism, it had prevented general war, and it had "won" the Cold War. The central question became whether the Alliance was still needed. In fact, the entire decade of the 1990s was one long period of uncertainty over NATO's role in the new world. Several alternatives were considered, including disbanding NATO and allowing national goals to take center stage once again in Europe. But a consensus emerged that without an alliance of like-minded nations, the world might again face the rise of national cultures and agendas, or small fights by groups whose purpose was antithetical to traditional Western civilization. That these challenges have not grown more significant in the past three decades is undoubtedly due to the decision to maintain valued institutions like NATO as one of the keepers of Western strategic culture. The value of the Alliance, and its self-perception, were enhanced by the recent memory of the end of the Cold War. As Julian Lindley-French has written, "NATO did nothing, but NATO meant everything. NATO won the Cold War."[10]

The sense of victory after the Cold War turned out to be short-lived. The 1990s proved to be a formidable period, as the Alliance reacted to new threats and dealt with internal debates over NATO's role in handling those new challenges. These challenges included conflicts in the Balkans, where the EU showed that it had not reached a level of maturity to conduct successful peacekeeping operations, requiring NATO to use military force for the first time in its history – and in Europe, no less. NATO's decision to enlarge its membership began in 1999 with three new member states, all former communist regimes.

At first, the enlargement of NATO did not reflect a deep-seated commitment to Western values or a common strategic culture on the part of the new members from Central and Eastern Europe. Their initial purpose for membership was primarily transactional, to receive protection against the possible return of Russia in their regions. It has taken over

20 years for some of these states to begin to adopt the cultural aspects of their membership. Simply working on the membership action plan, ensuring interoperability between weapons systems, and the like, does not create a shared culture. But working together in the halls of headquarters, witnessing how the older members behave and make decisions, and getting closer to the United States and the major European powers have all had a slow and perceptible influence on the enculturation of the new members that have become obvious to their allies and NATO analysts. As Seth Johnston has written,

> although states maintain considerable power over international institutions, the latter have autonomous capacities and can play consequential roles in facilitating their own adaptation.... Institutional actors – that is, those representing NATO itself, not necessarily its member states – have played underappreciated and consequential roles in facilitating adaptation.[11]

In this view he echoes David Yost, who wrote that "[f]or NATO, aspiration has been the compass providing direction in the face of recurrent setbacks and recalcitrant realities."[12] Furthermore, Wallace Thies has emphasized NATO's "self-healing tendencies" and its ability to overcome internal challenges and disagreements to move forward successfully. As he has written, "What sets NATO apart from so many previous alliances is *not* the absence of disagreements among its members but the ability to act in concert *despite* disagreements among its members."[13]

The peak of the 1990s challenges to NATO actually appeared in 2002–2003, as the United States faced pushback from several of its major European allies in its bid to lead a coalition invasion of Iraq. This led to yet another of the historically recurring crises of confidence within the Alliance, as the members split over whether to support the US operation. All the uncertainties over NATO's mission in the post-Cold War era came to a head in this loss of confidence in American leadership. At the time it seemed to create a nearly insurmountable rift between long-term allies. As William Hay and Harvey Sicherman put it, "If the Atlantic alliance, neé the West, was not physically breaking up, then it seemed on the verge of a collective nervous breakdown."[14] And yet within a year, the crisis had passed. The general commitment to common values and the desire to maintain the Alliance proved more important to all participants than did this single crisis. A common strategic culture informed the members that they had weathered such crises previously and would undoubtedly face them again in the future. But what mattered was to avoid allowing such a moment to devolve into a dispute that threatened the core basis on which the institution was built. In Hay and Sicherman's view,

> this episode illustrated what has always been true of the West and of the alliance itself. It is a nervous, high-strung operation of various proud states and sensitive peoples, yet sufficiently aware of common interests and common destiny to overcome international crises.[15]

Second Adaptation: The Return of an Existential Threat to Europe

Following the Iraq crisis, the Alliance returned to its new path of enlargement and out-of-area operations for the next decade without facing a major challenge. It focused on the development of deployable, expeditionary forces for the new types of fights that it was undertaking in faraway lands like Afghanistan; on the recruitment of more partner nations who were proving to be valuable team members for NATO operations; on continuing to enlarge the Alliance without unduly antagonizing its new strategic partner, Russia; and on reorganizing and downsizing the NATO structure and staff to reflect the more benign international environment. Member states, too, were busy: reducing their stock of heavy combat equipment left over from the Cold War; improving interoperability with allied equipment and procedures; and reducing the level of overall defense expenditure – part of the "peace dividend" following the end of that era. There were flareups and new operations during this time frame, notably internal debate over the purpose, conduct, and participation in the air campaign over Libya in 2011, but generally, the Alliance was on a comfortable track that was reflected in the 2010 Strategic Concept which emphasized that NATO had no enemies and that Europe was not threatened. As that document stated:

> While the world is changing, NATO's essential mission will remain the same: to ensure that the Alliance remains an unparalleled community of freedom, peace, security and shared values ... [the] Alliance thrives as a source of hope because it is based on common values of individual liberty, democracy, human rights and the rule of law, and because our common essential and enduring purpose is to safeguard the freedom and security of its members. These values and objectives are universal and perpetual, and we are determined to defend them through unity, solidarity, strength, and resolve.[16]

This complacency regarding European security changed in early 2014, a year that proved to be a game-changer for Europe and NATO. In retrospect, Western analysts of Russia should not have been surprised by Moscow's behavior in Ukraine, given the way Vladimir Putin had telegraphed his unhappiness with the Western international system, Russia's desire to be perceived as a great power, its supposed need for a sphere of influence in its near abroad, previous Russian military intervention in Georgia, and the like. Yet the military operations Russia undertook to capture Crimea and foment armed insurrection in the Donbas region of eastern Ukraine in the spring of 2014 were a seismic shock to the Alliance. Over the next several years, some member states of NATO proved hesitant to respond for fear of overreacting to Russia's newly aggressive foreign policy, hoping instead for a return to "normal" relations that had characterized the previous 20 years. Meanwhile, Russia continued to ramp up the pressure with large military exercises on NATO's borders, nuclear saber-rattling against specific NATO national capitals, a major air campaign in Syria to support the regime of President Assad, and increasingly aggressive air and naval maneuvers in and around the Allied territory. Rebecca Moore and Coletta summarized the "u-turn" in the evolution of the Alliance brought about by Russia's intervention in Ukraine and the Alliance's response:

> [I]t dragged NATO's attention back to Europe and the specific means by which Article 5 collective defense guarantees are achieved, abruptly clarified the distinction between NATO partners and members, reinvigorated the debate over NATO enlargement, prompted a reassessment of the synergy among multiple identities that NATO has established since its inception in 1949, and exposed fault lines within the Alliance regarding principal threats to the allies as well as conditions under which member states are willing to use force.[17]

It took two years and two NATO summits before the Alliance returned to a consensus position that Russia was no longer a trusted strategic partner, that there would be no going back to business as usual, and that the Alliance needed to return to its original focus on collective defense as its primary mission. Not every state wanted to follow this path, but a commitment to common defense allowed it to move on. It was not until the Warsaw Summit in July 2016 that the most uncertain of Western allies came to the realization that Moscow was no longer the country that Europe had learned to live with since the end of the Cold War. The summit called for significant new military responses to Russia along NATO's eastern frontier, and an overall commitment to old-fashioned collective defense – the original mission of the Alliance.

Simultaneously, to make matters more complicated, multiple new challenges arose in the south of Europe. As a result of failed and failing states in the Middle East and North Africa (MENA) region, Europe suddenly faced an uptick in terrorism, a huge influx of illegal migrants, and the rise of a new and particularly heinous adversary, the Islamic State of Iraq and Syria (ISIS). In short, NATO's southern flank was now perceived by some allies to be as dangerous as Russia. Several of the Mediterranean members of the Alliance pushed for new policies that would recognize those threats as equal to the one posed by the return of a nuclear-armed great power adversary to the east. In response, the Alliance announced a "360-degree" policy for threat awareness, while in fact still prioritizing the Russian challenge. This showed the power of language and symbology, and the lengths to which an organization will go to keep its members satisfied and tied to the underlying culture of the institution.

NATO as an organization knew it had to respond to the new threats, and that it needed solidarity and cohesion to do so. The indecisive members were eventually convinced of the need for decisive and unified action in response to the unexpected rise of these multiple new challenges from several directions. The Alliance stood firm, showing its resolve to ensure the security of its members in its summit declarations, in its decisions at the quarterly meetings of foreign and defense ministers, and in its operational actions. In particular, NATO accepted the vital need to return to its core mission: collective defense. The language in the Warsaw Summit declaration was some of the strongest since the end of the Cold War. It took the leadership of the United States and the transatlantic wing of the Alliance to make the point that a unified position was crucial to ensuring credible deterrence in this time of crisis. Appeals were made to the underlying values of the Alliance, and the need to stand together against the forces of nationalism and autocracy.[18] NATO's Warsaw Summit Communique made clear that:

> NATO's essential mission is unchanged: to ensure that the Alliance remains an unparalleled community of freedom, peace, security, and shared values, including individual liberty, human rights, democracy, and the rule of law. We are united in our commitment to the Washington Treaty, the purposes and principles of the Charter of the United Nations (UN), and the vital transatlantic bond.[19]

This renewed commitment to a Western alliance based on a strategic culture of shared common values also gave the Alliance the strength and patience to get through the following four years of uncertainty over America's commitment. Within six months of the successful Warsaw Summit, another development once again divided NATO Europe from the United States: the election of Donald Trump as president. Trump had stated throughout his electoral campaign that he did not like or trust NATO, that the United States might not honor its security commitment to certain countries if they came under attack, and that all members needed to pay the United States what amounted to protection money. These were views that diminished the values of NATO and questioned its validity. Trump carried these ideas into the White House and his first meetings with his peers in Europe. His leadership style emphasized "the deal," a transactional approach that put little stock in a sense of Western civilization based on values as the underlying bedrock of the Alliance. As one analyst wrote, "Trump abandoned the Jeffersonian West of liberty, multilateralism, and the rule of law, in favor of an ethno-religious-nationalist West," making him "the first non-Western president of the United States."[20] The Alliance was aghast at such language and beliefs, and what troubled other members most was Trump's unwillingness to accept the underlying strategic culture of an alliance of democratic nations that had more invested than simply combat brigades or defense budgets.

This was, in some ways, a parallel crisis of leadership not unlike the 2003 loss of confidence in the United States during the period of the Iraq invasion. Like that earlier period, a combination of the binding element of a strong NATO strategic culture, as well as a willingness to wait out the unpopular US president in favor of a successor who held a more transatlantic perspective, proved successful in weathering this political storm.

Can Strategic Culture Explain These Two Adaptations?

There are multiple alternative explanations as to why NATO was able to successfully adapt itself during the two transition periods reviewed above: basic military necessity (a realist approach); organization process theory, which emphasizes institutional resistance to change; or the adoption of the Harmel Report in 1967, which paved the way for the Alliance to have an alternative mission beyond simple military security.[21]

Beneath all these alternatives, however, remains the fact that the Alliance and its member states have continued their commitment to shared values, to the desire to have close allies on whom they can depend in times of crisis, and to maintaining a consensus on the purpose of the Alliance, even as those purposes have morphed and evolved. There are, quite frankly, few tangible signs of a political or strategic culture within the Alliance beyond its aspirational words in the treaty and in its regular summit communiqués. Yet that cultural underpinning is always there, behind the scenes, understood and respected by nearly all members. As Alastair Iain Johnston wrote, one purpose of symbols and language in developing a strategic culture "has to do with the creation and perpetuation of a sense of in-group solidarity directed at would-be adversaries."[22] Ernest Bormann has argued that "a political community first needs to exist as a 'rhetorical community,' bound together by myths and language which underscore the uniqueness of the community."[23] It appears to many analysts that Western democracies have developed a shared sense of identity such that "liberal democracy tags a state as one which shares key domestic political values with other democracies," and thereby "perceive each other as members of a larger in-group or community."[24]

The concept of a strategic culture defining the "West" is based on the fundamental concept of liberty and of governments created by the consent of free individuals. There has always been a spiritual aspect to Western culture, but modern societies are defined above all by their adherence to democracy. As Daniel Mahoney has written, the essential preconditions for democratic self-government include

> the territorial nation-state, the moral and intellectual inheritance of classical and biblical civilization, the mores that restrain individual self-assertion and allow free people to live together under the rule of law, and the civic spirit that leads free peoples to defend the "precious acquisition"... that is political liberty.[25]

Despite its slippery definition and vagaries, there are times when culture may prove to be the more dominant explanatory variable.[26] As David Haglund has written, strategic culture is too "obtuse" to be an actual theory.[27] But one does not always need a theory to explain what happens in the world. One's own life experiences are equally valid, and those shared by a group are even stronger. Recognizing the commitment to a set of values that are potentially universal, and that surpass the needs or wants of a nation-state, is one of those realities that is more easily understood than explained.

The challenge for political scientists, of course, is finding empirical evidence to support the concept of strategic culture. Many have employed elite analysis, based on the assumption that a nation's elites best reflect – and use in their public discourse and documents – those cultural aspects that represent their society and which their citizens value.[28] The consensus is that culture is an important but "fuzzy" independent or intervening variable in explaining why a nation-state, or an international institution, makes the decisions that it does. Strategic culture serves as a brake, or a conscience, on decisionmakers as they ponder the best action in a situation. As Benjamin Zyla has written, "Social actors reproduce norms and structure by reflexively basing their actions on their acquired knowledge, habits, and routines."[29] Such deep-rooted feelings are carried across time and, in the case of institutions, across national boundaries. Cultures can change, but it normally takes a seismic event in the international system, or the history of a state, to effect such a change.

Another explanation may reflect the concept of subcultures, an idea first explained by Gabriel Almond and Sidney Verba.[30] As utilized more recently by Zyla, sub-strategic cultures can explain the differences between the normative and operational outlooks of the EU and NATO, the two major security institutions in Europe.[31] But the concept of subcultures can also apply to the internal dynamics of an international organization. As other analysts have noted, it is likely that "Europe is not a unitary actor in international relations," lacking "a clearly defined single strategic culture."[32] But NATO's membership is larger than Europe, and it has both a strategic culture of its own as well as several subcultures. The most obvious way to define those subcultures can be seen in the division between "old" and "new" Europe – the older members of the Alliance, and those new member states that joined after the end of the Cold War. These two categories of sub-strategic cultures might also be linked to the transatlantic or Europeanist perspectives that dominate each group.[33]

Old Europe is best exemplified by France, which is the leading Europeanist nation. It supports NATO, but Paris is more interested in a stronger EU, one with its own European military and foreign policy establishment that could supplement or one day perhaps even supplant NATO. This is based to a significant degree on its concern that Europe cannot rely indefinitely on the United States and its security guarantees, and that the continent must take on more responsibility for its own defense. Furthermore, the EU as an institution has a broader mandate than NATO. Rather than focusing solely on security and defense, it also considers civilian missions to improve the lives and livelihood of people on Europe's periphery. Despite having nearly identical normative values and interpretations of future challenges as the EU, NATO is "a regionally confined military alliance with a primary reason d'état of ensuring the physical safety of its member states."[34]

The United Kingdom is often cited as the foremost Atlanticist state. It believes in the enduring value of the transatlantic link and in the leadership role of the United States in that relationship. It knows that Europe cannot provide for its own security without the support of the United States and Canada. These feelings are equally strong in many of the newer member states of the Alliance, especially those former members of the Soviet Union or the Warsaw Pact. Thus Poland, Romania, and the Baltic States are staunch supporters of the Atlanticist perspective. One can argue that this is a transactional position determined by recent history and the need for a strong protector, but it may also reflect an underlying strategic cultural perspective and belief that NATO "represents an institutionalization of the transatlantic security community based on common values and collective identity of liberal democracies."[35]

According to one recent study, Atlanticism and Europeanism share a continuum and can shift along that continuum as world events dictate. But the study's authors have no doubt that these two perspectives reflect alternative strategic subcultures whose presence can be found in all the member states. They point out that in a major poll by the German Marshall Fund in 2013 (just prior to the second case examined in this chapter), 58% of Europeans polled felt that NATO was still essential, and of those, 56% stated that this is because "NATO is an alliance of democratic countries," vice only 15% who felt it was because "there are still major threats that endanger our country."[36] This reflects a strong cultural component to the support for the Alliance in European societies.

The fact that subcultures are present in an Alliance of 30 nation-states is not surprising. That the views of the subcultures should become more apparent in times of relative quiet is also unsurprising. One author has likened the American perspective (one also embraced by the Atlanticists) as one that sees the world in terms of Hobbesian realism, ready and willing to use force to maintain stability, defend national and international interests, and advance human rights when possible. The Old Europeans, by contrast, take a more benign view of the world, and seek to achieve a Kantian peace on the continent and beyond. Living under the umbrella of American security guarantees for three generations has given them the luxury of building a network of institutions for multilateral problem-solving using soft power as their first choice.[37]

Challenges to Strategic Culture

In recent years, the core beliefs of Western strategic culture have come under pressure by the rise of nationalist parties in some European states, and a concomitant decline in democracy, freedom of the press, the rule of law, and other aspects of what are considered the values enshrined in Western documents, including the Washington Treaty. As of this writing, Hungary, Poland, and Turkey are no longer considered full democracies, and other nations, including Bulgaria, Italy, France, and the United States face domestic challenges from nationalist political movements that challenge Western culture and Alliance cohesion.[38] The EU took the unusual step at its June 2021 summit meeting of publicly criticizing President Viktor Orban of Hungary, and suing that country in international court for the way in which Budapest has flaunted EU regulations and overturned liberal Western ideals in its national laws and decrees.[39] Democracy remains fragile in several countries in southeastern Europe, including Greece, Slovenia, Montenegro,

Albania, and North Macedonia.[40] In addition, Hungary and Turkey are challenging Alliance cohesion with their deepening political and economic linkages to Russia, despite Russia's illegal invasion of Ukraine. Russian aggression against its neighbor posed significant challenges to Alliance cohesion and might have led to a potential fracturing of the Alliance's commitment to a common set of values based on the Western canon. Yet as of this writing, the Alliance has remained as strong and cohesive as at any point since the end of the Cold War. Its response to the rise of an existential threat on its borders was not only pragmatic but further buttressed its commitment to a common strategic culture.

Based on historical experience, it is likely that the strength of such beliefs and norms will bring those nations that are struggling democracies back into the Atlantic community. To facilitate such a return to a cohesive community of values, two European experts have recommended that the Alliance sign a new political pledge in which all states recommit to upholding transatlantic values. Indeed, the "NATO 2030" document published in December 2020, and endorsed at the June 2021 NATO Summit called for such a pledge in the form of a code of good conduct.[41] This would include not only a commitment to uphold transatlantic values such as democracy, individual liberty, and the rule of law but also some principles of good behavior, such as not blocking Alliance business for national issues and consulting allies on matters of common concern before making national decisions.[42] In addition, these principles suggest that NATO as a collective body needs to more closely monitor and assess each member's commitment to democracy, and actively intercede when necessary to pressure states who are falling behind to return to the fold.[43] Taking such a tough course of action entails some risks, of course, and will be very difficult for an institution that operates by consensus and caution when it comes to other states' behavior. But as Ellehuus and Morcos say, "The risks of ignoring NATO's internal strains far outweigh the benefits of addressing them."[44]

In a study examining Western strategic culture in NATO, Jordan Becker has proposed that there is an inverse relationship between the strength of strategic culture and the commitment to equitable burden sharing within the Alliance, as expressed in a nation's defense spending.[45] If this were true, it might seem to undermine the hypothesis in this chapter that there is, in fact, a strategic culture within the Alliance. However, it is equally possible that this actually proves the existence of such a culture. Since NATO does have strategic subcultures, unequal burden-sharing commitments are to be expected. Indeed, when the United States has aggressively pushed for more spending by its European allies, it has actually undercut the culture and affected cohesion within the Alliance.[46] The world witnessed this cause-and-effect relationship during the Trump administration, with the results validating Becker's hypothesis. Allied opinion of America's leadership role fell to record low levels when President Trump pushed the allies for additional spending, which seemed to reflect America's abandonment of the strategic culture to which all members had pledged their allegiance. The annual Munich Security Conference in 2020 went so far as to propose the concept of "Westlessness" as its theme, reflecting "the decay of the Western project," due in part to this approach by the American president.[47]

Upon entering office in early 2021, President Joe Biden made it his mission to restore faith in the underlying values embedded within the political and strategic culture of the Alliance, the concept of transatlanticism, and the leadership of the United States, all to a grateful sigh of relief from Europe. The honeymoon period between Biden and the European allies was diminished to some extent by the end of NATO involvement in

Afghanistan in August 2021, but overall his leadership style was a welcome change. And the allies seemed genuinely pleased with Biden's resolute but restrained approach to Russian behavior in the war in Ukraine, as well as US leadership of responses by the West.

Of course, a new American leader cannot resolve all the issues facing the Alliance. A major issue remains with Ukraine's troubles with Russia since 2014, and NATO's responsibility to intervene – or the lack thereof. As a partner, rather than a member state, the commitments made to Ukraine are clear and circumscribed. It does not warrant protection under Article 5 of the Washington Treaty. However, as one analyst has asked, if

> NATO takes seriously the idea of a security community that is not circumscribed by historical claims or spheres of influence, it has no choice but to contemplate what it owes partners that have made significant contributions to that larger security community.[48]

This unfortunately leaves the reader with a major unanswered question: is NATO a security Alliance of cooperative nation-states that only wants to protect a geographical space? Or is it truly a strategic culture that is willing to fight for an idea? And if the latter, how far will it go to pursue that goal?

Conclusion

Discussion of strategic subcultures within the Alliance may seem to contradict the hypothesis that there is a dominant strategic culture that motivates NATO and its members. Yet the experience of two major crises in the Alliance's recent history proves otherwise. It is apparent that there is a normative Western culture that recognizes the importance of befriending like-minded friends and allies that are fellow democracies, all subscribing to the values of liberty, freedom, and the rule of law. This is just as important for new member states as for the older members, even if the newer members have less experience in delivering those attributes to their citizenry. Such a commitment is mandated by Article 10 of the Washington Treaty, which requires any prospective member state to subscribe to those principles.[49] Wanting to join simply to have a military ally that can improve one's own security is not enough.

Because of the different backgrounds of current Alliance members, however, there is an obvious and understandable distinction between the older and newer members of the Alliance. This split approximates to a substantial degree the parallel differentiation between those states who emphasize transatlantic relations with North America, and those that prefer an approach to security that is more Eurocentric. These subcultures provide many of the divisive issues that concern NATO in normal times, but all of these concerns seem to fall by the wayside during a crisis when allied cohesion is required – as was shown in the two cases of Alliance adaptation examined in this chapter. As Douglas Porch has written,

> One of the endemic problems of the West... is that it requires constant defending, and that it holds together better when it can define a common agenda and strategy as it did during the Cold War. But disagreements about eligibility for membership in the Western club and what constitutes a proper Western response to threats are the rule more than the exception.[50]

Is there a Western strategic culture? Yes. But it must be inferred through the behavior and decisions made by NATO and its member states; it is very difficult to prove empirically. Its existence is seen in the efforts undertaken by the Alliance in two of its most recent critical junctures: transforming itself following the end of the Cold War and the disappearance of the adversary that was the focus of its attention during that conflict, and adapting its direction to new realities after the return of that adversary a quarter century later.

Strategic culture is a significant unseen binding force for the members of the Alliance, often only recognized following a crisis, or found in regularly published official statements from Brussels. On a daily basis, one is more cognizant of the interactions of strategic subcultures, which create so much of the ongoing clamor inside this international organization of 31 nation-states. But as the two cases in this chapter have shown, when the chips are down, and the challenges become existential, NATO's members have always rallied around their commitment to the institution that best serves their interests as a collection of free, democratic states. This is the ultimate basis for the existence of the Alliance. Strategic culture provides the bedrock for understanding why this is so.

Notes

1 Dean Acheson, *Present at the Creation: My Years in the State Department* (New York: W.W. Norton, 1967). Also Robert Kagan, *The Jungle Grows Back: America and Our Imperiled World* (Washington, DC: Brookings Institute, 2018).
2 "Brussels Summit Communique, issued by the Heads of State and Government participating in the meeting of the North Atlantic Council in Brussels, 14 June 2021," Press Release (2021)086, NATO Headquarters, 14 June 2021, para. 2.
3 John McCain, quoted in Ishaan Tharoor, "Can Biden Save the 'West?'" *The Washington Post,* 9 June 2021, https://www.washingtonpost.com/world/2021/06/09/biden-west-europe-trip/.
4 Stephen Schuker, "A Sea Change in the Atlantic Economy?" in William Hay and Harvey Sicherman (eds.), *Is There Still a West? The Future of the Atlantic Alliance* (Columbia: University of Missouri Press, 2007), p. 97.
5 Schuker, p. 100.
6 Douglas Porch, "Conflict within Cooperation: Western Security Relations, Past and Present," in Hay and Sicherman, p. 156.
7 Jeannie Johnson, Kerry Kartchner, and Jeffrey Larsen (eds.), *Strategic Culture and Weapons of Mass Destruction: Culturally Based Insights into Comparative National Security Policymaking* (New York: Palgrave Macmillan, 2009), p. 9
8 According to Seth Johnston, critical junctures "constitute the challenges that threaten institutional endurance. By disrupting institutional stability, these conditions can make institutional change more likely." Seth Johnston, *How NATO Adapts: Strategy and Organization in the Atlantic Alliance since 1950* (Baltimore, MD: Johns Hopkins, 2017), p. 3.
9 For background on extending stability as a NATO mission, see Rebecca Moore, *NATO's New Mission: Extending Stability in Post-Cold War World* (Santa Monica: ABC-Clio, 2007); and Jeffrey Larsen and Kevin Koehler, "Projecting Stability to the South: NATO's 'New' Mission?" in Eugenio Cusumano and Stefan Hofmaier (eds.), *Projecting Resilience across the Mediterranean* (Cham, Switzerland: Palgrave Macmillan, 2020), pp. 37–62.
10 Julian Lindley-French, *The North Atlantic Treaty Organization: The Enduring Alliance* (New York: Routledge Global Institutions, 2007), p. 59.
11 Seth Johnston, *How NATO Adapts: Strategy and Organization in the Atlantic Alliance since 1950* (Baltimore, MD: Johns Hopkins, 2017), p. 59.
12 David Yost, *NATO's Balancing Act* (Washington, DC: US Institute of Peace, 2014), p. 1.
13 Wallace Thies, *Why NATO Endures* (New York: Cambridge University Press, 2009), p. 296, quoted in Yost, p. 21.

14 Hay and Sicherman, p. 2.
15 Hay and Sicherman, p. 3.
16 See *Active Engagement, Modern Defence: Strategic Concept for the Defence and Security of the Members of the North Atlantic Treaty Organization* (Brussels: NATO Headquarters, November 2010), p. 5, https://www.nato.int/nato_static_fl2014/assets/pdf/pdf_publications/20120214_strategic-concept-2010-eng.pdf.
17 Rebecca Moore and Damon Coletta, "Introduction," in Moore and Coletta (eds.), *NATO's Return to Europe: Engaging Ukraine, Russia, and Beyond* (Washington, DC: Georgetown University Press, 2017), p. 3.
18 See "Warsaw Summit Communique issued by the Heads of State and Government participating in the Meeting of the North Atlantic Council in Warsaw," 9 July 2016, https://ccdcoe.org/uploads/2018/11/NATO-160709-WarsawSummitCommunique.pdf.
19 "Warsaw Summit Communique," para. 2.
20 Michael Kimmage, *The Abandonment of the West: The History of an Idea in American Foreign Policy* (New York: Basic Books, 2020).
21 The Harmel Report called for defense and dialogue as the dual-track basis of NATO strategy – the requirement to actually talk with the adversary, in addition to ensuring the Alliance had sufficiently sized and robust defense capabilities. This represented the beginnings of arms control negotiations, the Helsinki process and its confidence and security-building measures, and so on. It remains NATO policy to this day. See "The Future Tasks of the Alliance: Report of the Council—The Harmel Report," 14 December 1967, https://www.nato.int/cps/en/natohq/official_texts_26700.htm. Also see Schuyler Foerster and Jeffrey A. Larsen, *NATO Strategy: Integrating Defense and Collaborative Security,* NDC Research Paper No. 18 (Rome: NATO Defense College, March 2021).
22 Alastair Iain Johnston, "Thinking about Strategic Culture," *International Security,* 19:4 (Spring 1995), p. 58.
23 Ernest Bormann, "Symbolic Convergence: Organizational Communication and Culture," in Linda Putnam and Michael Pacanowsky (eds.), *Communication and Organizations: An Interpretative Approach* (Beverley Hills, CA: Sage, 1983), pp. 100–106.
24 Johnston, pp. 60–61.
25 Daniel Mahoney, "Humanitarian Democracy and the Postpolitical Temptation," in Hay and Sicherman, p. 27.
26 Johnson, Kartchner, and Larsen, p. 10.
27 David Haglund, "What Good Is Strategic Culture?" in Johnson, Kartchner, and Larsen, p. 27.
28 Including Benjamin Zyla, "Overlap or Opposition? EU and NATO's Strategic (Sub-) Culture," *Contemporary Security Policy* (December 2011), pp. 667–687, http://dx.doi.org/10.1080/13523260.2011.623066; Joshua Becker and Edmund Malesky, "The Continent or the 'Grand Large?' Strategic Culture and Operational Burden-Sharing in NATO," *International Studies Quarterly,* 61 (2017), pp. 163–180; and Heiko Biehl, Bastian Giegerich, and Alexandra Jonas (eds.), "Introduction," *Strategic Cultures in Europe: Security and Defence Policies across the Continent* (Springer 2013), pp. 7–16.
29 Zyla, pp. 670–671.
30 Gabriel Almond and Sidney Verba, *The Civic Culture: Political Attitudes and Democracy in Five Countries* (Princeton, NJ: Princeton University Press, 1963), pp. 27–31.
31 See Zyla.
32 Kerry Longhurst and Marcin Zaborowski (eds.), *Old Europe, New Europe, and the Transatlantic Security Agenda* (London: Routledge, 2005), p. 17.
33 Paul Luif, "The Strategic Cultures of 'Old' and 'New' Europe," *The International Spectator,* 41:2 (2006), pp. 109–112, http://dx.doi.org/10.1080/03932720608459420.
34 Zyla, p. 675.
35 Thomas Risse-Kappen, "Collective Identity in a Democratic Community," in Peter Katzenstein (ed.), *The Culture of National Security* (New York: Columbia University Press, 1996), quoted in Becker and Malesky, p. 165.
36 Quoted in Becker and Malesky, p. 166.
37 Stephen Shuker, "A Sea Change in the Atlantic Economy?" in Hay and Sicherman, p. 91.

38 World Justice Project, *WJP Rule of Law Index 2017–18*, quoted in Dick Zandee, "The Future of NATO: Fog over the Atlantic?" 18 December 2018, https://www.clingendael.org/person/dick-zandee.
39 "Thousands March in Hungary Pride Parade," PBS Newshour, 24 July 2021, https://www.pbs.org/newshour/world/thousands-march-in-hungary-pride-parade-to-oppose-lgbt-law.
40 Rachel Ellehuus and Pierre Morcos, "NATO Should Finally Take Its Values Seriously," *War on the Rocks*, 9 June 2021, https://warontherocks.com/2021/06/nato-should-take-its-values-seriously/.
41 *NATO 2030: United for a New Era, Analysis and Recommendations of the Reflection Group appointed by the NATO Secretary General*, 25 November 2020, https://www.nato.int/nato_static_fl2014/assets/pdf/2020/12/pdf/201201-Reflection-Group-Final-Report-Uni.pdf.
42 Ellehuus and Morcos.
43 Ellehuus and Morcos.
44 Ellehuus and Morcos.
45 Jordan Becker, "Strategic Culture and Burden Sharing in NATO: False Friends?" unpublished paper, US Military Academy, 2012.
46 Becker, 2012.
47 Tharoor.
48 Rebecca Moore, "The Purpose of NATO Partnership," in Moore and Coletta, p. 182.
49 *The North Atlantic Treaty*, 4 April 1949, https://www.nato.int/nato_static/assets/pdf/stock_publications/20120822_nato_treaty_en_light_2009.pdf.
50 Porch in Hay and Sicherman, p. 159.

Selected Bibliography

Acheson, Dean. *Present at the Creation: My Years in the State Department* (New York: W.W. Norton, 1967).

Active Engagement, Modern Defence: Strategic Concept for the Defence and Security of the Members of the North Atlantic Treaty Organization (Brussels: NATO Headquarters, November 2010), https://www.nato.int/nato_static_fl2014/assets/pdf/pdf_publications/20120214_strategic-concept-2010-eng.pdf.

Almond, Gabriel, and Sidney Verba. *The Civic Culture: Political Attitudes and Democracy in Five Countries* (Princeton, NJ: Princeton University Press, 1963).

Becker, Joshua, and Edmund Malesky. "The Continent or the 'Grand Large?' Strategic Culture and Operational Burden-Sharing in NATO," *International Studies Quarterly*, 61 (2017), pp. 163–180.

Biehl, Heiko, Bastian Giegerich, and Alexandra Jonas (eds.), *Strategic Cultures in Europe: Security and Defence Policies across the Continent* (Springer, 2013).

Ellehuus, Rachel, and Pierre Morcos. "NATO Should Finally Take Its Values Seriously," *War on the Rocks*, 9 June 2021, https://warontherocks.com/2021/06/nato-should-take-its-values-seriously/.

Foerster, Schuyler and Jeffrey A. Larsen. *NATO Strategy: Integrating Defense and Collaborative Security*, NDC Research Paper No. 18 (Rome: NATO Defense College, March 2021).

Hay, William, and Harvey Sicherman. *Is There Still a West? The Future of the Atlantic Alliance* (Columbia, MO: University of Missouri Press, 2007).

Johnson, Jeannie, Kerry Kartchner, and Jeffrey Larsen (eds.), *Strategic Culture and Weapons of Mass Destruction: Culturally Based Insights into Comparative National Security Policymaking* (New York: Palgrave Macmillan, 2009).

Johnston, Alastair Iain. "Thinking about Strategic Culture," *International Security*, 19:4 (Spring 1995), p. 58.

Johnston, Seth. *How NATO Adapts: Strategy and Organization in the Atlantic Alliance since 1950* (Baltimore, MD: Johns Hopkins, 2017), p. 3.

Kagan, Robert. *The Jungle Grows Back: America and Our Imperiled World* (Washington, DC: Brookings Institute, 2018).

Kimmage, Michael. *The Abandonment of the West: The History of an Idea in American Foreign Policy* (New York: Basic Books, 2020).

Larsen, Jeffrey, and Kevin Koehler. "Projecting Stability to the South: NATO's 'New' Mission?" in Eugenio Cusumano and Stefan Hofmaier, eds., *Projecting Resilience across the Mediterranean* (Cham, Switzerland: Palgrave Macmillan, 2020), pp. 37–62.

Lindley-French, Julian. *The North Atlantic Treaty Organization: The Enduring Alliance*, 2nd ed. (New York: Routledge Global Institutions, 2015).

Longhurst, Kerry, and Marcin Zaborowski (eds.), *Old Europe, New Europe, and the Transatlantic Security Agenda* (London: Routledge, 2005).

Luif, Paul. "The Strategic Cultures of 'Old' and 'New' Europe," *The International Spectator*, 41:2 (2006), pp. 109–112, 10.1080/03932720608459420.

Moore, Rebecca and Damon Coletta (eds.), *NATO's Return to Europe: Engaging Ukraine, Russia, and Beyond* (Washington: Georgetown University Press, 2017).

Moore, Rebecca. *NATO's New Mission: Extending Stability in Post-Cold War World* (Santa Monica: ABC-Clio, 2007).

NATO 2030: United for a New Era, Analysis and Recommendations of the Reflection Group appointed by the NATO Secretary General, 25 November 2020, https://www.nato.int/nato_static_fl2014/assets/pdf/2020/12/pdf/201201-Reflection-Group-Final-Report-Uni.pdf.

Risse-Kappen, Thomas. "Collective Identity in a Democratic Community," in Peter Katzenstein (ed.), *The Culture of National Security* (New York: Columbia University Press, 1996).

Tharoor, Ishaan. "Can Biden Save the 'West?'" *The Washington Post*, 9 June 2021, https://www.washingtonpost.com/world/2021/06/09/biden-west-europe-trip/.

"The Future Tasks of the Alliance: Report of the Council—The Harmel Report," 14 December 1967, https://www.nato.int/cps/en/natohq/official_texts_26700.htm.

The North Atlantic Treaty, 4 April 1949, https://www.nato.int/nato_static/assets/pdf/stock_publications/20120822_nato_treaty_en_light_2009.pdf.

Thies, Wallace. *Why NATO Endures* (New York: Cambridge University Press, 2009).

Yost, David. *NATO's Balancing Act* (Washington: US Institute of Peace, 2014).

Zandee, Dick. "The Future of NATO: Fog over the Atlantic," 18 December 2018, at https://www.clingendael.org/person/dick-zandee

Zyla, Benjamin. "Overlap or Opposition? EU and NATO's Strategic (Sub-) Culture," *Contemporary Security Policy*, December 2011, pp. 667–687, 10.1080/13523260.2011.623066.

8

NATURE, EMERGENCE, AND LIMITATIONS OF AN EU STRATEGIC CULTURE[1]

Christoph O. Meyer

Introduction

The discussion over whether and what kind of strategic culture the European Union (EU) has or needs arose after the decision of the 1999 Cologne Summit to create a European Security and Defense Policy (ESDP, or CSDP post-2009). Already then, Paul Cornish and Geoffrey Edwards[2] argued that a functioning European defense and security policy needed to be underpinned by a common strategic culture. They defined such a culture at the time as "the institutional confidence and processes to manage and deploy military force as part of the accepted range of legitimate and effective policy instruments, together with general recognition of the EU's legitimacy as an international actor with military capabilities."[3] Two years later, the EU's first-ever Security Strategy said more about what kind of strategic culture was needed, namely one "that fosters early, rapid, and when necessary, robust intervention."[4] While the EU's Global Strategy of 2016 does not mention "strategic culture" at all, the EU's High Representative/Vice-President Josep Borrell said in his investiture hearings that "[w]e need shared strategic culture and empathy to understand the different points of view."[5] This goal was further reflected in a call for a "common strategic culture" in the 2022 Strategic Compass, the EU's version of a security and defense review.[6]

To use the concept of strategic culture in the context of a non-state political system with both supranational and intergovernmental characteristics has from the beginning raised questions about definition, measurement, feasibility, and desirability. Conceptual questions are intrinsically tied to more optimistic or pessimistic assessments of whether and to what extent the EU currently has or could develop a shared strategic culture related to the use of force and/or the pursuit of security more broadly. More pessimistic analyses have tended to take the strategic cultures of France and the United Kingdom as explicit or implicit benchmarks and question whether the EU could ever "do strategy"[7] or point to the structural limitations arising from the "strategic cacophony" among its members.[8] They highlight what they see as large ideational differences and incompatibilities between smaller nonaligned and neutral countries in Europe, those with historically ingrained pacifism, members worried about the Russian threat to their territorial integrity and political sovereignty, and, finally, states with a high willingness to use force for a wide range of

 DOI: 10.4324/9781003010302-10

objectives.[9] More optimistic assessments question the appropriateness and narrowness of some of the conceptualizations of strategic culture when applied to the EU and stress a more holistic or post-national understanding of security.[10] They also see evidence of increasing "strategic culture-building" via Brussels-based institutions and learning from EU missions on the ground,[11] as well as cultural convergence among some member states in some specific areas.[12]

This chapter elaborates first on the conceptual challenges in studying the strategic culture of a supranational identity such as the EU and, particularly, how it relates to the strategic cultures of its member states. Second, it proceeds to examine the evidence in relation to the evolution, drivers, and nature of an EU strategic culture visible through developments in Brussels and learning from EU missions and operations. Third, it looks at the main differences and similarities in national strategic cultures as well as trends toward convergence or divergence. The chapter concludes with a brief discussion of the outlook for strategic culture building and its importance.

Conceptualizing EU and European Strategic Culture

This chapter conceptualizes strategic culture as a contributing but rarely sufficient causal factor in explaining foreign, security, and defense policy. It shapes and limits options of what is appropriate, but does not determine specific outcomes or decisions. It informs a given community in a semiconscious way on how to chart a rough trajectory in matters of international security, which paths to avoid as too risky, and which ends to pursue in foreign and security policy. Culture, whether at the level of society, political elites, or military organizations, transcends official strategy documents and planning papers even if the process and outcomes of strategy writing help to constitute, reaffirm, and occasionally also challenge and change ideas supported by such cultures. The value of the concept lies in its high permanence and deep roots among foreign policy communities regardless of party affiliation, as it is rooted in collective identities, memories, and lessons learnt from history as well as a country's geopolitical location. It is particularly useful in situations of high uncertainty and crisis when political actors tend to rely more on their conventional analytical prisms and the instruments they trust, rather than having time to question the adequacy of their preexisting worldviews and beliefs about what works.

Strategic culture can be defined as "the socially transmitted, identity-derived norms, ideas, and habits of mind that are shared among the most influential actors and social groups within a given political community, which help to shape a ranked set of options for a community's pursuit of security and defense goals."[13] This grounds strategic culture in modernist social constructivist theorizing and explicitly allows for the potential of incremental change and contradictions among its component beliefs, ideas, and norms. These include, first, goals for which the use of force is considered legitimate; for instance, territorial defense, defense of human rights, preemptive action against threats, or cultural and territorial expansion. Second, the degree to which force has to be domestically and/or internationally authorized in order to be considered legitimate; for example, with or without United Nations (UN) mandate, peer support, parliamentary approval, or consent of constitutional courts. Third, the willingness and ways in which force is to be used; in particular, the tolerance to risks arising for a country's own troops and civilians as well as foreign nationals, or the acceptability of using nuclear weapons. And finally, norms relating to the way in which a state should cooperate with other states and/or alliances, covering the spectrum from neutrality, acting within and through entities like the EU, or using force together with or in support of specific foreign actors such as the United States. This conceptualization overlaps with and cuts across alternative dimensions such as broader foreign policy and security ambitions of the state.[14]

The semantic difference between a European versus EU strategic culture is not a minor one even if some authors use them as synonyms. Not all European states are members of the EU, including, since 2020, the United Kingdom – one of the most militarily capable states. Furthermore, EU members have in the past participated in many military operations outside the EU framework, be it within the North Atlantic Treaty Organization (NATO), under the UN, or within "coalitions of the willing," thus creating potentially a convergence among military organizations regardless of framework. Furthermore, strategic culture could describe common or convergent ways of thinking among national foreign and defense policy elites in Europe about the common threats they face, which lessons to learn from international crises, and how to best cooperate to prevent future crises and counter threats. To talk of a European rather than an EU strategic culture may be appropriate insofar as the EU is heavily reliant on its member states for the provision of personnel, equipment, and often also leadership as "framework nations" in CSDP military operations. Furthermore, decisions in the EU's security and defense policy require unanimity and thus provide each and every member state a veto right. European strategic culture in this conception could be pictured as the area where EU strategic culture and the 27 member state strategic cultures intersect like a Venn diagram. Given that these differences are multidimensional and rather large on some dimensions, the ideational intersection would appear rather small. In reality, some member states and their strategic cultures matter more than others for decisions about the launch of CSDP operations and the EU's overall direction in defense matters.

Alternatively, one could follow authors like Alessia Biava[15] who stressed that the EU is not a state and should therefore not be compared to a statist conceptualization of strategic culture. Instead, she argues for taking seriously the EU's own broader understanding of security and its multilateral approach to achieving it and measuring progress against the EU's own objectives. Rather than emphasizing strategic culture as being narrowly concerned with the use of force against state or non-state adversaries, Biava stresses that an appropriate conceptualization of EU strategic culture recognizes the equal weight given to military and civilian instruments and approaches and that all missions are subject to a Europeanized planning process. In her conception, the EU already has a strategic culture, which is evolving over time and made every day in Brussels institutions when EU and member state officials meet regularly. Their daily interactions develop a shared sense of purpose and a set of formal and informal norms of working together, in part through socialization pressures in key committees and also through the cultivating role of central supranational bodies such as the EU Commission and the External Action Service. National officials operate in a context shaped by the overarching objective to achieve enough unity to make the EU's common policies work, including in security and defense. National officials acting as seconded officials or as representatives on committees appreciate that the EU has its own hierarchy of interests and threats and its own mechanisms, processes, and instruments to pursue and manage them. Therefore, the EU has achieved a degree of "actorness" in foreign, security, and defense affairs[16] and decisionmaking is in reality more "transgovernmental" rather than purely "intergovernmental."[17] Furthermore, this actorness is partially constituted by the emergence of a common strategic culture community of practitioners, politicians, officials, military, and civilian staff, which can be empirically studied by analyzing officials' documents, but also by interviewing or observing practitioners as they navigate the international and security landscape facing the EU. Through an empirical approach, one could determine whether and to what extent the EU's strategic culture differs from those of its member states. It is then a separate question of whether and to what extent such differences are also incompatibilities and under what conditions the EU is able to act, learn, and adapt in this area.

Drivers and Characteristics of EU Strategic Culture

The EU is still a relatively young and evolving political system compared to that of many states. While European integration had historical precursors, the distinct method through which it has been pursued since the 1950s and the shape it has assumed since then are in many ways unique by European and arguably also by international standards. The EU as an organization has its own identity, history, and indeed narratives of why it was created, the values it holds dear, and the purposes it serves or ought to serve externally. The early origins of the EU can be found in the collective lessons-learned process experienced by initially continental European countries who together concluded that future wars in Europe are best avoided through a process of promoting an "ever closer union" among "the peoples of Europe." This is also visible in the ideas enshrined in the EU's de facto constitution, the EU's so-called Treaties (on the EU and the functioning of the EU). In particular, Article 21 in the Treaty on European Union (TEU), talks about objectives and values guiding EU "external action," emphasizing, inter alia, human rights, development, environmental sustainability, and to "preserve peace, prevent conflicts and strengthen international security, in accordance with the purposes and principles of the United Nations Charter, with the principles of the Helsinki Final Act and with the aims of the Charter of Paris, including those relating to external borders."[18] Hence, the EU is frequently described as a "peace project" and was awarded the 2012 Nobel Peace Prize for its past achievements, including its efforts to promote stability, reconciliation, and human rights in Eastern Europe. It is mainly on the back of this identity that Manners argued that the EU has been in many respects, and indeed should be, a normative power as it promotes core liberal norms through a multilateral and peaceful approach.[19] The ensuing debate between scholars, however, showed that the EU was in fact more self-interested, hard-nosed, and indeed representative of a normal actor in foreign affairs[20] than a purely normative power, and that its role varied vis-à-vis different regions of the world.[21] Nevertheless, the goals of peacebuilding and conflict prevention remain central to the EU's identity and its legitimating narrative in foreign affairs. These values also underpin at a programmatic level many strategic documents, resolutions, and conclusions emanating from the EU central institutions.

While the use of force remained for decades "outsourced" to NATO, the erstwhile Western European Union (WEU), and individual member states, beginning in the late 1990s the EU made a significant shift toward taking hard security, defense, and use of force more seriously and raising its level of ambition as part of the Global Power Europe discourse.[22] This shift emanated from what has been described as transnational mediatized crises learning during the Yugoslav secession wars of the early 1990s and in particular the Kosovo War of 1998/1999.[23] It resulted in the 1998 Saint-Malo rapprochement between France and the United Kingdom, which provided the political will and unity necessary to equip the EU with the autonomy to act in security and defense matters where the United States was not involved. Subsequently, the Treaty of Amsterdam transferred to the EU the so-called Petersberg tasks of the WEU, leading to the WEU's eventual dissolution in 2011. The original Petersberg tasks were broadened by the Treaty of Lisbon, especially in light of the rise of jihadist terrorism, and are currently listed in Article 43 of the Treaty on European Union, including:

> ... joint disarmament operations, humanitarian and rescue tasks, military advice and assistance tasks, conflict prevention and peace-keeping tasks, tasks of combat forces in crisis management, including peace-making and post-conflict stabilisation. All of these may contribute to the fight against terrorism, including by supporting third countries in combating terrorism in their territories.[24]

This new policy and the efforts spearheaded by the EU's first High Representative for Foreign and Security Policy, Javier Solana – a former NATO general secretary – to create the structures and capabilities to make it a reality has led some observers to talk about a clash of the EU's civilian identity with a martial mindset and warnings of a creeping militarization.[25] In reality, though, the ESDP/CSDP turned out to be initially a far less drastic shift in culture than critics feared and advocates hoped. An overwhelming majority of ESDP/CSDP missions were civilian rather than military in nature and the military operations launched under the EU's auspices were generally limited in aspiration, duration, and, particularly, risk-taking. The EU missed many of its capability targets and could not muster the political will to deploy new instruments such as the small-scale EU battle groups to prevent mass atrocities in cases like the Democratic Republic of the Congo (DRC) in 2008 and Libya in 2011. The EU was buffeted from 2009 onward by a series of internal and external crises, which fed Euroskepticism and populism and led to significant cuts in national defense spending to balance budgets after the Financial Crisis. Member state support for a stronger foreign, security, and defense policy substantially waned after Solana left office, even though the Lisbon Treaty strengthened the legal basis for the policy and his office. For instance, it created the solidarity[26] and mutual assistance[27] clauses, the latter invoked for the first time by France after the terrorist attacks in Paris in 2015.

In the wake of the migration crisis, the war in Syria, the rise of the Islamic State of Iraq and Syria (ISIS) and, above all, the 2014 Crimea crisis following Russia's annexation of the peninsula, some of Europe's leaders and the EU's new High Representative Federica Mogherini pushed at the programmatic level for a reinvigoration of the CSDP. They argued for a stronger operational role for the EU in providing security and a somewhat less idealistic approach to foreign policy labeled "principled pragmatism." The first result was a new "Global Strategy" published just after the United Kingdom's vote to leave the EU in 2016. The election of US President Donald Trump in the same year, who had once labeled NATO "obsolete" and the EU a "foe," gave added impetus to the preexisting goal of achieving "strategic autonomy" for Europe. This was defined in the November 2016 European Council conclusions as "the capacity to act autonomously when and where necessary and with partners wherever possible." Subsequently, the Lisbon Treaty provisions for permanent structured cooperation (PESCO) were activated to allow a group of 25 member states to enter legally binding commitments to invest in a series of defense capability projects and make them available for EU military operations, with only Denmark and Malta remaining outside. New funding instruments were proposed and eventually agreed upon such as the

European Defense Fund and the European Peace Facility to allow the development of new capabilities for the EU and to support security forces in partner countries. Yet, even though strategic autonomy has been the "agreed language" of the EU since at least 2016,[28] different understandings and indeed disagreements between "Atlanticist" and "Europeanist" member states have bubbled up to the surface. For instance, German and Polish defense ministers criticized decoupling from the United States as an "illusion" in late 2020 and were rebuffed by French President Macron.[29] Russia's full-scale war of conquest and perpetration of mass atrocities crimes in Ukraine has not only accelerated existing momentum for the EU to better support and coordinate military spending and capabilities but also to use for the first time the European Peace Facility to purchase lethal aid for a third country with more than €3.5 billion going to support Ukraine as of 2023.

Institutional shifts within the EU have also contributed to the shaping of its strategic culture. The 1999 Cologne Summit decided to create new dedicated committees and structures in the area of security and defense, including the Political and Security Committee composed of national ambassadors, the EU Military Committee, and the EU Military Staff, including a Situation Centre. As a result, for the first time in the history of European integration, military officers became an integral and permanent part of the EU's institutional setup. This was followed by further institutional engineering, such as the creation of the European Defense Agency in 2004 and most notably the European External Action Service in 2010. The latter is composed of two-thirds permanent EU staff and one-third seconded officials and diplomats. The EU's new High Representative is a Vice President of the Commission, the EU's High Representative for Foreign Affairs, the Head of the EEAS and European Defense Agency, as well as the chair of the Foreign Affairs Council. At a more operational level, the United Kingdom's Brexit vote in 2016 created the political opportunity for the EU to move ahead with the creation of an operational headquarters for civilian and eventually also military operations with the EU Military Staff, the so-called Civilian and Military Planning and Conduct Capabilities (CPCC and MPCC, respectively). The MPCC is also meant to become the main headquarters for the new EU Rapid Deployment Capacity, and by doing so fix one of the shortcomings of the EU battlegroup concept.[30]

Parts of the new institutional machinery for the CSDP and Common Foreign and Security Policy (CFSP), such as Council working groups,[31] and particularly the Political and Security Committee, became drivers of socialization and peer pressure to create not only common norms on how to work together but also a degree of convergence around the values and threats to the EU and how to best tackle them.[32] Similarly, the research found some evidence of the emergence of a transgovernmental community of intelligence professionals, in part driven by the deepening of EU civilian and military intelligence cooperation.[33] At the same time, progress toward building a shared identity or "esprit de corps" among the EU's External Action Service has been slow, not least because of its hybrid nature and the reservations of member states.[34] Furthermore, EU foreign policy has become more politicized and some of its member states are governed by populist and right-wing parties. States like Poland, Hungary, and the Czech Republic have become more openly skeptical and less vulnerable to peer pressure.[35] The combined result is that committees like the Political and Security Committee may have reached their limits of what persuasion and socialization can achieve. They may struggle particularly with regard to constitutive norms about the identity and social purpose of the organization, rather than merely adherence to procedural norms around appropriate ways of professionally working together.[36]

The third, and arguably most important, driver of EU strategic culture building is the EU's 37 missions since 2001, 25 of which were predominantly civilian and 12 of which were military operations. At the time of writing in early 2023, the EU has 22 missions ongoing, of which 9 are military with a total of 4,000 personnel involved. Of the nine military missions most of them are training and capacity building with only three "executive operations."[37] Your author recently assessed these missions and operations on behalf of the European Parliament, and the findings[38] underlined the substantial heterogeneity of missions in terms of their mandate and particularly their size. The EUFOR Althea military mission in Bosnia involved 7,000 personnel at its peak, while the European Union Border Assistance Mission to Rafah (EU BAM Rafah) was staffed by only 16. Generally, EU military operations have been of low intensity. Its most robust engagement was the Artemis mission in the DRC in 2003, although one should not underestimate the potential risk that was involved in places like Bosnia. This contrasts with other operations that member states engaged in through other frameworks such as NATO in Afghanistan. EU missions are authorized by the UN and upon invitation of the host country. What is generally true of the EU's behavior in external affairs is particularly true for the domain of using force, as Matlary argues when stating that the EU's strategic culture is incapable of "coercive diplomacy."[39] Kiovula argues that the emergent EU military ethos is not geared toward high-intensity combat situations and emphasizes instead the mastery of a broader range of civilian skills.[40] EU operations have so far suffered relatively few casualties and could be deemed broadly successful, albeit against quite limited objectives, usually related to training and advice. Its maritime missions including NAFVOR Atalanta against the threat of piracy around the Horn of Africa and, potentially, EUFOR Sophia against people trafficking have placed a somewhat greater emphasis on promoting EU *realpolitik* interests, rather than human rights or ensuring stability in its neighborhood. Michael E. Smith writes that missions are more determined by the "availability of supply," rather than the "nature of external demand."[41] One might argue that a Russian victory in Ukraine is too severe to tolerate for most European countries, but most of the responses have been coordinated through the NATO framework with the EU launching a mission to enhance the capacity of the Ukrainian armed forces through training and the provision of weapons (EUMAM Ukraine).

The missions pursued to date have achieved some learning effects among the participating EU bodies, institutions, and member state practitioners at the working level in terms of speeding up the process of planning, deciding, resourcing, deploying, and implementing such missions.[42] While the EU is still criticized as being too slow – particularly in terms of speedily and fully generating the forces committed – the interplay of the Political and Security Committee, the EU Military Committee and Staff, the Committee for Civilian Aspects of Crisis Management (CIVCOM), the Crisis Management and Planning Directorate, and the Civilian Planning and Conduct Capability (CPCC) has improved. As a result, the EU has "learned to proceed more quickly through the process, from drawing up a first Crisis Management Concept (CMS) and developing the Operational Concept (CONOPS), to creating a detailed operational plan (OPLAN) and achieving a Council decision to launch an operation."[43] Apart from building institutional and operational capacities, the missions and operations have also become better integrated into broader foreign policy strategies and thus align increasingly better with the EU's comprehensive or integrated approach to conflict and instability.[44] Finally, the EU is trying to foster strategic culture convergence through other bodies such as setting up and strengthening the European Union Institute for Security Studies (EUISS) think tank, the creation of the European Security and Defense College (ESDC) in 2005, and the shared training offered through the Joint EU Intelligence School (JEIS), one of the projects launched under PESCO. The 2017 European intervention initiative launched by France, while formally outside the EU frame, also feeds into a convergence of strategic culture through its focus on four areas: strategic foresight and intelligence sharing, scenario development and planning, support for operations, and lessons learnt and doctrine.

European Strategic Culture through (Member) State Convergence?

One of the earliest attempts to categorize member states according to different dimensions of their strategic cultures was provided by Howorth,[45] who distinguished seven divergences in strategic cultures: allied vs. neutral, Atlanticist vs. Europeanist, professional power projection vs. conscript-based territorial defense, nuclear vs. non-nuclear, preference for military vs. civilian instruments, large vs. small states, and weapon system providers vs. consumers. These divergences do not align neatly with groups of countries and can often cut across them. The most frequently used category is the difference between Europeanist and Atlanticist countries, which was invoked to explain intra-European divisions over the US-led invasion of Iraq in 2003.[46] The focus on this dimension is not surprising given that most small-*n* comparisons of strategic cultures involved the EU's then most capable member states, namely France and the United Kingdom.[47] Both countries are nuclear powers with a high willingness to employ military force even in risky situations. Additionally, both pursue an ambitious vision of their countries' role in international security, support the UN, and support both humanitarian and strategic ends for the use of force. Yet, the two states drew opposite lessons from the historic Suez crisis: the United Kingdom pulled more closely toward the United States and France pulled further away. Since the late 1990s, there has been evidence of these differences narrowing due to, inter alia, generational change in France, convergence in threat perceptions, and trust among military professionals. The high point of this development was arguably the Anglo-French Lancaster House Treaty of 2010. Yet, the United Kingdom's choice of a hard Brexit in 2020, powered by animosity toward the EU and plans for revising the Anglosphere, has heightened the long-standing cultural gulf between France and the United Kingdom, as was evident in the furor over the AUKUS Treaty between the United Kingdom, United States, and Australia in 2021, negating an earlier submarine deal between France and Australia. Despite this spat, there is evidence, particularly post-February 2022, that France is trying to reintegrate the United Kingdom into European security and defence cooperation, for instance, through the European Political Community and the European intervention initiative.

No doubt strategic cultures differ also with regard to other dimensions, most notably on whether force could be used outside of self-defense and stopping mass atrocities, such as the toppling of undemocratic regimes with poor human rights records, but also including the general willingness to deploy troops into risky situations or expose them to risks through less restrictive rules of engagement. While the Afghanistan mission was a major departure and difficult experience for the German armed forces after the abolition of conscription, it is not clear that the country's political elites or indeed the society at large have become any more willing to deploy military force into risky situations. The German culture of restraint or reticence is at best changing very gradually, although the Russian invasion of Ukraine in 2022 has triggered deep soul-searching about the purpose of the use of force and led to a commitment to boost military spending and reform structures after the Chancellor's "Zeitenwende" speech. In fact, there appears to be evidence of countries with traditionally high willingness to use force becoming more restrained, giving a stronger role to parliament for authorizing the use of force with significant consequences as was seen in the British parliament's rejection of the intervention in Syria after the Assad regime used chemical weapons. The fallout from the interventions in Iraq, Libya, and recently, Afghanistan may lead to an incremental convergence toward greater restraint. The other two trends discernible in public opinion data are that neutral (Austria, Ireland) or nonaligned states (Finland, Sweden) have been becoming more supportive of a European role in defense even if levels of support are still lower than the EU average, particularly for Ireland and Austria.[48] Furthermore, there appears to be some evidence of Europeans placing a greater emphasis on being independent from the United States, no doubt influenced by the Trump administration's term in office. Where European states appear to be rather similar is in their general preference of civilian over military instruments, support for the goals of peacekeeping and conflict prevention, the importance of UN authorization, and a generally high level of support for a stronger European role in defense and foreign policy, even if the issue of the creation of a European army is more contentious.[49]

Furthermore, we see convergence at the level of military organizations and their relationships with the societies in which they are embedded.[50] One argument is that national strategic cultures are changing through intensified transnational links and interactions among national militaries, affecting particularly but not only the higher echelons.[51] This is due to both intensified training exercises and interactions on military missions. Mérand argued that the gradual and accumulative effect of both training and operations is the deprioritization of the central military norm of patriotism – dying for one's country.[52] The emergence of transnational military links is reinforced by a strong pan-European trend toward a professionalization of the armed services,[53] including a notable shift away from conscription and toward all-volunteer forces. These trends support Anthony King's argument that "a more common, though not unified European military culture may become more discernible."[54] The Crimea experience and especially Russia's war in Ukraine have led to major reassessments of the threat environment and the priority given to territorial defense over expeditionary warfare, particularly in countries bordering the Baltic Sea. Sweden and Finland submitted applications to join NATO, while Denmark ended its long-standing opt-out from the EU's Security and Defence Policy after a two-thirds majority of Danes supported this move in a referendum in 2022. Yet, the migration crisis, the rise of ISIS, and the security implications of climate change have also heightened threats from the south, making agreement among Northern, Southern, and Eastern European members potentially more difficult.

There are few larger-*n* comparative studies of strategic cultures. Apart from a study of five European and six non-European security cultures,[55] the most relevant study by Biehl et al. compares 27 strategic cultures in the EU, plus Turkey, based on individual expert views. They concluded that there was clearly no "single" European strategic culture.[56] However, as the editors rightly say, one should not retreat to a position that sees all countries' strategic cultures as unique and incompatible but instead try to identify clusters or groups of countries that have more in common with each other. Therefore, they identify three groups:[57] first, relatively small and often neutral states that see security as a manifestation of their statehood, such as Austria, Ireland, Portugal, Malta, or Cyprus. These states are often willing to punch above their weight in terms of ambitions in international security policy, but are rather reluctant to use military force and prefer the EU as the framework partner. The second group is composed of states that see security policy as part of international bargaining and are oriented toward the EU, except for the purpose of collective defense via NATO. This group tends to be moderately cautious in using force for crisis management operations. It includes Germany, Italy, and Spain, but also many smaller countries such as Belgium or the Czech Republic. The final group comprises countries very keen to protect or project state power. These are typically led by executives facing relatively weak constraints to the deployment of force, have a significant willingness to do so despite risks, and are typically strong advocates of either NATO or the EU. Members of this group include France and (historically) the United Kingdom, but also the Netherlands, Poland, Greece, Sweden, and Denmark. No doubt, one can conceive of alternative clusters, for instance involving countries that feel highly threatened by Russia or those who inhabit a similar geostrategic space, such as the countries bordering the Baltic Sea or countries with similar societal characteristics such as the Nordics.

Most of the literature on Europe's strategic cultures was published before the Russian annexation of Crimea in 2014, the migration crisis of 2015, the United Kingdom's Brexit vote in 2016, the election and 2017–2021 administration of US President Trump, the chaotic 2021 retreat of NATO from Afghanistan, and, crucially, Russia's invasion of Ukraine in 2022. Russia's war against Ukraine is still ongoing at the time of writing and given its severity, scale, and scope could have greater implications for societal change and lesson-learning within the EU compared with previous surprises and depending on whether it escalates further and its eventual outcome. While this is not yet an event that comes close to the collective societal experiences of military defeats or victories for most European countries, it has already triggered significant strategic shifts, policy, and spending decisions that will have lasting consequences.

Conclusion

The discussion above combines perspectives on the strategic culture of the EU and on those of member states. It is possible to identify elements of an evolving EU-specific strategic culture that is distinct from those of its member states, yet is also clearly influenced and constrained by their strategic cultures. While the basic features remain fairly stable, one can see elements of change with regard to some dimensions of strategic culture at both the EU and national levels, caused by learning from operational experiences and experiences of strategic surprise in Europe's neighborhood and broader security environment in the last ten years. The EU has so far managed to avoid a war of necessity of the scale of Yugoslavia where the US initially declined to be involved. This may well change in the future and the conflict may look different from previous engagements, particularly in relation to hybrid and cyber warfare activities. The growing military capacities and increasingly assertive behavior of China, the Indo-Pacific tilt of the United States, coupled with the experience of the COVID-19 pandemic has underlined that European security is about far more than protection against kinetic attacks. The hallmark of the EU's emergent strategic culture is precisely its capability and drive to harness civilian tools for the purpose of safeguarding not just EU citizens' security but also the human security of foreign nationals. At the same time as the EU is forging ahead with the rather heterogenous grouping of 25 member states in PESCO, cultural compatibility continues to underpin cooperation between individual member states on specific military capabilities, such as Benelux air policing, the Lancaster House Treaty between France and the United Kingdom, or the cooperation of France, Germany, Belgium, Spain, and Luxembourg in Eurocorps. It is clear that no single national strategic culture can be the blueprint or point of convergence for the rest of the EU. Indeed, this very idea embodied in the EU's first security strategy of 2003 has fallen out of fashion in favor of greater sensitivity to persistent intra-European differences and how they can be managed in creative ways. It is, however, also the case that with the departure of the United Kingdom from the EU and Russia's war on Ukraine, much will depend on the direction and speed of cultural change in Germany, the growing assertiveness of the Nordic and Eastern European countries in the aftermath of February 2022, and the direction of United States policy, especially after the 2024 Presidential election.

Notes

1 NB: This chapter was principally written before Russia's 2022 invasion of Ukraine.
2 Paul Cornish and Geoffrey Edwards, "Beyond the EU/NATO Dichotomy: The Beginnings of a European Strategic Culture," *International Affairs* 77, no. 3 (2001): 587–603.
3 Cornish and Edwards, "Beyond the EU/NATO Dichotomy," 587.
4 European Council, *A Secure Europe in a Better World: European Security Strategy*, 2003.
5 European Parliament, "Hearing with High Representative/Vice President designate Josep Borrell," European Parliament News Press Releases, 7 October 2019, https://www.europarl.europa.eu/news/en/press-room/20190926IPR62260/hearing-with-highrepresentative-vice-president-designate-josep-borrell.
6 Council of the European Union, *A Strategic Compass for Security and Defence*, Brussels, 2022, https://www.eeas.europa.eu/sites/default/files/documents/strategic_compass_en3_web.pdf.
7 Andrew Cottey, "A Strategic Europe," *JCMS: Journal of Common Market Studies* 58, no. 2 (2020): 276–291.
8 Hugo Meijer and Stephen G. Brooks, "Illusions of Autonomy: Why Europe Cannot Provide for Its Security If the United States Pulls Back," *International Security* 45, no. 4 (2021): 7–43.
9 Julian Lindley-French, "In the Shade of Locarno: Why European Defence Is Failing," *International Affairs* 78, no. 4 (2002): 789–811; Sten Rynning, "The European Union: Towards a Strategic Culture?" *Security Dialogue* 34, no. 4 (2003): 481; Adrian Hyde-Price, "European Security, Strategic Culture, and the Use of Force," *European Security* 13, no. 4 (2004): 323–343.
10 Alessia Biava, Margriet Drent, and Graeme P. Herd, "Characterizing the European Union's Strategic Culture: An Analytical Framework," *JCMS: Journal of Common Market Studies* 49, no. 6 (2011): 1227–1248; Janne Haaland Matlary, "When Soft Power Turns Hard: Is an EU Strategic Culture Possible?" *Security Dialogue* 37, no. 1 (2006): 105–121.
11 Alessia Biava, "The Emergence of a Strategic Culture within the Common Security and Defence Policy," *European Foreign Affairs Review* 16, no. 1 (2011): 41–58; Michael E. Smith, *Europe's Common Security and Defence Policy: Capacity-Building, Experiential Learning, and Institutional Change*, Cambridge: Cambridge University Press, 2017.
12 Christoph O. Meyer, *The Quest for a European Strategic Culture: A Comparative Study of Strategic Norms and Ideas in the European Union*, Basingstoke: Palgrave, 2006; Bastian Giegerich, *European Security and Strategic Culture*, Baden-Baden: Nomos, 2006.
13 Meyer, *The Quest for a European Strategic Culture.*
14 Heiko Biehl, Bastian Giegerich, and Alexandra Jonas, eds., *Strategic Cultures in Europe: Security and Defence Policies Across the Continent*, Berlin: Springer, 2013, 13–16.
15 Biava, "The Emergence of a Strategic Culture within the Common Security and Defence Policy."
16 Charlotte Bretherton and John Vogler, *The European Union as a Global Actor*, 2nd ed., Abingdon: Routledge, 2005.

17 Ana E. Juncos and Karolina Pomorska, "Manufacturing Esprit de Corps: The Case of the European External Action Service," *Journal of Common Market Studies* 52, no. 2 (2014): 302–319.
18 *Consolidated Version of Treaty on European Union,* Official Journal of the European Union, 2012, https://eur-lex.europa.eu/resource.html?uri=cellar:2bf140bf-a3f8-4ab2-b506-fd71826e6da6.0023.02/DOC_1&format=PDF, Article 21, paragraph 2c.
19 Ian Manners, "European Union 'Normative Power' and the Security Challenge," *European Security* 15, no. 4 (2006): 405–421.
20 Raffaella A. Del Sarto, "Normative Empire Europe: The European Union, Its Borderlands, and the 'Arab Spring,'" *JCMS: Journal of Common Market Studies* 54, no. 2 (2016): 215–232; Alexander Warkotsch, "The European Union and Democracy Promotion in Bad Neighbourhoods: The Case of Central Asia," *European Foreign Affairs Review* 11, no. 4 (2006): 509–525.
21 Natalia Chaban, Ole Elgstrom, and Martin Holland, "European Union as Others See It," *European Foreign Affairs Review* 11, no. 2 (2006): 245–262; Henrik Larsen, "The EU as a Normative Power and the Research on External Perceptions: The Missing Link," *JCMS: Journal of Common Market Studies* 52, no. 4 (2014): 896–910.
22 James Rogers, "From 'Civilian Power' to 'Global Power': Explicating the European Union's 'Grand Strategy' through the Articulation of Discourse Theory," *Journal of Common Market Studies* 47, no. 4 (2009): 831–862.
23 Meyer, *The Quest for a European Strategic Culture.*
24 *Consolidated Version of Treaty on European Union,* Article 43, paragraph 1.
25 Ian Manners, "Normative Power Europe Reconsidered: Beyond the Crossroads," *Journal of European Public Policy* 13, no. 2 (2006): 182–199.
26 *Consolidated Version of the Treaty on the Functioning of the European Union,* Official Journal of the European Union, 2012, https://eur-lex.europa.eu/resource.html?uri=cellar:2bf140bf-a3f8-4ab2-b506-fd71826e6da6.0023.02/DOC_2&format=PDF, Article 222.
27 *Consolidated Version of Treaty on European Union,* Article 42, paragraph 7.
28 Josep Borrell, "Why European Strategic Autonomy Matters," HR/VP Blog, 3 December 2020, https://eeas.europa.eu/headquarters/headquarters-homepage/89865/why-european-strategic-autonomy-matters_en.
29 Nicole Koenig, "Time to Go Beyond the Meta-Debate on EU Strategic Autonomy in Defence," *Hertie School: Jacques Delor Center,* Policy Brief, 4 December 2020, https://bit.ly/3DjOHw3.
30 Christoph O. Meyer, T van Osch, and Yt. Reykers (2022), The EU Rapid Deployment Capacity: This time, It's for Real?, In-Depth Analysis for SEDE-Subcommittee of the European Parliament, EP/EXPO/SEDE/FWC/2019-01/Lot4/1/C/14
31 Juncos and Pomorska, "Manufacturing Esprit de Corps."
32 Meyer, *The Quest for a European Strategic Culture*; Simon Duke, "Linchpin COPS: Assessing the Workings and Institutional Relations of the Political and Security Committee," *EIPA Working Paper* (5), 2005, http://aei.pitt.edu/5914/.
33 Mai'a K. Davis Cross, "A European Transgovernmental Intelligence Network and the Role of IntCen," *Perspectives on European Politics and Society* 14, no. 3 (2013): 388–402.
34 Pierre Vimont, "Esprit De Corps: Has the EEAS Missed Something?" *European Foreign Affairs Review* 26, no. 1 (2021): 13–24; Ana E. Juncos and Karolina Pomorska, "Playing the Brussels Game: Strategic Socialisation in the CFSP Council Working Groups," *European Integration Online Papers* 10, no. 11 (2006): 1–17.
35 Patrick Müller, Karolina Pomorska, and Ben Tonra, "The Domestic Challenge to EU Foreign Policy-Making: From Europeanisation to de-Europeanisation?" *Journal of European Integration* 43, no. 5 (2021): 519–534.
36 Anna Michalski, and August Danielson, "Overcoming Dissent: Socialization in the EU's Political and Security Committee in a Context of Crises," *JCMS: Journal of Common Market Studies* 58, no. 2 (2020): 328–344.
37 EEAS, "Military and Civilian Missions and Operations," 2023. https://www.eeas.europa.eu/military-and-civilian-missions-and-operations_en

38 Christoph O. Meyer, "CSDP Missions and Operations," in *Briefing for SEDE-Subcommittee of the European Parliament*: European Parliament, 2020.
39 Matlary, "When Soft Power Turns Hard: Is an EU Strategic Culture Possible?"
40 Tommi Koivula, "Towards an EU Military Ethos," *European Foreign Affairs Review* 14, no. 2 (2009): 171–190.
41 Smith, *Europe's Common Security and Defence Policy*, 16.
42 Smith, *Europe's Common Security and Defence Policy*; Biava, "The Emergence of a Strategic Culture within the Common Security and Defence Policy."
43 Meyer, "CSDP Missions and Operations," 9; A. I. Xavier, and J. Rehrl, "How to Launch a CSDP Mission or Operation," in Jochen Rehrl, ed., *Handbook on CSDP Missions and Operation: The Common Security and Defence Policy of the European Union*, 3rd ed., Vienna: Austrian Ministry of Defence and Sports, 2017, 78–82.
44 Smith, *Europe's Common Security and Defence Policy*, 18.
45 Jolyon Howorth, "The CESDP and the Forging of a European Security Culture," *Politique Européenne* 8, no. 4 (2002): 88–108.
46 Olaf Osica, "A Lesson in Politics: Poland and the Iraq Conflict," in *Yearbook of Polish Foreign Policy 2003*, Warsaw: Ministry of Foreign Affairs, 2004, 412–410; Piotr Buras and Kerry Longhurst, "The Berlin Republic, Iraq, and the Use of Force," *European Security* 13, no. 3 (2004): 215–245.
47 Meyer, *The Quest for a European Strategic Culture*; Giegerich, *European Security and Strategic Culture*.
48 Matthias Mader, Francesco Olmastroni, and Pierangelo Isernia, "The Polls—Trends: Public Opinion Toward European Defense Policy and NATO: Still Wanting It Both Ways?" *Public Opinion Quarterly* 84, no. 2 (2020): 551–582.
49 Mader, Olmastroni, and Isernia, "The Polls—Trends: Public Opinion Toward European Defense Policy and NATO."
50 Anthony King, "Towards a European Military Culture?" *Defence Studies* 6, no. 3 (2006): 257–277; Frédéric Mérand, *European Defence Policy: Beyond the Nation State*, Oxford: Oxford University Press, 2008.
51 Koivula, "Towards an EU Military Ethos"; Mérand, *European Defence Policy: Beyond the Nation State*; King, "Towards a European Military Culture?"
52 Frédéric Mérand, "Dying for the Union?" Military Officers and the Creation of a European Defence Force," *European Societies* 5, no. 3 (2003): 253–282; Mérand, *European Defence Policy: Beyond the Nation State*.
53 Bastian Giegerich and Alexander Nicoll, "European Military Capabilities," Strategic Dossier, London: International Institute of Strategic Studies, 2008.
54 King, "Towards a European Military Culture?" 273.
55 Emil Joseph Kirchner and James Sperling, *National Security Cultures: Patterns of Global Governance*, New York: Routledge, 2010.
56 Biehl, Giegerich, and Jonas, *Strategic Cultures in Europe*, 391.
57 Biehl, Giegerich, and Jonas, *Strategic Cultures in Europe*, 388–394.

Selected Bibliography

Biava, Alessia. "The Emergence of a Strategic Culture within the Common Security and Defence Policy." *European Foreign Affairs Review* 16, no. 1 (2011): 41–58.

Biava, Alessia, Margriet Drent, and Graeme P. Herd. "Characterizing the European Union's Strategic Culture: An Analytical Framework." *JCMS: Journal of Common Market Studies* 49, no. 6 (2011): 1227–1248.

Biehl, Heiko, Bastian Giegerich, and Alexandra Jonas, eds. *Strategic Cultures in Europe: Security and Defence Policies Across the Continent*. Berlin: Springer, 2013.

Bretherton, Charlotte, and John Vogler. *The European Union as a Global Actor*, 2nd ed.. Wiesbaden: Routledge, 2005.

Buras, Piotr, and Kerry Longhurst. "The Berlin Republic, Iraq, and the Use of Force." *European Security* 13, no. 3 (2004): 215–245.

Chaban, Natalia, Ole Elgstrom, and Martin Holland. "The European Union as Others See It." *European Foreign Affairs Review* 11, no. 2 (2006): 245–262.
Cornish, Paul, and Geoffrey Edwards. "Beyond the EU/NATO Dichotomy: The Beginnings of a European Strategic Culture." *International Affairs* 77, no. 3 (2001): 587–603.
Cottey, Andrew. "A Strategic Europe." *JCMS: Journal of Common Market Studies* 58, no. 2 (2020): 276–291.
Cross, Mai'A. K. Davis. "A European Transgovernmental Intelligence Network and the Role of IntCen." *Perspectives on European Politics and Society* 14, no. 3 (2013): 388–402.
Del Sarto, Raffaella A. "Normative Empire Europe: The European Union, Its Borderlands, and the 'Arab Spring.'" *JCMS: Journal of Common Market Studies* 54, no. 2 (2016): 215–232.
Giegerich, Bastian. *European Security and Strategic Culture*. Baden-Baden: Nomos, 2006.
Giegerich, Bastian, and Alexander Nicoll. "European Military Capabilities." *Strategic Dossier*. London: International Institute of Strategic Studies, 2008.
Howorth, Jolyon. "The CESDP and the Forging of a European Security Culture." *Politique Européenne* 8, no. 4 (2002): 88–108.
Hyde-Price, Adrian. "European Security, Strategic Culture, and the Use of Force." *European Security* 13, no. 4 (2004): 323–343.
Juncos, Ana E., and Karolina Pomorska. "Manufacturing Esprit de Corps: The Case of the European External Action Service." *JCMS: Journal of Common Market Studies* 52, no. 2 (2014): 302–319.
Juncos, Ana E., and Karolina Pomorska. "Playing the Brussels Game: Strategic Socialisation in the CFSP Council Working Groups." *European Integration Online Papers* 10, no. 11 (2006): 1–17.
King, Anthony. "Towards a European Military Culture?." *Defence Studies* 6, no. 3 (2006): 257–277.
Kirchner, Emil Joseph, and James Sperling. *National Security Cultures: Patterns of Global Governance*. New York: Routledge, 2010.
Koivula, Tommi. "Towards an EU Military Ethos." *European Foreign Affairs Review* 14, no. 2 (2009): 171–190.
Larsen, Henrik. "The EU as a Normative Power and the Research on External Perceptions: The Missing Link." *JCMS: Journal of Common Market Studies* 52, no. 4 (2014): 896–910.
Lindley-French, Julian. "In the Shade of Locarno? Why European Defence Is Failing." *International affairs* 78, no. 4 (2002): 789–811.
Mader, Matthias, Francesco Olmastroni, and Pierangelo Isernia. "The Polls—Trends: Public Opinion toward European Defense Policy and NATO: Still Wanting It Both Ways?" *Public Opinion Quarterly* 84, no. 2 (2020): 551–582.
Manners, Ian. "European Union 'Normative Power' and the Security Challenge." *European Security* 15, no. 4 (2006): 405–421.
Manners, Ian. "Normative Power Europe Reconsidered: Beyond the Crossroads." *Journal of European Public Policy* 13, no. 2 (2006): 182–199.
Matlary, Janne Haaland. "When Soft Power Turns Hard: Is an EU Strategic Culture Possible?" *Security Dialogue* 37, no. 1 (2006): 105–121.
Meijer, Hugo, and Stephen G. Brooks. "Illusions of Autonomy: Why Europe Cannot Provide for Its Security If the United States Pulls Back." *International Security* 45, no. 4 (2021): 7–43.
Mérand, Frédéric. "Dying for the Union? Military Officers and the Creation of a European Defence Force." *European Societies* 5, no. 3 (2003): 253–282.
Mérand, Frédéric. *European Defence Policy: Beyond the Nation State*. Oxford: Oxford University Press, 2008.
Meyer, Christoph O. *The Quest for a European Strategic Culture: A Comparative Study of Strategic Norms and Ideas in the European Union*. Basingstoke: Palgrave, 2006.
Meyer, Christoph O. "CSDP Missions and Operations." In *Briefing for SEDE-Subcommittee of the European Parliament*. European Parliament, 2020.
Michalski, Anna, and August Danielson. "Overcoming Dissent: Socialization in the EU's Political and Security Committee in a Context of Crises." *JCMS: Journal of Common Market Studies* 58, no. 2 (2020): 328–344.
Müller, Patrick, Karolina Pomorska, and Ben Tonra. "The Domestic Challenge to EU Foreign Policy-Making: From Europeanisation to de-Europeanisation?" *Journal of European Integration* 43, no. 5 (2021): 519–534.

Rogers, James. "From 'Civilian Power' to 'Global Power': Explicating the European Union's 'Grand Strategy' Through the Articulation of Discourse Theory." *JCMS: Journal of Common Market Studies* 47, no. 4 (2009): 831–862.
Rynning, Sten. "The European Union: Towards a Strategic Culture?" *Security Dialogue* 34, no. 4 (2003): 479–496.
Smith, Michael E. *Europe's Common Security and Defence Policy: Capacity-Building, Experiential Learning, and Institutional Change*. Cambridge: Cambridge University Press, 2017.
Vimont, Pierre. "Esprit De Corps: Has the EEAS Missed Something?." *European Foreign Affairs Review* 26, no. 1 (2021): 13–24.
Warkotsch, Alexander. "The European Union and Democracy Promotion in Bad Neighbourhoods: The Case of Central Asia." *European Foreign Affairs Review* 11, no. 4 (2006): 509–525.

9

IN DEFENSE OF THE NATIONAL PROFILE

Strategic Culture Then and Now

Jeffrey S. Lantis

The theory of strategic culture has advanced significantly through several "generations" of scholarship in international security studies, with the nation-state as the primary unit of analysis. Foundations for this work can be found in military and diplomatic history, sociology, and studies of comparative political culture. At its core is the principle that historical experiences and culture help define national strategic orientations that favor certain military-security policies over others. Trinkunas defines strategic culture as "the culturally and historically derived predispositions that exist in particular states and their elites concerning the use and effectiveness of the employment of force as an instrument of national policy."[1] Numerous studies contend that exploring these foundations may help us to explain and even predict security policy patterns.[2]

This chapter begins by surveying the scholarly literature on strategic culture, with particular attention to the geopolitical contexts in which it was formulated and how the theory has adapted with the times. It then highlights specific actors and institutions at the national level that constitute strategic culture. Next, the chapter draws links between theory and contemporary case examples. It addresses themes including geopolitical foundations of national strategic culture in historical experiences and narratives, the concept of "keepers" of strategic culture and the power of ideas, and continuity versus change. While this chapter ostensibly offers a defense of the national profile and the state level of analysis for study (focusing on key actors with the monopoly of the legitimate use of force), it also recognizes new insights and research at the intersection of strategic studies, constructivism, and culturalism.

A Brief History of Strategic Cultural Studies

The study of strategic culture is deeply rooted. Early work by Thucydides, Homer, Tacitus, Herodotus, and Sun Tzu articulated the concept of powerful connections between culture, military strategy, and warfighting by nation-states. Clausewitz later described war as a test of both "moral and physical forces" of states.[3] In the twentieth century, Liddell Hart's *The British Way in Warfare* (1932) examined traditional strategy preferences as a function of

DOI: 10.4324/9781003010302-11

culture and historical experience. National character studies based on anthropological models were advanced in the twentieth century, including early work on Japanese and German strategic choices.[4] Greatly influenced by sociology and anthropology, those studies identified the roots of state behavior in culture, language, religion, customs, and socialization.

While national character studies came under criticism and scrutiny after World War II, scholars advanced and refined strategic cultural approaches to explain great power orientations during the Cold War. For example, Weigley's *The American Way of War* (1973) provided a sweeping cultural and historical treatment of US military experiences and argued that those factors led the nation to adopt strategies of attrition and annihilation. Geertz defined culture as "an historically transmitted pattern of meanings embodied in symbols, a system of inherited conceptions expressed in symbolic form by means of which men communicate, perpetuate, and develop their knowledge about and attitudes towards life."[5] These patterns, he asserted, lead to distinct behaviors. Snyder set the tone for a new generation of studies by drawing connections between Soviet strategic culture, beliefs, attitudes, and socialization and the Soviet state's development of a nuclear strategy.[6] This challenge to the prevailing neorealist paradigm offered a compelling new way of thinking. Similarly, Gray (1981) studied distinctive national styles, with "deep roots within a particular stream of historical experience," that characterized strategy-making in countries like the United States and the Soviet Union. He defined strategic culture as "modes of thought and action with respect to force, which derives from perception of the national historical experience, from aspirations for responsible behavior in national terms," and even from "the civic culture and way of life."[7]

The study of strategic culture continued to expand in the post-Cold War era. For example, Johnston's *Cultural Realism: Strategic Culture and Grand Strategy in Chinese History* (1995) advanced a new ontology that argued that the strategic culture of a nation-state was measurable and could offer specific, testable "predictions about strategic choice."[8] The goal of this and other "third-generation" studies was to develop more rigorous models linking actors, political-military culture, and conditions to strategic decisions. Similar works also addressed the gap between prevailing neorealist models and state behavior, as demonstrated in Berger's (1998) treatment of Germany and Japan's "anti-militarist political-military cultures."[9] Scholars argued that cultural variables could explain the lag between structural change and alterations in state behavior and perhaps even account for why some countries appear to act "irrationally" and suffer the consequences of failing to adapt to the constraints of the international system.[10]

Constructivism also informed strategic cultural studies in the post-Cold War era, with scholarship identifying strategic culture as manifest on cognitive, evaluative, and expressive levels.[11] Katzenstein's *The Culture of National Security: Norms and Identity in World Politics* (1996) set the tone for a deeper recognition of the cognitive foundations of national security. Wendt argued that state identities and interests can be seen as "socially constructed by knowledgeable practice."[12] New constructivist work claimed to recognize the importance of "inter-subjective structures that give the material world meaning," including norms, culture, identity, and ideas on state behavior or on international relations more generally. The constructivist research program devotes particular attention to identity formation, with connections to organizational process, history, tradition, and culture. It supports the argument that national identities are "social-structural phenomena" that provide a "logic of appropriateness" regarding policy choices.[13] Farrell suggested that contemporary work in strategic cultural research might best be understood as a merger of two relevant streams of scholarship – culturalism, as derived from comparative politics (and sociological and anthropological studies), and constructivism from international relations theorists.[14]

For much of the twentieth century, the progression of strategic culture through several generations could be characterized as state-centric and primarily focused on great powers in the Global North. The strategic cultural theory offered reflexive considerations of the power of states, but this work was also infused with ontological tension. Even as it was advanced by Snyder, Gray, and others during the Cold War as an alternative to neorealism, many studies at the time remained, at their core, focused on materialist dimensions of power. States remained the primary unit of analysis, and theorists sought to add dimensionality to explanations of their actions to achieve power and influence in relation to rivals. However, strategic culture presented an inherent ontological challenge to realist assumptions in its contention that the state should not be evaluated as a unitary rational actor and that diverse levels exist in a political community: at the level of state, society, and the individual. These arguments existed in various forms for decades (even centuries) but have only become realized at their fullest potential with the end of the Cold War and the infusion of constructivist arguments about the evolution of ideational frames to understand contemporary security competition around the world. Today, a more ontologically reflexive approach allows scholars to expand applications to more diverse case studies and issue areas, with promising results.

Strategic Cultural Studies beyond the Global North

More scholars came to recognize the potential value of strategic cultural interpretations over time, and applications of the theory proliferated beyond the subject of great powers. For example, studies of comparative strategic cultures in nation-states in the Asia-Pacific contended that neorealist models of structural transformations should be supplemented with strategic cultural perspectives. In their study *Strategic Cultures in the Asia-Pacific Region* (1999), Booth and Trood defined strategic culture as "a distinctive and lasting set of beliefs, values and habits regarding the threat and use of force, which have their roots in such fundamental influences as geopolitical setting, history and political culture."[15] Subsequent comparative studies of Asia-Pacific strategic cultures also updated and informed this approach.[16] In addition, comparative strategic culture projects have examined the potential for regional groupings to develop communal associations or even common foreign and security policies, including the Association of Southeast Asian Nations and the North Atlantic Treaty Organization.[17]

Scholars have also applied strategic culture to analyze an increasingly diverse range of countries and different issue areas. Fonseca and Gamarra's edited volume (2017) examined the historical foundations, actors, and institutions that are significant in shaping strategic cultures in many Latin American countries.[18] More contemporary works explored the pressures of globalization, strategic culture in relation to dynamics such as regional integration, and counterterrorism. Further studies have examined links between Indonesian strategic culture and maritime security,[19] comparative strategic cultures in sub-Saharan Africa,[20] and strategic culture in Nigeria.[21] Additional works have examined the utility of strategic cultural models in studies of counterterrorism, deterrence, and foreign policy decisionmaking.[22] More recently, scholars have begun to expand applications of strategic culture into new realms as well, including cyber warfare strategies.[23]

Finally, it is important to note that while the development of strategic culture theory and its broader applications globally have created positive momentum in the field, the scholarship remains avowedly "self-conscious" about potential limitations and criticisms. Advancements in this field are often accompanied by critical reflections, analysis, and contextualization. Almond and Verba's work on political culture (1963) both added to and challenged assumptions of prior national character studies that were deeply historiographical and anthropological in nature.[24] Later, Klein (1988) challenged the distinction between political rhetoric and reality and argued against a discursive approach in the literature.[25] In the 1990s, scholars who advanced the third generation of strategic culture scholarship questioned the ontological and epistemological foundations of earlier work. Their goal was to establish testable hypotheses for the study of strategic culture that could reasonably address the gap between structural realist expectations and actual state behavior. These critiques addressed conceptual challenges for strategic cultural models, including the argument that traditional frameworks were tautological. These differences spurred significant intergenerational debates,[26] but scholars remained dedicated to the development of rich models, testable and generalizable conclusions, and the cumulation of knowledge in strategic cultural studies.[27]

Dimensions of Strategic Culture

This section surveys the actors, factors, and conditions at the nation-state level of analysis that often help constitute strategic culture. This includes attention to historical traditions and narratives and geopolitical concerns, but also ideational factors that can impact military-security policy decisionmaking.[28]

First, the birth and evolution of states, along with the powerful historical narratives that result from these developments, help shape strategic cultural identities. Every nation has historical experiences and narratives that are used to legitimize its political institutions and processes.[29] Some countries emerged from troubled pasts with colonialism and continue to struggle with its economic and political legacies. Others adopted a frontier narrative of contest, expansion, and power. Countries may have experienced revolutionary struggles that continue to define them, such as overcoming colonialism or overthrowing powerful elites. These strains come through in the strategic cultures of countries that are otherwise seemingly quite different, such as the United States and Venezuela. The net result of these experiences is that countries confront different strategic problems with varying material and ideational resources, and scholars argue that these translate into unique orientations and strategic responses. As Gray argued, individual historical circumstances help define the "prisms" through which states define geostrategic challenges and potential solutions.[30]

Next, strategic cultural studies argue that physical, geographical, and geopolitical factors influence the development of strategic culture. Among the most basic of these factors that help shape strategic thought are climate, national geography, and natural resources.[31] For example, the location of a country and its proximity to other powerful states could influence strategic thinking.[32] Countries that have established borders through conflict with neighboring states often continue to live in perpetual insecurity, while others may enjoy relative geographic isolation and abundant resources.[33] The cases of Israel and South Korea are illustrative here: facing a very difficult history, coupled with significant challenges to its legitimacy and territorial integrity, Israeli strategic culture adapted a more offensively oriented and aggressive program that would strike out against potential threats from seemingly every direction (including preemptive attacks).[34] Similarly, South Korean strategic culture has been shaped, in part, by decades of experience operating in the shadow of China and in the contested Asian security sphere.[35]

Third, strategic culture can be seen as an extension of political-cultural studies based on the assertion that state populations hold dominant values, attitudes, and beliefs regarding the political system. The foundation of this approach lay in early works by Weber and others that examined the relationship between the Protestant Ethic and the rise of capitalism in Europe.[36] This work was later informed by the national character studies, based on the premise that political-cultural characteristics can be compared across states. Almond (1956) proposed a comparative frame for this type of study, and Almond and Verba's *The Civic Culture: Political Attitudes and Democracy in Five Nations* provided a more rigorous approach to the study of aggregate values and their influence in shaping state behavior.[37] Individuals were socialized to support prevailing cultural values and traits, which in turn helped fuel consideration of foreign and security policy options. In this sense, political cultures were considered deeply rooted and well established, and pervasive influences on state behavior.[38]

Next, political institutions – including political parties, legislatures, and military organizations – also constitute key actors in strategic culture and influence policy preferences at the state level. Political parties adopt and maintain prevailing cultural narratives, and they serve as important actors and vehicles for legitimacy in the political process. Increasing attention is being given in political sociology studies to the role of parties in helping to shape and maintain dominant national narratives.[39] Studies contend that party organizations can be especially significant actors or vehicles by which groups translate preferences into dominant narratives and how experiences, in turn, help shape political discourse at the national level.[40] More broadly, domestic political attitudes toward the use of force vary significantly among states similarly situated in the international system. In many ways, these both reflect, and are reflected by, prevailing strategic preferences.

Military organizational culture can be an especially important influence on strategic culture.[41] Scholars contend that military organizations adopt unique approaches to strategic concerns, and that their experiences, innovations, and adaptations are manifested in strategic culture. These studies have foundations in past work, including Liddell Hart's pathbreaking study of "ways of war," while others address difficulties in civil-military relations over time.[42] Bloomfield and Nossal's treatment of Australian and Canadian strategic culture centers on the importance of military organizations and leadership, while Chafetz, Abramson, and Grillot's study of Ukrainian and Belarusian strategic choices also highlights how distinct military-security cultures may shape policies.[43] Many of these approaches are anchored in the study of military traditions and transformations, with attention to military organizational structure, concepts of war, and responses to future challenges.[44]

Scholars also identify political leaders and elites as important agents in broader strategic dialogues.[45] Leaders often seek to maintain or defend the common cultural narrative, playing the role of "keepers" of strategic culture.[46] But framing elites as keepers of strategic culture also highlights ideational dimensions of strategic culture, informed by constructivist treatments of agency. Scholars have investigated the question of stewardship of strategic culture, which can involve coalition- and consensus-building efforts by specific political players[47] or even the strategic "use" of culture to achieve preferred political objectives. Berger's work suggests that strategic culture is best characterized as a "negotiated reality" among elites.[48] Leaders pay respect to deeply held convictions such as multilateralism and historical responsibility, but they may also choose when and where to stake claims of strategic cultural traditions – they decide whether and how to consciously move beyond previous boundaries of acceptability in foreign policy behavior. This highlights the possibility that strategic cultures may be less static than often assumed, and instead evolve over time as political leaders respond to their assessment of changes in threats presented by both the internal and external environments.

New Research Areas: Revisiting the National Profile

The national profile remains the dominant frame for strategic cultural studies today, but as noted earlier, one of the hallmarks of this research program is a certain reflexivity – a willingness to acknowledge and debate successes and limitations.[49] This section surveys some of the more robust streams of research that revisit the national profile and suggest areas for further development through research and investigation.

At the broadest level, scholars continue to debate the question of continuity versus change in national strategic cultures. Strategic cultural studies have traditionally focused on the continuity of state behavior. Snyder defined strategic culture as "a set of semi-permanent elite beliefs, attitudes, and behavior patterns socialized into a distinctive mode of thought."[50] Strategic culture is seen as deeply rooted in historical experience and geopolitical circumstances, and its legacy often defines decades of strategic thought and action. As Duffield has argued,

> [t]he overall effect of national security culture is to predispose societies in general and political elites in particular toward certain actions and policies over others. Some options will simply not be imagined... some are more likely to be rejected as inappropriate or ineffective than others.[51]

Strong historical experiences become layered into narratives and decisionmaking processes, effectively filtering any future learning that might occur.

However, contemporary studies have emphasized more dynamic qualities of strategic culture, arguing that if historical memory, political institutions, and multilateral commitments shape strategic culture it might seem logical to also accept that security policies will evolve over time. In previous work on the subject, your author has argued that at least three factors can cause "strategic cultural dilemmas" and produce changes in security policy.[52] First, external shocks may fundamentally challenge existing beliefs and undermine past historical narratives. Informed by insights from punctuated equilibrium theory in comparative public policy,[53] dramatic or jarring events that undermine state security may serve to shake up and shift power within communities, and enable actors to build support for new strategic behavior patterns. Second, systemic changes or new demands may also create situations where primary tenets of strategic thought in a state come into direct conflict with one another (strategic cultural *dissonance*).[54] Third, as noted earlier, elites play a special role in strategic cultural continuity and change. Here, greater scrutiny is needed of the question of the "keepers" or stewards of strategic culture – and of whether strategic culture is truly cause or effect. While leaders pay respect to deeply held convictions associated with strategic culture, foreign policy analysis scholarship suggests that leaders often seek legitimation for preferred policy courses, which may or may not conform to traditional cultural boundaries.[55]

A second area of advancement in the literature focuses on competition among strategic subcultures in pluralist societies, demonstrating a strong ontological break with neo-realism. Over time, strategic cultural studies have come to accept the premise that the more contestation of dominant strategic cultural narratives, the better. This is at the same time encouraging for the future of the discipline and also potentially unsettling for some traditionalists. Johnston explicitly defined culture as "shared assumptions and decision rules that impose a degree of order on individual and group conceptions of their relationship to their social, organizational or political environment."[56] But contemporary work suggests that primary assumptions may not always be shared, rather elites or groups of elites adopt their own understandings of strategic cause and effect.[57] A sociological approach, for example, rejects static interpretations and instead sees strategic culture as fairly continuous (re)articulation and (re)interpretation (of political institutions, education, economic development, and so forth).[58] In fact, Johnston himself recognized competing strategic subcultures in studies of Chinese policy development. Thus, newer models informed by both constructivism and political sociology suggest that most countries maintain more of a fragile equilibrium that defines strategic culture.[59] Notably, some have contended that attention to strategic subcultures effectively helps "solve" the debate over continuity versus change in strategic culture. Continuity reflects the dominance of a particular subculture over others, but it does not preclude the development of latent groups or marginalized perspectives that could later come to the fore.

Third, scholars have revisited connections between strategic culture and deterrence policies related to weapons of mass destruction initiated by Snyder in the context of the Cold War. Manwaring argued that the old "nuclear theology" in strategic studies needed to be replaced "with broad, integrated, and long-term culturally oriented approaches."[60] Similarly, strategic culture provides insight into contemporary explorations of "tailored deterrence." This approach involves integrating the power of cultural symbols and meanings in cross-national communication strategies related to deterrence and dissuasion. Similarly, the Pentagon's 2006 Quadrennial Defense Review articulated a US plan to incorporate tailored deterrence, defined as a "context specific and culturally sensitive" conception of deterrence strategy. Bunn argued that this represented "a shift from a one-size-fits-all notion of deterrence toward more adaptable approaches suitable for advanced military competitors, regional weapons of mass destruction states, as well as non-state terrorist networks."[61]

Some scholars have applied strategic culture to study groups beyond the nation-state, including international and regional organizations, as well as violent non-state actors. For example, Cornish and Edwards (2005) explored the potential development of a European Union strategic culture in the face of myriad security challenges.[62] Others have highlighted how strategic culture helps shape nation-state approaches to counterterrorism.[63] Some assert that it is also possible that terrorist organizations develop their own strategic cultures. Last (2020) developed promising case studies of the strategic cultures of Islamist Jihadist terror networks, including Al-Qaeda in the Islamic Maghreb (AQIM) and Al-Qaeda central.[64] His study compared the ideas and behaviors of the groups to challenge the notion of Al-Qaeda as a monolithic movement. If one assumes that violent non-state actors are founded on core values and beliefs, have identifiable leaders and administrative structures, and may even control territory, then conceivably they are capable of formulating a strategic culture that, in turn, helps shape the structure and process of decisions related to the use of force.[65]

Finally, recent studies have explored applications of strategic culture to understand new realms and issue areas that transcend nation-states, including cyber warfare strategies.[66] Many have drawn analogies between the advent of cyber warfare today and the impact that nuclear weapons had on international security in the twentieth century. Scholars contend that if we approach strategic culture as an integrated system of symbols that helps to establish strategic preferences, then these frames might also help to account for cybersecurity and cyberattacks. Links between strategic culture orientations of nation-states and cyber activities might help explain actions ranging from the use of the Stuxnet worm to the SolarWinds hack. Contemporary studies have linked Russian strategic culture and cyber warfare, as well as India and other south Asian nations, the United States, China, and Iran.[67] In addition, it is notable that studies of the cyber realm also transcend the traditional boundaries of the nation-state – providing a richer and more expansive picture of a globalized world inhabited by significant actors with power and capabilities, including non-state actors with the capacity to impact global affairs.

Conclusion

The study of strategic culture has enriched our understanding of nation-state behaviors for decades in social science. Strategic cultural studies have become more ontologically reflexive in nature, greatly informed by constructivism and the richness resulting from broader theory applications. This chapter has primarily focused on the actors, factors, and conditions in nation-states that help to define strategic culture, as well as links to military and security policy decisionmaking. Substantial empirical research demonstrates the significance of strategic cultural orientations grounded in historical experience, capabilities, interests, and geopolitical circumstances. For example, the strategic cultures of great powers like China and the United States appear to be fairly predictably linked to foreign and defense policy – as do those of other actors ranging from European countries to nations in sub-Saharan Africa. In short, the evidence provides a strong "defense" of the national profile.

At the same time, though, this work does not preclude advancements in related areas of study. To the contrary, there appears to be complementarity between these approaches and the range of other interesting levels and domains of exploration taking place in the literature. The foundations for examination of strategic subcultures, for example, has its foundations in the core concept of identity group cultural profiles. The study of strategic cultures of various actors "beyond the state," from alliance structures to violent non-state actors, remains rooted in fundamental conceptualizations of the power of experience, circumstances, and authority structures reminiscent of the state. Finally, this work also provides a strong foundation for promising areas of theoretical advancement in studies of the strategic cultural foundations of behavior in new realms like cyberspace and orbital and outer space.

Notes

1 Harold Trinkunas, "Understanding Venezuelan Strategic Culture," Research Paper #11, Applied Research Center and Western Hemisphere Security Analysis Center (2008), p. 1.

2 Alastair Ian Johnston, "Thinking about Strategic Culture," *International Security* 19, no. 4 (1995): 32–64; Colin S. Gray, "Out of the Wilderness: Primetime for Strategic Culture," *Comparative Strategy* 26, no. 1 (2007): 1–20.
3 Karl von Clausewitz, *On War*, ed. and trans. by Michael Howard and Peter Paret (Alfred A. Knopf, 1993).
4 Ruth Benedict, *The Chrysanthemum and the Sword* (New York: Houghton Mifflin, 1946).
5 Russell F. Weigley, *The American Way of War: A History of United States Military Strategy and Policy* (Indiana University Press, 1993), p. 42.
6 Jack Snyder, *The Soviet Strategic Culture: Implications for Nuclear Options*, R-2154-AF (Santa Monica, CA: RAND Corporation, 1977), p. 8.
7 Colin S. Gray, "National Style in Strategy: The American Example," *International Security* 6, no. 2 (1981): 35.
8 Johnston, "Thinking about Strategic Culture," p. 1.
9 Thomas U. Berger, *Cultures of Antimilitarism: National Security in Germany and Japan* (Baltimore, MD: Johns Hopkins University Press, 1998).
10 Michael C. Desch, 1998. "Culture Clash: Assessing the Importance of Ideas in Security Studies," *International Security* 23, no. 1 (1998): 143.
11 Jeffrey S. Lantis, "Strategic Culture and National Security Policy," *International Studies Review* 4, no. 3 (2002): 87–113.
12 Alexander Wendt, "Anarchy Is What States Make of It: The Social Construction of Power Politics," *International Organization* 46, no. 2 (1992): 392.
13 Stephen M. Saideman "Thinking Theoretically about Identity and Foreign Policy," in Shibley Telhami and Michael Barnett, eds., *Identity and Foreign Policy in the Middle East* (Ithaca, NY: Cornell University Press, 2002), p. 169.
14 Theo Farrell, "Constructivist Security Studies: Portrait of a Research Program," *International Studies Review* 4, no. 1 (2002): 49–72.
15 Ken Booth and Russell Trood, eds., *Strategic Cultures in the Asia-Pacific Region* (London: Macmillan Press, 1999), p. 8; Andrew Scobell, *China's Use of Military Force: Beyond the Great Wall and the Long March* (Cambridge: Cambridge University Press, 2003).
16 Jeffrey S. Lantis, "Redefining the Nonproliferation Norm? Australian Uranium, the NPT, and the Global Nuclear Revival," *Australian Journal of Politics and History* 57, no. 4 (2011): 543–561; Ashley J. Tellis, ed., "Strategic Culture," in *National Security, and Policymaking in the Asia-Pacific*. https://carnegieendowment.org/2016/10/27/strategic-culture-national-security-and-policymaking-in-asia-pacific-pub-66166; *Alex* Burns and Ben Eltham. "Australia's Strategic Culture: Constraints and Opportunities in Security Policymaking," in *Strategic Cultures and Security Policies in the Asia-Pacific* (London: Taylor & Francis, 2016), pp. 22–45.
17 Peter Schmidt and Benjamin Zyla, "European Security Policy: Strategic Culture in Operation?" *Contemporary Security Policy* 32, no. 3 (2006): 484–493; P. Cornish and G. Edwards, "The Strategic Culture of the European Union: A Progress Report," *International Affairs* 81, no. 4 (2005): 801–820. Mily Ming-Tzu Kao, *Strategic Culture of Small States: The Case of ASEAN* (Tempe, AZ: Arizona State University, 2011); Jordan Becker and Edmund Malesky, "The Continent or the 'Grand Large'? Strategic Culture and Operational Burden-Sharing in NATO," *International Studies Quarterly* 61, no. 1 (2017): 163–180.
18 Eduardo Gamarra and Brian Fonseca, *Haitian Strategic Culture*. Applied Research Center, Latin American and Caribbean Center, Florida International University (Miami, FL: 2009).
19 Muhamad Arif and Yandry Kurniawan, "Strategic Culture and Indonesian Maritime Security," *Asia & the Pacific Policy Studies* 5, no. 1 (2018): 77–89.
20 Kevin Keasbey Frank, "Strategic Culture in Sub-Saharan Africa: The Divergent Paths of Uganda and Tanzania." PhD diss., The University of Southern Mississippi (2017), author's copy.
21 N. Kalu, *Political Culture, Change, and Security Policy in Nigeria* (London: Routledge, 2018).
22 Edward D. Last, *Strategic Culture and Violent Non-state Actors: A Comparative Study of Salafi-jihadist Groups* (Routledge, 2020); Fredrik Doeser, "Strategic Culture, Domestic Politics, and Foreign Policy: Finland's Decision to Refrain from Operation Unified Protector." *Foreign Policy Analysis* 13, no. 3 (2017): 741–759; Muhammad Shoaib Pervez, "Strategic Culture Reconceptualized: The Case of India and the BJP," *International Politics* 56, no. 1 (2019): 87–102.

23 A.M. Williams, "Strategic Culture and Cyber Warfare: A Methodology for Comparative Analysis," *Journal of Information Warfare* 19, no. 1 (2020): 113–116; Martti Kari and Katri Pynnöniemi, "Theory of Strategic Culture: An Analytical Framework for Russian Cyber Threat Perception," *Journal of Strategic Studies* 7, no. 1 (2019): 1–29.
24 Gabriel Almond and Sidney Verba, *The Civic Culture: Political Attitudes and Democracy in Five Nations* (Princeton, NJ: Princeton University Press, 1963).
25 Bradley S. Klein, "Hegemony and Strategic Culture: American Power Projection and Alliance Defense Politics." *Review of International Studies* 14, no. 1 (1998): 133–149.
26 Colin S. Gray, "Strategic Culture as Context: The First Generation of Theory Strikes Back," *Review of International Studies* 25, no. 1 (1999): 49–69; Alan Bloomfield, "Time to Move On: Reconceptualizing the Strategic Culture Debate." *Contemporary Security Policy* 33, no. 3 (2012): 437–461.
27 Christoph Meyer, *The Quest for a European Strategic Culture: Changing Norms on Security and Defence in the European Union* (Palgrave, 2006); Jeffrey S. Lantis, "American Strategic Culture and Transatlantic Security Ties," in Kerry Longhurst and Marcin Zaborowski, eds., *Old Europe, New Europe and the Transatlantic Security Agenda* (London: Routledge, 2005).
28 Jeffrey S. Lantis and Darryl Howlett, "Culture and National Security Policy," in John Baylis, James Wirtz, Eliot Cohen, and Colin S. Gray, eds., *Strategy in the Contemporary World* (Oxford: Oxford University Press, 2006): 116–141.
29 Kerry Longhurst and Marcin Zaborowski, eds., *Old Europe, New Europe and the Transatlantic Security Agenda* (Routledge, 2005); Colin S. Gray, "Strategic Culture as Context: The First Generation of Theory Strikes Back," *Review of International Studies* 25, no. 1 (1999).
30 Colin S. Gray, "The American Revolution in Military Affairs: An Interim Assessment," *The Occasional*, Wiltshire, UK: Strategic and Combat Studies Institute (1997), p. 28.
31 Dmitry Adamsky, *The Culture of Military Innovation: The Impact of Cultural Factors on the Revolution in Military Affairs in Russia, the US, and Israel* (Palo Alto, CA: Stanford University Press, 2010).
32 Nina Graeger and Halvard Leira, "Norwegian Strategic Culture after World War II: From a Local to a Global Perspective," *Cooperation and Conflict* 40, no. 1 (2005): 45–66; Henriki Heikka, Henrikki, "Republican Realism: Finnish Strategic Culture in Historical Perspective" *Cooperation and Conflict* 40, no. 1 (2005): 91–119.
33 Jeffrey S. Lantis and Andrew A. Charlton, "Continuity or Change? The Strategic Culture of Australia," *Comparative Strategy* 30, no. 4 (2011); Alan Bloomfield and Kim Richard Nossal, "Towards an Explicative Understanding of Strategic Culture: The Cases of Australia and Canada," *Contemporary Security Policy* 28, no. 2 (2007): 286–307.
34 Shaiel Ben-Ephraim, "From Strategic Narrative to Strategic Culture: Labor Zionism and the Roots of Israeli Strategic Culture," *Comparative Strategy* 39, no. 2 (2020): 145–161.
35 Jiyul Kim, "Strategic Culture of the Republic of Korea." *Contemporary Security Policy* 35, no. 2 (2014): 270–289.
36 Max Weber, *The Protestant Ethic and the Spirit of Capitalism*, translated by Talcott Parsons (New York: Charles Scribner's Sons ([1905] 1958).
37 Gabriel Almond and Sidney Verba, *The Civic Culture: Political Attitudes and Democracy in Five Nations* (Princeton University Press, 1963).
38 Gabriel Almond and Sidney Verba, eds., *The Civic Culture Revisited* (Cambridge, MA: Harvard University Press, 1980).
39 Jeff Goodwin, James M. Jasper, and Francesca Polletta, eds. *Passionate Politics: Emotions and Social Movement* (University of Chicago Press, 2015); Giselinde Kuipers, "Cultural Narratives and Their Social Supports, Or Sociology as a Team Sport," *The British Journal of Sociology* 70, no. 3 (2019): 708–731.
40 Aída Díaz de León, "The Politics of the Past and the Fragmentary Present," *Sites of Memory in Spain and Latin America: Trauma, Politics, and Resistance* (Spring, 2015).
41 Jeffrey W. Legro, "Military Culture and Inadvertent Escalation in World War II," *International Security* 18, no. 4 (2005): 108–142.
42 Peter J. Katzenstein and Nobuo Okawara, "Japan's National Security: Structures, Norms, and Policies," *International Security* 17, no. 4 (1993): 84–118; Elizabeth Kier, "Culture and Military Doctrine: France between the Wars," *International Security* 19, no.4 (1995): 65–93.

43 Glenn Chafetz, Hillel Abramson, and Suzette Grillot, "Culture and National Role Conceptions: Belarussian and Ukrainian Compliance with the Nuclear Nonproliferation Regime," in Valerie Hudson, ed., *Culture and Foreign Policy* (Lynne Rienner Publishers, 2007).
44 Theo Farrell, "Constructivist Security Studies: Portrait of a Research Program," *International Studies Review* 4, no. 1 (2002): 49–72; Kier, "Culture and Military Doctrine"; Adamsky, *The Culture of Military Innovation.*
45 Jack Snyder, *The Soviet Strategic Culture: Implications for Nuclear Options*, R-2154-AF. (Santa Monica, CA: RAND Corporation, 1977); Alastair Iain Johnston, "Thinking about Strategic Culture"; Thomas Hutzschenreuter, Ingo Kleindienst, and Claas Greger, "How New Leaders Affect Strategic Change Following a Succession Event: A Critical Review of the Literature," *The Leadership Quarterly* 23, no. 5 (2012): 729–755.
46 Jeffrey S. Lantis, "Strategic Culture and National Security Policy," *International Studies Review* 4, no. 3 (2002): 87–113.
47 John S. Duffield, *World Power Forsaken: Political Culture, International Institutions, and German Security Policy After Unification* (Palo Alto, CA: Stanford University Press, 1999); Jeffrey S. Latis, "Strategic Culture: From Clausewitz to Constructivism," *Strategic Insights* 10, no. 4 (2005): 24–37.
48 Thomas U. Berger, *Cultures of Antimilitarism: National Security in Germany and Japan* (Baltimore, MD: Johns Hopkins University Press, 1998).
49 Alan Bloomfield, "Time to Move On: Reconceptualizing the Strategic Culture Debate," *Contemporary Security Policy* 33, no. 3 (2012): 437–461.
50 Snyder, *The Soviet Strategic Culture,* p. 8.
51 Duffield, *World Power Forsaken,* p. 766.
52 Jeffrey S. Lantis, "Strategic Culture and National Security Policy," *International Studies Review* 4, no. 3 (2002): 87–113.
53 Frank R. Baumgartner, B.D. Jones, and P.B. Mortensen, "Punctuated Equilibrium Theory: Explaining Stability and Change in Public Policymaking," *Theories of the Policy Process* 8, no. 1 (2014): 59–103.
54 Lantis, "Strategic Culture: From Clausewitz to Constructivism."
55 Heikka, "Republican Realism."
56 Johnston, "Thinking about Strategic Culture," p. 34.
57 Brendon O'Connor and Srdjan Vucetic, "Another Mars–Venus Divide? Why Australia Said 'Yes' and Canada Said 'Non' to Involvement in the 2003 Iraq War," *Australian Journal of International Affairs* 64, no. 5 (2010): 526–548; Fredrik Doeser, "Strategic Culture, Domestic Politics, and Foreign Policy: Finland's Decision to Refrain from Operation Unified Protector," *Foreign Policy Analysis* 13, no. 3 (2017): 741–759; Fredrik Doeser and J. Eidenfalk, "Using Strategic Culture to Understand Participation in Expeditionary Operations: Australia, Poland, and the Coalition against the Islamic State," *Contemporary Security Policy* 40, no. 1 (2019): 4–29.
58 Justin Massie, "Making Sense of Canada's 'Irrational' International Security Policy: A Tale of Three Strategic Cultures," *International Journal* 64, no. 3 (2009): 625–645.
59 Iver B. Neuman and Henrikki Heikka, "Grand Strategy, Strategic Culture, and Practice: The Social Roots of Nordic Defence," *Cooperation and Conflict* 40, no. 1 (2005): 11–26. Bloomfield, "Time to Move On: Reconceptualizing the Strategic Culture Debate."
60 Ibid.
61 Elain M. Bunn, "Can Deterrence Be Tailored?" National Defense University, Institute for National Strategic Studies (Washington DC: 2007).
62 P. Cornish and G. Edwards, "The Strategic Culture of the European Union: A Progress Report," *International Affairs* 81, no. 4 (2005): 801–820.
63 Doeser and Eidenfalk, "Using Strategic Culture to Understand Participation in Expeditionary Operations"; W.Y. Rees and R.J. Aldrich, "Contending Cultures of Counterterrorism: Transatlantic Divergence or Convergence?" *International Affairs* 81, no. 5 (2005): 905–923.
64 Edward D. Last, *Strategic Culture and Violent Non-state Actors: A Comparative Study of Salafi-jihadist Groups* (London: Routledge, 2020).
65 Tamir Libel, "Rethinking Strategic Culture: A Computational (Social Science) Discursive-Institutionalist Approach," *Journal of Strategic Studies* 43, no. 5 (2020): 686–709.

66 A.M. Williams, "Strategic Culture and Cyber Warfare: A Methodology for Comparative Analysis," *Journal of Information Warfare* 19, no. 1 (2020): 113–116.
67 See, for example, Martii J. Kari and Katri Pynnöniemi. "Theory of Strategic Culture: An Analytical Framework for Russian Cyber Threat Perception," *Journal of Strategic Studies 46*, no. 1 (2019): 1–29; Muhammad Shoaib Pervez, "Strategic Culture Reconceptualized: The Case of India and the BJP," *International Politics* 56, no. 1 (2019): 87–102; Ashley J. Tellis, "Strategic Culture," in *National Security, and Policymaking in the Asia-Pacific. 2016.* https://carnegieendowment.org/2016/10/27/strategic-culture-national-security-and-policymaking-in-asia-pacific-pub-66166.

Select Bibliography

Adamsky, Dmitry. 2010. *The Culture of Military Innovation: The Impact of Cultural Factors on the Revolution in Military Affairs in Russia, the US, and Israel.* Stanford, CA: Stanford Security Studies.

Almond, Gabriel, and Sidney Verba. 1963. *The Civic Culture: Political Attitudes and Democracy in Five Nations.* Princeton, NJ: Princeton University Press.

Becker, Jordan, and Edmund Malesky. 2017. "The Continent or the 'Grand Large'? Strategic Culture and Operational Burden-Sharing in NATO." *International Studies Quarterly* 61(1): 163–180.

Ben-Ephraim, Shaiel. 2020. "From Strategic Narrative to Strategic Culture: Labor Zionism and the Roots of Israeli Strategic Culture." *Comparative Strategy* 39(2): 145–161.

Benedict, Ruth. 1946. *The Chrysanthemum and the Sword.* Boston, MA: Houghton Mifflin.

Berger, Thomas U. 1998. *Cultures of Antimilitarism: National Security in Germany and Japan.* Baltimore, MD: Johns Hopkins University Press.

Bloomfield, Alan. 2012. "Time to Move On: Reconceptualizing the Strategic Culture Debate." *Contemporary Security Policy* 33(3): 437–461.

Chafetz, Glenn, Hillel Abramson, and Suzette Grillot. 2007. "Culture and National Role Conceptions: Belarussian and Ukraining Compliance with the Nuclear Nonproliferation Regime." In Valerie Hudson, ed., *Culture and Foreign Policy.* Boulder, CO: Lynne Rienner Publishers.

Clausewitz, Karl von. 1993. *On* War, ed. and trans. by Michael Howard and Peter Paret. New York: Alfred A. Knopf.

Cornish, P. and Edwards, G. 2005. "The Strategic Culture of the European Union: A Progress Report." *International Affairs* 81(4): 801–820.

Desch Michael C. 1998. "Culture Clash: Assessing the Importance of Ideas in Security Studies." *International Security* 23(1): 141–170.

Doeser, Fredrik. 2017. "Strategic Culture, Domestic Politics, and Foreign Policy: Finland's Decision to Refrain from Operation Unified Protector." *Foreign Policy Analysis* 13(3): 741–759.

Doeser, F. and Eidenfalk, J. 2019. "Using Strategic Culture to Understand Participation in Expeditionary Operations: Australia, Poland, and the Coalition against the Islamic State." *Contemporary Security Policy* 40(1): 4–29.

Duffield, John S. 1999. *World Power Forsaken: Political Culture, International Institutions, and German Security Policy After Unification.* Stanford, CA: Stanford University Press.

Farrell, Theo. 2002. "Constructivist Security Studies: Portrait of a Research Program." *International Studies Review* 4(1): 49–72.

Gray, Colin. 1981. "National Style in Strategy: The American Example." *International Security* 6(2): 35–37.

Gray, Colin S. 2007. "Out of the Wilderness: Primetime for Strategic Culture." *Comparative Strategy* 26(1): 1–20.

Johnson, Jeannie L. Johnson, Kerry M. Kartchner, and Jeffrey A. Larsen, eds., 2009. *Strategic Culture and Weapons of Mass Destruction: Culturally Based Insights into Comparative National Security Policymaking.* New York: Palgrave Macmillan.

Johnston, Alastair Iain. 1995. "Thinking about Strategic Culture." *International Security* 19(4): 32–64.

Kao, Mily Ming-Tzu. 2011. *Strategic Culture of Small States: The Case of ASEAN.* Arizona State University.

Katzenstein, Peter J. Katzenstein. ed., 1996. *The Culture of National Security: Norms and Identity in World Politics*. New York: Columbia University Press.
Klein, Bradley S. 1998. "Hegemony and Strategic Culture: American Power Projection and Alliance Defense Politics." *Review of International Studies* 14(1): 133–149.
Lantis, Jeffrey S., and Darryl Howlett. 2006. "Culture and National Security Policy," in John Baylis, James Wirtz, Eliot Cohen, and Colin S. Gray, eds., *Strategy in the Contemporary World*. Oxford: Oxford University Press.
Lantis, Jeffrey S. 2002. "Strategic Culture and National Security Policy," *International Studies Review* 4(3): 87–113.
Last, Edward D. 2020. *Strategic Culture and Violent Non-state Actors: A Comparative Study of Salafi-jihadist Groups*. London: Routledge.
Massie, Justin. 2009. "Making Sense of Canada's 'Irrational' International Security Policy: A Tale of Three Strategic Cultures." *International Journal* 64(3): 625–645.
Meyer, Christoph. 2006. *The Quest for a European Strategic Culture: Changing Norms on Security and Defence in the European Union*. London: Palgrave.
Rees, W. Y., and R. J. Aldrich. 2005. "Contending Cultures of Counterterrorism: Transatlantic Divergence or Convergence?" *International Affairs* 81(5): 905–923.
Scobell, Andrew. 2003. *China's Use of Military Force: Beyond the Great Wall and the Long March*. New York: Cambridge University Press.
Snyder, Jack. 1977. *The Soviet Strategic Culture: Implications for Nuclear Options*, R-2154-AF. Santa Monica: RAND Corporation.
Weigley, Russell Frank. 1977. *The American Way of War: A History of United States Military Strategy and Policy*. Bloomington, IN: Indiana University Press.
Williams, A. M. 2020. "Strategic Culture and Cyber Warfare: A Methodology for Comparative Analysis." *Journal of Information Warfare* 19(1): 113–116.

10
STRATEGIC CULTURE AT THE NON-STATE ACTOR LEVEL

Edward D. Last

Introduction

The concept of strategic culture emerged during the Cold War to analyze the ideational basis of states' strategic behavior, beginning with the Soviet Union,[1] followed by the United States[2] and a host of other countries. More recently strategic culture has been applied to supranational actors such as the European Union and the North Atlantic Treaty Organization, sub-state actors such as the different branches of the US military,[3] including the Marines,[4] and violent non-state actors (VNSAs), particularly Al-Qaeda[5] as well as Islamic State.[6] Your author's own work compares the strategic culture of Al-Qaeda with that of its franchise, Al-Qaeda in the Islamic Maghreb (AQIM), and how AQIM's strategic culture combines Al-Qaeda's ideas and practices with more local influences.[7]

Keepers of Strategic Culture

Previously, VNSAs were not considered to have strategic culture or to be subject to analysis through a strategic cultural lens.[8] There is debate in the literature on who the keepers of strategic culture are. While the state is usually the referent group, is it the state per se, its institutions, the military, or the individual armed services that possess strategic culture?[9] Strategic culture may not reside with the state broadly, but rather with the politico-military decisionmakers of foreign and defense ministries and professionals of the armed services, who design and implement strategy. Or is it to be found within the broader society or the nation? In fact, societies generally contain multiple cultural groupings or subcultures, which may form around religious or other institutions.

What of non-nation-states containing multiple national, ethnic, or tribal divisions?[10] Since such states have varying degrees of national unity it is important to understand the extent to which these various subcultures inhabit a shared national culture and belief system. What is the dominant culture and how does it relate to the elites in power? As such, there may not only be subcultures but "strategic subcultures" within a state, as Snyder suggested of the Soviet Union: splits between the civilian government and the military.[11]

And finally, what about VNSAs? If sub-state actors such as the branches of the military can have strategic culture, there is no reason paramilitary organizations or VNSAs cannot. Any collective able to threaten and use violence can have strategic culture.[12] Just as state

DOI: 10.4324/9781003010302-12

security forces must adopt strategic doctrines to tackle terrorist or criminal gangs, these VNSAs also develop their own strategic cultures of confrontation or avoidance of state militaries and/or police. Sovereignty and statehood are not prerequisites for strategic culture.[13] Neither is the possession of territory.[14] J.M. Long even suggests that strategic cultural analyses of VNSAs may be easier than those of states.[15]

Long goes into great detail about how the nation is a more appropriate unit for strategic cultural analysis than the state. States, he argues, are only a useful unit of analysis for strategic culture to the extent that they constitute a community with a shared "national narrative" and are "coterminal with that nation and its narrative."[16] Similarly, Howard argues that Bin Laden's conception of the *umma* as an Islamic community is akin to a nation.[17]

A nation, Long contends, is a "protean" amorphous concept that he characterizes as "a group of people who strongly identify with an overarching, shared cultural narrative," adopting the notion of strategic culture as narrative. Thus, while they are not nations, what Long refers to as "violent non-state, national actors" (VNSNAs), or armed groups whose goals are bound to particular national territories, may be appropriately analyzed like "nations" at least to the extent that they share a national context or "heritage."[18] This is particularly so in the case of nationalist, secessionist, and national liberation movements, such as Irish nationalists and Basque separatists, the Liberation Tamil Tigers of Eelam (LTTE), the Moro Islamic Liberation Front in the Philippines, Chechen secessionists, or Hamas. This may also be true of local branches of Al-Qaeda and Islamic State, which may operate within single nations. However, Al-Qaeda and Islamic State franchises, such as AQIM, have increasingly become regional entities operating across borders.[19] Going further, Long claims transnational armed groups, or "violent non-state, non-national actors" (VNSNNAs) – primarily the original Al-Qaeda, and maybe the international jihadist movement more broadly – can also be compared to nations:

> ... the appeal that Salafi-jihadists make is that the bond of religion trumps state identification. This replicates the pattern of early Islam, wherein the forefathers claimed that loyalty to the *ummah*, the Islamic community, was to supersede *asabiyya*, loyalty to the kinship group.[20]

> What is clear is that al-Qaida does have pronounced characteristics that are appropriate to a nation: a deep sense of shared religious history, strong in-group/out-group demarcation, shared narrative of contemporary events, and – in an important way – a shared language.[21]

So, while Al-Qaeda is not a nation and has no state, it is, according to Long, nation-like to the extent that it shares a core narrative. Al-Qaeda and other VNSAs may embrace their narrative more closely even than nations, for their narrative is more ideological than national. Long argues that Al-Qaeda members "have been intentional about embracing a shared narrative that gives ... a thick account of the world."[22] One is born into a nation, but chooses an ideology and joins an armed group. While Al-Qaeda has ambitions of creating nations of Muslims under *sharia* law – and despite Islamic State's self-proclaimed and ill-fated "Caliphate" – these Salafi-Jihadists have desires to become like the Islamic empires of yore, "nations" in a more traditional sense, and restore Islamic power to its (perceived) former glory before the rise of the West.

> For al-Qaida, the organization is simply a tangible expression of this larger *ummah* [the Islamic nation]. And the *ummah* is not so much transnational as it is that nation that is trans-global.... The brilliance of Osama and others is in the crafting of a religious narrative that gives a thick account of this nation, the *ummah*, and thus makes it a cultural reality for which men and women are willing to die.[23]

Whether VNSAs are nation-like is largely trivial. Ultimately, for Long, whether an actor has a strategic culture boils down to whether it has a shared narrative. Pierman adds three more criteria: a sense of group identity (which is what shared narratives provide), "a leadership structure," and "a culturally relative and logical means-ends thought process."[24]

An organization need not be a nation to possess a coherent strategic narrative, but merely a strategic actor with shared purposes, ideologies, goals, and grievances. Any collective able to employ violence as part of a strategy may therefore have strategic culture. It is appropriate to see Al-Qaeda's strategic culture as a system of ideas, an ideology, or a religiopolitical doctrine, expressed in its narrative and practices, appealing to an "imagined community" of Sunni Muslims.[25] Their particular reading of Islamic scripture and history, as well as a succession of key texts by later religious ideologues, form the basis of Al-Qaeda's narrative or ideology and consequently its strategic culture.

However, Long's suggestion that strategic culture may be more easily applied to VNSAs than states seems unfounded. Strategic cultural approaches contend that the internal dynamics of an organization, whether a state, sub-state, or VNSA, are at least as important as the external environment in which they operate for understanding why they behave as they do. It is through their internal culture that they interpret the world around them.

Like states, VNSAs also have strategic cultures predisposing them "to fight certain wars in certain ways."[26] However, conceptions of state strategic cultures make certain assumptions that may not apply to VNSAs. VNSAs lack many of the institutions, formal structures, established rules, or extant policies present in states, including bureaucratic and legal mechanisms which hinder change and restrain certain behaviors. Decisionmaking is frequently dominated by a single or small group of individuals.

Violent Non-State Actors

VNSAs are armed groups that are not controlled by or attached to a state, but use violence to achieve political and/or economic goals. The literature on strategic culture and VNSAs focuses on politically motivated actors, jihadist organizations in particular, but it is quite possible it could be applied to groups primarily motivated by profit. That is not to say these political actors do not also dabble in profit-making activities. Indeed, jihadist groups such as AQIM are notorious for kidnapping for ransom and are purportedly involved in smuggling.[27]

Narrative Approaches

The first paper to employ an explicitly strategic cultural approach to analyze VNSAs, entitled *Strategic Culture, Al-Qaida, and Weapons of Mass Destruction* by Jerry Mark Long, was issued by the US Defense Threat Reduction Agency.[28] Two years later, Long's work was included alongside additional articles by James Smith and Thomas Johnson in an

Institute for National Security Studies (INSS) Occasional Paper entitled *Strategic Culture and Violent Non-State Actors.*[29] Here, Smith argues that strategic culture is a useful tool that "can and must be adapted" and applied for the analysis of terrorist, insurgent, and other armed groups.[30] Similarly, Long argues that strategic cultural approaches provide greater insight into VNSA behavior than calculations of "maximum utility" by placing them within their historical, religious, and cultural contexts.[31] What may seem irrational – suicide bombing for instance – makes sense when placed in its strategic cultural context.

Long, Richard Shultz, and Russell Howard all employ narrative approaches to analyze the strategic culture of Al-Qaeda.[32] Likewise Thomas Johnson analyzes the narratives within *shabnamah*, or "night letters," used by the Taliban to spread their message, instruct their supporters, and intimidate others to follow their demands.[33] These written tracts were pasted in prominent public places after dark and relied on the literate within the village to read them aloud. This technique has its origins in Persian and Afghan history and was used by the mujahideen during the Soviet-Afghan War (1979–1989) to coordinate protests and resistance. The Taliban employed it to threaten death to would-be collaborators with the government or International Security Assistance Forces (ISAF). Although the purpose of the night letters was to intimidate the populace, they also provide insight into the Taliban's worldview.[34]

In his report on Al-Qaeda's strategic culture, Shultz[35] defines strategic culture as "a state or non-state actor's shared beliefs and modes of behavior, derived from common experiences and narratives, which shape ends and means for achieving national security objectives. These beliefs and modes of behavior give strategic culture its core characteristics and constitute the framework through which capabilities are organized and employed."[36] This conception is problematic unless it is agreed that a non-state actor constitutes a "nation," as discussed earlier in the chapter, and can thus possess "national" security objectives.

Within the same work, however, Shultz offers another way to conceptualize strategic culture, arguing that narrative/strategic culture serves to define first the issues or grievances to be resolved, second the preferred solution to these issues, and finally the means by which to bring about this resolution.[37] This set of ingredients is referred to elsewhere as a strategic narrative.[38] Authors examining Al-Qaeda's internal narratives, however, do not always agree on their strategic interpretation. Garrett Pierman argues that Al-Qaeda's strategic culture and grand strategy constitute a cosmic struggle between good and evil, Islam and disbelief, and ultimately the annihilation of nonbelievers, making a negotiated settlement impossible.[39] Long argues, however, that "blind slaying" of unbelievers is not the strategic goal.[40] Indeed, Al-Qaeda has consistently couched its jihad in defensive terms. While the group does feel that God is on its side, this metaphysical conception is combined with concrete grievances and goals which underpin its strategic culture: the expulsion of Western forces and an end to Western support for "apostate" regimes in Muslim countries, revolution to replace these regimes with Islamic states ruled according to *sharia* law and Salafi principles, and ultimately the restoration of the Caliphate.

Your author's own work on Al-Qaeda central and its franchise, AQIM, combines narrative and practice approaches, conceiving of strategic culture as the interaction between strategic narratives and strategic practices.[41] Like Howard, that work adopts a comparative approach to describing and analyzing the strategic culture of Al-Qaeda. It is not compared to states, however, but to another VNSA in AQIM.

Al-Qaeda's strategic narrative has its basis in Salafism, an Islamic doctrine strictly following the example of the *salaf*, or pious predecessors – the Prophet Muhammad and his companions – and the Qur'an, while rejecting all later "innovations" as deviations from Islam to be shunned in order to restore Islam as it was revealed. The literature identifies Sayyid Qutb as a key source of Al-Qaeda's strategic culture. In *Milestones*, Qutb called for violent revolution against Muslim governments he judged apostate to be replaced by Islamic states.[42]

Shultz's report implies that a strategic studies curriculum as taught in military colleges is integral to strategic culture.[43] Indeed, national military colleges are an important institution for the transmission of strategic culture. For what Shultz has referred to as Al-Qaeda and associated movements (AQAM), such a literature, which has become known as "jihadi strategic studies," exists online.[44] Key authors in this literature include Abu Bakr Naji, Abu Ubayd al-Qurashi, and Abu Musab al-Suri, who have greatly influenced Al-Qaeda's strategic culture by outlining its worldview or "strategic narrative" and explaining its strategy. Since this strategic discourse is comparable to, and even influenced by, that conducted within state militaries addressing similar issues, it is reasonable to argue that Salafi-Jihadist groups have a strategic culture.[45] Indeed, their strategic analysis of how the world is and what to do given those circumstances is integral to the strategic narratives and strategic cultures of Salafi-Jihadist violent non-state actors (SJVNSAs). However, the strategic culture of individual SJVNSAs can be seen as nested within a broader Salafi-Jihadist, perhaps "Al-Qaedist," strategic culture in line with Shultz's concept of Al-Qaeda and associated movements. Rather than analyzing "AQAM" collectively, your author's research approach treats AQIM as a distinct actor with its own idiosyncratic strategic culture, distinct from but overlapping with that of Al-Qaeda central.[46]

There is much debate in the literature over whether strategic culture refers solely to ideational elements that inform strategy or if it also includes behavior or "practices" themselves.[47] Your author is of the latter opinion. However, like Long, Howard, and Thomas Johnson, Shultz's strategic cultural approach focuses primarily on the narrative elements of the Salafi-Jihadist strategic cultures. While he briefly touches upon the operational and tactical levels in discussing jihadi strategic discourse, Shultz does not mention any Al-Qaeda operations other than 9/11, arguing that their primary method is strategic communication and disseminating "operational knowledge."[48]

Jihadist Ways of War

"Ways of war" is more or less synonymous with strategic culture in the literature, but with a longer pedigree dating back to the interwar period in B.H. Liddell Hart's *The British Way in Warfare* (published in 1932).[49] However, unlike the works mentioned previously, ways of war approaches tend to emphasize strategic behavior and operational practice over the ideas behind the strategy. In *The Afghan Way of War*, Robert Johnson examines Afghanistan's tradition of guerrilla resistance to foreign occupation and intertribal warfare by sub-state and non-state groups, such as the Taliban.[50] The book is primarily a military history from the Anglo-Afghan Wars, through the Soviet-Afghan War, the Civil War that followed the Taliban rise to power, their subsequent ouster by the Western-backed Northern Alliance, and resort to insurgency. However, Johnson argues there is no timeless Afghan way of war, but one which has evolved and been adapted by different actors to changing circumstances.[51]

In contrast to Robert Johnson's history of multiple violent non-state and sub-state actors in Afghanistan, in *The Caliphate at War* Ahmed Hashim discusses the

"Jihadist way of war" of a single VNSA, Islamic State of Iraq and Syria (ISIS), in Iraq.[52] Hashim argues that ISIS "developed and implemented a way of warfare that has been unique within the jihadist world." Hashim distinguishes ISIS's way of war from a general Islamic way of war proposed by others,[53] and his approach focuses on the practices rather than the narrative of the Islamic State. Some scholars have noted that for a while, as a territorial entity, ISIS could be described as a "para-state," but soon reverted to violent non-state actorhood as its territory disintegrated. Although it claimed to be *the* Islamic state, and mimicked statehood in many ways, ISIS was not recognized as such by international society.[54] Similarly, AQIM held territory in northern Mali for a still briefer time.

While Hashim is careful to distinguish the way of war of this specific SJVNSA from notions of a broader Islamic or Arab way of war,[55] and Johnson notes the mutable nature of the Afghan way of war and its evolution over time, others have fallen into reductionism.[56] For instance, Peter Layton claims that there is a new "Arab way of war" exemplified by the "deliberate targeting of civilians," which does not distinguish between civilians and the military.[57] Yet there is nothing uniquely "Arab" about targeting civilians as represented by the German and Anglo-American aerial bombing campaigns of World War II, or ethnic cleansing during the Bosnian War. Calling this an "Arab" way of war is mistaken. Additionally, Layton does not distinguish between Arab state militaries and VNSAs. The attacks on civilians in question were not perpetrated by Arab states, but by Salafi-Jihadists whose ideology was not confined to Arabs. Layton presses the point by arguing that these attacks were perpetrated on behalf of Arab or Muslim societies and while the governments of these societies may not have *openly* supported these "assassins," the attackers relied on the complicity of that society. It is worth pointing out, however, that Salafi-Jihadists despise these "apostate" governments. Further, Salafi-Jihadism is an ideology that cannot be assumed to have majority support. Indeed, their violent acts frequently kill members of their own societies. Layton's notion of complicity is reminiscent of Al-Qaeda's claim that the electorate must be complicit with the alleged crimes of Western governments.

Like Layton, Bacevich claims that a "new Islamic Way of War" has emerged during the past 40 years as Muslim forces have abandoned conventional battle for asymmetric strategies of terrorism and guerrilla warfare to frustrate Western conventional advantages.[58] Again, this fails to distinguish between Muslim state militaries and Islamic VNSAs. There are many examples of Muslim countries fighting conventional wars in the past 40 years. Iraq fought a conventional war against Iran, and two admittedly one-sided wars with Western forces: the 1991 Gulf War and the beginning of the 2003 Iraq War before the ensuing insurgency. Moreover, the adoption of asymmetric warfare to counter a more powerful threat is not a uniquely Islamic practice, but a logical strategy.[59]

Raymond Ibrahim laments the lack of study of "Islam's war doctrines" for understanding contemporary jihadists' strategies, because of their divine, timeless nature.[60] Yet, this primordialist conception of culture fixed in some ancient past is inappropriate for studying Salafi-Jihadists.[61] Strategic cultures or "ways of war" evolve over time as demonstrated by Johnson[62] and Hashim.[63] Furthermore, VNSA strategic cultures tend to change more rapidly than those of states.[64] Indeed, SJVNSAs cannot be seen to have a single, immutable strategic culture. All will differ somehow. While influenced by the past, through religion, legend, and military traditions, SJVNSAs are not wholly defined by these and adapt traditional ideas to the exigencies of the present. They frequently reinterpret Islamic concepts creatively.[65] Sometimes it seems their practices are driving their theology when expedient, albeit framed in Islamic terms, rather than vice versa.[66]

While they claim theirs is *the* one true Islamic way of war, the strategic culture of the various iterations of Al-Qaeda in Iraq/ISIS[67] is both "a product of contemporaneous conditions" and "a particular and idiosyncratic interpretation of an Islamic way of war as perceived by Salafi-jihadists."[68] Neither the Al-Qaeda movement nor Salafi-Jihadism more broadly has a single, uniform strategic culture. Rather, different nodes of Al-Qaeda's network of franchises and affiliates, and now rivals, have their own strategic cultures nested within a broader Salafi-Jihadist strategic culture. Salafi-Jihadism is an "epistemic community" where participants endlessly debate theology, strategy, and politics.[69]

As Hashim argues "... there is more than just one jihadist way of war ... Islamist militants have known for a long time that a jihadist way of war in a particular theater of operations cannot mimic the jihadist way of war" elsewhere.[70] For states and VNSAs alike, strategic cultures and ways of war are influenced by their local contexts; affected by climate and geography, both physical and human; history, technology, and resources, both finances and materiel; and the size and strength of their forces versus those of their enemies. Tawhid wal-Jihad, as the group was originally known, recognized that the deserts of Iraq would require a different strategy from the mountains of Afghanistan they had just departed. According to Hashim, Zarqawi's band of jihadists opted for what they called "unrestricted warfare," unrestrained by moral, legal, or religious norms. It is, therefore, better to treat SJVNSAs as separate entities when undertaking strategic cultural analysis.[71]

In both Hashim's work and your author's, the jihadist way of war consists of both ideational and behavioral elements. We agree with Shultz that jihadist strategic cultures have been influenced by non-Islamic sources.[72] In *The Caliphate at War*, Hashim discusses the ideology and strategy of the group's founder, Abu Musab al-Zarqawi, and the key influence of Abu Abdullah al-Muhajir on him.[73] Unable to confront Western forces directly, the group that would become Al-Qaeda in Iraq sought to provoke sectarian conflict by attacking Shia Iraqis. Hashim traces the evolution of ISIS's way of war from 2003 to 2017 from their use of suicide bombers and vehicle or suicide vehicle-borne improvised explosive devices (VBIEDs/SVBIEDs), which primarily targeted Shia civilians and Western forces in Iraq, to the production and deployment of up-armored SVBIEDs on a massive scale as a battlefield weapon against Iraqi and Syrian security forces. One might add to these the use of drones for reconnaissance and as guided weapons; the professionalization of forces into a "quasi-conventional" army equipped with armored vehicles and artillery seized from Iraqi and Syrian depots; and expansion into Syria, creating a short-lived "Caliphate" spanning the Iraqi-Syrian border. ISIS's state-building was assisted by the corruption, incompetence, and poor leadership of the Iraqi military, which lacked the professionalism, firepower, and backup of recently withdrawn American land forces.

The ISIS way of war incorporated terrorism, guerrilla hit-and-run attacks, and semi-conventional warfare used in parallel.[74] Hashim characterizes this as hybrid warfare, employing irregular strategies of terrorism and guerrilla war alongside more conventional, mechanized warfare; and adaptability to alternate back and forth between these strategies as the exigencies of war and the strategic environment changed.[75] Mao Zedong conceived of strategy in a similar way, as a "continuum" progressing from guerrilla to conventional warfare, recognizing conventional and irregular forces could complement each other, as his forces demonstrated fighting the Japanese and Chinese Nationalist armies between 1937 and 1949.[76]

As ISIS's "Caliphate" began to crumble from 2015 onward in the face of dwindling finances, Western airpower, and regrouped Iraqi forces, Kurdish Peshmerga, and Shia militias, it returned to terrorism. ISIS engaged in suicide bombings against Shia civilians and conducted attacks outside Iraq and Syria, both across the Middle East and in Europe, most notably at the Bataclan in Paris. Air strikes forced ISIS to abandon semi-conventional mechanized maneuvers. Hashim traces how ISIS's strategy evolved from terrorism, to guerrilla warfare, to more conventional warfare, along the Maoist continuum and back again as it was overwhelmed by coalition airpower and Iraqi forces.[77]

Similarity and Difference in VNSA Strategic Cultures

The sources of VNSA strategic cultures are like those of states and may include historical experiences, ideology, religion, and geography, among other influences. Though they are not sovereign states, VNSAs and para-states are still highly territorial and frequently express a claim to a country, region, or lost empire. VNSA strategies and strategic cultures are deeply influenced by the topography of the territories in which they operate.

Howard's report examines the strategic cultures of three states – China, Iran, and North Korea – with at least superficially similar historical influences and therefore similar narratives to those of Al-Qaeda, a violent non-state actor.[78] All contain themes of past greatness, recent humiliation by and distrust of foreign powers, and self-reliance. Martyrdom is an important theme for both Al-Qaeda and Iran, influenced by Islam. All have authoritarian tendencies; though, according to Howard, since Al-Qaeda is not a state but a VNSA which members must choose to join, rebellion and opposition are not so problematic. Yet, the experience of AQIM and more particularly its predecessor, the Armed Islamic Group or GIA, shows that VNSAs can be more prone to factionalism, and sometimes violent disputes and purges, than states. GIA leaders Jamal Zitouni and Antar Zouabri brooked no dissent, purging and liquidating less extreme elements, causing the group to splinter. Among VNSAs, factions that disagree with their leadership frequently secede from the group. This is what led to the formation of the Salafist Group for Preaching and Combat, or GSPC, which broke away from the GIA and subsequently rebranded as AQIM in 2007. The GSPC's rebranding created further divisions, leading its former leader to quit the group. Indeed, secession is often the simplest way to resolve disagreements and effect strategic cultural change.[79] Similarly, as Howard concedes, the group formerly known as Al-Qaeda in Iraq seceded from the Al-Qaeda network due to disagreements with Ayman al-Zawahiri over its strategy, following its rebranding as the Islamic State of Iraq and ash-Sham (ISIS) and its move into Syria. According to Howard, Al-Qaeda as well as China, Iran, and North Korea have the strategic advantage over democratic states of swifter and bolder decisionmaking, because of their authoritarian nature without the need for popular support and lacking competing centers of power.[80] Yet, as we have seen, dissent remains a problem within VNSAs.

VNSAs differ from states in a number of other ways relevant to strategic culture. Their compartmentalized nature can lead to miscommunication and ill-discipline. These factors alongside their lack of bureaucratic decisionmaking procedures and the institutional restrictions typically found within states mean that VNSAs' strategic cultures tend to be more prone to sometimes rapid change than even the most authoritarian states. Contrary to Long's assertion that VNSA strategic cultures may be easier to identify, they are, perhaps, more elusive.[81] As clandestine organizations, the survival of VNSAs depends on their ability to adapt rapidly to changing circumstances and outwit the forces of the state and other adversaries. Moreover, VNSAs are prone to frequent personnel and leadership change. Their members are either lost in combat against state security forces or other non-state adversaries, or are purged or deposed due to factionalism within.

Conclusion

While strategic culture has traditionally been used to analyze states, there is no reason why strategic culture cannot be applied to VNSAs. Since statehood or nationhood is not a prerequisite, the literature employing cultural approaches in analyzing VNSAs is growing. The majority of existing strategic cultural analyses have been confined to the study of narrative with a few scholars, such as Robert Johnson and Ahmed Hashim, focusing on the behavioral element or "ways of war" of VNSA strategic cultures. Your author's own approach tries to combine these ideational and behavioral elements into a more comprehensive strategic cultural framework for studying VNSAs. In a world where non-state armed groups proliferate, strategic cultural approaches show promise for understanding what they believe and why they do what they do.

Notes

1 Jack L. Snyder, *The Soviet Strategic Culture: Implications for Limited Nuclear Operations*, Santa Monica: RAND Corporation 1977, http://www.rand.org/pubs/reports/2005/R2154.pdf
2 Colin S, Gray, "National Style in Strategy: The American Example," *International Security* 6:2 (1981), pp. 21–47.
3 Thomas G. Mahnken, "US Strategic Culture and Organizational Subcultures," in *Strategic Culture and Weapons of Mass Destruction*, Jeannie L., Johnson, Kerry M., Kartchner and Jeffrey A., Larsen, eds. (Basingstoke: Palgrave Macmillan, 2009), pp. 69–84.
4 Jeannie L. Johnson, *The Marines, Counterinsurgency, and Strategic Culture: Lessons Learned and Lost in America's Wars* (Washington DC: Georgetown University Press, 2018).
5 Russell D. Howard, *Strategic Culture: JSOU Report 13–8*, Joint Special Operations University, MacDill Airforce Base, 2013; Jerry Mark Long, "Does Al Qaeda Have a Strategic Culture?" in *Strategic Culture and* Weapons *of Mass Destruction*, Jeannie L., Johnson, Kerry M., Kartchner, and Jeffrey A., Larsen, eds. (Basingstoke: Palgrave Macmillan, 2009), pp, 201–218; Richard H. Shultz, "Strategic Culture and Strategic Studies: An Alternative Framework for Assessing al-Qaeda and the Global Jihad Movement," *JSOU Report 12–4*, Joint Special Operations University, MacDill Airforce Base, 2012.
6 Ahmed Hashim, *The Caliphate at War: The Ideological, Organisational and Military Innovations of Islamic State* (London: Hurst, 2018).
7 Edward D. Last, *Strategic Culture and Violent Non-State Actors: Comparing Jihadi Groups* (Abingdon: Routledge, 2020).
8 Shultz, "Strategic Culture and Strategic Studies," pp. 1–2.
9 Mahnken, "US Strategic Culture and Organizational Subcultures."
10 Alan Macmillan, Ken Booth, and Russell B. Trood, "Strategic Culture," in *Strategic Cultures in the Asia-Pacific Region*, Ken Booth and Russell B. Trood, eds. (Basingstoke: Macmillan, 1999), pp. 3–25.
11 Snyder, *The Soviet Strategic Culture.*
12 Kerry Longhurst, "The Concept of Strategic Culture," in *Military Sociology: The Richness of a Discipline*, Gerhard Kümmel and Andreas D, Prüfert, eds. (Baden-Baden: Nomos, 2000), p. 306.
13 Last, *Strategic Culture and Violent Non-State Actors.*
14 Garrett Pierman, "The Grand Strategy of Nonstate Actors: Theory and Implications," *Journal of Strategic Security* 8:5 (2015), pp. 69–78.
15 Jeffrey S. Lantis and Howlett, Darryl, "Strategic Culture," in *Strategy in the Contemporary World: An Introduction to Strategic Studies*, John Baylis, James J., Wirtz and Colin S., Gray, eds.,

3rd ed. (Oxford: Oxford University Press, 2010), pp. 84–103; and Long, "Does Al Qaeda Have a Strategic Culture?"

16 Long, "Does Al Qaeda Have a Strategic Culture?"

17 Russell D. Howard, *Strategic Culture: JSOU Report 13–8,* Joint Special Operations University, MacDill Airforce Base, 2013, p. 55.

18 Jerry Mark Long, *Strategic Culture, Al-Qaida, and Weapons of Mass Destruction,* Science Applications International Corporation (SIAC) prepared for the Defense Threat Reduction Agency, 2006; Long, "Does Al Qaeda Have a Strategic Culture?"

19 Last, *Strategic Culture and Violent Non-State Actors.*

20 Long, *Strategic Culture, Al-Qaida, and Weapons of Mass Destruction*, p. 203.

21 Long, "Does Al Qaeda Have a Strategic Culture?"

22 Jerry Mark Long, Email correspondence with Edward Last, Received 19 May 2011.

23 Long, *Strategic Culture, Al-Qaida, and Weapons of Mass Destruction*, p. 203.

24 Garrett Pierman, "The Grand Strategy of Nonstate Actors: Theory and Implications," *Journal of Strategic Security* 8:5 (2015), pp. 69–78.

25 Benedict Anderson, *Imagined Communities: Reflections on the Origin and Spread of Nationalism* (London: Verso, 2006).

26 James Burk, "Introduction," in *How 9/11 Changed Our Ways of War*, James Burk, ed. (Stanford, CA: Stanford University Press, 2013), p. 2.

27 See Last, *Strategic Culture and Violent Non-State Actors.*

28 Long, *Strategic Culture, Al-Qaida, and Weapons of Mass Destruction.*

29 James M. Smith, Jerry Mark Long, and Thomas H. Johnson, "Strategic Culture and Violent Non-State Actors: Weapons of Mass Destruction and Asymmetrical Operations Concepts and Cases," *INSS Occasional Paper 64* United States Air Force, Institute for National Security Studies, 2008. A third iteration of Long's article, from *Strategic Culture and Weapons of Mass Destruction*, edited by Johnson, Kartchner, and Larsen, will be cited henceforth.

30 Smith, Long, and Johnson, "Strategic Culture and Violent Non-State Actors."

31 Long, "Does Al Qaeda Have a Strategic Culture?" p. 202.

32 Long, "Does Al Qaeda Have a Strategic Culture?"; Shultz, "Strategic Culture and Strategic Studies"; and Howard, *Strategic Culture: JSOU Report 13–8,* pp. 55–63.

33 Thomas H. Johnson, "The Taliban Insurgency and Its Tribal Dynamics: An Analysis of Shabnamah (Night Letters)," *Small Wars & Insurgencies* 18:3 (2007), p. 28.

34 Johnson, "The Taliban Insurgency and Its Tribal Dynamics," pp. 58–59.

35 Shultz, "Strategic Culture and Strategic Studies," p. 5.

36 Shultz cites Roy Godson and Richard H., Shultz, "Glossary of Key Terms," in *Adapting America's Security Paradigm and Security Agenda,* Roy Godson and Richard H., Shultz, eds. (Washington, DC: National Strategy Information Center, 2010), p. 35.

37 Shultz, "Strategic Culture and Strategic Studies," p. 11.

38 Lawrence Freedman, "The Transformation of Strategic Affairs," *The Adelphi Papers* 45:379 (2006) pp. 22–23; Alister Miskimmon, Ben O'Loughlin, and Laura Roselle, *Strategic Narratives: Communication Power and the New World Order* (Abingdon: Routledge, 2013), pp. 4–7.

39 Pierman, "The Grand Strategy of Nonstate Actors."

40 Long, "Does Al Qaeda Have a Strategic Culture?"

41 Last, *Strategic Culture and Violent Non-State Actors.*

42 Long, "Does Al Qaeda Have a Strategic Culture?"; Shultz, "Strategic Culture and Strategic Studies," p. 14.

43 Shultz, "Strategic Culture and Strategic Studies," pp. 27–29.

44 Brynjar Lia and Thomas Hegghammer, "Jihadi Strategic Studies: The Alleged Al-Qaida Policy Study Preceding the Madrid Bombings," *Studies in Conflict & Terrorism* 27:5 (2004), pp. 355–375.

45 Shultz, "Strategic Culture and Strategic Studies," pp. 30–41.

46 Last, *Strategic Culture and Violent Non-State Actors.*

47 Colin S., Gray, "Strategic Culture as Context: The First Generation of Theory Strikes Back," *Review of International Studies* 25:1 (1999), pp. 49–69; Alastair I. Johnston, "Thinking About Strategic Culture," *International Security* 19:4 (1995), pp. 32–64; and, Alastair I. Johnston, "Strategic Cultures Revisited: Reply to Colin Gray," *Review of International Studies* 25:03 (1999), pp. 519–523.

48 Shultz, "Strategic Culture and Strategic Studies," pp. 22–23.
49 Lawrence Sondhaus, *Strategic Culture and Ways of War* (Abingdon: Routledge, 2006).
50 Robert Johnson, *The Afghan Way of War: How and Why They Fight* (New York: Oxford University Press, 2011).
51 Johnson, *The Afghan Way of War*, pp. 7, 36.
52 Hashim, *The Caliphate at War*.
53 Hashim, *The Caliphate at War*, pp. 289–290.
54 Katherine E. Brown, "Violence and Gendered Politics in the Proto-State 'Islamic State'," in *Revisiting Gendered States: Feminist Imaginings of the State in International Relations*, S. Parashar, J.A. Tickner, and J. True, eds. (Oxford: Oxford University, 2018), pp. 174–190. ISIS was, temporarily, a para-state having gained "de facto independence from the home country and aspir[ing] to the status of a full-fledged state, but are not recognized by the international community" (as cited in Bartosz H. Stanislawski, Katarzyna Pełczyńska-Nałęcz, Krzysztof Strachota, Maciej Falkowski, David M. Crane, and Melvyn Levitsky, "Para-States, Quasi-States, and Black Spots: Perhaps Not States, but Not 'Ungoverned Territories,' Either," *International Studies Review* 10:2 (2008), p. 371); and, Harmonie Toros and Filippo Dionigi, "International Society and Islamist Non-State Actors the Case of the Islamic State Organization," in *The Anarchical Society at 40: Contemporary Challenges and Prospects*, H. Suganami, M. Carr, and A.R.C. Humphreys, eds. (Oxford: Oxford University Press, 2017), pp. 145–161.
55 Hashim, *The Caliphate at War*.
56 Johnson, *The Afghan Way of War*.
57 Peter Layton, "The New Arab Way of War," *US Naval Institute Proceedings*, 129:3 (2003).
58 Andrew J., Bacevich, "The Islamic Way of War," *The American Conservative*, 11 September 2006.
59 Jeremy Black, "Determinisms and Other Issues," *The Journal of Military History* 68:4 (2004), p. 1217.
60 Raymond Ibrahim, "Studying the Islamic Way of War," *National Review*, 11 September 2008.
61 Patrick Porter, *Military Orientalism: Eastern War through Western Eyes* (London: Hurst, 2009), p. 12.
62 Johnson, *The Afghan Way of War*.
63 Hashim, *The Caliphate at War*.
64 Last, *Strategic Culture and Violent Non-State Actors*.
65 Shiraz Maher, *Salafi-Jihadism: The History of an Idea* (Oxford: Oxford University Press, 2016), pp. 65–66.
66 Last, *Strategic Culture and Violent Non-State Actors*.
67 This group has rebranded numerous times first appearing in Iraq in 2003 as Tawhid wal-Jihad led by Abu Musab al-Zarqawi, rebranding as Al-Qaeda in Iraq, then the the Islamic State of Iraq, before breaking away from Al-Qaeda to become the Islamic State of Iraq and ash-Sham (ISIS) in 2014 and finally just Islamic State.
68 Hashim, *The Caliphate at War*, pp. 289–291.
69 Thomas Hegghammer, "Introduction: What Is Jihadi Culture and Why Should We Study It?" in *Jihadi Culture: The Art and Social Practices of Militant Islamists*, Thomas Hegghammer, ed. (Cambridge: Cambridge University Press, 2017), p. 10.
70 Hashim, *The Caliphate at War*.
71 Hashim, *The Caliphate at War*; Last, *Strategic Culture and Violent Non-State Actors*.
72 Hashim, *The Caliphate at War*, p. 294.
73 Hashim, *The Caliphate at War*, pp. 126–141.
74 Hashim, *The Caliphate at War*.
75 Hashim, *The Caliphate at War*, pp. 23–24.
76 Peter R. Mansoor, "Introduction: Hybrid Warfare in History," in *Hybrid Warfare: Fighting Complex Opponents from the Ancient World to the Present*, Williamson Murray and Peter R. Mansoor, eds. (Cambridge: Cambridge University Press, 2012), p. 5.
77 Hashim, *The Caliphate at War*, pp. 288–370.
78 Howard, *Strategic Culture: JSOU Report 13-8*, pp. 65–81.
79 Last, *Strategic Culture and Violent Non-State Actors*.

80 Howard, *Strategic Culture: JSOU Report 13–8*, p. 81.
81 Lantis and Howlett, Darryl, "Strategic Culture"; and, Last, *Strategic Culture and Violent Non-State Actors.*

Selected Bibliography

Bacevich, Andrew J. "The Islamic Way of War." *The American Conservative*. 11 September, http://www.theamericanconservative.com/articles/the-islamic-way-of-war/ Accessed 12/6/2018, 2006.

Brown, Katherine E. "Violence and Gendered Politics in the Proto-State 'Islamic State'" in *Revisiting Gendered States: Feminist Imaginings of the State in International Relations*, eds. S. Parashar, J.A. Tickner, and J. True, Oxford: Oxford University Press pp. 174–190, 2018.

Hashim, Ahmed. *The Caliphate at War: The Ideological, Organisational and Military Innovations of Islamic State*. London: Hurst, 2018.

Hegghammer, Thomas. "Introduction: What Is Jihadi Culture and Why Should We Study It?" in *Jihadi Culture: The Art and Social Practices of Militant Islamists*, ed. Thomas Hegghammer, Cambridge: Cambridge University Press, 2017.

Ibrahim, Raymond. "Studying the Islamic Way of War." *National Review*, 11 September, 2008. https://www.nationalreview.com/2008/09/studying-islamic-way-war-raymond-ibrahim/

Johnson, Robert. *The Afghan Way of War: How and Why They Fight*. New York: Oxford University Press, 2011.

Johnson, Thomas H. "The Taliban Insurgency and Its Tribal Dynamics: An Analysis of Shabnamah (Night Letters)." *Small Wars & Insurgencies* 18:3 (2007), pp. 317–344.

Last, Edward D. *Strategic Culture and Violent Non-State Actors: Comparing Jihadi Groups*, Abingdon: Routledge, 2020.

Layton, Peter "The New Arab Way of War." *US Naval Institute Proceedings* 129:3 (2003). https://www.usni.org/magazines/proceedings/2003/march/new-arab-way-war

Lia, Brynjar and Thomas Hegghammer "Jihadi Strategic Studies: The Alleged Al-Qaida Policy Study Preceding the Madrid Bombings." *Studies in Conflict & Terrorism* 27:5 (2004), pp. 355–375.

Long, Jerry Mark. *Strategic Culture, Al-Qaida, and Weapons of Mass Destruction*. Washington, DC: SIAC prepared for the Defense Threat Reduction Agency, 2006.

Long, Jerry Mark. "Does Al Qaeda Have a Strategic Culture?" in *Strategic Culture and Weapons of Mass Destruction*, eds. Jeannie L. Johnson, Kerry M. Kartchner, and Jeffrey A. Larsen, eds. Basingstoke: Palgrave Macmillan, 2009. pp. 201–218.

Maher, Shiraz. *Salafi-Jihadism: The History of an Idea*. Oxford: Oxford University Press, 2016.

Pierman, Garrett. "The Grand Strategy of Nonstate Actors: Theory and Implications." *Journal of Strategic Security* 8:5 (2015), pp. 69–78.

Smith, James M. "Strategic Culture and Violent Non-State Actors: Concepts and Templates for Analysis" in *Strategic Culture and Violent Non-State Actors: Weapons of Mass Destruction and Asymmetrical Operations Concepts and Cases*, INSS Occasional Paper 64, eds. James M. Smith, Jerry Mark Long, and Thomas H. Johnson. Colorado: USAF Institute for National Security Studies, 2008. pp. 1–13.

Stanislawski, Bartosz H., Katarzyna Pełczyńska-Nałęcz, Krzysztof Strachota, Maciej Falkowski, David M. Crane, and Melvyn Levitsky. "Para-States, Quasi-States, and Black Spots: Perhaps Not States, but Not 'Ungoverned Territories' Either." *International Studies Review* 10:2 (2008), pp. 366–396.

Toros, Harmonie and Filippo Dionigi. "International Society and Islamist Non-State Actors The Case of the Islamic State Organization" in Th*e Anarchical Society at 40: Contemporary Challenges and Prospects*, eds. H. Suganami, M. Carr, and A.R.C. Humphreys. Oxford: Oxford University Press, 2017.

PART III

State of the Scholarship

National Strategic Culture Resources

11
AMERICAN STRATEGIC CULTURE IN THE ERA OF GREAT POWER COMPETITION[1]

Jeannie L. Johnson

Introduction

The field of strategic culture examines the ways in which national public culture and the various organizational cultures of a country's defense, intelligence, and diplomatic institutions impact the formation of its foreign and security policy. The national culture of the United States combines features of identity, preferred templates for action, distinctive values, and perceptions of the world shaped by the American experience, and the subculture of the US military expands upon and reinforces many dominant American traits while introducing some of its own. This combination of strategic culture traits exists because it has rewarded the American community, or at least significant subcommunities, with success in the formative moments of the national experience. Making sense of these salient traits within American strategic culture requires a thorough study of the United States' unique history – and yields valuable insights on the future contexts of great power competition for which United States preferred strategic action templates are, and are not, likely to be an effective match.

This chapter offers a frank assessment of US strategic culture directed at an audience of its own constituents and participants – US decisionmakers, military and national security professionals, and scholars of the American strategic landscape. The effectiveness of contemporary US adversaries in a renewed era of great power competition is enhanced by their studied understanding of American strategic culture – the ways in which both American national culture and the organizational cultures of the US security community advance or inhibit innovative thinking; the range of policy actions perceived to be both effective and permissible; the order of action warfighters default to in approaching an enemy; and the acts below the threshold of war that are likely to stymie US institutions that are left without a clear script for action. Near-peer competitors to the United States have already chalked up significant wins in the cyber domain, information operations, proxy warfare, and in expanding their own spheres of influence by exploiting weaknesses in the playbook of the American national security enterprise.[2] In this light, American decisionmakers, operators, and scholars especially stand to benefit by rising to Sun Tzu's edict to "know thyself."

DOI: 10.4324/9781003010302-14

The particular features of American strategic culture discussed in this chapter represent habits of mind and behavior in the American way of life that have been identified with consistency by scholars, ethnographers, historians, and foreign observers of the American condition over the last one hundred years.[3] The United States is characterized by a large and diverse population and houses myriad subcultures with distinct identities, norms, value orientations, and perceptions of the world. Some of the resulting cultural inclinations tend to weigh more heavily than others in American foreign and security policy decisionmaking and in the habits of American warfighters across combat theaters. It is important to note that while a given trait may qualify as being persistently within the American repertoire, its influence on thinking and behavior is likely to ebb and flow in response to contextual factors. This is key. Cultural influences do not provide a clear-cut script for action, *but they do tend to bound our beliefs about the range of effective and appropriate options available in a given situation.* For the purposes of military planning, the result can be strategically suboptimal: security practices and tactics that are a clear match with organizational or national identity and practice are privileged over potentially more effective policies that fall outside the strategic culture mainstream.

This chapter does not attempt to be comprehensive in cataloging the various features of American strategic culture nor the full historical contexts from which they derive. Rather, the purpose is to critically examine a range of cultural traits that have proven particularly problematic when applied to US operations abroad and are likely to be exploited by great power competitors in the present and immediate future. Illuminating these traits serves at least three key purposes:

1 Studied self-awareness makes it less likely that dedicated adversaries will be able to exploit blind spots in American security thinking and action.
2 Recognizing gaps in the US strategic culture repertoire – skills and competencies in which the United States has underinvested as a consequence of investing in preferred modes of action – may prompt budget and training shifts to address those gaps.
3 Clearly identifying the cultural roots of some preferred security practices may prompt strategic planners to reexamine their effectiveness and suitability with increased scrutiny before applying them to the field of great power competition.

Three scenarios within the arena of great power competition will provide useful models for examining key aspects of American strategic culture thinking and practice: future war with a near-peer power; proxy conflict played out by, with, and through third-party partner forces; and authoritarian advances in ideological warfare, including those made possible through the use of sophisticated digital technologies.

Preparing for Future War: Battling a Near-Peer Power

American strategic culture is rife with raw enthusiasm for preparing for a conventional war. It is, however, a very particular vision of conventional war. A significant amount of scholarly work has documented the US preferred way of war as one favoring wars of annihilation, against conventional enemies, accomplished in short time frames by employing lavish firepower in "an aggressive hunt for the main body of the foe."[4] The several prominent voices who have challenged this characterization focus on the US military's wealth of historic experience with unconventional ways of war and the ability of its fighting force to adapt and competently execute them.[5] These irregular warfare experiences, however, have not shifted the conventional preferences that dominate US strategic culture.[6]

The US Army has considerably more practice with irregular warfare than the conventional sort but has maintained a steady march toward primacy in conventional form and function since its frontier days. Even while fighting Native Americans whom they considered to be "masters of guerilla warfare," the US Army neglected to cultivate doctrine, training, or any professional literature that would reflect lessons learned in its combat against the continent's indigenous people.[7] Internal battles with local tribes were dismissed as "beneath the soldiers' vocation."[8] As the Army's irregular fights at home and abroad multiplied, the frequency of practice did little to dispel the general disdain toward low-threshold, often inconclusive missions. Robert Cassidy, in his historical study of US counterinsurgency and peacekeeping experiences, observes: "it is somewhat ironic, revealing, and disquieting that an institution with more history and experience fighting irregular conflicts of limited intensity than total wars without limits, would have its core culture so profoundly influenced by Sherman, Upton, and the World War II experience."[9]

Preference for conventional conflict is not terribly unusual for a national military – there is a significant advantage in being able to recognize one's enemy by virtue of a uniform. Conventional conflict is particularly attractive to the United States as a superpower, however, as it plays to key US strengths in material kit, logistical dominance, and technological superiority.[10] Preference notwithstanding, future US conflicts with great power competitors are likely to be muddied with nontraditional elements of hybrid warfare and be fought with local partners across unfamiliar territory. Moving military kit across great distances is a proven American strength, but effectively partnering with locals is not. As Eliot Cohen quipped, "[the] American proficiency at imparting technical skills is matched only by American insensitivity to local conditions."[11]

Reliance on Technological Overmatch

Technological superiority has long been the signature of modern American combat form, despite the strategic warnings against an overreliance on technology that have been sounding for some time.[12] Williamson Murray cautions that "[t]he greatest danger for the United States in the coming century is that the American military will possess self-satisfied, intellectually stagnant cultures that believe they have found the technological lodestone."[13] It is interesting that Murray made his claim in the same year that the now infamous Millennium Challenge 2002 exercise seemed to prove it. Set up as a war game to test some of the technologies designed to support the Pentagon's network-centric approach to warfare, the game was won almost before it began by the inventive low-tech tactics of Lieutenant General Paul Van Riper. Leading the team representing the enemy, Van Riper killed his own radio communications in favor of motorcycle messengers and light signals, then fitted patrol boats, pleasure boats, and small aircraft with missiles and explosives. His preemptive kamikaze attack neutralized 16 Navy ships and "killed" thousands. For all intents and purposes, the game was over. Frustrated by Van Riper's low-tech and unsportsmanlike methods, the game's orchestrators called for a scripted do-over in which Van Riper's enemy force was forced to follow a set protocol, which allowed it to be decimated by high-tech US forces.[14] The means of war privileged in the Millennium

Challenge 2002 serve as an extreme example of the American preference for dominating through technology rather than human ingenuity – a vulnerability that could be exploited by near-peer competitors that are already actively leveraging warfighting "shortcuts" to close the gap between the military strength of the United States and their own forces.

Furthermore, American technological superiority could soon be eroded as a distinct advantage. Near-peer powers are quickly catching up to some of the Pentagon's most sophisticated assets, which means that even proxy fights are likely to be fought with their sponsors' far more advanced weaponry. In addition, the entry threshold for military technology continues to drop, enabling traditionally weak or non-state actors (including lone-wolf individuals) to punch at an increasingly higher technological weight.[15] Even America's most sophisticated technological platforms, like the F-35, are vulnerable to the advancing cyber expertise of a growing number of belligerents. The networked cyber systems that provide the F-35 Lightning II's much-vaunted data-rich view of the battlefield also render this platform vulnerable to indirect hacks through one of its supporting systems.[16]

The growing utility of additive manufacturing (3D printing) also erodes some of the logistical advantages traditionally wielded by the United States and will complicate future strategies to interdict weapons. In not-so-future conflict areas and civilian zones, a significant number of small and light armaments are likely to be manufactured in real time and on-site.[17] In this increasingly technologically leveled arena, the contender that best understands the human terrain across which it is fighting will possess advantages in increasing friction for its enemy and decreasing it for itself. Cross-cultural competence skills, informed by an understanding of regional history, will be key warfighting assets.

Underinvesting in Cultural Competence

Despite these stakes, a deeply internalized commitment to the education and training necessary to achieve proficiency in cultural analysis will not come easily to the American defense establishment, in large part because it is not valued in the wider national community that it serves and from which its personnel are drawn. Americans have inherited a number of habits of mind that successfully advanced the prosperity of a young immigrant nation but have led to an undervaluing of cultural and historical analysis within US strategic culture. A forward-leaning optic and near dismissal of the past is a habit of mind documented across the American population from its earliest days. D.W. Brogan, a British observer of predominant American culture writing in 1944, argues that American pioneers came by this trait honestly. Very little of their survival depended on a study of Old World history. The forward-looking and problem-solving approach required in the making of America took "extraordinary energies" and cultivated a peculiarly American attitude, which refrains from looking back.[18] Writing 20 years later, Stanley Hoffman was more blunt, arguing that the American obsession with progress has resulted in a perspective that is the equivalent of "historical virginity."[19] Americans value novelty over tradition and often point out the change in their local settings as markers of progress.[20] A quip attributed to the quintessential American problem solver, Henry Ford, sums up this mentality: "history is more or less bunk. It's tradition. We don't want tradition. We want to live in the present and the only history that is worth a tinker's dam is the history we make today."[21] Contemporary scholars of American strategic culture and ways of war argue that "ahistoricism" continues as a serious deficit within US decisionmaking circles.[22] An American public that is largely uninterested in historic knowledge tends not to prioritize it when electing officials.[23] Public figures are rewarded instead for experience-based common sense and forward-looking ambition.[24]

Americans live in comfortable ignorance about other lands due to a combination of both cultural insularity and native enthusiasm about the American way of life. Samuel Huntington argued in 1957 that insular thinking is deeply rooted in American philosophies of liberalism, which focus inward on domestic affairs and largely ignore the foreign sphere.[25] Oliver Lee argues that this trend has continued to the present day. The American brand of individualism focuses first on the self, then in concentric circles moving outward – family, local community, and to some extent the nation. Little interest is reserved for the wider world beyond.[26] It is perhaps no surprise, then, that research conducted by Edward C. Stewart and Milton J. Bennett, authors of the seminal classic *American Cultural Patterns,* finds that most Americans possess "a cultivated ignorance of other nations," harboring biases that assume mutual similarity and gross oversimplifications.[27] This mindset is not a product of malice, nor is it terribly unique to Americans, but it is particularly consequential when it dominates the foreign and security policy decisionmaking of a world superpower.

Looking forward, US forces may reasonably anticipate being faced with fighting a near-peer power across theaters in which the adversary has already made significant inroads with the population. Understanding the depth and breadth of American disinterest in other cultures and recognizing it as a byproduct of historic national experience may enable strategic planners to identify and override instincts that put cultural training first on the chopping block when defense budgets experience stress. The national default settings that consistently under-prioritize cultural education and training in the US defense establishment are not only poor strategic practice but they are also a potentially serious liability in the face of looming conflict with culture-smart adversaries.

By, With, and Through: Great Power Competition and Proxy Warfare

Competition short of war – known variously as gray zone warfare, hybrid warfare, political warfare, and strategic competition among other labels – is intrinsically revisionist, whether employed by great power competitors or by non-state actors. These actors seek to shift the status quo in ways that would harm US interests without triggering a direct confrontation with American military forces.[28] Some of the core takeaways for advancing and protecting US interests in gray zone competition echo the points of the previous section: investments in understanding local contexts and culture will yield strategic advantages. The relationships the United States forges with local partners will be key to achieving US objectives. Leaders within the US special operating community are clear about the population-centric nature of gray zone mission sets already underway,[29] and pursue these almost exclusively by, with, and through partner forces.[30]

In addition to prioritizing language and culture training, US strategic planners and warfighters engaged in gray zone warfare will also benefit from a clear-eyed recognition of a few US strategic cultural predispositions – including a hyper-orientation to problem-solving, a devotion to effort optimism, and an obsession with quantitative metrics of success – that are likely to impact the success of relationships with local security forces and the host populations from which they are drawn.

Americans as Human Do-ings

In the early 1990s, a Russian student – fresh from the recent collapse of the Soviet Union – arrived at the college campus in Utah where your author was a junior instructor. After several weeks of enthusiastic welcoming and well-intended congratulations from her American peers ("How does it feel to be free?!"), she volunteered to give a comparative presentation on Russian and American culture. Her presentation began with the blunt statement, "Americans are not human beings." She went on: "Americans are human *do-ings*. They don't know how to *be*." Her pithy, jarring, and strikingly insightful assessment of American culture tracks with the observations of historians and ethnographers, and echoes primary themes across US military doctrine.

Problem-solving is key to American identity – being a problem solver is both a requirement for most occupations and an admired personal trait. For Americans, it is also perceived to be the primary purpose of human activity.[31] The military puts this impulse on steroids. The bias for action championed by US Marines is a bias in favor of problem-solving. John Boyd's OODA (observe, orient, decide, act) loop tempo, which forms the core of US Marine warfighting doctrine, is the ability to solve problems at a faster rate than the adversary in order to gain the initiative. Within this doctrine, overreaction can be forgiven – "errors by junior leaders stemming from overboldness are a necessary part of learning." But inaction cannot: "[o]n the other hand, we should deal severely with errors of inaction or timidity."[32]

As human do-ings, most Americans are comfortable with trial and error as a learning method and tend to venerate heroes who epitomize the innovative and action-oriented problem-solving of the frontier past.[33] An American's sense of identity is rooted in their occupation. For Americans, one is what one does.[34] With a value orientation firmly fixed on demonstrated activity, Americans tend to exude "busy-ness" as a status symbol. The fast pace of American life ("hurry sickness") is not new to this century nor to the last one. As early as the 1830s, Alexis De Tocqueville remarked on the excessive rushing about in American life.[35] Two hundred years later, Americans remain incessantly on the move.[36] Marines capitalize on the American hurry and take it further. Tempo and raw speed in the delivery of lethal effects is a core attribute of Marine combat doctrine: "speed is a weapon."[37]

The American preference for fast action means that tasks that require patience, restraint, and caution run counter to preferred American instincts. Working at the pace of local forces to execute key operations or with local officials responsible for civic action projects, the delivery of humanitarian aid, or the establishment of functioning governance can tax the patience of American diplomats and military forces. In response, Americans typically set a deadline. Americans possess near-mythical regard for deadlines as a means of increasing efficiency and accelerating progress, and they expect others to do the same.[38] When the threat of a deadline is insufficient to fast-track local action, the result is often a breach in the local relationship in favor of the efficiency of American forces doing it themselves. When the achievement of US strategic objectives involves local sustainment of security measures or public services, the preferred American pace tends to backfire.

The action orientation of US forces may also result in blind spots for military intelligence. Americans assign status based on demonstrated personal achievement and look for it in local leaders, a habit of mind that may cause them to overlook the import of key influencers within relationship-based societies who derive status from other sources, including family ties, religious position, or knowledge of local history.[39] In addition, impatience with the time required to research and assess the complex sociocultural angles of problem sets can leave American intelligence officers and planners easy prey to peddlers of single-solution concepts.[40]

Effort Optimism and the Engineering Fix

The action orientation of US problem solvers is fueled by an attractive trait within the American ethos – an unsinkable optimism. Expectations for success permeate American life and US national security documents.[41] Brogan points out that an outsized belief in one's own abilities and the possibility of success against long odds were the survival tools of the continent's earliest European settlers; intrepid enough to brave life in the New World, their optimism became a national brand among the young country's dominant groups. Within the "religions of economic and political optimism" he observed, "dissent, especially continuous pessimistic crabbing" was "near to treason."[42] Scholars across the decades have continued to note this theme, citing "effort optimism" – the belief that through hard work one can achieve anything – as a key American value.[43]

Colin S. Gray points out the implications for foreign and security policy: "[i]t is quintessentially American to be optimistic and to believe that all problems can be solved, if not today, then tomorrow, and most probably by technology."[44] The result, Gray cautions, is an American formula that can substitute optimism for hard-nosed analysis: "the problem-solving faith, the penchant for the engineering fix, has the inevitable consequence of leading US policy, including its use of armed force, to attempt the impossible."[45] Leonard Mason's review of a significant body of anthropological work on American culture provides support to Gray's claim. Accustomed to a history of success in mastering the geographic environment, Americans "are equally confident that undesirable social conditions can be remedied just as easily and are confused when such proves not to be the case."[46]

In the gray zone warfare context, applying an effort-optimism engineering fix to the training of partner security forces can result in extreme frustration on the part of US forces and a concomitant souring of the relationship with local partners. American cognitive patterns emphasize linear thinking and causal chains in which situations are diagnosed as a series of isolated, solvable problems rather than as a web of complex dynamics and relationships.[47] This orientation creates an exaggerated sense of control over the environment and a perception of independence from the decisions and actions of others.[48] Operating through this lens, US forces tend to fixate on their own training efforts as key to crafting effective local fighting forces rather than examining native "will to fight" factors over which they may have limited control.[49] Myopic analysis of this sort – fueled by can-do spirit – does not help alleviate frustration with local force performance nor produce the insights necessary to build or repair strategically important relationships.[50] Strategic and operational leaders may further be tempted to assume that increased US effort or resources offer a direct, linear solution to winning the loyalties, or at least shared interest, of local partners. Unfortunately, this blind spot is only compounded by the American proclivity for measuring operational success, including engagement with proxy forces, in numbers.

Obsession with Quantification

The compelling need to quantify the world, and experiences within it, is deeply rooted in the American psyche. Stewart and Bennett point out that the "[c]riteria that define success and failure [in the United States] are statistically measured, as are amounts of work, levels of ability, intelligence, and quality of performance." Americans find comfort in quantification because behavior that is quantified becomes objectified and is perceived as amenable to human control.[51]

In the foreign and security policy arena, measuring the successful growth of strategic relationships, the stabilization of fraught societies, and progress toward political objectives can be difficult, so Americans often default to the aspects that can be most easily counted. Even in population-centric warfare – where relationships are key – enemy dead and weapons confiscated are two typical measures of operational success. Pressed for other measurements, US forces often report on American *inputs* – the number of hygiene kits distributed, lengths of road built, cash distributed, and patrols run – treating them as successes in their own right, as evidence of strategic progress whether or not these inputs have significantly advanced the political goals they are meant to achieve. The same pattern is repeated for partner security training. Reports to military superiors emphasize the countable aspects of foreign internal development: number of local forces participating, hours spent on marksmanship, drills run, and certificates of completion signed.

Moral and mental forces – the cognitive realm – are at the center of twenty-first-century contests of influence. Aspects of the cognitive realm are difficult to measure, and the complexity of the task may tempt commanders to continue to default to superficial indicators – American material inputs or the efforts of its own forces – when assessing the health of partner relationships or the efficacy of US-supplied training. Without a concerted effort to forge new assessment tools, American habits of practice are destined to remain dangerously sophomoric and unreliable as strategic indicators.[52]

Great Power Ideological Competition: Authoritarian Ambitions

A third frontier over which the future of great power competition is likely to play out is in the contest of ideology, as the Western model of democratic governance faces increasing pressure from the proliferation of authoritarian and proto-authoritarian regimes. Strategic competitors solidifying their own unilateral rule at home will continue to seek to exploit domestic turmoil abroad and leverage political and economic upheaval in order to expand their own spheres of influence through well-developed weaponized narrative and psychological warfare campaigns undermining Western and Western-leaning governments.[53] Regimes seeking to stabilize governance at home may be attracted to the promise of advanced foreign surveillance systems designed to augment state control of citizen behavior. American cultural presets – particularly an evangelical enthusiasm for American-style democracy and a deep faith in the power of material generosity to accomplish strategic goals – represent dangerous blind spots that may prevent US decisionmakers and strategic planners from identifying or properly weighting great power competitors' advancing efforts to promote authoritarianism abroad and undermine the foundations of liberal democratic governance across US allies and partners.

Blind Spots in the American Zeal for Democracy

Although messianic tendencies toward exporting American political processes and values have been a hallmark of US foreign policy in the twenty-first century, early twentieth-century Americans were not so sure about their portability. Europeans were viewed as "not yet ready" for American-style democracy and Central and South Americans were regarded as not sufficiently "civilized."[54] As the American experiment in democracy gained momentum in power and status, its population came to see its virtues as universal and its adoption around the globe as inevitable.[55] Moved by this perceived eventuality, it became part of the American default setting to proselytize the American model at every opportunity.[56] However naively and overly simplistically Americans may view foreign cultures, Americans tend to have an honest desire to extend to others those virtues prized in their own culture: a democratic setting that advances individual worth, justice, and fair play, and offers the chance to realize the American dream – individualism, equal opportunity, and the right to pursue happiness.[57]

For Americans, it is difficult to conceive of a future in which authoritarianism emerges as the dominant international model. American policymakers and the public alike regard movements toward democratic governance to be an ineluctable and natural process of political evolution, one which will continue to mend relations between nations and result in international prosperity and stability. The "normal" march forward for humanity is believed to be advancement toward a better life for a consistently expanding percentage of the global population.[58] The basic sentiments that underpin this American view are captured in academic literature as "modernization theory," which held sway as a dominant paradigm in the 1950s and 1960s and made a modest comeback in the social sciences of the 1990s. As noted by Francis Fukuyama, "If one were to sum up the Americanized version of modernization theory, it was the sunny view that all good things went together: economic growth, social mobilization, political institutions, and cultural values all changed for the better in tandem."[59] Although modernization theory has fallen into ill repute within the scholarship of the academy, the basic tenets of this theory remain alive in the mental models of American leaders and much of the population. Moves away from, rather than toward, democratic practice are perceived as out of the natural order and are difficult for Americans to anticipate, consider probable, and adequately prepare toward.

One of the consequences of these combined assumptions is an ethnocentric tendency toward viewing the populations of undemocratic regimes as "underdeveloped Americans" who have been stymied in some way in their natural progression.[60] Americans believe that if liberated from their dictators and given the material resources to build a new life, the natural inclination of most people would be to gravitate toward democracy. American support for wobbling partner governments, therefore, has relied heavily on American material generosity and the deployment of its military force: two approaches that are not without their successes in history, but are likely to fall short in an era of technology-boosted ideological incursions against democracy.

Material Fixes to Immaterial Problems

Leaning on its default setting for the engineering fix, US policymakers tend to perceive struggling partner governments as a *problem to be fixed*, or, if the government has been pushed out in a regime change, an *item to be built*. US nation-building efforts have surged forward with native optimism, undiluted by a knowledge of history that would reveal the limits of a foreign power in attempting to do so. Consequent failures tend to be explained in the American mind as products of insufficient resources or insufficient American effort.[61]

Limited in their understanding of other cultures, Americans tend to lean on instincts of material generosity to forge relationships and provide support to governing structures. Drawing from the rich economic treasure of the United States, diplomatic and military personnel dispense humanitarian aid, build schools, and improve infrastructure in order to advance democracy and local life. Anthropologists Edward and Mildred Hall hold up the United States as genuinely distinctive in this regard, declaring that American material generosity is "matched by no other country we know."[62] Through gifts of resources and infrastructure Americans aim to win friends over to the virtues of democracy and accelerate the social and economic progress thought to be required for stability. Brogan, among others, tracks the origins of the American fixation on material advancement to the Protestant Reformation and its spread through American religious circles. The result was a cultural conflation of prosperity and virtue.[63] Belief in the positive morality of material success propelled the American economy forward, enabling robust economic growth and unparalleled logistical excellence. Americans typically measure their own national health through daily tracking of fiscal and material indicators.

Material investments are often deeply appreciated by the populations who receive them and in some critical cases have achieved the strategic objective of shoring up democracy. The material investments of the Marshall Plan were not insignificant in the effort to strengthen fragile democracies in post-World War II Europe, a model that continues to resonate with Americans today. The US experience since that era, however, has resulted in some negative lessons learned concerning the savvy application of material resources. Resources aimed specifically at the population – medical services, food supplies, hygiene kits, soccer balls, and the like – produce sincere gratitude when addressing a local need. A lush dispersal of funds and projects, however, can have unintended and counter-strategic effects. In some cases, the more generous the dispersal of goods, the more destabilizing the effects may be. A sudden injection of resources can create destabilizing winner-loser dynamics, fuel corruption, inadvertently supply the underground economy that feeds disruptive actors, destroy the fledgling profitability of local businesses, and even create perverse incentives among key actors to maintain a status quo of insecurity in order to remain on the receiving end of a steady flow of funds.[64]

Furthermore, when these investments are made without regard for local preferences or the ability of local resources to sustain the project beyond a US force presence, they stand as a testament to wasted effort, or worse, harm local dignity and well-being and have a souring effect on US relationships. A Marine captain of the Vietnam era offered sage insight concerning the overwhelming material gifts he saw being dispersed around him: "generosity which cannot be returned breeds hostility, not affection."[65] The fixation on winning local gratitude rather than amplifying local dignity can also lead the US population to become disillusioned and resentful when they believe their material gifts and well-meaning efforts are not appreciated.[66] These negative sentiments can become vulnerabilities exploited at home by political opposition, often resulting in a swell of public opinion to cut aid short and abandon the nation to its own devices.[67]

American confidence in the self-evident benefits of its own governing paradigm combined with a historically unfounded certainty regarding democracy's inherent stability may inhibit US public investment in the research, personnel, technologies, and defense focus necessary to protect democratic governance at home and provide something beyond token material support to partners abroad. Defense professionals pursuing public diplomacy and information operations on behalf of US interests acknowledge that their mission is understaffed, underbudgeted, and regarded by leadership as peripheral to more serious US defense objectives.

In considering policy that might counter the growing threats to democracy, it is worth noting that the record of human history indicates a strong preference for stability and predictable living over concerns about personal liberty. Regimes struggling with domestic chaos or bouts of violence – even those who are friends of the United States – may welcome foreign technological advances in digital surveillance of their populations as a means of getting their houses in order. The application of dataveillance authoritarianism by new national customers in Africa, the Middle East, and beyond may not meet with the stiff resistance the United States expects if effective at subduing disruptions to everyday life. The normalization of this type of technology has implications for international norms and the long-term viability of the democratic model. As China in particular emerges as both the peddler and the trendsetter in the tools of digital governance, its leadership in artificial intelligence and the digital age may pose a significant challenge to US global leadership.

Conclusion

Competition and the possibility of conflict with great power adversaries are unlikely to play directly to American strengths. It is being waged against shrewd adversaries who are well acquainted with American advantages in warfighting as well as the deficits and vulnerabilities that history has opened to view. Recognizing US cultural presets with more clarity will provide advantages in thinking strategically about civic as well as military action on the ground, prioritizing meaningful measures of progress toward long-term strategic goals, and avoiding the pitfalls of overestimating the salience, or even existence, of US technological overmatch vis-à-vis near-peer adversaries. Harvesting the ready-made lessons which might be learned from the last two decades of US warfighting and building on these in order to successfully advance the local relationships with partners and allies that are critical to US global success will require investments in history and culture – best practices currently outside the comfort zone of American strategic culture.

Perhaps most importantly, the United States must lean forward in anticipating and countering efforts to undermine the society of democratic nations that form the scaffolding of the post-World War II international order. US global leadership depends on its ability to model the advantages of democracy and to effectively convey strength and resilience to partner nations that are being attacked or actively courted by US competitors. The American model remains an experiment; one that requires consistent care and investment both at home and on behalf of the wide circle of US partners and allies abroad.

Notes

1 The content for this chapter was adapted from Jeannie L. Johnson, "Fit for Future Conflict?: American Strategic Culture in the Context of Great Power Competition," *Journal of Advanced Military Studies* 11, no. 1 (Spring 2020): 185–208, https://doi.org/10.21140/mcuj.2020110109. Reprinted by permission.

2 Linda Robinson et al., *Modern Political Warfare: Current Practices and Possible Responses* (Santa Monica, CA: Rand Corporation, 2018).
3 Jeannie L. Johnson, *The Marines, Counterinsurgency, and Strategic Culture: Lessons Learned and Lost in America's Wars* (Washington, DC: Georgetown University Press, 2018).
4 Russell F. Weigley, *The American Way of War: A History of United States Military Strategy and Policy* (Bloomington, IN: Indiana University Press, 1973); Victor Davis Hanson, *Carnage and Culture: Landmark Battles in the Rise of Western Power* (New York: Anchor Books, 2001), 22. Both Hanson and Echevarria note that this is not an attribute unique to the United States. Antulio J. Echevarria II, *Toward an American Way of War* (Carlisle, PA: Strategic Studies Institute March 2004), 2; Robert Cassidy, *Counterinsurgency and the Global War on Terror: Military Culture and Irregular War* (Stanford, CA: Stanford University Press, 2008), 103, 115; James Kurth, "Iraq: Losing the American Way," *American Conservative* 15 March 2004, http://www.theamericanconservative.com/articles/iraq-losing-the-american-way/; Thomas G. Mahnken, "US Strategic and Organizational Subcultures," in *Strategic Culture and Weapons of Mass Destruction: Culturally Based Insights into Comparative National Security Policymaking*, eds. Jeannie L. Johnson, Kerry M. Kartchner, and Jeffrey A. Larsen (New York: Palgrave Macmillan, 2009), 74; Colin S. Gray, *Irregular Enemies and the Essence of Strategy: Can the American Way of War Adapt*? Monograph (Strategic Studies Institute, 2006), 11.
5 Antulio J. Echevarria II, *Reconsidering the American Way of War: US Military Practice from the Revolution to Afghanistan* (Washington, DC: Georgetown University Press, 2014); Brian M. Linn and Russell F. Weigley, "'The American Way of War' Revisited," *Journal of Military History* 66, no. 2 (April 2002): 50133, https://doi.org/10.2307/3093069; and Max Boot, *The Savage Wars of Peace: Small Wars and the Rise of American Power* (New York: Basic Books, 2002).
6 Mahnken, "US Strategic and Organizational Subcultures," 73.
7 Robert Cassidy, *Peacekeeping in the Abyss: British and American Peacekeeping Doctrine and Practice after the Cold War* (Westport, CT: Praeger, 2004), 93.
8 Deborah D. Avant, "The Institutional Sources of Military Doctrine: Hegemons in Peripheral Wars," *International Studies Quarterly* 37, no. 4 (December 1993), 415, https://doi.org/10.2307/2600839; see also John A. Nagl, *Learning to Eat Soup with a Knife: Counterinsurgency Lessons from Malaya and Vietnam* (Chicago, IL: University of Chicago Press, 2005), 44.
9 Cassidy, *Peacekeeping in the Abyss*, 98.
10 Phillip S. Meilinger, "American Military Culture and Strategy," *Joint Force Quarterly* 46, no. 3 (2007), 81; Gray, "Irregular Enemies and the Essence of Strategy," 46; Williamson Murray, "An Anglo-American Strategic Culture?," (paper prepared for the symposium "Democracies in Partnership: 400 Years of Transatlantic Engagement," Williamsburg, VA, 18–19 April 2007), 152; Jeffrey Record, "The American Way of War: Cultural Barriers to Successful Counterinsurgency," CATO Institute, Policy Analysis no. 577 (1 September 2006), 5.
11 Eliot A. Cohen, "Constraints on America's Conduct of Small Wars," *International Security* 9, no. 2 (Autumn 1984): 169.
12 Colin S. Gray, "British and American Strategic Cultures" (paper prepared for the Jamestown Symposium "Democracies in Partnership: 400 Years of Transatlantic Engagement," Williamsburg, VA, 18–19 April 2007), 49; Theo Farrell, "Strategic Culture and American Empire," *SAIS Review* 25, no. 2 (Summer/Fall 2005): 8; Lawrence Sondhaus, *Strategic Culture and Ways of War*, (London: Routledge, 2006), 60–61; Mahnken, "US Strategic and Organizational Subcultures," 74; Sam C. Sarkesian, *America's Forgotten Wars: The Counterrevolutionary Past and Lessons for the Future* (Westport, CT: Greenwood Press, 1984), 5; Peter Warren Singer, *Insurgency in 2030* (Washington, DC: New America, 2019).
13 Williamson Murray, "Does Military Culture Matter?," in *America the Vulnerable: Our Military Problems and How to Fix Them*, eds. John F. Lehman and Harvey Sicherman (Philadelphia, PA: Foreign Policy Research Institute, 2002), 151.
14 Joe Galloway, "Rumsfeld's War Games," Military.com, 26 April 2006; Julian Borger, "Wake-up Call," *The Guardian*, 5 September 2002.
15 Singer, *Insurgency in 2030*.
16 Dreyton Schafer, *Cyber Threats: Technically Savvy American Aviation's Blind Spot*, monograph prepared for the Center for Anticipatory Intelligence, (Spring 2019), accessed on 17 January 2020 at https://cai.usu.edu/files/studentpaper-schafer.pdf

17 Singer, *Insurgency in 2030.*
18 D.W. Brogan, *The American Character* (New York: Alfred A. Knopf, Borzoi Books, 1944), 5.
19 Stanley Hoffman, *Gulliver's Troubles, Or the Setting of American Foreign Policy* (New York: McGraw Hill, 1968), 110. See also Robin Williams, "Values and Modern Education in the United States," in *Values in America*, ed. Donald Barrett (Notre Dame, IN: University of Notre Dame Press, 1961), 66; and Gary Althen, *American Ways: A Guide for Foreigners in the United States* (Yarmouth, ME: Intercultural Press, 1988), 10–11.
20 Brogan, *The American Character*, 136; Cora Du Bois, "The Dominant Value Profile of American Culture," *American Anthropologist* 57, no. 6 (December 1955), 1233, https://doi.org/10.1525/aa.1955.57.6.02a00130; Hellmut Lotz, "Myth and NAFTA: The Use of Core Values in US Politics," in *Culture and Foreign Policy*, ed. Valerie Hudson (Boulder, CO: Lynne Rienner Publishers, 1997), 79; Maryanne Kearney Datesman, Joann Crandall, and Edward N. Kearny, *American Ways: An Introduction to American Culture*, 4th ed. (White Plains, NY: Pearson Education, 2014), 105; Edward C. Stewart and Milton Bennett, *American Cultural Patterns: A Cross-Cultural Perspective,* rev. ed. (Boston, MA: Intercultural Press, 1991), 142–143; and Williams, "Values and Modern Education in the United States," 66.
21 Henry Ford interview in the *Chicago Tribune*, 25 May 1916.
22 Murray, "An Anglo-American Strategic Culture?," 157. Murray makes the important point that ahistoricism is not a uniquely American trait: "[T]here are few military organizations that possess a culture that encourages the study of even the recent past with any thoroughness." W. Murray, "Does Military Culture Matter?," 140.
23 Sarkesian, *America's Forgotten Wars*, xii.
24 Stewart and Bennett, *American Cultural Patterns*, 158; Lotz, "Myth and NAFTA," 80.
25 Samuel Huntington, *The Soldier and the State: The Theory and Politics of Civil–Military Relations* (Cambridge, MA: Belknap Press, an imprint of Harvard University Press, 1957), 149.
26 Oliver M. Lee, "The Geopolitics of America's Strategic Culture," *Comparative Strategy* 27, no. 3 (2008): 276, https://doi.org/10.1080/01495930802185627.
27 William Kincade, "American National Style and Strategic Culture," in *Strategic Power: USA/USSR*, ed. Carl G. Jacobsen (London: Macmillan, 1990), 13; Stewart and Bennett, *American Cultural Patterns*, 11.
28 James J. Wirtz, "Life in the 'Gray Zone': Observations for Contemporary Strategists," *Defense & Security Analysis* 33, no. 2 (2017): 106–114, https://doi.org/10.1080/14751798.2017.1310702.
29 James E. Hayes III, "Beyond the Gray Zone: Special Operations in Multidomain Battle," *Joint Forces Quarterly*, no. 91 (4th Quarter 2018): 64–65; Robinson, et al., *Modern Political Warfare*, 60–66; Joseph L. Votel et al., "Unconventional Warfare in the Gray Zone," *Joint Forces Quarterly*, no. 80 (1st Quarter 2016): 101–109.
30 Hayes, "Beyond the Gray Zone," 61.
31 Stewart and Bennett, *American Cultural Patterns*, 32, 37, 68.
32 *Warfighting*, MCDP 1 (Washington, DC: Department of the Navy, 1997), 57–58.
33 Stewart and Bennett, *American Cultural Patterns*, 69, 155; and Datesman et al., *American Ways*, 85–86.
34 Italics added. Stewart and Bennett, *American Cultural Patterns*, 76.
35 Datesman et al., *American Ways*, 106.
36 Du Bois, "The Dominant Value Profile of American Culture," 1234; and Leonard Mason, "The Characterization of American Culture in Studies of Acculturation," *American Anthropologist* 57, no. 6 (1955): 1268–1269, https://doi.org/10.1525/aa.1955.57.6.02a00160.
37 *Warfighting*, 39–40.
38 Mason, "The Characterization of American Culture in Studies of Acculturation," 1268; Stewart and Bennett, *American Cultural Patterns*, 74; Edward T. Hall and Mildred Reed Hall, *Understanding Cultural Differences: Germans, French, and Americans* (Yarmouth. ME: Intercultural Press, 1990), 140–141.
39 Dima Adamsky, *The Culture of Military Innovation: The Impact of Cultural Factors on the Revolution in Military Affairs in Russia, the US, and Israel* (Stanford, CA: Stanford University Press, 2010), 75.
40 Colin S. Gray, "Out of the Wilderness: Prime Time for Strategic Culture," in *Strategic Culture and Weapons of Mass Destruction: Culturally Based Insights into Comparative National Security*

Policymaking, eds. Jeannie L. Johnson, Kerry M. Kartchner, and Jeffrey A. Larsen (New York: Palgrave Macmillan, 2009), 224; Mason, "The Characterization of American Culture in Studies of Acculturation," 1268; Stewart and Bennett, *American Cultural Patterns*, 30–31; Datesman et al., *American Ways*, 106.

41 Maj. Russell A. Moore, USMC, "*Strategic Culture – How It Affects Strategic 'Outputs*'" (Quantico, VA: Marine Corps War College, Marine Corps University, Marine Corps Combat Development Command, 1998).

42 Brogan, *The American Character*, 32–34, 74.

43 Du Bois, "The Dominant Value Profile of American Culture"; Datesman, et al., *American Ways*, 85–86; Stewart and Bennett, *American Cultural Patterns*, 75.

44 Gray, "British and American Strategic Cultures," 45.

45 Gray, "Irregular Enemies and the Essence of Strategy," 33.

46 Mason, "The Characterization of American Culture in Studies of Acculturation," 1269; see also Datesman et al., *American Ways*, 135.

47 Hoffman, *Gulliver's Troubles*, 111; Kent Lam et al., "Cultural Differences in Affective Forecasting: The Role of Focalism," *Personality and Social Psychology Bulletin* 31, no. 9 (2005): 1296–1309, https://doi.org/10.1177/0146167205274691; see also Adamsky, *The Culture of Military Innovation.*

48 Stewart and Bennett, *American Cultural Patterns*, 69; and Claude S. Fischer, *Made in America: A Social History of American Culture and Character* (Chicago: University of Chicago Press, 2010), 210.

49 Ben Connable et al., *Will to Fight: Returning to the Human Fundamentals of War* (Santa Monica, CA: Rand Corporation, 2019).

50 Lauren Mackenzie and Kristin Post, "Relationship Repair Strategies for the Military Professional: The Impact of Cultural Differences on Expectations and Applications," *Marine Corps University Journal* 10, no. 1 (2019), 128–141

51 Stewart and Bennett, *American Cultural Patterns*, 127.

52 Ben Connable, in a thoughtful piece written for the Rand corporation, offers a selection of alternative assessment models that are both qualitative and quantitative in nature and leverage the wealth of lessons learned across recent US experience with population-centric warfare. Ben Connable, *Embracing the Fog of War: Assessment and Metric in Counterinsurgency* (Arlington, VA: Rand Corporation, 2012).

53 Robinson, et al., *Modern Political Warfare*; Wirtz, "Life in the 'Gray Zone,'" 109.

54 Marcus Cunliffe, "Formative Events from Columbus to World War I," in *American Character and Foreign Policy*, ed. Michael P. Hamilton (Grand Rapids, MI: William B. Eerdmans, 1986), 9.

55 Stanley Hoffman, *Gulliver's Troubles*, 111; and Michael J. Williams, *On Mars and Venus: Strategic Culture as an Intervening Variable in US and European Foreign Policy* (London: Lit Verlag, 2006).

56 Mahnken, "US Strategic and Organizational Subcultures," 71. Phillip Meilinger notes this trend as it applies to Mexico in 1847, Cuba and the Philippines in 1898, Europe in 1918, Germany and Japan after World War II, and in Korea 1950, Vietnam 1955, Iraq 2003, and Afghanistan 2001, "American Military Culture and Strategy," 81; Morell Heald, "Foreign Relations, American Style," in *American Character and Culture in a Changing World: Some Twentieth-Century Perspectives*, ed. John A. Hague (Westport, CT: Greenwood Press, 1979), 197–198; and Theo Farrell, "America's Misguided Mission," review of *Democracy by Force: US Military Intervention in the Post-Cold War World*, by Karin von Hippel, *International Affairs* 76, no. 3 (2000): 3, https://doi.org/10.1111/1468-2346.00153.

57 Sarkesian, *America's Forgotten Wars,* 14–15; and Lotz, "Myth and NAFTA," 81.

58 Brogan, *The American Character*, 65.

59 Francis Fukuyama, "Samuel Huntington's Legacy: Why His Works on World Order – Political and Otherwise – Are Still Relevant Today," *Foreign Policy* (6 January 2011), accessed on 24 April 2020 at https://foreignpolicy.com/2011/01/06/samuel-huntingtons-legacy/

60 Edward T. Hall, *The Silent Language* (New York: Premier Books, 1963), 9; and Althen, *American Ways*, xvi.

61 Stewart and Bennett, *American Cultural Patterns*, 75.

62 Hall and Hall, *Understanding Cultural Differences*, 153.

63 Brogan, *The American Character*, 67; Du Bois, "The Dominant Value Profile of American Culture," 1235.
64 Edwina Thompson, "Winning 'Hearts and Minds' in Afghanistan: Assessing the Effectiveness of Development Aid in COIN Operations" (report prepared on Wilton Park Conference, vol. 1022, 2010), 11–14.
65 R.E. Williamson, USMC, "A Briefing for Combined Action," *Marine Corps Gazette*, March 1968, 43.
66 Stewart and Bennett, *American Cultural Patterns*, 108; Hoffman, *Gulliver's Troubles*, 101–102.
67 Minxin Pei, Samia Amin, and Seth Garz. "Building Nations: The American Experience," in *Nation-Building: Beyond Afghanistan and Iraq*, ed. Francis Fukuyama (Baltimore, MD: Johns Hopkins University Press, 2006), 68; Hall and Hall, *Understanding Cultural Differences*, 152; and Lee, "The Geopolitics of American's Strategic Culture," 280.

Selected Bibliography

Adamsky, Dima. *The Culture of Military Innovation: The Impact of Cultural Factors on the Revolution in Military Affairs in Russia, the US, and Israel* (Stanford, CA: Stanford University Press, 2010).
Boot, Max. *The Savage Wars of Peace: Small Wars and the Rise of American Power* (New York: Basic Books, 2002).
Brogan, D. W. *The American Character* (New York: Alfred A. Knopf, Borzoi Books, 1944).
Cassidy, Robert. *Peacekeeping in the Abyss: British and American Peacekeeping Doctrine and Practice after the Cold War* (Westport, CT: Praeger, 2004).
Echevarria II, Antulio J. *Reconsidering the American Way of War: US Military Practice from the Revolution to Afghanistan* (Washington, DC: Georgetown University Press, 2014).
Gray, Colin S. "British and American Strategic Cultures" (paper prepared for the Jamestown Symposium "Democracies in Partnership: 400 Years of Transatlantic Engagement," Williamsburg, VA, 18–19 April 2007).
Heald, Morell. "Foreign Relations, American Style," in *American Character and Culture in a Changing World: Some Twentieth-Century Perspectives*, ed. John A. Hague (Westport, CT: Greenwood Press, 1979).
Johnson, Jeannie L. *The Marines, Counterinsurgency, and Strategic Culture: Lessons Learned and Lost in America's Wars* (Washington, DC: Georgetown University Press, 2018).
Mahnken, Thomas G. "US Strategic and Organizational Subcultures," in *Strategic Culture and Weapons of Mass Destruction: Culturally Based Insights into Comparative National Security Policymaking*, eds. Jeannie L. Johnson, Kerry M. Kartchner, and Jeffrey A. Larsen (New York: Palgrave Macmillan, 2009).
Sarkesian, Sam C. *America's Forgotten Wars: The Counterrevolutionary Past and Lessons for the Future* (Westport, CT: Greenwood Press, 1984).
Williams, Michael J. *On Mars and Venus: Strategic Culture as an Intervening Variable in US and European Foreign Policy* (London: Lit Verlag, 2006).
Weigley, Russell F. *The American Way of War: A History of United States Military Strategy and Policy* (Bloomington, IN: Indiana University Press, 1973).

12
RUSSIAN STRATEGIC CULTURE
A Critical Survey of the Literature

Mette Skak

Old habits die hard. This saying captures well why strategic culture matters. Strategic culture refers to security policy habits of mind, notably on the level of decisionmakers, and to habitual behavior.[1] Within the discipline of international relations, strategic culture is often treated as part of the constructivist turn as it is an approach that challenges neo-realism.[2] But in reality, the study of strategic culture is a child of Cold War Sovietology, more specifically the pioneering work on the so-called *Operational Code of the Soviet Politburo* by the eminent Russian-born sociologist Nathan C. Leites.[3] In other words, the field is rooted in empirical research into twentieth-century Russian – i.e., Soviet – security policy mentality, as when Jack Snyder invented the term "strategic culture" in 1977, he intended it to capture the peculiar nature of Soviet nuclear doctrine.[4] Insights about Russia and the Soviet Union are often cited to show what strategic culture is all about, namely national security policy idiosyncrasies borne by local scope conditions and historical experiences with war versus peace, victory and defeat, and so forth.[5]

Accordingly, this chapter will begin by devoting some space to Leites' operational code approach along with other clues to contemporary Russian strategic culture taken from Cold War Sovietology. The chapter then comments upon selected contributions to the study of post-Soviet Russian strategic culture by distinguishing between classical military *way of war studies* of contemporary Russia and studies of *broader strategic culture,* i.e., analyses of Russian nonmilitary strategic culture, including intelligence culture. Why such a holistic approach to strategic culture is warranted in the case of post-Soviet Russia follows later. Another holistic principle to be followed is argued by one veteran from the field of strategic culture, Colin S. Gray. According to Gray, perceptions cannot be separated from behavior as both analytical dimensions go into what the concept of culture refers to, namely *context*, which in turn implies complex causality.[6] Then there is holism in the sense of addressing both ends and means when analyzing strategic culture.[7] Concerning ends, Alastair Iain Johnston rightly argues that the purpose of strategic culture analysis is to identify the grand strategy of a given actor.[8] But Gray has a point when stressing the analysis of means, such as a given actor's preference for military versus nonmilitary means, their willingness to resort to espionage, and so on.[9] Lastly, the chapter offers your author's own theoretical and methodological position when studying today's Russian strategic culture.

 DOI: 10.4324/9781003010302-15

On the Operational Code Findings of Leites: Insights from Cold War Sovietology

In 1946, American diplomat George Kennan offered critical insight into Bolshevik strategic culture when distinguishing between the official and unofficial levels of Soviet diplomacy in his famous Long Telegram. Here Kennan defined "official" Soviet diplomacy as actions undertaken formally by the Soviet government, for instance in order to further the prestige of the Soviet Union, whereas "unofficial level" diplomacy referred to the subterranean plane of actions for which the Soviet government would deny responsibility.[10] Kennan elaborated by speaking about "a concealed Comintern" operating through foreign communist parties, seeking to help Moscow to pursue its more controversial yet vital goals abroad. Clearly, Kennan thereby wanted to draw attention to the Soviet unofficial diplomacy as possibly the truly operational Soviet diplomacy.

Operational code analysis refers to the art of distilling rules and norms for political action from the pattern of actual action, strategies, and key explicit statements. Hence, it is much about making the implicit explicit or, drawing upon the Dutch sociologist Joris Van Bladel, distilling the *mentality* that decisionmakers have internalized as their own guide to action as opposed to their use of instrumentalized outward ideology.[11] Leites' aim was to reconstruct the operational code of conduct that motivated the Soviet Politburo at the time of Stalin. Leites' method was to, first, cite rules explicitly stated by Lenin or Stalin; second, to infer rules easily recognized by Bolsheviks; and, lastly, to imply otherwise operational rules. He organized his findings into 20 thematic subchapters and stressed that rules were often contradictory and imperfect, which only added to their significance. Stalin himself would seem to violate Rule 16 – a rule tabooing "adventurism"[12] – when he "applied" Rule 5 from the subchapter on the exertion of pressure. Leites held Stalin and the whole Soviet Politburo to have grossly miscalculated the American countermeasures of cold warfare, for instance in the Korean War (1950–1953).[13]

What is striking is the uniquely militant zero-sum mindset of Stalin and his fellows. The notion of win-win games was alien to them as was the earnest pursuit of compromise in order to overcome security dilemmas. Lenin was obsessed with war and revolution, not coexistence as such. His morality was one of class struggle, which led him to recognize all means of struggle.[14] This holism is a clue to grasping post-Soviet Russian strategic culture, argues Oscar Jonsson in his highly recommendable monograph.[15] Only the Party was bestowed with agency, whereas all other actors were "forced" into their parts in history.[16] Hence, Bolshevik ontology represented what Gray terms a culture of *high context.*[17] By this he means a bias toward conspiracy theorizing, of seeing plots and subplots or connections between events where other observers see none, to quote from Leites.[18] This is another clue to what characterizes the Kremlin today.[19]

Leites documents the deeply Social Darwinian mentality of Lenin and Stalin by quoting the latter in 1931: "Those who fall behind get beaten. But we do not want to be beaten (…) [about Russia:] All beat her – for her backwardness, for military backwardness."[20] Those phrases were repeated by the Russian President Vladimir V. Putin upon the Beslan massacre in 2004 and later when arguing the case for comprehensive armament.[21] Stalin's own conclusion – "You are mighty, therefore you are right" – reflects the twenty-first-century Russian approach to world politics as an arena for great powers like Russia.[22] Despite certain fundamental differences between contemporary Russia and the Soviet Union, there is Cold War continuity between the operational code of the Politburo and that of Putin's Kremlin.[23] Take the myth of Russia as a besieged fortress (*osazhdyonnaya krepost'*) which drove Soviet Cold War policy and now drives Russian cybersecurity thinking as established by the Finnish scholar Martti J. Kari.[24]

As for the pioneering study on strategic culture by Snyder, he found the war years of 1941–1945 to represent the formative era for the Brezhnev generation of Soviet decisionmakers. Also on this account, there is continuity to the Putin generation as exactly those years of the Great Fatherland War (*Velikaya Otechestvennaya Voina*) have become the foundational myth of post-Soviet Russia as seen in the 2020 amendments to the constitution. Snyder concluded that Soviet nuclear doctrine of the 1970s was in the hands of the Soviet military brass, not the civilian academic experts.[25] Although the Soviet Army was not Bonapartist,[26] its professionalism was taken seriously by the post-Stalin Politburo. After Mikhail S. Gorbachev came to power, analysts discussed how much of a revolution his liberal institutionalist New Political Thinking brought to the strategic culture. On this topic, John Glen together with Henrikki Heikka drew optimistic conclusions,[27] whereas Stephen S. Blank and Robert G. Hermann both argued the superficiality of the Gorbachev revolution.[28] Similarly, the in-depth analysis of the rise, fall, and reprise of Soviet-Russian military interventionism by Andrew Bennett concluded that post-Soviet Russian decisionmakers were unlearning what their Soviet predecessors learned from their deeply tragic war in Afghanistan (1979–1988), his post-Soviet case being Russia's brutal war against Chechen secessionism.[29] Putin's presidency was born in the Second Chechen War (1999–2009).

So one conclusion to draw from the super-ideological Soviet epoch and Western Sovietology is that Russian strategic culture was and is a strategic culture of militarism, albeit without Bonapartism.[30] It was a militarism driven by civilian hawks like Lenin and Stalin, later the "chief ideologue" Mikhail Suslov, and the boss of the Soviet secret police (the KGB) Yuri Andropov.[31] Similarly, Heikka's incisive analysis of the Soviet "cult of the offensive" points to militarism as a generic operational code for the Soviet elite.[32] In continuation of this, the Russian military analyst Aleksandr Golts exposes the astonishing revival of militarism in Putin's Russia, including the diffusion of militarism into Russian civil society.[33] Czarist Russia proudly fought offensive wars, and both the Winter War against Finland (1939–1940) and the Korean War count as Stalin's wars. As for the myth about the Russian fear of invasion from the West, Russia defeated both Napoleon and Hitler and hence ought to have a "cult of the defensive."

Lastly, virtually all works on Russian strategic culture stress the continuity between Czarist Russia, the Soviet Union, and Putin's Russia concerning the Russian identity of being a great power, including a geopolitical myth of entitlement to spheres of influence in the post-Soviet neighborhood. As pointed out by the historian Graeme P. Herd in his fine study of Putin's operational code, this turns the United States as a superpower into Russia's strategic benchmark.[34] Inspired by his Russian colleague Sergei Medvedev, Herd considers the pursuit of *being feared* a staple of the Russian conception of its great power self, and hence a staple of Russian strategic culture.[35] As will be argued below, the powerful Ukrainian defense against the 2022 Russian invasion undermines this Russian yearning for awing the surrounding world into submission.

Works on the Russian Way of War

When analyzing the Russian "way of war" one must examine both *deeds* – the Russian military practice as manifested in armed warfare, military exercises, and weapons procurement – and *words*, i.e., works on Russian military thought including nuclear doctrine. The Russian full-scale invasion of Ukraine starting on 24 February 2022 and the ongoing warfare at this time of writing serve as a point of departure for this analysis. It is the largest military confrontation in Europe since World War II, entailing huge Russian and Ukrainian casualties, material destruction, and waves of inhabitants fleeing Ukraine. The conflict is another case of offensive war waged by Russia and thus follows the pattern of militarism and aggression within Russian strategic culture. Indeed, there is almost consensus among analysts on interpreting the war as caused by Putin's peculiar imperialist approach to Ukraine.[36] In other words, the decision to invade reflects Putin's own Social Darwinism as displayed in his deliberations on "colonies" versus truly sovereign states like Russia.[37] Putin's likely perception that Ukraine was moving in the European direction of consolidated democracy and rule of law and receiving military training from North Atlantic Treaty Organization (NATO) powers in support of this process appears to have mobilized the Russian president's operational code of acting in defense of his own authoritarian regime's security. The section following this one will elaborate on this point.

Nevertheless, the Russian Army's poor performance on the ground in Ukraine in 2022 represented a paradox and surprised many military experts. The original *Blitzkrieg* intended to conquer Ukraine and seize Kyiv within three days in order to topple the popularly elected President Volodymyr Zelensky soon failed.[38] The supposedly sophisticated arms and equipment of Russian soldiers often turned out to be worthless, and their morale was low and dropped even further when confronted with the Ukrainian Army's resolve to fight back. To take one example, there was a 64-kilometer-long Russian convoy of armed vehicles heading toward Kyiv that ended up as just a traffic jam – maybe due to Ukrainian attacks against the vehicles leading the convoy; otherwise due to vehicle breakdowns making ambush easy and opening the opportunity for desertions.[39] Russian elite paratroopers were squandered, and many Russian generals had to enter the frontline and became prey for Ukrainian snipers. The *Blitzkrieg* war plan was built on poor intelligence, leading many analysts to theorize about possible groupthink surrounding Putin as the key decisionmaker.[40] Others insist on Putin's own delusion about Ukraine as the catalyst for this disastrous approach.[41] Either way, the war has exemplified strategic cultural idiosyncracies and faulty authoritarian decisionmaking habits dying hard – contrary to decisionmaking based on a serious assessment of strengths and weaknesses.[42] Hence Russia's decision to invade Ukraine vindicates the key analytical insight about Russian and Soviet strategic culture offered by the aforementioned veteran in the field, Colin Gray, that some strategic cultures may turn out to be dysfunctional or self-destructive.[43]

Although the outcome of the war remains unknown at this time of writing, it stands to reason that Russia has been severely weakened militarily by this self-inflicted strategic nightmare. Contrary to its wishes, Russia is now being feared less, and it is as isolated as ever because of the uniquely drastic economic sanctions hitting Russia. In one important respect, however, Russia remains a terrifying power – namely in terms of atrocities committed against civilians along with the utter material destruction characterizing Russian warfare. The Cold War historian Mark Kramer considers this brutality against Ukrainians to be a legacy from Russia's Soviet past – which also happened to include both looting and deportations, especially in Stalin's time.[44] Kramer stresses the Russian willingness to use its own soldiers as cannon fodder as another feature of Russian and Soviet military culture and hence as an operational expression of today's Russian strategic culture.

Before the 2022 Russian invasion of Ukraine, the aforementioned argument by Bennett about Russian military interventionism was vindicated by the Russian interventions in Georgia (2008), Ukraine (2014–), and Syria (2015–).[45] What has been highlighted as a novelty, however, is the accompanying military reform and comprehensive technological modernization of Russia's armed forces. One much-acclaimed analysis of this ostensible Russian "revolution in military affairs" is the monograph by the British-German expert Bettina Renz, who emphasized the "unprecedented political will at the highest level" driving the post-2008 reform.[46] But Russia's surprisingly poor performance in Ukraine has led analysts to point to problems of corruption and import dependence on Western high-technology military parts.

The air campaigning in support of the Assad regime in Syria was a pioneering intervention outside Russia's post-Soviet "near abroad" and a successful one at that.[47] It may thus have fed into Putin's hubris toward Ukraine. Russia has also held multiple spectacular military exercises and practiced surprise "snap" maneuvers as well as engaged itself in upgrading old Soviet military bases, including in the Russian Arctic. On one occasion, Russian fighter jets trained conducting a surprise attack on Stockholm, the capital of (then) neutral Sweden. Danish airspace above the island Bornholm, which falls under NATO's security guarantee, was also targeted. This new pattern of Russian assertiveness brought the seasoned French analyst Bruno Tertrais to warn about Russia's "reckless and dangerous military provocations... its violations of arms-control and disarmament treaties, and its temptation to play the nuclear card as a tool for political coercion...."[48] Likewise, Cynthia Roberts invoked the term brinkmanship to capture the escalation inherent in the Russian annexation of Crimea.[49] Moreover, the scenario constructed for the subsequent large-scale Zapad-2017 military exercise by the Russian and Belarusian commanders suggests that so-called colour revolutions – beforehand perceived by Russian decision-makers to be instigated by Western powers – are becoming the operational *casus belli* for the Kremlin. Thus, Russian Defense Minister Sergei Shoigu used the term to motivate the Russian intervention in Syria in the context of the Arab Spring.[50]

On this point, it is worth invoking the strategic cultural analysis of current Russian practices by Steven R. Covington, who stresses the importance of the colour revolutions for the Russian military.[51] Like Renz, he reminds us of the overlapping roles of Russia's Internal Troops and Armed Forces and cites the nationwide "anti-Maidan" type of military exercise in 2015, presumably drawing lessons from the popular rising against Ukrainian President Viktor Yanukovich the year before.[52] The founding of a new Russian National Guard, *Rosgvardiya,* by Putin in 2016 is another most telling fact as it enjoys more powers than the Russian Army.[53] Equally important, according to the powerful contribution by Pavel Shchelin, the Russian National Security Strategy of 31 December 2015 essentially turned *regime security* into national security.[54] This resonates with Covington who stresses the uneasy mix of defensive and offensive features within Russian strategic culture, leading him to conclude that Russia displays an outright destabilizing national military strategy.[55] As for Zapad-2017, this military exercise indicated Kremlin hypersensitivity to any NATO initiative toward Belarus. The next year's Vostok-2018 was remarkable for its involvement of a minor contingent of Chinese People's Liberation Army soldiers, signaling Sino-Russian amity. By contrast, Tsentr-2019 was seen by Russia analysts in the West as marking some distancing from China's One Belt, One Road initiative that stretches into Central Asia.

What also characterizes contemporary conventional Russian warfare is the use of special forces – "little green men" without Russian insignia, in the case of the 2014 Crimea annexation – along with the integration of non-kinetic information warfare with kinetic warfare and the resort to private military companies like the Wagner Group, facilitating a "lean" or smart application of force in Syria.[56] In addition, the initial period of war is considered decisive. Altogether, this very holism is turning into a pitfall for Russian military and civilian decisionmakers. If they cannot distinguish defensive from offensive measures – and they blur the borders between war and peace as argued by Jonsson – they cannot distinguish war from peace, stresses Covington.[57] In short, they cannot falsify either, but end up securitizing everything in a hopeless Soviet-like armaments and mobilization race to the bottom. Some Russian officers seem to realize this dangerous outcome. Drawing on Keir Giles' observation about the ambiguous and possibly lowered Russian threshold for war – colour revolution as *casus belli* – Jonsson emphasizes that "Russia already considers itself to be in a state of war,"[58] if only a non-kinetic war with the liberal Western powers. Specifically, he points to the IT or digital revolution as a driver for the Russian pursuit of information war[59] against the West, along with the fear of colour revolution in Russia as an additional driver. For all of the above reasons, then, strategic culture analyses of Russia must be equally holistic.

As for Russian non-conventional warfare and nuclear doctrine, several positions can be identified: on the one hand, scholars arguing the offensive nature of the Russian approach, represented by scholars like Blank and Nikolai N. Sokov; and, on the other hand, scholars arguing its defensive nature like Olga Oliker and Tertrais. At issue is the possible lowering of the nuclear threshold – the Russian *casus belli* for nuclear war. The offensive position cannot be easily dismissed as even defensivists emphasize the disturbingly ambiguous Russian nuclear thinking. Thus, Blank has a point when arguing that Russian decisionmakers turn conventional and nuclear weapons into a whole and that their strategy is one of escalation control.[60] But Tertrais insists that one should not fear Russia "for the wrong reasons" and he fairly persuasively debunks myths about the lowering of Russia's nuclear threshold through the notorious de-escalation doctrine.[61] Among other things, he stresses that it was only ex post the annexation of Crimea and the onset of secessionist war in Donbas that Putin in 2015 publicly declared to have been ready to activate the Russian nuclear forces if the West had intervened in Ukraine. One remarkable twist to the debate between defensivists versus offensivists is the interpretation advanced by Pyotr Topychkanov. He shows how Russia's non-compliance with the Intermediate-Range Nuclear Forces treaty may be rooted in fear of intermediate short-range missiles deployed "on the arch that spans from North Korea to Israel including Pakistan, India, Iran" and ultimately China, as specified in 2007 by Sergei Ivanov, the Presidential Administration Chief of Staff at the time and a close associate of Putin.[62]

A truly original, instrumental interpretation of today's Russian nuclear doctrine is offered by the eminent expert on Russian strategic culture, Dima Adamsky. He shows how adept the Russian Orthodox Church was when offering its services to the country's demoralized and poorly funded armed forces following the collapse of the Soviet Union,[63] thereby turning itself into a *keeper*[64] of Russian strategic culture – devoid of Soviet atheism, however. The title of Adamsky's monograph is *Russian Nuclear Orthodoxy*, by which he means the following new mantra in Moscow: "to stay Orthodox, Russia should be a strong nuclear power" and, conversely, that "to stay a strong nuclear power, Russia should be Orthodox."[65] Today, each leg of the Russian nuclear triad (i.e., its land-launched nuclear missiles, its nuclear-missile-armed submarines, and strategic aircraft with nuclear bombs and missiles) has its patron saint, Orthodox icons appear on nuclear platforms, and the war in Syria was declared a holy battle by the church. The problem with this theocratization is the possible nuclear crisis instability and loss of escalation control through a loss of restraint by those in charge of pushing the nuclear button.[66] Reviewers of Adamsky's work concluded that while one should not exaggerate the operational significance of Russian nuclear orthodoxy, it represents an entirely novel, and most likely lasting, twist to Russian strategic culture.

Selected Works on Broader Russian Strategic Culture

Below follows a brief review of other instructive works on Russian strategic culture at the level of Kremlin decisionmakers divided into *ends* – i.e., the Russian grand strategy – and *means* preferred by the Russian security policy elite. A case in point regarding the former is the intriguing comparison between Chinese and Russian strategic culture written by the American sociologist and Russian-speaking Sinologist Gilbert Rozman, the only caveat about his analytical venture being the misleading term "national identities." He identifies several overlapping cultural features uniting Putin and the Chinese President Xi Jinping.[67] For both, the culminating year of the Kosovo War (1999) was a turning point, and they share an acute fear of colour revolutions and actively balance against US power. Rozman rightly considers the Putin regime to be beholden to the legacy of communism when defining Russia's foreign and security interests in defiance of Western interests and liberal values. He concludes in terms of a shared grand strategy, indeed an operational code of regime security, as the utmost priority for their increasingly authoritarian political systems: "a strong correspondence in regime interests" whose essence is that they are "regimes fearful of values associated with the international community,"[68] meaning liberal values about the integrity of the individual, the rule of law, and so on. But Putin's operational code is arguably more revisionist than Xi's and China is obviously much more of a geo-economic power.

Still, there is something to be said for geo-economics and Kremlin strategic culture judging from the pioneering essay on Putin-era Russian strategic culture by the ex-Central Intelligence Agency (CIA) veteran Fritz W. Ermarth. He argued a possible civilizing effect upon Russian militarism from the Kremlin's utter dependence on exporting energy to the world market.[69] Whereas Czarist Russia saw itself as having only two reliable allies – its army and fleet – nowadays it is oil and gas, Ermarth quipped. However, because of that the Finnish scholar Veli-Pekka Tynkkynen perceives the Kremlin to be caught in a new vicious circle of hydrocarbon culture and climate denialism, preventing Russia from taking on the necessary green transformation of its economy[70] – another argument about a deeply dysfunctional strategic culture.[71] One Russian sociologist perceives Russia's war against Ukraine to be a proxy war against the "greening" of Europe's energy consumption.[72] In any event, Russia is clearly warring against Ukraine's status as a geo-economic great power concerning the export of food. As for the Russian grand strategy of upholding great power status in world affairs, it is of immense importance that the surrounding world granted Russia the Soviet Union's permanent seat in the United Nations Security Council upon the Soviet collapse,[73] although few Russians care to take this fact into consideration. Hereby the Kremlin received veto power over security affairs all over the globe. Moreover, Russia inherited the entire Soviet arsenal of nuclear weapons.

In terms of means used by Russia to pursue its grand strategy, once again holism applies as was advised by Lenin as operational code for the Soviet Politburo. As exposed in great, entertaining detail in a monograph by Thomas Rid, post-Soviet Russia has reinvented the huge inventory of Soviet active measures[74] (*aktivnye meropriyatiya*) roughly corresponding to what is known as covert action within the CIA. Indeed, what characterizes Russian strategic culture in practice is the frequent resort to subversive measures carried out by the powerful contemporary Russian secret services, for instance the military intelligence branch GRU under the General Staff. A case in point covered by Rid is the Russian interference via social media platforms, troll farms, and other mechanisms of weaponized narrative in the US presidential election in 2016 against the Democratic candidate Hillary Clinton – a case leading to the arrest of 12 GRU agents among other things.[75] The advantage associated with avoiding overt operations for the benefit of covert ones is the in-built plausible deniability as the generic strategic cultural professional ethos of intelligence agencies. This option of protecting oneself against retaliation becomes even more powerful within the cyber and IT digital domains – two key arenas for Russian active measures in the twenty-first century.[76]

In other words, Russia benefits from the so-called attribution problem: the difficulty in tracking the perpetrator behind cyberattacks, of which some of the most disruptive like NotPetya and the SolarWinds attack have been linked to Russia. Michael Kofman uses the colorful Russian term *nabeg,* or raiding/brigandry, about this novel Russian way of war.[77] Another pertinent analyst of the Russian approach to cyber technology argues that whereas Russia did not invent today's computer technology, Russians are typically the first to theorize its possible use – and abuse.[78] This pattern matches the findings of Adamsky who examined how the Soviet military brass approached the original Revolution in Military Affairs within NATO.[79] To this, one could add the aggressive postmodern media policy of the Kremlin within Russia as vividly described by the British-Russian TV journalist Peter Pomerantsev.[80] These deliberations about the preference for means and methods drawn from the Russian intelligence community lead to the conclusion that Russian strategic culture is also a pervasive and Manichean *intelligence* culture.[81] Your author belongs to this approach.[82]

Consensus prevails on the point that Putin's coming into power dramatically advanced the power of the various specialized Russian intelligence branches – occasionally misnamed *siloviki* or "men of power," including the military. Better used is the term *chekisty,* which is the precise Russian term as it harkens back to Lenin's dreaded counterespionage body known under the acronym Cheka. Putin happily employs the term "chekist" about himself and his colleagues. They actively cherish the legacy of the Cheka boss Dzerzhinsky, for example in the context of *Rosgvardiya*, the Russian National Guard.[83] So to be more precise, Russian strategic culture is a *culture of counterintelligence* in the ominous sense of linking the domestic struggle against terrorism, extremism, and colour revolutions with presumable designs by Western intelligence agencies, for instance, concerning Russian opposition figures like Aleksei Navalny.[84] Hereby the signature Russian high-context strategic culture plays out as the conviction that hostile foreign secret services are immensely powerful, cunning, and successful at that – an obvious syndrome of mirror imaging.

The Putin leadership's cherishing of the memory from its old KGB boss Andropov is also no coincidence as his less brutal repression and gentler Soviet times fit into the formative youth experience of the Putin age cohort, as argued in a lengthy study of the mentality or operational code of Putin's power coalition.[85] However, the author of this fine study views

Putin's regime as a case of plain authoritarianism, whereas your author here deems it more precise to label Putin's Russia as a counterintelligence state, a concept used by the intelligence historian Robert W. Pringle about the Andropov era.[86] The term was invented by another academic intelligence analyst, Michael J. Waller, who wrote: "the counterintelligence state is characterized by a large elite force acting as watchdog of security. This apparatus is not accountable to the public and enjoys immense political powers."[87] The concept that captures his definition is autonomy or lacking checks and balances on the intelligence organs. Some might counterargue that the uniquely strong Russian office of the president turns this institution into the true principal; a role Putin insists upon fulfilling, as powerfully argued by Herd.[88] But if anything, Putin personifies the idiosyncratic Russian intelligence culture and cannot be considered above its peculiar groupthink and obsession with security, nor can another of his key chekist decisionmakers, hawk and confidante Nikolai Patrushev.[89]

In other words, your author here takes a methodological cue from Lantis and Howlett, who emphasize the need to identify who are the *keepers* of strategic culture.[90] This could be pushed further into a discussion of the tricky but vital *agent-principal* problem within any strategic culture. On this account, positions differ as several analysts, for good reasons, insist on the Russian General Staff – notably General Valery Gerasimov – as a keeper bordering on principal or agenda-setter for Russian strategic culture, in relation to whom all others are mere agents or executors of policy. This exemplifies Jonsson's themes drawing on Covington.[91] True, Covington calls the General Staff the brain of the army, but later on adds that *Rosgvardiya* is intended to strengthen the overlap between the military and the country's internal security.[92] He portrays the outcome of this "Bermuda triangle" consisting of Putin, *Rosgvardiya*, and the General Staff as one that "can produce extremely distorted views of political and military reality."[93] Hence Covington concludes his analysis in terms of "the unrealities of President Putin's worldview."[94] Still others equally insist on the loss of control by Putin, the principal, to his autonomous yet mutually jealous intelligence agencies.[95] The bizarre decisionmaking and execution of the invasion of Ukraine in 2022 discussed above underscores the need for further research into agent-principal dynamics within Russian security policy.

Concluding Thoughts

Admittedly, the above brief inquiry into who are the keepers of today's Russian strategic culture by trying to distinguish principals from mere agents proceeding from counterintelligence state theory may seem to end where any analysis of Soviet nuclear strategy was deemed to end according to Gray – namely without having detected a "smoking gun."[96] Even so, taking on the issue is just as vital for the field of strategic culture as raising the issue of operational *casus belli* concerning conventional and nuclear war. In other words, any study of strategic culture ought to be not only holistic – i.e., open-minded to what actually matters – but also fairly focused and not shying away from security policy relevance, on the contrary. This academic ethos more or less permeates all works commented upon above, although some of them may have drawn premature conclusions about Russia overcoming its deeply militaristic and Manichean Soviet strategic culture. Accordingly, once Leites reconstructed the operational code inherent in Leninism and Stalinism,[97] he paved the way for something akin to a cumulative research effort whose collective finding is the many Soviet atavisms in twenty-first-century post-Soviet Russian strategic culture.

One such Soviet atavism is the Kremlin's Leninist approach to agency, even if it is no longer only the Party that is bestowed with agency. Nowadays, it is foreign intelligence services and their US principals that are considered almighty, whereas angry citizens taking to the streets are considered devoid of capacity for acting on their own in defense of their interests – devoid of agency, in short. In other words, the field rightly stresses elements of continuity at the expense of change in the Kremlin's strategic culture. Another Soviet atavism is found at the level of grand strategy or ends, namely the emphasis on regime security, not broad national security. On top of this comes a relentless pursuit of great power status; a grand strategy that turns the superpower into Russia's strategic benchmark, the Kremlin's Cold War mindset notwithstanding. What is novel is the Russian nuclear orthodoxy and the overlap with Chinese strategic culture at the level of presidential operational codes. Last but not least, the means to achieving these grand strategic ends have been technologically updated to the digital era for purposes of aggressive cyber and information warfare. This represents another Soviet atavism: the willing resort to today's inventory of active measures. It is a strategic culture resting on a surviving Cold War mentality. All such Soviet atavisms matter as they make Russian strategic culture – whether in the military "way of war" sense or more broadly – different from Western strategic cultures. Ultimately, they matter for the existential issue of war or peace with Russia.

Notes

1 Jack L. Snyder, *The Soviet Strategic Culture: Implications for Limited Nuclear Operations*. A Project AIR FORCE report prepared for the United States Air Force, *Rand,* Santa Monica, R-2154-AF September 1977: 8.
2 John Glenn, Darryl Howlett, and Stuart Poore (Eds.), *Neorealism Versus Strategic Culture*. Aldershot & Burlington, VT: Ashgate, 2004.
3 Nathan Constantin Leites, *The Operational Code of the Politburo*. Rand Corporation, Santa Monica, 1951/2007. https://www.rand.org/pubs/commercial_books/CB104-1.html
4 Snyder, *Soviet Strategic Culture*. See also Johnson, Jeannie L., Kerry M. Kartchner, and Jeffrey A. Larsen (2009), *Strategic Culture and Weapons of Mass Destruction: Culturally Based Insights Into Comparative National Security Policy Making*. New York: Palgrave Macmillan.
5 Colin S. Gray, "Strategic Culture as Context: The First Generation Theory Strikes Back," *Review of International Studies,* Vol. 25, No. 1, 1999: 49–69. See also Jeffrey S. Lantis and Darryl Howlett, "Strategic Culture," in John Baylis, James J. Wirtz, and Colin S. Gray (Eds.), *Strategy in the Contemporary World*, 3rd ed. Oxford: Oxford University Press, 2013, 76–95.
6 Gray, "Strategic Culture as Context."
7 Mette Skak, "Russian Strategic Culture: The Generational Approach and the Counterintelligence State Thesis," in Roger E. Kanet (Ed.), *Routledge Handbook of Russian Security*, London: Routledge, 2019, 109–118.
8 Alastair Iain Johnston, "Thinking about Strategic Culture," *International Security,* Vol. 19, No. 4, 1995: 36–43.
9 Gray, "Strategic Culture as Context," 68.
10 George Kennan, "Telegram, George Kennan to George Marshall ['Long Telegram']," 22 February 1946. Harry S. Truman Administration File, *Elsey Papers*: 8, 11 ff.
11 Joris Van Bladel, "The Dual Structure and Mentality of Vladimir Putin's Power Coalition. A Legacy for Medvedev," *FOI Swedish Defence Research Agency,* FOI-R-2519-SE, May 2008.
12 Leites, *Operational Code*, 17.
13 Leites, *Operational Code,* 76.
14 Leites, *Operational Code,* 7. See also Stephen J. Blank, "Class War on a Global Scale. The Leninist Culture of Political Conflict," in idem et al. (Eds.), *Conflict, Culture, and History. Regional Dimensions.* Alabama: Air University Press, Maxwell Air Force Base, 1993, 1–55.

15 Oscar Jonsson, *The Russian Understanding of War: Blurring the Lines between War and Peace.* Washington, DC: Georgetown University Press, 2019.
16 Leites, *Operational Code,* 1.
17 Gray, "Strategic Culture as Context," 56.
18 Leites, *Operational Code,* 3.
19 Mette Skak, "Russian Strategic Culture: The Role of Today's *Chekisty,*" *Contemporary Politics,* Vol. 22, No. 3, 2016: 324–341; Skak, "Russian Strategic Culture: The Generational," 2019.
20 Leites, *Operational Code,* 79.
21 Vladimir V. Putin, "Being Strong: National Security Guarantees for Russia," *Rossiiskaya Gazeta,* 20 February 2012. http://www.voltairenet.org/article172934.html
22 Michael Kofman, "Raiding and International Brigandry: Russia's Strategy for Great Power Competition," *War on the Rocks,* 14 June 2018. https://warontherocks.com/2018/06/raiding-and-international-brigandry-russias-strategy-for-great-power-competition/
23 Graeme P. Herd, "Putin's Operational Code and Strategic Decision-Making in Russia," in Roger E. Kanet (Ed.), *Routledge Handbook of Russian Security.* London: Routledge, 2019, 17–29.
24 Martti J. Kari, "Russian Strategic Culture in Cyberspace: Theory of Strategic Culture – A Tool to Explain Russia's Cyber Threat Perception and Response to Cyber Threats," *JYU Dissertations 122, University of Jyväskylä,* 2019. http://urn.fi/URN:ISBN:978-951-39-7837-2
25 Snyder, *Soviet Strategic Culture.*
26 Brian D. Taylor, *Politics and the Russian Army: Civil-Military Relations, 1689–2000.* Cambridge, UK: Cambridge University Press, 2003.
27 John Glenn, "Russia," pp. 173–203 in John Glenn, Darryl Howlett, and Stuart Poore (Eds.), *Neorealism Versus Strategic Culture.* Aldershot & Burlington, VT: Ashgate, 2004, 173–203; Henrikki Heikka, *Beyond the Cult of the Offensive: The Evolution of Soviet/Russian Strategic Culture and Its Implications for the Nordic-Baltic Region.* Helsinki: Ulkopoliittinen Instituutti & Institut für Europäische Politik, 2000.
28 Blank, "Class War on a Global Scale"; Robert G. Hermann, "Identity, Norms and National Security: The Soviet Foreign Policy Revolution and the End of the Cold War," in Peter J. Katzenstein (Ed.), *The Culture of National Security: Norms and Identity in World Politics.* New York: Columbia University Press, 1996, 271–326.
29 Andrew Bennett, *Condemned to Repetition? The Rise, Fall, and Reprise of Soviet-Russian Military Interventionism, 1993–1996.* Massachusetts: MIT Press, 1999.
30 Fritz W. Ermarth, "Russian Strategic Culture in Flux. Back to the Future?," In Kerry M. Kartchner, Jeffrey A. Larsen, and Jeannie L. Johnson (Eds.), *Strategic Culture and Weapons of Mass Destruction: Culturally Based Insights Into Comparative National Security Policy Making.* New York: Palgrave Macmillan, 2009, 85–96.
31 Robert W. Pringle, "Andropov's Counterintelligence State," *International Journal of Intelligence and Counterintelligence,* Vol. 13, No. 2, 2010: 193–203.
32 Hikka, *Beyond the Cult of the Offensive.*
33 Alexandr Golts, *Military Reform and Militarism in Russia.* The Jamestown Foundation, 2019.
34 Herd, "Putin's Operational Code," 25f.
35 Graeme P. Herd, *Understanding Russian Strategic Behavior: Imperial Strategic Culture and Putin's Operational Code.* Routledge: Contemporary Security Studies, 2022, 34–35. See also Sergei Medvedev, *The Return of the Russian Leviathan.* Cambridge, UK: Polity Press, 2022, 90 ff.
36 Fiona Hill and Angela Stent, "The World Putin Wants: How Distortions About the Past Feed Delusions About the Future," *Foreign Affairs,* September/October, accessible at: https://www.foreignaffairs.com/russian-federation/world-putin-wants-fiona-hill-angela-stent.
37 Putin, "'Taking It Back': Vladimir Putin Likens Self to Peter the Great," *Al Jazeera,* 9 June 2022, accessible at: https://www.aljazeera.com/news/2022/6/9/putin-likens-self-to-tsar-hints-at-russian-territory-expansion. See also the Social Darwinian and premature announcement of victory that was soon removed by Russia from the internet: Petr Akopov, "Nastuplenie Rossii i novogo mira," *RIA Novosti,* 26.02.2022, accessible at https://web.archive.org/web/20220226051154/https:/ria.ru/20220226/rossiya-1775162336.html.
38 Jack Watling and Nick Reynolds, "Operation Z: The Death Throes of an Imperial Delusion." *Royal United Services Institute,* Special Report, 22 April 2022.

39 Lawrence Freedman, "Why War Fails: Russia's Invasion of Ukraine and the Limits of Military Power." *Foreign Affairs,* July/August 2022, accessible at: https://www.foreignaffairs.com/articles/russian-federation/2022-06-14/ukraine-war-russia-why-fails. See also Watling and Reynolds, "Operation Z."
40 Kimberly Marten, "President Putin's Rationality and Escalation in Russia's Invasion of Ukraine, *PONARS Eurasia Policy Memo* no. 756, 9 March 2022. Accessible at: https://www.ponarseurasia.org/president-putins-rationality-and-escalation-in-russias-invasion-of-ukraine.
41 Freedman, "Why War Fails," See also Hill and Stent, "The World Putin Wants."
42 See also Watling and Reynolds, "Operation Z."
43 Gray, "Strategic Culture as Context,": 65f.
44 Mark Kramer, "Russia Is Repeating Its Brutal History in Ukraine," *Cognoscenti.* Commentary, 4 May 2022. https://www.wbur.org/cognoscenti/2022/05/04/russians-invasion-ukraine-overtones-wwii-mark-kramer.
45 Bennett, *Condemned to Repetition?*
46 Bettina Renz, *Russia's Military Revival.* Cambridge, UK: Polity Press, 2018, 63.
47 Michael Kofman and Matthew Rojansky, JD, "What Kind of Victory for Russia in Syria?," *Military Review,* March–April, 2018: 6–23, https://www.armyupress.army.mil/Journals/Military-Review/Online-Exclusive/2018-OLE/Russia-in-Syria/.
48 Bruno Tertrais, "Russia's Nuclear Policy: Worrying for the Wrong Reasons," *Survival,* Vol. 60, No. 2, 2018: 42.
49 Cynthia A. Roberts, "The Czar of Brinkmanship: A Classic Cold War Strategy Makes a Comeback in the Kremlin," *Foreign Affairs* (essay?) Monday, 5 May 2014.
50 Sergey Shoigu, "Minister: Russian Operation in Syria Stopped Chain of Colour Revolutions in Middle East," 21 February 2017. *TASS, Russian News Agency.* https://tass.com/politics/932137. See also Jonsson, *The Russian Understanding of War.*
51 Steven R. Covington, "The Culture of Strategic Thought Behind Russia's Modern Approaches to Warfare." *Belfer Center for Science and International Affairs at Harvard Kennedy School, Paper,* October 2016: 23f.
52 Renz, *Russia's Military Revival,* 120.
53 Aleksandr Golts, "Росгвардия поднимает генштаб" (Rosgvardia gets the upper hand over the General Staff, *The New Times,* No. 19–20 (446), 05.06.2017. https://newtimes.ru/articles/detail/116432
54 Pavel Shchelin, "Russian National Security Strategy: Regime Security and Elite's Struggle for 'Great Power' Status," *Slovo,* Vol. 28, No. 2, Spring, 2016: 85–105.
55 Covington, "The Culture of Strategic Thought Behind Russia's Modern Approaches," 43.
56 Kofman and Rojansky, "What Kind of Victory for Russia in Syria?"
57 Covington, "The Culture of Strategic Thought"; Jonsson, *The Russian Understanding of War.*
58 Jonsson, *The Russian Understanding,* 15–16.
59 Ibid., 4.
60 Blank, "Reflections on Russian Nuclear Strategy," in Roger E. Kanet (Ed.), *Routledge Handbook of Russian Security,* 2019, 154–168.
61 Tertrais, "Russia's Nuclear Policy: Worrying for the Wrong Reasons."
62 Petr Topychkanov, "Is Russia Afraid of Chinese and Indian Missiles?," *Carnegie Moscow Center,* 3 November 2014.
63 Dima Adamsky, *Russian Nuclear Orthodoxy: Religion, Politics and Strategy,* Stanford, CA: Stanford University Press, 2019.
64 Lantis and Howlett, "Strategic Culture," 96f.
65 Adamsky, *Russian Nuclear Orthodoxy,* 235.
66 Adamsky, *Russian Nuclear Orthodoxy,* 10.
67 Gilbert Rozman, *The Sino-Russian Challenge to the World Order: National Identities, Bilateral Relations and East Versus West in the 2010s.* Washington, DC: Woodrow Wilson Center Press & Stanford University Press, 2014, 274.
68 Rozman, *The Sino-Russian Challenge to the World Order,* 275. See also Herd, "Putin's Operational Code."
69 Ermarth, "Russian Strategic Culture in Flux: Back to the Future?"

70 Veli-Pekka Tynkkynen, *The Energy of Russia: Hydrocarbon Culture and Climate Change.* Cheltenham, UK & Northampton, MA: Edward Elgar, 2019.
71 Gray, "Strategic Culture as Context," 65f.
72 Alexey Levinson, "Why Is the Kremlin So Committed to This War?," 14 June 2022, *Meduza.* https://meduza.io/en/feature/2022/06/14/why-is-the-kremlin-so-committed-to-this-war.
73 Yehuda Z. Blum, "Russia Takes Over the Soviet Union's Seat at the United Nations," *European Journal of International Law,* Vol. 3, No. 2, August 1992, 354–361.
74 Thomas Rid, *Active Measures: The Secret History of Disinformation and Political Warfare.* London: Profile Books, 2020.
75 Rid, *Active Measures. The Secret History of Disinformation and Political Warfare,* 377–409.
76 Jonsson, *The Russian Understanding of War,* 93–123.
77 Kofman, "Raiding and International Brigandry: Russia's Strategy for Great Power Competition," *War on the Rocks,* 14 June 2018. https://warontherocks.com/2018/06/raiding-and-international-brigandry-russias-strategy-for-great-power-competition/
78 James J. Wirtz, "Cyber War and Strategic Culture: The Russian Integration of Cyber Power into Grand Strategy," Chapter 3 in Kenneth Geers (Ed.), *Cyber War in Perspective: Russian Aggression against Ukraine.* NATO CCD COE Publications, Tallinn, 2015. Available as pdf: https://ccdcoe.org/uploads/2018/10/Ch03_CyberWarinPerspective_Wirtz.pdf
79 Dima Adamsky, *The Culture of Military Innovation. The Impact of Cultural Factors on the Revolution in Military Affairs in Russia, the US, and Israel.* Stanford, CA: Stanford University Press, 2010.
80 Peter Pomerantsev, *Nothing Is True and Everything Is Possible: Adventures in Modern Russia.* London: Faber & Faber, 2015.
81 Kimberly Marten, "The 'KGB State' and Russian Foreign Policy Culture," *Journal of Slavic Military Studies,* Vol. 30, No. 2, 2017, 131–151; Kimberly Marten, "The Intelligence Agencies and Putin: Undermining Russia's Security?" in Roger E. Kanet (Ed.), *Routledge Handbook of Russian Security,* Abingdon: Routledge, 2019, 192–202; Brian D. Taylor, *State Building in Putin's Russia: Policing and Coercion after Communism.* Cambridge, UK: Cambridge University Press, 2011.
82 Skak, "Russian Strategic Culture: The Role of Today's *Chekisty";* Skak, "Russian Strategic Culture: The Generational Approach."
83 Aleksandr Golts, "Росгвардия поднимает генштаб" (Rosgvardia Gets the Upper Hand over the General Staff, *The New Times,* No. 19–20 (446), 05.06.2017. https://newtimes.ru/articles/detail/116432
84 Paul Goble, "Putin's Speech to Chekists Highlights His Increased Focus on Domestic Opponents and Foreign Enemies," *Window on Eurasia,* Staunton, 25 February 2021. http://windowoneurasia2.blogspot.com/2021/02/putins-speech-to-chekists-highlights.html
85 Van Bladel, "The Dual Structure and Mentality of Vladimir Putin's Power Coalition."
86 Pringle, "Andropov's Counterintelligence State."
87 Quoted from Skak, "Russian Strategic Culture: The Generational Approach," 114.
88 Herd, "Putin's Operational Code."
89 Skak, "Russian Strategic Culture: The Role Of Today's *Chekisty";* Herd, "Putin's Operational Code"; Mark Galeotti, "New National Security Strategy Is a Paranoid's Charter," *The Moscow Times,* 5 July 2021. https://.www.themoscowtimes.com/2021/07/05/new-national-security-strategy-is-a-paranoid-charter-a74424
90 Lantis and Howlett, "Strategic Culture."
91 Jonsson, *The Russian Understanding of War,* 18.
92 Covington, "The Culture of Strategic Thought," 3 vs. 24.
93 Ibid., 24.
94 Ibid., 46.
95 Mark Galeotti, "Putin's Hydra: Inside Russia's Intelligence Services," *European Council on Foreign Relations,* 11 May 2016. See also Herd, "Putin's Operational Code"; and Marten, "The Intelligence Agencies and Putin."
96 Jonsson, *The Russian Understanding,* 18.
97 Leites, *Operational Code.*

Selected Bibliography

Adamsky, Dima. *The Culture of Military Innovation. The Impact of Cultural Factors on the Revolution in Military Affairs in Russia, the US, and Israel.* Stanford, CA: Stanford University Press, 2010.

Adamsky, Dima. *Russian Nuclear Orthodoxy: Religion, Politics, and Strategy*. Stanford, CA: Stanford University Press, 2019.

Bennett, Andrew. *Condemned to Repetition? The Rise, Fall, and Reprise of Soviet-Russian Military Interventionism, 1993–1996*. Massachusetts: MIT Press, 1999.

Blank, Stephen J. "Class War on a Global Scale: The Leninist Culture of Political Conflict," pp. 1–55 in idem et al. (eds.) *Conflict, Culture, and History. Regional Dimensions*. Alabama: Air University Press, Maxwell Air Force Base, 1993, pp. 1–55.

Blank, Stephen J. "Reflections on Russian Nuclear Strategy," in Roger E. Kanet (ed.) London: Routledge Handbook of Russian Security, 2019, pp. 154–168.

Covington, Steven R. "The Culture of Strategic Thought Behind Russia's Modern Approaches to Warfare. *Belfer Center for Science and International Affairs at Harvard Kennedy School, Paper*, October 2016.

Ermarth, Fritz W. "Russian Strategic Culture in Flux: Back to the Future?," in Kerry M. Kartchner, Jeffrey A. Larsen and Jeannie L. Johnson, *Strategic Culture and Weapons of Mass Destruction: Culturally Based Insights Into Comparative National Security Policy Making*. New York: Palgrave Macmillan, 2009, pp. 85–96.

Galeotti, Mark. "Putin's Hydra: Inside Russia's Intelligence Services," *European Council on Foreign Relations*, 11 May 2016.

Glenn, John. "Russia," in John Glenn, Darryl Howlett and Stuart Poore (eds.). *Neorealism Versus Strategic Culture*. Aldershot & Burlington, VT: Ashgate, 2004, pp. 173–203.

Golts, Alexandr. *Military Reform and Militarism in Russia*. The Jamestown Foundation, 2019.

Heikka, Henrikki. *Beyond the Cult of the Offensive: The Evolution of Soviet/Russian Strategic Culture and Its Implications for the Nordic-Baltic Region*. Helsinki: Ulkopoliittinen Instituutti & Institut für Europäische Politik, 2000.

Herd, Graeme P. "Putin's Operational Code and Strategic Decision-Making in Russia," in Roger E. Kanet (ed.). *Routledge Handbook of Russian Security*. London: Routledge, 2019, pp. 17–29.

Herd, Graeme P. *Understanding Russian Strategic Behavior: Imperial Strategic Culture and Putin's Operational Code*. Routledge: Contemporary Security Studies, 2022.

Hermann, Robert G. "Identity, Norms and National Security: The Soviet Foreign Policy Revolution and the End of the Cold War," in Peter J. Katzenstein (ed.). *The Culture of National Security. Norms and Identity in World Politics*. New York: Columbia University Press, 1996, pp. 271–326.

Hill, Fiona and Angela Stent. "The World Putin Wants: How Distortions About the Past Feed Delusions the Future." *Foreign Affairs*. September/October 2022. https://foreignaffairs.com/russian-federation/world-putin-wants-fiona-hill-angela-stent

Jonsson, Oscar. *The Russian Understanding of War: Blurring the Lines between War and Peace*. Washington, DC: Georgetown University Press, 2019.

Kramer, Mark. "Russia Is Repeating Its Brutal History in Ukraine," *Cognoscenti. Commentary*. 4 May 2022. https://www.wbur.org/cognoscenti/2022/05/04/russians-invasion-ukraine-overtones-wwii-mark-kramer

Leites, Nathan Constantin. *The Operational Code of the Politburo*. Santa Monica: Rand Corporation, 1951/2007. Available at https://www.rand.org/pubs/commercial_books/CB104-1.html

Marten, Kimberly. "The 'KGB State' and Russian Foreign Policy Culture," *Journal of Slavic Military Studies*, Vol. 30, No. 2 (2017), pp. 131–151.

Marten, Kimberly. "The Intelligence Agencies and Putin: Undermining Russia's Security?," in Roger E. Kanet (ed.). *Routledge Handbook of Russian Security*. London: Routledge Publishers, 2019, pp. 192–202.

Pomerantsev, Peter. *Nothing Is True and Everything Is Possible: Adventures in Modern Russia*. London: Faber & Faber, 2015.

Pringle, Robert W. "Andropov's Counterintelligence State," *International Journal of Intelligence and Counterintelligence*, Vol. 13, No. 2 (2010), pp. 193–203.

Renz, Bettina. *Russia's Military Revival*. Cambridge, UK: Polity Press, 2018.

Rid, Thomas. *Active Measures: The Secret History of Disinformation and Political Warfare*. London: Profile Books, 2020.
Rozman, Gilbert. *The Sino-Russian Challenge to the World Order: National Identities, Bilateral Relations and East Versus West in the 2010s*. Washington, DC: Woodrow Wilson Center Press & Stanford University Press, 2014.
Shchelin, Pavel. "Russian National Security Strategy: Regime Security and Elite's Struggle for 'Great Power' Status," *Slovo*, Vol. 28, No. 2 (Spring 2016), pp. 85–105.
Skak, Mette. "Russian Strategic Culture: The Role of Today's *Chekisty*," *Contemporary Politics*, Vol. 22, No. 3 (2016), pp. 324–341.
Skak, Mette. "Russian Strategic Culture: The Generational Approach and the Counterintelligence State Thesis," in Roger E. Kanet (ed.). *Routledge Handbook of Russian Security*. London: Routledge, 2019, pp. 109–118.
Snyder, Jack L. *The Soviet Strategic Culture: Implications for Limited Nuclear Operations*. A Project AIR FORCE report prepared for the United States Air Force, *Rand*, Santa Monica, R-2154-AF, September 1977.
Taylor, Brian D. *Politics and the Russian Army: Civil-Military Relations, 1689–2000*. Cambridge, UK: Cambridge University Press, 2003.
Taylor, Brian D. *State Building in Putin's Russia: Policing and Coercion after Communism*. Cambridge, UK: Cambridge University Press, 2011.
Tertrais, Bruno. "Russia's Nuclear Policy: Worrying for the Wrong Reasons," *Survival*, Vol. 60, No. 2 (2018), 33–44.
Tynkkynen, Veli-Pekka. *The Energy of Russia: Hydrocarbon Culture and Climate Change*. Cheltenham, UK & Northampton, MA: Edward Elgar, 2019.
Van Bladel, Joris. "The Dual Structure and Mentality of Vladimir Putin's Power Coalition: A Legacy for Medvedev," *FOI Swedish Defence Research Agency*, FOI-R-2519-SE, May 2008.
Watling, Jack and Nick Reynolds. "Operation Z: The Death Throes of an Imperial Delusion." *Royal United Services Institute*, Special Report, 22 April 2022.
Wirtz, James J. "Cyber War and Strategic Culture: The Russian Integration of Cyber Power into Grand Strategy," Chapter 3 in Kenneth Geers (ed.). *Cyber War in Perspective: Russian Aggression against Ukraine*. Tallinn: NATO CCD COE Publications, 2015. Available at: https://ccdcoe.org/uploads/2018/10/Ch03_CyberWarinPerspective_Wirtz.pdf

13
CHINA'S IDENTITY THROUGH A HISTORICAL LENS[1]

Neil Munro

Introduction

The purpose of this chapter is to give a simple description of China's identity by summarizing salient features of its history and relating them to current issues in great power competition. It does not characterize China's strategic culture in general terms, a task that has been ably attempted elsewhere.[2] The chapter does not take an either/or position in the debates over whether China's strategic culture is offensive or defensive, realist or driven by ideas. Whether it is offensive or defensive depends on the situation, and the conclusions of practitioners who use the strategic culture concept suggest that, while realism is the core of Chinese foreign policy, ideas are the flesh that covers the bones.[3]

The importance of understanding China has never been greater, particularly for military and diplomatic leaders of the world's preeminent power, the United States. In part, this is due to rising tension in the bilateral relationship, where terms like "strategic competition" and "rivalry" increasingly displace "partnership" or "cooperation." In part, this is due to the West's relative ignorance of China, compared to China's understanding of the West. Popular Western understandings of China are tainted by the influence of previous generations of writers who, in the service of various imperial projects, constructed the East as exotic, effeminate, and dangerous.[4] This leads to two common mistakes. The first is to demonize China, regarding everything Chinese with suspicion, skepticism, fear, or mistrust. The second is to idealize it, treating Chinese knowledge as a source of special insights and taking too seriously some of the things that the Chinese like to say about themselves, such as "China seeks a harmonious world." In part, the need to understand China better comes from the brute fact of China's rise: its gross domestic product (GDP), which is second only to that of the United States, or bigger if one measures it in purchasing power parities; and its military capabilities, which while still less impressive than those of the United States and Russia are on a rising trajectory.[5]

This chapter takes a strategic culture approach.[6] The author is concerned with describing the key historical events that formed China's identity. Identity is defined as "[the] nation-state's view of itself, comprising the traits of its national character, its intended regional and global roles, and its perceptions of its eventual destiny."[7] The international relations approach closest to strategic culture is constructivism, which problematizes the formation and transformation of state interests and provides explanations

DOI: 10.4324/9781003010302-16

for them in terms of historical processes of identity formation.[8] As Jeannie L. Johnson points out, "Values weighed by a rational actor in a cost/benefit analysis are often ideational as well as material and cannot be accurately assessed without a substantive knowledge of the actor's preferences."[9] Therefore, being equipped with a rational mind and a set of internationally transferable assumptions about state behavior is often insufficient. Strategists need to ground such assumptions in a deep understanding of the identity of the actor.

China's identity is the outcome of a series of tensions emerging from its history. China has risen as a great power in the modern world after taking several wrong turns and experiencing what it describes as a "Century of Humiliation." Along the way, tensions have emerged between feelings of superiority and inferiority, between the needs for development and equality, between demands for freedom and order, and between China's territorial ambitions and geopolitical reality. John Gerard Ruggie suggests a conception of time as "different temporal forms that bring deeper and wider 'presents' into view" and a conception of space as a "social construct that people, somehow, invent ... [and which] generates emergent properties of its own."[10] Seen in this light, China has its unique history, but there is no mysterious essence that one must have spent decades in the country to grasp.

The structure of this chapter is chronological, following the broad outlines of Chinese history over the past century and a half, before opening out into a discussion of current geopolitical issues and concluding with a characterization of the tensions underlying China's identity.

China's Inferiority-Superiority Complex

China is driven to be an overachiever. Iver B. Neumann writes that "if Russia had an inferiority complex towards Europe in 1991, a quarter-century down the road that has been inverted into a superiority complex."[11] Neumann's starting point is that all states have a need for recognition and that citizens' beliefs determine the grounds on which recognition may be sought. China is different from Russia in that instead of a kind of defensive pride, China continues to have feelings of both superiority and inferiority simultaneously. Psychologists define "subjective overachievement" as the co-occurrence of self-doubt and anxiety over performance, which drives an individual to exert extra effort, leading to better results than expected.[12] Like the straight-A high school student who lacks popularity but works harder than their peers and eventually ends up with a much better income, China has made it in the material sense.

However, China's feeling of superiority does not rest on GDP alone. As its diplomats never tire of reminding foreign journalists, Chinese civilization is 5,000 years old. It is an exaggerated claim, since not much is known about the first 2,000 years, and there were several long periods when China was split into multiple states or ruled by foreign dynasties. Nevertheless, there is a remarkable degree of cultural continuity, owing in part to the use of ideograms, which make even very ancient texts intelligible.[13] Admiral Zheng He's voyages around Southeast Asia and the Indian Ocean at the beginning of the fifteenth century demonstrated China's interest in the outside world. His ships dwarfed the one Columbus would sail to America 90 years later. Crucially, however, the purpose of the voyages was to "placate and moralize" rather than trade and conquer, and the Sino-centric tribute system demanded that Zheng He should confer gifts from the Ming emperor, thus gaining face and establishing obligation, rather than demanding trade or other concessions.[14]

Confucianism, China's traditional system of ethics, values social stability through hierarchy, and therefore what mattered in international relations was the establishment of a pecking order. The maps of the world prepared by Jesuits at the Ming court 200 years later prove that at the highest level, at least, China's rulers were aware of the size and shape of the major continents, even if the zest for expensive voyages had faded.[15] However, the next dynasty, the Qing, turned to a policy of active self-isolation, motivated by the fear that southern China, which had seen large-scale rebellions in support of the Ming, would become too prosperous if allowed to trade freely, creating alternative power centers.[16] Security concerns thus led the Qing to restrict foreign trade to just one guild, known in English as the Cohong (from the Chinese *Gonghang*), based in Guangzhou (Canton), in the far south of the country.

Humiliation and Glory

China's sense of inferiority comes from the Century of Humiliation beginning with the First Opium War from 1839 to 1842. Provoked by Chinese attempts to curtail the trade in opium, the British sent a fleet of 42 ships, including HMS *Nemesis*, Britain's first oceangoing iron warship. The Chinese had only swords, spears, primitive muskets, and seventeenth-century cannon with which to repel attacks by long-range naval artillery. The fact that they fortified Guangzhou while leaving other ports vulnerable showed a basic lack of understanding of how to fight wars at sea. The British had the ability to transport troops quickly along China's coast and the steam-powered *Nemesis* was able to maneuver in the shallow waters of Chinese rivers. The Qing dynasty's lack of preparation and strategic ignorance were not fully analyzed in China until 1995, when Mao Haijian published *Tianchao de Bengkui* (*Collapse of the Celestial Empire*).[17] The outcome of the war forced the Qing to abolish the Cohong, open five ports to international trade, accept permanent diplomatic envoys, pay an indemnity, cede Hong Kong in perpetuity, provide extraterritoriality for British subjects, fix import tariffs, and provide a most-favored nation clause to Britain.[18] Whatever Britain received, the United States and France also demanded.

The First Opium War set a pattern: the presentation of unreasonable demands, swift violence from the foreign powers, and the signing of an unequal treaty obliging the Chinese to make concessions and pay reparations. The Second Opium War (1857–1860), the Sino-French War (1884–1885), the Sino-Japanese War (1894–1895), the suppression of the Boxer Rebellion (1899–1901), and the Second Sino-Japanese War (1937–1945) formed a continuing series of aggressions in the Chinese mind, all aimed at stripping China of its sovereignty and pillaging its wealth. In this period, modernization and industrialization were thrust on China by foreigners who saw the economic potential and wanted a piece of it, treating the Chinese as a colonized people. The sign "No dogs and no Chinese allowed!" which appears in Bruce Lee's 1972 film *Fist of Fury* may not have existed in the form it appears in the film, but for the first 60 years of its existence until 1928, Huangpu Park in Shanghai did have regulations banning the admission of Chinese unless they were police or servants accompanying a foreigner, as well as bans on dogs and bicycles.[19]

The end of China's civil war put an end to such humiliation, a turning of the tables best symbolized by People's Liberation Army (PLA) artillery crippling HMS *Amethyst* (F116) as the ship made its way up the Yangtze to relieve another British ship at Nanjing in the summer of 1949. Mao Zedong's speech to the Chinese People's Political

Consultative Conference in September that year – including the famous sentence "the Chinese people have stood up!" – celebrated victory over the Japanese, the European imperialist powers, and the Nationalist Kuomintang (KMT) party, but it also warned of the need for continuing vigilance against "reactionaries."[20] Thus, the "liberation" did not end internal strife, which continued hand in hand with the construction of the People's Republic of China (PRC). Mao envisaged a united front under the leadership of the working class, but in reality, the Chinese Communist Party (CCP) stood above all classes and Mao stood above the CCP.[21] From that point on, the CCP identified itself with China and its propaganda conflated the two. De jure and in practice, the PLA was and remains the CCP's army.

Development versus Equality

Deng Xiaoping's verdict on the founder of the PRC, echoing Mao's verdict on the Soviet Union, was that he was 70% good and 30% bad. In the same statement, Deng also said that China would never do to Mao what the Soviet Union had done to Stalin. The refusal to completely repudiate past leaders is an important feature of CCP ideology, keeping the party anchored to its past and limiting the range of possible futures. The "30%" is a terse admission of the suffering that Mao had inflicted to build a basic command economy. Through the Great Leap Forward, Mao tested two great *idées fixes*: that man's will rather than objective social and economic laws is the most important force in history, and that the undeveloped consciousness of the peasants conferred an advantage because their minds were like a blank sheet of paper. Mao failed to consider overreporting, a side effect of his absolute power, which meant that grain harvest statistics were inaccurate and too much food was taken out of the countryside to fund industrialization. Compounded with natural disasters, the Great Leap Forward caused a famine costing about 30 million lives between the spring of 1959 and the end of 1961.[22]

After a decisive break with the Soviet Union, which was perceived as taking too soft a line with the West, Mao applied the same idea of blankness to the youth, turning them into Red Guards and using them to attack the political and social elites, whom he perceived as corrupt and wavering in ideological commitment.[23] In the Cultural Revolution, thousands of intellectuals and officials were beaten to death and millions of city dwellers were sent into the countryside to work on farms. When Red Guard factions started fighting one another, Mao called in the army to restore order. After Mao's death, his wife, Jiang Qing, and three of her henchmen took the blame for the Cultural Revolution and were put on trial. In 1981, the so-called Gang of Four were all given long sentences and China made a decisive break with Mao's extreme leftism.

The debate within the CCP on Maoist ideas had focused on whether "relations of production" (class struggle) or "productive forces" (industrialization and technology) were the priority in building Communism. Mao's view was that fixing relations of production came first. When the Central Committee passed a resolution in 1958, attempting to soft pedal the Great Leap Forward, warning against "impetuous actions" and "utopian dreams" and reasserting that building Communism would take considerable time and could only be done after developing the productive forces, Mao was annoyed and the next year those who disagreed with him, including the Defense Minister Marshall Peng Dehuai, were purged as members of an "anti-Party clique."[24] Deng's reevaluation of Mao meant the return to power of those holding to more orthodox interpretations of Marxism-Leninism.

However, the world in 1979 did not look the same as the world in 1959. Undemocratic but capitalist states in East Asia had started their ascent to industrialized status.[25] In June 1981, the *People's Daily* carried an article entitled "Principal Problems of the Soviet Economy" in which the economist Lu Nanquan pointed out that while huge investment had helped build a sound industrial base, overreliance on this method of economic growth had reduced economic efficiency, resulting in sluggish economic growth.[26] A new assessment of Marxist-Leninist orthodoxy had begun.

Reform and Opening Up, as Deng's policies became known, delivered what China craved – rapid development and, at last, respect on the international stage. It was a case of "crossing the river by feeling for the stones," as the CCP did not have an established blueprint. Hence, Deng was praised for pragmatism and a gradual, decentralized approach whereby policy ideas were tried out in small areas before being scaled up. This created a pro-reform constituency, including enterprises and regions where policies had worked, and the non-state sectors of the economy demonstrated innovation and took up the slack when the state sector was eventually downsized.[27] The most important change in the early years was the introduction of the Household Responsibility System, which was a euphemism for de-collectivization: family farms replaced the people's communes.[28] Township and village enterprises and private enterprises began to account for a steadily increasing share of the value of industrial output.[29] Deng was happy to humor UK Prime Minister Margaret Thatcher and US President Ronald Reagan when they lauded him as a "market reformer." Yet, in ideological terms, he was far from liberalism, as his reinterpretation of Marxism involved the assertion that China was in the "primary stage of socialism."[30] In this stage, China would remain a dictatorship under the leadership of the CCP and its focus would be on economic development.

Freedom versus Order

Political and economic liberalism diffused into China, and a rift developed between those who wanted to move more quickly on the economy and even experiment with political reform and hardliners who wanted to stick closely to orthodox Marxism-Leninism. Protests broke out after the death of General Secretary Hu Yaobang in April 1989, a reformer, over a perceived failure by the party leadership to mourn him properly. Events escalated as students occupied Tiananmen Square and began to make diverse demands. At the end of May, Mikhail Gorbachev made an untimely visit, the first Sino-Soviet summit since the 1961 split, further increasing the pressure on the hardliners. On the night of 3–4 June, Deng gave the order to clear the square by force. To Western media, who were in the city to cover the summit, the narrative was clear: a pro-democracy movement had been crushed. Western governments applied sanctions and investors pulled out. Deng defended himself by saying it was a "counter-revolutionary rebellion" that was "bound to happen and was independent of man's will."[31] The "6–4 Incident," as the Chinese call it, showed the limits of political liberalization but also brought marketization into question. The collapse of Communism in Eastern Europe at the end of the year and the disintegration of the Soviet Union two years later stimulated deep reflection on what had gone wrong.[32] Deng's southern tour in 1992 bolstered his position against conservatives, and he was able to convince the CCP that rapid development was their only means of salvation.

From the crucible of these events, a mentality combining cynicism, materialism, and nationalism emerged among Chinese elites in the 1990s. Materialism was the obverse of

Communist ideology and reflected the zeitgeist of the previous decade.[33] Cynicism was a response to corruption resulting from the "commodification" of state power, disappointment with the outcomes of 1989, and loss of belief in Communism.[34] Chinese propagandists like to frame the growth of nationalism in this period as a reaction to repeated provocations by Western powers, specifically US talk about "containing" China, attempts to spread democracy through "peaceful evolution," and memories of the Century of Humiliation. Indeed, nationalistic books like *China Can Say "No!"* had huge commercial success.[35] However, the CCP also encouraged state-led nationalism, for example, through a "patriotic education campaign" in schools and universities.[36] Nationalism began to replace Communism as the basis for social solidarity.

Current Geopolitical Tensions

Chinese nationalism has a popular dimension. Citizens protested in 1999 against the accidental bombing by North Atlantic Treaty Organization (NATO) forces of the Chinese embassy in Belgrade, again in 2001 after a US spy plane collided with a Chinese jet near Hainan, causing the death of the Chinese pilot, and in 2005, 2010, and 2012 against Japan over various issues. The 2005 and 2012 protests included attacks on property and individuals. Official commemoration of past humiliation at the hands of foreign powers draws mass participation but also sometimes arouses skepticism.[37] There is little evidence that nationalism has ever gotten out of the CCP's control, or that the regime has ever felt pressured to modify its diplomatic stances in response to popular pressure. Participation in protest activity is predicted by social network diffusion.[38]

After their victory over the KMT in 1949, the CCP set out to build the *Zhonghua minzu* or "Chinese people" with the Han majority at its core. This involved exoticizing 55 ethnic minorities in order to assimilate them, to "recognize ethnic diversity into irrelevance" by conferring autonomous status on titular minority regions and various privileges on minorities, while simultaneously depriving them of their ability to self-organize.[39] This "first generation" ethnicity policy came under criticism after the Soviet collapse because it was perceived to have "politicized" ethnicity.[40] Protests in Tibet and Xinjiang, provoked by economic inequality and religious and identity issues, reinforced the regime's perception that the first generation policy was not working. In 2009, clashes between members of the Uighur nationality and Han Chinese in Xinjiang's capital, Urumqi, convinced the CCP that a new approach was needed for this region. Even though violence was perpetrated by both sides, the authorities blamed the Uighurs and resorted to totalitarian methods of suppression involving mass internment, intensified surveillance, indoctrination, and restrictions on religious practice. The solution found by the regime is tantamount to cultural genocide. Uighurs are included in the *Zhonghua minzu* but at the same time prevented from feeling part of it.[41]

Officially known as the Republic of China (ROC), Taiwan is the rump regime established by the KMT after they fled the mainland in 1949. The PRC regards it as a renegade province. China's Anti-Secession Law of 2005 commits China to pursue peaceful reunification, but, according to Article 8, in the event of "secession" or if the "possibilities of a peaceful reunification should be completely exhausted" China will use "non-peaceful means and other necessary measures to protect China's sovereignty and territorial integrity."[42] Although the adoption of the law is sometimes portrayed as a threat, US-based scholar Suisheng Zhao argues that, on the contrary, it seeks to balance emotional pressures with national interests.[43] War is thus the last resort to be used only after every other means has been tried. A factor preventing war is the ambiguous position of the United States. The Taiwan Relations Act (1979) does not commit the United States to defend the island, but it does allow the United States to sell arms to it or defend it if the US president so decides. In 1992, representatives of the CCP and KMT reached a consensus recognizing the principle of "one China," but they shelved the question of which regime, the ROC or the PRC, should constitute the state. China's interpretation of the principle is that Taiwan should eventually join the PRC under a "one country, two systems" arrangement analogous to Hong Kong. Cai Ingwen, Taiwan's president since 2016 from the Democratic Progressive Party (DPP), has not accepted the consensus as a basis for relations with China.

China claims almost the whole of the South China Sea (Figure 13.1) and pursues its claims with "creeping assertiveness," a strategy combining negotiation with occupation.[44] It has built runways and fortifications on disputed atolls, pouring concrete over coral reefs that took thousands of years to grow, and used "maritime militias" to coerce other countries' vessels into leaving the area. In 2009, China referred to the South China Sea as a "core interest," a term used for Taiwan, Xinjiang, and Tibet. It has claimed the status of an archipelagic state so that it can treat the South China Sea as an internal sea; it applies an expansive interpretation to the land features that can be used as the basis for claiming territorial seas and an exclusive economic zone, and it claims the right to regulate military activities within these areas.[45] When in 2016 the Philippines won an arbitration ruling under the UN Convention on the Law of the Sea (UNCLOS) supporting its claim to part of the Spratly Islands, China refused to recognize the arbitration court, even though it is an UNCLOS signatory. China's protestations that the South China Sea islands form part of its "historic territory" do not stand up to scrutiny: indeed, when Chinese nationalists first began to agitate for sovereignty over the Spratlys and the Paracels in the first few decades of the twentieth century, there was confusion between the two archipelagos.[46] There are questions over what China hopes to achieve in the South China Sea, but it appears to some military observers to be part of a wider strategy aimed at neutralizing US deterrence against an operation to retake Taiwan.[47]

China's approach to the Diaoyu (Senkaku) Islands dispute is similar. It has created an Air Defense Identification Zone (ADIZ) surrounding the islands, requiring civilian aircraft to identify themselves, and regularly sends an enlarged coast guard fleet to patrol the area. China believes Japan's claim to the islands is based on the Treaty of Shimonoseki, which ended the first Sino-Japanese War in 1895, and therefore the islands should have been returned by Japan after World War II.[48] Japan believes that the islands were part of the Ryuku Kingdom, which was annexed by Japan in 1879, and therefore have nothing to do with World War II. Since the United States and Japan have a mutual defense pact, the United States could be obliged to defend the islands if China were to try to take them by force.

Conclusion: China's Identity

China repeats that it does not wish to be a hegemon, at least not on a global scale. The logic of the so-called Thucydides Trap is that when a rising power challenges the existing hegemon, conflict occurs more often than not.[49] Scholars have pointed to the dangers that

Figure 13.1 Competing Claims in the South China Sea.

Source: SVG version of the South China Sea claims map by Voice of America, 2012.

emotions can bring to a power transition: an overconfident, ambitious China makes a strategic blunder, or an insecure, even paranoid US overreacts to a provocation.[50] Other scholars have argued that the United States has less to fear and can even benefit from China's rise.[51] Be that as it may, China is preparing for conflict and has the second largest military budget in the world. Moreover, at 1.9% of GDP, its defense spending is both easily affordable and rapidly growing.

China wants security and respect within its existing borders, the opportunity to flourish as a key player in the global economy, plus Taiwan, the Diaoyu Islands, and control over the South China Sea. The concept of *geo-body* is useful in understanding the nature and extent of China's ambitions – it refers to the constructed homeland, which is "not merely space or territory ... [but] a component of the life of nation ... a source of pride, loyalty, love, passion, bias, hatred, reason, unreason."[52] Tibet, Taiwan, Hong Kong, and Xinjiang are all part of China's geo-body and China will go to war to defend its claims to them. It is doubtful, however, whether the CCP would risk a war with a major power over any territories that lie beyond its geo-body. It has yet to sink a US or allied vessel engaged in freedom of navigation patrols in disputed territorial waters, though the possibility cannot be excluded.

China's ideas about international order today reflect its status as the largest economy in the world, measured in terms of purchasing power parities. It is no longer interested in promoting worldwide revolution, but it does want to change those rules of the game that it perceives as being to its disadvantage. Given its approach to unresolved territorial disputes, it is reasonable to conclude that China views the world as anarchic and makes realist calculations about what other states might do. It tends to project the traditional Confucian view that respect for hierarchy is the best guarantee of stability: small countries should know their place. However, applying Confucian ideas to international relations requires also that powerful countries live up to the ideals of "true kingship" (*wangquan*) as opposed to "hegemony" (*baquan*) by showing benevolence to lesser powers and taking their responsibilities seriously.[53] China is keen to claim the mantle of legitimacy for its actions by framing them in terms of its own view of international order, one that is distinct from and superior to the liberal world order defended by the West.

While some Chinese might regard Confucian ethics as an overly idealistic basis on which to conduct foreign policy, values remain important. The 18th CCP Congress in 2012 delivered a "five in one" development strategy, focusing on economic, political, cultural, and social development as well as building an "ecological civilization."[54] The 19th Congress in 2017 renewed the commitment to green growth and recognized China's responsibility to the "community with a shared future for mankind," which was widely interpreted as a commitment to take climate change seriously.[55] Changes such as these, which are written into the CCP constitution, represent strategic decisions taken at the highest level.

Traditional Chinese ideas challenge Western assumptions in other ways. Qin Yaqing argues that Western international relations theory is based on individual rationality, whereas China practices "relationality," which assumes that international actors base their actions on relations. Relations are logically prior to rational calculations, whether these be instrumental or normative; contrasting elements are mutually inclusive, not wholly separate, like yin and yang; and hence the natural state of the world is harmony, not conflict.[56] While these propositions might seem abstruse, they inform judgments about what is right and what is rational. Berating Chinese negotiators, as the US Secretary of State Anthony J. Blinken did in Alaska in March 2021 at the first face-to-face high-level talks after President Joe Biden's election, shows a lack of concern for the relationship, and therefore seems irrational. China, by contrast, is scrupulous in attention to protocol and never fails to roll out the red carpet for visiting leaders of even the smallest powers. This helps it win support from other developing countries when it faces diplomatic confrontation with the West.

Russia, whom the Chinese call "the fighting nation," has played different roles in Chinese history, but must now be seen as an ally of China. US foreign policy pushed these two countries closer together, through NATO expansion, the development of missile defense systems, promotion of democracy abroad, and denial of Chinese and Russian aspirations to great power status.[57] Russia's "strategic partnership" with China is a "constructive engagement and positive-sum cooperation, based on shared political, security and economic interests."[58] Among these interests, security is paramount. Russia is now seen as a reliable partner for China in the struggle to make the world safe for authoritarianism. China has given rhetorical support to Russia's 2022 invasion of Ukraine, blaming the West for provoking and fueling the conflict, but it has so far refrained from helping Russia militarily, making new Belt and Road Initiative deals with Russia, or systematically busting US sanctions. Despite some recent American rhetoric, China is not bent on exporting its own model of government. China feels comfortable with authoritarian powers and finds them easier to deal with, but perhaps unlike Russia, it has no messianic streak driving the export of its ideology.

China seeks to enlarge its influence but has a limited appetite for responsibility. While the hegemon is answerable to the international community for everything that happens, and worries about losing its position, the great power with limited responsibility can walk away from problems where the stakes are low. China seeks absolute control over its own geo-body, but beyond those boundaries, it has not been prepared to make great sacrifices for its vision of global order. Arguably, this is a more favorable position than hegemony.

China's identity has been formed by contradictory drives: feeling at once inferior and superior, meeting the needs of development and the desire for equality, assuaging demands for freedom and ensuring order, and bridging the gap between China's geo-body and geopolitical realities. It is only by keeping such tensions in mind that we can hope to understand how its leaders are likely to behave under pressure and to avoid the twin errors of underestimating or overestimating China's strength and the scale of its ambition. China takes great pride in its recent accomplishments, seeing them as a vindication of its choices and confirmation of its values. It believes that its destiny is to dominate East Asia, and through that to play a leading role in the world. The challenge for the United States today is to find a balance between moderating and accommodating that ambition without sacrificing its own values and political influence.

Notes

1 The content for this chapter was adapted from Neil Munro, "China's Identity through a Historical Lens," *Journal of Advanced Military Studies: Special Issue on Strategic Culture* (January 2022): 35–48, https://doi.org/10.21140/mcuj.2022SIstratcul003. Reprinted by permission.

2 Alastair Iain Johnston, *Cultural Realism: Strategic Culture and Grand Strategy in Chinese History* (Princeton, NJ: Princeton University Press, 1995); Huiyun Feng, *Chinese Strategic Culture and Foreign Policy Decision-Making: Confucianism, Leadership and War* (London and New York: Routledge, 2007).

3 Christopher A. Ford, "Realpolitik with Chinese Characteristics: Chinese Strategic Culture and the Modern Communist Party State," in *Strategic Asia 2016–2017: Understanding Strategic Cultures in the Asia-Pacific*, Ashley J. Tellis, Alison Szalwinski, and Michael Wills, eds. (Washington, DC: The National Bureau of Asian Research, 2016), 29–62.

4 Edward W. Said, *Orientalism* (New York: Vintage Books, 2003).

5 Purchasing power parities (PPP) adjust GDP for differences in prices between countries, since, for example, a dollar in China can buy more than a dollar in the United States. Samuel E. Fleischer, *Measuring China's Military Might* (New York: Nova Science Publishers, 2010).

6 Beatrice Heuser, "Beliefs, Culture, Proliferation and Use of Nuclear Weapons," *Journal of Strategic Studies* 23, no. 1 (2000): 74–100, https://doi.org/10.1080/01402390008437779; and Jeannie L. Johnson, *Strategic Culture: Refining the Theoretical Construct* (Washington, DC: Defense Threat Reduction Agency, 2006).
7 Johnson, *Strategic Culture*, 11.
8 John Gerard Ruggie, "What Makes the World Hang Together? Neo-Utilitarianism and the Social Constructivist Challenge," *International Organization* 52, no. 4 (Fall 1998): 855–885; and Alexander Wendt, "Anarchy Is What States Make of It: The Social Construction of Power Politics," *International Organization* 46, no. 2 (Spring 1992): 391–425.
9 Johnson, *Strategic Culture*, 3.
10 Ruggie, "What Makes the World Hang Together?," 875.
11 Iver B. Neumann, "Russia's Europe, 1991–2016: Inferiority to Superiority," *International Affairs* 92, no. 6 (November 2016): 1382–1383 https://doi.org/10.1111/1468-2346.12752.
12 Kathryn C. Oleson et al., "Subjective Overachievement: Individual Differences in Self-Doubt and Concern with Performance," *Journal of Personality* 68, no. 3 (2000): 493, https://doi.org/10.1111/1467-6494.00104.
13 Chi Li, *The Beginnings of Chinese Civilization* (Singapore: Springer Singapore, 2021).
14 Qian Zheng, "Solving a Riddle: Why Did Zheng He Sail to the Western Seas?—China's Own Vision of World Order," *China's Ethnic Groups* 3, no. 2 (2005): 24–29.
15 Cheng Fangyi, "Pleasing the Emperor: Revisiting the Figured Chinese Manuscript of Matteo Ricci's Maps," *Journal of Jesuit Studies* 6, no. 1 (2019): 31–43, https://doi.org/10.1163/22141332-00601003.
16 Wang Gungwu, *Anglo-Chinese Encounters since 1800: War, Trade, Science and Governance* (Cambridge, UK: Cambridge University Press, 2003), 15.
17 Gungwu, *Anglo–Chinese Encounters*, 18.
18 James L. Hevia, *English Lessons: The Pedagogy of Imperialism in Nineteenth Century China* (Durham, NC: Duke University Press, 2003), 5.
19 Robert A. Bickers and Jeffrey N. Wasserstrom, "Shanghai's 'Dogs and Chinese Not Admitted' Sign: Legend, History and Contemporary Symbol," *China Quarterly* 142, no. 46 (June 1995): 444–466, https://doi.org/10.1017/S0305741000035001.
20 Timothy Cheek, *Mao Zedong and China's Revolutions: A Brief History with Documents* (New York: Palgrave, 2002), 125–127.
21 Maurice Meisner, *Mao Zedong: A Political and Intellectual Portrait* (Cambridge, UK: Polity Press, 2007), 110–112.
22 Vaclav Smil, "China's Great Famine: 40 Years Later," *British Medical Journal* 319, no. 7225 (1999): 1619, https://doi.org/10.1136/bmj.319.7225.1619.
23 Roderick MacFarquhar and Michael Schoenhals, *Mao's Last Revolution* (Cambridge, MA.: Belknap Press, an imprint Harvard University Press, 2006), 5–6.
24 Meisner, *Mao Zedong*, 151–157.
25 Robert Wade, *Governing the Market: Economic Theory and the Role of Government in East Asian Industrialization* (Princeton, NJ: Princeton University Press, 1990).
26 Gilbert Rozman, *The Chinese Debate about Soviet Socialism, 1978–1985* (Princeton, NJ: Princeton University Press, 1987), 92.
27 Jinglian Wu, *Understanding and Interpreting Chinese Economic Reform* (Singapore: Thomson–Southwestern, 2005), 72–73.
28 Gregory C. Chow, *China's Economic Transformation* (Malden, MA: Blackwell Publishers, 2007), 49.
29 Wu, *Chinese Economic Reform*, 66.
30 Maria Hsia Chang, "The Thought of Deng Xiaoping," *Communist and Post-Communist Studies* 29, no. 4 (1996): 383.
31 Robert F. Ash, "Quarterly Chronicle and Documentation," *China Quarterly* 120, no. December (1989): 919.
32 Neil M. I. Munro, "Democracy Postponed: Chinese Learning from the Soviet Collapse," *Journal of Current Chinese Affairs*, no. 4 (2008): 31–63, https://dx.doi.org/10.2139/ssrn.1345959.
33 Orville Schell, *To Get Rich Is Glorious: China in the Eighties* (London: Robin Clark, 1985).

34 David L. Wank, *Commodifying Communism Business, Trust, and Politics in a Chinese City* (Cambridge, UK: Cambridge University Press, 1999).
35 Qiang Song et al., *Zhongguo Keyi Shuo Bu [China Can Say No]* (Beijing: Zhonghua gongshang lianhe chuban she, 1996).
36 Suisheng Zhao, "A State-Led Nationalism: The Patriotic Education Campaign in Post-Tiananmen China," *Communist and Post-Communist Studies* 31, no. 3 (September 1998): 287–302, https://doi.org/10.1016/S0967-067X(98)00009-9.
37 Robert D. Weatherley and Ariane Rosen, "Fanning the Flames of Popular Nationalism: The Debate in China over the Burning of the Old Summer Palace," *Asian Perspective* 37, no. 1 (January–March 2013): 53–76, https://doi.org/10.1353/apr.2013.0004.
38 Min Zhou and Hanning Wang, "Participation in Anti-Japanese Demonstrations in China: Evidence from a Survey on Three Elite Universities in Beijing," *Journal of East Asian Studies* 16, no. 3 (November 2016): 391–413, https://doi.org/10.1017/jea.2016.21.
39 Thomas S. Mullaney, "How China Went from Celebrating Ethnic Diversity to Suppressing It," *The Guardian*, 10 June 2021.
40 Ma Rong, "A New Perspective in Guiding Ethnic Relations in the 21st Century: 'De-Politicization' of Ethnicity in China," *Procedia Social and Behavioral Sciences* 2, no. 5 (2010): 6831–6835, https://doi.org/10.1016/j.sbspro.2010.05.034.
41 David Tobin, *Securing China's Northwest Frontier: Identity and Insecurity in Xinjiang* (Cambridge, UK: Cambridge University Press, 2020).
42 Anti-Secession Law (Full text) (03/15/05). Embassy of the People's Republic of China in the United States of America. http://www.china-embassy.org/eng/zt/999999999/t187406.htm, 9 December 2021.
43 Suisheng Zhao, "Conflict Prevention across the Taiwan Strait and the Making of China's Anti-Secession Law," *Asian perspective* 30, no. 1 (2006): 92.
44 Ian James Storey, "Creeping Assertiveness: China, the Philippines and the South China Sea Dispute," *Contemporary Southeast Asia* 21, no. 1 (April 1999): 95–118.
45 Oriana Skylar Mastro, "How China Is Bending the Rules in the South China Sea," *The Interpreter*, 17 February 2021.
46 Bill Hayton, "The Modern Origins of China's South China Sea Claims: Maps, Misunderstandings, and the Maritime Geobody," *Modern China* 45, no. 2 (2019): 127–170, https://doi.org/10.1177/0097700418771678.
47 Capt David Geaney, "China's Island Fortifications Are a Challenge to International Norms," DefenseNews, 17 April 2020.
48 Erica Strecker Downs and Phillip C. Saunders, "Legitimacy and the Limits of Nationalism: China and the Diaoyu Islands," *International Security* 23, no. 3 (Winter 1998–1999): 125.
49 Graham T. Allison, *Destined for War: Can America and China Escape Thucydides's Trap?* (London: Scribe Publications, 2018).
50 Biao Zhang, "The Perils of Hubris? A Tragic Reading of 'Thucydides' Trap' and China–US Relations," *Journal of Chinese Political Science*, no. 24 (2019): 129–144, https://doi.org/10.1007/s11366-019-09608-z.
51 Donald Gross, *The China Fallacy: How the U.S. Can Benefit from China's Rise and Avoid Another Cold War* (New York: Bloomsbury, 2012).
52 Thongchai Winichakul, *Siam Mapped: A History of the Geo-Body of a Nation* (Honolulu: University of Hawaii Press, 1994), 17.
53 Xuetong Yan et al., *Ancient Chinese Thought, Modern Chinese Power* (Princeton, NJ: Princeton University Press, 2013).
54 Hu Jintao, "Report at 18th Party Congress" (Beijing, 8 November 2012).
55 Xi Jinping, "Report at 19th Party Congress" (Beijing, 18 October 2017).
56 Yaqing Qin, *A Relational Theory of World Politics* (Cambridge, UK: Cambridge University Press, 2018).
57 Tao Wenzhao and Xu Shengwei, "The US Factor in Post-Cold War China–Russia Relations," *International Politics* (2020): https://doi.org/10.1057/s41311-020-00211-1.
58 Bobo Lo, "The Long Sunset of Strategic Partnership: Russia's Evolving China Policy," *International Affairs* 80, no. 2 (March 2004): 295.

Selected Bibliography

Allison, Graham T. *Destined for War: Can America and China Escape Thucydides's Trap?* London: Scribe Publications, 2018.

Bickers, Robert A., and Jeffrey N. Wasserstrom. "Shanghai's 'Dogs and Chinese Not Admitted' Sign: Legend, History and Contemporary Symbol." *China Quarterly* 142, no. 46 (June 1995): 444–466, 10.1017/S0305741000035001.

Chang, Maria Hsia. "The Thought of Deng Xiaoping," *Communist and Post-Communist Studies* 29, no. 4 (1996): 377–394.

Cheek, Timothy. *Mao Zedong and China's Revolutions: A Brief History with Documents.* New York: Palgrave, 2002.

Downs, Erica Strecker, and Phillip C. Saunders. "Legitimacy and the Limits of Nationalism: China and the Diaoyu Islands." *International Security* 23, no. 3 (Winter 1998–1999): 114–146.

Feng, Huiyun. *Chinese Strategic Culture and Foreign Policy Decision-Making: Confucianism, Leadership and War.* London and New York: Routledge, 2007.

Fleischer, Samuel E. *Measuring China's Military Might.* New York: Nova Science Publishers, 2010.

Ford, Christopher A. "Realpolitik with Chinese Characteristics: Chinese Strategic Culture and the Modern Communist Party State." In *Strategic Asia 2016–2017: Understanding Strategic Cultures in the Asia-Pacific*, Ashley J. Tellis, Alison Szalwinski, and Michael Wills, eds. Washington, DC: The National Bureau of Asian Research, 2016.

Gross, Donald. *The China Fallacy: How the U.S. Can Benefit from China's Rise and Avoid Another Cold War.* New York: Bloomsbury, 2012.

Gungwu, Wang. *Anglo-Chinese Encounters since 1800: War, Trade, Science and Governance.* Cambridge, UK: Cambridge University Press, 2003.

Hayton, Bill. "The Modern Origins of China's South China Sea Claims: Maps, Misunderstandings, and the Maritime Geobody." *Modern China* 45, no. 2 (2019): 127–170, 10.1177/009770041 8771678.

Hevia, James L. *English Lessons: The Pedagogy of Imperialism in Nineteenth Century China.* Durham, NC: Duke University Press, 2003.

Johnston, Alastair Iain. *Cultural Realism: Strategic Culture and Grand Strategy in Chinese History.* Princeton, NJ: Princeton University Press, 1995.

MacFarquhar, Roderick, and Michael Schoenhals. *Mao's Last Revolution.* Cambridge, MA: Belknap Press, an imprint Harvard University Press, 2006.

Meisner, Maurice. *Mao Zedong: A Political and Intellectual Portrait.* Cambridge, UK: Polity Press, 2007.

Munro, Neil M. I. "Democracy Postponed: Chinese Learning from the Soviet Collapse." *Journal of Current Chinese Affairs*, no. 4 (2008): 31–63, 10.2139/ssrn.1345959.

Qin, Yaqing. *A Relational Theory of World Politics.* Cambridge, UK: Cambridge University Press, 2018.

Rong, Ma. "A New Perspective in Guiding Ethnic Relations in the 21st Century: 'De-Politicization' of Ethnicity in China." *Procedia Social and Behavioral Sciences* 2, no. 5 (2010): 6831–6845, 10.1016/j.sbspro.2010.05.034.

Rozman, Gilbert. *The Chinese Debate about Soviet Socialism, 1978–1985.* Princeton, NJ: Princeton University Press, 1987.

Said, Edward W. *Orientalism.* New York: Vintage Books, 2003.

Smil, Vaclav. "China's Great Famine: 40 Years Later." *British Medical Journal* 319, no. 7225 (1999): 1619–1921, 10.1136/bmj.319.7225.1619.

Storey, Ian James. "Creeping Assertiveness: China, the Philippines and the South China Sea Dispute." *Contemporary Southeast Asia* 21, no. 1 (April 1999): 95–118.

Tobin, David. *Securing China's Northwest Frontier: Identity and Insecurity in Xinjiang.* Cambridge, UK: Cambridge University Press, 2020.

Weatherley, Robert D., and Ariane Rosen. "Fanning the Flames of Popular Nationalism: The Debate in China over the Burning of the Old Summer Palace." *Asian Perspective* 37, no. 1 (January–March 2013): 53–76, 10.1353/apr.2013.0004.

Wenzhao, Tao, and Xu Shengwei. "The US Factor in Post-Cold War China–Russia Relations." *International Politics* (2020): 1–23. 10.1057/s41311–020–00211–1.

Wu, Jinglian. *Understanding and Interpreting Chinese Economic Reform*. Singapore: Thomson–Southwestern, 2005.
Yan, Xuetong, et al. *Ancient Chinese Thought, Modern Chinese Power*. Princeton, NJ: Princeton University Press, 2013.
Zhang, Biao. "The Perils of Hubris? A Tragic Reading of 'Thucydides' Trap' and China–US Relations." *Journal of Chinese Political Science*, no. 24 (2019): 129–144, 10.1007/s11366-019-09608-z.
Zhao, Suisheng. "A State-Led Nationalism: The Patriotic Education Campaign in Post-Tiananmen China." *Communist and Post-Communist Studies* 31, no. 3 (Sep 1998): 287–302, 10.1016/S0967-067X(98)00009-9.
Zhao, Suisheng. "Conflict Prevention across the Taiwan Strait and the Making of China's Anti-Secession Law." *Asian Perspective* 30, no. 1 (2006): 79–94.
Zheng, Qian. "Solving a Riddle: Why Did Zheng He Sail to the Western Seas?—China's Own Vision of World Order." *China's Ethnic Groups* 3, no. 2 (2005), 24–29.

14
THE STRATEGIC CULTURE OF THE UNITED KINGDOM

Alastair Finlan

Introduction

The concept of strategic culture has gained significant intellectual purchase in the United Kingdom. It has sparked rich, occasionally contentious, and fruitful intellectual discussions across numerous academic disciplines, from international relations and its subfield, strategic studies, to military history and war studies since the 1970s. The fascination with strategic culture is perhaps unsurprising. Looking critically through a lens at the history of the present-day United Kingdom reveals much. For a small island nation located within the European continent, it has been a remarkably active and proactive global actor in military and security issues/interventions around the world for the last 300 years. For a time, its colonial empire spanned the Atlantic and Indian Oceans and, on two occasions, in the twentieth century, it was a central player in the most destructive wars in human history. Notwithstanding the decline and retreat from empire after World War II, the United Kingdom has continued its predilection to be a major influencer in world affairs in the twenty-first century. From Afghanistan to Iraq, from the Global War on Terror to "Global Britain,"[1] the United Kingdom has applied dynamic military and diplomatic force with varying degrees of success in international relations from Europe to the Pacific. Strategic culture matters in the United Kingdom because of its seemingly unquenchable thirst and explicit preference for military, political, and security influence, not just in Europe, but in Asia, Africa, the Americas, the Arctic, and the Antarctic. The puzzle for historians, political scientists, and social scientists is how to characterize the strategic culture of the United Kingdom today and how to identify its key and enduring elements that provide a glimpse of future directions.

The Academic Discussion

The fascination with ideas that could be considered as coming under the broad description of strategic culture today has long roots in the United Kingdom. For example, in the 1930s the well-known journalist and strategist Basil Liddell Hart published his famous book, *The British Way in Warfare.*[2] In addition to being one of the most famous defense commentators of his time, Liddell Hart also directly and indirectly affected generations of scholars and their doctoral students in the UK with his work in this area during the Cold War. Notwithstanding these shoots of early interest, it was the work of the American scholar

DOI: 10.4324/9781003010302-17

Jack Snyder,[3] who first used the term "strategic culture" in 1977, which provoked a sustained interest in this concept within the field of international relations and more specifically in one of its policy-orientated subfields, strategic studies. Snyder's excellent work lit an intellectual fire in the imagination of many scholars in the United Kingdom during the Cold War to excavate this fresh concept more fully. It is remarkable that the first wave of interest in strategic culture attracted two young academics who would become recognized scholarly titans in their respective areas of expertise.

The first was Ken Booth with his extraordinary and quite radical book, *Strategy and Ethnocentrism*,[4] published in 1979, which encouraged strategists to look beyond the existing and quite tired parameters of the strategy debates in the bipolar world – from zero-sum games to rational actors – to embrace the emancipating cultural pathway. The interest in the cultural turn in strategy was acutely felt in Booth's alma mater, the famous Department of International Politics at Aberystwyth University, and stimulated significant work in this area by staff and graduate students over several decades, including John Baylis and Kristan Stoddart,[5] Alan Macmillan,[6] Colin McInnes,[7] Kerry Longhurst,[8] and Alastair Finlan.[9] Today, *Strategy and Ethnocentrism*, while still in print, is not so well known to contemporary students, but remains a staple for scholars with an interest in the origins of strategic culture in the United Kingdom. In contrast, the other better-known proponent of strategic culture, Colin S. Gray, still stands (notwithstanding his untimely death in 2020) at the forefront of the strategic culture debate. Gray's lifetime of work in strategic studies is unparalleled in terms of its sustained focus over a long period of time, development of thought, and sheer insight into the concept. His international engagement with strategic culture in the United States and the United Kingdom encompassed path-leading publications in the top journals in the field and started with his idea of a "national style in strategy."[10] Interestingly, but often overlooked, it was Booth's initial work that greatly influenced Gray's first major publication in this area. Gray's continuous intellectual dialogue that deepened and developed his ideas on the topic over the years stimulated and inspired other generations of scholars to take the cultural pathway in international relations. His efforts culminated in perhaps the most famous debate on strategic culture[11] between himself, as a representative of the first generation, and the most well-known proponent of the third generation, Alastair Iain Johnston.

This closely followed elegant and erudite discussion between two "tier one" scholars, to use a description taken from another important research area to Gray (Special Operations Forces or SOF), greatly added to the broader understanding of the concept in the wider field of international relations. It also increased the visibility of Gray's idea of culture as context. The debate itself, which many scholars felt Gray had won, also fanned more discussion in Britain. Stuart Poore, for example, followed it up with an insightful rejoinder[12] and later, with colleagues at Southampton University (another center of expertise on strategic culture in the United Kingdom) John Glenn and Daryl Howlett, produced a fine addition to the literature entitled *Neorealism Versus Strategic Culture* in 2004.[13] Other voices have also been influential in the discussion. Theo Farrell, for example, has been one of the most prolific scholars in the area of culture, security, and strategy in the last two decades, which makes him broadly a contemporary of the third generation of scholarship with his most significant contribution being his book *The Norms of War: Cultural Beliefs and Modern Conflict* published in 2005.[14] Kerry Longhurst also made an excellent contribution to the conversation in international relations on strategic culture with her book *Germany and the Use of Force* in 2004.

Across other disciplines, notably military history and war studies, conversations on aspects of strategic culture have also occurred that have an interesting connection to Basil Liddell Hart because he helped the founder of the Department of War Studies at King's College London, Michael Howard, set up this radical new experiment in higher education in the United Kingdom. Liddell Hart also mentored one of the department's most dedicated teachers, Brian Bond,[15] who was a prolific producer of successful PhD students (50 of them). The remarkable Bond stable of PhD students has flourished in higher education in the United Kingdom and includes intellectual luminaries such as Alex Danchev[16] and David French.[17] Several would go on to further knowledge in the area of the British way in warfare and even Liddell Hart's ideas on the subject.

It is impossible to discuss the evolution of the Department of War Studies without mentioning the interlinkage with Oxford University that provides a regular flow and interflow of staff and students. The nexus between King's and Oxford has always been intimate and was cemented by Michael Howard's appointment as the Chichele Professor of the History of War in 1977, who also wrote *The British Way in Warfare: A Reappraisal* in 1975.[18] In the twenty-first century, this focus on strategic culture and ways of war was greatly strengthened by the appointment of Hew Strachan as Chichele Professor between 2002 and 2015. Strachan's contribution to understanding and critically engaging with the strategy of the Global War on Terror is frankly unrivaled and his thoughts on strategic culture are captured in his outstanding collection of thoughts in *The Direction of War: Contemporary Strategy in Historical Perspective* published in 2013.[19]

UK Strategic Culture in the Twenty-First Century

It is more problematic to interrogate, define, and describe UK strategic culture in the twenty-first century than it was in the last century because – surprisingly – the referent object has changed. When Booth and Gray focused their intellectual lens on Britain, their subject was one nation, centrally governed from Westminster. Today, the United Kingdom is composed of four distinct political entities (England, Northern Ireland, Scotland, and Wales) with quite different national outlooks on important issues related to security and national defense. The primacy of England with the British Parliament in London in security and strategy continues *for now* (dependent on the thorny question of full independence in the devolved nations) and its armed forces remain a UK-wide composition. Interestingly, the 2016 Brexit referendum that ultimately took the United Kingdom out of the European Union may well be the catalyst for the future break-up of the United Kingdom, which is exactly the opposite of the Vote Leave campaign's dream to return Britain back to its pre-European Union glory days. Brexit's repercussions could herald a return to a situation not seen since the Middle Ages with an independent England, Wales, and Scotland with profound consequences for the idea of a unified UK strategic culture.

Defining a specific strategic culture is not easy due to all the potential variables that can influence or constitute it at a particular moment of time in a state. A good starting point, to begin with, is to look toward academic discussions on the topic. The baseline of all definitions of what strategic culture is usually follows the inspiration of Snyder who described it as "the sum total of ideas, conditioned emotional responses, and patterns of habitual behavior that members of a national strategic community have acquired through instruction or imitation and share with each other."[20] This definition captures the characteristic wide-angle lens of the early work on strategic culture that drew in an enormous number of potential variables that could shape this phenomenon in a state. It was the inclusion of the "patterns of habitual behavior" that led to Johnston's famous critique of Snyder's work as being "mechanically deterministic explanations of strategic choices."[21] Put more simply, linking attitudes to behavior is problematic because they do not necessarily correlate. In fairness to Snyder, there is no mention of attitudes in his definition, but rather a nod to expectations which can be defined as norms. Johnston's laudable methodological rigor in his research has made the issue of behavior and strategic culture a sensitive one, suggesting that scholars need to take care of this deterministic pitfall.

It is perhaps more useful when attempting to articulate a UK strategic culture in the twenty-first century to draw upon subsequent work that built upon Snyder's initial foundational excavation of the concept. Much insight can be taken from Gray's ideas of a distinct "national style" and "culture as context" that emphasize how people are steeped in their nation's history and geography. In this way, they are encultured strategic actors who affect group outlook and self-image/identification in a state. Equally, from Gray's great interlocutor on strategic culture, Johnston, the ideas of "assumptions" and recognizable "preferences"[22] provide more tangible direction to tease out a distinctive and identifiable UK strategic culture today. Focusing explicitly on the time period 2000–2022 permits a framework to look closely at several aspects of UK strategic culture that clearly stand out as cherished elements with the British strategic community, regardless of political affiliation.

The Special Relationship

The Atlantic alliance between the United States and the United Kingdom is the most important security relationship to the latter since World War II. It is impossible to discuss UK strategic culture from 1941 to 2022 without acknowledging the so-called Special Relationship that gained a new significance in the twenty-first century. The cultural context is significant in view that the United States was a former British colony, so the historical links are strong and on two occasions in the twentieth century the United States intervened decisively to aid Britain in its time of urgent need in two world wars. By the twenty-first century, the relationship between the two countries deepened markedly from a preference (a choice from many alternative options) to an assumption (an unquestioned and accepted truth). This marks a significant development in UK strategic culture that can be shown empirically in the last 20 years. A noticeable feature of the Global War on Terror initiated by George W. Bush in the aftermath of the horrendous attack of 11 September 2001 has been the sheer intimacy of the unwavering partnership and support of the United Kingdom to the United States in the major military efforts in Afghanistan, Iraq, Libya, and Syria. The extent to which Britain has been in lockstep with the United States included participating in an internationally legally dubious invasion of Iraq in 2003, and, more recently, meekly following the catastrophic departure of the United States from Afghanistan in 2021.

In past times, British political elites would not hesitate to express their differences of opinion or have another preference on security issues, such as the refusal to be involved in the Vietnam War.[23] Nevertheless, the new generation of political elites over the last two decades from left to right, from Tony Blair to Boris Johnson, have displayed a remarkably consistent policy that now has the status of an assumption to work closely and subordinately with their more powerful ally. Blair, for example, despite contributing the second largest coalition force in the Iraq War, was not even consulted when Bush preemptively started the war with an attempt to kill Saddam Hussein.[24] Johnson, an American dual citizen by birth (who only renounced his citizenship in 2016)[25] displayed little influence in the decision to withdraw US forces from Afghanistan that precipitated a feverish international crisis after the Taliban swept to power again in August 2021. The military partnership with the United States reached a remarkable milestone in 2021 when the new British aircraft carrier, HMS *Queen Elizabeth*, set sail on a long deployment to the Pacific with eight UK F-35 aircraft and ten US Marine Corps F-35 aircraft.[26] This milestone can be seen as perhaps taking the Atlantic alliance to another level of cooperation and partnership, but the superior number of American aircraft on a state-of-the-art British fifth-generation aircraft carrier carries much potent symbolism.

The Nuclear Deterrent

Showing the concrete manifestation of beliefs, preferences, and assumptions in a state in terms of its defense and security posture, from the ideational level to the physical form, is perhaps one of the hardest challenges for researchers to do empirically. It is widely acknowledged that strategic culture is a "nebulous concept"[27] as a result of the difficulties in identifying clear markers for its existence. Borrowing from military culture studies and the broader field of organization theory, it is possible to outline or map a strategic culture by highlighting the choice of specific defense systems. Throughout history, nations have exhibited choices of weaponry that suit their style of warfare. In ancient times, Sparta placed an emphasis on armored infantry in a phalanx using six-foot spear technology. Alexander the Great developed pike and phalanx warfare with extraordinary success. The Romans employed legions with the short stabbing sword at the heart of their style of warfare. In the modern age, the choice of weaponry mirrors, to borrow a phrase from Gray, a "national style"[28] and exudes how a nation orientates itself toward warfare. The possession of the nuclear deterrent has been a staple of UK strategic culture since the 1950s. From the 1960s onward, it increasingly became synonymous with American-supplied submarine-launched ballistic missiles, first in the form of the Polaris system and then in the 1980s as the new Trident system. In the twenty-first century, the Blair government quietly started the process of looking into renewing this system, which gained Parliamentary approval in 2016.[29]

The existence of the Trident nuclear weapons system is arguably one of the most visible aspects of the Special Relationship. The United States has never shared this cutting-edge technology with another partner in international relations. Moreover, the independence of the British strategic nuclear weapons is questionable on several levels. First, the missiles are manufactured by and leased from the United States and represent a technology pathway that proved to be beyond the financial means for the United Kingdom to pursue alone since the 1960s. Second, the current warhead design closely follows an American design, the W76.[30] An unexpected leak also showed UK involvement in the proposed new replacement warhead, the W93.[31] Third, the launching of the weapon has traditionally been dependent on American satellites. Only the submarines are genuinely British. The cost of this weapons system is exorbitant for the British state (the replacement is currently estimated to be £31 billion[32] but with running and infrastructure costs significantly higher) and possesses profound consequences for UK conventional forces. In the 1980s when Britain purchased Trident II under Margaret Thatcher, her defense secretary placed the cost in the Royal Navy's budget,[33] arguing that as it was carried by a naval platform, it therefore was a naval system. The reality was this savvy financial sleight of hand (Trident was a national defense system) saw a rapid deterioration of the Royal Navy's surface fleet from the 1980s–2000 as the service struggled to pay for the nuclear cuckoo in the budget nest.

What does the United Kingdom's strategic nuclear deterrent mean from a strategic culture perspective?[34] It is the supreme symbol of great power status in international relations. The United Kingdom remains one of the few members of the exclusive club of nations that possess nuclear weapons of mass destruction. More so, its possession of the most advanced submarine-based weaponry places it in a very small grouping of nations with this seemingly invulnerable component of the nuclear triad of technology (that in other states includes air-launched weapons, land-based missiles, and submarine-launched missiles). The United Kingdom has only the latter system, but it has surprisingly and quite recently indicated a desire to increase its stockpile of warheads. The United Kingdom's status as a nuclear weapons state also feeds into its cultural context and cherished history as one of the "big three" victors of World War II, as well as one of the largest colonial empires in the nineteenth and early twentieth centuries. The idea that Britain was (is) a great power is deeply embedded in the British political elites of the major two parties. This perhaps explains, when looking at all the prime ministers in the United Kingdom since 2000, how little the deviation is between them with regard to the renewal of the nuclear deterrent, notwithstanding the huge financial costs of the next generation of weapons and platforms.

Embracing Sea Power

One of the major challenges of strategic culture involves the intellectual process of attempting to produce generalizations about a state and its defense orientation and preferences based on its history. A broad interpretation of the United Kingdom's history would suggest that sea power has been at the heart of UK strategic culture since the defeat of the Spanish Armada in 1588, the Battle of Trafalgar in 1805, the two world wars, and the Falklands Conflict in 1982. This is what Colin Gray described as "maritime primacy"[35] and a tracing of historical patterns adds great strength to this idea. A glimpse at the shape of the Royal Navy in 2021, with two fifth-generation aircraft carriers, advanced F-35 stealth aircraft, state-of-the-art Astute nuclear-powered submarines, and Type 45 air defense destroyers would reinforce this perception as a preference toward an investment in sea power. In essence, the British state is consistently a maritime power. Nevertheless, these descriptions come with caveats. They do not indicate the complex and shifting relationship of the British state with sea power. In 1981, for example, Margaret Thatcher initiated one of the most significant reductions of the size of the Royal Navy in modern times with the disposal of two aircraft carriers, both of its specialist amphibious assault ships, and nine frigates/destroyers (from 59 to around 50).[36] A snapshot of UK strategic culture in 1981 would have indicated that naval power did not matter so much to the state in view that it was willing to abandon entire roles (amphibious assault and out-of-area capability) in order to pay for the nuclear deterrent. This would suggest that the nuclear deterrent was more important than sea power. Interestingly, these drastic cuts were mitigated, not by a change of heart by political elites in Westminster, but rather from the fallout of the Falklands Conflict, an unexpected, but not unanticipated, short war between British and Argentine forces on the fringes of the Antarctic. This closely fought campaign (by no means an easy or certain victory for the United Kingdom) reinforced the need for sea power. Nevertheless, the victory in the South Atlantic did not radically change the gradual diminution of British sea power.

A cursory glance at the state of the United Kingdom's sea power in 2012, at the height of the Global War on Terror, would also provide a sense of a very different strategic culture to the situation in the early 2020s. At that point, the Royal Navy had no aircraft carriers, no strike air power, a reduced collection of frigates and destroyers – around 18 in number[37] – and many ships laid up at port due to a shortage of fuel because the navy's budget was under strain. The focus of the state in 2012 was very much on pumping huge amounts of money into its land forces, which were struggling to extract themselves in Iraq and bogged down in a horrendous insurgency in Afghanistan that would continue for another two years. It would be easy to claim that land power with air power in a support role was the characteristic of UK strategic culture with sea power as a wasting asset in this period. Two elements need to be taken into account, however, when attempting to create an academically credible generalization about UK strategic culture. The first is that change is often slow with regard to defense technologies, especially complex systems such as modern aircraft, main battle tanks, and warships. It took the United Kingdom ten years to produce the new generation of aircraft carriers from contract to commission, so the situation in 2012 represented an interregnum when old capabilities were being discarded while new capabilities were being built. This fact necessitates that a strategic culture lens adopts a comprehensive view that takes into account not just cultural artifacts such as the weapon systems on view, but also the investments in weaponry to come.

British sea power was materially weak in 2012, but its direction clearly followed that of previous generations, reinforcing Gray's idea of "maritime primacy." Furthermore, the United Kingdom decided, as it has in the past, to fight at the highest levels of naval warfare and its provision of fifth-generation forces (ships and aircraft) demonstrates this preference. The second factor encompasses the subcultural dimension of strategic culture and one element in particular: that of military culture. No highly developed state has a homogenous military culture. Instead, they possess several distinctive military cultures represented variously by the air force, army, navy, marines, special forces and, more recently, space force, all in competition with each other for resources and primacy. It can be said that on occasion, a state's strategic culture reflects the color of a service branch to a greater extent than the others. During the Falklands Conflict, the Royal Navy dominated UK strategic culture, not only out of necessity as only naval forces could transport all the military power from the North Atlantic to the extreme edge of the South Atlantic Ocean but also because of the critical advisory nexus with chastened political elites who were completely out of their expertise and comfort zones (the age before the "Iron Lady" emerged after the victory). In the Persian Gulf War, the Royal Air Force had the most influence on the command arrangements based in the United Kingdom. The Global War on Terror will very much be seen as a time of British Army influence that may carry significant costs to the service in view of the woeful outcomes in both Afghanistan and Iraq.

The subcultural dimension, of which military culture is just one aspect, indicates that strategic culture is dynamic. It is regularly affected by powerful competitor military organizations seeking material gain and influence that challenge the idea of immutability or a lack of change in strategic culture. Instead, this points toward the notion that strategic culture has significant scope for identification during specific periods with a particular military culture within a state. This strengthens another argument of Gray that a "strategic personality"[38] can be discerned from a cultural lens focused on the security and strategy dimensions of a state.

Defining the strategic culture of a state is intellectually exacting, but the existence of certain concrete facets, such as the impact of geography[39] on its history and strategy, makes the process a little more quantifiable. For the United Kingdom, its geography or island status, detached by a small stretch of water from the European continent, has profoundly influenced its strategic culture for hundreds of years and more. The debate between a continental commitment or a maritime strategy[40] is an ancient one that has echoed passionately in the castles of kings to the corridors of power in Number 10 Downing Street today. Europe has been and always will be the United Kingdom's most vital security issue. From the time of Roman ascendency to the catastrophic invasion of the Normans in 1066, the progenitors of the modern United Kingdom have long understood that splendid isolation from European power politics is not a viable option. Over time, the United Kingdom intervened in Europe with significant military forces to alter the balance of power (Napoleonic Wars, World War I, and World War II) and defend its vital interests.

In the modern age, this timeless facet of UK strategic culture has been greatly affected and complicated by the Brexit debate that has attempted to offer a fanciful, often surreal, excursion from the actual strategic reality rooted in the geography and history of the United Kingdom. The role of social media, particularly that stemming from a hostile external actor such as Russia, to create a mass communication influence (an influence operation) on the British public has been raised at the highest levels of government as a possibility,[41] but there is no inclination on behalf of recent governments to follow up this potential unorthodox security threat to British interests. Remarkably, the Intelligence and Security Committee of Parliament that raised this matter in a 2020 report also revealed that none of the top security organs or agencies in the British state saw it as their responsibility to protect "democratic processes"[42] that would profoundly affect vital strategic relationships and the economic security of the state. It is a decision that previous spymasters of the state such as Francis Walsingham, who protected Elizabeth I from external threats emanating from the European continent, would have found perhaps mystifying. In the short term, the immediate impact of Brexit on the United Kingdom's strategic culture is limited in view of the deep-rooted stabilizing elements that ensure incremental and not radical change is the norm.

In the twenty-first century, notwithstanding the rhetoric of Brexit, the United Kingdom's membership of the NATO alliance means that it remains committed to the defense of Europe. The state's role within NATO has been a stable component of UK strategic culture since the late 1940s and it is highly unlikely that this aspect will change in the near future. The question is more about how it fulfills its obligation. In the Cold War, the United Kingdom had thousands of troops permanently stationed in Germany. The vast majority of those bases and troops were either shut, withdrawn, or handed back to German authorities between 2010 and 2020.[43] This shift places the United Kingdom back in the same situation as it was in 1914 and earlier. In the event of a major crisis in Europe involving military forces, UK troops would have to be moved into theater by air, sea, or railway using the Channel Tunnel. This reality also returns the United Kingdom to the same logistical challenges of moving large concentrations of military forces and equipment from the home island to the continent. The old discussions on the merits and pitfalls of continental commitment versus a maritime strategy will be very much alive for the foreseeable future in UK strategic culture.

Conclusion

The influence of Snyder's idea of strategic culture continues to maintain a prominent position in academic debates in the United Kingdom. It is a rare scholar whose work provokes four decades' worth of discussion in a remote and now "isolated" part of Europe. Strategic culture found a home in the United Kingdom due to the pioneering work of Booth and Gray as well as a deep abiding interest in intellectual circles in ways of war that predated Snyder's thoughts by nearly half a century. Together, two extraordinary intellectual sponsors of strategic culture combined with a receptive environment have generated significant wattage in the development of an idea that continues to spark discussions and interest in the present day. Strategic culture is one of the big concepts in international relations in the United Kingdom. Mapping and defining UK strategic culture is not easy in view of the constraints of a short chapter. Much is missing such as the idea of limited liability, limited wars, irregular warfare, colonial warfare, and the development of unorthodox/unconventional warfare pioneered by units (invented in the United Kingdom) at the forefront of the use of force in the early twenty-first century, commonly known as special forces. Instead, this chapter's focus has been on the most significant aspects of UK strategic culture, including the extraordinary (for the United Kingdom) Special Relationship, the nuclear deterrent, the refreshed focus on sea power, and the continental commitment in the strange half-light of Brexit. The immediate focus is the chaotic retreat from Afghanistan that perhaps will have less impact on UK strategic culture than US strategic culture because this is the fourth time the United Kingdom has departed from Afghanistan in haste. The social memory of other disasters in Afghanistan is well embedded in British culture, but not so in the United States, which may provoke more serious effects in political and military culture. The future for UK strategic culture is likely to see an emphasis on elements of the armed forces that were largely untouched by the stain of defeat in Afghanistan, such as the Royal Navy, whose state-of-the-art and powerful aircraft carriers symbolize the appropriated Brexit vision of "Global Britain" that reinforces the critical relationship with the United States and flies the British flag around the world.

Notes

1 "Global Britain" is a policy that has been developed by several Conservative governments in the United Kingdom, but it is most closely associated with former Prime Minister Boris Johnson.
2 Basil Liddell Hart, *The British Way in Warfare* (London: Faber & Faber, 1932).
3 Jack L. Snyder, "The Soviet Strategic Culture: Implications for Limited Nuclear Operations," R-2154-AF (Santa Monica: Rand, 1977).
4 Ken Booth, *Strategy and Ethnocentrism* (New York: Holmes & Meier, 1979).
5 John Baylis and Kristan Stoddart, *The British Nuclear Experience: The Roles of Beliefs, Culture and Identity* (Oxford: Oxford University Press, 2014).
6 Alan Macmillan, "Strategic Culture and National Ways in Warfare: The British Case," *The RUSI Journal*, Vol. 140/5 (1995), pp. 33–38.
7 Colin McInnes, "The British Army's New Way in Warfare: A Doctrinal Misstep?," *Defense and Security Analysis*, Vol. 23/2 (2007), pp. 127–141.
8 Kerry Longhurst, *Germany and the Use of Force: The Evolution of German Security Policy 1990–2003* (Manchester: Manchester University Press, 2004).
9 Alastair Finlan, *The Royal Navy in the Falklands Conflict and the Gulf War: Culture and Strategy* (London: Frank Cass, 2004) and idem, *Contemporary Military Culture and Strategic Studies: US and UK Armed Forces in the 21st Century* (Abingdon: Routledge, 2013).

10 Colin S. Gray, "National Style in Strategy: The American Example," *International Security*, Vol 6/2 (1981), pp. 21–47.
11 Colin S. Gray, "Strategic Culture as Context: The First Generation of Theory Strikes Back," *Review of International Studies*, Vol. 25/1 (1999), pp. 49–69; Alastair Iain Johnston, "Strategic Cultures Revisited: Reply to Colin Gray," *Review of International Studies*, Vol. 25/3 (1999), pp. 519–523.
12 Stuart Poore, "What Is the Context? A Reply to the Gray-Johnston Debate on Strategic Culture," *Review of International Studies*, Vol. 29/2 (2003), pp. 279–284.
13 John Glenn, Darryl Howlett, and Stuart Poore (eds), *Neorealism Versus Strategic Culture* (Aldershot: Ashgate Publishing, 2004).
14 Theo Farrell, *The Norms of War: Cultural Beliefs and Modern Conflict* (Boulder, CO: Lynne Rienner Publishers, 2005).
15 Brian Bond, *Liddell Hart: A Study of His Military Thought* (London: Cassell, 1977).
16 Alex Danchev, *Alchemist of War: The Life of Basil Liddell Hart* (London: Weidenfeld & Nicolson, 1998).
17 David French, *The British Way in Warfare 1688–2000* (London: Unwin Hyman, 1990).
18 Michael Howard, *The British Way in Warfare: A Reappraisal* (London: Jonathan Cape, 1975).
19 Hew Strachan, *The Direction of War: Contemporary Strategy in Historical Perspective* (Cambridge: Cambridge University Press, 2013).
20 Snyder, "The Soviet Strategic Culture," p. 8.
21 Alastair Iain Johnston, *Cultural Realism: Strategic Culture and Grand Strategy in Chinese History* (Princeton, NJ: Princeton University Press, 1995), p. 6.
22 Ibid, pp. ix–x.
23 The British Prime Minister Harold Wilson kept the United Kingdom out of the Vietnam War.
24 See Alastair Finlan, *Contemporary Military Strategy and the Global War on Terror: US and UK Armed Forces in Afghanistan and Iraq 2001–2012* (New York: Bloomsbury, 2014), p. 133.
25 Patrick Wintour, "Boris Johnson among Record Number to Renounce American Citizenship in 2016," *The Guardian*, 9 February 2017.
26 Sam LaGrone, "U.K., U.S. F-35Bs Launch Anti-ISIS Strikes from HMS Queen Elizabeth," *USNI News*, 22 June 2021.
27 Booth, *Strategy and Ethnocentrism*, p. 14.
28 Gray, "National Style in Strategy," p. 21.
29 "MPs Vote to Renew Trident Weapons System," *BBC News*, 19 July 2016. The vote was 472 for renewal and 117 against.
30 Hans M. Kristensen and Matt Korda, "United Kingdom Nuclear Weapons, 2021," *Bulletin of the Atomic Scientists*, Vol. 77/3 (2021), p. 156.
31 Ibid.
32 Ibid.
33 Finlan, *The Royal Navy in the Falklands Conflict and the Gulf War*, p. 50.
34 For a discussion, see Baylis and Stoddart, *The British Nuclear Experience: The Roles of Beliefs, Culture and Identity*.
35 See Colin S. Gray, "British and American Strategic Cultures," paper delivered at the Jamestown Symposium (2007), p. 34.
36 Finlan, *The Royal Navy in the Falklands Conflict and the Gulf War*, p. 31.
37 *United Kingdom Defence Statistics 2012*, Chapter 4 (London: MoD, 2013), n.p.
38 Gray, "British and American Strategic Cultures," p. 16.
39 Ibid, p. 17.
40 See John Baylis, "The Continental Commitment versus a Maritime Strategy," in idem, *British Defence Policy: Striking the Right Balance* (New York: Palgrave Macmillan, 1989) pp. 19–29.
41 Intelligence and Security Committee of Parliament, *Russia*, HC 632 (21 July 2020), 12.
42 Ibid, p. 10.
43 Mattha Busby, "British Army Hands Back Last Headquarters in Germany," *The Guardian*, 22 February 2020.

Selected Bibliography

Baylis, John, and Kristan Stoddart. *The British Nuclear Experience: The Roles of Beliefs, Culture and Identity* (Oxford: Oxford University Press, 2014).

Booth, Ken. *Strategy and Ethnocentrism* (New York: Holmes & Meier, 1979).

Farrell, Theo. *The Norms of War: Cultural Beliefs and Modern Conflict* (Boulder, CO: Lynne Rienner Publishers, 2005).

Glenn, John, Darryl Howlett, and Stuart Poore (eds.). *Neorealism Versus Strategic Culture* (Aldershot: Ashgate Publishing, 2004).

Gray, Colin S. "National Style in Strategy: The American Example," *International Security*, Vol. 6/2 (1981), pp. 21–47.

Gray, Colin S. "Strategic Culture as Context: The First Generation of Theory Strikes Back," *Review of International Studies*, Vol. 25/1 (1999), pp. 49–69.

Gray, Colin S. "British and American Strategic Cultures," paper delivered at the Jamestown Symposium (2007).

Howard, Michael. *The British Way in Warfare: A Reappraisal* (London: Jonathan Cape, 1975).

Johnston, Alastair Iain. "Strategic Cultures Revisited: Reply to Colin Gray," *Review of International Studies*, Vol. 25/3 (1999), pp. 519–523.

Liddell Hart, Basil. *The British Way in Warfare* (London: Faber & Faber, 1932).

Longhurst, Kerry. *Germany and the Use of Force: The Evolution of German Security Policy 1990–2003* (Manchester: Manchester University Press, 2004).

Macmillan, Alan. "Strategic Culture and National Ways in Warfare: The British Case," *The RUSI Journal*, Vol. 140/5 (1995), pp. 33–38.

Poore, Stuart. "What Is the Context? A Reply to the Gray-Johnston Debate on Strategic Culture," *Review of International Studies*, Vol 29/2 (2003), pp. 279–284.

Snyder, Jack L. "The Soviet Strategic Culture: Implications for Limited Nuclear Operations," R-2154-AF (Santa Monica: Rand, 1977).

Strachan, Hew. *The Direction of War: Contemporary Strategy in Historical Perspective* (Cambridge: Cambridge University Press, 2013).

15
FRENCH STRATEGIC CULTURE
International Engagement in the Name of National Autonomy

Maria Hellman

French strategic culture could be summed up in the two statements: "doing foreign policy with military means," and inconsistent behavior stemming from "a tension between being a country of human rights per se and at the same time being a country resting on self-reliance and independence."[1] The first statement points to the acceptance of a strong and costly military with large deployments abroad, whose presence is aimed at defending international norms and liberal democratic values. The second statement refers to the pursuit of contradicting strategies prioritizing national defense interests while simultaneously presenting itself as a multilateral security actor.[2] In the words of Irondelle and Schmitt, France tends to "punch above its weight in the international system."[3]

This chapter presents key contributions within the strategic culture literature on France structured according to themes included in the scholarly work in the field. The concept of strategic culture generally tends to invite a focus on long-term, national perspectives stressing stability rather than change. France's strategic culture, however, seems particularly coherent and stable over time with few scholarly debates over its content, intentions, driving forces, and consequences for France's role in the world. These trends are reflected in the literature on the subject. However, the conformity might also stem from a relatively limited number of epistemologies, perspectives, and approaches to the research of the strategic culture of France, which have been perhaps less apt to reveal features but rather consequences and effects of the strategic culture. The second part of the chapter aims to identify some such gaps in the literature and to suggest questions and alternative approaches for future studies.

Key Themes from the Literature on French Strategic Culture

History and Past Experiences

In addition to prominent works on irregular and regular warfare strategies, there is an equally strong strand of scholarship on France's historical experiences of wars and conflicts. Several of these works are longitudinal and stretch over long time periods.[4] The historical experiences often aim to explain present-day thinking about warfare and strategy

 DOI: 10.4324/9781003010302-18

as well as France's role in the world and its present defense policies. Some works date back to the Westphalian wars, Louis XIVth's reign, and the 1870s,[5] others center on the de Gaulle era from World War II onward,[6] and yet others on the time period from the end of the Cold War.[7] A recurring theme in these historical accounts is analyses of how France gained its great power status, made a substantial imprint on civil society in foreign places by way of its *mission civilisatrice* ("civilizing mission") and later on – although in a modified version – through *la francophonie* (the global French-speaking community). Special attention is given to the legacy of General Charles de Gaulle. The mark by de Gaulle runs deep in French strategic culture, not only because of his role as a war hero leading the resistance movement against the Nazi occupation during World War II but also for having brought France out of the humiliation of defeat and reinstalling a sense of national pride reclaiming France as a great power. The introductory phrase of de Gaulle's memoirs has become famous: "la France ne peut être la France sans la grandeur" (France cannot be France without greatness).[8]

Gaullism (i.e., the conservative French nationalist policies and principles associated with de Gaulle dating back to the resistance movement during World War II) is the key factor used to explain France's strong emphasis on national autonomy, which was made manifest, for example, in its withdrawal from the North Atlantic Treaty Organization's (NATO) integrated command structure in 1966. The intent of de Gaulle and his followers was to declare France as a great power and to carve out what was termed "the third way" for an independent France between the American and Soviet superpowers. In line with these strategies, France acquired the role of a key global diplomat in defense of republican values of liberty and equality in which peace and prosperity could be achieved without the recourse to the use of force.[9] In Moïsi's words: "France regards itself as carrying a universal message about 'the rights of man,' liberty, equality and fraternity."[10] De Gaulle's strategy was continued by later presidents and became known as French exceptionalism.[11]

Colonialism established France as a world power with French territories and consequently citizens across the globe. Even today, France has over two million citizens living outside of the *hexagone* (mainland European France). French language, values, traditions, and ways of life were exported as universal liberal values. Nevertheless, the mission to disseminate French (liberal) cultural values was paralleled by continuous manifestations of military might, a strong defense, and a dependence on nuclear deterrence.[12] Many of these features have lived on, adapting to new times and to changing global security situations. This is true also for the special kind of relationship that France cultivated with the Third World,[13] where it kept up close diplomatic relations with especially African state leaders. As Chafer writes, the support given by African troops during World War II reinforced the significance of Franco-African relations, but it was also a firmly held belief in France that its great power status rested on close links with Africa.[14]

France's Continuing Ambitions for Great Power Status

Despite its declining military and economic strength, defeats in three consecutive wars, and the loss of its empire in colonial conflicts and wars, France has retained its great power ambitions and continues to offer a view on global security challenging that of the United States. France's great power status is heavily reliant on nuclear deterrence,[15] but equally important for its aspirations for world influence is a sense of responsibility for peace and prosperity for French citizens in all of France's regions and territories across the globe, and for human and civil rights in France's spheres of interest. However, as Luis Simón points out, "France's so-called axis of strategic priority (Northern Africa, the Sahel, Levant, Horn of Africa/Red Sea and the Gulf) is characterized by mounting instability and increasing penetration by external powers."[16]

Writings about France's great power ambitions most often center on ways in which French leadership has successfully managed to achieve and maintain a great power role despite lacking matching material and economic resources. Studies show and argue how active and skilled diplomatic activities in key international institutions, of which permanent membership of the UN Security Council is by far the most significant, have afforded the middle power France the position of a great power.

Love-Hate Relationship with the United States

The relationship with the United States is another major theme in the literature on French strategic culture, and for good reason. Although at heart an ally, it has often been argued that France needs the United States as a polar opposite to understand and cultivate its national identity and to affirm its position in the world. However, French anti-American sentiments are not fictitious but have often been spurred by conflicts between the two states over trade issues, international status, influence in international organizations, strategies for military interventions, or the political and cultural dominance of the United States. Many writers refer to the competition for world influence as the key driving force behind the antipathy. Moïsi writes: "between the lines of France's carefully worded diplomatic statements one can discern a distinct distaste for America's oft-proclaimed sole-superpower status, or on matters of culture, where France is always the first to denounce the American 'cultural imperialism.'"[17]

There is a striking consistency over time in the instability of the relationship between France and the United States. Depending on the chosen time frame, some studies come to emphasize how France has sometimes let go of its exceptionalism and approached the United States to invest more heavily in the transatlantic link.[18] Studies covering other time periods show a greater reluctance from France to let go of its Gaullist tradition.[19] The frustration over US unilateralism has been a recurring feature of this troubled Franco-American relationship. Uncertainty has been added in recent years as to what the US defense commitments to Europe are. The United States has confirmed its continued support for European security, while at the same time prioritizing the Pacific and Southeast Asia. This triggers the old French distrust of the United States and the Gaullist idea of a great France taking lead position in a Europe independent of America.[20]

Another significant theme in the literature on Franco-American security relations deals with the competition between the two states on global security strategies and conflict resolutions. In these cases, France often represents the European Union (EU). Some argue that France aids the United States through burden-sharing in global security matters, such as by engaging in operations fighting jihadism in the Sahel region in Africa. In several conflicts, France has promoted diplomacy and negotiations where the United States has wanted a quick and massive military intervention.[21] However, in times of crisis France has repeatedly been a loyal US ally. This was clearly demonstrated in the wake of the 9/11 terror attacks when France immediately expressed support for the United States, first through diplomatic channels and then by offering and providing military support in the US attacks on Afghanistan and Al-Qaeda. On 13 September 2001, *Le Monde* even published an editorial with the now famous phrase "Nous sommes tous Américains" ("We are all Americans").[22]

A Strong Proponent of Europeanization and Multilateralism

As one of the founding fathers of what is today the EU and with ambitions to increase European independence from the United States in regard to security and defense, France took a lead position early on. After the end of the Cold War, the drive to develop the European Security and Defense Policy (ESDP) was intensified and France has been clear in its ambitions to promote Europe as a military power. Literature on France's role in the Europeanization of defense and security policy is as rich as the Europeanization of French defense policy. There is a shared theme in the treatment of the French foreign and security policy stance vis-á-vis Europe, characterized by a long-standing tension between a profound belief in multilateralism and an equally profound belief in national autonomy.[23] The constant hesitancy in the EU on whether and how it should be turned into a proper strategic actor to be reckoned with globally serves as a hindrance to France's engagement. Terpan writes: "the more the EU emerges as a global power, the more France will be willing to include – or integrate – its defense policy in a European collective framework."[24]

Despite a strong undercurrent of national autonomy, French strategic culture in reference to defense and military matters is deeply integrated into and overlapping with the EU's strategic culture.[25] With Brexit taking the United Kingdom out of the EU, European strategic culture is likely to be less divisive and feature fewer conflicts between French and EU interests – but it is also a European strategic actor which is weakened globally,[26] not the least if the United Kingdom intensifies its collaboration with the United States outside of NATO.

France and the North Atlantic Treaty Organization

The tension between national independence and multilateralism has been more strongly noted regarding France's participation in NATO than in the EU and the UN, and in part, this has been due to Franco-American tensions, which have been extensively covered in the literature.[27] With France reentering NATO's integrated command structure in 2008, it has been said that old traditions connected to French strategic culture are now abandoned and that the yoke of de Gaulle is finally lifted. Other scholars argue that "the Gaullist consensus on NATO was not as widely shared as commonly thought and that party preferences matter in order to understand France's NATO policy, thus adding a welcome degree of granularity in the study of this troubled relationship."[28] This strand of research has thus criticized what has been seen as a rather rigid view of French NATO membership as well as France's relations with the United States. Schmitt argues that "[t]here is a growing gap between the rhetoric of French security policy, emphasizing 'autonomy' and 'sovereignty' out of habits from the Cold War, and the actual security practices showing a gradual embedding within the transatlantic security structures."[29]

Strong Executive Presidential Powers on Matters of Defense and Security: La Grande Muette

The French constitution, laws, and regulations have a strong influence on the state's strategic culture. Scholars have written extensively on France's fifth constitution and have explained its aims to ensure stability, political consensus on national security issues, and efficient governance. This was done by the creation of a strong executive presidential power, close civil-military relations at the top level of government, and a limited legislative power.[30]

The French constitution is unique in that it gives the president the sovereign right to decide on an armed intervention against a foreign country without consultations with or consent from the parliament or the government.[31] Moreover, matters concerning security and defense issues are managed by a close circle of political leaders and military advisors, the so-called Restricted Security Council (le Conseil de Défense Restreint). This system, in which the military and the president decide on decisive defense and security issues such as military interventions, is often referred to as *la grande muette* (the great mute) since it is excluded from the public and from other bodies of the government. Officially it is called the *domaine réservé* – the area preserved for the exercise of presidential powers.[32] The president is continuously involved in military matters and meets with his military advisors, the Chief of Staff and the Special Chief of Staff, on an almost daily basis. Because of the lack of transparency in this part of the exercise of power, there are few studies on how the actual decisionmaking is done.

French Civil-Military Relations

With the advent of cybersecurity and information warfare such as disinformation, French civil-military relations have become less strict and the role of the military more difficult to define. Experiences from the Afghanistan mission, the terror attacks in Paris in 2015 (and Nice in 2016), and external developments in global security thinking have also brought changes to France's military culture. The military is strictly forbidden to make any political statements other than when called to act as experts in military matters.[33] However, the separation between the political and the military spheres as stipulated in the constitution of the Fifth Republic has increasingly come into question. Several spectacular events have taken place that confirm the cleavages which seem to be emerging in France's previously cohesive and consensual strategic culture on this account.

In 2017, after President Emanuel Macron had taken office, the tension between the military and the government intensified over defense spending and as a result of disagreements over the defense budget the French chief of staff resigned. When the critical voices against French engagement in Afghanistan grew louder, a few military high-ranking officers joined the public debate. Although the government gave reprimands, the military has continued to get involved in political discussions.[34] In April 2021, a group of military officers and soldiers – 20 of whom were retired generals – wrote an open letter which was published in the French far-rightist journal *Valeurs actuelles*,[35] arguing that France is on the verge of civil war and that perils of religious extremism are mounting. The armed forces minister, Florence Parly, responded that this was a breach of the political neutrality and loyalty that the military is obliged to respect.[36]

Yet another point of contention that shakes the traditional relationship between the military and the political spheres is caused by the fight against terrorism in France. Following the attacks in Paris and Nice in 2015 and 2016, the French government deployed 13,000 troops in France, increased the defense budget, and gave more autonomy to the military. Bove et al. argue from a critical standpoint that "[t]hese examples illustrate greater military intervention in politics following terrorist attacks, which in turn can change the civil-military equilibrium, and may hinder civilian control over the military and undermine the quality of democracy."[37]

Public Support for the French Military

The significance of the military in French strategic culture and the approach to the military by the elites is paralleled by considerable respect for the military by the public[38] and a traditionally widespread acceptance of the use of military force in international operations in defense of human security and civil liberties.[39] However, the changing role of the military can be seen in parallel shifts in public support for these external operations (OPEX), which seems to show less stability over time.

During the Afghanistan mission, support for the legitimacy of French participation dropped from 66% in 2001[40] to a majority of the French population being against a continuation of the mission in 2008,[41] and the opposition against a continuation of French participation kept increasing in the following years.[42] This change in attitude has been attributed to the lack of a coherent and convincing strategic narrative over time. To begin with, the mission was motivated by defending democratic values and human rights and aiding the United States in the fight against terrorism – both of which resonated with French interests – but later, fighting an American war became the argument for those opposing the mission. President Nicolas Sarkozy's strategy to collaborate with the United States in order to secure American support in case of future needs and to pave the way for France's reintegration into NATO's command structure did not appeal to either the opposition or to the French public.[43]

On-the-Ground Operations

Most scholars writing on French strategic culture and France's role in the world agree on the notion of a French exceptionalism in international relations and global security. France is an atypical international state actor in many respects with seemingly contradictory ambitions. However, research shows that in regard to military strategies and operations, there might be less of a French exceptionality and more of a compatibility with other countries' armed forces.

Relations with the United States are a case in point. Military interventions by the two powers are differently designed with the French being smaller in scale, adapted to the specific area, and underpinned by in-depth cultural competence and sensitivity in contrast to the American interventions based on massive force and high technology arms. Even so, the collaboration and trust between the United States and French armed forces have shown to be characterized by mutual respect. They have appeared relatively unaffected by political tensions between the two countries' governments, even when these have been at their worst.[44] Tenenbaum, studying irregular warfare doctrines, attests to convergences between French forces and their NATO allies and suggests that this might be due to "a political, strategic, and possibly cultural community leading most Western liberal democracies toward similar answers to irregular warfare."[45] Other studies point to American admiration of the French armed forces in Africa, praising them for their tactics and effectiveness in the fight against jihadist movements.[46]

Key Questions and Areas for Future Research

The Impact of External Changes: US-China-Russia Dynamics and Their Spheres of Interest

It is safe to say that foreign powers' perceptions of other nations have a bearing on the status and reputation of the latter. This in turn is relevant for influence and power within multilateral institutions, for relations with allies, and for regional leadership roles. Future studies on French strategic culture would do well to incorporate insights and perspectives from studies on international status building and work to connect these with security studies. Case studies of international crises might be fruitful to analyze. How did, for example, the French decision to block a UN Security Council resolution on the military intervention in Iraq in 2002/2003 affect France's international reputation long term and what were the consequences for the legitimacy of and domestic support for French strategic culture? A more recent question concerns what gains in international status France has won through the anti-jihadist operations in the Sahel region, of which it is the lead nation, and what future strategies might arise as a result of this. Analysts have argued that these missions have been beneficial for France's status in the eyes of the United States. At the same time, connections have been made between the terror attacks in Paris and its external operations.

Dependency on a strong international reputation does not stand in contrast to an independent approach to global security in the French case. Although rarely demarcating its distance to or difference from the United States during international crises, France is known to represent an independent and often alternative perspective on global security: how it is to be maintained and defended. The willingness in the past decades by many states, especially Western ones, to engage in international multilateral humanitarian interventions gave further impetus to the French strategy. With troops stationed around the world and with a constitution that enables decisions about military interventions in foreign countries to be taken quickly, the era of intervention warfare has served France well and strengthened its position in the world. Furthermore, the multinational humanitarian interventions have been grounded on the defense of human rights and international law, both of which are cornerstones of French strategic culture. The question is to what extent France's great power status is dependent on the willingness of the international community to engage in humanitarian military interventions.

France's belief in multilateralism is profound and its reliance on the EU is strong. If France is to have a voice in the world and retain influence and image as a great power it must retain its influence in the UN Security Council, but it must also work through the EU. Several scholars and practitioners note that the EU member states face similar security challenges and threats – ranging from cyber threats and disinformation from China and Russia to military build-ups in these countries and the looming risk of conflict in the wake of Russia's 2022 invasion of Ukraine – and therefore need to align and work to establish a European strategic culture.[47] In this line of thinking much has been written on the French role in the EU and in the ESDP, and the literature in the field makes few distinctions between the EU security strategy and that of France. However, there are exceptions to the similarities in French and European strategic interests which call for critical questions. How are French national security interests negotiated into the EU talks about the European security strategy and in reference to different issue areas, and what is the view and response from other EU member states on the French position? France's stance on China in reference to trade agreements is an example of such a question, where national autonomy and strife for multilateralism are juxtaposed.[48] The interface between the French and the EU strategic culture post-Brexit also calls for more systematic analyses and theoretical models to increase the understanding of the development of the EU strategic culture and France's influence in the Union.

The Blurring of Boundaries between the External and the Internal Spheres of Security

The interplay between the EU and France on security and defense matters can be seen as part of a bigger question about the blurring of external and internal boundaries regarding international relations, France's position in the world, and its multilateral engagements. Several incidents and recent crises have shown that external and internal violent conflicts and threats are connected. Threats from within are connected to military engagements abroad. This raises questions about how France's strategic culture might be affected by the interconnections between external and internal security and threats. Future studies are needed to increase the understanding of these linkages.[49] The blurring of the boundaries between internal and external security concerns could be seen to reflect new challenges in defining the national interest. France is a particularly interesting case since its long tradition of military engagements in foreign countries has brought insecurity to the *hexagone* several times before in modern history. Future work might ask if the latest attacks on French soil will lead to a rethink of national security or if this is perceived by the military and political elites as one of the somber parts of the strategic culture which France has accepted as part of its legacy and is reflective of the mission to disseminate universal values.

Such research should also call attention to the influence of domestic party politics on France's strategic culture. Will leftist and rightist-centrist parties seek to downplay France's multilateral argument and its engagement within NATO, and in order to appeal to the national conservative electorate stress autonomy and independence to a greater extent? Or will a lessened public support for multilateral cooperation and OPEX push French political leaders to promote a stronger position for France in Europe and an increasing national defense budget? What is the impact of anti-Islamist and anti-immigration sentiments on the traditional values which are part of the French strategic culture?

Civil-Military Relations and Political-Military Relations

The blurring of the boundaries between internal and external aspects of strategic culture is also connected to the increasingly tense relations between the military and the political sphere. The separation of the political sphere and the military remains a principle deeply rooted in the Fifth Republic and when it has been breached by senior military officers making political statements these have been severely reprimanded and lost their titles. However, different political developments such as defense budget restraints, the installment of a state of emergency, and the deployment of 10,000 French soldiers inside the *hexagone* following the Paris terror attacks in 2015 have caused unusual turbulence in the political-military relations. More research is needed to study what changes the interactions between the military and the political leadership are undergoing, how these are connected to the threat perceptions and the fight against terrorism, and what bearing these connections might have on the French strategic culture in the long run.

Another aspect of the changing political-military relations is that which concerns collaboration between civilian and military staff in military operations. French strategic culture shows a heavy emphasis on military aspects[50] but in light of the evolving political attitudes toward the military and the domestic situation in France post-2015, it might be expected that civil tasks assigned to the military increase in scope and significance. This is likely to have an impact on the role of the military profession and add civilian aspects to the concept of strategic culture. Such a development would also be in line with other security developments, such as cyber defense, which also depends on civil-military cooperation. More studies are needed to analyze the links between French military culture and strategic culture.

Public Opinion and the Sociology of Soldiering

Much of the literature on French strategic culture is state-centered, but with changes in views on security and the need to legitimize military spending, operations, and so on, more attention should be given to public perceptions and opinions.[51] It cannot be taken for granted that the French strategic culture is nationally shared. Longitudinal studies would be fruitful as well as comparative work to learn whether support is changing over time dependent on the type of crisis, mandate through UN Security Council resolutions or through the EU, international collaboration, and so on. Little is also known about the connections between the public image of the French defense forces and the support for military operations.

It might be suggested that the legitimacy of military operations is tied to perceptions of the military force. The French military has been a highly respected group of citizens and a corps who used to enjoy privileges in exchange for risking their lives in combat. However, with the economic crisis in France, the privileges and benefits for French soldiers and officers have been radically reduced. The social perception of the military culture and the status of the soldier and the officer need to be studied and related to public opinion of military operations and the defense forces.[52]

Assessing the Literature on French Strategic Culture: A Call for Critical Perspectives and Narrative Theory

Reviewing the literature in the field, there is a remarkable consensus on how the nature and character of French strategic culture are presented and explained. Most writers stress the continuity over time and the link between France's national self-image and French strategic culture dating back to the end of World War II or even longer.

As outlined above, previous research is rich in the field of history, with historical accounts shaping the current strategic culture and with President de Gaulle as the key figure. The historical analyses tie in with the construction and playing out of national self-images, France's role in Europe and its security concerns in regard to both national independence and European independence, and its love-hate relationship with the United States. Across these themes there are values, shared and referred to by most scholars, relating to France's identity as a staunch defender of multilateralism, human rights, and the dissemination of liberal values for peace and prosperity, first and foremost seen through its engagements in the UN Security Council, and through which national independence as well as global influence is maintained – the latter being disproportionately large in reference to the size of the armed forces and the economic strength of the country.

Despite the very concept of strategic culture involving long-term and stable values rather than short-term and changing values, there is reason to identify gaps within the literature related to changes in global security thinking, changing dynamics between the global powers, domestic circumstances, and to ways in which strategic cultures are being studied. To some extent, a strategic culture depends on national consensus. At a general level, there has to be a shared understanding and acceptance of the basic tenets of what that culture is. However, with the domestic problems facing France and with the way conflict and global security politics are becoming increasingly unpredictable and fast-changing,[53] this consensus might be more difficult to maintain and uphold. French citizens are known to distrust their leadership, but with the economic decline and increasing tensions in society, strong populist movements, recent terror attacks, and a resurgent threat of war in Europe there is even less certainty about future French public attitudes toward a large and expanding defense budget, a strong military presence abroad, and continued great power ambitions.

With this in mind, the concept of strategic culture might be more useful if not only applied for descriptive purposes but also made to serve as an analytic model and an explanatory variable. Reconceptualizing the concept of strategic culture might contribute to new ways of seeing security work. If the concept is expanded to include also internal threats and political violence this can serve to highlight adaptions and changes needed for improved resilience, conflict prevention, and post-conflict management. In other words, more focus should be given to the interplay between international status and the construction of strategic culture. How do these factors mutually reinforce one another and how might a weakened status impact on the perceptions of strategic culture and the use of strategic culture as a legitimizing tool in external affairs?

The existing scholarship on French strategic culture is dominated by geopolitics and realism. Although there are also studies from a constructivist perspective focused on national self-images and soft power mechanisms, for example, more studies are needed that take a critical and post-positivist perspective. This would also contribute to a more multidisciplinary view. One inroad to exploring the strategic culture from a critical perspective would be to integrate the use of narrative theory.[54] This implies viewing the articulation of strategic culture as storytelling in which messages and images are constructed and take on meaning, allowing for not only the construction of consensual but also contradictory and conflictual stories. A narrative approach might show how strategic cultures come into being, how they evolve or remain unchanged, and how they become powerful tools in exercising international influence and attaining domestic legitimacy.[55]

Furthermore, narrative theory points to the salience of collective identity and identity formation and how these link to power positions of actors in a system. This gives deeper insights into the cultivation of French national self-images and can include different actor perspectives – political leaders and the military as well as journalists, experts, and citizens. Narratives are also key to legitimatize interventions and other actions that stem from strategic culture. One might argue that without a convincing strategic narrative, the military or security operation will not succeed and the legitimacy of the operation will decrease.

Expanding the studies of French strategic culture to epistemologies other than positivism will lead to a broadening of the use of theories and methodologies. There are, for example, few theory-driven case studies drawing on ontological security or securitization.[56] Generally, insights about aspects of French strategic culture would be enhanced with more empirical analysis of primary sources, in particular using qualitative studies. Discourse-, framing-, and narrative analyses exploring the inner workings of government and military agencies, views on the military profession, France's diplomatic work in the UN Security Council and the EU, and the content of destabilizing disinformation campaigns are examples of such studies.[57] However, access to primary data is often limited. The closed circle around the president where major security decisions are taken makes any such analysis difficult, if not impossible. Studies are more fruitfully conducted drawing on public records, news media, and social media material.

In sum, the field of French strategic culture shows significant unity with few or no debates on what the culture contains, how it developed, and how it is being played out in different contexts. This might be due to the French case and the way in which the studies on French strategic culture have evolved, but it might partly be due to the understanding of strategic culture which directs focus on deep values and attitudes not prone to quick and frequent changes.

Notes

1 Maria Hellman, "Assuming Great Power Responsibility: French Strategic Culture and International Military Operations," in *European Participation in International Operations. The Role of Strategic Cultures,* ed. by Malena Britz (Champagne: Palgrave Macmillan, 2016), 23–48 referring to Bastien Irondelle and Olivier Schmitt, "France," in *Strategic Cultures in Europe: Security and Defence Policies Across the Continent* ed. by Heiko Biehl, Bastian Giegerich, and Alexandra Jonas (Wiesbaden: Springer VS, 2013), 125–137.
2 Hellman, "Assuming Great Power Responsibility: French Strategic Culture and International Military Operations."
3 Bastien Irondelle, and Olivier Schmitt, "France," 125.
4 See, for example, Frédéric Bozo, *La Politique Étrangère de la France depuis 1945* (Paris: Flammarion, 2012).
5 See Bertrand Badie, "French Power-Seeking and Overachievement," in *Major Powers and the Quest for Status in International Politics. Global and Regional Perspective,* ed. by T. Volgy, R. Corbetta, K.A. Grant, and R.G. Baird (New York: Palgrave Macmillan, 2011), 97–114.
6 Philip H. Gordon, *A Certain Idea of France: French Security Policy and the Gaullist Legacy* (Princeton, NJ: Princeton University Press, 1993); Stanley Hoffmann, "La France dans le monde 1979–2000," *Politique étrangère* 65, no. 2 (Summer) (2000): 307–317; Pierre Lellouche, *Légitime défense. Vers une Europe en sécurité au XXIème siècle* (Paris: Editions Patrick Banon, 1996).
7 Alice Pannier and Olivier Schmitt, *French Defence Policy Since the End of the Cold War* (London: Routledge, 2021); Pernille Rieker, *French Foreign Policy in a Changing World. Practising Grandeur* (Cham: Palgrave Macmillan, 2017).
8 Charles de Gaulle, *Mémoires de guerre: l'appel 1940–1942* [War Memoirs: The Call to Honour 1940–1942] (Paris: Librarie Plon, 1954), 1.
9 Charlotte Wagnsson, *Security in a Greater Europe: The Possibility of a Pan-European Approach* (Manchester: Manchester University Press, 2008).
10 Dominique Moïsi, "The Trouble with France," *Foreign Affairs* (May–June 1998): 95.
11 Olivier Schmitt, "The Reluctant Atlanticist: France's Security and Defence Policy in a Transatlantic Context," *Journal of Strategic Studies* 40, no. 4 (2017): 463–474. https://doi.org/10.1080/01402390.2016.1220367; Maurice Vaïsse, *La Puissance ou l'influence? La France dans le Monde depuis 1958* (Paris: Fayard, 2009); Thierry Tardy, "France: Between Exceptionalism and Orthodoxy," in *Global Security Governance: Competing Perceptions of Security in the 21st Century,* ed. by Emil J. Kirchner and James Sperling (London: Routledge, 2007), 25–45.
12 François Géré, "Faute de frappe ou l'érosion de la stratégie française de dissuasion nucléaire," *Revue Défense Nationale* 7, no. 782 (2015): 176–185.

13 Claude Wauthier, Quatre présidents et l'Afrique: De Gaulle, Pompidou, Giscard d'Estaing, Mitterrand: quarante ans de politique africaine (Paris: Seuil, 1995).
14 Tony Chafer, "From Confidence to Confusion: Franco-African Relations in the Era of Globalisation," in *France on the World Stage: Nation State Strategies in the Global Era*, ed. by Mairi Maclean and Joseph Szarka (Basingstoke: Palgrave Macmillan, 2008), 37.
15 Lucien Poirier and François Géré, *La réserve et l'attente: l'avenir des armes nucléaires* françaises (Paris: Economica, 2001).
16 Luis Simón, "The Spider in Europe's Web? French Grand Strategy from Iraq to Libya," *Geopolitics* 18, no. 2 (2013): 403), DOI: 10.1080/14650045.2012.698336
17 Dominique Moïsi, "The Trouble with France," *Foreign Affairs* (May–June 1998): 94.
18 Alice Pannier, "From One Exceptionalism to Another: France's Strategic Relations with the United States and the United Kingdom in the Post-Cold War Era," *Journal of Strategic Studies* 40, no. 4 (2017): 475–504; Olivier Schmitt, "The Reluctant Atlanticist: France's Security and Defence Policy in a Transatlantic Context," *Journal of Strategic Studie*s 40, no. 4 (2017): 463–474. https://doi.org/10.1080/01402390.2016.1220367
19 Benjamin Leruth, "Gaullism as Doctrine and Political Movement," in *The Routledge Handbook of French Politics and Culture* ed. by Marion Demossier, David Lees, Aurélien Mondon, and Nina Parish (Abingdon: Routledge, 2019), 36–44.
20 See Frédéric Charillon, *La France peut-elle encore agir sur le monde*? (Paris: Armand Collin, 2010).
21 See Frédéric Bozo, *A History of the Iraq Crisis. France, the United States and Iraq 1991–2003* (New York: University of Colombia, 2016). Olivier Kempf (interview) refers to the dispute between the United States and France in the run-up to the Iraq invasion in 2003 in Hellman, "Assuming Great Power Responsibility: French Strategic Culture and International Military Operations."
22 Jean-Marie Colombani, "Nous sommes tous Américains," *Le Monde,* 13 September 2001.
23 Anand Menon, "From Crisis to Catharsis: ESDP after Iraq," *International Affairs* 80, no. 4 (2004): 631–648.
24 Fabien Terpan, "The Europeanization of the French Defence Policy," Conference paper at the EU in International Affairs Conference, 2008.
25 See Pernille Rieker, "From Common Defence to Comprehensive Security: Towards the Europeanization of French and Foreign Security Policy," *Security Dialogue* 37, no. 4 (2006): 509–528.
26 Nicole Gnesotto, *Faut-il Enterrer la Défense Européenne?* (Paris: La Documentation Française 2014).
27 Sten Rynning, Changing Military Doctrine: Presidents and Military Power in Fifth Republic France, 1958–2000 (Westport: Praeger, 2002); Bertrand Badie, "French Power-Seeking and Overachievement"; Annick Chizel and Stéfanie von Hlatky, "From Exceptional to Special? A Reassessment of France-NATO Relations since Reintegration," *Journal of Transatlantic Studies* 12, no. 4 (2014): 353–366.
28 Oliver Schmitt, "The Reluctant Atlanticist: France's Security and Defence Policy in a Transatlantic Context," *Journal of Strategic Studies*, 40, no. 4 (2017): 471 referring to Stephanie Hoffmann, *European Security in NATO's Shadow: Party Ideologies and Institution Building* (Cambridge: Cambridge University Press, 2013).
29 Olivier Schmitt, "The Reluctant Atlanticist: France's Security and Defence Policy in a Transatlantic Context," 463.
30 Bastien Irondelle, "The Fifth Republic at Fifty: Logics and Dynamics," Conference paper presented at the APSA Annual Meeting, Chicago (29 August 2007).
31 Anthony Forster, *Armed Forces and Society in Europe* (Basingstoke, UK: Palgrave, 2006).
32 Bastien Irondelle, "Europeanization without European Union? French Military Reforms 1991–1996," Paper delivered to the ECSA Seventh Biennial International Conference (31 May–2 June, Madison, 2001), 14.
33 Samy Cohen, "Le pouvoir politique et l'armée," *Pouvoirs* 125, no. 2 (2008): 19–28.
34 See, for example, writings by Vincent Desportes, a highly distinguished general who resigned after having expressed critique against the French Afghanistan mission. Vincent Desportes, "La guerre en Afghanistan et la France: un bien lointain conflit," *Revue Défense Nationale* 740 (2007): 52–60.

35 "Pour un retour de l'honneur de nos gouvernants: 20 généraux appellent Macron à défendre le patriotisme," *Valeurs Actuelles*, 21 April 2021. https://www.valeursactuelles.com/politique/pour-un-retour-de-lhonneur-de-nos-gouvernants-20-generaux-appellent-macron-a-defendre-le-patriotisme/
36 Florence Parly (@florence_parly), "Tweet," 25 April 2021, https://twitter.com/florence_parly?lang=en.
37 Vincenzo Bove, Mauricio Rivera, and Chiara Ruffa, "Beyond Coups: Terrorism and Military Involvement in Politics," *European Journal of International Relations* 26, no. 1 (2020): 264.
38 Barbara Jankowski, "War Narratives in a World of Global Information Age: France and the War in Afghanistan," *IRSEM Paris Paper* 8 (2013), 13.
39 Bastien Irondelle and Olivier Schmitt, "France," in *Strategic Cultures in Europe: Security and Defence Policies Across the Continent*, ed. by Heiko, Biehl, Bastian Giegerich, and Alexandra Jonas (Wiesbaden: Springer VS, 2013), 133.
40 Sondage IPSOS, "Les Français et le conflit en Afghanistan," Sondage (2001). https://www.ipsos.com/fr-fr/les-francais-et-le-conflit-en-afghanistan.
41 The polls range between 55% and 68% of the public being against a prolonged French participation. Sondage IFOP pour Dimanche Ouest France, "Les Français et l'envoi de troupes supplémentaires en Afghanistan," 4 April 2008.
42 Ronald Hatto, "French Strategic Narratives, Public Opinion, and the War in Afghanistan," in *Strategic Narratives, Public Opinion, and War: Winning Domestic Support for the Afghan War*, ed. by Beatrice De Graaf, George Dimitriu, and Jens Ringsmose (London: Routledge, 2015), 157–176, 162.
43 Hatto, "French Strategic Narratives, Public Opinion, and the War in Afghanistan."
44 Hellman, "Assuming Great Power Responsibility: French Strategic Culture and International Military Operations."
45 Élie Tenenbaum, "French Exception or Western Variation? A Historical Look at the French Irregular Way of War," *Journal of Strategic Studies* 40, no. 4 (2017): 572.
46 Cizel and von Hlatky, Annick Chizel, Annick and Stéfanie von Hlatky, "From Exceptional to Special? A Reassessment of France-NATO Relations since Reintegration," *Journal of Transatlantic Studies* 12, no. 4 (2014): 353–366; Michael Shurkin, "France's War in Mali. Lessons for an Expeditionary Army," Rand Report (Santa Monica: Rand Corporation, 2014); Anand Menon, "France Displaces Britain as Key US Military Ally," Agence France Press 19 March 2015.
47 Strategic Culture from European Union Institute for Security Studies, a report from a seminar on 18 June 2021.
48 See Alister Miskimmons, Ben O'Loughlin, and Ben Zeng, *One Belt, One Road, One Story? Towards an EU-China Strategic Narrative* (Basingstoke: Palgrave Macmillan, 2021).
49 Studies on threat perceptions in connection to national security are very limited in the French case. The White Papers 2008 and 2013 outlining the French security and defense strategies have few if any mention of threat images and perceptions. President of the French Republic, *White Paper on Defence and National Security* (Paris: Odile Jacob, 2008); President of the French Republic, *White Paper. Defence and National Security* (Paris: Odile Jacob, 2013).
50 Chiara Ruffa, "What Peacekeepers Think and Do: An Exploratory Study of French, Ghanian, Italian, and South Korean Armies in the United Nations Interim Force in Lebanon," *Armed Forces and Society* 40, no. 2 (2013): 199–225. DOI: 10.1177/0095327X12468856
51 See Jankowski, "Opinion Publique et Armées á l'Epreuve de la Guerre en Afghanistan" One exception to this is Ronald Hatton, "French Strategic Narratives, Public Opinion, and the War in Afghanistan."
52 See Chiara Ruffa, "Military cultures and force employment in peace operations," *Security Studies* 26, no. 3 (2017): 391–422.
53 See Miskimmons, O'Loughlin, and Zeng, *One Belt, One Road, One Story? Towards an EU-China Strategic Narrative.*
54 See, for example, Beatrice De Graaf, George Dimitriu and Jens Ringsmose, *Strategic Narratives, Public Opinion, and War: Winning Domestic Support for the Afghan War*, ed. by Beatrice De Graaf, George Dimitriu, and Jens Ringsmose (London: Routledge, 2015).

55 Olivier Schmitt, 2018 "When Are Strategic Narratives Effective? The Shaping of Political Discourse through the Interaction between Political Myths and Strategic Narratives," *Contemporary Security Policy* 39, no. 4 (2018): 487–511. DOI: 10.1080/13523260.2018.1448925

56 There are comparative studies on securitization in which France is included. See, for example, Philippe Bourbeau, *The Securitization of Migration. A Study of Movement and Order* (London: Routledge, 2011).

57 Massimo Flore, "Understanding Citizens' Vulnerabilities (II): From Disinformation to Hostile Narratives. Case Studies Italy, France and Spain," Luxembourg: Technical Report, European Commission, 2020.

Selected Bibliography

Badie, Bertrand. "French Power-Seeking and Overachievement." In *Major Powers and the Quest for Status in International Politics. Global and Regional Perspective*, edited by T. Volgy, R. Corbetta, K.A. Grant, and R.G. Baird, 97–114. New York: Palgrave Macmillan, 2011.

Bove, Vincenzo, Mauricio Rivera and Chiara Ruffa. "Beyond Coups: Terrorism and Military Involvement in Politics," *European Journal of International Relations* 26, no. 1 (2020): 263–288.

Bozo, Frédéric. *La Politique Ètrangère de la France depuis 1945*. Paris: Flammarion, 2012.

Charillon, Frédéric. *La France peut-elle encore agir sur le monde?* Paris: Armand Collin, 2010

Cohen, Samy. "Le pouvoir politique et l'armée," *Pouvoirs* 125 (2008): 19–28.

De Gaulle, Charles. *Mèmoires de guerre: l'appel 1940–1942* [War Memoirs: The Call to Honour 1940–1942]. Paris: Librarie Plon, 1954.

Drévillon, Hervé and Olivier Wieviorka. eds. *Histoire militaire de la France: de 1870 à nos jours*. Paris: Perrin, 2018.

Durieux, Benoit. *Clausewitz en France. Deux siécles de réflexions sur la guerre 1807–2007*. Paris: Economica, 2008.

Géré, François. "Faute de frappe ou l'érosion de la stratégie française de dissuasion nucléaire," *Revue Défense Nationale* 7, no. 782 (2015): 176–185.

Gnesotto, Nicole. *Faut-il Enterrer la Dèfense Europèenne?* Paris: La Documentation Française, 2014.

Gordon, Philip H. *A Certain Idea of France. French Security Policy and the Gaullist Legacy*. Princeton, NJ: Princeton University Press, 1993.

Guisnel, Jean and Bruno Tertrais. *Le président et la bombe; Jupiter à l'Élysée*. Paris: Odile Jacob, 2016.

Hatton, Ronald. "French Strategic Narratives, Public Opinion, and the War in Afghanistan." In *Strategic Narratives, Public Opinion, and War. Winning domestic support for the Afghan War*, edited by Beatrice De Graaf, George Dimitriu, and Jens Ringsmose, 157–176. London: Routledge, 2015.

Hellman, Maria. "Assuming Great Power Responsibility: French Strategic Culture and International Military Operations." In *European Participation in International Operations. The Role of Strategic Cultures*, edited by Malena Britz, 23–48. Basingstoke: Palgrave Macmillan, 2016.

Hoffmann, Stanley. "La France dans le monde 1979–2000," *Politique ètrangère* 2 (2000): 307–317.

Irondelle, Bastien and Sophie Besancenot. "France. A departure from exceptionalism?" In *National Security Cultures. Patterns of Global Governance*, edited by J. Kircher Emil and Sperling James, 21–41. London: Routledge, 2010.

Irondelle, Bastien and Olivier Schmitt. "France." In *Strategic Cultures in Europe. Security and Defence Policies across the Continent*, edited by Biehl Heiko, Bastian Giegerich, and Alexandra Jonas, 125–137. Wiesbaden: Springer VS, 2013.

Meunier, Sophie. "Anti-Americanisms in France." *French Politics, Culture & Society* 23, no. 2 (Summer 2005): 126–141.

Moïsi, Dominique. "The Trouble with France." *Foreign Affairs* (May–June 1998): 94–98.

Pannier, Alice. "From One Exceptionalism to Another: France's Strategic Relations with the United States and the United Kingdom in the Post-Cold War Era." *Journal of Strategic Studies* 40, no. 4 (2017): 475–504.

Pannier, Alice, and Olivier Schmitt. Fr*ench Defence Policy since the End of the Cold War*. London: Routledge, 2021.
Rieker, Pernille. *French Foreign Policy in a Changing World: Practicing Grandeur*. Cham: Palgrave Macmillan, 2017.
Ruffa, Chiara. "Military Cultures and Force Employment in Peace Operations." *Security Studies* 26, no. 3 (2017): 391–422.
Rynning, Sten. *Changing Military Doctrine: Presidents and Military Power in Fifth Republic France, 1958–2000*. Westport: Praeger, 2002.
Schmitt, Olivier. "The Reluctant Atlanticist: France's Security and Defence Policy in a Transatlantic Context." *Journal of Strategic Studies* 40, no. 4 (2017): 463–474. 10.1080/01402390.2016.1220367
Tardy, Thierry. "France: Between Exceptionalism and Orthodoxy." In *Global Security Governance: Competing Perceptions of Security in the 21st Century*, edited by Emil J. Kirchner and James Sperling, 25–45. London: Routledge, 2007.
Tenenbaum, Élie. "French Exception or Western Variation? A Historical Look at the French Irregular Way of War." *Journal of Strategic Studies* 40, no. 4 (2017): 554–576. 10.1080/01402390.2016.1220368.

16

SWEDISH EXCEPTIONALISM AND STRATEGIC CULTURE[1]

Swedish Nuclear Decisionmaking

Harrison Menke

Introduction

Sweden offers unique insights into the study of strategic culture. When looking through a prism of realism or idealism, Swedish decisions can be contradictory and confusing. To better understand Sweden and its unique decisionmaking process, a deeper examination of distinct sociopolitical factors is required. This chapter will first highlight historical influences on Swedish strategic culture. It will then examine the key factors and functions shaping that culture. Second, the chapter will assess the impact of Swedish strategic culture on two key case studies: Sweden's decision to pursue and abandon nuclear weapons and its decision not to join the Treaty on the Prohibition of Nuclear Weapons. From these examples, this chapter posits the existence of two subcultures within Swedish strategic culture: Constructive Neutralists and Traditional Neutralists. In so doing, this chapter seeks to identify a distinctive pattern of decisionmaking that can be applied to Swedish foreign policy more broadly. Finally, this chapter concludes with a brief review of the key contributions within the strategic culture literature on Sweden, as well as some evaluations of the strengths and gaps within the body of literature and recommendations for future areas of research for scholars and students studying the strategic culture of Sweden.

Background: Historical Influences on Swedish Strategic Culture

A study on any nation-state's strategic culture should begin with an assessment of important historical factors that inform collective perceptions. Sweden has an extensive history that spans centuries. For example, from 800 to 1050 AD Swedish Vikings actively roamed across the Baltic region. Sweden played a decisive role in the Thirty Years' War, when Gustav Adolphus landed in Pomerania and defeated imperial German troops. Gustav's actions fundamentally altered the European political landscape and cemented Sweden's status as a great power.[2] From its zenith during the seventeenth and eighteenth centuries, a series of disastrous defeats slowly eroded Sweden's power.[3] In its final defeat by the Russian Empire, Sweden was forced to accede to the humiliating Treaty of Fredrikshamn. Sweden yielded Finland, an integral buffer since at least the fourteenth

DOI: 10.4324/9781003010302-19

century representing nearly a third of Swedish territory and a fourth of its population. This marked the definitive loss of Sweden's great power status, both geographically and culturally.[4]

The loss of Finland birthed the modern Swedish state.[5] The great sense of national injury, both real and perceived, compelled Sweden to reexamine its self-identity and create a new national frame of reference. To circumvent the intellectual void, young nationalists and romanticists within the *Götiska Förbundet* movement revitalized bygone myths and traditions from the Viking Age and Sweden's Gothic origin. These analysts leveraged certain characteristics and values that were held to be specifically Swedish to stake out a future course.[6] This resulted in novel ideational formulations about what constituted the Swedish nation. The "New Sweden" explicitly eschewed nostalgia for the Imperial Age.[7] For example, an early 1900s study of rural Swedish families found that because the flag was associated with royalty and militarism, it was avoided as a national symbol.[8] Rather, Sweden's past was engineered in a more utilitarian context, emphasizing ethnic and civic nationalism that cemented a bond across the entire *Folk* (nation).[9] Freedom, independence, and solidarity were central themes and seen as particularly typical of Sweden.[10] These traits were reflected in the philosophical teachings of Johan Gottfried Herder as well as Erik Gustaf Geijer's influential poems, *Manhem* (an Old Norse term for Sweden), *Odalbonden* (The Yeoman Farmer), and *Vikingen* (The Viking).[11]

The second key historical source of strategic culture was the nation-building impact of the Social Democratic Party (SDP). When the SDP came to power in the 1930s, Sweden was poor, rural, and faced an emigration epidemic.[12] To combat these challenges, the SDP pursued an ambitious social agenda known as the *Folkhem* project, literally meaning "the people's home." Central to the *Folkhem* ideology is that welfare and social entitlement are inalienable rights. At its foundation, *Folkhem* represents a symbiotic social relationship wherein greater equity is afforded to both the privileged and the disadvantaged alike to ensure social cohesion.[13] *Folkhem* eventually transcended the domestic arena, wherein core principles such as egalitarianism, humanism, justice, and consensus were applied to all aspects of Swedish activity.[14] Indeed, this ideology became so deeply embedded in national discourse and societal development that it formed an integral part of Swedish national identity.[15] While some ideals and institutions may have preceded the welfare state, they were nonetheless assimilated under the SDP's platform and became powerful artifacts within the national cognition.[16]

Swedish Exceptionalism and Its Subsidiary Principles

The discussion above demonstrates that shared norms, values, and ideals have been central to engineering a particular and distinct vision of *Svenskhet,* or "Swedishness." While the preceding discussion outlines causes and motivations, a deeper exploration is necessary to identify the relevant manifestations in Sweden's strategic culture. This section aims to classify and evaluate the sources of Sweden's strategic culture, expressed here as Swedish exceptionalism.

At the foundation of Swedish cognition is a common acceptance of Sweden's differences, superiority, and national mission.[17] Scholars tend to consider the publication of Marcus Childs' 1936 book *Sweden: The Middle Way* as the originator of the idea of Sweden as a special state.[18] With the world facing global conflict, Childs portrayed Sweden as an effective compromise between communism and capitalism, or a "middle way." Indeed, the

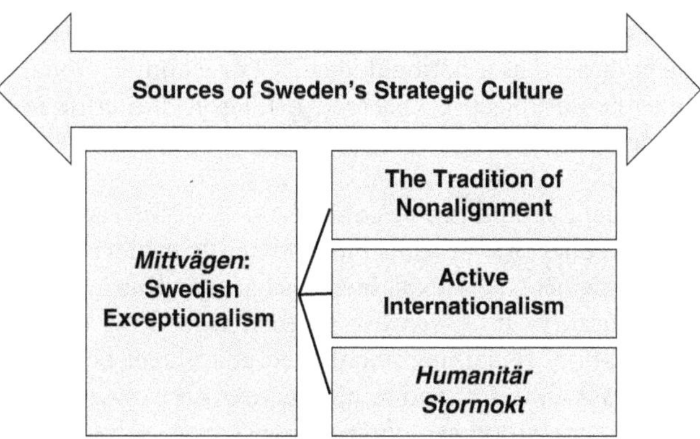

Figure 16.1 Sources of Swedish Strategic Culture.

Source: Author's Creation.

Swedes have long perceived themselves as being in the middle of primitive great power politics, thus necessitating an alternative model. As noted by Arne Ruth, however, this notion generated new conceptions that extended well beyond political balancing.[19]

Indeed, *Mittvägen* (the middle way) can better be characterized as a deeply rooted belief system centered on Sweden as a unique exception in the world. Scholars have rightly assumed that this self-perception is based on dominant values and beliefs in domestic society. For example, Peter Lawler suggests Scandinavian exceptionalism is partly driven by "distinctive values, including that of solidarity."[20] Similarly, Annika Berman posits that Sweden's self-narrative is that of "an outward-looking internationalist state whose commitments to justice and equality are not confined to co-nationals."[21] Sweden's exceptional status has become a commonly held perception, reinforced by public figures.[22] Swedish officials regularly refer to their efforts toward pioneering or leading progress globally as extraordinarily progressive.[23] As a result, the collective sense of exceptionalism constitutes a mythology that is the very backbone of the national consciousness.[24]

Within the *Mittvägen* construct, three mutually supporting core principles or ideals can be extrapolated (see Figure 16.1). These include nonalignment, internationalism, and moralism/humanitarianism. Each will be briefly described in turn.

The Tradition of Nonalignment

The oldest and perhaps most influential element within *Mittvägen* is the tradition of nonalignment. Sweden prides itself on over 200 years of peace and a legacy of neutrality. After 1809, Sweden was a small state surrounded by great military and economic powers coveting Sweden's strategic geographic position. Weakened and unwilling to return to militarism, new formulations were envisioned to secure Sweden's security, thus resulting in a commitment to nonalignment.[25] This approach was perceived to be more enlightened and represented a rejection of crude and brutal power politics that had characterized Sweden's imperial past. To be sure, nonalignment was "gradual practice rather than conscious design."[26] For example, Britain and France used the Swedish island of Gotland to project power into the Baltic Sea against Russia during the Crimean War as Sweden weighed joining the conflict.[27] Yet, as Ann-Sofie Dahl noted, nonalignment offered Sweden a useful means to achieve national objectives and eventually grew to become an integral part of Swedish life embraced as a national ideal.[28] For example, nonalignment is believed to enable the state to be an objective observer, mediator, and critic in international politics.[29] This detachment is an intrinsic element of "Swedishness" and foundational to Swedish exceptionalism.[30]

However, nonalignment should not be confused with disinterest. Swedish officials often default to the enduring belief that peace is most assuredly achieved through freedom from political-military entanglements. As explained by Barbara Kunz, underpinning this argumentation suggests causality. It is *because* Sweden is neutral that it has managed to avoid war; by extrapolation, remaining nonaligned guarantees peace in the future.[31] This national myth continues to manifest and reinforce Swedish views. In this way, the Swedish approach can better be understood as *militära alliansfrihet*, or freedom from alliances. On one hand, autonomy was argued to generate stability and peace. As noted by former Swedish Prime Minister Per Albin Hansson in 1939,

> Friendly with all other nations and strongly linked to our neighbors, we look on no one as our enemy. There is no place in the thoughts of our people for aggression against any other country, and we note with gratitude, the assurances from others that they have no wish to disturb our peace, our freedom, or our independence.[32]

Given the uncertainty inherent within alliance constructs, Swedish analysts tend to view membership as destabilizing and risky.[33] However, Sweden has acted in its own self-interest when needed. During both World War I and World War II, Sweden supported different coalitions thought to be aligned with Swedish interests. According to historian Mikael af Malmborg, Sweden's position "cannot be captured with the simple dichotomies of 'neutral', or 'aligned'."[34] While Sweden seeks to be independent on the world stage and disentangled from military alliances, it does not seek to be strictly impartial.

Active Internationalism

According to Fred Halliday, internationalism exhibits two distinct modalities: liberal and hegemonic. Liberal internationalism entails an optimistic belief that processes of integration can result in greater cooperation, peace, and prosperity.[35] This notion is strongly linked with multilateralism, in that establishing the global community requires mechanisms that rely on cooperation between sovereign entities.[36] Sweden's approach is clearly distinguished from hegemonic internationalism, which uses asymmetric and unequal belligerence to transform the status quo. Hegemonic internationalism is closely associated with militarism. In both cases, however, a common theme of presupposed superiority of status is present. John A. Agnew explains that internationalist states understand themselves as having a special role and mission in the world to pioneer their enlightened beliefs and practices globally.[37] Similarly, Swedes believe themselves predestined to spread their unique approach to the rest of the world.[38]

The tenets described above are emblematic of and unmistakable in Swedish foreign policy. Swedish activism emphasizes equality, self-determination, and respect for a strong international legal order. Embedded within this approach is the belief that universal values, supported by interstate consensus, are necessary conditions for international peace and development.[39] This encourages greater assistance to developing nations and mediation between rival powers, as well as vocal criticism against more powerful states. Swedish strategic culture assumes that its principles of nonalignment and morality provide Sweden with an independent voice in international politics.[40] As noted by Ann-Sofie Nilsson,

> Only neutrality provides Sweden with the opportunity, and room, to act successfully as an independent third force in world affairs.... As such a third force, Sweden has the ability to intervene, diplomatically and politically, in situations in which other actors would be enjoined from taking action either by reason of prior commitment or their alignment elsewhere.[41]

This enables Sweden to distinguish its foreign policy from other countries, thereby reinforcing an image of exceptionalism.[42]

Humanitär Stormakt

Tying each of the above principles together is a strong preference to act in accordance with moral and humanitarian values. Such values are framed as being Sweden's global duty to be seen on the right side of history. The central concept is *värdegrund* (common value foundation). Kennert Orlenisu defines *värdegrund* as certain basic principles, such as equal value among individuals, which cannot be ignored.[43] Sweden has pursued primarily universalist norms to better the world regarding social and economic rights. In practice, the concept of *värdegrund* serves as a guideline for politically acceptable narratives and acts as a useful "trump card" when disputes emerge in forming national consensus.[44] Consequently, "equality at home and justice abroad have come to be regarded as complementary and mutually supporting values."[45] This is considered so exceptional that foreign observers have branded Sweden a *humanitär stormakt,* or humanitarian superpower.[46]

Functions of Swedish Strategic Culture

Underpinning each principle within Swedish exceptionalism is a series of foundational attributes (see Figure 16.2). These attributes can be grouped into four often overlapping elements: identity, values, norms, and perceptual lens. Each element will be examined sequentially. It should be noted that there are likely more attributes within Swedish strategic culture, but for the purposes and focus of this chapter, only those attributes which were considered pertinent to Swedish nuclear decisionmaking were explored.

Swedish Identity

As the discussion above demonstrates, Sweden's national identity is a project that has continued to evolve based on emerging needs. As noted by Erik Örjan Emilsson, there is often a lack of consensus about national identity and what it means to be Swedish.[47] For example, some Swedes still cling to the notion of an ethnically, religiously, and racially homogenous population. However, Sweden has never been as homogenous as these narratives contend.[48] Indeed, *Folkhem* envisions Swedish nationality based on core values and ideals rather than physical qualities.[49] This approach has deepened due to the unprecedented rate of immigration to Sweden in recent years. Ideals center on

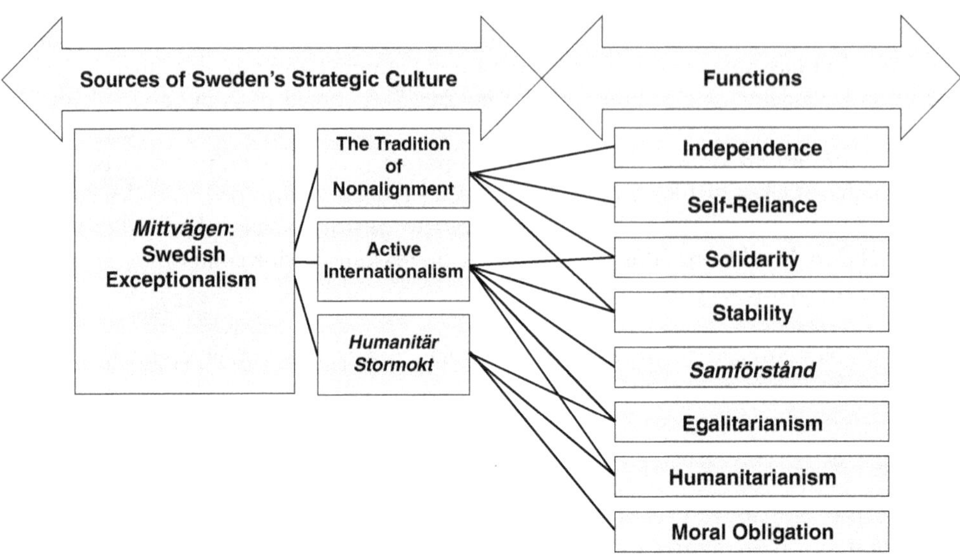

Figure 16.2 Functions of Swedish Strategy.
Source: Author's Creation.

egalitarianism, respect for universal human rights and democratic institutions, and modernity (e.g., aggressively championing gender and economic rights). Swedes pride themselves on overlooking brutish delineations based on sectarian biases and accepting a modern state based on commonly accepted values and norms. This has reaffirmed and strengthened Sweden's exceptional position and unique contribution to international affairs.

Religion is more recessed and muted in Swedish identity formulation than in some other states. Only 6% of the 62% of the population who are members of the Church of Sweden attend services regularly, while 65% of church members come in contact with the church through one or more of the rituals of baptism, marriage, and death during a year.[50] According to Gustafsson Lundberg and David Thurfjell, Swedes do not consider themselves as religious: "Instead they tend to perceive the other (Muslims, Jews, or Hindu) as religious, while being blind to their own Christian Protestant heritage."[51] Here, the religious other is associated with an outdated way of thinking, while Swedish secular and rational reasoning is depicted as well-informed and enlightened. However, religion probably contributed to the acceptance of certain values and norms. Thurfjell refers to the Swedish worldview as Protestant humanism, a position he explains as a combination of "post-materialism, secular rationalism, relativism and, not least, individualism."[52] He concludes that the Swedish Lutheran faith and long-standing Nordic traditions evolved together, streamlining Sweden's individualistic identity.

Finally, Sweden views itself as an influential member of the international arena that operates from a distinctly rational and moral perspective. These values (to be addressed below) help distinguish Sweden from other nation-states, particularly great powers prone to militarism and violence. Indeed, the moral undertones within Swedish identity suggest that Sweden is always on "the side of good," whereas activities by others are much more dubious. As noted by Einhorn, only Nordic cooperation and cohesion transcends this view.[53] Indeed, *Norden* (Nordic) was often seen to characterize progressivism, egalitarianism, secularism, and modernism, whereas Europeanization represented backwardness characterized by Catholic values.[54] *Norden* helps explain Sweden's hesitancy toward broader identification with the rest of Europe that persisted until after the Cold War.

Swedish Values

A central theme of freedom and independence runs throughout Swedish values. Geijer emphasized freedom as a special characteristic of Scandinavian people in his influential lectures during the early 1800s. Geijer argued that the Swedish Theory of Love could be applied to international relations, in which respect and mutual trust require a degree of independence afforded by all parties involved.[55] Sweden values mutual respect (ensured by an adequate degree of independence) to ensure no actor is overlooked, ignored, or dominated. This concept is also reflected in the importance Swedes attach to sovereignty, both at the individual and state level, which underpins Sweden's nonalignment posture.

An equally important influence has been the acceptance of universal values and egalitarianism, which have come to be regarded as mutually supportive values and are applied globally.[56] Sweden's humanitarian superpower status imbues a collective consciousness that it is duty bound to objectively do right. Implicit is the assumption that the interests and demands of states can be satisfied within a global rules-based framework rooted in universal justice to facilitate harmony. Sweden views itself as working for the good of the international community even at the expense of immediate state interests. Here, Sweden's universalist predisposition is closely tied to solidarity with developing nations. In Archer's words, "As a neutral small state, Sweden has a legacy of supporting international law, international organizations, and multilateralism to safeguard the independence and security of small states in the international system."[57] By aligning its engagement toward small-state solidarity, Sweden can improve conditions in the international system.

Finally, Sweden exhibits a strong preference for consensus, bridge-building, and cooperation. The Swedish ethnologist Åke Daun explicitly captures the importance of consensus as "conflict avoidance" among the traits of the "Swedish mentality."[58] Consensus implies the resolution of conflicts – either temporarily or indefinitely – by a compromise between inevitable differences between values and goals.[59] However, Rosenberg argues that the Swedish word for common understanding, *samförstånd*, tends to obfuscate the distinction between consensus and compromise. Consequently, Sweden's view on consensus is flexible in application to ensure confrontation is avoided or diminished.

Swedish Norms

Given the above discussion within identity and values, Sweden is clearly oriented toward cooperative mechanisms. However, this should not be taken to mean that Sweden dismisses the utility of violence in certain circumstances. Indeed, Sweden regularly highlights the need for a strong military. The distinction between Sweden and other states rests on its self-perceived humanitarian mandate.[60] This includes upholding the values and ideals articulated by international institutions, through force when necessary. This orientation in turn shapes an attitude that compels Stockholm to forcefully oppose activities deemed ethically intolerable, such as genocide. [61] This altruistic perspective underscored Sweden's sense of justified duty during support to military operations in the former Yugoslavia, Afghanistan, and Libya.[62]

Sweden's support of Operation Unified Protector is a useful example to help identify the normative factors that underpinned Swedish decisionmaking. In a study by Fredrik Doeser, four overlapping elements had the greatest impact on the decision to participate in military operations: core tasks, operational mandate, willingness to use force, and organizational framework.[63] From this framework, two overlapping themes played a particularly important role. First, a strong moralistic element was present that under certain conditions Sweden has a moral obligation to use military force. As noted by then Defense Minister Sten Tolgfors, "Sweden helps others ... because it is the right thing to do."[64] Here, the use of force should give moral weight to the inherent equality of all human beings and objectively improve a situation. As noted by a Foreign Ministry official, "We perceived that we could do something useful and good."[65] Second, the role of an internationally sanctioned mandate was critical. As noted by then Prime Minister Fredrick Reinfeldt: "I think we have a moral obligation to act when international society is united around an operation."[66] Implicit is the necessity of acting by, with, and through legally recognized international institutions. This provides the basis for legitimacy required to escalate up to and past the threshold of armed conflict. As stated by then Foreign Minister Carl Bildt shortly before the UN Security Council agreed on Resolution 1973, "We want to stop tanks, artillery and massacres. That demands a clear UN mandate.... We hope to help people in need and we want to support the UN."[67] The UN directive, therefore, was instrumental to ensure Sweden acted within internationally defined rules.

Swedish Perceptual Lens

On the international stage, the most durable component of Swedish strategic culture is Sweden's self-perception as a small, influential state. Sweden is a relatively weak state surrounded by powerful neighbors. As argued by Ole Elgström, this part of the Swedish cognitive system formed the basis for Swedish nonalignment decisions.[68] At the core of Sweden's security perceptions is its view of Russia. Since the 1600s, a durable feature of the Swedish worldview was the image of Russia as the natural archenemy. Centuries of mistrust and suspicion have reinforced this image. For example, Swedish King Oscar I described Russia as an enemy to the whole liberal world.[69] During World War I, Swedish King Gustaf and Knut Wallenberg, then Swedish foreign minister, told German ambassadors that Sweden would never stand with Russia.[70] Similar hardline views are echoed today.[71] Michael Nordlind contends that demonization of Russia has prohibited unbiased analysis and inherently confines Moscow to the role of the enemy.[72] Valentin Sevéus goes further, stating that current Swedish attitudes can be explained by anti-Russian indoctrination that has historically prevailed in Sweden.[73]

Despite the enduring threat posed by Russia, Swedish leaders assumed that any attack on Sweden would likely occur as part of a broader conflict.[74] For example, Russian threats during the 1860s were contextualized by the Crimean War with Britain, France, and the Ottoman Empire. During the Cold War, it was widely held that the North Atlantic Treaty Organization (NATO), or at least the United States, would come to Sweden's aid in case of attack.[75] This was due in large part to Sweden's strategic location; Soviet forces in Sweden could disrupt the European balance of power and more effectively isolate West Germany, Denmark, and Norway. Thus, implicit within Swedish defense planning was a belief that US and Swedish defense interests were interconnected. As noted by Göransson, "After a week into hostilities, the Supreme Commander expects Sweden to receive military help."[76] At least until the recent decision to join the NATO Alliance (negotiations unfolding at the time of this writing), Swedish planning probably assumed that although Sweden intended to be neutral in the event of war, it would be so "on the NATO side."[77]

Manifestations of Sweden's Strategic Culture: Nuclear Decisionmaking

The following section applies the study of strategic culture to two key Swedish decisions: the decision to pursue and ultimately abandon nuclear weapons and the decision not to join the Treaty on the Prohibition on Nuclear Weapons (TPNW). These case studies will be used to determine whether there is a discernable pattern related to Swedish nuclear weapons choices.

Decision to Pursue Then Abandon Nuclear Weapons

Shortly after the atomic bombings of Hiroshima and Nagasaki, Sweden set about pursuing a small nuclear program. The growing Soviet threat led the Supreme Commander, the Air Force Commander, and the Minister of Defense to favor the acquisition of nuclear weapons. Proponents reasoned that a critical component of neutrality was the state's ability to deter opponents from attack. Vulnerability to nuclear threats, therefore, mandated that Sweden be self-sufficient in providing for its own security, including the acquisition and possible employment of nuclear weapons.[78] Sweden planned on around 100 tactical nuclear weapons fixed to torpedoes, missiles, and nuclear-capable aircraft to complement conventional operations in destroying points of embarkation to deny an amphibious invasion.[79] In addition to military factors, national pride through advancements in science were also visible. The SDP's 1959 nuclear weapon committee report emphasized that "the demand for a Swedish atomic weapon has been motivated primarily by the desire to see Sweden keep up with technical developments in weaponry."[80]

Changes both domestically and externally ultimately slowed Sweden's nuclear ambitions. The Swedish nuclear program was initially tight-knit and secretive; however, resource limitations exposed the program to Parliamentary oversight and public debate. Leading public objections was *Aktionsgruppen Mot Svensk Atombomb* (The Action Committee against Swedish Nuclear Weapons), an influential group which found favor with sympathetic SDP members.[81] Moreover, Swedish leaders assumed that Sweden tacitly and implicitly was under the US nuclear umbrella despite its neutral status.[82] Indeed, Sweden covertly cooperated with NATO, primarily the United States.[83] At first, Sweden delayed, or decided not to decide, a final decision on acquiring nuclear weapons. Yet eventually, as a result of these realities, Sweden abandoned its nuclear pursuits.

Underpinning Swedish decisionmaking was an ongoing competition over the appropriate hierarchy of Swedish cultural values. On one hand, proponents used the tradition of nonalignment, sovereignty, and independence to gain influence for their cause.[84] Nuclear weapons ensured *handlingsfrihet*, or freedom of action. Simultaneously, opponents leveraged equally compelling normative ideations to criticize the military's nuclear pursuit. For example, SDP members argued that Sweden should address the root cause of conflict through international relations rather than defense capabilities.[85] Multilateral initiatives, namely the Nuclear Nonproliferation Treaty (NPT), became viewed as a more ethical tool to moderate nuclear risks. According to Maria Rost Rublee, a high degree of trust in international treaties exhibited by Sweden and its corresponding desire not to harm the NPT negotiations led to Swedish nuclear forbearance.[86] In each instance, a strong moral undertone is clear. Then Foreign Minister Östen Undén outright characterized the possession of nuclear weapons as immoral due to the damage their use would inflict on innocent victims.[87] Former Swedish Prime Minister Olof Palme reaffirmed this view, noting in a 1985 interview, "I became more moral as the years went on."[88]

Ultimately, military necessity was bounded by a web of values, tradition, and ideology. While there were objective reasons for pursuing nuclear weapons, the endeavor could not overcome core tenets of Swedish strategic culture. Indeed, even the preferred process – delay – reflected the Swedish principles of compromise, caution, and consensus. In typical Swedish fashion, an internal agreement was formed in which the military would receive increased funding for conventional forces and a de facto maintenance of *handlingsfrihet*, should security conditions worsen in a way that might necessitate revisiting the nuclear option. In exchange, Sweden would forgo nuclear weapons and instead leverage international cooperative frameworks to lessen risks to the security environment.

Decision to Support but Not Join the TPNW

The TPNW is an internationally legally binding document to prohibit nuclear weapons as an initial step toward their total elimination.[89] Sweden was among the 122 UN Member States to vote in favor of its adoption in 2017. For TPNW advocates, the current step-by-step approach to disarmament outlined in the NPT has been unsuccessful because it does not solve the fundamental issue that is preventing nuclear abolition – global perception on the utility of nuclear deterrence and the need to make greater movement toward nuclear abolition. Thus, the TPNW seeks to delegitimize and stigmatize entrenched views on the acceptability of nuclear violence and build momentum toward elimination through public opinion and activism.[90] The TPNW offered a fresh approach to disarmament in which Sweden could play a leading role. The former Swedish Foreign Minister Margot Wallström even penned an op-ed in Sweden's largest newspaper, acknowledging her ambition to see Sweden sign the treaty.[91] However, in 2019, Sweden declined to sign the TPNW, coming as a surprise to TPNW advocates.[92] Some observers suggested that Sweden had succumbed to intense US pressure.[93]

Yet a closer examination reveals more complex motivations. An official government study identified several concerns with the TPNW. In addition to more technical aspects, two core arguments can be identified. First, the TPNW was assessed to rival rather than complement foundational nonproliferation and arms control frameworks, such as the NPT. For Sweden, disarmament is a step-by-step process to ensure stability post-nuclear demobilization; thus military risks must be addressed comprehensively.[94] Underpinning this line of thought is the desire to maintain consensus. By ignoring legitimate security concerns, the TPNW risks disrupting the fragile compromise between nuclear and nonnuclear states established by the NPT. Acceding and promoting the TPNW would mean common consensus on future nuclear issues would be harder to reach.[95] As noted in the report, "Sweden will be perceived in Europe as part of a small minority in a serious security policy disagreement between proponents and opponents of the Treaty... Swedish expertise in the area of nonproliferation and disarmament will suffer."[96] Sweden's stance would open the country up to charges of hypocrisy contrary to Sweden's exceptional self-image.

Second, the TPNW would complicate cooperation with nuclear-armed states and alliances. While Wallström argued that Sweden should make its own policy assessments as an "alliance free country," it was equally clear that accession would remove future freedom of action, such as joining NATO – a nuclear alliance.[97] Implicit in this argumentation is the value placed on extended nuclear deterrence in maintaining stability and solidarity in the face of current and emerging strategic risks.[98] Beyond erasing any possible nuclear assurances, Swedish consent would be perceived, as Lundin notes, as a "fundamental criticism of the strategic doctrine subscribed to by almost all of Sweden's neighbors and partners in NATO. In this context, Sweden would no longer be perceived as like-minded."[99]

The TPNW ultimately failed to accommodate the strong cultural and ideational factors embedded within Swedish cognition and decision style. Indeed, the treaty largely ignored core Swedish principles such as solidarity, stability, independence, and consensus through internationally recognized mechanisms. By supporting it, Sweden would not be acting in a morally defensible, exceptional manner.

Summary of Swedish Nuclear Decisionmaking

What do Swedish nuclear decisions reveal? Certainly, the central importance of consensus is visible. Each decision involved a compromise to suit all parties' interests. Regarding the TPNW, Sweden's rejection was paralleled with continued, if not increased, activism through the NPT. The decision to abandon the pursuit of nuclear weapons was coupled with increased funding for conventional forces. Conceptually, the illustrations reaffirm the powerful role ideational factors have on Swedish cognition. Even compelling reasons for military advances must align with Sweden's unique cultural principles. Thus, Sweden demonstrates a case in which normative attitudes bound the space in which military and political decisions function. Values, therefore, are often prioritized over interests.

Interestingly, the concept of solidarity probably played the most dynamic role. For example, during the NPT negotiations solidarity with nonnuclear and small states was stressed. Within the TPNW context, solidarity was mainly understood as unanimity with Western military powers. Both instances suggest the concept is malleable. In other words, solidarity can be adapted to support any argument. Appreciation for how solidarity is articulated and defined provides a useful indicator as to how Sweden may respond in future decisions.

The case studies also shed light into the dualism inherent in Swedish decisionmaking. While Sweden is an outspoken advocate of nonproliferation and disarmament, it appears to accept – perhaps implicitly – the value of nuclear deterrence. In addition to assuming tacit security guarantees by the United States, Swedes assess that stability depends on the presence of conventional and nuclear forces to dissuade aggression. However, the benefits afforded by nuclear deterrence cannot be divorced from international obligations. Developments by the nuclear powers that exceed these agreements such as the NPT will continue to be vigorously opposed. Given this posture, being simultaneously in favor of nuclear deterrence and nuclear disarmament is not a conceptual problem from a Swedish perspective.[100]

While the Swedish approach is not internally contradictory, multiple cultural factors are in direct competition. Groups tend to organize around certain elements that are then pitted against the other. The pattern of disagreement appears to indicate the existence of at least

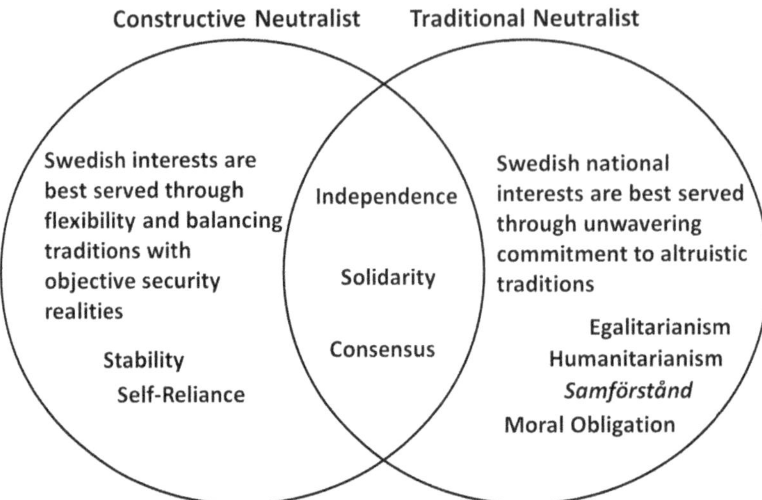

Figure 16.3 Subcultures in Swedish Strategic Culture.

Source: Author's Creation.

two subcultures (see Figure 16.3). While each shares a common foundation – that Sweden should be nonaligned and independent – they differ on how best to achieve said objectives. The two groups can be differentiated as traditional neutralists and constructive neutralists. Traditional neutralists emphasize altruistic traditions and are reluctant to make any compromises about Sweden's foreign policy orientation. Oppositely, constructive neutralists are more flexible in their interpretations and focus more on issues of national interest. This is not to suggest constructive neutralists disavow the moralistic mandate or that traditional neutralists dismiss practical policy options. Rather, the variance exists in the relative weight placed on specific values by each group.

Future Research in Swedish Strategic Culture

This chapter has leveraged Swedish decisionmaking on nuclear weapons policy as a case study to illuminate insights that can be gained from utilizing a strategic culture framework. The extant body of literature on this topic provides a useful foundation for the study of Swedish strategic culture. Taken collectively, it explores a range of important factors, to include historic, social, and political, among others. However, current scholarship tends to focus on specific elements of Swedish strategic culture rather than the comprehensive whole.[101] This results in only fragmentary insights regarding Swedish decisions. The lack of English language resources further compounds the problem, inhibiting a deeper understanding of Swedish strategic choices.

The body of literature on Sweden's strategic culture also suffers from three additional gaps. First, research efforts typically do not employ a systematic framework nor a standard definition of strategic culture. Case studies tend to utilize improvised methodologies tailored to meet the needs of a specific study.[102] The failure to maintain commonality can generate confusion and sometimes contradictory analyses. Second, research frequently merges disparate elements in ways that neglect critical aspects unique to Sweden. Sweden is regularly assimilated into broader groupings that are either regional (Nordic, Scandinavian, and European) or functional (small state) in nature. In other cases, Sweden is presented as a monolithic actor that can be explained simply by understanding a single political party or leader.[103] While these structures are useful in some cases, they often miss key ideational influences that shape Swedish decisionmaking. Finally, and perhaps most importantly, Swedish studies often discount the influence of strategic culture on leadership decisions. For example, some analysts suggest structural realism is more relevant to deciphering Swedish strategic choices, particularly on nuclear issues. Here, Swedish nonalignment and moral policies are a means to compensate for military disadvantages and preserve a regional balance of power. Consequently, these analysts argue that Swedish neutrality is not idealistically motivated but generated from pragmatic self-interest common among weak states,[104] and that Swedish decisions are simply a reflection of the reality imposed by realism.[105] To be sure, strategic culture is not a simple or magical solution to difficult problems. Yet it does provide a useful mechanism to better identify the unique influences on leadership decisionmaking. In some cases, these factors are just as palpable as realist concerns. Dismissing the impact of strategic culture, therefore, precludes a more comprehensive assessment necessary to explain strategic behavior.

Consequently, the requirement for greater study – not less – devoted to strategic culture is needed. With regard to Sweden, perhaps the most important contribution for scholars and students will be to ensure future research is accessible to the broadest audience possible. This is particularly important given the current scarcity of English-language resources. Translated products would offer added opportunities to strengthen the current body of literature. In addition, future research might also consider further examination of the two subcultures identified in this chapter. As noted, Sweden is frequently treated as a uniform entity; disaggregating particular groups (perhaps initially along political party lines) would prove beneficial in advancing the study and understanding of Sweden's strategic culture. The current literature, while a useful starting point, reveals seams and gaps that can and should be moderated by additional strategic culture research.

Notes

1 The views expressed are those of the author and do not reflect the official policy or position of the Defense Threat Reduction Agency, the Department of Defense, or the US government.
2 Hans Brems, "Sweden: From Great Power to Welfare State," *Journal of Economic Issues*, Vol. 4, No. 2/3, June–September 1970, pp. 12–13.
3 Ibid.
4 Dag Blanck, "The Transnational Viking: The Role of the Viking in Sweden, the United States, and Swedish America," *Journal of Transnational American Studies*, Vol. 7, No. 1, 2016, p. 4.
5 Petri Karonen, "The Peace Treaty of Fredrikshamn and Its Aftermath in Sweden and Finland," *Sjuttonhundratal: Nordic Yearbook for Eighteenth-Century Studies*, Vol. 7, 2010, pp. 168–183.
6 Blanck, 2016, p. 3.
7 Erik Örjan Emilsson, "Recasting Swedish Historical Identity" (Göteborg: Centrum för Europaforskning, University of Gothenburg, 2009), p. 193.
8 Orvar Löfgren, "The Nationalization of Culture," *Ethnologia Europaea*, Vol. 19, No. 1, 1989, pp. 38–39.
9 Ibid.
10 Ibid., p. 6.

11 See Erik Gustaf Geijer, *Vikingen*, translated by Kristina Andersson Bicher, *The Harvard Review*, 18 October 2017, https://harvardreview.org/content/the-viking/.
12 Franklin D. Scott, "Sweden's Constructive Opposition to Emigration," *The Journal of Modern History*, Vol. 37, No. 3, September 1965, pp. 310–312.
13 Fredrik Sunnemark, "Who Are We Now Then? The Swedish Welfare State in Political Memory and Identity," *Kultura: International Journal for Cultural Researchers*, No. 5, 2014, pp. 7–16.
14 Mikko Kuisma, "Social Democratic Internationalism and the Welfare State after the 'Golden Age,'" *Cooperation and Conflict*, Vol. 42, No. 1, March 2007, p. 12.
15 Arne Ruth, "The Second New Nation: The Mythology of Modern Sweden," *Daedalus*, Vol. 113, No. 2, Spring 1984, p. 71.
16 Göran Rosenberg, "The Crisis of Consensus in Postwar Sweden," in Nina Witoszek and Lars Trägårdh (eds.), *Culture and Crisis: The Case of Germany and Sweden* (New York: Berghahn Publishing, 2002), pp. 173–174.
17 Eliassen Restad, *American Exceptionalism: An Idea That Made a Nation and Remade the World* (London: Routledge, 2015), p. 17.
18 Marquis William Childs, *Sweden: The Middle Way* (New Haven: Yale University Press, 1936).
19 Ruth, 1984, pp. 90–92.
20 Peter Lawler, "Scandinavian Exceptionalism and European Union," *Journal of Common Market Studies*, Vol. 35, No. 4, December 1997, p. 568.
21 Annika Bergman, "Co-Constitution of Domestic and International Welfare Obligations: The Case of Sweden's Social Democratically Inspired Internationalism," *Cooperation and Conflict*, Vol 42, No. 1, 2007, p. 1.
22 Elin Lager, "Swedish Exceptionalism in Foreign Policy Discourse: An Analysis of the Swedish Government's Statements of Foreign Policy 2002–2018," Stockholm University, 2020.
23 Ibid.
24 Ruth, 1984, p. 92.
25 Christine Agius, *The Social Construction of Swedish Neutrality: Challenges to Swedish Identity and Sovereignty* (Manchester: Manchester University Press, 2006).
26 Joseph Board, "Reinterpreting the Bombast," *Washington Quarterly*, Autumn, 1998, p. 189.
27 Clive Ponting, *The Crimean War: The Truth Behind the Myth* (New York: Random House, 2011).
28 Ann-Sofie Dahl, "To Be Or Not to Be Neutral: Swedish Security in the Post-Cold War Era," in Efraim Inbar and Gabriel Sheffer (eds.), *The National Security of Small States in a Changing World* (London, UK: Routledge, 1997), pp. 175–197 .
29 Ulf Bjereld, "Critic or Mediator? Sweden in World Politics, 1945–1990," *Journal of Peace Research*, Vol. 32, No. 1, 1995, p. 23-33. Also see Douglas Brommesson."Nordicness" in Swedish Foreign Policy – From Mid Power Internationalism to Small State Balancing?" *Global Affairs*, Vol. 4, No. 4–5, 2015, pp. 23–33, 292–295, 397–399, https://www.tandfonline.com/doi/full/10.1080/23340460.2018.1516116.
30 Barbara Kunz, "Sweden's NATO Workaround: Swedish Security and Defense Policy against the Backdrop of Russian Revisionism," IFRI, *Focus Strategique*, No. 64, November 2015, https://www.ifri.org/sites/default/files/atoms/files/fs64kunz_0.pdf.
31 Ibid.
32 Georg Holmin, "The Defense of Neutrality," in *Sweden: A Wartime Survey* (New York: The American-Swedish Exchange, 1942), p. 24.
33 Sven Hirdman, "Ten Myths and Misunderstandings about Swedish Security Policy," *Dagen Nyheter*, 19 June 2015, https://www.dn.se/debatt/tio-myter-och-missforstand-om-svensk-sakerhetspolitik/.
34 Mikael af Malmborg, Neutrality and State-Building in Sweden (Basingstoke: Palgrave, 2001), p. 151
35 Fred Halliday, "Three Concepts of Internationalism," *International Affairs*, Vol. 64, No. 2, Spring 1988, p. 187.
36 Cecelia Lynch, "The Promise and Problems of Internationalism," *Global Governance*, Vol. 5, 1999, p. 87.
37 John A. Agnew, "An Excess of'National Exceptionalism': Towards a New Political Geography of American Foreign Policy," *Political Geography Quarterly*. Vol. 2, No. 2, 1983, p. 151.

38 Ann-Sofie Dahl, "Sweden: Once a Moral Superpower, Always a Moral Superpower?" *The International Journal*, Vol. 61, No. 4, Autumn 2006, p. 898.
39 Marie Demker, "A Magic Moment in Swedish Foreign Policy: Voting 'YES' to Algerian Self-Determination in 1959," *Cooperation and Conflict*, Vol. 33, No. 2, June 1998, pp. 135–137.
40 Ole Elgström, "Introduction: Sweden's International Relations," in Jon Pierre (ed.), *The Oxford Handbook of Swedish Politics* (Oxford: Oxford University Press, 2016), p. 432.
41 Ann-Sofie Nilsson, "Swedish Social Democracy in Central America: The Politics of Small-State Solidarity," *Journal of Interamerican Studies and World Affairs*, Vol. 33, No. 3, Autumn 1991, p. 171.
42 Douglas Brommesson, "Normative Europeanization: The Case of Swedish Foreign Policy Reorientation," *Conflict and Cooperation*, Vol. 45, No. 2, June 2010, pp. 229–231.
43 Greg Simons and Andrey Manoilo, "Sweden's Self-Perceived Global Role: Promises and Contradictions," *Research in Globalization*, Vol. 1, December 2019, https://www.sciencedirect.com/science/article/pii/S2590051X19300085#bb0095.
44 Ibid.
45 Ruth 1984, p. 71.
46 Dahl, 2006, p. 896.
47 Emilsson, 2009, p. 191.
48 Tomas Stenbäck, "Swedish Belief and Swedish Tradition: The Role of Religion in Sweden Democrat Nationalism," Akademin För Utbildning Och Ekonomi, 2020. https://www.diva-portal.org/smash/get/diva2:1458910/FULLTEXT01.pd, p. 20.
49 Jansson, 2017, p. 86.
50 Stenbäck, 2020, p. 18.
51 Ibid., p. 19.
52 Per Ewert, "Secularizing the Church of Sweden: By Politics Alone," *Transatlantic Blog*, Acton Institute, 10 May 2019, https://www.acton.org/publications/transatlantic/2019/05/10/secularizing-church-sweden-politics-alone.
53 Eric S. Einhorn, "Just Enough ("Lagom") Europeanization: The Nordic States and Europe," *Scandinavian Studies*, Vol. 74, No. 3, Fall 2002, pp. 275–277, 284–285 .
54 Kuisma, 2007.
55 Emilsson, 2009, p. 196.
56 Ruth, 1984, p. 71.
57 Clive Archer, "Conflict Prevention in Europe: The Case of the Nordic states and Macedonia," *Cooperation and Conflict*, Vol. 29, No. 4, December 1994, pp. 367–386 and Annika Bergman, "Co-Constitution of Domestic and International Welfare Obligations: The Case of Sweden's Social Democratically Inspired Internationalism," *Cooperation and Conflict*, Vol. 42, No. 1, March 2007, pp. 73–99.
58 Åke Daun, *Svensk Mentalitet, ett Jamforande Perspektiv* (Stockholm: Rabén & Sjögren, 1989), p. 102. Cited in Göran Rosenberg, "The Crisis of Consensus in Postwar Sweden," Lund University, 2002, p. 170.
59 Ibid.
60 Dahl, 2006.
61 Ulf Bjereld, "Sweden – A Moral Superpower?" *The Nordic Page*, 1 April 2015, https://www.tnp.no/magazine/?s=Moral+Superpower.
62 Fredrik Doser, "Sweden's Libya Decision: A Case of Humanitarian Intervention," *International Politics*, Vol. 51, 2014, p. 208.
63 Fredrik Doeser, "Finland, Sweden, and Operation Unified Protector: The Impact of Strategic Culture," *Comparative Strategy*, Vol. 35, No. 4, 2016, pp. 287–291.
64 Sten Tolgfors,"Socialdemokraterna Försvårar Sveriges Internationella Ansvarstagande," *Svenska Dagbladet*, 11 May 2011.
65 Doeser, 2014.
66 Wolfgang Hansson, "Sverige Skickar åtta Gripenplan: Men de får Inte Attackera Markmål," *Aftonbladet*, 29 March 2011.
67 Doeser, 2001.
68 Elgström, 2016.
69 Ibid.

70 Ryszard M. Czarny, *Sweden: From Neutrality to International Solidarity* (Springer International Publishing, 2018), p. 31.
71 Speech by Minister for Defence Peter Hultqvist on Northern European Security, Key Note Speech at John Hopkins University, Washington, DC, 17 May 2017, https://www.government.se/speeches/2017/05/speech-by-minister-for-defence-peter-hultqvist-on-northern-european-security/.
72 Michael Nordlind, "The New Russian Hatred and Its Function," *Motpol*, 23 November 2014, https://motpol.nu/nordanvind/2014/11/23/det-nya-rysshatet-och-dess-funktion/.
73 Gregory Simons, Andrey Manoilo, and Philipp Trunov, "Sweden and the NATO Debate: Views from Sweden and Russia," *Global Affairs*, Vol. 5, No. 4–5, 13 April 2019, https://www.tandfonline.com/doi/full/10.1080/23340460.2019.1681014.
74 Magnus Christiansson, "Solidarity and Sovereignty: The Two-Dimensional Game of Swedish Security Policy," *The Quarterly Journal*, Vol. 10, No. 1, Winter 2010.
75 Discussion with Kerry Kartchner, 23 April 2021.
76 Ott Ummelas, "Sweden's NATO Skepticism Endures as Russia Flexes Its Muscles," *Bloomberg News*, 10 January 2021, https://www.bloombergquint.com/onweb/sweden-shuns-nato-as-stability-outweighs-worries-about-russia
77 Kunz, 2015.
78 Alexandre Debs and Nuno P. Monteiro, *Nuclear Politics: The Strategic Causes of Proliferation* (Cambridge: Cambridge University Press, 2017), pp. 181–182.
79 Paul M. Cole, *Sweden without the Bomb: The Conduct of a Nuclear-Capable Nation without Nuclear Weapons* (Santa Monica: RAND Publishing, 1994), pp. 74–80. Also see Thomas Jonter, "Getting Rid of the Swedish Bomb," *Physics Today*, Vol. 72, No. 9, 2019, https://physicstoday.scitation.org/doi/10.1063/PT.3.4293.
80 Ibid., p. 6.
81 Ibid., p. 23. Also see Lars van Dassen, "Sweden and the Making of Nuclear Non-Proliferation: From Indecision to Assertiveness," Swedish Nuclear Power Inspectorate Report, March 1998, https://inis.iaea.org/collection/NCLCollectionStore/_Public/29/032/29032967.pdf, p. 14.
82 Rebecca Davis Gibbons, "Supply to Deny: The Benefits of Nuclear Assistance for Nuclear Nonproliferation," *Journal of Global Security Studies*, Vol. 5, No. 2, April 2020, p. 9.
83 Mikael Holmström, *The Hidden Alliance: Sweden's Secret Ties to NATO* (Atlantis, 2015: Stockholm, Sweden) and Tsukamoto Katsuya, Kudo Toyoko, and Sue Shuji, "Why Do States Remain Nonnuclear? Effects of Extended Deterrence on Nuclear Proliferation," National Institute for Defense Studies (Japan) Security Reports, No. 10, December 2009, p. 62.
84 Debs and Monteiro, 2017.
85 Paul M. Cole, "Atomic Bombast: Nuclear Weapon Decisionmaking in Sweden 1945–1972," Occasional Paper No. 26, Stimson Center, April 1996, https://www.stimson.org/1996/atomic-bombast-nuclear-weapon-decisionmaking-sweden-1945-1972/.
86 Maria Rost Rublee, *Nonproliferation Norms: Why States Choose Nuclear Restraint* (Athens: University of Georgia Press, 2009), p. 180.
87 Dassen, 1998, p. 11.
88 Steve Coll, "Neutral Sweden Quietly Keeps Nuclear Option Open," *The Washington Post*, 25 November 1994, https://www.washingtonpost.com/archive/politics/1994/11/25/neutral-sweden-quietly-keeps-nuclear-option-open/754e8f39-b158-4ec5-812c-63592ac1889d/.
89 United Nations, Treaty on the Prohibition of Nuclear Weapons (English), https://www.un.org/disarmament/wp-content/uploads/2017/10/tpnw-info-kit-v2.pdf.
90 Nick Ritchie, "The Treaty on the Prohibition of Nuclear Weapons: Delegitimizing Unacceptable Weapons," in Shatabhisha Shetty and Denitsa Raynova (eds.), *Breakthrough or Breakpoint? Global Perspectives on the Nuclear Ban Treaty,* European Leadership Network, Global Security Special Report, December 2017, p. 51.
91 "Swedish Foreign Minister Wants to Sign the Ban," ICAN, 2017, https://www.icanw.org/swedish_foreign_minister_wants_to_sign_the_ban.
92 "A Big Step Backwards: Why Sweden's Decision Not to Sign TPNW Damages Its Reputation as a Leader on Disarament," International Campaign to Abolish Nuclear Weapons, 2019, https://www.icanw.org/a_big_step_backwards_why_sweden_s_decision_not_to_sign_tpnw_damages_its_reputation_as_a_leader_on_disarmament.

93 Christopher Woody, "Mattis Reportedly Threatened Sweden with Retaliation over Signing a Nuclear Weapons Ban," *Business Insider*, 5 September 2017, https://www.businessinsider.com/mattis-threatened-sweden-over-a-nuclear-weapons-ban-treaty-2017-9.
94 Lars-Erik Lundin, 'Inquiry into the Consequences of a Swedish Accession to the Treaty on the Prohibition of Nuclear Weapons," Ministry of Foreign Affairs, Sweden, 2019, p. 42.
95 Emil Dall, "Sweden's Choice: NATO or the Nuclear Ban?" *Commentary*, RUSI, 22 September 2017, https://rusi.org/commentary/sweden%E2%80%99s-choice-nato-or-nuclear-ban.
96 Lundin, 2019, p. 44.
97 Margot Wallström, "A World Free of Nuclear Weapons Is Possible," *Svenska Dagbladet*, 25 August 2017, https://www.svd.se/wallstrom-en-varld-fri-fran-karnvapen-ar-mojlig.
98 Lundin, 2019, p. 45.
99 Ibid., p. 45.
100 Cole, 1994, p. 135.
101 The most notable and most cited exception is Gunnar Åselius, "Swedish Strategic Culture after 1945," *Cooperation and Conflict*, Vol. 40, No. 1, 2005, pp. 25–44. However, this document tends to focus more exclusively on the evolving culture within the Swedish military.
102 For example, see Doeser's methodology.
103 Mitchell Reiss, *Without the Bomb: The Politics of Nuclear Nonproliferation* (New York: Columbia University Press, 1988).
104 Board, 1998, p. 239.
105 See Cole, 1994; Jan Prawitz, *From Nuclear Option to Non-Nuclear Promotion: The Swedish Case* (The Swedish Institute for International Affairs, 1995); Eric Arnett, "Norms and Nuclear Proliferation: Sweden's Lessons for Assessing Iran," *Nonproliferation Review*, Winter 1998, pp. 32–43.

Selected Bibliography

Agius, Christine. *The Social Construction of Swedish Neutrality: Challenges to Swedish Identity and Sovereignty*. Manchester: Manchester University Press, 2006.

Bergman, Annika. "Co-Constitution of Domestic and International Welfare Obligations: The Case of Sweden's Social Democratically Inspired Internationalism." *Cooperation and Conflict*, Vol. 42, No. 1, 2007.

Bjereld, Ulf. "Sweden – A Moral Superpower?" *The Nordic Page*, 1 April 2015. https://www.tnp.no/magazine/?s=Moral+Superpower.

Blanck, Dag. "The Transnational Viking: The Role of the Viking in Sweden, the United States, and Swedish America." *Journal of Transnational American Studies*, Vol. 7, No. 1, 2016.

Brommesson, Douglas. "Nordicness" in Swedish Foreign Policy – From Mid Power Internationalism to Small State Balancing?" *Global Affairs*, Vol. 4, No. 4–5, 2015, https://www.tandfonline.com/doi/full/10.1080/23340460.2018.1516116.

Brommesson, Douglas. "Normative Europeanization: The Case of Swedish Foreign Policy Reorientation." *Conflict and Cooperation*, Vol. 45, No. 2, June 2010, pp. 224–244.

Childs, Marquis William. *Sweden: The Middle Way*. New Haven: Yale University Press, 1936.

Christiansson, Magnus. "Solidarity and Sovereignty: The Two-Dimensional Game of Swedish Security Policy." *The Quarterly Journal*, Vol. 10, No. 1, Winter 2010, pp. 1–24.

Czarny, Ryszard M. *Sweden: From Neutrality to International Solidarity*. Cham, Switzerland: Springer International Publishing, 2018.

Dahl, Ann-Sofie. "Sweden: Once a Moral Superpower, Always a Moral Superpower?" *The International Journal*, Vol. 61, No. 4, Autumn 2006, pp. 895–908.

Doeser, Fredrik. "Finland, Sweden, and Operation Unified Protector: The Impact of Strategic Culture." *Comparative Strategy*, Vol. 35, No. 4, 2016.

Doser, Fredrik. "Sweden's Libya Decision: A Case of Humanitarian Intervention." *International Politics*, Vol. 51, 2014, pp. 196–213.

Elgström, Ole. "Introduction: Sweden's International Relations." *The Oxford Handbook of Swedish Politics*, edited by Jon Pierre. Oxford: Oxford University Press, 2016.

Emilsson, Erik Örjan. "Recasting Swedish Historical Identity." *Göteborg: Centrum för Europaforskning.* University of Gothenburg, 2009.
Geijer, Erik Gustaf. *Vikingen.* Translated by Kristina Andersson Bicher. *The Harvard Review*, 18 October 2017, https://harvardreview.org/content/the-viking/.
Hentilä, Seppo. "The Origins of the Folkhem Ideology in Swedish Social Democracy." *Scandinavian Journal of History*, Vol. 3, No. 1–4, 1978.
Holmin, Georg. "The Defense of Neutrality." In *Sweden: A Wartime Survey.* New York: The American-Swedish Exchange, 1942.
Holmström, Mikael. *The Hidden Alliance: Sweden's Secret Ties to NATO.* Stockholm: Atlantis, 2015.
Karonen, Petri. "The Peace Treaty of Fredrikshamn and Its Aftermath in Sweden and Finland." *Sjuttonhundratal: Nordic Yearbook for Eighteenth-Century Studies*, Vol. 7, 2010, pp. 168–183.
Kunz, Barbara. "Sweden's NATO Workaround: Swedish Security and Defense Policy against the Backdrop of Russian Revisionism." *IFRI, Focus Strategique*, No. 64, November 2015. https://www.ifri.org/sites/default/files/atoms/files/fs64kunz_0.pdf.
Kuisma, Mikko. "Social Democratic Internationalism and the Welfare State after the 'Golden Age.'" *Cooperation and Conflict*, March 2007.
Lager, Elin. "Swedish Exceptionalism in Foreign Policy Discourse: An Analysis of the Swedish Government's Statements of Foreign Policy 2002–2018." Stockholm University, 2020.
Nilsson, Ann-Sofie. "Swedish Social Democracy in Central America: The Politics of Small-State Solidarity." *Journal of Interamerican Studies and World Affairs*, Vol. 33, No. 3, Autumn 1991, pp. 196–200.
Rosenberg, Göran. "The Crisis of Consensus in Postwar Sweden." Lund University, 2002.
Ruth, Arne. "The Second New Nation: The Mythology of Modern Sweden." *Daedalus*, Vol. 113, No. 2, Spring 1984, pp. 53–96.
Simons, Greg and Manoilo, Andrey. "Sweden's Self-Perceived Global Role: Promises and Contradictions." *Research in Globalization*, Vol. 1, December 2019. https://www.sciencedirect.com/science/article/pii/S2590051X19300085#bb0095.
Stenbäck, Tomas. "Swedish Belief and Swedish Tradition: The Role of Religion in Sweden Democrat Nationalism." Akademin För Utbildning Och Ekonomi, 2020. https://www.diva-portal.org/smash/get/diva2:1458910/FULLTEXT01.pdf.
Sunnemark, Fredrik. "Who Are We Now Then? The Swedish Welfare State in Political Memory and Identity." *Kultura: International Journal for Cultural Researchers*, No. 5, 2014, pp. 7–16.

17

THE STRATEGIC CULTURE OF THE FEDERAL REPUBLIC OF GERMANY[1]

Thomas Biggs

Germany is a noteworthy case for the study of strategic culture because of the radical cultural change the nation has undergone in the past century. The unique situation Germany finds itself in has provided no shortage of books and articles addressing aspects of the nation's strategic culture, even when this literature is not explicitly focused on the concept of "strategic culture" itself. Recent scholarship offering insight into Germany's strategic culture does not use this specific terminology but rather focuses on individual aspects of German culture that are significant at the strategic level. There is excellent coverage of Germany in the literature as it relates to defense, remembrance, international relations, and domestic politics. Naturally, in the field of strategic culture, defense receives the most coverage, followed by international relations.

The modern Federal Republic of Germany is often noted by commentators for its general unwillingness to use force on the international stage. Germany has shown itself to be a nation of progressive politics: it has been reluctant to take part in military interventions, has pursued an aggressive environmentalist policy, and has opened its doors to migrants from Middle Eastern conflicts. The recent rise of the Alternative für Deutschland (AfD) as one of the largest opposition parties in the Bundestag and increased pockets of neo-Nazi activity have caused rising concern for German officials, given that the government has spent almost three-quarters of a century rebuilding the country's postwar image. Germany's modern strategic culture as a whole formed as a reaction to the strategic culture the state developed in the years before and during World War II.

In a 2006 SAIC study, Jeannie L. Johnson defined strategic culture as:

> [T]hat set of shared beliefs, assumptions, and modes of behavior, derived from common experiences and accepted narratives (both oral and written), that shape collective identity and relationships to other groups, and which determine appropriate ends and means for achieving security objectives.[2]

The strategic culture framework as defined by Johnson consists of four dimensions for investigating a state or group: identity, values, norms, and perceptual lens.[3] How a state

DOI: 10.4324/9781003010302-20

views itself, what ideas it gives priority to, what actions it views as acceptable and unacceptable, and how it sees the world around itself will all factor significantly into any security decision the state or group makes. This chapter will follow this approach to strategic culture analysis, investigating elements of current German identity, values, norms, and perceptions, and will review the literature supporting this investigation. It will also explore manifestations of Germany's strategic culture, providing examples of how each of these cultural dimensions have influenced and continue to shape its strategic decisions. Following this review of Germany's strategic culture, the chapter will address the key contributions within the strategic culture literature on Germany, the strengths and gaps in this existing body of literature, and the areas of research that could be investigated by future scholars of German strategic culture.

Historical Background

A brief, high-level overview of German history reflects that the modern unified German nation-state is a relatively recent development in history. Germanic tribes led by Hermann (Arminius) defeated the Romans at Teutoburg Forest in 9 AD, and a Germanic kingdom came into being after the splitting of Charlemagne's realm in 843 AD. But through most of the past two millennia the region that is now Germany was a loose association of small kingdoms, sharing the same language but differing in culture and politics. For example, the Mark of Brandenburg, situated around Berlin, was primarily agricultural in its economy and was governed by a traditional monarchy. In contrast, the city-state of Hamburg maintained a fiercely republican system of governance and built itself from the wealth of maritime trade and fishing. To this day, both of these areas (now federal states) retain distinct civic cultures. Germany came to value Christian religious freedom, as it was the birthplace of the Reformation (1517) and the main battlefield of the brutal Thirty Years' War (1618–1648) that began over religion and killed a devastating portion of its population.[4]

Napoleon's conquest of the German states in 1806 accelerated the development of German nationalism. Rebels from all the German states, and eventually armies led by the Kingdom of Prussia, fought against and defeated France. In 1848, there was a failed revolutionary attempt to create a unified German-speaking state due to intense policy disagreements between the delegates from each of the states.[5] Unionist Otto von Bismarck commented on these events in 1862 when he became minister-president of the Kingdom of Prussia: "Not through speeches and majority decisions will the great questions of the day be decided – that was the great mistake of 1848 and 1849 – but by iron and blood."[6] "Iron and blood" would define Germany's outlook for the next 80 years. Militarism and faith in the army, already core tenets of Prussian culture, became key parts of German identity. War and preparedness for war would also foster the rigid conformity and national unity that would be needed to bind the disparate principalities together.

Prussia led the German states in three wars from 1864 to 1871, in which they decisively defeated Denmark, Austria, and France. Prussia and its allies captured territory from all three, greatly expanding the reach of the German realm. Upon the defeat of France in January 1871, the leaders of the German states agreed to the formation of the German Empire, with Prussian King Wilhelm I as its Kaiser and Bismarck as its Chancellor. The establishment of Germany as an influential military force upset the established balance of power in Europe. Alsace-Lorraine, a predominantly German-speaking region conquered by the Kingdom of France in 1639 during the Thirty Years' War, was annexed by the new German nation-state as one of its spoils of victory. France remained bitter about the loss of this rich agricultural and industrial territory for years afterward, and other European states viewed Germany with suspicion.[7] Military competition and a complex network of alliances ultimately led to World War I, which ended with the defeat of Germany and the fall of the empire to a left-wing revolution. The defeat was caused largely by material shortages imposed by the British blockade of Germany and the failure of several key offensives by the German army. Far-right German nationalists, however, blamed the defeat on far-leftist and Jewish betrayal.[8] General bitterness about the war and the need to blame someone for Germany's problems opened the door to the rise of the National Socialist German Workers' Party, or the Nazis.[9]

Nazi leadership under Adolf Hitler led a destructive campaign of political and racial persecution that targeted Jews, Sinti and Romani, Slavic peoples, homosexual and bisexual men, and the disabled, among others. German and East European Jews suffered the greatest under what came to be known as the Holocaust, with around six million Jews ultimately murdered by the Nazis. World War II was initiated by the Nazi regime and ended with what can be described as the complete destruction of Germany. Every major city was mostly in ruins by 1945, and the Allied occupation completely dismantled almost all institutions of the old Germany. The land was split between the Western Allies in the West of Germany and the Soviets in the east. Prussia, deemed by the Allies to be the source of Germany's problematic militarism, was dissolved in 1947.[10] Germans were expelled from territories granted to Poland and Czechoslovakia by the Soviets and resettled in the Allied occupation zones or elsewhere. Two separate German states were established in 1949 from the occupation zones. The Federal Republic of Germany (West Germany), established with the approval of the Western Allies, ultimately became the current government of modern united Germany, while the German Democratic Republic (GDR, or East Germany) was a Soviet-aligned communist state. The GDR became widely known in the West for its militaristic posture and its intrusive intelligence service, the Stasi, which invaded the private lives of its citizenry. East Germany collapsed in the upheavals of 1989, and on 3 October 1990, it was integrated into the Western Federal Republic.

Today's reunified Germany has assumed a political and economic leadership role in the European Union (EU). It rejects militarism, although it has taken part in several North Atlantic Treaty Organization (NATO) missions since the Yugoslav Wars in the 1990s. Germany has become increasingly diverse in the past three decades, with 12.8% of the population identifying as ethnicities other than German in 2020.[11] People of Turkish heritage make up the largest non-German ethnic group in the country, at 1.8% of the population.[12] Given these demographic shifts, the German population – and in turn German identity – is more diverse than ever before.

Sources of Germany's Strategic Culture

The sources of a state's strategic culture include the specific historical and cultural experiences that have forged its identity, values, norms, and perceptual lenses. Events in recent history will have a significant impact on these dimensions, but as reflected in the historical background offered above, events hundreds of years ago can still exert a profound impact. The following section will further elaborate on specific sources of Germany's strategic culture that are of enduring significance.

The consequences of Nazism have left a profound mark on Germany. Even as of 2023, Berlin has not grown back to its pre-World War II population of over four million, falling about a half million short.[13] The Jewish population of Germany is still only about a fifth of what it was before the Holocaust and composes less than a percentage point of the country's population.[14] Preventing an event like the Holocaust from happening again is at the core of modern Germany's strategic culture. Germany's dominant culture shies away from all forms of nationalism. Users on internet forums discussing Europe have pointed out that, outside of football season, most Germans tend to keep their patriotism in check. The national flag is used sparingly by citizens and is not often seen on display, even in government offices.[15] Germany's restrictive attitudes toward its military, the Bundeswehr, and its intelligence agencies are intended to prevent a reversion to the excesses of the Wehrmacht and Gestapo. The ultimate keeper of Germany's strategic culture is its "Basic Law," which is equivalent to a constitution. This 1949 document enshrined the founding principles of the Federal Republic of Germany and exemplifies Germans' dedication to human rights for all people, which outweighs both duties to the state and any national pride. The very first sentence of the document is "Human dignity shall be inviolable."[16]

The legacies of both World War II and the Cold War still hang over Germany's military. American and Soviet forces were stationed in West and East Germany, respectively, throughout the Cold War. Had World War III started during those tense years, the first shot would likely have been fired in Germany. West Germany made the controversial decision in 1955 to establish a permanent military force (the Bundeswehr) and join NATO in order to face the Soviet threat to its territory. This decision was also made because of pressure from the Western Allies, who wanted to pull their own troops from Germany for service in other strategic locations.[17] This rearmament was controversial among the West German public, as the memory of the Wehrmacht being an agent of the Nazi regime was not at all distant. The West German government forbade the Bundeswehr from taking part in any mission outside of NATO territory in Europe during the Cold War, insisting that it exist strictly for the defense of Germany from foreign attack and nothing else.[18]

Because of the need to maintain a standing army for defense against the Soviet threat, the Bundeswehr instituted conscription for German men[19] with the added intent to "counteract anti-democratic political ambitions of the officer corps of the armed forces."[20] The postwar government believed Germans had learned a lesson in the 1920s and 1930s when the Reichswehr (Weimar Germany's military) was able to distance itself from the democratic civilian government and became a breeding ground for Nazi sympathies, and the government hoped that peacetime conscription of civilians from all walks of life in the enlisted ranks would prevent this from happening again.[21] Men who were opposed to military service on moral or religious grounds were allowed to instead perform *Zivildienst* ("civil service"), working in locations such as hospitals or farms in peaceful service to the state.[22] Conscription persisted until 2011 when the Merkel government deemed that there was no longer any threat to Germany that merited it.[23] Deployment of Bundeswehr troops outside of NATO territory was first permitted by the German Constitutional Court in 1994, just before Germany's involvement in NATO's Operation Deliberate Force in Bosnia.[24]

Germany's Strategic Cultural Identity

Identity in Johnson's strategic culture framework relates to how a state views itself and consists of the important traits in a state's "national character." These traits can include who is considered a member of the nation-state, sources of pride in the state, shared historical narratives, and views on what the state is destined to become. It is important to understand Germany's identity with regard to its strategic culture because of how drastically Germany's identity has shifted in the past century and how this sense of identity influences the other three dimensions of German strategic culture.

As exemplified by its Basic Law, Germany prides itself on its democratic values. Germany knows it is the economic leader of the EU and one of the most influential countries in the world. Germany is the largest economy in the EU and the fourth largest globally as of 2021, with a gross domestic product of 4.26 trillion USD.[25] Germany is famous worldwide for its leadership in science and industry, hosting renowned corporations such as BioNTech, Bayer AG, ThyssenKrupp, and Mercedes-Benz.

As noted previously, Germans are known to be reluctant to show patriotism due to its associations with Nazism. Pride in the German ethnicity is effectively forbidden due to the Nazis' fixation with race. Language is the primary unifying factor of the German people. The mainstream culture is proud of the country's religious and ethnic diversity, brought on by historical and current migrations. Germans tend to view themselves more as "German Europeans" than solely as Germans and most desire to be prominent members of a European community, collaborating respectfully with neighbors that were enemies in the recent past.[26] The West German identity that manifested in 1949 and still exists in Germany today centers around "Christianity and the cult of German idealism associated with Goethe, Schiller, and the classical literary tradition."[27] Germany styles itself as a cultured land of "*Dichter und Denken*" ("Poets and Thinkers"). Militarism and the glorification of war have been expunged from any mainstream sense of German identity.

Because of two world wars, the Holocaust, and 40 years of communist rule in half of the country, many Germans (37.8%) are atheist or otherwise nonreligious.[28] Christianity is the largest practiced religion, with Roman Catholicism accounting for 27.7% and Protestant sects for 25.5% of the population. Though religion has no discernible influence on government affairs, all Christian holidays, from Easter and Christmas to Ascension Day and Pentecost, are state holidays for all working Germans. About 5.1% of Germans follow Islam, reflecting the significant Turkish community and immigrants from other Middle Eastern states. As noted above, the German Jewish community makes up less than 1% of the population.[29]

German Strategic Culture Values

Values are the ideations and physical materials that a state sees as most important and which are "selected over others," per Johnson. Values are often seen in what form of government a state maintains, the state's views on justice, a state religion if one is officially declared, and what the state promotes in international dialogues. Germany's values, like its identity, have undergone major alterations in the postwar era. Some of these alterations were imposed by the occupying Allies; others, however, were revivals of deep-rooted German values from earlier in the nation's history.

Germany values being a democracy after living under monarchy and dictatorship for so many years. This value is evident in its flag, the black-red-gold republican standard flown by revolutionaries in 1848. The colors of this flag originated during the Napoleonic Wars, known as the *Befreiungskriege* or "Wars of Liberation." The colors were worn on the uniforms of the Lützow Free Corps, a Prussian Army unit consisting of volunteers from around the German kingdoms, because the colors were cheap and easily available at the time.[30] Because of the war, the colors became associated with German liberty and were adopted by republicans and nationalists alike. The black-red-gold standard became the flag of the German "Weimar" Republic in 1919 but was thrown out by the Nazis 14 years later in favor of the black-white-red flag of the old German Empire. West and East Germany both readopted the black-red-gold flag in 1949 as a rejection of Nazism and a commitment to republican ideals. Germany's *Bundesadler* or "Federal Eagle," displayed on the country's seal and state flag, shows a long continuity in symbolism: an eagle has represented Germany since the era of the Holy Roman Empire, and the use of the eagle in the Federal Republic of Germany, even after use of an eagle by the Imperial and Nazi regimes, shows a degree of respect for the shared history of the German states.[31]

Germany strongly values economic security. The experience of hyperinflation in 1923 is recent enough to still influence German economic policy.[32] Cash is still used widely by average Germans, as opposed to credit cards and other more contemporary forms of currency. Economic security also influences Germany's attitude toward the Euro, as Germany is the host of the European Central Bank. Germany did not print Euros to stave off the late 2000s financial crisis because of the effect that doing the same had in 1923: an extremely devalued currency and unbearable losses of assets for Germans across social classes.[33]

Germany is determined to protect human rights, civil liberties, and religious freedom. These concepts are also enshrined in the text of the "Basic Rights" section of the Basic Law. In relation to this, "coming to terms with the past" (*Vergangenheitsbewältigung*) is a significant part of German culture. The German government has conducted a massive undertaking in recognizing the injustices committed by the Nazi regime and to a somewhat lesser extent the East German regime. Holocaust education is required in every German public school. State-sponsored memorials and interpretive centers have been erected at every former concentration camp and many other sites associated with Nazism around Germany. Berlin's "Topography of Terror," an exhibition covering the rise of the Nazi police state erected in the ruins of the Gestapo and Schutzstaffel (SS) headquarters, is one of the best-known examples of these *Gedenkstätten* ("memorial sites").[34]

On the Hofstede scale, which provides a framework for cross-cultural comparison, Germany scores 35 in power distance, 67 in individualism, 66 in masculinity, 65 in uncertainty avoidance, 83 in long-term orientation, and 40 in indulgence.[35] Germany scores low in power distance because it is a democracy and its government very much depends on the consent of the governed. Open communication between the government and citizenry is expected; this is a norm as much as it is a value. Germany's democratic tradition is also reflected in its high score for individualism – it is a state where "loyalty is based on personal preferences."[36] Germany is a "masculine" society because of its value of competition. This can be considered a possible holdover from the Prussian traditions of rigidness and efficiency, but whatever its source Germans value performance: the Hofstede scale describes the population as "living in order to work."[37] Germany is uncertainty avoidant, preferring deductive rather than inductive reasoning "in line with the philosophical heritage of Kant, Hegel and Fichte" (some of the famous *Denken*).[38] Germany has a very high score in long-term orientation, and given its history of war, destruction, economic crisis, reconstruction, and so forth it is no surprise that Germans show adaptability and prepare themselves to endure a range of future challenges.[39]

Norms of German Strategic Culture

Johnson defines norms as the "accepted and expected modes of behavior within a state." Essentially, these are the actions that a state's populace will generally consider acceptable or unacceptable. German culture reflects particularly strong views in this field that significantly influence strategic conduct.

Modern Germany views peace as natural and war as an aberration, in stark contrast to its Prussian, Imperial, and Nazi predecessors. This view is even in contrast with the West and East Germany of the Cold War – both were always prepared for a ground war to break out on their soil due to the positioning of American and Soviet military forces, respectively, within them. Since reunification, the Bundeswehr has taken part in NATO military missions in Afghanistan and in the waters off the Horn of Africa, well outside the "European NATO territory" originally permitted by the government. This is a significant development from the Bundeswehr of the Cold War, and it has not come without controversy among the German public much like rearmament did. Because of this, the German Bundestag maintains a "strict prerogative" over the deployment of German troops outside of NATO states and has considerable control over the use and withdrawal of troops in such theaters.[40] Limitations on Germany's use of force are expected by the population. Germany opposed the 2003 invasion of Iraq because it ran counter to what it saw as a legitimate operation for its military, alongside perceptions of the war itself being a violation of international law and too risky an operation to undertake.[41]

Germany has two civilian intelligence services, the Bundesnachrichtendienst (BND) (Federal Intelligence Service; foreign intelligence) and the Bundesamt für Verfassungsschutz (BfV) (Federal Office for the Protection of the Constitution; domestic intelligence), both also subject to governmental limitations. The BND, for example, is not allowed to store any of the communications it collects and analyzes unless it specifically pertains to an immediate threat.[42] The BfV reflects the Basic Law's value of human rights through its orientation against violent extremist right-wing and left-wing groups.[43]

Extremism itself is regarded as an aberration in Germany, or that is at least what the state works to achieve. Most far-right parties in Germany, including anything remotely related to Nazism, are banned, and several far-left parties are banned as well. Per Article 21 of the Basic Law, parties that "by reason of their aims or the behavior of their adherents, seek to undermine or abolish the free democratic basic order or to endanger the existence of the Federal Republic of Germany" are deemed "unconstitutional."[44] This manifested early on in West Germany's history with the banning of the Nazi-influenced Socialist Reich Party in 1952 and the Communist Party of Germany in 1956.[45] The AfD has escaped this fate since it has been able to present itself as a populist party that (somewhat) successfully downplays its more extreme followers and does not openly call for disestablishment of the German democracy.[46] This has prevented legal action from taking place in spite of the party's discriminatory ideology. Like most other states in the present world, Germany will have to answer the difficult question of how to contend with extremist groups that operate within the bounds of the law and potentially find their way into government positions.

Germany's Perceptual Lenses

Perceptual lenses relate to how a state views its situation and the world around it. These are informed by the state's experiences or lack of experiences. Johnson states that behavior is always influenced by a perception of reality instead of reality itself, and this maxim also proves true with the behavior of states. What a state perceives as true will determine what security decisions it makes, regardless of what other states might think or perceive. Germany's perceptual lenses influence what the state presently sees as its most pressing security issues.

Germany is most concerned about non-state actor terrorism, far-right movements, and Russian militancy as threats to its security. Its experience with extremism in the past makes it especially concerned about the global rise of right-wing populism, including within Germany. The migrant crisis of the 2010s has brought an increase in small-scale terrorist attacks to Germany, and the country is struggling to balance its acceptance of large numbers of migrants with its counterterrorism efforts. Germany views Russia as its greatest conventional security threat[47] and faces the challenging task of balancing its opposition to Russian aggression with its reliance on Russian natural gas while it seeks energy alternatives. The Russian invasion of Ukraine, beginning in February 2022, prompted dramatically increased defense spending from Germany, underscoring how significant the threat is that Russia presents to the stability of the modern European system Germany helped build. Germany is also keeping an eye on China's rising economic prowess, realizing its potential as a security threat but working with Beijing on trade matters.

Germany has become increasingly wary of the United States in recent years, as this important ally started to withdraw from a leadership role in European defense and global security under the Trump administration. Germany has enjoyed a close partnership with the United States in the realm of foreign policy since the United States was instrumental in setting up West Germany's military and intelligence services, which still collaborate closely with their US counterparts.[48] Even after disagreements like the 2003 Iraq War and the revelation in 2013 that the National Security Agency and BND were spying on each other's governments, Germany has continued cooperation with the United States in counterterrorism operations that began after 9/11, sharing the United States' threat perception about Islamist extremist terrorism. The Trump administration, however, caused more strain on the two countries' relationship than either of these incidents. Norbert Röttgen, chair of the Bundestag's Foreign Affairs Committee, remarked in October 2020:

> The four years of the Trump presidency have meant that everything, very fundamentally, has been called into question. The very existence of NATO, the predictability of US foreign policy. It has been a disruption which we haven't seen since World War Two.[49]

The Trump administration sharply criticized the trade surplus between the two countries and threatened to move US troops out of Germany because of NATO commitment disagreements. Germany is working to raise the percentage of its GDP that it spends on defense to 2%, the target for all NATO members, but was still under target as of 2020.[50] This could notably rise in the wake of Russia's 2022 invasion of Ukraine, but such shifts may sit uneasily with other elements of Germany's strategic culture. For years, Germany has perceived that it is meeting its commitment to NATO in other ways, including participating in NATO combat operations and hosting the US bases that have been vital to missions in Iraq and Afghanistan.[51] It maintains that continued security cooperation with the United States is good for the country and relations have warmed again with the Biden administration's renewed dialogue and cooperation with Europe and NATO. Nonetheless, Trump's presidency and the continued presence he has maintained in American politics have certainly caused an erosion of trust that may never be fully reversed.

Manifestations of Germany's Strategic Culture

All four of Johnson's dimensions of strategic culture influence the decisions a state makes, both domestically and internationally. These dimensions can also manifest in practices of the state's civilian populace, related or unrelated to governmental affairs. This section provides examples of how Germany's unique strategic culture has been reflected in recent history.

The clearest manifestation of German strategic culture is in its foreign policy. Germany has worked to become a "good neighbor" in Europe after being feared by its neighbors for so long and seeks to reconcile with and incorporate them into a shared European community. Bringing countries like the Czech Republic and Poland, which suffered greatly under Nazi occupation, into the EU and "making borders porous" has been part of Germany's way of righting the wrongs of the past.[52] Germany has collaborated significantly with France in particular, going so far as to establish a shared military battalion that conducts maneuvers in both Germany and France.[53]

Germany provides Israel with significant monetary and material military aid as a major part of its efforts to make rapprochement with the Jewish people. Shared commemoration of the Holocaust by the two states is a frequent occurrence, manifested in August 2020 with a ceremonial flyover of Dachau by Israeli and German air force pilots.[54] Germany also officially recognizes the Palestinian Authority and has provided millions of dollars in civil aid to it since 1993, reflecting Germany's desire for a two-state solution to the Israel-Palestine conflict.[55]

Germany's strategic culture manifests in the restrictions it places on its armaments. Because of its own destruction by conventional weapons during World War II and the fear of being a primary target in the event of a nuclear attack during the Cold War, Germany is averse to nuclear weapons. It is a non-nuclear weapons state, although it is protected by the US nuclear umbrella and hosts US nuclear weapons.[56] On the world stage, Germany has taken a leading role in nuclear disarmament efforts. Germany's Federal Foreign Office has long called for the international abolition of tactical nuclear weapons platforms (which during the Cold War had been stationed in both East and West Germany), occasionally putting it at odds with its neighbors and the United States.[57]

Germany's primary counterterrorism unit, GSG 9, is a police unit and is not tied to the military. Upon the unit's creation in 1972 in response to the Munich Olympics massacre, the German government had concerns about it being associated with Nazi "special" units like the SS.[58] The Basic Law also forbids the Bundeswehr from being used in combat on German soil during peacetime. Germany eventually did establish a military special forces unit within the Bundeswehr, the Kommando Spezialkräfte (KSK), in 1996. This decision was a reaction to the deaths of German civilians in the Rwandan crisis, which GSG 9 was unauthorized to respond to. The event highlighted Germany's need for its own force that could carry out special operations in NATO missions overseas.

In 2020, Germany disbanded part of the KSK because of the discovery of far-right extremist members in leadership positions within the unit.[59] The incident came after multiple previous instances of White supremacist paraphernalia being found in Bundeswehr barracks, which were also dealt with by firings and reorganizations.

Strengths and Gaps in the Strategic Culture Literature

The discussions above have highlighted several key contributions within the strategic culture literature on Germany. A few further seminal works from the existing body of literature merit a brief discussion. Julian Junk and Christopher Daase's chapter on Germany in *Strategic Cultures in Europe: Security and Defence Policies Across the Continent* provides an excellent overview of German strategic culture around defense and is one of few sources highlighted in this review that explicitly uses the label of strategic culture; all other works cover some cultural dimension without direct reference to the concept of strategic culture itself. Poultney Bigelow's two-volume *History of the German Struggle for Liberty* provides an account of the origins of the German nation-state in the nineteenth century and clarifies the importance of modern German symbols. This work is an excellent starting point for researching the origins of the democratic movement in the nineteenth century, the importance of democracy to Germany, and how all of these have manifested in the modern Federal Republic.

Karlfried Knapp's 1995 article "What's German? Remarks on German Identity" explores the idiosyncratic behaviors of Germans and their attitudes toward patriotism, Germany's role in Europe and the world, and the use of national symbols. Jan-Werner Müller's 2014 article "Germany's Two Processes of 'Coming to Terms with the Past'" covers the role that memory plays in modern Germany, and Stephen Brockmann's 2002 article "Germany as Occident at the Zero Hour" narrates how German culture was effectively reset in 1945, a year known as *Stunde Null* ("Zero Hour") in German historiography. These articles examine the impact of Germany's twentieth-century history and the way this still impacts domestic decisionmaking.

In the domain of military affairs, Wilfrid von Bredow's 1992 article "Conscription, Conscientious Objection, and Civic Service: The Military Institutions and Political Culture of Germany, 1945 to Present," Kerry Longhurst's 2004 book *Germany and the Use of Force*, and Sophia Becker's 2013 article "Germany and War: Understanding Strategic Culture Under the Merkel Government" give researchers a vivid picture of Germany's present attitudes about the use of its military at home and abroad, and the historical roots of its military policies, including conscription and involvement in wars since the 1990s. Tuomas Forsberg examines Germany and the Iraq War as a specific case study in his 2005 piece "German Foreign Policy and the War on Iraq: Anti-Americanism, Pacifism or Emancipation?" and Beatrice Heuser supplies a standout work examining the cultural roots of Germany's nuclear aversion in her 1998 book *Nuclear Mentalities? Strategies and Belief in Britain, France, and the FRG.*[60]

The existing body of literature is very strong in the exploration of security and military affairs, but few utilize the dedicated concept of strategic culture or associated frameworks for assessment. Similarly, there are many works about Germany's history and civic culture, but few build the bridge directly connecting how these elements relate to German strategic culture. These gaps within the literature present ample opportunities for future scholars to advance the dedicated study of Germany's strategic culture.

Another area of great significance to Germany that could be built out further in the literature is the economic domain. Wolf D. Gruner's 2017 article "Is the German Question – Is the German Problem Back? The Role of Germany in Europe from an Historical Perspective," and Stephen Brown's 2011 short piece "If Germans Are Stubborn on Euro, Blame Weimar" explore cultural dimensions shaping German economic policy and the reverse effect of economics on culture. In historical studies, the impact of the Weimar era and the inflation of the *Papiermark* receives the most coverage, but more could be written about the impact of other events and ideas in German economic history on the state's strategic culture, including the ascension of the Social Democratic Party in the late nineteenth century, the extensive privatization of the Nazi regime, Konrad Adenauer's famous "Rhine Capitalism" model, and the communist economic system of East Germany.

Additional future studies in German strategic culture could investigate Germany's ongoing involvement in NATO missions in Kosovo, the Horn of Africa, and Afghanistan, providing new case studies on the influence of culture on the country's defense posture and conduct. Germany's evolving response to the resurgent threat of Russian aggression likewise merits focus. The shape of German politics after 2021, when Merkel stepped down from her lengthy chancellorship, offers much interesting ground for assessing how German strategic culture will shape and be shaped by successive chancellors. Germany's relationship with the United States, with ties strengthening again under the Biden presidency, is also likely to evolve according to Germany's perception of what is arguably Europe's second "special relationship" spanning the Atlantic. Across the eventful 2020s and beyond, Germany will remain a fascinating case study for any scholar of history and policy, and there is space for future scholars of strategic culture to make valuable contributions in this domain.

Notes

1 Special thanks to Dr. Kerry Kartchner for introducing me to strategic culture and assisting me with the arrangement of this chapter, and to Jörn Beissert, former Counselor for Asia and Security Policy at the German Embassy in Washington, DC, for interviewing with me and my partner Helena Doms for a project in 2018.

2 Jeannie L. Johnson, "Strategic Culture: Redefining the Theoretical Construct," Science Applications International Corporation (2006), 5.

3 Johnson, 14. See also Jeannie L. Johnson and Matthew T. Berrett, "Cultural Topography: A New Research Tool for Intelligence Analysis," *Studies in Intelligence* 55, no. 2 (2011): 1–22.

4 Quentin Outram, "The Demographic Impact of Early Modern Warfare," *Social Science History* 26, no. 2 (2002): 248.

5 Wilhelm Cleven, "The Failure of a Revolution France, Germany and The Netherlands in 1848: A Comparative Analysis" (Utrecht University, 2008), 38.

6 Jeremiah Riemer, trans., "Excerpt from Bismarck's 'Blood and Iron' Speech (1862)," German History in Documents and Images, last modified 17 December 2003, http://germanhistorydocs.ghi-dc.org/sub_document.cfm?document_id=250&language=english.

7 Fabrice Jesne and Eric Schnakenbourg, "The European Balance of Power," Sorbonne Université Digital Encyclopedia of European History, last updated 22 June 2020, https://ehne.fr/en/encyclopedia/themes/europe-europeans-and-world/organizing-international-system/european-balance-power.

8 "Antisemitism in History: World War I," United States Holocaust Memorial Museum, https://encyclopedia.ushmm.org/content/en/article/antisemitism-in-history-world-war-i.

9 "Background: Life before the Holocaust," Learning Voices of the Holocaust, British Library, last updated 5 January 2006, https://www.bl.uk/learning/histcitizen/voices/testimonies/life/backgd/before.html.

10 US State Department of State, "Territorial Reorganization Inside Germany: Abolition of the State of Prussia," in *Germany, 1947–1949: The Story in Documents*, US Department of State, US Government Printing Office (Washington, DC, 1950), 151.
11 The World Factbook (Washington, DC: Central Intelligence Agency, 2020), https://www.cia.gov/library/publications/the-world-factbook/geos/gm.html.
12 CIA, The World Factbook, 2020.
13 CIA, The World Factbook, 2020.
14 CIA, The World Factbook, 2020.
15 Karlfried Knapp, "What's German? Remarks on German Identity." *Institut fur Anglistik/Amerikanistik* (1995), 223–224, http://webdoc.sub.gwdg.de/edoc/ia/eese/articles/knapp/10_95.html.
16 Christian Tomuschat et al., trans. "Basic Law for the Federal Republic of Germany: I. Basic Rights, Article 1, (1)," Gesetze im Internet, Bundesministerium der Justiz und für Verbraucherschutz, last modified 28 March 2019, https://www.gesetze-im-internet.de/englsch_gg/index.html.
17 Sophia Becker, "Germany and War: Understanding Strategic Culture under the Merkel Government," Institut de Recherche Strategique de L'Ecole Militaire, Paris Papers (2013) 16.
18 Becker, 2013, 16.
19 Kerry Longhurst, *Germany and the Use of Force* (United Kingdom: Manchester University Press, 2004), 120.
20 Wilfried Von Bredow, "Conscription, Conscientious Objection, and Civic Service: The Military Institutions and Political Culture of Germany, 1945 to Present," *JPMS: Journal of Political and Military Sociology* 20, no. 2 (Winter 1992): 291, https://search.proquest.com/docview/1303263304/fulltextPDF/516F6F04AB674E83PQ/1?accountid=13158.
21 Bredow, 1992, 291.
22 Bredow, 1992, 292.
23 Becker, 2013, 69.
24 Becker, 2013, 16.
25 World Bank, "GDP (current US$) – Germany," World Bank National Accounts Data, and OECD National Accounts Data Files, last updated 2021, https://data.worldbank.org/indicator/NY.GDP.MKTP.CD.
26 Knapp, 1995, 223.
27 Stephen Brockmann, "Germany as Occident at the Zero Hour," *German Studies Review* 25, no. 3 (2002): 480, doi:10.2307/1432597.
28 CIA, The World Factbook, 2020.
29 CIA, The World Factbook, 2020.
30 Poultney Bigelow, *History of the German Struggle for Liberty*, Vol. 2 (Harpers & Brothers, 1896), 97–98, https://books.google.com/books?id=IawBAAAAYAAJ&pg=PA110&source=gbs_toc_r&cad=3#v=onepage&q&f=false
31 German Bundestag, "The Federal Eagle," National Symbols, last modified 2010, https://www.bundestag.de/en/parliament/symbols/eagle.
32 Wolf D. Gruner, "Is the German Question – Is the German Problem Back? The Role of Germany in Europe from an Historical Perspective," *Rivista Di Studi Politici Internazionali*, Nuova Serie, 84, no. 3 (335) (2017): 365.
33 Stephen Brown, "If Germans Are Stubborn on Euro, Blame Weimar," *Reuters*, 10 November 2011, https://www.reuters.com/article/eurozone-germany-history/if-germans-are-stubborn-on-euro-blame-weimar-idUSL5E7MA17O20111110.
34 Jan-Werner Müller, et al., "Germany's Two Processes of 'Coming to Terms with the Past'—Failures, After All?" in *Remembrance, History, and Justice: Coming to Terms with Traumatic Pasts in Democratic Societies* (Budapest; New York: Central European University Press, 2015), 214.
35 "Country Comparison: Germany," Hofstede Insights. Last updated 2020. Accessed on 31 October 2020. https://www.hofstede-insights.com/country-comparison/germany/.
36 "Country Comparison," 2020.
37 "Country Comparison," 2020.
38 "Country Comparison," 2020.

39 "Country Comparison," 2020.
40 Julian Junk and Christopher Daase, "Germany," in *Strategic Cultures in Europe: Security and Defence Policies Across the Continent*, ed. Heiko Biehl, Bastian Giegerich, and Alexandra Jonas (Potsdam: Springer VS, 2013), 142.
41 Tuomas Forsberg, "German Foreign Policy and the War on Iraq: Anti-Americanism, Pacifism or Emancipation?" *Security Dialogue* 36, no. 2 (2005): 219–220, http://www.jstor.org/stable/26298888.
42 Mark Lowenthal, "Foreign Intelligence Services: Germany," in *Intelligence: From Secrets to Policy*, 7th ed. (CQ Press, 2017), 532.
43 Lowenthal, "Foreign Intelligence Services," 2017, 533.
44 Christian Tomuschat et al., trans. "Basic Law for the Federal Republic of Germany: I. Basic Rights, Article 21, (2)," Gesetze im Internet, Bundesministerium der Justiz und für Verbraucherschutz, last modified 28 March 2019, https://www.gesetze-im-internet.de/englisch_gg/index.html.content/uploads/2008/09/Frida-Tronnberg.pdf.
45 Angela Bourne and Fernando Casal Bertoa, "Germany No Longer Bans Extremist Parties. But Which European Democracies Do, and Why?," Democratic Audit, 25 September 2017, https://www.democraticaudit.com/2017/09/25/germany-no-longer-bans-extremist-parties-but-which-european-democracies-do-and-why/.
46 Bourne, "Germany No Longer Bans Extremist Parties," 2017.
47 Jörn Beissert (former Counselor for Asia and Security Policy at the German Embassy in Washington, DC) in discussion with the author, October 2018.
48 Lowenthal, "Foreign Intelligence Services," 2017, 532.
49 Jenny Hill, "US Election 2020: Why It Matters So Much to Germans," BBC News, 14 October 2020, https://www.bbc.com/news/election-us-2020-54522984.
50 Hill, "US election 2020," 2020.
51 Beissert interview, 2020.
52 Lily Gardner Feldman, "The Principle and Practice of 'Reconciliation' in German Foreign Policy: Relations with France, Israel, Poland and the Czech Republic," *International Affairs* 75, no. 2 (1999): 338.
53 Feldman, "The Principle and Practice of 'Reconciliation' in German Foreign Policy," 335.
54 "German and Israeli Jets Mark First Joint Flyover, Honor Holocaust Cictims," Deutsche Welle, 18 August 2020, https://www.dw.com/en/german-and-israeli-jets-mark-first-joint-flyover-honor-holocaust-victims/a-54605587.
55 Paul Belkin, "Germany's Relations with Israel: Background and Implications for German Middle East Policy," Library of Congress. Foreign Affairs, Defense, and Trade Division, Congressional Research Service, 2007, 11.
56 Patricia M. Lewis, et al., *Four Emerging Issues in Arms Control, Disarmament, and Nonproliferation: Opportunities for German* Leadership, Report, James Martin Center for Nonproliferation Studies (CNS), 2009, 25.
57 Lewis, *Four Emerging Issues*, 2009, 28.
58 Doron Zimmermann, "Between Minimum Force and Maximum Violence: Combating Political Violence Movements with Third-Force Options," *The Quarterly Journal* 4 (2005): 52–53.
59 "'Toxic Leadership Culture': Germany Shakes up Elite Army Force over Far-Right Links," Thelocal.de, 30 June 2020, https://www.thelocal.de/20200630/germany-to-partly-dissolve-elite-force-over-far-right-links-minister.
60 Beatrice Heuser, *Nuclear Mentalities? Strategies and Belief in Britain, France, and the FRG* (New York: St. Martin's Press, 1998).

Selected Bibliography

Becker, Sophia. "Germany and War: Understanding Strategic Culture under the Merkel Government." Institut de Recherche Strategique de L'Ecole Militaire, Paris Papers, 2013.

Belkin, Paul. "Germany's Relations with Israel: Background and Implications for German Middle East Policy." Library of Congress. Foreign Affairs, Defense, and Trade Division, Congressional Research Service, 2007.

Bredow, Wilfried Von. "Conscription, Conscientious Objection, and Civic Service: The Military Institutions and Political Culture of Germany, 1945 to Present." *JPMS: Journal of Political and Military Sociology* 20, no. 2 (Winter, 1992): 289–303. https://search.proquest.com/docview/1303263304/fulltextPDF/516F6F04AB674E83PQ/1?accountid=13158.

Brockmann, Stephen. "Germany as Occident at the Zero Hour." *German Studies Review* 25, no. 3 (2002): 477–496. Accessed October 1, 2020. doi:10.2307/1432597.

Brown, Stephen. "If Germans Are Stubborn on Euro, Blame Weimar." *Reuters*, 10 November 2011. https://www.reuters.com/article/eurozone-germany-history/if-germans-are-stubborn-on-euro-blame-weimar-idUSL5E7MA17O20111110

Forsberg, Tuomas. "German Foreign Policy and the War on Iraq: Anti-Americanism, Pacifism or Emancipation?" *Security Dialogue* 36, no. 2 (2005): 213–231. http://www.jstor.org/stable/26298888.

Gardner Feldman, Lily. "The Principle and Practice of 'Reconciliation' in German Foreign Policy: Relations with France, Israel, Poland and the Czech Republic." *International Affairs* 75, no. 2 (1999): 333–356.

Gruner, Wolf D. "Is the German Question – Is the German Problem Back? The Role of Germany in Europe from an Historical Perspective." *Rivista Di Studi Politici Internazionali*, Nuova Serie, 84, no. 3 (335) (2017): 341–373.

Jesne, Fabrice, and Eric Schnakenbourg. "The European Balance of Power." Sorbonne Université Digital Encyclopedia of European History. Last updated June 22, 2020. https://ehne.fr/en/encyclopedia/themes/europe-europeans-and-world/organizing-international-system/european-balance-power.

Johnson, Jeannie L. "Strategic Culture: Redefining the Theoretical Construct." Science Applications International Corporation, 2006.

Junk, Julian, and Christopher Daase. "Germany." In *Strategic Cultures in Europe: Security and Defence Policies Across the Continent*, edited by Heiko Biehl, Bastian Giegerich, and Alexandra Jonas, 139–152. Potsdam: Springer VS, 2013.

Knapp, Karlfried. "What's German? Remarks on German Identity." Institut fur Anglistik/Amerikanistik, 1995. http://webdoc.sub.gwdg.de/edoc/ia/eese/articles/knapp/10_95.html.

Lewis, Patricia M., Dennis M. Gormley, Miles A. Pomper, Lawrence Scheinman, Stephen I. Schwartz, Nikolai N. Sokov, and Leonard S. Spector. *Four Emerging Issues in Arms Control, Disarmament, and Nonproliferation: Opportunities for German Leadership*. Report. James Martin Center for Nonproliferation Studies (CNS), 2009. 25–49.

Longhurst, Kerry. *Germany and the Use of Force*. United Kingdom: Manchester University Press, 2004.

Müller, Jan-Werner. "Germany's Two Processes of 'Coming to Terms with the Past'—Failures, After All?" In *Remembrance, History, and Justice: Coming to Terms with Traumatic Pasts in Democratic Societies*, edited by Tismaneanu Vladimir and Bogdan C. Iacob, 213–236. Budapest; New York: Central European University Press, 2015.

Pokrandt, Christine. "Germany's Lingering Identity Crisis." Cultural Diplomacy, 2020. http://www.culturaldiplomacy.org/pdf/case-studies/germanys-lingering-id.pdf.

18

REINTERPRETING THE (RELATIVELY) IMMUTABLE FEATURES OF THE NORTH KOREAN STRATEGIC CULTURE

Steve S. Sin

For the concept of strategic culture to be useful as an analytical approach, one must first define the features of strategic culture that are observable for the state that is being analyzed. As with the debate and evolution of the concept of strategic culture, what features to include in the analyses have also been a topic of heated discussion. The consensus in the literature on this topic seems to be that one must include those features that prove fruitful in answering the research questions while also being able to be clearly observed by a third party. This consensus means that the features of strategic culture that an analyst focuses on could potentially (and really should) differ based on the questions being asked. For example, some analysts may choose to focus mainly on a state's military aspects related to war and its leadership's approach to the use of force, while others may choose to examine the political aspects of a nation's life more broadly.

The features chosen for analysis can be divided into categories, which will allow one to conduct cross-temporal and/or cross-sample comparisons. The features can be divided into those that are not change-prone and those that are mutable – either over time or those that are currently undergoing changes – as well as those that are ideational and material. Examples of features that can be classified as not change-prone are geography, historical experiences, myths, and symbols – though the interpretation of the ideology, as well as the interpretation and importance of specific myths and symbols, can change over time as society continues to evolve. Examples of mutable features are those such as political and military structure, technological development and adoption, key texts that inform actors' strategies, and societal norms.

(Relatively) Immutable Features of North Korean Strategic Culture

This chapter first explores major features of the North Korean strategic culture categorized as being relatively immutable by scholars of strategic culture. These include its geography, formative historical experiences, myths, and symbols, including the creation myth of the

DOI: 10.4324/9781003010302-21

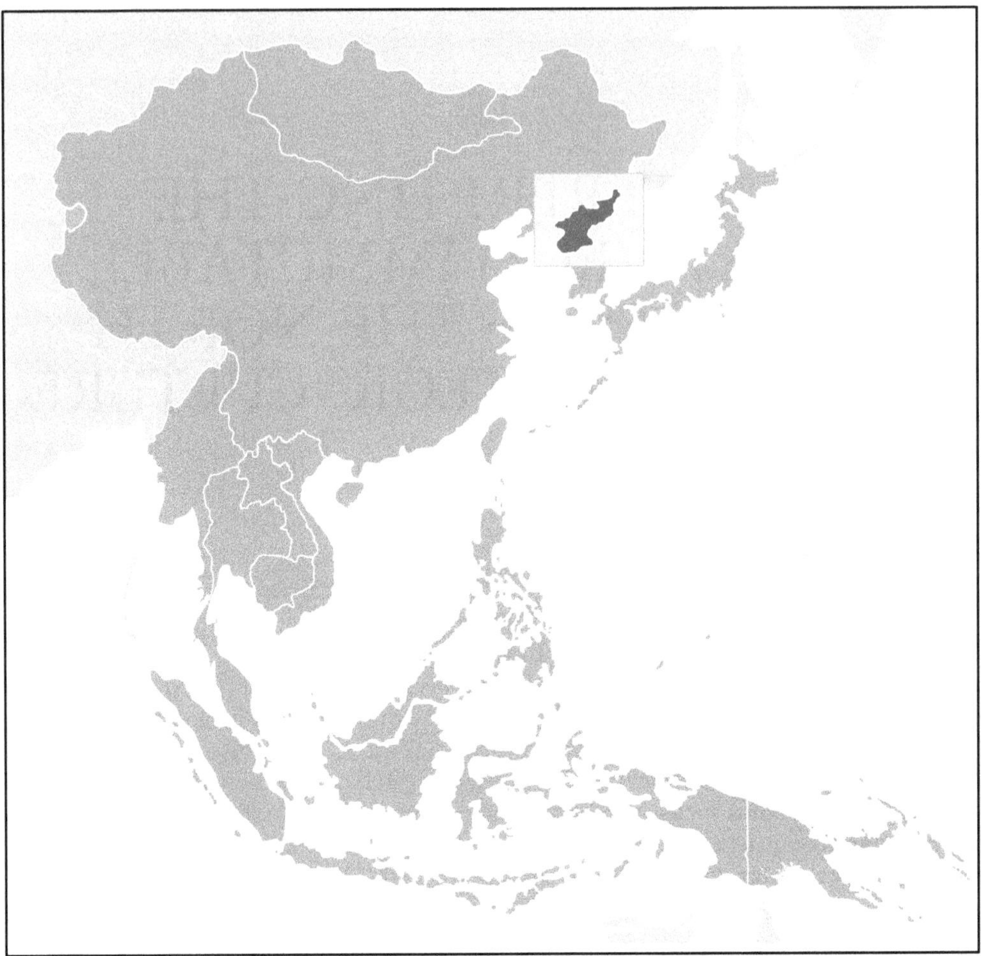

Figure 18.1 North Korea Location Map.
Source: CIA Factbook 2021.

Kim family, and the use of political ideology to entrench that myth and exert control over the North Korean people. The chapter concludes by advancing the argument that Kim Il-sung used a mutable feature of the strategic culture – namely through the establishment of a societal norm in the form of a political ideology – to entrench an immutable feature in North Korean society and exert control through it.

Geography

The Korean Peninsula is a relatively small yet prominent peninsula that protrudes for approximately 1,000 kilometers southward from the northeast portion of the Asian continental landmass (Figure 18.1). The peninsula, together with its many islands located in the south, forms a nearly complete land bridge between China and Japan. The southern tip of the peninsula is only a little over 40 kilometers from the nearest Japanese territory of Tsushima Island.

The Korean peninsula shares land borders with China and Russia to the north, with the main island of the Japanese archipelago (Honshu) located approximately 200 kilometers to its southeast across the East Sea (Sea of Japan). Approximately 190 kilometers to its west, across the West Sea (Korea Bay in North Korea and Yellow Sea in China), is China's Shandong Peninsula. Korea's east coast is, unsurprisingly, bordered by the East Sea (Sea of Japan) and its south coast by the South Sea. The peninsula has 8,460 kilometers of coastline, with 2,495 kilometers of it belonging to North Korea.

For North Korea, the Amnok (Yalu) and Duman (Tumen) rivers form its 1,433 kilometers of land borders shared with China and Russia. Of these, 1,416 kilometers are shared with the Chinese provinces of Liaoning and Jilin, while just 17 kilometers are shared with Russia. Additionally, the portion of the land border with China that includes Mount Paektu (Paektu-san in Korean or White Head Mountain) is yet to be clearly demarcated. North Korea's land border with South Korea (also known as the Republic of Korea) is a 238-kilometer Demarcation Line that runs through the middle of a 4-kilometer-wide Demilitarized Zone (DMZ) between the North and South Koreas that went into effect with the armistice of the Korean War in 1953. Additionally, North Korea claims territorial waters 12 nautical miles from its shores and exercises its claim of an exclusive economic zone 200 nautical miles from its shores. The total land area of North Korea is 120,410 square kilometers, which is approximately 55% of the Korean Peninsula's land area, or by comparison slightly larger than the US state of Virginia (Figure 18.2).[1]

The geographical location and characteristics of the Korean Peninsula have made it exceptionally strategically important and explain a great deal about the historically unremitting interests – which continue to the present day – of its neighbors as well as the United States to influence Korea's political, military, economic, and cultural evolutionary directions.

Formative Historical Experiences

Owing largely to its strategic geographical location, the Korean Peninsula has experienced over 900 foreign invasions throughout its recorded history, including the Mongol invasions during the thirteenth century, the Japanese invasion in the sixteenth century, and the Ching invasions in the seventeenth century.[2] The contention over the Korean Peninsula, however, became much more heightened with the arrival of Western powers in East Asia in the nineteenth century. The ingress of the Western powers caused the erosion of the long-established Chinese-centric regional order[3] in the Korean Peninsula and exposed it to the pressures of foreign great power rivalries.

Japan became the ultimate beneficiary of the great power rivalries of the late nineteenth and early twentieth centuries when it came to control over the Korean Peninsula. When Japan beat China in 1895 in the First Sino-Japanese War, the Treaty of Shimonoseki removed China's suzerainty over Korea, allowing Japan to gain increased influence in the peninsula. After winning the Russo-Japanese War a decade later, Japan forced Korea to sign the Japan-Korea Treaty of 1905, in which Korea became Japan's protectorate. Although the Emperor of Korea requested the United States to intervene under the "good offices" clause of the United States-Korea Treaty of 1882, the United States denied the request and supported Japan's colonization of Korea. The United States supported Japan's move to exert control over Korea because of its desire to check the Russian influence in East Asia as well as the Japanese promise that Japan would not make any aggressive moves

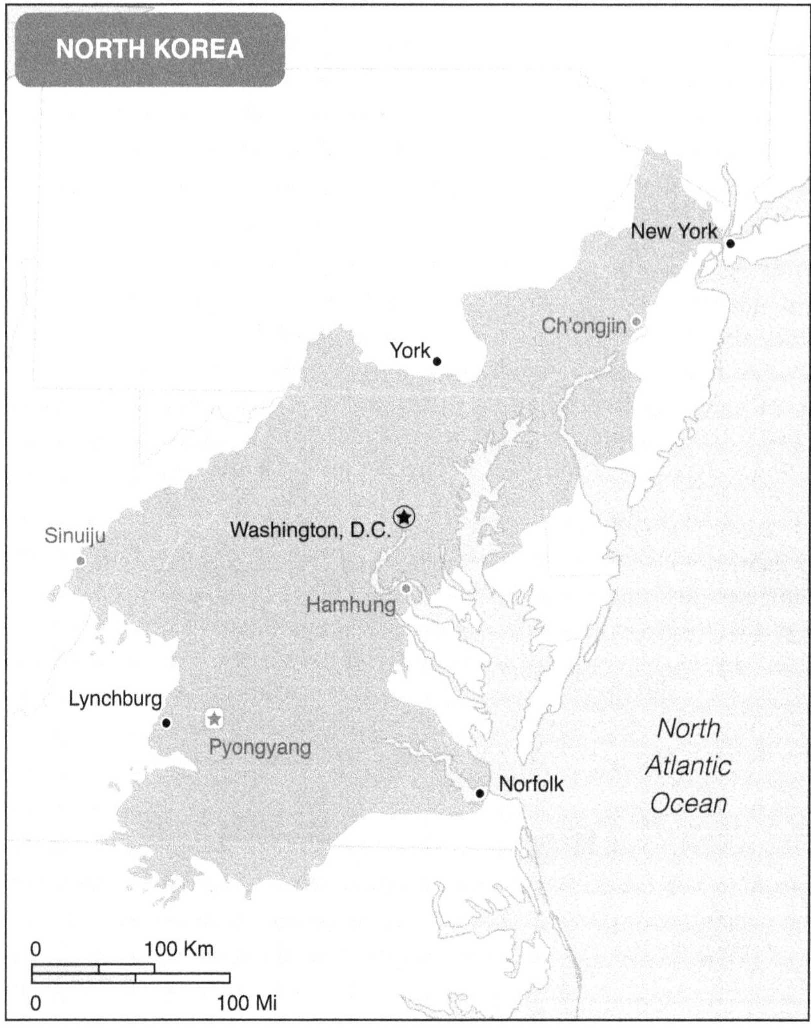

Figure 18.2 North Korean Land Area Comparison.
Source: CIA Factbook 2021.

against the Philippines in exchange for its control over Korea. The Taft-Katsura Agreement of July 1905 affirmed a similar understanding between Japan and the United States.[4] Around the same time, Great Britain also recognized Japan's suzerainty over Korea in return for gaining exclusive rights over Singapore. In 1907, the Emperor of Korea secretly dispatched envoys to the Hague Peace Conference to appeal for assistance from the international community; however, no country in attendance would support the envoys. In July of that year, Japan forced Korea to sign the Japan-Korea Treaty of 1907 and gained full control of Korean domestic affairs. Finally, in 1910, Japan consolidated its control over Korea by imposing the Japan-Korea Annexation Treaty.[5]

Japanese rule over the Korean Peninsula continued until the end of World War II. After the Japanese defeat in 1945, Korea regained its independence; however, not as a unified peninsula. Per the Yalta Conference agreements, the Soviet Union accepted the Japanese surrender in the part of the Korean Peninsula north of the 38th parallel, while the United States accepted the surrender south of that line. This arrangement divided the Korean Peninsula into two spheres – dominated by the Soviet Union in the north and the United States in the south – and eventually led to the creation of the two Koreas that persist to the present day.

These experiences with foreign powers, embodied in unequal treaties and international settlements that ignored Korean positions, planted in the Korean collective psyche a deep distrust and hatred of foreigners – especially against the Japanese and the "Westerners." Additionally, these bitter experiences and severe consequences made Koreans extremely sensitive about their national security and foreign interference or dominance in their domestic and international affairs. Finally, the geopolitical influences and the international security episodes described above have come to serve as the foundation of the North Korean worldview and the political ideology that undergirds its strategic culture.

Myths and Symbols

National identity is the fundamental ideational feature of any country's strategic culture, and national identity influences how and why the people of that country hold certain perceptions about themselves and others. One of the best ways to glean a country's national identity is through the examination of its myths and symbols. A country's myths and symbols are both spatial and temporal. They are spatial because they form complex multidimensional and multilayered puzzle pieces where each piece represents a specific myth and symbol that applies to a specific group or segment of the society, and these pieces come together to form the myths and symbols for the entire nation. Second, they are temporal because they are born out of (and change based on) the country's history: from its founding, through the shared historical experiences of its people, to its modern-day occurrences. It is in this way that national myths and symbols are closely tied to the nation's historical experiences and establish themselves as the unifying "narratives of the people."

What is important to note is that while the interpretation of these myths and symbols will certainly evolve over time, the core narratives surrounding them tend to be rather immutable. For all states, educational institutions primarily serve as the mechanism to consistently propagate these core national narratives. Choosing which national stories are told in educational institutions and how they are told, therefore, is an integral part of the formation and sustainment of a country's strategic culture.

North Korea is no exception to this. In fact, North Korea's educational system is a prime example that demonstrates the relationship between a country's myths and symbols, education, and strategic culture. For North Korea, the most important historical education – thus the most important set of myths and symbols – is the period of Japanese colonization and the triumph of the Korean freedom fighters (who later became the first generation of political and military leaders for North Korea) over the Japanese that led to the eventual liberation of the Korean Peninsula. The national stories told about this period and the contributions made by the Korean freedom fighters (especially Kim Il-sung) form the basis for the North Korean strategic culture.

Creation of the Myth of Kim Il-sung: The Symbol of North Korea

Leader worship in North Korea is unlike any that one can observe in other former Soviet bloc states and other dictatorships, and it is at the core of the legitimacy for three generations of the Kim family's hereditary rule over North Korea. In North Korea, Kim Il-sung – the state's first leader – is a mythical figure who lives in portraits and statues, music, art, literary works, and folklore. The creation of the Kim Il-sung myth began under the patronage of the Soviet Union. At the end of World War II, Japan had been defeated, and the Korean people were looking for a hero to lead them after years of Japanese colonial rule. As the war ended, North Korea – the territory north of the 38th parallel – was governed by the Soviet military under the control of T. F. Stykov of the Maritime Province Military District and Major General Lebedev of the 25th Soviet Army. For a more permanent governance solution, the Soviet Union wanted to establish a socialist regime that it could control easily.[6] To accomplish this, Soviet leaders needed a suitable ethnic Korean military figure loyal to the Soviet Union. Soviet leaders found that Kim Il-sung was that man – the most suitable and loyal person available for the job.[7] It is widely believed outside of North Korea, in fact, that Kim Il-sung had actually never set foot in the Korean Peninsula before 1945 – September 1945 to be exact – when he was selected by the Soviet Union to be the new leader of the Korean people living in the Northern occupation zone of the Korean Peninsula.[8]

In propping up the pro-Soviet government and Kim Il-sung, the Soviet Union carefully crafted the mythical stories of Kim Il-sung. As Stalin's cult of personality was at its peak in the Soviet Union at the end of World War II, the Soviet Union transliterated its cult of personality model to North Korea, replacing Stalin with Kim Il-sung. To bolster Kim Il-sung's legitimacy, every event, every poster, and every portrait in North Korea had the exclamation, "Hurrah, General Kim Il-sung!" alongside "Hurrah, Stalin!" and Kim Il-sung was always portrayed together with Marx, Lenin, and/or Stalin. The Soviet Union also planned and executed a delicately scripted propaganda campaign to elevate Kim Il-sung to a mythical guerilla hero of the Korean people who was instrumental in leading, planning, orchestrating, and conducting a war of liberation against the occupying Japanese forces in Korea and China. According to the propaganda, Kim Il-sung first set foot in the Korean Peninsula in the 1930s to conduct raids against the Japanese, and eventually moved his guerilla headquarters to Korea in 1937 to begin his war of attrition against the Japanese forces.[9]

Of course, a mythical cult of personality figures cannot be complete without an accompanying story of one's humble beginnings. As the story went – crafted by Soviet and North Korean propagandists – Kim Il-sung was born on 12 April 1912 in Pyongyang as the first child to peasant parents, but departed to Manchuria, China, soon after his birth when his father, a staunch anti-imperialist and an anti-Japanese fighter, moved the family there. It is while there that Kim Il-sung became an active pro-communist nationalist fighter against the Japanese. Through the guidance and education of his parents, the narrative continues, Kim Il-sung was able to grow as a nationalist fighter who deeply cared for the Korean people and their independence from the imperial forces. After the death of his father in 1926 and of his mother in 1932, Kim Il-sung redoubled his efforts to liberate the Korean people, became a guerrilla leader, and returned to Korea in 1937 to take the fight to the Japanese occupation forces.[10]

Through the deployment of the Soviet-style cult of personality propaganda campaign that employed the narrative of a returning hero of national liberation, the Soviet Union was able to establish the legend of Kim Il-sung and place it as the core of the North Korean national myth: around which all other national myths and symbols revolve, and from which the North Korean regime derives its legitimacy. In fact, Kim Il-sung and later his son and successor Kim Jung-Il "refined, expanded, and reinterpreted" the stories surrounding the myth of Kim Il-sung as needed over the years to justify their legitimacy and their policy decisions. The cult of personality originating around Kim Il-sung is used today by his grandson, Kim Jong Un, as he continues to assert his legitimacy as the rightful leader of North Korea, navigate North Korea's delicate political landscape, and cajole the masses that North Korea is stronger than ever.[11]

Use of Political Ideology to Entrench the Myth and Exert Control

While the Kim Il-sung myth has been instrumental in providing the Kim family political legitimacy and making the North Korean people accept the hereditary transfer of power by the Kim family, it has been paired with the systematic cultivation of a set of social norms that defined accepted and expected behavioral patterns for the society in order for the consecutive Kim regimes to maintain power and control. Kim Il-sung sought to develop controlling social norms flavored with his own variant of Marxist-Leninist political ideology. These would mimic Soviet levels of social control but create a system that was uniquely North Korean and distinguishable from Soviet-style political ideology.[12] In doing so, Kim Il-sung developed the *Juche* ideology. This term first appeared in a speech Kim Il-sung gave in 1955; it was not until 1965, however, that Kim Il-sung outlined the four fundamental principles of the *Juche* ideology: (1) self-reliance in ideology (*Juche*); (2) political self-reliance (*Jaju*); (3) economic self-reliance (*Jarip*); and (4) self-reliant defense (*Jawi*).[13] In 1982, Kim Jong-Il formalized and consolidated the *Juche* ideology principles into three pillars – independence in politics, a self-sufficient economy, and self-reliant defense – and adopted them as the official guidelines for North Korea's domestic and foreign policies.[14]

The most noticeable element of the *Juche* ideology is its anti-imperialist characteristics, harkening back to the common Korean historical experiences of the late nineteenth and first half of the twentieth centuries. As such, the self-reliant defense principle of the *Juche* ideology was prioritized.[15] In line with this focus, Kim Il-sung instituted mandatory military service for most North Koreans and structured North Korea's production around defense and heavy industries. Additionally, in keeping with the *Juche* ideology's anti-imperialist characteristics, Kim Il-sung used the principle of self-reliant defense to legitimize the institutionalization of mass mobilization. Ostensibly this was instituted to repel any future imperialist invasions of North Korea and to support the nation's efforts to eject any foreign influencers from the Korean Peninsula, but in reality, it was a perfect vessel for the Kim regime to exert control over the North Korean society. The priority given to the military and the institutionalization of mass mobilization effectively militarized the entire North Korean society, and this was – and still is today – an instrumental part of the mechanism of control and maintenance of power for the Kim regime, now three generations in.

Given the proximity between the *Juche* ideology and North Korea's anti-imperialist nationalism, operating within the frame of the *Juche* ideology means one is inherently hostile to foreign power influence over the Korean peninsula. From this perspective, the US military presence in South Korea makes the United States a clear and present danger that the Kim regime can use to convince the North Korean society that North Korea needs to continue on the path of self-reliant defense (by way of continued weapons development programs, nuclear or otherwise) even in the face of hardship and steep opportunity costs. The same imperative is argued for the North Korean population to strictly adhere to the *Juche* ideology – and by proxy, remain loyal to the Kim regime.

Evaluating the Literature on North Korea's Strategic Culture

Current literature on North Korean strategic culture can be divided into two general categories. First, a body of work explores the sources of North Korea's strategic culture, and the impact North Korea's strategic culture has on its threat perceptions, in order to extricate implications for regional as well as global security. The works that belong to this category tend to analyze the origins of various components that collectively make up the North Korean strategic culture in order to determine how those individual origins come together to form a cohesive worldview, leading to the North Korean regime's and elites' perceptions of threat about the actors and environment around them.[16]

The second general category of works, and perhaps the more voluminous, attempts to extrapolate North Korea's nuclear ambitions and prospects for denuclearization utilizing the framework of strategic culture as an analytic method. This body of work analyzes North Korea's behaviors surrounding its nuclear weapons development in order to understand the regime's motivations behind its continued nuclear weapons development despite clear and extraordinary disincentives associated with such pursuits. This body of work also seeks to determine the prospects for North Korean denuclearization and avenues available for the United States and the international community to encourage or induce North Korea to walk the path of verifiable denuclearization.[17]

While not a large part of the overall literature on the North Korean strategic culture, some of the other bodies of work on the subject include examination of the relationship between the North Korean leadership and its strategic culture,[18] as well as works that examine the relationship between the prospects of reunification and North Korea's strategic culture.[19] Relevant to the strategic culture literature but not always explicitly included in this grouping are works on the North Korean leadership myths and North Korean identity. These works provide rich historical backgrounds and the processes of identity and myth creation, evolution, and entrenchment[20] that are essential to understanding how a regime can exploit these processes to lay the foundations for the development and sustainment of a strategic culture that is instrumental to a regime's continued preservation of power and control.

There are four primary trends that can be noticed as one reviews this large and diverse body of work on North Korea's strategic culture. First, a key strength of the current literature is that the amount of research and knowledge accumulated on the various components and features that make up North Korea's strategic culture is vast. Countless scholars and students of North Korea have poured over primary and secondary sources for decades to fit together the puzzle pieces that, in aggregate, form the complex mosaic of the North Korean strategic culture. This literature has an appreciation of the Korean historical experiences that influence the worldviews of the Korean people, and one is also able to glean an understanding of the uniquely North Korean narratives related to Korean historical experiences through the current literature. Furthermore, the literature effectively captures the Kim family's mechanisms of control – in the forms of political ideology; political, military, and educational institutions; as well as arts and literature – over the North Korean society. Overall, one is able to gain a sense of how the regime is able to employ each lever of national control to achieve the ultimate goal of regime sustainment and survival for the Kim family.

The second noticeable trend in the literature is the treatment of North Korea's strategic culture as a constant, meaning that the literature tends to employ North Korea's strategic culture as a constant frame and applies that frame to the situation that is being analyzed. This is a shortcoming in the literature. To be sure, the methodology of applying strategic culture as the framework for analyzing a particular situation is sound, and this is not the issue here. Rather, the problem is that the literature tends to treat North Korea's strategic culture itself as a constant, providing little room for the strategic culture to evolve as the Kim regime itself, the environment in which the regime finds itself both domestically and internationally, and the North Korean society all evolve. More specifically, given this tendency, the literature does not adequately capture the impact that changes in leadership characteristics may have on the features that undergird the regime's strategic culture, thus having an impact on the overall North Korean strategic culture. For example, a number of works have examined how North Korea's strategic culture influenced the state's behaviors during and after the Cold War; however, these works generally do not address how much of the differences in behavior were due to the fact that Kim Il-sung was in charge during the Cold War and operated within his worldview as informed by his internalization of the strategic culture, while Kim Jong-Il assumed power starting in 1994 and operated within his worldview as informed by his own internalization of the strategic culture.

Third, there is a need for an additional exploration of the relationships among various components of the North Korean strategic culture. While further investigation in this area may not yield a new formulation for understanding the overarching North Korean strategic culture, it may yield a deeper understanding of the mechanisms that are informing the strategic culture; thus providing more nuanced insight into how external actors should approach and/or respond to North Korea. For example, it is widely accepted in the literature that much of the North Korean strategic culture is predicated upon the common Korean historical experiences of the late nineteenth and the first half of the twentieth centuries. It is also widely accepted that the *Juche* ideology was heavily informed by Kim Il-sung's experience during this historical period. But what if this common historical experience was not the driver for these political ideology and strategic culture formulations but rather a mechanism that Kim Il-sung selected to justify the political ideology and entrench these elements of strategic culture into the North Korean society? One might argue that Kim Il-sung recognized that appealing to the North Koreans' shared historical experiences, in particular their hatred of Japan and Western powers, represented the lowest common denominator narrative around which he might rally North Koreans in order to justify his own flavor of Marxist-Leninist political ideology and enable his social controls over North Korean society.[21] Interestingly, even if it is true that the anti-imperialist nationalism inherent in the North Korean strategic culture was a vessel created by Kim Il-sung (and reinforced by Kim Jong-Il and Kim Jong Un) to justify his ideologies, policies, and control over North Korea, one's broad understanding of North Korea's strategic culture may not be any different than the literature's current understanding. However, an appreciation of these differences would provide a more nuanced set of insights into the motivations behind and the potential spectrum of North Korea's behaviors, which could lead to the development of new policy avenues for approaching contemporary situations with North Korea and attempts to influence the regime's behavior.

Finally, and somewhat related to the point mentioned above, is the common assertion in the current literature that Confucianism had a foundational role in the development of the *Juche* ideology and North Korea's strategic culture in general. Scholars who advance this assertion point to the ideas of loyalty, filial piety, and the hierarchal composition of society that exist in Confucianism, and can be seen embedded in traditional Korean culture and identity, as the basis for their argument. In essence, the argument is that Kim Il-sung had a certain understanding of Korean culture and Korean identity which was greatly influenced by Confucianism; thus the ideology he developed incorporates significant elements of Confucian thought and informs what we understand as North Korea's strategic culture in general. Joseph Bermudez's 2009 work on North Korea's strategic culture exemplifies how scholars who advance this assertion interpret the role of Confucianism in the formulation of the *Juche* ideology and the application of North Korea's strategic culture:

> Korean society within both the ROK [South Korea] and DPRK [North Korea] has a strong underpinning of Confucian philosophy. One notable aspect of this is the stringent authoritarian hierarchal rules. Within government organizations this is expressed by the fact that subordinates will rarely, if ever, disagree with their superiors. In fact, they are encouraged not to. Therefore, if a superior is known to possess a particular view on a subject, their subordinates – whether they believe the view to be correct or not – will tend to work new information into that view. These authoritarian rules are also manifested in deep institutional loyalty that results in a frequent refusal to share information and detrimental inter-agency competition. While such submission to superiors and institutional loyalty are witnessed in some form throughout the business, military, and intelligence communities in the world, it is quite evident within the DPRK [North Korea] they are taken to extremes under the umbrella of Chuche [Juche].[22]

Similar to the shortcoming noted above in examining the relationships among various components of the North Korean strategic culture, the literature only points to the influence that Confucianism could have had on Kim Il-sung as he developed the *Juche* ideology. As such, the literature fails to consider the potential that Confucianism was in fact a convenient device Kim Il-sung employed to justify his ideology's logic and formulation in order to increase his own legitimacy. Before even considering the nuances and implications behind whether the culture influenced Kim Il-sung or Kim Il-sung appropriated the culture to buttress his legitimacy, however, the existent literature is incomplete, to say the least, in its understanding of the three primary Confucian thoughts – loyalty, filial piety, and societal relationship – that it claims have influenced the formulation of North Korea's strategic culture. The literature has the tendency to interpret these characteristics of the Confucian thoughts as being restrictive and/or repressive by mirror imaging Western values (as seen in the passage above) and ignoring the fact that Confucianism explicitly states that all relationships are reciprocal. It also fails to account for the fact that the Chosun dynasty – the Korean dynasty that adopted Confucianism as its core value and governance system – institutionalized a number of government and civil society institutions specifically designed to check and devolve the king's authority. For example, the literature focuses on the aspect of loyalty whereby a subordinate must be loyal to his or her superior. While this is very true, the Confucian teachings on loyalty also clearly state that the superior must reciprocate his or her loyalty to the subordinate through the virtue of propriety. In fact, Confucius makes clear that loyalty is not about one being subservient to authority, and that loyalty begins from the top, not vice versa:

> If the lord directs his minister with li [rules of propriety], the minister will serve his lord with loyalty.[23]

Likewise, filial piety is also a bidirectional concept in Confucianism, much like loyalty (and unlike the assertion often made in the existent literature). While it is true that filial piety demands that one must respect parents, ancestors, and the hierarchies within society, it also demands that one should not blindly obey. In fact, the Confucian concept of filial piety clearly defines that blind obedience is contrary to the value of filial piety and one must provide advice and argue respectfully if those in authority are wrong. It further states that one must still respect those in authority even if they do not listen to one's advice or argument.[24] Furthermore, the requirement for reciprocity extends beyond loyalty and filial piety. It is the basis for all social networks and relations.[25] The idea of reciprocity being the foundation for proper administration of a nation is seen in the following conversation between Confucius and Duke Jing of Qi:

> Duke Jing of Qi asked Confucius about governance. Confucius replied, "Let the ruler be ruler, ministers ministers, fathers fathers, sons sons." The Duke said, "Excellent! Truly, if the ruler is not ruler, ministers not ministers, fathers not fathers, sons not sons, though I possess grain, will I be able to eat it?"[26]

Given that Confucianism emphasizes good governance and social harmony resulting from all members of society fulfilling the duties and responsibilities given to each individual, the fact that Kim Il-sung adopted only those parts of Confucian teachings that provided him legitimacy to control the North Korean society lends credibility to the argument that (1) Kim Il-sung had a very incomplete understanding of Confucianism and the Confucian concepts himself; or (2) Kim Il-sung hijacked Confucianism as a device for societal control. Unfortunately, the current literature cannot provide a clear assessment on this front. Moreover, given that the current literature only has a limited and imprecise understanding of the relationship between Confucianism and North Korea's strategic culture, it is difficult to assess how and if traditions of Confucianism in the North Korean society will influence North Korea's decisionmaking in the future. Finally, the Kim family has ruled North Korea with an iron fist for close to 75 years; thus, it is not clear if the North Korean society even adheres to any portion of the Confucian traditions it had at the founding of North Korea, and if so, how the societal understanding and interpretations of those traditions have evolved over time. For these reasons, this would be a fruitful area of future exploration that can lead to further development of the strategic culture framework for analyzing North Korean decisionmaking.

Conclusion

Applying the lens of strategic culture to analyze a state's potential behavior and decisionmaking creates room for one to identify specific national tendencies that have roots in the state's perceptual lenses, through which various collectives within the state receive and process information. Notwithstanding the expansive degree of knowledge accumulated on the various components and features of North Korea's strategic culture in the current body of the literature, there are several important areas of future research on this topic that should be endeavored.

First is the expansion of the North Korean strategic culture research into a wider range of security-related topic areas. As mentioned previously, most of the literature on North Korea's strategic culture deals with the implications of the state's behaviors and activities on regional and global security, and this discussion is dominated by North Korea's continued development of nuclear weapons. Although this is understandable and is an indisputably important subject, more research is warranted in other security-related areas such as emerging technology (including North Korea's use of cyber warfare), climate change, and new/renewable energy. Given that sanctions have curtailed most emerging and new/renewable energy technologies from entering North Korea, it would be a worthwhile research agenda to examine the propensity of the Kim regime to develop and/or adopt these technologies in the future given its strategic culture norms, values, and perceptual lens. It would also be interesting to examine how North Korea might choose to address the consequences of climate change in the future, which will have security implications reaching beyond its borders. It would be especially interesting to determine the conditions under which North Korea would consider climate change as a topic for its domestic and/or foreign policies – not only because of the intrinsic importance of the topic of climate change but also because one might be able to gain insights into the conditions needed to change North Korea's strategic culture, what that potential change might look like, and what level of flexibility exists within the current strategic culture to accommodate shifts to newly declared security "priorities."

Related to this, another area of future research would be to examine changes in North Korea's strategic culture over time, and how each generation of the Kim dynasty has impacted North Korea's strategic culture (or, perhaps, how North Korea's strategic culture has affected the three Kim rulers to date differently). This research agenda would allow one to trace past changes in perceptual lenses and potentially anticipate future ones, which could contribute to more enhanced and nuanced analyses of North Korean behavior.

Finally, more research is needed on how the North Korean leadership, the elites, and the society interpret shared immutable features of the strategic culture. Whether these immutable features are foundations of the strategic culture or simply vessels appropriated to justify it (and thus the legitimacy of the regime), the interpretations of what they mean and represent change with the passage of time, changes in leadership, makeup of the elites, shifts in society, and so forth. It is, therefore, extremely important for one to understand exactly how shared immutable features, such as national myths, symbols, and history, are being interpreted or reinterpreted. A clear understanding of this, including which narratives are being elevated or devalued and which narratives are being created or eliminated, will better enable analysts to detect potential changes that could inform North Korea's outward behavior.

Applying the lens of strategic culture to analyze a state's decisionmaking and potential behaviors provides tremendous value by allowing the analyst to better understand that state's motivations. An analytic method, however, is only as good as the assumptions and the data used to build the model utilized to conduct the analysis. In the area of North Korea's strategic culture, previous scholarship has focused on building a solid model that can be used for analysis. It is now time to update and sustain the model by reevaluating and expanding the datasets and testing whether long-standing assumptions still hold.

Notes

1 Central Intelligence Agency, "North Korea," World Factbook (Central Intelligence Agency, 15 December 2021), https://www.cia.gov/the-world-factbook/countries/korea-north/.
2 Chang-il Ohn, *Hamminjok Jeonjaengsa* [War History of Korea] (Jipmundang, Seoul, Korea, 2001).
3 For a detailed explanation of the China-centric world order and China's relations with its neighbors during this period, see Key-Hiuk Kim, *The Last Phase of the East Asian World Order: Korea, Japan, and the Chinese Empire, 1860–1882* (University of California Press, Berkley, CA, 1980), 1–38.
4 Ki-Jung Kim, "Theodore Roosevelt's Image of the World and United States Foreign Policy toward Korea, 1901–1905," *Korea Journal* 35, no. 4 (1995): 39–53.
5 Hilary Conroy, *The Japanese Seizure of Korea, 1868–1910* (University of Pennsylvania Press, Philadelphia, PA, 2016), 494–496; Chong Ik Eugene Kim and Han-Kyo Kim, *Korea and the Politics of Imperialism, 1876–1910* (University of California Press, Berkley, CA, 1967), 53.
6 Kwang Seo Kee, "North Korean Policy of the Soviet Union during 1945–1947 Represented in the Documents of Russian Archives," *Journal of History and Culture* 29 (2005): 6.
7 It is believed, according to some sources, that Kim Il-sung held the rank of a Major in the Soviet military at the time of his selection for the mission to play the part of the revolutionary leader to lead the Korean people [see: Dae-Sook Suh, *Kim Il-sung: The North Korean Leader* (Columbia University Press, New York, 1988), 55–73].
8 Hokkanen, Jouni. *Pohjois-Korea – Siperiasta itään* [*North Korea – East of Siberia*] (Johnny Kniga, Helsinki, Finland, 2016), 125.
9 Bong Baik, *Kim Il Sung: Biography I* (Mishihara, Tokyo, Japan, 1969), 363–382.
10 Suh, *Kim Il-sung*, 3–11; Baik, *Kim Il Sung*, 11–78.
11 Kim Jung Un aligned himself closely with Kim Il-sung's cult of personality by mimicking Kim Il-sung's hair and fashion styles, and attitudes presented during speeches, venues visited, and events participated.
12 Kim Il-sung wanted a uniquely North Korean political ideology to reverse some of the strong Soviet influence inherent in the North Korean system as it was the instrumental force that created North Korea. He also wanted to increase his legitimacy as his "own person" in the eyes of the North Korean elites and the populace rather than being seen as a proxy of the Soviet Union.
13 Kim, Il-sung. *Kim Il Sung Jeojakjip* [Collected Works of Kim Il-sung], Volume 19 (Chosunrodongdang Chulpansa, Pyongyang, North Korea, 1981), 306.
14 Kim, Jong-Il. *Kim Jong-Il Seonjip* [Kim Jong Il Selected Works], Volume 9 (Chosunrodongdang Chulpansa, Pyongyang, North Korea, 1997), 459.
15 Myoung Seob Kim, "The North Korean Nuclear Problem and the Geopolitics of Northeast Asian Six Party Talks: A Historical Reflection and Forecast," in *Nuclear Internal Policy* (Institute for Far Eastern Studies, Kyungnam University, Seoul, Korea, 2012), 237.
16 See, for example, Joseph S. Bermudez, "North Korea and the Political Uses of Strategic Culture," in *Strategic Culture and Weapons of Mass Destruction,* Jeannie L, Johnson, Kerry M. Kartchner, and Jeffrey A. Larsen, eds. (Palgrave Macmillan, New York, 2009), 189–200; Dae-Ho Byun, *North Korea's Foreign Policy: The Juche Ideology and the Challenges of Gorbachev's New Thinking* (Research Center for Peace and Unification of Korea, Seoul, Korea, 1991); Kang Choi, "Korea: A Tradition of Peace – The Danger of War," in *Strategic Cultures in the Asia-Pacific Region* (Palgrave Macmillan, London, UK, 1999); Yong-Pyo Hong, "North Korea's Strategic

Culture and Threat Perception: Implications for Regional Security Cooperation," *Korea Observer* 42, no. 1 (2011): 95–115; Leif-Eric Easley, "North Korean Identity as a Challenge to East Asia's Regional Order," in *Japan and Asia's Contested Order* (Palgrave Macmillan, Singapore, 2019), 109–144.

17 See, for example, Hyo Jong Son, "Nuclear Dilemma of North Korea: Coexistence of Fear and Ambition – North Korea's Strategic Culture and Its Development of Nuclear Capability," *The Korean Journal of Defense Analysis* 29, no. 2 (2017): 195–211; Shane Smith, "North Korea's Strategic Culture and Its Evolving Nuclear Strategy," in *Crossing Nuclear Thresholds: Leveraging Sociocultural Insights into Nuclear Decisionmaking* (Palgrave Macmillan, New York, 2018), 227–250; Cameron Chansong Lee, "Understanding the Strategic Culture of Nuclear Decisions," *SAIS Review of International Affairs* 39, no. 2 (2019): 197–200; Manseok Lee and Lee Sangmin. "North Korea's Choice of a Nuclear Sstrategy: A Dynamic Approach," *Defense & Security Analysis* 36, no. 4 (2020): 377–397.

18 See, for example, Merrily Baird, "Kim Chong-il's Erratic Decisionmaking and North Korea's Strategic Culture," in *Know Thy Enemy: Profiles of Adversary Leaders and Their Strategic Cultures*, Barry R. Schneider and Jerrold M. Post, ed. (US Air Force Counterproliferation Center, Maxwell Air Force Base, AL, 2003), 109–140.

19 See, for example, Sachio Nakato, "North Korean Unification Strategy: Strategic Culture and Future Prospects," in *One Korea: Visions of Korean Unification,* Tae-Hwan Kwak and Seung-Ho Joo, ed. (Routledge, New York, 2016), 43–63.

20 See, for example, In Ae Hyun, "Analyzing the Structure of the North Korean Leader Myth and Creating the Kim Jong Un Myth," *Journal of Peace and Unification* 5, no. 1 (2015): 69–107; Hye-sin Yun, "The Kim Il-sung Discourse as a Modern Myth," *North Korean Review* 17, no. 2 (2021): 76–92.

21 Specifically, Kim Il-sung invoked hatred against Japan for its occupation of the Korean Peninsula and hatred against the United States for betraying Korea in favor of allowing Japan to counter Russia and receiving a security guarantee for the Philippines, an American colonial dependency until 1946; and the United Kingdom for unilaterally recognizing Japan's suzerainty over Korea (which was never within its rights to do so) in return for the United Kingdom's exclusive rights over Singapore.

22 Bermudez, "North Korea and the Political Uses of Strategic Culture," 195.

23 Ba Yi, Book III, 3.19, In *The Analects of Confucius* (An Online Teaching Translation, 2015: https://chinatxt.sitehost.iu.edu/Analects_of_Confucius_(Eno-2015).pdf), 12. Chinese text: 定公問:「君使臣，臣事君，如之何?」孔子對曰:「君使臣以禮，臣事君以忠。」

24 Wei-Ming Tu, *Centrality and Commonality: An Essay on Confucian Religiousness* (State University of New York, Albany, NY, 1989), 794.

25 Stephen Feuchtwang, "Chinese religions," in *Religions in the Modern World: Traditions and Transformations* (Routledge, London, UK, 2016), 143–172.

26 Yan Yuan, Book XII, 2.11, In *The Analects of Confucius* (An Online Teaching Translation, 2015: https://chinatxt.sitehost.iu.edu/Analects_of_Confucius_(Eno-2015).pdf), 62. Chinese text: 齊景公問政於孔子。孔子對曰：「君君，臣臣，父父，子子。」公曰：「善哉！信如君不君，臣不臣，父不父，子不子，雖有粟，吾得而食諸?」

Selected Bibliography

Baik, Bong. *Kim Il Sung: Biography I*. Mishihara, Tokyo, Japan, 1969.

Baird, Merrily. "Kim Chong-il's Erratic Decisionmaking and North Korea's Strategic Culture." *Know Thy Enemy: Profiles of Adversary Leaders and Their Strategic Cultures*, edited by Barry R. Schneider and Jerrold M. Post, 109–140. US Air Force Counterproliferation Center, Maxwell Air Force Base, AL, 2003.

Bermudez, Joseph S. "North Korea and the Political Uses of Strategic Culture." In *Strategic Culture and Weapons of Mass Destruction*, edited by Jeannie L. Johnson, Kerry M. Kartchner, and Jeffrey A. Larsen, 189–200. Palgrave Macmillan, New York, 2009.

Booth, Ken. "The Concept of Strategic Culture Affirmed." In *Strategic Power: USA/USSR*. Edited by Carl G. Jacobsen, 121–128. Palgrave Macmillan, London, UK, 1990.

Byun, Dae-Ho. *North Korea's Foreign Policy: The Juche Ideology and the Challenges of Gorbachev's New Thinking*. Research Center for Peace and Unification of Korea, Seoul, Korea, 1991.

Choi, Kang. "Korea: A Tradition of Peace – The Danger of War." In *Strategic Cultures in the Asia-Pacific Region*. Edited by Ken Booth and Russell Trood. Palgrave Macmillan, London, UK, 1999.

Easley, Leif-Eric. "North Korean Identity as a Challenge to East Asia's Regional Order." In *Japan and Asia's Contested Order*, edited by Yul Sohn and T. J. Pempel, 109–144. Palgrave Macmillan, Singapore, 2019.

Hong, Yong-Pyo. "North Korea's Strategic Culture and Threat Perception: Implications for Regional Security Cooperation." *Korea Observer* 42, no. 1 (2011): 95–115.

Hyun, In Ae. "Analyzing the Structure of the North Korean Leader Myth and Creating the Kim Jong Un Myth." *Journal of Peace and Unification* 5, no. 1 (2015): 69–107.

Kim, Il-sung. *Kim Il Sung Jeojakjip* (Collected Works of Kim Il-sung), Volume 19. Chosunrodongdang Chulpansa, Pyongyang, North Korea, 1981.

Kim, Jong-Il. *Kim Jong-Il Seonjip* (Kim Jong Il Selected Works), Volume 9. Chosunrodongdang Chulpansa, Pyongyang, North Korea, 1997.

Lee, Manseok, and Sangmin Lee. "North Korea's Choice of a Nuclear Strategy: a Dynamic Approach." *Defense & Security Analysis* 36, no. 4 (2020): 377–397.

Nakato, Sachio. "North Korean Unification Strategy: Strategic Culture and Future Prospects." In *One Korea: Visions of Korean Unification*, edited By Tae-Hwan Kwak, and Seung-Ho Joo, 43–63. Routledge, New York, 2016.

Smith, Shane. "North Korea's Strategic Culture and Its Evolving Nuclear Strategy." In *Crossing Nuclear Thresholds: Leveraging Sociocultural Insights into Nuclear Decisionmaking*, edited by Jeannie L. Johnson, Kerry M. Kartchner, and Marilyn J. Maines, 227–250. Palgrave Macmillan, New York, 2018.

Son, Hyo Jong. "Nuclear Dilemma of North Korea: Coexistence of Fear and Ambition – North Korea's Strategic Culture and Its Development of Nuclear Capability." *The Korean Journal of Defense Analysis* 29, no. 2 (2017): 195–211.

Yun, Hye-sin. "The Kim Il-sung Discourse as a Modern Myth: The Classical Mythic Format and Ego-Oriented Modernity." *North Korean Review* 17, no. 2 (2021): 76–92.

19
THE STRATEGIC CULTURE OF SOUTH KOREA

Briana Marie Stephan[1]

South Korea: Introduction and Background

South Korea, also known as the Republic of Korea (ROK), is a collectivist nation that values hierarchy and structure, upholds social values and religious ideals, and prioritizes a sense of duty and commitment to family. Additionally, South Korea maintains a long-term orientation related to societal goals, economic advancement, and national security.[2] The nation prides itself on strategic thought (forward thinking) and Confucian ideology (respect for self, upholding justice, and acting in a beneficent manner), which filters through every aspect of society. This chapter will walk through the more salient aspects of South Korea's identity, values, norms, and perceptual lens that define its unique strategic culture, which comprises a number of complex and interdependent variables that have been shaped by the nation's long history, religious and moral ideologies, and aspirations for economic success.

The aim of this chapter is to identify themes that are widely understood about South Korean strategic culture, which is defined here as "shared beliefs, assumptions, and modes of behavior, derived from common experiences and accepted narratives (both oral and written), that shape a collective identity and relationships to other groups, and which determine appropriate ends and means for achieving security objectives."[3] A country's strategic culture impacts and influences how it will conduct business with foreign nations. Lessons learned from war, political engagements, military strategy and execution of those strategies, economic affairs, and treatment of its citizens shape the way in which a nation will cooperate broadly on matters pertaining to national security, economic policy, and other endeavors. Misperception and lack of understanding pertaining to a country's strategic culture has been and will continue to be a contributing factor to failed attempts at developing shared international policies and alliances.

Lastly, this chapter will provide a brief capture of the current strategic culture literature and provide key points for future analysis and study. It should be noted that the bulk of the scholarly literature examined in this chapter reflects the state of play in 2019, and as noted in the concluding section, the geopolitical events and shifts that have transpired since then have continued to shape South Korea's collective culture and perception of national security issues.[4]

 DOI: 10.4324/9781003010302-22

South Korea's Cultural Identity

Questions of cultural identity generally speak to questions relating to conception and sense of self, often as distinct from others. South Korea's cultural identity revolves around a number of key elements. For example, South Korea's past experience with war and colonization as well as the outcomes of those conflicts has contributed to and further solidified South Korea's sense of nationalism, which is deeply rooted in its ancient *minjok* heritage,[5] meaning the belief that Koreans were born from a "pure bloodline" and have a distinct culture that is separate from other Asian nations. This nationalism is a foundation that encourages adherence to Korean cultural norms and obedience to roles, responsibilities, and duties to family, community, and country. Next are influences of religion, desires for economic prosperity, and creating national stability. Among other variables, South Korean leadership firmly believes that elevating the country's economic status is inextricably linked to an increase in national pride and a sense of belonging among its people.

Identity Through History and Religion

A driving force behind South Korea's identity is its military and political history, which was also greatly impacted by religion. Prior to the sixteenth century, three ruling Korean kingdoms (Goguryeo/Goryeo, Baekje, and Silla) were expansionistic and in constant conflict with one another. These kingdoms were eventually overthrown by the Chosŏn dynasty, which introduced a neo-Confucian religion, a need for civil moral order, and "a minimal standing army and dependence on militias for defense."[6] Invasions from external groups seeking territorial expansion occurred in the sixteenth century followed by Japanese and Manchu Qing invasions in the seventeenth century. Power struggles over the peninsula continued into the nineteenth and twentieth centuries. Two wars fundamentally changed the geopolitical atmosphere in Asia in this period: the Sino-Japanese War (1894–1895), which dislocated China from its long-standing role as the center of gravity in East Asia; and the Russo-Japanese War (1904–1905), which crushed Russia's influence in the Far East and cemented Japan's rise as the great power in East Asia during the first half of the twentieth century.[7] Japan annexed South Korea in 1910 and South Korea was then subject to Japanese rule until the end of World War II in 1945. The mistreatment of Koreans under Japanese rule was particularly impactful. In the words of Kim:

> The memory of how Korea [in general] became an unwilling and ineffectual object of great power struggle and the humiliation of losing its sovereignty to Japan, whom it has for centuries regarded as inferior, remains seared in the minds of Koreans, north and south alike.[8]

South Korea's elders, the country's leadership, and members of the National Assembly – all of whom function as "keepers of strategic culture" – have repeated and reinforced the narratives of South Korea's dark colonial chapter for decades, passing them down to each succeeding generation. These narratives emphasize the importance of Korean independence and the mass suffering that would ensue if that independence were ever taken. Also cultivated is a national narrative of heroic martial heritage, which is meant to tie the nation together and instill pride in maintaining the Korean legacy.

Finally, the state's most significant conflict of the twentieth century, the Korean War, had a tremendous impact on the strategic culture of modern South Korea. The Korean War – also known as the "Forgotten War" in the West, the "Fatherland Liberation War" by North Koreans, or the "Six-Two-Five" in South Korea (coined for the day it started)[9] – made clear the ongoing need for US allied support and for forging relations with the United Nations (UN) and other international partners. Since the settlement of the 1953 Korean War armistice, modern South Korea has built a distinctive identity that remains influenced, in part, by ongoing North Korean threats. Multiple rounds of negotiations, diplomacy efforts, initiatives, and peace talks have made progress in some areas, but have not resulted in a formal peace treaty to end the war. Further, North Korea continues to advance and openly test its abilities as a nuclear weapons state, despite UN sanctions for nuclear and ballistic missile testing violations on numerous occasions (2006, 2009, 2013, early 2016, late 2016, and 2017).[10] Geographically, the 1953 armistice essentially isolated South Korea, making it comparable to an island cut off from the mainland by its land border with North Korea and forced into an uncomfortable arrangement in which it shares critical trade and industry waterways with China and Japan.

Identity Through Economy

In the 1960s, President Park Chung Hee set forth to "transform the South Korean mindset [after the Korean War] from one of dejection and victimization to one of hope and glory imbued with a sense of purpose and pride in their role in building a new nation of wealth and power."[11] Each leader since that time has played a part in helping shape the culture and rebuilding the South Korean people's collective identity from the atrocities of Japanese colonial rule – though some were more successful than others. South Korean leaders have encouraged citizens to be "the masters of their own destiny" and in so doing make the nation a nexus of economic accomplishment.[12] South Korea aims to be the strategic linchpin[13] of the region with a more prominent position in regional, international, and humanitarian missions (e.g., UN missions). Over the course of several decades, South Korea has become much more than a country of purely agricultural value (debunking the initial shortsightedness of some Western nations) and has since paved a path to being an industry leader in exports of textiles, machinery, steel, chemicals, and modern technologies, including electronics, communication systems, and computer systems.[14]

South Korea's Values

Values can be physical or ideological constructs. For example, nations may place significant value on natural resources, sovereignty, and freedom of religion, from among other priorities. For South Korea, much of what is valued is derived from its cultural history as a collectivist nation that values the greater good of the people as a whole over the individual.

Religious Values

South Korea greatly values and appreciates religious freedom. In South Korea, Confucianism, Buddhism, and Christianity are the most common belief systems and there is a strong tie to spirituality among the South Korean people. This can be seen through many of the ceremonial and religious offerings provided to deities or spirits prior to holidays, special events, or celebrations. South Koreans tend to be firm believers in the cultural notion that leaders are obligated according to the "Mandate of Heaven." Originally this inferred that that political leadership was an extension of God and a conduit for God's will.[15] Modern interpretations focus on the second aspect that leaders must serve the people well in order to prove themselves and maintain a legitimate claim to power. This was evidenced by President Park Geun-hye's impeachment in 2017 after South Korean citizens felt she had shamed the title, her father's name, and the nation as a whole.[16] President Park had lost the faith of the public, and therefore the right to continue in power. Thus, there is a cultural system of checks and balances held even at the highest levels of power as individuals are held accountable for their actions and must be deemed worthy of maintaining their roles within the community – a feature that speaks directly to Confucian values.

Political and Economic Values

South Korea's consolidation of a democratic political system in the final decades of the twentieth century is highly valued by its public as a monumental step toward modernization, enhancing trade, building strategic international partnerships, and improving how the state is viewed among the international community.[17] For South Korea, democracy equates to relevancy. South Korea values its strategic military and economic alliance with the United States but has also placed significant effort since the 1980s on enhancing its own industries. The state has focused on building connections to trade markets, expanding trade agreements with China, India, and other countries, and considers economic success, alongside democracy, to be key to staying relevant in the global circuit.[18] South Korea further recognizes the importance of forging new relationships that could bring funding and resources into the country. In keeping with the state's focus on securing its place in the global economy, South Korea greatly values its nuclear energy program. In 2019, South Korea was ranked the fifth-largest nuclear energy producer in the world, with 24 operational nuclear reactor units,[19] and ranks second in the Asia region (behind Japan) for nuclear energy consumption.

National Security Values

As a nation surrounded by adversaries and competitors, missile threats, and strategic uncertainty, the South Korean state's focus on security must balance a complex range of variables. Notwithstanding its heavy focus on civil nuclear energy, South Korea remains a Nuclear Nonproliferation Treaty (NPT) signatory (signed in 1968 and ratified in 1975) and is also a Comprehensive Test Ban Treaty (CTBT) signatory and Nuclear Suppliers Group (NSG) member. It is a strong supporter and promoter of peaceful nuclear energy use and, despite a belligerent neighbor to the north, has not publicly signaled any intention to revisit the development of nuclear weapons. South Korea pursued them at one point during the Cold War and was the only country to do so while under the protections of the US nuclear umbrella's security guarantees.[20]

South Koreans find value and nostalgia in the idea of reunification with the North, but most realize that reunification entails more than just the physical aspect of terminating the demilitarized zone and combining the two countries; it would also have tremendous social, economic, political, cultural, and national security implications.[21] There are a variety of concerns with the reunification of the North and the South, such as the monetary costs associated with melding two vastly economically different countries, increases in national debt, the potential loss of democratic standing or challenges in enforcing democracy, and the severity of economic impacts and redefinition of economic goals.

South Korea's Norms

Cultural norms are often understood to be the standards that individuals, groups, and communities live by. Norms are sets of shared expectations and rules that ultimately guide individual and group behavior in different social settings. Norms naturally vary across cultures, which plays a role in many cultural misunderstandings and prejudices. This is especially critical to understanding why international policies and frameworks for security are not always successful when attempting to align the behavioral expectations of different countries. For South Korea, accepted norms across facets of religion, education, law and order, politics, and strategy deeply impact how the state engages with the international community.

Religious and Social Norms

Confucianism plays a governing role in South Korean society. Religious norms shape morals, set behavioral expectations, and play a role in establishing boundaries in relationships between individuals and groups. Some of the strongest Confucian ideals can be seen in the daily lives of South Koreans: the prioritization of the family or collective society over that of the individual guides the understanding and sense of responsibility that each individual should contribute to the common good.[22] This expectation has carried over to business settings as well, with individuals in the workplace expected to treat coworkers with respect, follow a hierarchical chain of command, and maintain harmonious work relationships.[23] As noted previously, Confucianism also plays a role in South Korean politics, determining the behaviors of leaders that will and will not be tolerated by the population and the red lines across which a leader may be believed to have lost the "Mandate of Heaven."

Norms of Law and Order

South Korea is a law-abiding nation that prefers to have structure and order in society and a chain of command that is understood and respected. Confucianism is also the basis of the legal system for the country, bounded by the principle of "do unto others as you would have others do unto you" nested in the Constitution of the Republic of Korea.[24] South Korea expects agreements to be upheld and promises kept by those they call allies or with whom they conduct business. This is partly evidenced by the numerous treaties and international forums the country participates in, such as the UN, NPT, CTBT, NSG, Proliferation Security Initiative (PSI), International Atomic Energy Agency (IAEA) protocols and safeguards, and so forth. Respecting law and order is a norm prioritized in the name of safeguarding the country and enhancing its domestic and international security posture.

National Security Norms

It is important to recognize that North Korean missile tests and threats have become common to some degree for the peninsula, but are still far from acceptable by South Korea, the United States, and allied nations' leadership. North Korea's weapons development and saber-rattling across the past decade have shown important advances in the state's capabilities and have effectively bolstered its credible nuclear deterrence strategy in the region.[25] In terms of its own defense and deterrence, South Korea relies on the United States for military assistance and has been fairly hesitant to take any military or political actions that the North may deem threatening or retaliatory in nature. President Moon begrudgingly agreed to the deployment of US-supplied Terminal High Altitude Area Defense (THAAD) systems to the peninsula after the 2017 missile tests by North Korea[26] despite the fact that many South Koreans protested this action and felt that it could provoke war with the North. Notwithstanding best efforts, sanctions, and governmental discussions, tensions remain constant between the North and the South. South Korea maintains its stance against the development of its own nuclear weapons, but its hedging posture suggests mounting insecurities related to future North Korea relations.

South Korea's Perceptual Lens

Every nation has a point of view related to its current situation and standing in the world. This perceptual lens, or the culturally informed filter through which reality is viewed, may or may not align with other states' perceived realities. Perception is influenced by historical and present-day experiences (or the lack thereof) and outcomes of situations impacting like-minded nations. What South Korea has perceived to be true related to its allied relations, place in the global economy, and the current state of its national security directly impacts how the state postures for present and future security challenges.

Perception of US Relations

South Korea understands the weight of its dependence on the United States for security and economic prosperity. The US-South Korea alliance is built on the legacy and enduring presence of complex regional security challenges and has not been spared in recent years from pressures in the Asia-Pacific amid the resurgence of great power competition.[27] While the country deeply values the alliance, South Korea is also keen on building more self-reliance on its security and economic framework and avoiding sole dependence upon coverage by the US military or its nuclear umbrella. As the United States and North Korea have traded dialogue and arranged summits at various intervals, the South has remained vigilant of the persistent threats related to North Korean nuclear weapons advancements and particularly the arsenal's growing "mobility, reliability, potency, precision, and survivability,"[28] all of which pose significant threats. South Korea sees the continuous tension with North Korea as a primary existential security threat, and despite the endurance of the US-South Korea alliance, it has questioned whether or not the United States would honor its pledge to assist the South if conflict broke out again on the peninsula. This doubt resurfaced when North Korea announced that it possessed weapons that could reach the US mainland,[29] and South Korea has experienced a heightened sense of insecurity related to the United States handling not only nuclear diplomacy with North Korea but its broader commitment to regional security in Asia. For now, South Korea demonstrates a steady commitment to the US relationship but has sought out more trade and economic relations with other powers, including China and India to help secure economic stability.[30]

To maintain a healthy partnership, the United States and South Korea will likely need to engage in more transparent conversations and co-led diplomatic efforts, especially on issues pertaining to North Korea and strategy development related to the North Korean nuclear weapons program.[31] South Korea has been part of extensive efforts to combat the nuclear threat including being a signatory to a Joint Declaration on the Denuclearization of the Korean Peninsula in 1992; creating the Sunshine Policy in the late 1990s to provide aid to the North in exchange for better relations; engaging in the Six-Party Talks in the mid-2000s; and assisting in the creation of many other actions and policies. Despite the failures of these efforts and North Korea's violations or withdrawals, South Korea still maintains hope for future relations with its "kin to the North."

Perception of Other State Relations

South Korea is leery of China's rise[32] and of Russia's weapons systems modernization and advancements, particularly in the wake of Russia's 2022 invasion of Ukraine – and fairly so, as the Korean War images burned so deeply into South Korean minds resulted from North Korean, Chinese, and Russian war efforts (some more active than others). Despite this, South Korea has come to appreciate the economic boost it can receive from improved relations with China. In Petrovsky's words:

> Simultaneously with release of the new national security strategy, the South Korean Foreign Ministry announced a reform of its organizational structure and the planned creation of a separate department to deal exclusively with relations with China. Until now, the U.S. was the only country to enjoy the privilege of having such a "personal" department there. This is a clear sign of China's growing importance to South Korea.[33]

Going forward, Seoul may have to find a way to walk a fine line between both Washington's and Beijing's interests as it attempts to build its economy and maintain security protections in the region.[34] South Korea's economic growth is projected to remain highly dependent on close trading relationships, which includes China and Japan – who alone account for approximately one-quarter of South Korean exports.[35]

Perception of Security and Strategic Alliances

The anxious atmosphere around North Korea's continued nuclear weapons program growth and the United States' growing frustrations with the Kim regime have left South Korea feeling sandwiched between maintaining the strategic alliance and trying to find a degree of stability with the North. South Korea's national security strategy released in December 2018 pushed for peaceful resolutions to the divide between the North and South,[36] presenting no drastic change from Seoul's approach to the numerous treaties and multinational engagements that the South has participated in over the years. President Moon then emphasized "security that ensures peace" and aimed to solidify the country's role in future negotiations related to denuclearization.[37] Seoul still maintains its position that a peace treaty will only be signed once denuclearization has been completed, which raises questions over whether strategies like the 2018 document are doomed to be little more than wish lists since the North seems highly unlikely to surrender its prized nuclear arsenal. Even if the North signals willingness to begin denuclearization at some future point, ironclad IAEA verification of the North's disarmament would present another steep challenge. The distrust that exists between North Korea and the rest of the world will be difficult to reconcile.

Assessing the State of Scholarship on South Korea's Strategic Culture

There is a significant body of literature related to contributing factors influencing Korean strategic culture (both North and South). The history of the Asian continent and the Korean peninsula as it pertains to historical battles and conquests is well documented and much of it examines perspectives of North Korea, South Korea, China, and Japan. The sources drawn on for this chapter primarily focused on the strained relationship between North and South Korea with the intent of capturing salient features of South Korean strategic culture resulting from years of conflict, the outcomes of those conflicts, political and military security strategy and treaties, and South Korea's growing dependence on international partnerships as it becomes more economically connected through international trade. There are, however, some important gaps in the strategic culture literature regarding the impact of more recent events on South Korean strategic culture and how unfolding contemporary issues may impact South Korean foreign policy and alliances.

Aside from the ever-present threat presented by North Korea and its nuclear capabilities, South Korea also faces complex strategic challenges related to China's economic prowess and its strategic ties to Russia; fraught tensions between Russia and the West; dependence on foreign powers for protection; challenges in fortifying an economic relationship with Japan; long-term ripple effects from the COVID-19 pandemic; intragovernmental abuse of power; and lack of inclusion in decisionmaking over political and economic issues in East Asia. Each of these issue areas is a strong candidate for further study utilizing a strategic culture lens. Two of these – the continuing response to the COVID-19 pandemic, and the global reaction to Russia's 2022 invasion of Ukraine – are explored in greater detail below.

Case Study: COVID-19 Public Health Crisis Response

Further scholarly work could examine South Korea's values and norms as they pertain to public health as a security issue – including dimensions ranging from the country's stance on health care, to developing national strategic plans to interdict spread of pathogens, to competing in the expanding medical supply chain market. The jarring global experience of the COVID-19 pandemic and the shortage of critical medical products has sparked debates related to varying countries' dependence on foreign partners for the development, procurement, transport, and delivery of raw materials, medical products, and medical countermeasures (MCMs). Reports and studies have been published in the wake of the pandemic related to nations' responses to the viral outbreak and their ability to provide much-needed care to citizens. Government and private sector stakeholders across the international community have had to work together to combat supply shortages and growing needs for therapeutics and MCMs. However, despite global lockdowns and travel restrictions, South Korea weathered the pandemic without implementing the harshest measures seen around the globe, including total economic lockdowns or universal mandates for citizens to stay at home, which significantly aided the South Korean population's ability to successfully navigate through the worst years of the crisis.[38]

South Korea's close proximity to China was a significant concern in terms of potential outbreaks early in the pandemic. However, despite waves of outbreaks, South Korea was able to keep the numbers of infections and deaths relatively low when compared to international counts, in part due to South Korea's communication strategy, mask mandates, quick closure of various entertainment facilities, and compliance of the South Korean population to requests from the Korean government and leadership.[39] Much like Western allies, South Korea set out to develop tests to help with diagnosing individuals and contact tracing capabilities for those who were infected. One interesting vein of future research would be a comparison of South Korea's approach to combatting COVID-19 with other states' responses to the pandemic, and to South Korea's own previous handling of other outbreaks in terms of developing strategic plans, working with the international community, and maintaining an open dialogue with the Korean people. According to one source:

> South Korea's approach [to COVID-19] ... depended on public buy-in and trust, which authorities were able to achieve, for the most part, through transparency and openness. In this regard, authorities learned from their experience with Middle East Respiratory Syndrome (MERS) in 2015. With MERS, they had withheld information to avoid creating panic among the public, but the resulting information vacuum was filled by rumor and misinformation.[40]

This assessment is important because public buy-in and trust were difficult to achieve for many of South Korea's counterparts, including the United States. Was this due to the United States being a more individualistic society versus a collectivist one? Additionally, South Korea directed enormous effort into thinking strategically about how to reinvigorate the economy after the pandemic, which resulted in the "Korean New Deal"[41] that aimed to "stimulate investments in advanced technology, upskilling Korean workers, and positioning the country to emerge from the pandemic as a leading player in the data economy and the green economy, rather than using government funding strictly to rebuild the economy."[42] A comparative analysis of this plan and other nations' COVID-19 relief plans may give further insight into the manifestations and implications of those countries' values, norms, and strategic direction. It may also be worth noting which industries South Korea invests in more heavily post-pandemic, and if any of these moves signal a strategic shift in partnerships or toward more in-country production and self-reliance.

Case Study: Response to Russia's 2022 Invasion of Ukraine

Just weeks before Russia launched its full-scale invasion of Ukraine in February 2022, Russian President Vladimir Putin and Chinese President Xi Jinping declared a "new era" in the global order and endorsed their respective territorial ambitions in Ukraine and Taiwan, unveiling a long-term agreement that challenged the North Atlantic Treaty Organization and the United States.[43] That agreement between Russia and China fell short of a formal alliance, but included nebulous pledges to stand "shoulder to shoulder" against the United States and its allies both ideologically and militarily to achieve their respective territorial ambitions and to cooperate on mutual interests in space, web governance, artificial intelligence, and climate issues. Politically, the document claimed that there is "no one-size-fits-all" type of democracy and heralded both forms of authoritarian rule in Moscow and Beijing as successful democracies.[44] Although the relationship between Russia and China has been clearly strained by the scale and intensity of Russia's war in Ukraine, the two powers remain in a complicated marriage of convenience that features varying levels of tacit and explicit support.

This complex dynamic offers important ground for strategic culture scholarship that assesses the implications for the United States and its allies – including South Korea – of Russia's war in Ukraine and its evolving partnership with China. Ripple effects through the global system from this crisis are likely to be vast and enduring, and may well hold significant implications for South Korea's future. How will South Korea's perceptions of the credibility of the US alliance evolve as it observes US policy toward Ukraine? Will North Korea take advantage of an era of growing international chaos to make a move against South Korea – and if so, what military and security norms might shift in South Korea in response to a renewed existential threat? How might cascading effects from great power competition between the United States, Russia, and China impact South Korea's alliance relationship with the United States and other aligned nations in East Asia – especially if China crosses the monumental threshold of a forced reunification with Taiwan? As the global security challenges of the 2020s continue to unfold, further utilization of the strategic culture paradigm to assess the impacts of these events on South Korea's strategic perceptions and decisionmaking can offer important insights to scholars and policymakers around the world.

Conclusion

At its simplest construct, strategic culture is the examination of cultural factors that affect strategic behavior. Therefore, a country's strategic culture takes into account aspects of history, military action, religion, geography, sociological demographics, political systems, economics, and the environment. All of these factors are interrelated and influence the cultural identity, norms, values, and unique perceptual lens that have shaped South Korean behavior. South Korea's strategic culture will continue to have an impact on the country's military, political, and economic conduct and goals. Manifestations of strategic culture are evident in South Korea's negotiating style and approach to international discussions, wartime strategy development with allies, including force management and modernization, crisis management, conflict resolution, and even setting economic milestones for success. Among the many factors, events, and cultural nuances that shape South Korean strategic culture, the most prominent manifestations of the nation's strategic culture going forward may be best divided into three pillars, captured by Kim: "attaining prosperity and strength as an enduring national purpose and objective; countering the existential North Korean threat; and maintaining strong alliance with the United States."[45]

South Korea's culture is indeed unique. Blended with a globally competitive economy, democratic government, and close alliance with the United States are elements of Confucian thought and moral constructs that can be seen throughout many, if not all, aspects of South Korean life – the state's political structure, business practices, and even its strategic approach to alliances with Western partners and industries. South Korea holds tightly to ancient practices and belief systems, yet sees the need for modern advantages of Western technological and military advancement. A complex political history has had a profound impact on the development of South Korea's insecurities related to the need for protection, its desire for relations with powerful international partners, and its drive to be seen as a regional stronghold for economic growth.

What is perhaps most fascinating about modern South Korea's strategic culture is that just 70 years ago the country was on the brink of devastation and economic ruin.[46] The Korean War left the country broken and battered through years of bloodshed, yet with determined political leadership and assistance from the international community, the South was resilient. South Korean leaders used a range of tools and campaigns – including the nation's own troubled historic period of authoritarian rule – to set South Korea on the road to significant national growth. South Korea set its sights on aligning economically and in some cases politically with countries that could help support a South Korean comeback. In this fashion, South Korean strategic culture has served "as a mechanism and process to uphold, enhance, and promote South Korea's vision of the ideal and authentic Korean nation and nationalism."[47]

Notes

1 Special thanks to Dr. Kerry Kartchner for introducing the author to the concept of strategic culture and to numerous professors for their mentorship during her Countering Weapons of Mass Destruction fellowship with the National Defense University.

2 Hofstede Insights, "Country Comparison: South Korea and United States," accessed 24 June 2019, https://www.hofstede-insights.com/country-comparison/south-korea,the-usa/

3 Jeannie L. Johnson, Kerry M. Kartchner, and Jeffrey A. Larsen, eds., *Strategic Culture and Weapons of Mass Destruction: Culturally Based Insights into Comparative National Security Policymaking*, Springer, 2009, 9.

4 This statement refers in particular to presidential and political elections, missile tests conducted by North Korea, economic crisis, the outbreak of COVID-19 that has left lasting impacts on partnerships and national and economic security around the globe, and Russia's 2022 invasion of Ukraine.

5 Gi-Wook Shin, "Korea's Ethnic Nationalism Is a Source of Both Pride and Prejudice, According to Gi-Wook Shin," *The Korea Herald*, 2 August 2006, https://aparc.fsi.stanford.edu/news/koreas_ethnic_nationalism_is_a_source_of_both_pride_and_prejudice_according_to_giwook_shin_20060802.

6 Jiyul Kim, "Strategic Culture of the Republic of Korea," *Contemporary Security Policy*, 35, no. 2 (2014): 5.

7 Kim, "Strategic Culture of the Republic of Korea," 5.

8 Kim, "Strategic Culture of the Republic of Korea," 6.

9 Lincoln Riddle, "The Korean War—A War of Many Names," *War History Online*, last modified 16 June 2017, https://www.warhistoryonline.com/korean-war/10-top-facts-the-korean-war-m.html?edg-c=1.

10 Mary Beth Nikitin, "North Korea's Nuclear and Ballistic Missile Programs," in Focus Report IF10472 (Version 16), Congressional Research Service, 1.

11 Kim, "Strategic Culture of the Republic of Korea," 5.

12 Kim, "Strategic Culture of the Republic of Korea," 3.

13 Katrin Katz and Victor Cha, "South Korea in 2011: Holding Ground as the Region's Linchpin," *Asian Survey*, 52, no. 1 (2012): 52–64.

14 Luis Suarez-Villa and Pyo-Hwan Han, "The Rise of Korea's Electronics Industry: Technological Change, Growth, and Territorial Distribution," *Economic Geography*, 66, no. 3 (1990): 273–292.

15 Philip J. Ivanhoe, "'Heaven's Mandate' and the Concept of War in Early Confucianism," in Hashmi H. Sohail and Steven P. Lee, eds., *Ethics and Weapons of Mass Destruction: Religious and Secular Perspectives*, Cambridge: Cambridge University Press (2004), 272.
16 Gi-Wook Shin and Rennie J. Moon, "South Korea's President Lost the Mandate of Heaven," *The Diplomat,* 21 November 2016, https://thediplomat.com/2016/11/south-koreas-president-lost-the-mandate-of-heaven/.
17 Chien-peng Chung, "Democratization in South Korea and Inter-Korean Relations," *Pacific Affairs*, 76, no. 1 (Spring 2003), 9–35.
18 Ivanhoe, "'Heaven's Mandate' and the Concept of War in Early Confucianism," 273.
19 "South Korea Is One of the World's Largest Nuclear Power Producers," US Energy Information Administration, 27 August 2020, https://www.eia.gov/todayinenergy/detail.php?id=44916#:~:text=In%202019%2C%20South%20Korea's%20nuclear,the%20country's%20total%20electricity%20generation.
20 Hans M. Kristensen and Robert S. Norris, "A History of US Nuclear Weapons in South Korea," *Bulletin of the Atomic Scientists,* 73 no. 6 (2017): 349–357.
21 Jong-Yun Bae, "South Korean Strategic Thinking toward North Korea: The Evolution of the Engagement Policy and Its Impact upon U.S.-ROK Relations," *Asian Survey*, 50 no. 2 (2010): 335–355.
22 Hofstede Insights, "Country Comparison: South Korea and United States."
23 Kwang-Ok Kim, "The Reproduction of Confucian Culture in Contemporary Korea: An Anthropological Study," in Tu Wei-Ming, ed., *Confucian Traditions in East Asian Modernity: Moral Education and Economic Culture in Japan and the Four Mini-Dragons*, Harvard University Press (1996), 220.
24 South Korea's government is composed of three branches – judicial, executive, and legislative – and is founded on Confucian principles and the Constitution of the Republic.
25 Nikitin, "North Korea's Nuclear and Ballistic Missile Program," 2.
26 Institute for Security and Development Policy (ISDP), "THAAD on the Korean Peninsula" (2017): 1–8, http://isdp.eu/content/uploads/2016/11/THAAD-Backgrounder-ISDP-2.pdf.
27 Youngshik Bong, "Continuity Amidst Change: The Korea-United States Alliance," in Michael Wesley, ed., *Global allies: Comparing US Alliances in the 21st Century,* Australian National University Press (2017): 57.
28 Nikitin, "North Korea's Nuclear and Ballistic Missile Program," 1.
29 Toby Dalton, Byun Sunggee, and Lee Sang Tae, "South Korea Debates Nuclear Options," Carnegie Endowment for International Peace (2016): 1.
30 Rajiv Kumar, "South Korea's New Approach to India," ORF Issue Brief 263, Observer Research Foundation (2018):1 –8.
31 Jenny Town, "Challenges of Negotiating with North Korea," *Stimson* (Issue Brief), 7 April 2021, https://www.stimson.org/2021/biden-review-challenges-of-negotiating-with-north-korea/.
32 Kumar, "South Korea's New Approach to India," 1–8.
33 Vladimir Petrovsky, "South Korea's New National Security Strategy: Priorities and Nuances," *Modern Diplomacy*, 29 December 2018, https://moderndiplomacy.eu/2018/12/29/south-koreas-new-national-security-strategy-priorities-and-nuances/.
34 Petrovsky, "South Korea's New National Security Strategy."
35 Human Rights Watch "World Report 2020: South Korea," accessed 19 February 2022 from https://www.hrw.org/world-report/2020/country-chapters/south-korea.
36 Petrovsky, "South Korea's New National Security Strategy."
37 Petrovsky, "South Korea's New National Security Strategy."
38 Paul Dyer, "Policy and Institutional Responses to COVID-19: South Korea," *Brookings*, 15 June 2021, https://www.brookings.edu/research/policy-and-institutional-responses-to-covid-19-south-korea/.
39 Dyer, "Policy and Institutional Responses to COVID-19: South Korea."
40 Dyer, "Policy and Institutional Responses to COVID-19: South Korea."
41 A copy of the *New Deal* can be downloaded from South Korea's Ministry of Economy and Finance at: https://english.moef.go.kr/pc/selectTbPressCenterDtl.do?boardCd=N0001&seq=4948#:~:text=The%20Korean%20New%20Deal%2C%20announced,employment%20and%20social%20safety%20net.

42 Dyer, "Policy and Institutional Responses to COVID-19: South Korea."
43 Robin Wright, "Russia and China Unveil a Pact against America and the West," *The New Yorker,* 7 February 2022, https://www.newyorker.com/news/daily-comment/russia-and-china-unveil-a-pact-against-america-and-the-west.
44 Wright, "Russia and China Unveil a Pact against America and the West."
45 Kim, "Strategic Culture of the Republic of Korea," 1.
46 Suarez-Villa and Han, "The Rise of Korea's Electronics Industry," 273.
47 Kim, "Strategic Culture of the Republic of Korea," 3.

Selected Bibliography

Bae, Jong-Yun. "South Korean Strategic Thinking toward North Korea: The Evolution of the Engagement Policy and Its Impact upon U.S.-ROK Relations." *Asian Survey*, 50, no. 2 (March/April 2010): 335–355. doi:10.1525/as.2010.50.2.335

Bong, Youngshik. "Continuity Amidst Change: The Korea–United States Alliance." *Global Allies: Comparing US Alliances in the 21st Century*, edited by M. Wesley (2017): 45–58. http://www.jstor.org.proxy.missouristate.edu/stable/j.ctt1sq5twz.7

Chung, Chien-peng. "Democratization in South Korea and Inter-Korean Relations." *Pacific Affairs* (2003): 9–35. http://www.jstor.org.proxy.missouristate.edu/stable/40023987

Dalton, Toby, Byun Sunggee, and Lee Sang Tae. "South Korea Debates Nuclear Options." (Carnegie Endowment for International Peace, 2016). https://carnegieendowment.org/2016/04/27/south-korea-debates-nuclear-options-pub-63455

Dyer, Paul. "Policy and Institutional Responses to COVID-19: South Korea." *Brookings* (2021), accessed 20 February 2022. https://www.brookings.edu/research/policy-and-institutional-responses-to-covid-19-south-korea/

Hayes, Peter, and Roger Cavazos. "Complexity and Weapons of Mass Destruction in Northeast Asia." In *Complexity, Security and Civil Society in East Asia: Foreign Policies and the Korean Peninsula*, edited by Peter Hayes and Kiho Yi, 261–318. Cambridge, UK: Open Book Publishers, 2015. http://www.jstor.org.proxy.missouristate.edu/stable/j.ctt16gzg36.9

Hofstede Insights. "Country Comparison: South Korea and United States." 2022. https://www.hofstede-insights.com/country-comparison/south-korea,the-usa/

Human Rights Watch. "World Report 2020: South Korea." Accessed 19 February 2021. https://www.hrw.org/world-report/2020/country-chapters/south-korea

Institute for Security and Development Policy. "THAAD on the Korean Peninsula" (2017), 1–8, accessed 17 July 2019, http://isdp.eu/content/uploads/2016/11/THAAD-Backgrounder-ISDP-2.pdf

Ivanhoe, Philip J. "Heaven's Mandate" and the Concept of War in Early Confucianism." In *Ethics and Weapons of Mass Destruction: Religious and Secular Perspectives*, pp. 270–276. Cambridge University Press, 2004.

Johnson, Jeannie L., Kerry M. Kartchner, and Jeffrey A. Larsen, eds. *Strategic Culture and Weapons of Mass Destruction: Culturally Based Insights into Comparative National Security Policymaking*. Springer, 2009.

Katz, Katrin, and Victor Cha. "South Korea in 2011: Holding Ground as the Region's Linchpin." *Asian Survey* 52, no. 1 (2012): 52–64.

Kim, Jiyul, "Strategic Culture of the Republic of Korea," *Contemporary Security Policy: Strategic Cultures and Security Policies in the Asia-Pacific* 35, no. 2 (2014): 1–22. https://www.academia.edu/16475213/Strategic_Culture_of_the_Republic_of_Korea_Contemporary_Security_Policy_35.2_2014_

Kim, Kwang-Ok. "The Reproduction of Confucian Culture in Contemporary Korea: An Anthropological Study." *Confucian Traditions in East Asian Modernity: Moral Education and Economic Culture in Japan and the Four Mini-Dragons* (1996): 220.

Kristensen, Hans M., and Robert S. Norris. "A History of US Nuclear Weapons in South Korea." *Bulletin of the Atomic Scientists* 73, no. 6 (2017): 349–357, DOI: 10.1080/00963402.2017.1388656

Kumar, Rajiv. "South Korea's New Approach to India." *ORF Issue Brief* 263 (2018): 1–8. https://www.orfonline.org/research/south-koreas-new-approach-to-india-45135/

Millett, Allan R. "Introduction to the Korean War." *The Journal of Military History* 65, no. 4 (2001): 921–935. doi:10.2307/2677623

Nikitin, Mary Beth D. "North Korea's Nuclear and Ballistic Missile Program." *Congressional Research Service* (2022): 1–2, accessed August 2022, https://crsreports.congress.gov/product/pdf/IF/IF10472.

Petrovsky, Vladimir. "South Korea's New National Security Strategy: Priorities and Nuances," *Modern Diplomacy* (29 December 2018): 1. https://moderndiplomacy.eu/2018/12/29/south-koreas-new-national-security-strategy-priorities-and-nuances/

Riddle, Lincoln. "The Korean War—A War of Many Names," *War History Online*, 16 June 2017, https://www.warhistoryonline.com/korean-war/10-top-facts-the-korean-war-m.html

Shifrinson, Joshua. "Security in Northeast Asia." *Strategic Studies Quarterly* 13, no. 2 (2019): 23–47, https://www-jstor-org.proxy.missouristate.edu/stable/26639672

Shin, Gi-Wook. "Korea's Ethnic Nationalism Is a Source of Both Pride and Prejudice, According to Gi-Wook Shin." *The Korea Herald* (2006), accessed 20 February 2022, https://aparc.fsi.stanford.edu/news/koreas_ethnic_nationalism_is_a_source_of_both_pride_and_prejudice_according_to_giwook_shin_20060802

Shin, Gi-Wook, and Rennie J. Moon. "South Korea's President Lost the Mandate of Heaven," *The Diplomat*, 21 November 2016. https://thediplomat.com/2016/11/south-koreas-president-lost-the-mandate-of-heaven/

Suarez-Villa, Luis, and Pyo-Hwan Han. "The Rise of Korea's Electronics Industry: Technological Change, Growth, and Territorial Distribution." *Economic Geography* 66, no. 3 (1990): 273–292.

20

JAPAN'S STRATEGIC CULTURE

An Interpretation of Identity

Michael Barron

Introduction

Japan's strategic culture is a product of its unique geography, climate, resources, history and experiences, demographics, myths, symbols, transnational norms, political structure, economic development, and relationships. Geopolitically, Japan is a stratovolcanic archipelago (a vast chain of 6,852 islands), and most of its 127 million people live on the six main islands with over 90% of the islands uninhabited, yet still conferring territorial integrity and strategic utility.[1] The country spans a longitudinal distance of over 2,000 kilometers, stretching from the colder northern islands bordering Russia further south into the tropical environs of Okinawa and the Ryukyu Islands chain, at some points located only 200 km from the Asian continent. Japan's strategic culture has been profoundly shaped by its geographic reality as a vulnerable set of islands that lacks natural resources and is challenging to defend, is adjacent to increasingly illiberal continental neighbors, and is buffeted by progressively contested maritime commons.

Despite its relatively small population size, Japan is home to one of the world's most robust economies: its gross domestic product is third in the world, and its per capita gross national product and level of technological, social, and educational development also rank among the highest.[2] Japan is home to a model and mature democratic government (albeit with a ceremonial/symbolic emperor) with an American-drafted Western-style constitution, adopting a broadly Western liberal style of government with comparable state governance institutions.

This set of facts and circumstances gives rise to three areas of strategic consideration. First, because Japan is exclusively an island nation with few natural resources, there are real limitations on its national strength and strategic depth in terms of self-sufficiency. The need for uninterrupted and secure trade routes and sources, especially for energy, is important. Second, this dynamic is compounded by Japan's relatively small and widely dispersed geographic area and comparably modest population size.[3] In light of these and various other indicators such as a moderately waning economy, population decline and its consequences, and relative strategic strength, Japan surpasses middle powers such as India, Brazil, and many individual European Union states but cannot rival other great powers, which effectively limits its global influence. Third, Japan is fundamentally a maritime nation, with over 85% of its commerce seaborne and nearly 55% of all its

DOI: 10.4324/9781003010302-23

foodstuffs from the ocean. This maritime identity gives rise to much of its security posture and policies.[4]

This concept of Japan as a maritime state is critical in understanding the recent evolution of Japan's strategic identity.[5] Japan's strong civilian economy and technological prowess have allowed it to rebuild post-World War II national wealth mainly through maritime trade, retaking markets in the United States, Europe, and most crucially throughout East and Southeast Asia.[6] Important to Japan's identity is not only how the country has been a long-standing regional leader but also how it has fostered deep economic integration within East as well as Southeast Asia. Japan has developed a strong reputation of being a fair, equitable, and stable trading partner, which has led to dense economic and business linkages in the area and Japan often being the driving force in the organization and crafting of trade partnerships and agreements.

Japan sits at the geopolitical juncture of a uniquely difficult set of regional and global security flashpoints, including an increasingly assertive and militarized China; a nuclear and ballistic missile belligerent North Korea; mercurial political relationships with South Korea; and unresolved territorial-island disputes with the People's Republic of China, Taiwan, South Korea, and Russia. Japan lies squarely within the so-called First Island Chain, an arc of Pacific archipelagos off the continental coast of East Asia, with all its compounding geopolitical challenges. In addition, because of its long-standing US security-military alliance Japan falls within the shadow of American interests and their unfolding hegemonic challenges in the region.[7] On defense, Japan is trying to find the optimum balance as it attempts to avoid entrapment while also feeling isolated, meaning that if it does not do enough to appease the United States as well as the international security community, it fears being pressured to assume a greater military burden for national and regional defense. Japan has been confronted with the risks of the significant asymmetry in its mutual defense treaty with the United States and its incorporation within the greater US regional security framework.

Against this backdrop, over the last two decades, a renewed commitment to Japan's maritime disposition as elemental to the nation's future strength has gained increasing traction and this core strategic focus has shaped many of Japan's recent defense reforms.[8] Redefining Japan as a maritime state in response to encroaching powers is salient now more than ever and the issue of claiming national space in the seas around Japan has become critical. The rapidly evolving regional security landscape and associated challenges to Japan's island and sea claims have Japan's leaders looking outward, committing once again to orienting to the sea and working collectively across political lines and domestic constituencies to connect domestic and foreign policies to general allied interests, international laws of the sea, and increasing Japan's hard military power and gray-zone warfare capabilities (including paramilitary assets such as an expanded set of Japanese Coast Guard missions).[9] Strongly supporting a free and open maritime commons and strategic waterways throughout the Indo-Pacific is a fundamental security concern and underpins a debate about how the island country's nature defines its strategic culture: "as inward-looking and tightly defined or as open-ended and engaged with the world."[10]

Facing these contemporary challenges is the question of why Japan's strategic culture matters.[11] The Japan-US security alliance is the cornerstone of the US Pacific security-military strategy and represents innumerable and invaluable American interests. Despite periodic political perturbations in the relationship, Japan remains one of the United States' strongest allies and Asia's most democratic and vibrant open society.[12] Against the background of an increasingly challenging security environment, strengthening the US-Japan alliance and increasing Japan's material capabilities and capacities across multiple new domains of deterrence (joint operations; intelligence, surveillance, and reconnaissance; space; cyber; and missile defense) have been Japanese elites' incremental policy ambitions and institutional goals – a new security agenda, illustrated in the shift from Yoshida to Koizumi to Abe Doctrines. How Japan internally effects these responsive changes, both to expand its own military capabilities and to provide more systemic contributions to the US-Japan alliance in proportion to its economic and diplomatic power, has taken on increasing urgency and politico-cultural salience.

Despite facing many realist security threats, Japan's cultural narratives and identity conceptions have been powerful political constraints. Japan has largely achieved its security objectives through a strategic culture and national strategic community drawn from perceptions of its post-Word War II historical experiences and its strong desire to be accepted as a peaceful, responsible, engaging, and economically leading nation in the eyes of the world. Its nuanced security relationship with the United States confers the added dimension of deterrence, and Japan is a classic case of how these "costs and benefits" are culturally determined values. Japan's strategic culture not only provides the essential context for understanding its evolving strategic policy dilemmas but it is also the defining characteristic shaping its military policy and armed forces.

How Japan's strategic culture shapes, influences, and bounds its military force and contextualizes key security and defense decisions will be examined in this chapter through a four-function model: collective identity, shared values, common norms, and perceptual lenses, in addition to action templates, which are the manifestations of this strategic culture. While values, norms, and the perceptual lens through which a nation-state holds its worldviews are important, this chapter examines the constructs and forces of identity as the primal strategic cultural determinant. This emphasis on the role of identity is consistent with Kartchner, who characterizes identity as the predominant cultural trait from which the other three domains are derived in this four-factor framework.[13] The chapter concludes with a recapitulation of the key contributions of the strategic culture literature on Japan, offers some comments on the strengths and gaps within that body of literature, and recommends future areas of research for scholars and students.

The Making of Modern Japan's Strategic Culture

Japan's domestic politics, policies, strategic identity, and culture are, for the intents and purposes of this chapter, to be considered as the most recent iteration of its "Third Modern Period," often characterized as the immediate post-World War II period to the present.[14] Japan is unique in that almost the entirety of its state structures, functions, and religious affiliations were abruptly reset following World War II in the wake of the war's destructive aftermath and subsequent US occupation. Japan's devastating defeat in World War II gave rise to a new and distinct strategic culture based on domestic antimilitarism and buttressed by a robust military alliance with the United States. The human, civil-social, and economic destruction of the war led to fervently antimilitary public opinion, prompting elected officials to reflect the public's commitment to these attitudes even in the face of contrarian views from political elites.[15] Another war legacy was a keenly felt mistrust of anything reminiscent of a secret state. Most significantly, the United States imposed upon Japan a constitution that was deliberately designed to restrict Japan from ever waging offensive war again and essentially bifurcated civil-military relations so that defense officials and military elites would not be part of any security policymaking process.[16]

Further, Japan is constitutionally and legally proscribed from possessing traditional armed forces, from the use of force except in self-defense, and from participation in foreign wars.[17] Japan is the only state in the world that does not use the term "military" to describe its defense organizations, and it is technically unable to meet the United Nations' (UN) decreed right to collective self-defense as an expression of recognized sovereignty. The essence of this constitutional prohibition and structural political reality is codified in what is known as Article 9, the so-called war-renouncing clause. Japan's constitution, which endures unchanged since its inception in 1947, remains a powerful and deeply realized social contract. For most Japanese people, Article 9 defines their national identity and is the cornerstone of Japanese strategic culture.[18] The nature and value of Article 9 remain aspirational for Japan in describing the nation it would like to embody in the world as a peace-seeking, pacifistic, antimilitaristic country determined not to repeat its "mistakes of the past" (the popular Japanese concept of peace-state or *heiwa kokka*).[19]

Maintaining Article 9 is an expression of those whose worldviews hold that this self-constrained domestic antimilitarism is an inviolate constitutional provision. This constitutionally derived pacifism has become part of Japan's national identity and culture, and as Ejima notes, "the relationship between the Japanese constitution as a pacifist text and the Japanese people's identity as peace loving people is clear."[20] The paramount view held by most Japanese people is that their nation's nonmilitary outlook is essential to the state's strategic culture, thereby creating a profound social understanding that the right to wage war is axiomatically forbidden.[21] Japanese society's broader sentiment of antimilitarism, centered around this strong attachment to Article 9 and a deep suspicion of using military power for security ends, has led Japan's elected officials to remain committed to an "exclusively defense-oriented posture" that upholds antimilitaristic prohibitions and abjures the right or exercise of collective self-defense.[22] This in theory includes potential entrapment in entangling alliances and associated issues such as caps on defense expenditures and military technology exports.

Lastly, Japan is profoundly unique among all nations in that it is the only country to have been subject to the use of nuclear weapons, which forms another core component of the country's identity and contemporary strategic culture. Japan's pacifism remains largely unchanged since 1947 and is the core message of its overt nonmilitary diplomacy.

Sources of Japan's Strategic Culture

Two historical continuities are also relevant to Japan's extant strategic culture: its historical Western inclinations, and, in trying to establish its rightful place in the world, wide subscription to the prevailing liberal international order. Going as far back as the sixteenth century when the Europeans tried to enter the region, the Japanese state was able and effective in repelling the European political-economic-military interventionists while simultaneously exploiting their new knowledge, commodities, and weapons technologies (guns and cannons).[23] Matsumura highlights how the Japanese people continued to be renowned for assimilating the West's know-how and technology while preserving their autonomy and culture. This was reintroduced at the end of World War II as Japan was liberally integrated into the prevailing international order of states for diplomacy and economic relationships, and through its US defense pacts was free to pursue economic recovery and reconstruction.[24] Until the recent rise of South Korea and Taiwan, and now China, Japan was by far the most economically developed and Westernized Asian nation and still holds one of the most robust stock exchanges, bond markets, and financial centers while maintaining high marks in all global economic indicators. It is a leading G-20 and G-7 country and for decades was the Organization for Economic Cooperation and Development's Eastern Hemisphere powerhouse. However, despite leading or collaborating with regional economic organizations such as the Asia-Pacific Economic Cooperation forum, the revised Trans-Pacific Partnership, and the Association of Southeast Asian Nations, Japan has never been able to translate its economic might into a comparable level of strategic clout.[25]

Since 1951, Japan has been party to a "grand strategic bargain" with the United States that comprises superpower military protection, forward-deployed forces and bases, and the nuclear deterrent umbrella. Japan is a strong subscriber to its role in international institutions and organizations such as the UN, along with its many attendant norms, conventions, and laws.[26] Throughout the Yoshida years that were focused on recovering from the aftermath of World War II and decolonization, Japan sought economic strength through security by regional diplomacy and economic cooperation. This so-called Yoshida Doctrine is not only historically important but also fundamental to understanding Japan's current security policies. Japan declared a strong internationalist focus in its foreign and security policy, and its exclusively defensive-oriented posture avoided exacerbating security dilemmas in the region, first in the Cold War and now primarily with China. The state still struggles with submitting to pressures from China, Russia, North Korea, and other East Asian states on historical issues, and has played a delicate balancing game between its strategic security and economic allies, not wanting to get too close to China nor stray too far from the United States.

Four key narratives shape Japan's strategic culture: its postwar pacifism, pride in postwar accomplishments, outward face to the world, and sensitivity to its populace. First, as noted above, Japan's modern strategic culture has been significantly shaped by the security identity of domestic antimilitarism and by its military alliances and relationships. This predominant thread in Japan's strategic culture is deeply rooted in nonmilitary approaches to its interactions with regional neighbors and the pacifist model that has become the basis of its regional identity.[27] The values and beliefs integral to this security identity are readily claimed by many Japanese political actors and exert a major influence on the state's strategic culture.[28] Critically, pacifist symbols and symbolic strategies are used by Tokyo's decisionmakers and policy elites to secure legitimacy and continually convey a sense of competency and responsiveness to domestic opinion.[29] One major consequence of this self-constrained pacifist discourse is that it induces a strong inertia in Japanese strategic decisionmaking.[30]

Second, Japan is still working within many of the constructs of the immediate post-World War II period, including Prime Minister Yoshida's unequivocal relinquishment of any semblance of militarism and assuming the costs of the prosecution of the war. Much of this political culture and civil-social contract was a strategy formed by a totally defeated power to navigate the circumstances of defeat.[31] Japan's devastating defeat incurred a deep sense of humiliation and shame, but unlike China, Japan has used this historical narrative to become a pacifist state with the intent of being a model international peacekeeping country, and has successfully utilized this narrative to reconfigure its foreign policies toward economic benefits along pacifist lines.[32] Japan is still struggling to overcome the encumbrance of needing to apologize for its colonial past and war-related activities and sometimes feels this has placed it in a subservient position in the face of US, Chinese, and other East Asian pressures.[33]

Third, Japan goes to great lengths to ensure that its international image is commensurate with its domestic antimilitarism and that it carries its strong image as a peace-loving Japan abroad. As Oros notes, "Japan ranked highest among countries surveyed in its image as a positive contributor to international society in 2012, and ... it has consistently ranked highly in the years the poll has been conducted."[34] This is reflected in the routine and extensive deployment of Japan's Self-Defense Forces (JSDFs) as generous providers of disaster relief and humanitarian assistance.[35]

Fourth, this pacifistic social contract is so deeply ingrained in the Japanese psyche that Japan's polity is extremely sensitive to domestic opinion on such issues. Japan hosts innumerable public and private polls, including key governmental agency-sponsored polls, and public opinion polls are among one of the Japanese government's prime sources of legitimacy and authority.[36] These polls are considered barometers of the balance between ideology and interests, and can be used to assess thresholds or opportunities for policy change from the national polity. This also confers an element of moral judgment from society whereby Japanese ruling elites must balance public judgments against national material interests. The Japanese legislative body, the Diet, has its own public polling machinery and this further serves as a source of civil society strength and limits the permissiveness of the legislature to reshape Japan's policies. This is an important moderating factor in Japan's strategic culture and is a very real political impediment to abruptly altering either its strategic culture or its security policy.[37] Politically, postwar Japan has always had a conservative government, as the Liberal Democratic Party (LDP) has been in power for all but about five of the last 70 years, making the party one of the most important of Japan's so-called keepers of strategic culture. While political elites shape most of Japan's security identity, domestic institutions and electoral politics present a strong force affecting its identity construction and offering predictions about the resilience of Japan's strategic culture.[38] To quote Oros, "many diverse and powerful actors shape Japan's strategic culture, making long-term change to the culture a lengthy and complicated process."[39]

In many ways, the keepers of Japan's strategic culture and security identity are primarily Japan's key political parties and elites, particularly the long-ruling LDP and the surrounding political and bureaucratic actors who model the state's strategic culture and routinely shape policy within its bounded rationality. A few key state institutions play subservient political "keeper" roles, such as Japan's Ministry of Defense (MOD) and Ministry of Foreign Affairs (MOFA), but the Cabinet Legislation Bureau (CLB) stands out as one of the main keepers or maintainers of the dominant strategic culture.[40] The CLB is charged with oversight of priority security issues such as constitutional interpretation, Article 9 policies, and collective self-defense, and is ultimately controlled by the elected government and headed by a Prime Minister appointee.[41] Consensus is a critically important part not only of the political process but also as a mechanism to keep the culture. Effectively, the Japanese Prime Minister must obtain the consent of other cabinet members to make decisions, and unanimity is usually the basis for managing the cabinet.[42]

While these institutions (JSDFs, MOD, MOFA, and CLB) play important roles in enacting changes to Japan's strategic culture, change is initiated and controlled by the ruling party.[43] However, the populace and public opinion have a considerable role as "minders" of the keepers, as their viewpoints constitute, in Oros' words, a political "hegemony of the dominant antimilitarist security identity of domestic antimilitarism at the mass level."[44] Japan is an extreme example of how popular culture and civil society act in an indispensable role in creating or maintaining a country's strategic culture. This poses a significant and ultimately limiting constraint for anyone seeking to change Japan's strategic culture – illustrated by the fact that even with the majority strength of the LDP Prime Minister Shinzo Abe resigned without formal amendment of Article 9.

Cracks Appearing in Japan's Post-World War II Sense of Identity

Among developed nations, Japan has one of the strongest monocultural identities. Japan has long struggled with being labeled one of the world's most xenophobic countries and has some of the most strident anti-immigration policies. One of the more critical economic exigencies related to Japan's rapidly changing demographics is its looming low-skilled and semi-skilled labor shortages, and Japan is currently wrestling with changing immigration practices. This policy shift also brings forth several considerations about Japan's regional diplomatic and economic relationships and may be a development vehicle for East and South Asia. Japan consciously attempts to preserve its cultural identity and stability through immigration control, and this fear of disruption impacts its foreign relations and sense of duty and responsibility to the rest of the world. Japanese ethnocentric considerations also shape understanding and views on other key national policies and represent the nexus of identities and interests.

Immigration serves as a test of how both material-based and identity-based concerns shape public views and how public attitudes influence the range of policy options considered and selected by officials.[45] Interestingly, Davidson and Peng have shown how an "identities approach to understanding anti-immigration sentiment has explanatory power in Japan," and that "material-based concerns can, when sufficiently strong, override identity based concerns" in a pragmatic manner.[46] This has two implications that may be applicable to other core national policy issues starting with security and defense. First, there are costs when Japan prioritizes ideations or cultural perceptions over interests; and second, there are cases where domestic support for issues such as remilitarization policy reform can be "non-ideological, considered, and conditional."[47] Conway warns that in today's changing security environment, strict pacifism may present a problem. In the innumerable polls of constitutional change, public opinion has been dynamic, public attitudes toward government decisions and policies often change in response to world events.[48]

Ethnically and racially, the Japanese population is 98% Japanese. The Yamamoto Japanese are the dominant group, the exceptions being the Ryukyuan Japanese (1.3 million) who are indigenous to the Ryukyu Islands chain and have different language and cultural subgroups, as well as very low populations of Ainu who inhabit the far northern Kuril Islands and the Russian island of Sakhalin.[49] This largely monoethnic population and resultant ethnic-identity formation carry an implicit assumption of uniformity, including values and norms, which despite the presence of a nontrivial population of minority peoples helps drive a sense of ethnonationalism.[50] During the era of Imperial Japan, this ethnic component of identity shaped the concepts of the Greater Asian Co-Prosperity Sphere and Pan-Asianism, which included the aims of removing European colonial influence from Asia and harnessing other Asians for the benefit of Japan.[51] Hagstrom and Gustafsson, quoting Takami, discuss this "hierarchic worldview" and an "associated sense of Japanese uniqueness" that has endured within the postwar pacifist state identity.[52] This worldview also derives from one of Japan's identities being that of an "economic great power/superpower/giant."[53] This self-perceived status and state identity conception of self, relative to others in the region and on the world stage, is currently the object of many escalating challenges in the Western Pacific arena.

Japan is a country with a long and rich cultural heritage and was able to preserve this inward-looking perception of itself while simultaneously looking outward to find its place in a trade/mercantile-based world and take the best of what the rest of the world had to offer. This approach to an economically oriented foreign policy led to oscillation between shutting out and embracing different facets of foreign cultures, and conceptions of Japanese national identity and norms continue to comprise both native contributions from the Japanese population and the adoption of imported ideas and practices.[54] This adaptation is exemplified by the identification of a modern Japanese state identity more resembling a Western state than its ancient antecedents and those of its neighbors, and this may suit Japan for its emerging geostrategic landscape in East Asia. Japan has historically been torn between maintaining traditions and pursuing Westernization, and since the Meiji era has entertained the conception of Japan as a bridge between the East and the West.[55] One of the core elements of Japan's identity is that it sits on the outskirts of Western civilization but has continued to thrive as an independent culture not overtaken by Western culture – a message the country conveys to other cultures.[56] This also helps form Japan's identity as a democratic state, emphasizing both the differences from undemocratic states and similarities with other democratic states. This messaging is a vital component of its foreign policy and alliance relationships and has taken on a greater tenor in recent expressions of its foreign policy.[57]

Japan's Identity-Focused Perceptual Lens: Responding to an Evolving Security Environment

It is critical to understand that Japanese strategic culture over the last 70 years has taken to an extreme the elements contained in its constitution. Despite the varying circumstances from inception to Japan's currently changing security environment, for the overwhelming majority of Japanese people, these elements have become essential to their view of the world and not something to be manipulated for expedient gain.[58] This view includes Japan's security relationship and guarantees from the United States as well as supporting the global system that has protected and benefitted Japan. The constitution confers Japan's place in the world as a model for democratic idealism and pacifism and taps deeply into popular aspirations in this vein.[59] Yet while most Japanese people view their nation's nonmilitary outlook as essential to its strategic culture, Hughes describes how modern Japan is confronted with a changing worldview where a more activist foreign policy is necessary to confront myriad threats and to allow for greater security cooperation with its partners.[60] Modern Japan's openness to the world and noninsularity may be an ideal vehicle for greater cooperation and alliances.

Japan's traditional modernity is undergoing a crucial test with the changing geopolitical environment and its roles and relationships in the world. It is being confronted with having to face a "reluctant realism" and has had to adroitly and incrementally address these burgeoning threats through legislative security change and institutional defense reforms, including the recent formation of both a Ministry of Defense and a National Security Council and the release of National Security Strategy documents modeled after the United States. Japan's "resentful realism" and balancing China's rise has in particular been a consuming act for Japan's ruling elite over the last two decades, as they have been forced to contemplate a future in which China will dominate the region economically, politically, and militarily.[61] Japan's outward-looking view of the world has been altered by the inconvenient truth of China's rise to power and the nuclearization of the Korean peninsula. Within the short span of Abe's most recent term, Japan witnessed North Korea develop, test, and launch over Japanese territory a spectrum of intermediate to long-range ballistic missiles, and the admitted North Korean kidnappings of Japanese people have only added to negative security perceptions. North Korea's nuclear weapons programs, ballistic missile capacities, and rogue state status genuinely represent an existential threat to Japan.[62]

Interestingly, despite Japan's deep economic interdependence with China and now South Korea, mutual public opinion and perception surveys continue to find significant mutual distrust and negative perceptions of each other.[63] Japan is very aware that modern Chinese nationalism is suffused with virulent anti-Japanese sentiment, and historical lingering unfortunately also heavily flavors its relationship with South Korea. Japan is additionally having to contend with America's strategic choices and what sustainable level of commitment the United States will provide vis-à-vis its China and Russia policies and its treaty and alliance partnerships with South Korea, Taiwan, and the Philippines. The state's threat perceptions also arise from the fact that Japan is a key international actor who under the current constitution can never amount to a full-fledged military power and assume a complementary international or allied role. Yet Japan is situated at the front door of the growing US-China, US-North Korea, and Russian tensions and has nationalistic territorial claims and contentions of its own across all of these regions.

Abe's revisionist nationalist right and his successors ignited political flames by increasing militarization over highly contested maritime territories and the near continuous "violation" of Japan's sovereignty by Chinese maritime vessels in the Senkaku Islands. These spatial assertions and the conceptualization of Japanese national territory giving rise to sea claims are the first in modern Japanese history. The focus of Japan's new strategic vision is the defense of these islands and the seas around them. Japan is critically dependent upon a free and open global economic system and sea lines of communication throughout all the Indo-Pacific commons, but the Chinese now operate with impunity in parts of Japan's air and sea space as they continue to push both Japan and the United States out. Toward the latter end, the Japanese have openly engaged diplomatically, economically, and now militarily with more than symbolic relationships with India, Australia, Singapore, Malaysia, Vietnam, and the Philippines. These changes are designed to give Japan strategic options and move away from a stand-alone pacifistic defense posture to a more dynamic one based on bilateral and multilateral relationships that bridge strong diplomatic, economic, and military means. These joint operations also give Japan a hedge against Chinese objections, since it is aware that the deployment of Japanese ships and planes to the South China Sea, for example, would be seen as highly provocative. Japan is highly in favor of the strategic rebalancing by the United States toward Asia, and Kafura reinforces Hagstrom and Gustafsson's contention that, most importantly, the majority of the Japanese public understands that the US-Japanese alliance is currently the only viable means of guaranteeing Japan's security. Positive public opinion polls and perceptions of

US-Japanese relationships are vital.[64] Yet this is not without complication, as the US forward-deployed force posture within Japan has also led to deep-seated tensions between the accommodating security keepers and the populace, especially with regard to the forward-staged ground forces in the increasingly strategic Okinawan island chain.

The complex and evolving security environment of East Asia lends credibility to an argument of a dual strategic culture emerging within Japan, one among the ruling and policy elites and another among the electoral populace. This dueling tension exists between a largely pacifistic population and a ruling elite that appropriately worries about the many noted emerging threats: North Korea's growing nuclear strike and blackmail capabilities, a militarily modernized and increasingly belligerent China, being caught in the middle of a Sino-US strategic competition, fear of abandonment as a capable US ally, and being exposed if the United States does not maintain either the military presence, capability, or political will to provide security. All of these present credible challenges to Japan's responsibility for its own security and ability to help maintain regional stability. Japan's strategic identity has both restrained and compelled it to address these new security challenges, with some elites favoring bolstering Japan's military posture, while public opinion polls continue to show that an overwhelming majority of the population favors preserving the laws that prevent such action.[65] Dudden summarizes this by noting that "Japan's strategic outlook has been shifting for roughly two decades from one of rather clear-cut dependence on the US protective shield to one of increasing autonomy."[66] Japan will need to forge its own way ahead with free agency – a path that is likely to include an independent defense policy addressing its regional security concerns while maintaining its alliance with the United States. However, any significant movement in the direction of collective defense, including offensive capabilities, will require Japanese civil society willfully edging away from or relinquishing the strength and value of its pacifistic identity.

State of the Scholarship on the Impact of Identity on Japanese Strategic Culture

Literature on Japan's strategic culture has highlighted the dominant strand of its pacifistic belief system and antimilitaristic postwar culture's influence on key Japanese actors.[67] Berger presciently examined the linkages between Japan's culture and military strategy in light of the prospect of rising regional powers and described pragmatic ways that Japan's burgeoning security needs could be met through more internationally structured frameworks while preserving many of the tenets of its cultural antimilitarism.[68] The bulk of the expanding body of literature on Japan's approaches to being confronted with a rapidly changing strategic environment has been primarily focused on the national ideations that either support or seek revision of its constitutional constraints, specifically Article 9.[69] The totality of strategic culture impacting national strategy and policymaking throughout all of the Asia-Pacific was perhaps best brought together by Tellis et al. in 2016 with a series of comprehensive surveys of how countries in the region think about and respond to their strategic environment through the lenses of their histories, belief systems, and social-civic institutions.[70] The much broader considerations of US policy toward Japan and the region at large – as well as a deeper understanding of how Japan's culture continues to shape its partnership with the United States, especially regarding balancing against China – reveal the tensions and potential effects of Japan's rising nationalism versus its purposed internationalism.[71] Tellis et al.'s work also examined the cultural dynamics that may favor change; a crucially important emerging area of inquiry as adaptation may become Japan's defining characteristic. Breaking out of the long-held mold of its stoic reputation of conservatism and risk aversion could be one of Japan's key politico-cultural adaptations.

The Japan case literature highlights how strategic culture is a powerful tool of security studies and as a formalized analytical model provides insights into the bridge between material and ideational expectations of state behavior.[72] The key contributions by Oros and Lantis question how Japan's enduring beliefs help to explain its foreign policy decisions and the limits of understanding its political activity, including the origins of political support for its security practices.[73] While Japan's security identity continues to broadly shape its national defense posture, its institutionalized antimilitarism is incrementally being challenged by ruling elites, including through the recent establishment of American models of a Ministry of Defense, National Intelligence Services, and a National Security Council. Work by Oros has shown how the trajectory of Japan's dominant strand of security identity suggests that it will continue to play a reluctant and restrained role in developing military power beyond some narrow scenarios.[74] However, the bulk of the relevant literature building Japan's strategic culture and security portfolio was written prior to a rapidly changing and increasingly foreboding international environment that has led to a resurgence of Japanese leadership driving political efforts aimed at increasing military activity near and abroad.[75] Further political inroads include the 2021 determination that Japanese defense spending is not limited to the 1% cap.[76] Lantis has provided a more refined depiction of the near-contemporary nexus of Japanese cultural predispositions and its burgeoning realist geopolitical security threats and opportunities,[77] but Oros agrees that Japan's postwar identity is "under siege" and the moral determinisms of peace and pacifism will increasingly converge upon the agendas of ambitious political actors.[78]

Conclusion

Japan is truly at a crossroads as a fundamental shift in East Asian geopolitics forces its hand, not only in flexing its muscles but also in stretching its constitutional limits to the extent possible through a series of increasingly aggressive constitutional reinterpretations and legislative reforms that now represent the single largest reconsideration of Japanese military policy post-World War II.[79] The horrors of World War II still continue to haunt policymakers in Tokyo, and Japan is keenly aware of this legacy and the suspicions that any military ambitions may have on its continental neighbors' threat perceptions and relationships. Japan's constitutionally decreed pacifistic ethos "forever renouncing the use of force as a means of settling international disputes" is a long-standing taboo that is being seriously reevaluated by Japan's ruling polity as they keep a close finger on the pulse of public outcry.

A key question for Japan and the United States is how Japan's strategic culture will fit the task of balancing against China, as well as confronting the Taiwan issue, especially with regard to Japan's determinations to either become more nationalistic and militarize or remain within a more staid and restrained international order, conforming to the ideational inheritances of pacifism and antimilitarism. In consideration of Japan's potential rearming, the tensions extant in Japan's security ideals and movements are summed up by Smith:

> The United States is largely seen as an advocate for greater action by the Japanese military, both in terms of its radius of operation and the latitude it has to cooperate with US forces on shared missions. But the US stake in Japan's constitution extends

> far beyond this narrow debate over how the Self-Defense Force operates. It is in the interest of the US to ensure that any changes Abe makes are fully supported by the Japanese people. Otherwise, any decision on collective self-defense would undermine confidence in the alliance if it was perceived as appeasing Washington rather than serving Japan's own interests. The Japanese people must support this evolving role for their military and remain confident that their government is willing only to use military force for the purpose of self-defense.[80]

A fourth generation of post-World War II Japanese have now inherited a set of enduring cultural and security identity attributes amid a growing awareness of pressing Asian geopolitical challenges. Japan's historical context and associated narratives will be tested by the structural expectations of a rapidly evolving environment and domestic policy responses that frame Japan-Asia relations. Areas for future research in this domain should include how much of Japan's strategic cultural characteristics and resultant foreign policy decisions have been situational. The question of conditionality is critical, as this provides a more exacting path of foreign policy decisionmaking such as fewer limits on national defense investments, expanded security alliances and multilateral agreements, and domestic political acquiescence. Studies in this vein may produce insights into how Japan's identity and physical survival are best ensured, as well as its more realist necessitation of "normalization" as a country.

Japan's latent nationalist feelings will need to be carefully balanced against its internationalist ambitions. While the culture of nationalism has not been pervasive in the contemporary period and there has not been much populist movement, Japan is far from immune to such impulses (consider Nippon Kaigi, Japan's largest ultranationalist non-governmental organization), especially if the country suffers from growing Chinese regional domination, the continuing erosion of its economic standing, and the downsides of globalization. Japan's adaptation to its situational reality will be strongly culturally determined, and while its sense of identity and self-respect is as a uniquely peaceful state, there may be enough cultural vestiges and determinants lingering to validate Ruth Benedict's interpretation of "the conditionality of their outlook on life ... being situational. If the world around them was peaceful, the Japanese would be committed to pacifism, but if the great powers were to gear themselves up for war again, Japan would soon revert to the old militarism."[81]

Notes

1 "The World Factbook: Japan," Washington, DC: Central Intelligence Agency, 2021, https://www.cia.gov/the-world-factbook/countries/japan/.
2 Christopher W. Hughes, "Japan's Grand Strategic Shift: From the Yoshida Doctrine to an Abe Doctrine?" Essay, In *Power, Ideas, and Military Strategy in the Asia-Pacific*, eds. Ashley J. Tellis, Allison Szalwinski, and Michael Wills (Seattle, WA: The National Bureau of Asian Research, 2017), 80; Alexis Dudden, "Two Strategic Cultures, Two Japans," *Strategic Asia 2016–2017*, 93–94, The National Bureau of Asian Research, 2016, https://www.nbr.org/publication/strategic-asia-2016-2017-understanding-strategic-cultures-in-the-asia-pacific/.
3 CIA, "The World Factbook: Japan."
4 CIA, "The World Factbook: Japan."
5 Hughes, "Japan's Grand Strategic Shift," 76.
6 Dudden, "Two Strategic Cultures, Two Japans," 101.
7 Hughes, "Japan's Grand Strategic Shift," 76–77.

8 Dudden, "Two Strategic Cultures, Two Japans," 99–103; Masahiro Matsumura, *The Japanese State Identity as a Grand Strategic Imperative* (The Brookings Institution Center for Northeast Asian Policy Studies, 2008), 1–2.
9 Ibid., 99.
10 Ibid., 101.
11 Andrew L. Oros, "Japan's Strategic Culture: Security Identity in a Fourth Modern Incarnation?" *Contemporary Security Policy 35*, no. 2 (2014): 229. doi: 10.1080/13523260.2014.928070.
12 Dudden, "Two Strategic Cultures, Two Japans," 99.
13 Kerry M. Kartchner, "Introduction: Sociocultural Approaches to Understanding Nuclear Thresholds," in *Crossing Nuclear Thresholds*, eds. Jeanne L. Johnson, Kerry M. Kartchner, and Marilyn J. Maines (Cham, Switzerland; Palgrave Macmillan, Springer International Publishing AG. 2018), 9–11.
14 Andrew L. Oros, "Japan's Strategic Culture: Security Identity in a Fourth Modern Incarnation?" *Contemporary Security Policy* 35, no. 2 (2014): 231–232, doi: 10.1080/13523260.2014.928070.
15 Ibid., 228, 232–233.
16 Masahiro Matsumura, "The Japanese State Identity as a Grand Strategic Imperative" (Brookings Institution, 2008), 1–28.
17 Oros, "Japan's Strategic Culture," 233.
18 Dudden, "Two Strategic Cultures, Two Japans," 92, 97.
19 Ibid., 98–99.
20 Akiko Ejima, "What Is the True Value of the Constitution of Japan? Japanese People's Perception of the Constitution," IACL-AIDC Blog (26 January 2021), https://blog-iacl-aidc.org/cili/2021/1/21/what-is-the-true-value-of-the-constitution-of-japan-japanese-peoples-perception-of-the-constitution.
21 Dudden, "Two Strategic Cultures, Two Japans," 110.
22 Hughes, "Japan's Grand Strategic Shift," 92.
23 Matsumura, "The Japanese State Identity," 7.
24 Ibid., 11–12.
25 Hughes, "Japan's Grand Strategic Shift," 80.
26 Ibid., 85–86.
27 Karl Gustafsson, Linus Hagström, and Ulv Hanssen, "Long Live Pacifism! Narrative Power and Japan's Pacifist Model," *Cambridge Review of International Affairs* 32, no. 4 (2019): 502–504. doi: 10.1080/09557571.2019.1623174.
28 Oros, "Japan's Strategic Culture," 234.
29 Matsumura, "The Japanese State Identity," 13.
30 Ibid., 25.
31 Hughes, "Japan's Grand Strategic Shift," 75, 96.
32 Tom French, "Narratives of Humiliation: Chinese and Japanese Strategic Culture," *ISN ETH Zurich* (2012), 1–3, https://www.files.ethz.ch/isn/188767/ISN_141042_en.pdf.
33 Hughes, "Japan's Grand Strategic Shift," 96.
34 Oros, "Japan's Strategic Culture," 232.
35 Ibid., 231–233.
36 Hughes, "Japan's Grand Strategic Shift," 103; Chalula Rininita Anindya, "The Evolving Security Policy of Japan and the Adherence to Antimilitarism Culture," *Global: Jurnal Politik International* 18, no. 2 (2016): 152–164, doi: 10.7454/globalv18i2306.
37 Anindya, "The Evolving Security Policy of Japan," 161–162.
38 Andrew L. Oros, "International and Domestic Challenges to Japan's Postwar Security Identity: 'Norm Constructivism' and Japan's New 'Proactive Pacifism," *The Pacific Review* 28, no. 1 (2014), doi: 10.1080/09512748.2014.970057.
39 Oros, "Japan's Strategic Culture," 235.
40 Ibid., 236–237.
41 Ibid., 236–238.
42 Kristian Gustafson, Philip H. J. Davies, and Ken Kotani, "A Reconstruction of Japanese Intelligence Issues and Prospects," Essay, In *Intelligence Elsewhere: Spies and Espionage Outside the Anglosphere* (Washington, DC: Georgetown University, 2013), 181–198.
43 Oros, "Japan's Strategic Culture," 238.
44 Ibid., 240–241, 243.

45 Jeremy Davison and Ito Peng, "Views on Immigration in Japan: Identities, Interests, and Pragmatic Divergence," *Journal of Ethnic and Migration Studies*, 47 no. 11 (2021): 2578–2579.
46 Ibid., 2578.
47 Chris Conway, "Forgive, Not Forget: The Way Ahead for Japan's Remilitarization," *The Oxford Blue*, 12 August 2021, https://www.theoxfordblue.co.uk/2021/08/12/forgive-not-forget-japans-remilitarisation/.
48 Ejima, "What Is the True Value of the Constitution of Japan?
49 "The World Factbook: Japan," Washington, DC: Central Intelligence Agency, 2021, https://www.cia.gov/the-world-factbook/countries/japan/.
50 Jeff Kingston, "The Ideology of Japanese Identity – MULTIETHNIC JAPAN, by John Lie," *The Japan Times*, 16 September 2001, https://www.japantimes.co.jp/culture/2001/09/16/books/book-reviews/the-ideology-of-japanese-identity.
51 Ibid.
52 Linus Hagström and Karl Gustafsson, "Japan and Identity Change: Why It Matters in International Relations," *The Pacific Review* 28, no. 1 (2014): 6, doi: 10.1080/09512748.2014.969298, 6.
53 Ibid., 12.
54 Matsumura, "The Japanese State Identity," "Japan," *Encyclopedia.com,* https://dir.md/arts/encyclopedias-almanacs-transcripts-and-maps/japan?cv=1&host=www.encyclopedia.com.
55 Shin'ichi Kitaoka, "'Japan's Identity: Neither the West Nor the East," The Japan Forum on International Relations, The Nippon Foundation, https://www.jfir.or.jp/e/special_study/seminar1/conversation.htm.
56 Ibid.
57 The White House, "U.S.-Japan Joint Leaders Statement: U.S.-Japan Global Partnership for a New Era." 16 August 2021, https://www.whitehouse.gov/briefing-room/statements-releases/2021/04/16/u-s-japan-joint-leaders-statement-u-s-japan-global-partnership-for-a-new-era/.; Sakura Murakami and Kiyoshi Takenaka, "Japan signals more active role on China's tough stand on Taiwan," *Reuters*, 5 October 2021, https://www.reuters.com/world/asia-pacific/biden-promised-us-commitment-defending-senkaku-islands-japan-pm-kishida-2021-10-05/.
58 Dudden, "Two Strategic Cultures, Two Japans," 104–105.
59 Ibid., 98.
60 Hughes, "Japan's Grand Strategic Shift," 83, 100–101.
61 Christopher W. Hughes, "Japan's 'Resentful Realism' and Balancing China's Rise," *Chinese Journal of International Politics* 9, no. 2 (2016): 109, https://doi.org/10.1093/cjip/pow004.
62 Hughes, "Japan's Grand Strategic Shift," 76, 88, 93.
63 "Poll Shows Rise to over 60% of Chinese Viewing Japan Unfavorably," *Nippon*, 17 November 2021, https://www.nippon.com/en/japan-data/h01154/.
64 Craig Kafura, "Public Opinion and the US-Japan Alliance at the Outset of the Trump Administration," Chicago Council on Foreign Affairs, 2017. https://globalaffairs.org/sites/default/files/2020-11/Public%20Opinion%20and%20the%20US-Japan%20Alliance%20at%20the%20Outset%20of%20the%20Trump%20Administration%20PDF%20Report.pdf
65 Dudden, "Two Strategic Cultures, Two Japans," 109.
66 Ibid., 109–110.
67 Thomas U. Berger, "From Sword to Chrysanthemum: Japan's Culture of Anti-militarism," *International Security* 17, no. 4 (1993): 119–150, doi: 10.2307/2539024.
68 Ibid.
69 Jeffrey P. Richter, "Japan's 'Reinterpretation' of Article 9: A Pyrrhic Victory for American Foreign Policy?" *Iowa Law Review* 101 (2016): 1223–1262; Institute for Security & Development Policy, "Amending Japan's Pacifist Constitution – Article 9 and Prime Minister Abe," 10 January 2019, https://isdp.eu/publication/amending-japans-pacifist-constitution/; Council on Foreign Relations, "Constitutional Change in Japan," 2021, https://www.cfr.org/japan-constitution/.
70 Ashley J. Tellis, "Strategic Culture, National Security, and Policymaking in the Asia-Pacific," Carnegie Endowment for International Peace, 27 October 2016, https://carnegieendowment.org/2016/10/27/strategic-culture-national-security-and-policymaking-in-asia-pacific-pub-66166.
71 Ibid.

72 Jeffrey S. Lantis, "Strategic Cultures and Security Policies in the Asia-Pacific," *Contemporary Security Policy* 35, no. 2 (2014): 166–186. doi: 10.1080/13523260.2014.927676.
73 Ibid.; Oros, "Japan's Strategic Culture," 227.
74 Ibid.
75 Miki Nose, "Japan Defense Spending Isn't Bound by 1% GDP Cap, Suga Says," *Nikkei Asia*, 31 August 2021, https://asia.nikkei.com/Politics/Japan-defense-spending-isn-t-bound-by-1-GDP-cap-Suga-says; Mari Yamaguchi, "Japan Cabinet OKs More Defense Funds Amid Potential Threats," *AP News*, 21 December 2020, https://apnews.com/article/technology-yoshihide-suga-north-korea-national-budgets-china-af175b243f0419e724eb7c7ef0efdbb8; Walter Russell Mead, "Tokyo Lobs the Ball Back at Beijing," *Wall Street Journal*, 26 July 2021, https://www.wsj.com/articles/japan-china-taiwan-nobuo-kishi-nakayama-xi-jinping-dpp-11627328357?mod=article_relatedinline.
76 Nose, "Japan Defense Spending" (2021).
77 Lantis, "Strategic Cultures and Security Policies," 166.
78 Oros, "Japan's Strategic Culture," 240.
79 Conway, "Forgive, Not Forget," 4.
80 Sheila A. Smith, "Reinterpreting Japan's Constitution," Council on Foreign Relations, 2 July 2014, https://www.cfr.org/blog/reinterpreting-japans-constitution.
81 Benedict R., *The Chrysanthemum and the Sword: Patterns of Japanese Culture* (First Mariner Books, 2005), xi–xii.

Selected Bibliography

Anindya, Chaula Rininta. "The Evolving Security Policy of Japan and the Adherence to Antimilitarism Culture." *Global: Jurnal Politik Internasional* 18, no. 2 (2016): 152–164. doi: 10.7454/global.v18i2.306

Berger, Thomas U. "From Sword to Chrysanthemum: Japan's Culture of Anti-militarism." *International Security* 17, no. 4 (1993): 119–150. doi:10.2307/2539024

Central Intelligence Agency. "The World Factbook: Japan." https://www.cia.gov/library/publications/the-world-factbook/geos/tw.html

Council on Foreign Relations. "Constitutional Change in Japan." 2021. https://www.cfr.org/japan-constitution/

Dudden, Alexis. "Two Strategic Cultures, Two Japans" Essay. In *Strategic Asia 2016–2017*, 93–94. The National Bureau of Asian Research (2016). https://www.nbr.org/publication/strategic-asia-2016-2017-understanding-strategic-cultures-in-the-asia-pacific/

French, Tom. "Narratives of Humiliation: Chinese and Japanese Strategic Culture." ISN ETH Zurich (2012). https://www.files.ethz.ch/isn/188767/ISN_141042_en.pdf

Gustafsson, Karl, Linus Hangström, and Ulv Hanssen. "Long Live Pacifism! Narrative Power and Japan's Pacifist Model." *Cambridge Review of International Affairs* 32, no. 4 (2019): 502–520. doi: 10.1080/09557571.1623174

Hagström, Linus, and Karl Gustafsson. "Japan and Identity Change: Why It Matters in International Relations." *The Pacific Review* 28, no. 1 (2014): 12. doi: 10.1080/09512748.2014.969298

Hughes, Christopher. "Japan's Grand Strategic Shift: From the Yoshida Doctrine to an Abe Doctrine?" Essay. In *Power, Ideas, and Military Strategy in the Asia-Pacific*, ed. Ashley J. Tellis., Allison Szalwinski, Michael Wills, Seattle (WA): The National Bureau of Asian Research (2017).

Institute for Security and Development Policy. "Amending Japan's Pacifist Constitution – Article 9 and Prime Minister Abe." 10 January 2019. https://isdp.eu/publication/amending-japans-pacifist-constitution/

Kingston, Jeff. "The Ideology of Japanese Identity – Multiethnic Japan, by John Lie." *The Japan Times*. 16 September 2001. https://www.japantimes.co.jp/culture/2001/09/16/books/book-reviews/the-ideology-of-japanese-identity/

Kitaoka, Shin'ichi. "'Japan's Identity and What It Means.' Japan's Identity: Neither the West Nor the East." *The Japan Forum on International Relations*. The Nippon Foundation. https://www.jfir.or.jp/e/special_study/seminar1/conversation.htm

Kotani, Ken. "A Reconstruction of Japanese Intelligence Issues and Prospects." Chapter 10 In *Intelligence Elsewhere: Spies and Espionage Outside the Anglosphere*, eds. Kristian Gustafson, Philip H. J. Davies, and Ken Kotani, Washington, DC: Georgetown University, 2013, 181–198.

Matsumura, Masahiro. "The Japanese State Identity as a Grand Strategic Imperative," *The Bookings Institution Center for Northeast Asian Policy Studies*, Washington, DC: The Brookings Institution, May 2008.

Oros, Andrew L. "Japan's Strategic Culture: Security Identity in a Fourth Modern Incarnation?" *Contemporary Security Policy* 35, no. 2 (2014): 229. doi: 10.1080/13523260.2014.928070.

Oros, Andrew L. "International and Domestic Challenges to Japan's Postwar Security Identity: 'Norm Constructivism' and Japan's New 'Proactive Pacifism'." *The Pacific Review* 28, no. 1 (2014): 130–160. doi:10.1080/09512748.2014.970057

Richter, Jeffrey P. "Japan's 'Reinterpretation' of Article 9: A Pyrrhic Victory for American Foreign Policy?" *Iowa Law Review* 101 (2016): 1223–1262.

Ruth, Benedict. *The Chrysanthemum and the Sword: Patterns of Japanese Culture*. First Mariner Books, 2005.

Smith, Sheila A. *Japan Rearmed. The Politics of Military Power*. Cambridge, MA: Harvard University, 2019, 1–19.

Tellis, Ashley J. "Strategic Culture, National Security, and Policymaking in the Asia-Pacific." Carnegie Endowment for International Peace. 27 October 2016. https://carnegieendowment.org/2016/10/27/strategic-culture-national-security-and-policymaking-in-asia-pacific-pub-66166

Tellis, Ashley J., Alison Szalwinski, and Michael Wills. "Strategic Asia 2017–18: Power, Ideas, and Military Strategy in the Asia Pacific," Washington, DC: National Bureau of Asian Research (2017).

21

EXAMINING THE STRATEGIC CULTURE OF SINGAPORE

Lion City, Poison Shrimp

Nathan Fedorchak

Introduction

Singapore's existence is something of a miracle. In just 70 short years it has transformed itself from a small, impoverished island at the tip of the Malay Peninsula to one of the richest and most strategically important countries in Asia. The country's rich strategic culture can help explain how it rose to such prominence and maintained its status without major military conflict. It is important to understand Singapore's strategic culture as Southeast Asia emerges as one of the most important regions in twenty-first-century geopolitics. Singapore is among the most important trade conduits in the world, sits near the strategically important Strait of Malacca, and has the most advanced and capable military in the region. Because of this, Singapore has the unenviable task of needing to balance its relations with both the United States and China as those two nations see their relationship deteriorate into a great power competition, much of which may play out on Singapore's doorstep. As Singapore considers its options, it will be necessary to understand the factors influencing its strategic decisionmaking.

This chapter will be organized as follows. First, a brief background on the history of Singapore will be provided and the sources of its strategic culture will be explored, as well as the geostrategic relevance of Singapore that warrants a closer examination of its strategic culture. Following that, using some elements of the Cultural Topography (CTops) method first introduced by Jeannie Johnson and Matthew Berrett,[1] this chapter will examine Singapore's identity and how it contributes to the nation's strategic culture. An examination of Singapore's strategic culture values as well as a discussion of the nation's norms will follow, particularly focusing on its balancing of international institutions with hard military power through a small state realist paradigm. Singapore's perceptual lens will then be examined, with a particular focus on the historical elements that have informed Singapore's perceptions of itself and its place in the world. Finally, beyond providing an overview and examination of Singapore's strategic culture, this chapter will conclude with a review of the literature on Singapore's strategic culture, assessing the strengths and weaknesses of the existing literature, and finally offering some recommendations for possible future inquiries to add to the existing academic literature.

DOI: 10.4324/9781003010302-24

A note on methodology may be appropriate at the outset. This chapter uses elements of the Cultural Topography Analytic Framework (CTAF) to assess the strategic culture of a given country.[2] Building on Johnson and Berrett's CTops method, CTAF was originally developed to help in assessing nuclear decisionmaking by potential adversaries and to develop appropriate policy responses. However, CTAF maintains great relevance for understanding security decisionmaking beyond the realm of weapons of mass destruction (WMD) as many of the same questions that are relevant in a WMD context are also relevant below that security threshold. One of the key steps in the CTAF process is conducting research to understand four primary elements of the subject in question: their identity, values, norms, and perceptual lens. While an investigation into a particular country is not necessarily limited to these four factors alone, when executed well they can provide a sufficient overview and understanding of a country such that well-informed policy decisions can be made.

Background and Sources of Singapore's Strategic Culture

Modern Singapore was founded as a colonial trading outpost of the British Empire in 1819 by Sir Stamford Raffles, a British statesman serving as the Lieutenant-Governor of Bencoolen (part of modern Indonesia). Following World War I, Singapore's strategic position became even more critical to the British Empire as demand for industrial goods from Southeast Asia like tin and rubber skyrocketed. When World War II came, Sir Winston Churchill declared that Singapore would be a fortress island, like the "Gibraltar of the East," and withstand Imperial Japanese aggression. Despite this, the Imperial Japanese Army launched an attack on Singapore in 1942, taking the city after just over one week of fighting, kicking off a brutal occupation that would last until 1945 when Singapore was once again reclaimed by the British. The memory of this capitulation and occupation remains a foundational source of Singaporean strategic culture, as will be discussed at greater length below.

After more than a century of colonial rule, Singapore became an independent country in 1959, briefly joined the Federation of Malaysia from 1963 to 1965, and became an independent nation once more when it was ejected from the Federation.[3] This split occurred as a result of intractable ethnic disputes stemming from Malaysian fears about Singapore's majority ethnic-Chinese population and accusations that Singapore was abusing and oppressing its Malay population.[4] These ethnic disputes culminated in the "Communal Riots" of 1964, when two separate race-based crimes escalated into riots that left dozens dead and resulted in the arrest or detention of thousands. These riots led directly to Singapore instituting a number of laws around racial harmony and increased emphasis in schooling on the dangers of racial division for national stability. These measures now serve as a critical source of national values and norms as reflected in the state's strategic culture.

Singapore has since grown into a multiethnic commercial republic of more than five million people and is one of the economic powerhouses of Asia, all in a landmass of just 281 square miles. Singapore has almost no natural resources of its own, making trade a critically important component of its national survival. This lack of defensive depth and reliance on the outside world for vital resources are key sources of the nation's strategic culture and place a high emphasis on international cooperation. Its citizens consistently have some of the highest incomes, highest levels of educational attainment, and highest health outcomes in the region.[5] Singapore is the largest destination for the transshipment of goods globally, operates one of the world's largest and busiest ports, and remains a financial powerhouse in the region.[6] The nation is strategically positioned astride the Strait of Malacca, one of the busiest shipping lanes in the world and a critical chokepoint for the entire region. These factors have all contributed to Singapore's meteoric rise and factored into how the nation views itself and its own security.

Singapore's Identity: The Lion City

Understanding identity is critical to understanding the strategic culture of any nation. Identity here can be understood to mean "character traits a given group assigns to itself, the reputation it pursues, the individual roles and statuses it designates to members, and the distinctions it makes between those people who are considered members of the group ('us'), and those who are not members of the group ('them')."[7] Any discussion of Singapore's strategic culture must start with a discussion of its identity.

Singapore's status as a young, multiethnic, multireligious republic has meant it does not traditionally draw its national identity from a shared racial or religious heritage. There are three major ethnic groups in Singapore: Chinese (74%), Malay (13%), and Indian (9%). During colonial rule, the British emphasized the racial divides between the three peoples as a way to maintain political power – a practice the postcolonial government was quick to reverse to reduce racial tension and enhance political stability.[8] It has also emphasized religious toleration and pluralism among the primary religions in the country (Buddhism, Christianity, Islam, Taoism, and Hinduism).[9]

Promoting multiculturalism is a key component of Singaporean security and a core value in the state's strategic culture. Singapore's 2004 National Security Strategy explicitly argued that a society built on tolerance and acceptance would be less conducive to the growth of homegrown threats such as radical Islamist terror.[10] The government instituted "Inter-Racial Confidence Circles" as part of its effort to encourage dialogues between citizens and foster a better understanding of the differences between cultures.[11] Singapore also celebrates "Racial Harmony Day" as a national holiday each July and has, since its earliest days as an independent nation, made racial harmony a key factor in Singaporean identity. Despite this state support for multiculturalism, there remains an official policy in Singapore's armed forces of keeping the largely Muslim Malay population out of sensitive high-level roles within the armed services[12] – a modern extension of the historical skepticism and initial exclusion of the Malay population's service in the Singaporean military which remains an enduring source of frustration.[13]

Singapore's government is a quasi-authoritarian mix of representative parliamentary democracy and autocratic one-party rule that strictly curtails civil liberties and individual freedoms. Lee Kuan Yew is considered to be the modern founder of Singapore, serving as Prime Minister from 1959 to 1990 and establishing the People's Action Party (PAP), the political party that has ruled the country for the entirety of its modern existence and continues to dominate elections.[14] Given its status as the dominant party for Singapore's entire modern history, the PAP and the Singaporean military have been the primary "keepers" of the nation's strategic culture – the elite individuals and institutions that shape and guide national strategic culture. Combined, these institutions have achieved remarkable civil-military fusion, leading to "full alignment with the 'values, interests, and national goals' of the civilian government."[15]

The PAP has attempted to develop symbols and common beliefs that Singaporeans can take pride in and unite behind. Some of these are the traditional symbols any nation uses to foster a sense of national pride, like a flag, a pledge of allegiance, and a national anthem.[16] Singapore takes its name from the phrase "Singa Pura," which means "Lion City," and the government chose a lion as the national symbol. This was in part due to the lion's representation of three national values – courage, excellence, and strength – and its "tenacious mien" symbolizing the nation's resolve to overcome any challenges.[17] In 1991, Singapore's Parliament adopted Five Shared Values as a focus for

> Singaporeans of all ethnic groups to embrace as the nation progressed into the 21st century. The aim of introducing the shared values was to help forge a Singaporean identity that would incorporate the various aspects of the nation's multicultural heritage with the attitudes and values that had contributed to Singapore's success.[18]

These values will be discussed in greater detail below but represent a prominent recognition by the government of the potential problem posed by a lack of a cohesive national identity and what it saw as counterproductive and growing materialism in the populace throughout the 1980s.[19]

More directly, the government's consistent rhetoric that Singapore is a nation under continuous threat with its national survival constantly at stake has done much to shape the nation's identity and inform its values.[20] Recognizing that its geographic, resource, and population constraints create a strong strategic disadvantage relative to other powers in the region, Singapore has long made its military strength a key piece of its identity.[21] This can be seen not only in internal identity-building efforts like the mandatory conscription for all citizens that acts as a social leveler and point of shared experience[22] but also in more outward-facing displays of national pride like the military parade on National Day.[23] The "Total Defence" doctrine discussed below also plays a role in ensuring that every Singaporean feels they have some duty to defend the country from both foreign and domestic threats.[24]

Finally, Singapore has developed a strong identity as a responsible actor on the world stage that can work well with others. This sense of identity stems from a strategic need to avoid antagonism as much as it does an understanding that the country may be able to leverage power most effectively as a negotiator. Examples of Singapore's role in this way abound, but perhaps the best recent example is its role as the host for a 2018 summit between then-US President Donald Trump and North Korea's Kim Jong Un.[25]

Singapore's Values: Stability through Unity

Values are defined within the CTAF framework as "deeply held beliefs about what is desirable, proper, and good that serve as broad guidelines for social life."[26] Understanding a nation's values allows an observer to understand the things on which a country will, at root, make decisions. A nation's values influence its norms of behavior and its perceptual lens, and can even influence its identity if its identity is bound up in the things it values.

As a relatively young country, Singapore has very intentionally set out to cultivate a shared set of values for its population. This deliberate cultivation of values is most prominently seen in the articulation of the "Five Shared Values" in 1991 that encapsulate the things Singaporeans hold dear and speak directly to Singapore's strategic culture. The five values are maintained to the present and include: "(1) Nation before community and society above self, (2) Family as the basic unit of society, (3) Community support and respect for the individual, (4) Consensus, not conflict, and (5) Racial and religious harmony."[27]

These "shared values" manifest themselves in much of Singapore's strategic cultural decisionmaking. For example, the value "nation before community and society above self" is best reflected in the mandatory military service[28] and the doctrine of "Collective Defence" (discussed below), in which every citizen is seen to have a critical role to play in the defense of society. The third and fourth values speak to the importance placed on societal stability and consensus building, which manifests itself in the way Singapore interacts with other nations. Singapore's pioneering role in founding the Association of Southeast Asian Nations (ASEAN),[29] its growth as a financial hub of Southeast Asia, and its willing participation in a variety of international nonproliferation regimes point to a preference for working together with other nations rather than taking more adversarial tacks or going it alone.

Lastly, it is clear from the "shared values" that Singapore values taking a long view in decisionmaking. All five shared values are designed to maintain internal regime stability over a long time horizon by minimizing racial and religious differences, keeping peace at home, and focusing instead on external threats to the nation. A concrete strategic example can be found in Singapore's understanding as early as the 1970s of the need for the United States to maintain a strategic presence in the Pacific to balance other large powers in the region and prevent the further spread of communism, ensuring the regional stability necessary for Singapore's success.[30] Singapore proactively engaged the Americans and pointed to communism spreading in Cambodia as a *raison d'être* for the Americans to continue to engage in the region, but also helped launch ASEAN as a hedge against the possibility that American engagement would not be enough or would not last.[31]

Singapore's Norms: Small State Realism

National norms are often built on values, and are here defined as "accepted, expected, or customary behaviors within a society or group."[32] This is an important area of study when examining strategic culture because it speaks to how keepers of strategic culture make decisions about national security.

Singapore's norms are in many ways defined by a small state realist tradition of power balancing and a Hobbesian perception deeply held by Lee Kuan Yew and subsequent leaders that "if there was no international law and order, and big fish eat small fish and small fish eat shrimps, [Singapore] wouldn't exist."[33] This understanding of how interstate politics works has led to the development of a Singaporean strategic culture that values cooperation and consensus building on the international stage to promote stability and security but also an understanding of hard power as the necessary element that underwrites those diplomatic efforts. Singapore's role as one of the five founding members of ASEAN reflects the emphasis it places on cooperation with the group's nine other member states to ensure regional peace and promote economic stability. Singapore maintains what has been described as "'ASEAN centrality' in regional economic, diplomatic, and security policy issues."[34] ASEAN considers itself "probably the most successful inter-governmental organization in the developing world today"[35] and its continued success is fundamental to Singapore's strategic culture in that regional stability and economic integration reduce the possibility of a conflict between the small island nation and its much larger neighbors.[36]

Singapore maintains some level of connection to both its former colonial rulers and fellow members of the former British empire, joining the Five Powers Defence Arrangements in 1971 along with the United Kingdom, Malaysia, Australia, and New Zealand. While not a formal treaty of mutual defense, the Arrangements do call for any of the five countries to consult the other four in the event of an attack to decide on the best course of action.[37] The five countries routinely hold joint training and readiness exercises, and Singapore hosted a fortieth-anniversary celebration and exercises in 2011.[38]

Singapore has also demonstrated a history of power balancing between great powers, always with regional stability in mind. It most obviously did this in building its security relationship with the United States and encouraging it to remain in the region after the Vietnam War when there was a mood of retrenchment in the United States but fear of the spread of communism abroad remained.[39] Communism was long viewed as a threat to Singapore's stability; therefore, working with the United States to make the conditions for communism's spread more difficult further contributed to the stability of the regime internally so it could focus on the external threats it saw as existential.[40] Singapore has continued to deepen its security partnership with the United States, supporting American efforts in the Global War on Terror that it saw as tremendously threatening to its security, especially as terror cells formed and committed violence in nearby Indonesia.[41] Singapore has also worked on US-led multilateral efforts, most notably the Proliferation Security Initiative (PSI) that aims to stop the use of international shipping for the proliferation of WMD.[42]

Singapore engages deeply in international institutions and the promotion of the international rule of law, particularly through the United Nations (UN). When Singapore was admitted to the UN in 1965, then Minister of Foreign Affairs S. Rajaratnam delivered a speech to the General Assembly outlining what Singapore saw as the essential value of the UN, perfectly encapsulating the international relations norms of the country. Rajaratnam noted the importance of sovereignty and collective security, saying:

> For [Singapore], the essentials of the [UN] Charter are the preservation of peace through collective security, promotion of economic development through mutual aid and the safeguarding of the inalienable right of every country to establish forms of government in accordance with the wishes of its own people.[43]

Since joining, Singapore has taken its role as a member state seriously and has contributed forces to 17 peacekeeping missions, served as a critical negotiator in the debate around the United Nations Convention on the Law of the Sea (UNCLOS) in 1982, and served as a nonpermanent member of the Security Council in the early 2000s.[44]

This eagerness to engage in international institutions does not mean that Singapore does little to provide for its own defense and relies solely on power balancing and international institutions to provide stability. In fact, the opposite is true as Singapore understands the necessity of credible hard power to undergird and give force to its diplomatic efforts. Due to its small size and lack of defensive depth, and fueled by the perception that it is constantly under threat, Singapore has developed a robust indigenous defense industry and placed high priority on naval and air power assets to try to push the battlefield as far from home as practicable. These defense policies were driven initially by the perceived threat from Malaysia and the British withdrawal from Singapore in 1960, which heightened the sense of vulnerability the country felt. As Banlaoi puts it,

> for Singapore, developing a strong domestic defence industry was a strong expression of national sovereignty. Developing a self-reliant defence industry was viewed as an important means to strengthen national political independence and as an integral part of a country's defence capability.[45]

This early investment has paid off as Singapore has the most advanced defense industrial base among ASEAN members and is proficient enough in weapons development and production that it exports a portion of what it produces and is among the top 20 largest arms industries worldwide.[46]

Despite this, there are some areas where Singapore goes abroad in search of acquiring cutting-edge military technologies. Singapore is a top 10 customer for American arms sales, importing nearly $4 billion worth of armaments between 2008 and 2018.[47] Following the purchase of F-16 fighter aircraft in prior decades, the US State Department approved the sale of 12 F-35 Joint Strike Fighters to Singapore in 2020 in a move that angered China and was indicative of both the close relationship between the United States and Singapore but also the shortcomings of Singapore's domestic defense industry and the country's recognition of the need for standoff capabilities to ensure its own defense.[48]

Singapore's Perceptual Lens: Becoming a Poison Shrimp

The perceptual lens is the "[filter] by which individual actors or members of groups determine 'facts' about others"[49] and serves as a critical tool for understanding why keepers of strategic culture will make certain decisions over others. It is often deeply rooted in history and contains some element of self-perception as well. Further, the perceptual lens incorporates the cognitive dimension of strategic culture, which Dima Adamsky describes in *The Culture of Military Innovation* as "the preferred collection of strategies to perceive, organize, and process information."[50]

Singapore's perception of the world around it is influenced by a wide array of factors, but principal among these are the state's strategic geographic position and lack of geographic depth, its relatively small population size, and its lack of abundant natural resources. This perception is largely a consequence of Singapore's historical experience with interstate violence, the seminal event being the Japanese occupation during World War II. In February 1942, the Imperial Japanese army swept down the Malay Peninsula, brushing aside British and Commonwealth resistance, and then crossed the Straits of Johor. The Japanese Army, despite being vastly outnumbered, quickly took the island[51] and kicked off a brutal occupation that lasted for nearly the duration of the conflict, which included atrocities committed toward civilians and prisoners of war alike.[52]

The experience of this attack, both in terms of how the Japanese army achieved it and how the British garrison capitulated, has resonated strongly in how Singapore perceives its security as dependent first and foremost on being self-reliant. The country celebrates "Total Defence Day" every February as a "solemn reminder ... of the price of failure to defend our country" and a reinforcing mechanism for the doctrine. The Total Defence doctrine was adapted from Switzerland and Sweden and was made official state policy in 1985.[53] Its primary purpose, then as now, is to serve as a deterrent against conventional war – if a nation goes to war with Singapore, it goes to war with all of Singapore. The "five pillars" of Total Defence represent the critical aspects of society: military defense, civil defense, economic defense, social defense, and psychological defense.[54] This full-society mobilization coupled with mandatory national service creates a kind of societal "strategic depth" that makes up in some way for the lack of geographic strategic depth. This strategy has been described as the "poison shrimp" approach, after Prime Minister Lee Kuan Yew's suggestion that Singapore become a "poisonous shrimp" in a "world where the big fish eats the small fish and the small fish eats shrimps."[55]

The relevance of Total Defence extends beyond its initial conception as a deterrent against conventional war. As Senior Minister of State for Defense Heng Chee How described in a 2019 speech,

> [t]he security environment today is complex and volatile. And it definitely goes beyond overt military conflict and outright war. The threats have expanded to terrorism, cyberattacks, disinformation and other subversive activities that lurk below the threshold of military conflict, and this is what we might now call hybrid threats.[56]

There is a growing sense within Singapore that these threats both from state and non-state actors represent a new and perhaps more immediate threat than conventional war. This stems in part from the state's experience with homegrown terrorism in the form of the regional terrorist organization Jemaah Islamiyah, a radical Islamist terror organization with members based in Singapore that plotted attacks on various American military facilities in the country following the 9/11 attacks.[57] It also stems from the steep rise in Jihadi terrorism throughout the region, particularly in Indonesia. Additionally, Singapore is uniquely vulnerable among other countries in the region to cyberattacks because of the emphasis it has put on technology through its Smart Nation program and its continued growth as a world high-tech and financial hub. Singapore added a sixth pillar to the Total Defence doctrine in 2019 – digital defense – which encourages citizens to consider how their actions online could impact others or compromise the security of the nation.[58]

The nations that Singapore has traditionally perceived as the greatest threats to its security are Malaysia and Indonesia. The memory of how the Japanese Army was able to invade Singapore and the acrimonious breaking away of Singapore from the Malaysian Federation in 1965 continues to color the relationship between Singapore and Malaysia.[59] Singapore went so far as to secretly work with the Israeli Defense Forces (IDF) to develop its military capability to hedge against a Malaysian invasion in the middle of the twentieth century.[60] IDF forces trained Singaporean forces, supplied them with tanks and airplanes, and provided advice about how to structure an effective military based on population and space constraints. This cooperation continues to the present day, as the two nations jointly developed an anti-tank missile in the early 2000s and their defense industries work closely in supplying third parties with defense systems.[61] Interstate tensions with Malaysia and Indonesia remain today. For example, Malaysia controls Singapore's water supply, which has been a frequent point of tension.[62] Former Indonesian President Bacharuddin Jusuf Habibie once derisively said he saw Singapore as a "little red dot" that was not viewed as a friend of Indonesia, a statement that continues to resonate and has become something of a rallying cry for Singaporeans.[63] This also stems from strong anti-Chinese sentiment in both Malaysia and Indonesia; Singapore's status as a majority-ethnic Chinese country historically forced it to be cautious in its dealings with these neighbors for this reason in addition to differences in size.[64] To be sure, the perception of the threat from these two countries has diminished with time – Singapore is a member of the Five Powers Defence agreement with Malaysia and works closely with both Indonesia and Malaysia as part of its membership in ASEAN. However, Singapore remains vigilant in the event relations deteriorate.

Singapore's evolving perception of China is increasingly complex, exacerbated by progressively assertive Chinese activities in the South China Sea. Singapore's economic reliance on China further complicates Singapore's position.[65] This economic intermingling has sometimes occasioned fear that China will leverage its power to intimidate and coerce Singapore, though this has not yet occurred and would doubtless be damaging for China as well.[66] Singapore's status as a predominantly ethnic-Chinese nation also colors its relations with China – Singapore has long harbored anxiety about being seen as a "third China" and was the last member of ASEAN to normalize relations with Beijing, only doing so in 1990.[67] In its role as an ASEAN member, Singapore actively shaped negotiations during the ASEAN-China Summit in 2018 that led to the conclusion on negotiations over a code of conduct for the South China Sea; a sign of Singapore's perception that China still believes in some semblance of a rules-based international order, even if that perception is changing.[68] Singapore has been subject to somewhat unique forms of harassment from China in the gray zone relative to its Southeast Asian peers, as when China delayed the shipment of armored personnel carriers bound for Singapore as they transited Hong Kong following training exercises in Taiwan or when Singapore's Prime Minister was not invited to a Belt and Road Initiative summit where China hosted nearly all other ASEAN leaders. In response to these tactics, Singapore has typically fallen back on appeals to the rules-based international order and the importance of free trade that has made both nations prosperous.[69] However, Singapore has acknowledged that there has been a shift in how China perceives the rules-based order, indicating a possible evolution in how it will interact with China in future disputes.[70] This was especially demonstrated by China's rejection of the Hague tribunal's ruling that China's claims to sections of the South China Sea, also claimed by the Philippines, were invalid.[71]

Singapore's relationship with the United States has gotten closer as China has sought to militarize the South China Sea – moves that have been perceived as the impetus for Singapore's aforementioned F-35 purchase.[72] Further, Singapore has extended American naval and air base use rights through 2035, updating the original agreement signed in 1990 and ensuring American presence in Singapore for the foreseeable future.[73] All of this suggests that Singapore perceives the need to keep the United States in the region to balance the China threat, just as it did during the Cold War, and that its perception of which country provides it the best security guarantees has not changed.

Assessing the Literature on Singapore's Strategic Culture

There is a dearth of literature on Singapore's strategic culture that takes a holistic approach and attempts to analyze the state's strategic culture as a whole. Mily Ming-Tzu Kao's study "Strategic Culture of Small States in ASEAN"[74] uses a qualitative and quantitative framework to try to ascertain which factors weigh most heavily on the strategic culture of a few case study states including Singapore, which differs from this chapter's use of the CTAF framework. Beyond that, most of the literature examined and incorporated in this chapter focuses on individual aspects of Singapore's culture or provides insights into one element or another of the nation's strategic culture.

The body of literature on Singapore's strategic culture is fairly small and of relatively recent vintage, with much of the most useful work being published beginning in the early 2000s. The body of literature that has grown up over this time has primarily focused on Singapore's identity and its perceptual lens. Ortmann's work examining Singapore's invention of its national identity[75] is perhaps the best of these as it focuses explicitly on the history of and reasons for the Singaporean government's development of a national identity and attempts, albeit in a somewhat limited way, to project forward how this national identity might evolve over time. Tan and Lew's examination of the role that the armed forces play in shaping national identity[76] is more recent and quite good, though it is somewhat limited by its examination of only the armed forces. Vasu and Ramakrishna's specific examination of identity in the context of counterterrorism efforts[77] is instructive, particularly when placed in the context of this other scholarship. If there is a drawback to the existing body of work on Singaporean identity, it is that it is somewhat dated and may fail to account for changing social dynamics in the country which could have a long-term impact on some of the core tenets of national identity, including what appears to be slowly eroding support for the PAP. Jun Yan Chang's recent work on Singapore's securitization of its vulnerability[78] is an excellent and concise examination of how the country perceives the world around it. Peng Lam Er has written two commendable papers[79] for the National Institute for Defense Studies Joint Research series, including one that speaks to how Singapore is attempting to navigate the US-China conflict amid a myriad of other security challenges. On the whole, the literature on Singapore's strategic culture mostly exists in silos focused on specific aspects of the strategic culture and lacks cumulativeness.

Due to the relative immaturity of the literature on Singapore's strategic culture, there are a number of potentially rich opportunities for future research. While individual studies exist examining Singapore's relationship with the United States and China individually, and a limited number of studies examine how it balances that relationship from a security standpoint, a thorough strategic cultural analysis of how Singapore may balance both relationships in the future is lacking. Such a study could incorporate historical analysis of how other small states have balanced relations with great powers in the past and how the strategic culture of those states compares with Singapore's, which could help illuminate what options Singapore may have and may pursue. Another avenue of study that could prove useful would be to build on the work of Lily Zubaidah Rahim's examination of Singapore and Malaysia and provide a strategic cultural comparison of Singapore, Malaysia, and Indonesia in light of China's rise, given the history of animosity between these countries and their respective economic relationships and strategic concerns with China.

Summary and Conclusion

Singapore's strategic culture reflects both its relative youth among the nations of the world and its small geographic size. That Singapore prizes diplomacy and cooperation is no surprise given its size, and its strategic culture will likely continue to be defined both by the sense of constant threat of encroachment by outside forces and fear of instability from within. It will have to increasingly find ways to balance its relationships with the United States and China, particularly as the relationship between the two superpowers becomes more acrimonious and threatens the regional stability that Singapore depends upon. There is some indication that Singapore's perception of China is changing and that it is recognizing the increasingly aggressive nature of the regime in Beijing – but there is little that Singapore is able to do about this given the deep economic interdependence between the two countries. It remains to be seen if the course of that contest of powers could cause Singapore's strategic culture to shift as the conditions that have allowed it to peacefully prosper change. That question is beyond the scope of this chapter, though it seems unlikely that the fundamentals of Singapore's strategic culture will change much given that the geopolitical position in the region is worsening rapidly. For now, the small island nation will likely continue to maintain a strategic culture that aims to be a friend to all and a strong proponent of international organizations and consensus building when possible, but a "poison shrimp" committed to "total defence" if it must.

Notes

1 Jeannie L. Johnson and Matthew T. Berrett, "Cultural Topography: A New Research Tool for Intelligence Analysis," *Studies in Intelligence* 55, no. 2 (2011): 1–22.

2 Jeannie L. Johnson and Marilyn J. Maines, "The Cultural Topography Analytic Framework," in Jeannie L. Johnson, Kerry M. Kartchner, and Marilyn J. Maines (eds.), *Crossing Nuclear Thresholds: Leveraging Sociocultural Insights into Nuclear Decisionmaking (*Palgrave MacMillan, 2018), 19, 29–60.

3 "The Singapore Exception," July 2015, *The Economist.* https://www.economist.com/special-report/2015/07/16/the-singapore-exception

4 Jamie Han, "Communal Riots of 1964," *Singapore InfoPedia.* Accessed May 2021. https://eresources.nlb.gov.sg/infopedia/articles/SIP_45_2005-01-06.html?s=1964+racial+riots

5 Ibid.

6 Ben Dovlen and Emma Chanlett-Avery, "U.S. Singapore Relations," Congressional Research Service *In Focus* last updated 30 July 2020. https://www.everycrsreport.com/files/2020-07-30_IF10228_daa171d5ee2b9f9c3ae4b632a29c6e3bee8b59ff.pdf.

7 Jeannie L. Johnson and Marilyn J. Maines, "The Cultural Topography Analytic Framework," in Jeannie L. Johnson, Kerry M. Kartchner, and Marilyn J. Maines (eds.), *Crossing Nuclear Thresholds: Leveraging Sociocultural Insights into Nuclear Decisionmaking (*Palgrave MacMillan, 2018), 19, 29–60. *CTAF Training Slides,* October 2017.

8 "Singaporean Culture – Core Concepts," *Cultural Atlas.* Accessed May 2021. https://culturalatlas.sbs.com.au/singaporean-culture/singaporean-culture-core-concepts#singaporean-culture-core-concepts.

9 Ibid.

10 Norman Vasu and Kumar Ramakrishna, "Countering Terrorism: Multiculturalism in Singapore," *Connections* 5, no. 4 (2006): 143–156. Accessed May 2021. http://www.jstor.org/stable/26323271.

11 Ibid.

12 Sean P. Walsh, "The Roar of the Lion City: Ethnicity, Gender, and Culture in the Singapore Armed Forces," *Armed Forces & Society* 33, no. 2 (January 2007): 265–285.

13 Alon Peled, "A Question of Loyalty: Ethnic Minorities, Military Service, and Resistance," 1993. Program on Nonviolent Sanctions & Cultural Survival. https://web.archive.org/web/20060906153229/http://www.wcfia.harvard.edu/ponsacs/seminars/Synopses/s93peled.htm

14 Dovlen and Chanlett-Avery. "U.S. Singapore Relations."

15 Issac Neo. "The Management of Threats in Singapore: Civil-Military Integration," *Singapore Policy Journal* 3 January 2020. https://spj.hkspublications.org/2020/01/03/the-management-of-threats-in-singapore-civil-military-integration/.

16 Stephan Ortmann, "Singapore: The Politics of Inventing National Identity," *Journal of Current Southeast Asian Affairs* 28, no. 4 (December 2009): 23–46. https://doi.org/10.1177/18681034 0902800402.
17 National Heritage Board. "The Lion Head Symbol." August 2021. https://www.nhb.gov.sg/what-we-do/our-work/community-engagement/education/resources/national-symbols/the-lion-head-symbol
18 "Shared Values Are Adopted – Singapore History". Accessed 15 February 2021. https://eresources.nlb.gov.sg/history/events/62f98f76-d54d-415d-93a1-4561c776ab97#:~:text=The%20five%20Shared%20Values%20that,5)%20Racial%20and%20religious%20harmony.
19 Ortmann, "Singapore: The Politics of Inventing National Identity," 23–46.
20 Rahim, Lily Zubaidah. *Singapore in the Malay World: Building and Breaching Regional Bridges* (Routledge, 2010), 78.
21 Fred Wel-Shi Tan and Psalm B.C. Lew, "The Role of the Singapore Armed Forces in Forging National Values, Image, and Identity," *Military Review* 97, no. 2 (March–April 2017): 9, https://www.armyupress.army.mil/Portals/7/military-review/Archives/English/MilitaryReview_2017430_art006.pdf
22 Ibid.
23 Ortmann, "Singapore: The Politics of Inventing National Identity," 23–46.
24 "Fact Sheet: Evolution and History of Total Defence over the Past 35 Years," 2019. https://www.mindef.gov.sg/web/portal/mindef/news-and-events/latest-releases/article-detail/2019/February/15feb19_fs2/!ut/p/z1/jZBLC8IwEIR_kewmto052qo10hofrcZcJILVglapD9BfbxDBQ7F2bwvfzO4MaFCgC3PPd-aanwpzsPtKe2sme8EQHTqWbkKwu0iSyPXHImUeLN-AnDohcZBGMp65FuC8k-CEYshAN9Hjj-liM30NoOvtR_8O2AZoGQfxDvTZXPetvMhOoCgSDmqw3ZQ3Uz5AETfbbghfZxdqX9K1pjP2Ab6tLUJmgYnwRnxOsN-uANVa_wU7H9NUPSOfi1y8AJn0npE!/dz/d5/L2dBISEvZ0FBIS9nQSEh/.
25 Yen Nee Lee, "White House Explains Why It Chose Singapore to Host Summit with North Korea," June 2018, *CNBC*. https://www.cnbc.com/2018/06/08/why-trump-and-kim-picked-singapore-for-meeting.html.
26 Johnson and Maines, "Cultural Topography Analytic Framework Methodology," 21.
27 "Shared Values Are Adopted – Singapore History."
28 Andrew T.H. Tan, "Singapore's Survival and Its China Challenge," *Security Challenges* 13, no. 2 (2017): 11–31. https://www.jstor.org/stable/26457716.
29 Tan, Eugene, "A Small State Perspective on the Evolving Nature of Cyber Conflict: Lessons from Singapore," *Prism.*
30 Ibid.
31 Ibid.
32 Johnson and Maines, "Cultural Topography Analytic Framework Methodology," 20.
33 Tan, "Singapore's Survival and Its China Challenge."
34 Cortez A. Cooper, II and Michael S. Chase, Regional Responses to U.S.-China Competition in the Indo-Pacific: Singapore. Santa Monica, CA: RAND Corporation, 2020. https://www.rand.org/pubs/research_reports/RR4412z5.html.
35 "History: The Founding of ASEAN." *Association of Southeast Asian Nations*. https://asean.org/asean/about-asean/history/.
36 Eugene Tan, "A Small State Perspective on the Evolving Nature of Cyber Conflict: Lessons from Singapore," *Prism*. 8, no. 3. https://ndupress.ndu.edu/Media/News/News-Article-View/Article/2054199/a-small-state-perspective-on-the-evolving-nature-of-cyber-conflict-lessons-from/.
37 Khoo How San, "The Five Power Defence Arrangements: If It Ain't Broke ..." *Pointer* 26, No. 4 (October–December 2000), https://web.archive.org/web/20040702143722/http://www.mindef.gov.sg/safti/pointer/back/journals/2000/Vol26_4/7.htm.
38 "Joint Press Statement on the 8th FPDA Defence Ministers' Meeting" November 2011 *Ministry of Defence*.
39 See Seng Tan, "Facilitating the US Rebalance: Challenges and Prospects for Singapore as America's Security Partner," *Security Challenges* 12, no. 3 (2016): 20–33. Accessed 20 March 2021. https://www.jstor.org/stable/26465596.
40 Bilveer Singh, "The Communist Threat in Perspective," *The Straits Times,* 5 January 2015, https://www.straitstimes.com/opinion/the-communist-threat-in-perspective.

41 Vasu and Ramakrishna, "Countering Terrorism: Multiculturalism in Singapore," 143–156.
42 "Disarmament." Ministry of Foreign Affairs. Accessed March 2021. https://www.mfa.gov.sg/SINGAPORES-FOREIGN-POLICY/International-Issues/Disarmament#:~:text=Singapore%20supports%20disarmament%20and%20the,strengthening%20international%20peace%20and%20security.&text=Effective%20non%2Dproliferation%20efforts%20require%20global%20and%20regional%20cooperation.
43 Lim Tin Seng, "Singapore Joins the United Nations," *Singapore Infopedia: A Singapore Government Agency Website*. Accessed April 2021. https://eresources.nlb.gov.sg/infopedia/articles/SIP_2019-04-12_104028.html#:~:text=This%20was%20an%20important%20milestone,a%20sovereign%20and%20independent%20state.&text=Since%20joining%20the%20UN%2C%20Singapore,participating%20in%20UN%20peacekeeping%20missions.
44 Ibid.
45 Geoffrey Till, Emrys Chew, and Joshua Ho, eds., "Globalization's Impact on Defence Industry in Southeast Asia," in *Globalization and Defence in the Asia-Pacific* (Taylor & Francis, 2009), Ch. 11, 201. https://books.google.com/books?id=Iy3abta6oGoC&pg=PA1&dq=Singapore+:+comprehensive+security+–+total+defence+/+Richard+A.+Deck&source=gbs_toc_r&cad=3#v=snippet&q=%22strengthen%20national%20political%22&f=false
46 Ibid.
47 Thomas C. Frohlich, "Saudi Arabia Buys the Most Weapons from the US Government. See What Other Countries Top List," 26 March 2019, *USA Today*. https://www.usatoday.com/story/money/2019/03/26/us-arms-sales-these-countries-buy-most-weapons-government/39208809/
48 Reuters Staff, "U.S. State Dept. Approves Sale of 12 F-35 Jets to Singapore," *Reuters*.
49 Johnson and Maines, "Cultural Topography Analytic Framework Methodology," 22.
50 Dima Adamsky, *The Culture of Military Innovation* (Stanford: Stanford University Press, 2010), 18.
51 Terry Stewart, "The Fall of Singapore," *Historic UK*. Accessed March 2021. https://www.historic-uk.com/HistoryofBritain/The-Fall-of-Singapore/
52 "Remembering Civilian Victims of Japanese Occupation," 16 February 2020, *The Straits Times*. https://www.straitstimes.com/singapore/remembering-civilian-victims-of-japanese-occupation
53 "Fact Sheet: Evolution and History of Total Defence over the Past 35 Years".
54 Ibid.
55 Stephen Kuper, "Taking a Closer Look at Singapore's 'Poison Shrimp' Defence Doctrine," *Defenceconnect.Com.Au*. 11 February 2020. https://www.defenceconnect.com.au/key-enablers/5555-taking-a-closer-look-at-singapore-s-poison-shrimp-defence-doctrine
56 Heng Chee How, "Speech by Senior Minister of State for Defence, Mr Heng Chee How, at the Total Defence Awards 2019," 17 October 2019. https://www.mindef.gov.sg/web/portal/mindef/news-and-events/latest-releases/article-detail/2019/October/17oct19_speech/!ut/p/z1/jZBBC4JAEIV_Uczspm57zApTtC1T2_YiJmZCqZR06Ne3RNAhMuc28L038x4okKDq7F6VWVc1dXbW-15ZKRPz2RINuhJmRHCaRJFv2is3ZhbsXoDYGA4xkPoiCE0NcD6JcE3RYaCG6PHHTHGYv-gdQ_fbevwO6AXoNZkEJqs2606iqjw1IioSDFHnXHIorSMKavCM8vbVFkZ_0U6rXNmRv4NNb4jANrF3L41uCi_EX8F3sv2jtJY7lw7e5W7lPYCVWww!!/dz/d5/L2dBISEvZ0FBIS9nQSEh/
57 Vasu and Ramakrishna, "Countering Terrorism: Multiculturalism in Singapore."
58 Eugene Tan, "A Small State Perspective on the Evolving Nature of Cyber Conflict: Lessons from Singapore."
59 Rahim, *Singapore in the Malay World: Building and Breaching Regional Bridges*.
60 Amnon Brazilai, "A Deep, Dark, Secret Love Affair: A Team of IDF Officers, Known as the 'Mexicans,' Helped Singapore Establish an Army. It Was the Start of a Very Special Relationship," *Haaretz.Com*. 16 July 2004. https://www.haaretz.com/1.4758973.
61 Ibid.
62 Tashny Sukumaran, "Singapore-Malaysia Relations Threaten to Boil over as Mahathir Makes Splash About Water Prices," 4 March 2019, *South China Morning Post*. https://www.scmp.com/week-asia/geopolitics/article/2188586/singapore-malaysia-relations-threaten-boil-over-mahathir-makes

63 "A Little Red Dot in a Sea of Green," 18 July 2015, *The Economist*. https://www.economist.com/special-report/2015/07/16/a-little-red-dot-in-a-sea-of-green.
64 Tan, "Singapore's Survival and Its China Challenge."
65 Ibid.
66 Lyle J. Morris, Michael J. Mazarr, Jeffrey W. Hornung, Stephanie Pezard, Anika Binnendijk, and Marta Kepe, *Gaining Competitive Advantage in the Gray Zone: Response Options for Coercive Aggression Below the Threshold of Major War*. Santa Monica, CA: RAND Corporation, 2019. https://www.rand.org/pubs/research_reports/RR2942.html.
67 Tan, "Singapore's Survival and Its China Challenge."
68 Tan Dawn Wei, "Asean, China Agree on Early Completion of Sea Code," 15 November 2018, *The Straits Times*. https://www.straitstimes.com/singapore/asean-china-agree-on-early-completion-of-sea-code.
69 Morris, Mazarr, et al., 120.
70 Morris, Mazarr, Hornung, Pezard, Binnendijk, and Kepe, *Gaining Competitive Advantage in the Gray Zone*.
71 Tom Phillips, Oliver Holmes, and Owen Bowcott, "Beijing Rejects Tribunal's Ruling in South China Sea Case," 12 July 2016, *The Guardian*. https://www.theguardian.com/world/2016/jul/12/philippines-wins-south-china-sea-case-against-china
72 Joseph Trevithick, "Singapore Moves Closer to Joining What China Calls the 'U.S. F-35 Friends Circle'," 18 January 2019. *The Drive*. https://www.thedrive.com/the-war-zone/26049/singapore-moves-closer-to-joining-what-china-calls-the-u-s-f-35-friends-circle
73 Charissa Young, "Key Pact on US Use of Air, Naval Bases in Singapore Renewed till 2035," 25 September 2015, *The Straits Times*. https://www.straitstimes.com/singapore/key-pact-on-us-use-of-air-naval-bases-in-spore-renewed-till-2035.
74 Mily Ming-Tzu Kao, "Strategic Culture of Small States: the Case of ASEAN" [Dissertation], 2011. https://repository.asu.edu/attachments/56880/content/Kao_asu_0010E_10849.pdf
75 Ortmann, "Singapore: The Politics of Inventing National Identity."
76 Tan and Lew, "The Role of the Singapore Armed Forces in Forging National Values, Image, and Identity."
77 Vasu and Ramakrishna, "Countering Terrorism: Multiculturalism in Singapore."
78 Jun Yan Chang, "Conscripting the Audience: Singapore's Successful Securitisation of Vulnerability," in *National Service in Singapore*, 1st ed. (World Scientific, 2019), 83–103.
79 Peng Lam Er, "Singapore Security Outlook 2017: Between a Rock and a Hard Place," National Institute for Defense Studies Joint Research Series No. 16. *National Institute for Defense Studies*, 2018, http://www.nids.mod.go.jp/english/publication/joint_research/series16/pdf/chapter06.pdf; Peng Lam Er, "Singapore's Security Outlook: The Immutability of History, Geography, and Demography?" National Institute for Defense Studies Joint Research Series No. 5. *National Institute for Defense Studies*, 2010, http://www.nids.mod.go.jp/english/publication/joint_research/series5/pdf/5-4.pdf.

Selected Bibliography

Chang, Jun Yan. "Conscripting the Audience: Singapore's Successful Securitisation of Vulnerability." *In National Service in Singapore*, 1st ed. 83–103. World Scientific, 2019.

Cooper, Cortez A. III, and Michael S. Chase. *Regional Responses to U.S.-China Competition in the Indo-Pacific: Singapore*. Santa Monica, CA: RAND Corporation, 2020.

Er, Peng Lam. "Singapore Security Outlook 2017: Between a Rock and a Hard Place." *National Institute for Defense Studies Joint Research Series* No. 16. National Institute for Defense Studies. 2018.

Er, Peng Lam. "Singapore's Security Outlook: The Immutability of History, Geography, and Demography?" *National Institute for Defense Studies Joint Research Series* No. 5. National Institute for Defense Studies. 2010. http://www.nids.mod.go.jp/english/publication/joint_research/series5/pdf/5-4.pdf

Harris, Emma J. "Report: The Security Evolution of Singapore: An Examination of Security and Counter-Terrorism Legislation." *International Institute for Counter-Terrorism* (ICT), 2016. Accessed April 2021. http://www.jstor.org/stable/resrep09454.

Hofstede Insights, "Country Comparison – Singapore" https://www.hofstede-insights.com/country-comparison/singapore/
How, Heng Chee. "Speech by Senior Minister of State for Defence, Mr Heng Chee How, at the Total Defence Awards 2019." MinDef Singapore, 2019.
Kao, Mily Ming-Tzu. "Strategic Culture of Small States: the Case of ASEAN." 2011, 69–109. https://repository.asu.edu/attachments/56880/content/Kao_asu_0010E_10849.pdf
Kuper, Stephen. "Taking a Closer Look at Singapore's 'Poison Shrimp' Defence Doctrine." 2020, Defenceconnect.Com.Au. https://www.defenceconnect.com.au/key-enablers/5555-taking-a-closer-look-at-singapore-s-poison-shrimp-defence-doctrine.
Leifer, Michael. *Singapore's Foreign Policy: Coping with Vulnerability*. London: Routledge, 2000.
Neo, Issac. "The Management of Threats in Singapore: Civil-Military Integration." January 2020. *Singapore Policy Journal*. https://spj.hkspublications.org/2020/01/03/the-management-of-threats-in-singapore-civil-military-integration/
Ortmann, Stephan. "Singapore: The Politics of Inventing National Identity." *Journal of Current Southeast Asian Affairs* 28, no. 4 (December 2009): 23–46. 10.1177/186810340902800402.
Peled, Alon. "A Question of Loyalty: Ethnic Minorities, Military Service, and Resistance." 1993. *Program on Nonviolent Sanctions & Cultural Survival*. https://web.archive.org/web/20060906153229/http://www.wcfia.harvard.edu/ponsacs/seminars/Synopses/s93peled.htm
Rahim, Lily Zubaidah. *Singapore in the Malay World: Building and Breaching Regional Bridges*. London: Routledge, 2009.
Raska, Michael. "Singapore's Approach to Military Modernization." Defense.info. 2019. https://defense.info/re-shaping-defense-security/2019/04/singapores-approach-to-military-modernization/
"Remembering Civilian Victims of Japanese Occupation." 2020. *The Straits Times*. https://www.straitstimes.com/singapore/remembering-civilian-victims-of-japanese-occupation
Seng, Lim Tin. "Singapore Joins the United Nations." *Singapore Infopedia: A Singapore Government Agency Website*. https://eresources.nlb.gov.sg/infopedia/articles/SIP_2019-04-12_104028.html#:~:text=This%20was%20an%20important%20milestone,a%20sovereign%20and%20independent%20state.&text=Since%20joining%20the%20UN%2C%20Singapore,participating%20in%20UN%20peacekeeping%20missions.
"Shared Values Are Adopted – Singapore History." Eresources.Nlb.Gov.Sg., 2021.
"Singapore – Current National Security Situation." Undated. Federation of American Scientists. https://fas.org/irp/nic/battilega/singapore.pdf.
"Singaporean Culture – Core Concepts." 2021. *Cultural Atlas*. https://culturalatlas.sbs.com.au/singaporean-culture/singaporean-culture-core-concepts#singaporean-culture-core-concepts
"Singapore's Foreign Policy: Disarmament." 2021. https://www.mfa.gov.sg/SINGAPORES-FOREIGN-POLICY/International-Issues/Disarmament#:~:text=Singapore%20supports%20disarmament%20and%20the,strengthening%20international%20peace%20and%20security
Singh, Bilveer. "The Communist Threat in Perspective." *The Straits Times*, January 2015. https://www.straitstimes.com/opinion/the-communist-threat-in-perspective
Special Report. "The Singapore Exception." 16 July 2015, *The Economist*. https://www.economist.com/special-report/2015/07/16/the-singapore-exception
Stewart, Terry. "The Fall of Singapore." *Historic UK*. https://www.historic-uk.com/HistoryofBritain/The-Fall-of-Singapore/
Tan, Andrew T. H. "Singapore's Survival and Its China Challenge." *Security Challenges* 13, no. 2 (2017): 11–31. https://www.jstor.org/stable/26457716.
Tan, Eugene. "A Small State Perspective on the Evolving Nature of Cyber Conflict: Lessons from Singapore." *Prism*. 8, no. 3. /a-small-state-perspective-on-the-evolving-nature-of-cyber-conflict-lessons-from/
Tan, Fred Wel-Shi and Psalm B. C. Lew "The Role of the Singapore Armed Forces in Forging National Values, Image, and Identity." *Military Review*. March–April 2017. https://www.armyupress.army.mil/Portals/7/military-review/Archives/English/MilitaryReview_2017430_art006.pdf
Tan, See Seng. "Facilitating the US Rebalance: Challenges and Prospects for Singapore as America's Security Partner." *Security Challenges* 12, no. 3 (2016): 20–33. https://www.jstor.org/stable/26465596.

Vasu, Norman and Kumar Ramakrishna. "Countering Terrorism: Multiculturalism in Singapore." *Connections* 5, no. 4 (2006): 143–156. http://www.jstor.org/stable/26323271.
Walsh, Sean P. "The Roar of the Lion City: Ethnicity, Gender, and Culture in the Singapore Armed Forces." *Armed Forces & Society* 33, no. 2 (January 2007): 265–285. 10.1177/0095327×06291854.
Yaacob, Abdul. "Singapore's Forward Defence Strategy Goes Naval." *East Asia Forum*. 2019, https://www.eastasiaforum.org/2019/12/21/singapores-forward-defence-strategy-goes-naval/.
Yew, Lee Kuan. "Why Singapore Is What It Now Is." *Estudios Internacionales* 40, no. 159 (2008): 171–176. http://www.jstor.org/stable/41391966.

22

INDIAN STRATEGIC CULTURE

A Review Essay

Muhammad Shoaib Pervez[1]

Introduction

Strategic culture is an essentially contested concept and has presented a conceptual conundrum due to an intense debate between its advocates and detractors regarding its potential usage in international politics. Without delving into this conceptual maze, as it is beyond the scope of this review essay on Indian strategic culture, this chapter will focus on just one working definition given by Ashley Tellis, who paraphrased Jeannie L. Johnson[2]:

> Strategic culture refers to those inherited conceptions and shared beliefs that shape a nation's collective identity, the values that color how a country evaluates its interests, and the norms that influence a state's understanding of the means by which it can best realize its destiny in a competitive international system.[3]

From this definition, this chapter takes the approach that understanding the role of strategic culture in a state's security discourse is a two-step process. The first step involves the exploration and identification of certain influential sociocultural norms of society or those norms which carry popular support and are instrumentally used by state elites in their discursive practices to gain public legitimacy. The second step involves studying the security discourse of a state formulated by the strategic behavior of its elites at critical junctions of time. This is a very simplistic explanation of strategic culture, considering the fact that there are four distinct generations or "waves" of strategic culture scholarship, including national identity conceptions (NICs) during the Cold War; critical explorations at the societal level and questioning the hegemony of state narratives; a more contemporary debate on causality versus contextuality of strategic culture between Alastair Iain Johnston and Colin Gray[4]; and fourth-generation scholarship focused in part on providing more structured methodological frameworks for strategic culture research. This chapter will not delve into all these debates as they fall beyond the scope of this discussion; however, it is sufficient to mention here that strategic culture, as a concept, enjoys a great scholarly lineage.

With regard to understanding Indian strategic culture, in addition to ongoing conceptual debates, there is another problem of identifying influential sociocultural norms in a

 DOI: 10.4324/9781003010302-25

huge country like India which is also a multiethnic, multilingual, and multireligious polity. Indian society is a very complex one divided into castes, dogmas, and sociocultural norms. It is not easy to dig out influential norms which prescribe or proscribe behavior of a state with a given identity.[5] Indian strategic culture as espoused by its elites is woven around an intermixing of complex mythologies, dogmas, norms, and an inherent pride in its civilizational history. It is very difficult to construct a monolithic conception of Indian strategic culture and what it entails. In plain words, India does not have any one repository of strategic culture which functions as the fountain spring of all of its strategies and security practices. The diversity of Indian culture is the essence of its elites' strategic thinking, and hence this review essay will be appreciative of the myriad forms of Indian strategic culture instead of aiming to identify any single form which does not exist at all.

There are certain unique aspects of the Indian political system which must be taken into consideration as they all have a bearing upon the state's strategic culture. First, India's massive billion-plus population is projected to overtake the current number one most populous country, China, in mid-2023. Second, it is one of those very few postcolonial states who have given ascendency to the political elites and have marginalized the military elites from all important matters of security. Almost all of the other postcolonial states who achieved independence in the second half of the twentieth century have seen a predominant rise of their military in all affairs of the state and specifically in the security domain. With little or no direct influence of a military-bureaucratic nexus in Indian strategic decision-making, the military and the bureaucratic elites are all subservient to the political elites. Third, the political elites in India are the real entrepreneurs of norms and they are the ones who have developed key facets of Indian strategic culture.

Fourth, the understanding of Indian strategic culture requires an appreciation of distinct political trends of Indian history. For example, from its independence in 1947 until 1964, the founding father of India, Prime Minister Jawaharlal Nehru, was at the helm of affairs. Nehru belonged to the Congress Party of India, which vehemently followed and advocated the "Ahimsa" (nonviolence) norms of Mahatma Gandhi, the spiritual father of India. This norm encapsulates nonviolence toward other powers and brotherhood toward everyone. This approach was in the formative phase of the nascent Indian state which at that time had championed itself as the beacon of hope for all postcolonial states struggling to sustain themselves after throwing away the colonial yoke. Hence India adopted a secular form of democracy in order to allay the anxieties of its minorities who were all fearful of Hindu domination. From 1964 until 1998, this tryst with a secularized form of democracy continued with a small interregnum (a brief period of rule of Hindu fundamentalist parties like Bharatiya Jana Sangh from 1975 to 1977). It was only in 1998 that the phenomenal rise of "Hindutva" (Hindu Raj or the rule of Hindu religion-based political parties in India in the late 1990s) led to the practical demise of secularism in India, and the dawn of the twenty-first century witnessed the progression of the Bharatiya Janata Party (BJP), a fundamental Hindu party aiming for the revival of Hindutva norms. The strategic culture of assertive Hindutva ideology led to distinct security practices of overt nuclearization, war rhetoric for the outside world, and coerced assimilative integrationist policies at home.

There is a need to understand these shifts in the political behavior of the Indian state before explaining the nuances of its strategic culture. It is remarkable to see the progress of Indian strategic culture from Ahimsa norms (nonviolence) to modern-day Hindutva ones (assertive Hindu identity). Yet as Tellis notes, "the dominance of civilians and general exclusion of the military from decision-making has hindered India from becoming a power that regularly uses the military as an instrument of policy."[6]

This chapter will not only recap the key literature on Indian strategic culture but will also explain the strengths and weaknesses of these scholarly works and offer recommendations for future research on Indian strategic culture. As a body of work, there are three distinct domains in the Indian strategic culture literature. One is the work written by policy experts working in think tanks for their respective governments; the second is the work done by pure academicians; and the third is done by scholars writing in the nuclear domain, especially after India went nuclear again in 1998 after a lull of 25 years (following the first "Peaceful Nuclear Explosion" [PNE] in 1974). Accordingly, this review chapter is divided into four distinct sections. The first section will synthesize various themes on Indian strategic culture as explained by think tank experts and tailor-made for policy formulation; the second section will develop a synthesized version of academic experts' works; and the third section will explain the key works on Indian nuclear strategy. The final section will conclude by elaborating on future trends or areas of research on Indian strategic culture. But first, this chapter will establish the philosophical embeddedness of Indian strategic culture.

The genealogy of Indian strategic culture's philosophical foundations is tied to the Sanskrit epics of Ramayana and Mahabharata. But the problem here is "there is not a single Ramayana or a single Mahabharata in India. Each version presents its own traditions and cultural values endemic to its own smaller circles."[7] This leads us to look toward Kautilya who wrote Arthashastra around the third century BC when Alexander the Great invaded India and defeated Raja Porus (327 BC). His advice to local Indian kings was based on a complex terminology in the local vernacular of sociocultural norms, which include among others the *mandala* system (a system of external policy relations through identification of friends and foes in the geostrategic area of a state) and the *matsya-nyaya system* (compromises and coercions in foreign decisionmaking) in order to govern the mosaic-like Indian society riddled into castes and tribes. Kautilya was a realist and his prescriptions to the kings and princes carry a Machiavellian taste. These prescriptions are akin to a realist account of state behavior but at the same time also offer tips for rapprochements as and when the ulterior motive of the king is served best through compromises. This "Kautilyan brand of realism seems to pervade the Indian policy of nonalignment which has been the cornerstone of Indian foreign policy and security policy since India's independence."[8] Zaman further explains the influence of Kautilyan thought through Indian adoption of nonalignment policy:

> Nonalignment was a strategy to stay away from the bloc conflicts, not global politics in its entirety. It was a strategy to use diplomatic or, when the situation permitted, military means to gain influence despite material weakness. Simply put, nonalignment was a low-risk strategy to gain influence on the cheap.[9]

Zaman is of the opinion that Kautilyan thought has a lingering impact on Indian elites' strategic thinking, as he notes:

> [T]he Kautilyan admonishment to be wary of depending on one group of "allies" also can be used to explain the fact that India has not dropped its allies in the developing world. Even while rubbing shoulders with the rich and powerful economies of the

North, India has been active in organizing groups of developing economies on trade issues and at times has been willing to deadlock negotiations rather than concede.[10]

Indian Strategic Culture Explained by Policy Experts

The seminal research on Indian strategic culture was done by George K. Tanham in a RAND research report, which explained that four principal factors worked in the backdrop of Indian strategies and security practices: "Indian geography; the discovery of Indian history by Indian elites over the past 150 years; Indian cultural and social structures and belief system; and the British rule."[11] Tanham explained Indian defensive posture as being due to the uniqueness of the Hindu belief system based on reincarnation of the human body into seven different forms, positing that pacifist Indian strategic culture is inherently linked to this belief system. Tanham's methodology is interpretive and decouples Indian strategic culture from the methodological rigors of rational approaches, which are the prime basis of analysis of strategies in Western governments. He stated:

> The complex [Indian] view of life makes the future appear uncertain and less subject to human manipulation than it does to a Westerner. Rational analysis, so vital to Western societies, has less influence in Indian society, as so many other factors play important or dominant roles. The acceptance of life as a mystery and the inability to manipulate events impedes preparation for the future in all areas of life, including the strategic.[12]

Hence on a methodological scale, Tanham's approach of interpretivism explains Indian strategy as the product of a complex Hindu way of life where there is an "absence of strategic thinking, defensive posture and a lack of an expansionist military tradition."[13] The prescriptions of Tanham carry a contemporary taste, which is obvious considering that his is a policy document written for the US State Department as a practitioner manual to understand Indian strategic culture for better policymaking. His study gives a pessimistic account of Indian "sub-optimal" performance.[14] Burgess offers an explanation for this pessimism in the Tanham report:

> Among the cultural and social structures and belief systems that Tanham cites for a lack of strategic vision is the dominance of Hinduism and belief in re-incarnation, which indefinitely lengthens the shadow of the future and reduces the sense of urgency among Indian leaders until crises [are] upon them. Also, the relative strength of Indian imperial states over the centuries in the South Asia subcontinent and the ability to assimilate invaders provide India's leaders with confidence that they will ultimately prevail.[15]

If India's pacifist attitudes are taken as an extension of Hinduism then the question here arises of how to explain the assertiveness and decisiveness in Indian strategy from 1998 onward once the Hindu fundamental party BJP came to power. The way the BJP used Hindutva norms as a deliberate strategy to reincarnate the strategic culture of India through its security practices brings out a lacuna in Tanham's analysis. There is no perpetual pacifism or assertiveness in Hinduism; it all depends on the articulation of professed norms of behavior which are tactfully employed by the elites for the securitization of their strategies. The discursive practices or "speech acts" of elites play a dominant role in this regard.[16] The pacifism or assertiveness in Hinduism needs a focus on actual elites' practices. In this vein, Pardesi's optimistic account gave the carte blanche of "offensive realism" as the future strategy of Indian behavior based on his rich analysis of five pan-Indian powers: the Mauryas, the Guptas, the Mughals, British India, and the Republic of India.[17]

Another interesting study on Indian strategic thinking is by Javed Hassan, a Brigadier and later Lieutenant General of the Pakistan Army, who in his book *India: A Study in Profile* states:

> The Indian obsession with the mythical past has given birth to a number of beliefs which most Indians (Hindus) hold as an article of faith. "The more myths one encounters, the more the basic theme seems to be reinforced ... (that) is the process of developing order out of chaos." The importance of these 'mythical' beliefs are; one, that in the distant past Mahabharata (Great Bharat) extended over an area from (including) parts of Iran and Afghanistan to Indonesia, that this India was always a united country ruled by Great Aryan Kings (named Bharatas) till the invaders from the north west came and destroyed this unity; two, that India must reclaim and reunite all the lost territories and emerge once again as a great Mahabharata, which is its destiny. This myth along with the aspiration was reflected in 1947 when India was named as "India that is Bharatavarsh" at the time of independence.[18]

As an interesting case study of Indian history and strategy, Hassan further notes that Indians as a race have never been migrated from their own territory, have often capitulated to foreign aggressors, and the "creed of Hinduism has been the dominant impulse of Indian civilization."[19]

There is almost a consensus in all think tank policy reports prepared by the intelligentsia in the West, especially in the United States, that Indian strategic culture is quite unique on many fronts. India has a vast expanse of territory, but at the same time, its underperformed security potential is due to the ascendency of civilian elites, an underfunded colonial-minded bureaucracy, and marginalization of military elites in important matters of state security.[20] In Burgess's assessment, "[d]omestic politics and outmoded bureaucracy also contribute to a phlegmatic and short-sighted strategic culture and to a foreign policy that does not serve a rising India well."[21] The next section will cover some of the scholarly academic works on Indian strategic culture.

Indian Strategic Culture Written by Academicians

On an academic front, the research on Indian strategic culture focuses on a theoretically informed empirical understanding of Indian state behavior. Different theoretical frameworks from realism to liberalism and social constructivism have been applied to understand Indian strategic culture and this has resulted in a plethora of research from indigenous Indian scholars and Westerners alike. India has developed a Monroe Doctrine type of "sphere of influence" around the Indian Ocean whereby the presence of outside powers in the Indian sphere of influence is strictly abhorred.[22] In Brewster's words, "[t]his is essentially an expression of South Asia as a single strategic unit, with India having a special role as the custodian of regional security."[23]

The journey of Indian strategic culture from secularism toward Hindutva was also interjected with *realpolitik*, as Ian Hall explains: "Indian strategic culture is informed by ideas taken from Hindu texts, nineteenth and twentieth-century religious [thought], and modernist thinkers. These ideas shape three traditions of strategic thought: Nehruvian tradition, *realpolitik*, and Hindu nationalism."[24] Kanti Bajpai has rephrased these three schools of strategic thought as "Nehruvian, neoliberalism, and hyperrealism" and goes on to draw common assumptions as well as different propositions of all three schools.[25] All three agree on the centrality of state sovereignty, state interests as defining features of world politics, and power both in soft and hard forms as the end of all means adopted by state survival.[26] The differences between these three schools are in the strategy toward foreign policy, where first the Nehruvian school asks for the mitigation of anarchy through cooperation at all levels, from people to people and from state to state. In other words, this school offers both bottom-up as well as top-down strategies for cooperation by emphasizing the goodness in human behavior. The neoliberalism school explains the efficacy of compliance toward international institutions for cooperation by asserting that "economic well-being is vital for national security ... [emphasizing] free market policies."[27] Hyperrealism focuses on the accumulation of power in order to be a hegemon and believes that only a strong military force can change things in the economic realm.[28] Bajpai elaborates on these three visions of Indian strategic culture toward its policies vis-à-vis China and Pakistan, but the general sweep of his arguments is not based on any deductive logic or isolation of strategic culture causality determining concrete policy actions, a core feature of the third wave of strategic culture scholarship.[29] Work in this vein would benefit from developing the causality or embeddedness of strategic culture behind Indian state behavior, and from further exploration of strategic culture's relationship to elite politics in India and the role played by the ideology of respective political parties.

This crucial question of how strategic culture develops at the elite level and where this can be found is addressed by your author in his work on the BJP and Indian strategic culture.[30] This work develops an intermediary step of elites' socialization by the norms of a political party who later work as state elites and develop security practices of the Indian state. It argues that the norms of strategic thinking are embedded in the political ideology of political parties.[31] For example, it is the BJP's norms of Hindutva that developed the strategic culture of "hyperrealism" with emphasis on the use of optimal power as the surest way of state survival. This was amply demonstrated during the second term of Indian Prime Minister Narendra Modi (2019 onward) through Indian state security practices both internally (e.g., scrapping of Article 370 on Kashmir) as well as externally (e.g., the use of surgical strikes in Pakistan to stop cross-border terrorism). The problem here is that elites' socialization as an intermediary step needed to be investigated by an in-depth case study of Indian political parties who came to power with a focus on their hierarchical structure and the process of socialization adopted by them. Your author's case study of Hindutva ideology exclusively focused on the Indian decision to become a nuclear power in 1998 and did not proceed with an in-depth analysis of political parties and strategic decisionmaking in India.[32]

According to Rodney Jones, "India has an omniscient patrician strategic culture," which originally in Sanskrit means "Bharat jagat guru," or "India: The World's Teacher."[33] Jones further explains five "philosophical and mythological foundations" of this type of strategic culture, which are sacredness of Indian identity, timelessness of perpetual goals, inherent great status of India, quest for truthful knowledge, and hierarchical structure of the world order.[34] Each of these embedded myths gave rise to unique elements of strategy devised by the Indian elites. For example, the instrumental implications of this mythical knowledge led to an "enigmatic" vision of India, the perpetual nature of self-interests, the "contradictory" nature of the world, power politics, security being "sedentary," and strategy being "assimilative."[35] There is one caveat in Jones' analysis and that is the generalizations made in line with the five mythical strands of Indian strategic culture. Further focused research needs to elaborate on each of these myths by explaining the distinct phases of Indian foreign policy, with particular attention as to what type of behavior is shown by the Indian state in line with this mythical knowledge. But Jones' idea of an omniscient-patrician strategic culture is quite thought-provoking and needs to be augmented with a fine-grained analysis of Indian strategy adopted during "crunch times" of decisionmaking by Indian elites.

Strategic Culture and Indian Nuclear Strategy

There is a third body of work on Indian strategic culture in the realm of its nuclear program. Here scholars have defined the strategy of nuclear decisionmaking by analyzing the foreign policy of India and all have developed their causal arguments on different sets of variables.[36] The application of Indian strategic culture over nuclear weapons has also been discussed in a number of articles. Rajesh Basrur explained:

> Indian nuclear strategy (and strategic culture) has retained elements of deterrence – from Nehru's minimal open door to Vajpai's pronouncements on credible minimal deterrence – and thereby stayed within the neorealist framework that emphasizes states' need to ensure security through the possession of military capability in an anarchic self-help system. In this sense, strategic culture as intermediate structure supplements rather than undermines the neorealist concept of system structure.[37]

On an academic note, the transformation in Indian strategic culture of idealism coupled with ambiguity over nuclearization is carried from Nehruvian thoughts (moorings of long-serving Indian Prime Minister Jawaharlal Nehru, 1947–1964) to Hindutva ones (ideology given by the BJP from the 1990s onward). This is very interesting as, from carrying the secularist credentials of its founding fathers, India has embraced an overt Hindutva ideology. Runa Das notes:

> In pursuing this line of strategic thinking, known as Nehruvianism, postcolonial India's political leaders were influenced by Gandhian nonviolence, which drawing from ancient Indian civilizational moorings, was considered by the Indian leadership as an alternative to the conflict-ridden strategic conceptions of the West.[38]

India tested its nuclear devices in 1998 after a lull of almost 25 years since the country's PNE in 1974 under Indira Gandhi, the Congress Party Prime Minister and the daughter of Nehru. The political horizon was completely changed in 1998 with the ascendency of the BJP, and the strategic culture also showed a transformation with more cultural values imbued into the security practices of the Indian state. Runa Das explained this strategic transformation from secularism to Hindutva as "[o]ne nation, one people, one culture underpin the BJP's construct of India as a Hindu rashtra.... Pitrabhoomi (fatherland), jati (bloodline) and Sanskriti (culture) are identified by the BJP as constituting the cultural boundaries of India as a nation."[39] This new Hindutva outlook of Indian strategic culture was aptly explained by the BJP's minister of External Affairs Jaswant Singh in his book *Defending India*:

> [T]o define Indian strategic culture one has to examine the very nature of India's nationhood: The very characteristic of its society; and the evolution of its strategic thought over the ages ... it is [mainly] a by-product of the political culture of a nation and its people ... this is where history and racial memories influence a nation's strategic thought.... Above all else, India is Hindu and Hindus thinks differently from non-Hindus ... that has given birth to a culture from which we hope to extract the essence of its strategic thought.[40]

On one hand, George Perkovich's analysis of Indian nuclear strategy is an excellent historical account of the discourse of nuclear strategy in which he explained the role played by the "strategic enclave" of Indian nuclear scientists as well as "the normative national identity interest in achieving major power status."[41] The focus of Perkovich's analysis is on "the normative identity-nuclear nexus as the prime motive behind the Indian decision to go nuclear" in 1998.[42] Itty Abraham, on the other hand, focused on the postcolonial mindset of Indian elites who are bent on achieving something great (acquired modernity) for India in the community of nations.[43] The achievement of the bomb is a conduit for this endeavor. The strategy of nuclear decisionmaking is linked here to the pro-bomb lobby within the Indian scientific community coupled with this mindset of Indian elites which helped India to formulate a distinct strategy for achieving its target of building this "indigenous" bomb.[44]

The approach of Abraham is hinged upon exploring the mindset of Indian elites through their discursive practices. In both Perkovich and Abraham's studies, the historical analysis is sociocultural, while Jacques Hymans' study is more of a psychological account of Indian elites' decisionmaking.[45] Hymans has linked the Indian elites' "national identity conceptions to nuclear proliferation,"[46] explaining the nuclear strategy of India through the "oppositional nationalist identity conceptions" (NICs) of BJP elites like the Indian prime minister Atal Behari Vajpayee. Oppositional nationalism is described by Hymans "as based on a stark black-white dichotomization of us against them." Hymans further developed his NIC model on two dimensions of solidarity and status: Solidarity referring to the content of national identity and status to the global standing of this identity in comparison to others.[47] Hymans developed this socio-psychological theory of NICs of Indian elites to explain nuclear strategic decision-making; however, it fails to give room to the political parties, which socialize these elites by inculcating values of specific NICs.

Chris Ogden in a similar vein explains the "security identities" of Indian elites, which he distinguishes from strategic culture and links to ideological norms while explaining the nuclear posturing of the Indian state,[48] but Ogden neglects to explain the cycle of norms through its emergence, cascade, and internationalization.[49] Your author has argued that the role of political parties as harbingers of strategic culture is very important, to the extent of framing Indian strategic culture as "an understanding of social practices of elites embedded in sociocultural ideology of their political party forming some falsifiable determinant of strategic behavior."[50] Ogden's type of research is a snapshot survey of Indian strategic culture as it exclusively focuses on nuclear decisionmaking with selective use of sociocultural norms of Indian society, but the analysis of strategic culture can be much better served if a holistic approach to foreign policy analysis is employed in all these case studies.

The impetus behind the Indian nuclear program is tied to certain contradictory "rationales" as explained by Ogden, which exists for "self-sufficiency" in security along with "repeated calls for universal disarmaments."[51] These contradictions in Indian security practices in the realm of nuclear proliferation can be better explained by Indian strategic culture. During the first three decades of Indian independence, Indian strategic culture under Nehruvian influence called for global norms of nuclear nonproliferation and at the same time aimed for nuclear reliance to make India "Mahan" (Grand) again.[52] In the last decade of the twentieth century (1998 onward) and still at the time of this writing, an obsessiveness toward Hindutva norms brought forward the strategy of over-nuclearization in Indian nuclear discourse. The shift in India's nuclear behavior from an ambiguous posture (1974–1998) to an overt one from 1998 forward was explained by Sumit Ganguly as due to three factors, one of them that "the fitful movement toward a nuclear weapons capacity closely followed the shifting calculations of Indian leaders, who responded to a mix of ideology (initially a force of restraint), statecraft, and domestic pressures reflecting security concerns."[53] This ideology and strategy nexus is explained through the strategic culture of India as connivance between political party ideology and elites' security practices at the helm of affairs.[54] Even the posturing problem in Indian nuclear strategy can be better explained through the lens of strategic culture: The distinct and changing posturing of Indian nuclear weapons is in line with the ideological commitments of its elites' strategic thinking. This is an area where future research in Indian strategic culture can be undertaken as most of the nuclear proliferation policies used for nuclear posturing are devised on a rational calculus with a positivist methodology.[55] A detailed case study of Indian nuclear posturing with an interpretive postpositivist methodology can reveal much as compared to the positivist framework.

Conclusion

Future studies on Indian strategic culture would benefit from an in-depth historical analysis of Indian strategic thinking by linking the practices of its elites with that of sociocultural norms of Indian society. A synthesis of comparative research on Indian political parties with that of cultural anthropology aimed at exploring sociocultural norms can be one area of future research. The role of Indian political parties in propagating norms of behavior as well as the socialization of elites has been consistently glossed over in almost all major works on Indian strategic culture. This is in spite of the fact that most of the time modern India is ruled by two major parties – the Congress Party and the BJP – with each professing contrasting norms of behavior for India's elites. This requires juxtaposing the modus operandi of these two parties, looking at their distinct ideologies, focusing on their methodologies used for elite socialization, and finally explaining the strategic thinking and practices of their elites.

On a normative note, in order to explore influential norms behind strategic thinking, there is a dire need to deconstruct the richness of the Indian polity by focusing on the diversity of people living in this state and what norms of behavior each of these communities allude to. At present, there is ample research on the predominant norms of the Hindu community, and seldom any research is done on the multicultural norms of Indian society. A media approach focusing on the propagation of norms is another area of research where the methodology of popular culture can help us understand the phenomenon of norms cascades.[56] The cycle of norms starts with their generation, moves to cascade or propagation, and finally to internalization (becoming second habit). In this chain, the middle tier of norms cascade or propagation is very important, as at this stage the norms receive popular acceptance and legitimacy. There is significant value in studying how such influential norms are being propagated in a vast country like India. In India there are a plethora of media channels and the Indian films industry (Bollywood) propagates such norms of behavior.[57] This is important because the critical shifts evident in Indian strategic culture from Ahimsa (nonviolence) to Hindutva (assertive Hindu identity) require an exploration of mechanism of norms propagation used to garner public support behind such shifts. The traditional yardstick of rational choice focusing on material capabilities and cost-benefit calculation of a state is inadequate as it cannot explain critical shifts in Indian strategic behavior. This requires a postpositivist interpretivist framework where a critical analysis of the security discourse of India can help bring to light the myriad forms of Indian strategic culture. A via-media approach helps bridge the gap between a rationalist and an interpretivist framework by incorporating the role played by identity and norms in constructing state interests.

To conclude, Indian strategic culture is at a crossroads of yet another major critical shift – a clear break from the past. The strategic culture of India constructed through Ahimsa norms (nonviolence toward other states) has been rejected by popular vote with the advent of the BJP on the national political horizon. The call made by present-day elites of the BJP aims for assertive security practices and overt nuclear posturing based on Hindutva norms, which is a break from the Indian founding fathers' ideology of Ahimsa and subsequent elites' practices of being flagbearers of the Non-Aligned Movement as well as spokesmen of nuclear ambiguity. This critical junction has presented a crossroads for Indian strategic culture, as either it will fully embrace the Hindutva ideology or it will move away from religious schisms to help India shine again. Further research utilizing the strategic culture paradigm can offer crucial insights as Indian elites face the herculean tasks of overcoming parochial political interests in order to embrace the challenges of this century.

Notes

1 *Acknowledgments*: I am thankful to Allah for giving me the strength to complete this task. I am also grateful for the prayers of my parents Shahida Pervez and that of my late father Wajeeh Uddin Pervez. I am also thankful to my beloved wife Sadia for her unwavering support through thick and thin and finally to my sons Muneeb and Moiz for giving me time to pursue my intellectual pursuits.
2 Jeannie L. Johnson, "Strategic Culture: Refining the Theoretical Construct," in *Science Applications International Corporation (SAIC)* (Defence Threat Reduction Agency, 2006).
3 Ashley J. Tellis, "Overview," in *Understanding Strategic Cultures in the Asia Pacific*, ed. Ashley J. Tellis, Alison Szalwinski, and Michael Wills (Seattle: The National Bureau of Asian Research, 2016), p. 5.

4 Alastair Iain Johnston, *Cultural Realism: Strategic Culture and Grand Strategy in Chinese History* (Princeton, NJ: Princeton University Press, 1995). Jeffrey S. Lantis and Darryl Howlett, "Strategic Culture," in *Strategy in the Contemporary World*, eds. John Baylis, James J. Wirtz, and Colin S. Gray (New York: Oxford University Press, 2013). Tellis, "Overview."
5 Peter J. Katzenstein, *The Culture of National Security: Norms and Identity in World Politics*, ed. John Gerard Ruggie, New Directions in World Politics (New York: Columbia University Press, 1996).
6 Tellis, "Overview," p. 2.
7 Ana Sinha, "Critical Analysis of Tanham Interpretation of Indian Strategic Culture" (MA thesis, University of Delhi, 2019).
8 Rashed Uz Zaman, "Kautilya: The Indian Strategic Thinker and Indian Strategic Culture," *Comparative Strategy* 25 (2006), p. 242.
9 Ibid.
10 Ibid., p. 243.
11 George K. Tanham, *Indian Strategic Thought: An Interpretive Essay* (Santa Monica, CA: RAND, 1992), p. V.
12 Ibid., p. 17.
13 Ibid.
14 Ibid.
15 Stephen Burgess, "India's Strategic Culture, Foreign and Security Policy, and Relations with the United States," in *American Political Science Association Convention* (2009), p. 10.
16 Ole Waever, "Securitization and Desecuritization," in *On Security*, ed. Ronnie D. Lipschutz (New York: Columbia University Press, 1995).
17 Manjeet Singh Pardesi, *Deducing India's Grand Strategy of Regional Hegemony from Historical and Conceptual Perspectives* (Singapore: Institute of Defence and Strategic Studies Nanyang Technological University, 2005).
18 Javed Hassan, *India: A Study in Profile* (Rawalpindi: Army Education Press GHQ, 1990), p. 11.
19 Ibid., p. 49.
20 Burgess, "India's Strategic Culture, Foreign and Security Policy, and Relations with the United States."
21 Ibid., p. 16.
22 David Brewster, "Indian Strategic Thinking about East Asia," *The Journal of Strategic Studies* 34, no. 6 (2011), pp. 825–852.
23 David Brewster, "Indian Strategic Thinking about the Indian Ocean: Striving towards Strategic Leadership," *India Review* 14, no. 2 (2015), pp. 221–237.
24 Ian Hall, "The Persistence of Nehruvianism in India's Startegic Culture," in *Understanding Startegic Culture in the Asia-Pacific*, eds. Ashley J. Tellis, Alison Szalwinski, and Michael Wills (Seattle: National Bureau of Asian Research, 2016), p. 141.
25 Kanti Bajpai, "Pakistan and China in Indian Strategic Thought," *International Journal* 62, no. 4 (2007), pp. 805–822.
26 Ibid.
27 Ibid., p. 808.
28 Ibid.
29 Johnston, *Cultural Realism: Strategic Culture and Grand Strategy in Chinese History*. Lantis and Howlett, "Strategic Culture."
30 Muhammad Shoaib Pervez, "Strategic Culture Reconceptualized: The Case of India and the BJP," *International Politics* 56, no. 1 (2019), pp. 87–102.
31 Ibid.
32 Ibid.
33 Rodney W. Jones, "India Strategic Culture and the Origins of Omniscient Paternalism," in *Strategic Culture and Weapons of Mass Destruction*, eds. Jeannie L. Johnson, Kerry M. Kartchner, and Jeffrey A. Larsen (New York: Palgrave Macmillan, 2009).
34 Ibid.
35 Ibid., p. 118.

36 Itty Abraham, *Making of the Indian Atomic Bomb: Science, Secrecy and the Postcolonial State* (London: Zed Books, 1998). George Perkovich, *India's Nuclear Bomb: The Impact on Global Proliferation* (Berkeley, CA: University of California Press, 1999). Jacques E.C. Hymans, *The Psychology of Nuclear Proliferation: Identity, Emotions, and Foreign Policy* (Cambridge: Cambridge University Press, 2006).
37 Rajesh M. Basrur, "Nuclear Weapons and Indian Strategic Culture," *Journal of Peace Research* 38, no. 2 (2001), p. 196.
38 Runa Das, "The Prism of Strategic Culture and South Asian Nuclearization," *Contemporary Politics* 15, no. 4 (2009), p. 399.
39 Ibid., p. 403.
40 Jaswant Singh, *Defending India* (New Delhi: Macmillan Press, 1999), pp. 2, 5.
41 Perkovich, *India's Nuclear Bomb: The Impact on Global Proliferation*, p. 452.
42 Muhammad Shoaib Pervez, *Security Community in South Asia: India-Pakistan* (New York: Routledge, 2013), p. 108.
43 Abraham, *Making of the Indian Atomic Bomb: Science, Secrecy and the Postcolonial State.*
44 Ibid.
45 Hymans, *The Psychology of Nuclear Proliferation: Identity, Emotions, and Foreign Policy.*
46 Ibid.
47 Ibid.
48 Chris Ogden, "Norms, Indian Foreign Policy and the 1998–2004 National Democratic Alliance," *The Round Table-The Commonwealth Journal of International Affairs* 99, no. 408 (2010), pp. 303–315. Chris Ogden, *Hindu Nationalism and the Evolution of Contemporary Indian Security: Portents of Power* (New Delhi: Oxford University Press, 2014).
49 Martha Finnemore, *National Interests in International Society*, ed. Peter J. Katzenstein, Cornell Studies in Political Economy (London: Cornell University Press, 1996).
50 Pervez, "Strategic Culture Reconceptualized: The Case of India and the BJP," p. 92.
51 Ogden, *Hindu Nationalism and the Evolution of Contemporary Indian Security: Portents of Power*, p. 153.
52 Jawaharlal Nehru, *The Discovery of India* (London: Meridian Book Limited, 1947). "Jawaharlal Nehru's Speeches" (Delhi: Publications Division, Ministry of Information and Broadcasting, Government of India, 1968). Mushirul Hasan, ed. *Selected Works of Jawaharlal Nehru* (Delhi: Jawaharlal Nehru Memorial Fund, 2006).
53 Sumit Ganguly, "India's Pathway to Pokhran Ii," *International Security* 23, no. 4 (1999), p. 170.
54 Pervez, "Strategic Culture Reconceptualized: The Case of India and the BJP."
55 Neil Narang, Erik Gartzke, and Matthew Kroeing, *Nonproliferation Policy and Nuclear Posture* (London: Routledge, 2017). Ibid. Vipin Narang, *Nuclear Strategy in the Modern Era: Regional Powers and International Conflict* (Princeton, NJ: Princeton University Press, 2014).
56 Finnemore, *National Interests in International Society.*
57 Pervez, *Security Community in South Asia: India-Pakistan.*

Selected Bibliography

Abraham, Itty. *Making of the Indian Atomic Bomb: Science, Secrecy and the Postcolonial State.* London: Zed Books, 1998.

Bajpai, Kanti. "Pakistan and China in Indian Strategic Thought." *International Journal* 62, no. 4 (2007): 805–822.

Basrur, Rajesh M. "Nuclear Weapons and Indian Strategic Culture." *Journal of Peace Research* 38, no. 2 (2001): 181–198.

Brewster, David. "Indian Strategic Thinking about East Asia." *The Journal of Strategic Studies* 34, no. 6 (2011): 825–852.

Brewster, David. "Indian Strategic Thinking About the Indian Ocean: Striving towards Strategic Leadership." *India Review* 14, no. 2 (2015): 221–237.

Burgess, Stephen. "India's Strategic Culture, Foreign and Security Policy, and Relations with the United States." In *American Political Science Association Convention*, 1–16, 2009.

Das, Runa. "The Prism of Strategic Culture and South Asian Nuclearization." *Contemporary Politics* 15, no. 4 (2009): 395–411.
Finnemore, Martha. *National Interests in International Society*. Cornell Studies in Political Economy. Edited by Peter J. Katzenstein. London: Cornell University Press, 1996.
Finnemore, Martha. "Norms, Culture, and World Politics: Insights from Sociology's Institutionalism." *International Organisation* 50, no. 2 (1996): 325–347.
Ganguly, Sumit. "India's Pathway to Pokhran Ii." *International Security* 23, no. 4 (1999): 148–177.
Hall, Ian. "The Persistence of Nehruvianism in India's Strategic Culture." In *Understanding Strategic Culture in the* Asia-Pacific, edited by Ashley J. Tellis, Alison Szalwinski, and Michael Wills, 141–167. Seattle: National Bureau of Asian Research, 2016.
Hasan, Mushirul, ed. *Selected Works of Jawaharlal Nehru*. Delhi: Jawaharlal Nehru Memorial Fund, 2006.
Hassan, Javed. *India: A Study in Profile*. Rawalpindi: Army Education Press GHQ, 1990.
Hymans, Jacques E.C. *The Psychology of Nuclear Proliferation: Identity, Emotions, and Foreign Policy*. Cambridge: Cambridge University Press, 2006.
Johnson, Jeannie L. "Strategic Culture: Refining the Theoretical Construct." In *Science Applications International Corporation (SAIC)* Defence Threat Reduction Agency, 2006.
Johnston, Alastair Iain. *Cultural Realism: Strategic Culture and Grand Strategy in Chinese History*. Princeton NJ: Princeton University Press, 1995.
Jones, Rodney W. "India Strategic Culture and the Origins of Omniscient Paternalism." In *Strategic Culture and Weapons of Mass Destruction*, edited by Jeannie L. Johnson, Kerry M. Kartchner, and Jeffrey A. Larsen, 117–136. New York: Palgrave Macmillan, 2009.
Katzenstein, Peter J. *The Culture of National Security: Norms and Identity in World Politics*. New Directions in World Politics. Edited by John Gerard Ruggie. New York: Columbia University Press, 1996.
Narang, Neil, Erik Gartzke, and Matthew Kroeing. *Nonproliferation Policy and Nuclear Posture*. London: Routledge, 2017.
Narang, Vipin. *Nuclear Strategy in the Modern Era: Regional Powers and International Conflict*. Princeton, NJ: Princeton University Press, 2014.
Narang, Vipin. "Strategies of Nuclear Proliferation: How States Pursue the Bomb." *International Security* 41, no. 3 (2017): 110–150.
Nehru, Jawaharlal. *The Discovery of India*. London: Meridian Book Limited, 1947.
Nehru, Jawaharlal. "Jawaharlal Nehru's Speeches." Delhi: Publications Division, Ministry of Information and Broadcasting, Government of India, 1968.
Ogden, Chris. *Hindu Nationalism and the Evolution of Contemporary Indian Security: Portents of Power*. New Delhi: Oxford University Press, 2014.
Ogden, Chris. "India: The (Accepted) Gatecrasher." In *Handbook of Nuclear Proliferation*, edited by Harsh V. Pant, 149–160. London: Routledge, 2014.
Ogden, Chris. "Norms, Indian Foreign Policy and the 1998–2004 National Democratic Alliance." *The Round Table-The Commonwealth Journal of International Affairs* 99, no. 408 (2010): 303–315.
Pardesi, Manjeet Singh. *Deducing India's Grand Strategy of Regional Hegemony from Historical and Conceptual Perspectives*. Singapore: Institute of Defence and Strategic Studies Nanyang Technological University, 2005.
Perkovich, George. *India's Nuclear Bomb: The Impact on Global Proliferation*. Berkeley: University of California Press, 1999.
Pervez, Muhammad Shoaib. *Security Community in South Asia: India-Pakistan*. New York: Routledge, 2013.
Pervez, Muhammad Shoaib. "Strategic Culture Reconceptualized: The Case of India and the BJP." *International Politics* 56, no. 1 (2019): 87–102.
Singh, Jaswant. *Defending India*. New Delhi: Macmillan Press, 1999.
Sinha, Ana. "Critical Analysis of Tanham Interpretation of Indian Strategic Culture." MA thesis, University of Delhi, 2019.
Tanham, George K. *Indian Strategic Thought: An Interpretive Essay*. Santa Monica, CA: RAND, 1992.

Tellis, Ashley J. "Overview." In *Understanding Strategic Cultures in the Asia Pacific*, edited by Ashley J. Tellis, Alison Szalwinski, and Michael Wills, 3–26. Seattle: The National Bureau of Asian Research, 2016.

Waever, Ole. "Securitization and Desecuritization." In *On Security*, edited byRonnie D. Lipschutz, 46–86. New York: Columbia University Press, 1995.

Zaman, Rashed uz. "Kautilya: The Indian Strategic Thinker and Indian Strategic Culture." *Comparative Strategy* 25 (2006): 231–247.

23
PAKISTAN'S MILITARY-CENTRIC STRATEGIC CULTURE
A Review Essay

Arshad Ali

Introduction

There are various debates over the concept of strategic culture and its role in explaining the strategic conduct of states. Many agree that the strategic culture of a nation is based on "its shared beliefs and assumptions [meaning that] the observer needs to immerse himself or herself in that nation's history, attitudes, and conduct."[1] For the purpose of this chapter examining Pakistani strategic culture, this study focuses on two definitions that explain military-centric strategic culture: First, that "strategic culture stands for the 'beliefs and assumptions that frame ... choices about international military behavior, particularly those concerning decisions to go to war, preferences for offensive, expansionist or defensive modes of warfare and levels of wartime casualties that would be acceptable.'"[2] Second, that "strategic culture [is] 'the set of attitudes and beliefs held within a military establishment concerning the political objective of war and the most effective strategy and operational method of achieving it.'"[3] These definitions are important in order to grasp the central politico-military and military-organizational aspects of Pakistan's strategic culture.

To understand Pakistani strategic culture, it is necessary to reflect on the security environment in which Pakistan came into being after the partition of the Indian subcontinent in August 1947. Since that point, Pakistan's strategic thinking has evolved in the context of hostility with India, security concerns on the Western Afghan borders, and the interests of global superpowers in Pakistan's geostrategic position during and after the Cold War.[4] For example, Pakistan fought three major wars and one limited war with India, including the Kashmir war right after independence in 1947. Moreover, Afghanistan had territorial claims over Pakistan's northwestern Pashtun areas along the "Durand Line" in Khyber Pakhtunkhwa and Balochistan provinces and opposed Pakistan's admission to the United Nations. Compounding this, both the former Soviet Union and the United States had significant interests in Pakistan due to its geostrategic location during the bipolar contest of the Cold War. Thus, Pakistan was born in a very difficult security circumstance where survival became the main priority of the country.[5]

This uneasy existence has shaped Pakistan's strategic compulsions by giving ascendency to the Pakistani military elites who developed the contours of the state's strategic culture.

DOI: 10.4324/9781003010302-26

The dominance of the military started right after the partition when General Ayub Khan became the first native Commander in Chief in 1951. Under General Ayub, the military gradually increased its influence in state affairs and ultimately took over the government, imposing the first martial law administration in 1958.[6] Since then, military elites have remained influential in all key strategic decisionmaking, including the issue of Kashmir, nuclear weapons, relations with Afghanistan and India, and participation in the Cold War and the Global War on Terror (GWOT).[7]

Despite being a parliamentary system, Pakistan has been governed by a troika of the prime minister, the president, and the army chief.[8] However, the prime minister has historically been very weak and does not have much say in the strategic affairs of the country. In fact, the army chiefs formed the core of Pakistan's strategic culture during their 33 years of direct rule where the primary task was to establish foreign and security policy for the very survival of the country. In the remaining period, the military elites ruled indirectly with the support of the president and by manipulating the parliamentary system. In the 1990s, the Pakistani president dismissed three elected governments – in 1990, 1993, and 1996 – with the support of the military elites to maintain their dominance in state affairs.[9] The military elites justified their upper hand on the grounds of the security deficit with India: perceiving that India was going to be a permanent enemy in the foreseeable future, they believed the country required "a strong military and permanent military preparedness" to maintain its security.[10] Thus, Pakistan's strategic culture has been almost exclusively defined by the military elites in the country.

In addition, the Pakistani military inherited the legacy of the British Army after the nation's independence in 1947. The British Army's strategic thinking, designed for colonial rule in India and the empire's interests in the South-Central Asian region, played a key role in the formation of Pakistan's strategic culture.[11] Christine Fair explains that Pakistan inherited British narratives about the subcontinent's defense, and particularly the idea of Afghanistan as a neutral buffer against Russian threats.[12] Furthermore, the development of Pakistan's nuclear program and the use of proxy war were added to Pakistan's strategic culture in the 1970s. Over time, military elites escaped civilian oversight of Pakistan's strategic assets and further elevated the role of asymmetric warfare in the country's strategic culture. Thus, carrying through to the 2020s, the Pakistani military has maintained heavily predominant control over the country's foreign and security policies and its strategic assets.

Against this backdrop, there are certain unique politico-military and military organizational aspects of Pakistan's strategic culture that are taken into consideration in this chapter. First, the partition of India was incomplete, so the Pakistani military had to complete the unfinished agenda of Kashmir. From 1947 to 1971, the Kashmir conflict was the main board of contest between India and Pakistan. Second, Pakistan adopted a military-centric national security approach by prioritizing military competition with India over nation-building. Third, the Pakistani military considers itself the guardian of territorial borders and Pakistan's ideological frontier, which provides the military prestige and position in the state and society. Fourth, the military has established financial and political autonomy to act independently in state affairs. Finally, the military considers the civilian leadership weak and too incompetent to protect the country's strategic interests. Thus, in the eyes of the military elites, Pakistan needs a "guided democracy" where the military elites take charge of all strategic decisionmaking with respect to the Kashmir conflict, Pakistan's nuclear program, participation in war, and relations with the United States, India, and Afghanistan.

This chapter will review the literature on the domineering role of the military elites in Pakistan's strategic culture and explore twin aspects of Pakistan's military-centric strategic culture: First, the organizational norms of the Pakistani army and how it functions as a unified chain of command; and second, the military's spillover effect as the specific organizational culture it has developed leads to its domination of all affairs of the state. In total, the chapter assesses what type of norms are disseminated by Pakistan's military elites in the name of state security as they hold the helm of state affairs.

The Culture of Pakistan's Military

Organizational Structure and Norms

Since its independence in 1947, Pakistan has adopted a military-centric security approach to protect its territorial integrity and national sovereignty. The Pakistani military elites developed their institutional mechanisms by resolving all the outstanding issues confronting the new country through military means and resources, and the military presented itself as the only institution capable of protecting Pakistan's national interests from external and internal threats. Interestingly, there was general acceptance of the military's ascending political role in the country. As a result, these elites governed the country directly through military rule for some time and indirectly for the remaining period with the support of the president and by manipulating democratic processes in the country.[13] Therefore, Pakistan's "culture of national security"[14] is directly influenced by military institutional norms, as the military is the only organized forum that is a consistent key player in strategic decisions of national importance.

At the structural level, the Pakistani military comprises three branches with a total of 650,000 active personnel, among which the army is the most politically assertive and largest service with 550,000 personnel. It is followed by the Pakistani air force and navy with 45,000 and 25,000 personnel, respectively.[15] The bulk of the forces, around 75%, come from the largest Punjab province, followed by 20% from Khyber Pakhtunkhwa. Sindh and Balochistan provinces are underrepresented in the military with 5% of the national forces coming from each, while the remaining forces come from other Pakistani administrative units.[16] The role of the military was defined for the first time in Article 245 of the 1973 constitution, outlining that the military is required to "defend Pakistan against external aggression or threat of war, and, subject to law, [must] act in aid of civil power when called upon to do so."[17]

From the vantage point of institutional norms, the military corps commanders play a prominent role in major strategic decisions in the country, especially in working closely with the Chief of Army Staff (COAS) – the most powerful position in the armed forces – to discuss key security and political matters, including foreign and security policies, imposing martial law administration, and domestic power and politics. According to Blom, "although the COAS is all-powerful, he needs the support of the CCC (Corps Commanders Conferences) for any major change in foreign policy."[18] Thus, the corps commanders have a significant role in the political affairs of the country. Furthermore, the military has the 10th Command structure known as Army Strategic Forces Command, which is responsible for holding key strategic assets, including nuclear weapons.[19] Interestingly, the Chairman of the Joint Chiefs of Staff Committee is the highest position in the military, nominally presiding over all three armed services, but in practice the Joint Chief is powerless, and such a position is offered to a senior military officer simply to become a four-star general.[20] Thus, the military elites have established a command-and-control structure that is effectively used in the political sphere of the country as well as the strategic domain.

Historical Factors Elevating the Military

The centrality of the Pakistani military in state affairs was historically driven by the fact that the civil and military bureaucracies were the only institutions that existed before the partition in 1947. Under the terms of the partition agreement, Pakistan got 33% of the military resources and only 17.5% of the financial resources from British India upon independence.[21] Pakistan had to establish nearly all other institutions from scratch to make a functional state.

It is important to note that Muhammad Ali Jinnah, the founder of Pakistan, wanted to have a democratic setup in Pakistan with a supportive and politically neutral military. In his speech at the Staff College Quetta to army officers, Jinnah admonished military elites "not to forget that the armed forces were the servants of the people and you do not make national policy; it is we, the civilians, who decide these issues and it is your duty to carry out these tasks with which you are entrusted."[22] Subsequently, the Pakistani leadership attempted to establish civilian-led institutions and agencies to govern the country. In June 1948, the government established the Defense Committee of the Cabinet, which was headed by the defense minister to manage and oversee the armed forces. However, this body failed to regulate the military and its functions.[23] Subsequently, the Pakistani military started its role as a watchdog in defense matters in 1952. General Ayub became a government cabinet member in 1951 and then took over as defense minister in 1954.[24] In this context, General Ayub presented himself as a prominent statesman to address political crises and establish a viable foreign and security policy during the Cold War. He played an instrumental role in formulating the alliance with the United States during this period, which facilitated subsequent US financial and military aid to Pakistan.

Interestingly, there are seldom any formal documents on national security doctrines or security strategies of Pakistan to date, as all decisions are highly personalized following the unique traits of the army chief in power. Under General Ayub's leadership, the military elites defined the strategic priorities and foreign policy directives of the country in the 1950s. Stephen Cohen noted in his article:

> From the day Pakistan was created the army has helped to establish internal order and protect Pakistan's permeable and often ill-defined borders; it used its power and special position within Pakistan to ensure that it received adequate weapons, resources, and manpower. Finally, it always regarded itself as the special expression of the idea of Pakistan, and a few officers have argued for an activist role in reforming or correcting the society when it fell below the standard of excellence set by the military.[25]

There were many factors driving the military's ascendent role which were set right after the partition in 1947. The Pakistan Muslim League was a popular independence movement among Indian Muslims, but it was unable to convert itself into a grassroots political party. Moreover, the political leadership became weak after the death of prominent leaders, including Jinnah in 1948 and Liaquat Ali, the country's first prime minister, in 1951.

After that point, the country witnessed continuous political turmoil, experiencing four governor generals and seven prime ministers in the first decade after independence. There was no change in the military command when General Ayub became the Commander in Chief in 1951.[26] Consequently, the military took advantage of the general political instability to establish its dominance in the political spheres of the country. In addition, the military got firsthand experience with government administration when martial law was imposed in Lahore after riots started against the Ahmadiyya sect and the civilian administration failed to maintain law and order. The martial law administration successfully restored stability and transferred the control of the city to civilian administration in a relatively short time. Hasan Askari noted the following implication of the martial administration on the state and society:

> First, the weaknesses and deficiencies of the political institutions and leaders were exposed – that they could not satisfactorily perform their primary duty of political and administrative management. Second, it gave the military firsthand experience of civilian affairs and the machinations of the political leaders – that some political leaders were involved with smugglers, hoarders and other criminal elements. Third, it created a strong impression in the public mind that the military could cope with a difficult situation even when the political leaders failed, thereby giving a boost to the Army's reputation as a task-oriented and efficient entity with a helpful disposition towards the people.[27]

Therefore, this administration of martial law left a deep imprint on the military's institutional culture. Since then, the military has viewed itself as highly competent and the most uncorrupt institution in the country, while perceiving the political leadership as weak and incompetent in dealing with state affairs. More importantly, it has become a military norm to act under domestic compulsion to establish state control in the country.

External and Internal Pressures Driving Military Predominance

The formation of the Pakistani military's culture was confronted by the constant challenge of India, which was larger in terms of territory, population, economy, and military power. Moreover, the conflict over Kashmir and other bilateral issues intensively shaped Pakistan's early foreign and security policy, consequently prioritizing an arms race with India over domestic nation-building. During the Cold War, the Pakistani government allocated extensive budgetary resources to the armed forces, with the defense expenditure recorded as 53.5% of the total government expenditure during the 1947–1983 period.[28] Furthermore, the United States assisted Pakistan in building its military capabilities after the formation of its alliance in the 1950s, offering $8.7 billion to Pakistan in economic and military aid between 1951 and 1959. This military alliance and high government expenditure helped restructure Pakistan's military capabilities and balance against India's military power. Thus, the perceived threat from India was used to advance the military institution at the expense of civilian institutions from 1947. This allocation of resources lent professional autonomy to the military, and its interests were prioritized in the state's strategic culture while the civilian leadership became subordinate.

While the Pakistani military was modernized according to its own requirements at the expense of civilian institutions and true nation-building, Pakistani military elites nonetheless became heavily involved in domestic affairs by establishing welfare and commercial enterprises using their political and administrative autonomy during military regimes. Since 1954, there has been vertical and horizontal growth in their business ventures. According to media reports, the military runs more than 700 military companies, 96 enterprises, and holds about 10% of the private-sector assets in the country.[29] Furthermore, the military has established public sector organizations that are engaged in various businesses such as transportation, construction, and telecommunication. The purpose behind the business ventures was to promote import-substitution industries and boost the military and private sectors to create employment opportunities in the country. Thus, over time the military has presented itself as a nation-building institution.[30] This has given rise to the Pakistani "military-industrial complex," which gives financial autonomy to the military to act independently.

Furthermore, the Pakistani military has established a monopoly over the definition of "national interests" and has subsequently presented its own institutional interests as national security interests, in particular by utilizing hyper-patriotism and religious ideology. Following the military coup in 1999 led by General Musharraf, a former army chief said that "anyone who did not support the coup was not a patriotic Pakistani."[31] Therefore, the military in Pakistan has been treated as an untouchable subject and any objective criticism of the military has been considered an anti-Pakistan activity and unacceptable maligning of the armed forces.

Despite Pakistan's parliamentary system, the military enjoys greater control over the political parties, media, the Inter-Services Intelligence (ISI), and paramilitary forces as well as the ministry of defense, finance, and foreign affairs. Many retired officers have been allocated to key positions in civil administrations, which provide them with control over other state institutions as well as the ability to run the government.[32] According to Stephen Cohen:

> In a country such as Pakistan, where for many years the armed forces have been at the helm of civil affairs, the influence of adulatory publicity on them cannot be overlooked. It appears to have affected them deeply enough to change their professional attitude and standards and to breed in them the unfortunate belief that armed forces could do no wrong.[33]

The military has justified its dominance on the grounds of the external security threat from India and internally weak and incompetent political leadership to deal with such threats. Thus, Pakistan's military-centric national security approach has played a major role in shaping the country's strategic culture because military elites have a veto within Pakistan's national security state that allows them to largely define the strategic culture of the country.

Gaps in the Strategic Culture Literature

This section has reviewed the extensive literature on Pakistan's strategic culture that explains the military's nearly exclusive role at the helm of strategic affairs. However, the role of Pakistani civilian leadership in strategic affairs has been largely ignored in the literature. After the 1971 India-Pakistan war, the civilian leadership of Z.A. Bhutto got a chance to influence the state's military-centric strategic culture; however, Bhutto continued with the military-centric strategic approach. Although he played an instrumental role in the country's nuclear program, it was later taken over by the military elites. Similarly, Prime

Minister Nawaz Sharif got the opportunity in his second term (1997–1999) to influence Pakistan's strategic approach, but he continued with the same military-centric strategy and conducted Pakistan's first nuclear test after India's 1998 nuclear tests due to the influence of the military. There is very limited academic debate on the role of civilian leadership in shaping Pakistan's strategic culture and little discussion on why even key civilian leaders have been unable to transform it.

The addition of the Pakistani nuclear program has further expanded the military's power and has escaped civilian control or oversight. The military elites have leveraged the secrecy of the nuclear program, which is managed through a closed system. Information relating to its operations and financial resources are kept confidential due to national security concerns, which has magnified the military's control over the program. The military elites' enduring belief that the civilian leadership is weak fuels the perception that civilian leaders could not handle the international pressure over Pakistan's nuclear program.[34] Benazir Bhutto said that she asked for a briefing on Pakistan's nuclear program after becoming prime minister, but she was denied. It was US Ambassador Robert Oakley who informed her about Pakistan's nuclear program.[35]

Some shift in the national dialogue took place after 2007, since which time Pakistan's strategic culture has been discussed and criticized by civilian leaders in the parliament, media, and civil society in public. These leaders have objectively questioned Pakistan's past strategic policies and their impact on the state and society and have demanded that the parliament has the right to decide the country's strategic interests and its foreign and security policies. However, the role of civil society in Pakistan's strategic culture is still significantly underexplored in the literature.

The Indian Factor in Pakistan's Strategic Culture

As discussed above, Pakistan has perceived or encountered a real threat from India since its inception in 1947. In the wake of the partition, the Indian leadership was of the view that the Pakistani state was weak and the country could not survive for long. According to India's first minister of Home Affairs Vallabhbhai Patel, "Pakistan was not viable and would soon collapse."[36] Consequently, the Pakistani leadership perceived that India wanted to undo the partition of India, and thereby revamp the old "Akhand Bharat" – reunified India.[37] Moreover, Kashmir became a main source of conflict and rivalry between India and Pakistan as an area that both countries saw as part of their respective territories following the partition.[38] This led to the 1947–1948 war between the two countries over the issue in which India occupied two thirds of Kashmir and Pakistani forces were defeated during their first war with India.[39]

After this formative conflict, Pakistan's military elites drew some conclusions that shaped formative elements in the state's strategic culture: First, that the partition of India was incomplete; second, that India was going to be an existential threat to Pakistan; and finally, that India sought to be a regional hegemon in South Asia.[40] From that point, the military viewed India to be a permanent external threat in the foreseeable future, and as a result the profile of the Pakistani military increased on the grounds of the security deficit with India. According to General Ayub Khan:

> Our aim must be to make India realize that it is not worth her while to maintain a hostile attitude towards us. India's military strength would always be greater than ours.

Our aim should be to build up a military deterrent force with adequate offensive and defensive power; enough, at least, to neutralize the Indian army. India can concentrate her forces against us without warning. We must, therefore, have a standing army to take the field at moment's notice. In our circumstances a territorial army has hardly any place; it would take too much time to mobilize and train such an army.[41]

The 1971 war between India and Pakistan changed the security dynamics of South Asia and shifted the strategic balance in favor of India. However, Pakistan did not accept Indian regional hegemony and considered asymmetric ways to balance the growing dominance of India in the region. The Pakistani leadership realized that nuclear deterrence was necessary for the territorial integrity and national sovereignty of the country. It was Z.A. Bhutto who proposed the nuclear program during the cabinet meetings in General Ayub's regime. In 1965, Bhutto famously said that "if India builds the bomb, we will eat grass and leaves for a thousand years, even go hungry, but we will get one of our own."[42] Bhutto noted in his book:

Pakistan's security and territorial integrity are more important than economic development.... All wars of our age have become total wars; all European strategy is based on the concept of total war; and it will have to be assumed that a war waged against Pakistan is capable of becoming a total war. It would be dangerous to plan for less and our plans should, therefore, include the nuclear deterrent ... our problem in its essence, is how to obtain such a weapon in time before the crisis begins.[43]

Huge strategic efforts and budgetary allocation were necessary to develop the nuclear program. In 1974, India carried out a nuclear test that provoked Pakistan to start developing its own nuclear program.[44] Consequently, Z.A. Bhutto got an opportunity to materialize his proposed nuclear program on the grounds of the strategic deficit and preventing future events like the 1971 debacle with India.[45] Since then, the development of the nuclear arsenal has become a major component of Pakistani strategic thinking. According to Pervez Hoodbouy and Zia Mian:

Overwhelmed by the power of the bomb, they saw it as magical; a panacea for solving Pakistan's multiple problems. They told themselves and their people that the bomb would bring national security, allow Pakistan to liberate Kashmir from India, bind the nation together, make its people proud of their country and its leaders, free the country from reliance on aid and loans, and lay the base for the long-frustrated goal of economic development.[46]

In addition, Taxila Heavy Industries and the Pakistan Aeronautical Complex in Kamra were established to rebuild, upgrade, and modernize the state's military capabilities in the 1970s. These facilities were used for the production of sophisticated weapons such as tanks, F-6s, the Mushshak and K-8 Karakoram trainer aircraft, armored personnel carriers, radar, and avionics equipment. Most recently, it has developed the JF-17 aircraft in a joint venture with China.[47] These measures have demonstrated Pakistan's resistance against Indian hegemony in South Asia, and the development of the nuclear program and other sophisticated weapons have further extended the military's control over strategic assets in the country – exempted from civilian influence or oversight.

In addition to the nuclear program, Pakistan has used jihadi groups in proxy wars to protect its strategic interests and balance Indian military power. Pakistan supported the Khalistan movement in Indian Punjab in the 1980s and the Kashmir uprising in 1989 by providing covert military aid.[48] This strategy of proxy support was effectively used by the Pakistani military elites against strong Indian military power, based in part on the assumption that India would not react beyond the "Line of Control" (LoC) in case of a proxy war in Kashmir because of Pakistan's nuclear capabilities and India's restraint policy after the Simla Accord in 1971. However, the Indian leadership has demonstrated a more aggressive posture after the Pulwama terror attacks in February 2019 and carried out airstrikes inside Pakistan beyond the LoC, violating the international border. This signals that India has changed its traditional maximum restraint strategy, which has significant impact on Pakistan's asymmetric warfare and nuclear deterrence strategies against India's military power.[49]

There is limited discussion on the 2019 Indian airstrikes in the context of Pakistan's strategic culture. Most of the literature on the Indian factor in Pakistan's strategic culture has focused on a historical perspective, covering Pakistan's traditional rivalry with India and arguing that the military has justified its dominance on the grounds of security shortfalls toward India. However, the recent developments in India, especially with the rise of Indian Prime Minister Narendra Modi and Hindu nationalism, have changed Pakistan's strategic outlook. India's decision to step away from its traditional restraint policy toward Pakistan by carrying out airstrikes in the wake of the 2019 Pulwama terror attack likely means that India has formed a new nuclear threshold in which it has created a strategic space for conventional war. Therefore, there is a need for fresh insights in the literature on the Indian factor in Pakistan's strategic culture in the era of Modi.

Ideological Factors in Pakistan's Strategic Culture

Religious-National Identity

Unlike Turkey, Indonesia, and Algeria, Pakistan won independence from the British colonial empire through a constitutional and political struggle in 1947. The Pakistani military had no direct role in the independence movement against the British empire; in fact, the Pakistani military was part of the "ex-colonial" army, not a "national liberation" army or a "post-liberation army."[50] Rather, the Pakistani military inherited the British legacy, which was designed for colonial rule in the subcontinent. Most of the military officers were commissioned and trained by the British military before joining the Pakistani armed forces after independence. Therefore, at the time of independence, these officers lacked military and political legitimacy which could be used to interfere in the affairs of the new nation-state.[51] Thus, the military had to redefine its role and identity according to the ideology of the new country.

Against this backdrop, the Pakistani military took its inspiration from the ideology of the independence movement known as "two nation theory." This ideology was based on the idea that Muslims and Hindus are two separate nations based on their religion; a premise on which the Muslim League demanded a separate homeland for the Indian Muslims. This ideology has defined Pakistan's political, social, military, and educational discourses after independence. Hasan Askari Rizvi noted that "Islam is an integral part of

Pakistan's strategic culture because it contributes to shaping societal dispositions and orientations of [its] policy makers."[52] Interestingly, the Pakistani military was the first to redefine itself on the basis of this ideology and presented itself as the sole guardian and defender of the country's ideological foundation and its territorial frontiers. However, the Pakistani military continued the practices of the British military and has taken pride in its British legacy of martial virtues and Western standards of professionalism in warfare. Therefore, the military adopted the religious ideology of the new Pakistani nation in combination with Western military professionalism.

In addition, the Pakistani military takes motivation from early Muslim warriors and the battles they fought against great empires such as Byzantium and Persia. John Keegan argues that "it was Islam itself, which lays so heavy an emphasis on the fight for the faith, that made them so formidable in the field."[53] The Pakistani military has viewed itself as a successor of the Muslim warrior tradition – fighting for the noble cause of Islam and subsequently conquering the great empires of their time. Through this ideological lens, the Pakistani military fostered the belief that "one Pakistani soldier is worth ten Indian soldiers."[54] Thus, the military has presented itself as a Muslim military force, which provides the institution with public support and legitimacy to have control in both state and society. According to former interior secretary Khalid Aziz:

> The Pakistan military strongly believed that if religious identity is imposed on Pakistan, they will be mentally ready to face any enemies who might attack Pakistan[,] especially India. Consequently, the religious narrative was propagated as the national identity by [the] military establishment.[55]

Asymmetric Warfare and Proxy Groups

As discussed earlier, the Pakistani military adopted asymmetric warfare strategies to achieve strategic parity against India in South Asia, providing covert support to jihadi groups as a "politico-military strategy" in the Afghanistan and Kashmir conflicts. After 1979, General Zia devised a strategy to wage jihad against the Soviet occupation of Afghanistan. The United States joined the bandwagon after seeing that this strategy was successful in countering Soviet forces in Afghanistan. The Central Intelligence Agency and ISI conducted the world's biggest covert operation to counter the Soviet Union during the 1980s; moreover, the ISI received a significant amount of financial aid from Western and Gulf countries – around $2 billion from the United States and a roughly equal amount from Saudi Arabia and other Gulf countries – for the Afghan resistance movement against the Soviet Union.[56] Ahmad Rashid writes:

> General Zia did not allow the CIA or any other foreign intelligence agency to aid the Mujahedin directly, enter Afghanistan, or plan the Mujahedin's battles and strategy. That became the prerogative of the ISI, which, with its newfound wealth and American patronage, had become a state within a state, employing thousands of officers in order to run what was now also Pakistan's Afghan war.[57]

Thus, the ISI became another key player under the influence of military elites in shaping Pakistan's strategic policymaking during the Soviet-Afghan War. Since then, the ISI has coordinated with the army elites to further expand military control in state affairs.

In addition, General Zia institutionalized religion in the armed forces. Many religious scholars were appointed to work along with the armed forces from the Jamaat-e-Islami and Deobandi religious groups. Moreover, Pakistan's northwestern region became the base point of a transnational jihadi resistance movement to fight against the Soviet forces in Afghanistan. As a result, an alliance was established between the military, militants, and mullahs (religious class), which served Pakistan's strategic interests during the Soviet-Afghan War. This alliance has brought a new class of military officers, which is "ideological in profession of belief and practice of both" as put by one Pakistani strategist. In this process, the Pakistani military viewed itself in a larger pan-Islamic context during General Zia's regime.[58] Thus, the Soviet-Afghan War added a new flavor to the Pakistani strategic narratives institutionalizing religion in the military and using jihad as a "politico-military strategy."[59]

Like the Soviet occupation of Afghanistan, the 9/11 terror attacks had a major impact on Pakistan's strategic culture. It was General Musharraf who formulated the strategy to join the GWOT. Musharraf apparently attempted to part ways with militant groups in Kashmir and Afghanistan, which led to a major rift between the military elites. Consequently, some key officers resigned from the military after Pakistan joined the GWOT. Subsequently, the Pakistani military conducted operations against select militant groups by differentiating on the basis of threats and strategic utilization. For example, Pakistan cracked down on groups that turned against the Pakistani state, including Tehrik-i-Taliban Pakistan, the Islamic Movement of Uzbekistan, and Al-Qaeda.[60] However, the Pakistani military thought it would overstretch the armed forces and evoke a possible blowback by taking on all the militant groups. Paul Staniland et al. explain that:

> the true crackdowns [by the Pakistani military] have only occurred when groups ideologically radicalized against the military and began making unacceptable political demands, rather than in response to outside pressure or a change in core military preferences. Pakistan's army appears entirely comfortable with a fractured monopoly of violence, as long as it functions on the military's political terms.[61]

Therefore, the military elites have utilized the militant groups based on their strategic use to secure Pakistan's interests in South Asia. However, Islam, which was applied as a unifying national force, took a very conservative turn during General Zia's regime, leaving deep imprints on Pakistan's state and society. This religiosity has put soldiers under the delusion that "no power on earth can subdue the valour of the Mujahidin."[62] The Pakistani military elites viewed the glory of Islam as being made manifest in the Pakistani military, which was born to lead and rule after the Soviet Union withdrew from Afghanistan. More importantly, the practice of asymmetric war and the use of proxy groups have further strengthened the role of the military in the state's strategic culture. It was the Pakistani military that planned and executed the Afghan War against the Soviet forces and more recently the GWOT, both with the support of the United States.

This review of the literature highlights studies that explain how the Pakistani military has redefined its role based on the "two nation theory" after the partition, and moreover how the military has used jihadi groups instrumentally to achieve its strategic interests in the region. The use of such proxy groups was an effective strategic tool during the Cold War, but it has become a very dangerous phenomenon after the emergence of radical Islam in the post-Cold War era. There is a need for further study on the ways proxy war in the form of jihadi groups transformed from an effective instrument of preference within Pakistani strategic culture to a dangerous phenomenon in the post-Cold War era, especially after the 9/11 terror attacks.

Conclusion

The core themes across the literature on Pakistan's strategic culture focus on the central role of the military elites in developing Pakistan's strategic culture after the state's independence in 1947. The Kashmir issue and rivalry with India have been used by the military to justify its near-exclusive control in state affairs on the pretext of Pakistan's security deficit. The Pakistani military redefined itself on the country's ideological right after its inception, and subsequently military elites presented themselves as guardians to protect the state's territorial integrity and ideological foundations, which has in turn provided position and prestige to the military in the Pakistani state and society. The military receives a large share of budgetary resources, which has granted it professional and financial autonomy and reinforced military elites' belief that the civilian leadership is too weak and incompetent to deal with state affairs. Hence, the military elites have established a powerful role and veto in defining Pakistan's strategic culture, and significantly, the military's role as a savior and guardian has been accepted in the country given the geostrategic global environment and domestic compulsion to protect the country's national interests.

This chapter considers a wide range of the literature on Pakistan's strategic culture that explores the predominant role of the military in state affairs; however, there is very limited discussion in the literature over the role of civilian leadership in building or influencing the military-centric strategic culture of the country. In a similar fashion, while there has been extensive scholarship on Pakistan's traditional rivalry with India in the context of Kashmir and the major wars fought by the two countries, further study is needed on Pakistan's response to India's increasingly aggressive strategic behavior under the leadership of Prime Minister Modi. Pakistan's adoption of asymmetric warfare strategies after 1971 to balance Indian military power and achieve its regional strategic interests has been well studied; however, the use of proxy groups has provoked severe blowback in the form of extremism, radicalization, and terrorism. As Pakistan has become increasingly isolated globally and put on the gray list of the Financial Action Task Force for these issues, the state faces an increasing need to reform its strategic culture and look for alternative strategic options to protect its national interests in the region and further afield. Further study is needed on the shifts in policy and paradigms that Pakistan's elites may adopt to navigate the complex security environment of the 2020s and beyond. These include how Pakistan may use its strategic location for regional connectivity by adopting an economic-driven foreign policy that could help restructure its weakening economy and international image, with a particular eye toward the potential impact of China's multibillion-dollar projects in the region under the Belt and Road Initiative from which Pakistan could only benefit if it transforms elements of its strategic thinking and norms.

Notes

1 Peter R. Lavoy, "Pakistan's Strategic Culture," *SAIC*, 31 October 2006, 5–6.
2 Peter Rosen's definition, as quoted in Anand V., "Revisiting the Discourse on Strategic Culture: An Assessment of the Conceptual Debates," *Strategic Analysis* 44, no. 3 (2020): 199–200.

3 Bradley S. Klien's definition, as quoted in Anand V., "Revisiting the Discourse on Strategic Culture," 199–200.
4 Sumita Kumar, "Pakistan's Strategic Thinking," *Strategic Analysis* 35, no. 3 (2011): 479.
5 Arshad Ali, Pakistan's National Security Approach and Post-Cold War Security: Uneasy Co-existence (Abingdon, Oxon: Routledge, 2021), 32–34.
6 Arshad Ali and Robert G. Patman, "The Evolution of the National Security State in Pakistan: 1947–1989," *Democracy and Security* 15, no. 4 (2019): 309–311.
7 Priyanka Singh, "Army: The Be-All or End-All of Pakistan Politics?," *Strategic Analysis* 39, no. 3 (2015): 319.
8 Shuja Nawaz, *Crossed Swords: Pakistan, Its Army, and the Wars Within* (Oxford University Press, 2009), 410–414.
9 Bennett O. Jones, *Pakistan: Eye of the Storm* (Yale University Press, 2002), 276–277.
10 Arshad Ali and Robert G. Patman, "The Evolution of the National Security State in Pakistan: 1947–1989," *Democracy and Security* 15, no. 4 (2019): 320.
11 Arshad Ali and Fazal Subhan, "Deconstructing the Myth of Pashtun as a Nation of Extremists and Warriors," in Muhammad Shoaib Pervez ed., *Radicalization in Pakistan: A Critical Perspective* (Routledge 2020), 130.
12 C. Christine Fair, *Fighting to the End: The Pakistan Army's Way of War* (Oxford University Press, 2014), 112–115.
13 C. Christine Fair and Shuja Nawaz, "The Changing Pakistan Army Officer Corps," *The Journal of Strategic Studies* 34, no. 1 (2011): 63.
14 Peter J. Katzenstein, *The Culture of National Security: Norms and Identity in World Politics* (Columbia University Press, 1996).
15 Ayesha Siddiqa, Military Inc. Inside Pakistan's Military Economy (Pluto Press, 2007), 71.
16 Ibid.
17 The Constitution of Islamic Republic of Pakistan, 1973, Article 245.
18 Paul Staniland et al., "Pakistan's Military Elite," *Journal of Strategic Studies* 43, no. 1 (2020): 74–76.
19 C. Christine Fair and Seth G. Jones, "Pakistan's War Within," *Survival* 51, no. 6 (2009): 163.
20 Paul Staniland et al., "Pakistan's Military Elite," *Journal of Strategic Studies* 43, no. 1 (2020): 74–76.
21 Arshad Ali, *Pakistan's National Security Approach and Post-Cold War Security: Uneasy Co-existence* (Abingdon, Oxon: Routledge, 2021), 35.
22 Roedad Khan, "Quaid's Visit to Staff College Quetta," *The Nation*, Pakistan, 22 November 2012.
23 Aqil Shah, *The Army and Democracy: Military Politics in Pakistan* (Harvard University Press, 2014), 35–36.
24 Kaushik Roy and Scott Gates, *Conventional Warfare in South Asia, 1947 to the Present* (Routledge 2011), xvii.
25 Stephen Cohen, "Pakistan: Army, Society, and Security," *Asian Affairs: An American Review* 10, no. 2 (1983): 1.
26 Arshad Ali, *Pakistan's National Security Approach and Post-Cold War Security: Uneasy Co-existence* (Abingdon, Oxon: Routledge, 2021), 37.
27 Hasan A. Rizvi, *Military State and Society in Pakistan* (Palgrave Macmillan, 2000), 78–79.
28 Hasan A. Rizvi, "Pakistan's Defence Policy," *Pakistan Horizon* 36, no. 1 (1983): 38.
29 Ron Moreau, "The Military's Long Reach," *Foreign Policy*, 12 October 2009, http://foreignpolicy.com/2009/10/12/the-militarys-long-reach/.
30 Ayesha Siddiqa, *Military Inc. Inside Pakistan's Military Economy* (Pluto Press, 2007), 50.
31 Ahmad Faruqui, "The Army and Democracy: Military Politics in Pakistan," *The RUSI Journal* 159, no. 6 (2014): 75.
32 Yunas Samad, "The Army and Democracy: Military Politics in Pakistan," *Commonwealth & Comparative Politics* 54, no. 1 (2016): 145.
33 Smruti S. Pattanaik, "Civil-Military Ccoordination and Defence Decision-Making in Pakistan," *Strategic Analysis* 24, no. 5 (2000): 940.
34 Shuja Nawaz, *Crossed Swords: Pakistan, Its Army, and the Wars Within* (Oxford University Press, 2009), 437.

35 Ibid., 415.
36 Nicholas Mansergh, "The Partition of India in Retrospect," *International Journal* 21, no. 1 (1966): 16.
37 Jaffrelot, C., Pakistan at the Crossroads: Domestic Dynamics and External Pressures (Columbia University Press, 2016).
38 Varun Vaish, "Negotiating the India-Pakistan Conflict in Relation to Kashmir," *International Journal on World Peace* 28, no. 3 (2011): 57–58.
39 Kaushik Roy and Scott Gates, Conventional Warfare in South Asia, 1947 to the Present (Routledge, 2011), x.
40 C. Christine Fair, "Pakistan's Strategic Culture Implications for How Pakistan Perceives and Counters Threats," *The National Bureau of Asian Research –NBR* special report no. 61, December 2016, 4–6.
41 Ayub Khan, *Friends Not Masters: A Political Autobiography* (New York: Oxford University Press, 1967), 47.
42 Yaqoob Khan, "Eating grass," *Express Tribune*, Pakistan, 24 January 2015, http://tribune.com.pk/story/826538/eating-grass/.
43 Zulfikar A. Bhutto, *Myth of Independence* (Oxford University Press, 1969), 152–153.
44 Alexander Evans, "Pakistan and the Shadow of 9/11," *The RUSI Journal* 156, no. 4 (2011): 66.
45 Pervez Hoodbhoy and Zia Mian, "Nuclear Fears, Hopes and Realities in Pakistan," *International Affairs* 90, no. 5 (2014): 1128.
46 Ibid., 1127–1141.
47 Muhammad N. Mirza et al., "Military Spending and Economic Growth in Pakistan," *Margalla Papers,* Islamabad XIX, no. 19 (2015): 169.
48 Alexander Evans, "Pakistan and the Shadow of 9/11," *The RUSI Journal* 156, no. 4 (2011): 66.
49 Abhinav Pandya, "The Future of Indo-Pak Relations after the Pulwama Attack," *Perspectives on Terrorism* 13, no. 2 (2019): 65–68.
50 Raymond A. Moore, "The Army as a Vehicle for Social Change in Pakistan," *The Journal of Developing Areas* 2, no. 1 (1967): 59.
51 Aqil Shah, *The Army and Democracy: Military Politics in Pakistan* (Harvard University Press, 2014), 31–32.
52 Runa Das, "The Prism of Strategic Culture and South Asian Nuclearization," *Contemporary Politics* 15, no. 4 (2009): 398–399.
53 John, Keegan, *A History of Warfare* (New York: Random House, 1993), 196.
54 Ahmad Faruqui, "Pakistan's Strategic Myopia," *The RUSI Journal* 145, no. 2 (2000): 52.
55 Arshad Ali, *Pakistan's National Security Approach and Post-Cold War Security: Uneasy Co-existence* (Abingdon, Oxon: Routledge, 2021), 154.
56 Husain Haqqani, "Pakistan: Between Mosque and Military," *Carnegie Endowment for International Peace* (2005): 85.
57 Grant Holt and David H. Gray, "A Pakistani Fifth Column? The Pakistani Inter-Service Intelligence Directorate's Sponsorship of Terrorism," *Global Security Studies* 2, no. 1 (2011): 58.
58 Ibid., 55–56.
59 Arshad Ali, *Pakistan's National Security Approach and Post-Cold War Security: Uneasy Co-existence* (Abingdon, Oxon: Routledge, 2021), 61.
60 Staniland, P. et al., "Politics and Threat Perception: Explaining Pakistani Military Strategy on the North West Frontier," *Security Studies* (2018): 2.
61 Ibid.
62 C. Christine Fair, *Fighting to the End: The Pakistan Army's Way of War* (Oxford University Press, 2014), 99.

Selected Bibliography

Arshad Ali, *Pakistan's National Security Approach and Post-Cold War Security: Uneasy Co-existence*. Abingdon, Oxon: Routledge, 2021.

Arshad Ali and Robert G. Patman, "The Evolution of the National Security State in Pakistan: 1947–1989." *Democracy and Security* 15, no. 4 (2019): 301–327.

Zulfikar A. Bhutto, *Myth of Independence*. Oxford University Press, 1969.
Stephen Cohen, "Pakistan: Army, Society, and Security." *Asian Affairs: An American Review* 10, no. 2 (1983): 1–26.
Runa Das, "The Prism of Strategic Culture and South Asian nuclearization." *Contemporary Politics* 15, no. 4 (2009): 395–411.
Alexander Evans, "Pakistan and the Shadow of 9/11." *The RUSI Journal* 156, no. 4 (2011): 64–70.
Christine Fair, *Fighting to the End: The Pakistan Army's Way of War*. Oxford University Press, 2014.
Christine Fair and Seth G. Jones, "Pakistan's War Within." *Survival* 51, no. 6 (2009): 161–188.
Christine Fair and Shuja Nawaz, "The Changing Pakistan Army Officer Corps." *The Journal of Strategic Studies* 34, no. 1 (2011): 63–94.
Husain Haqqani, "Pakistan: Between Mosque and Military." *Carnegie Endowment for International Peace*, 2005.
Pervez Hoodbhoy and Zia Mian, "Nuclear Fears, Hopes and Realities in Pakistan." *International Affairs* 90, no. 5 (2014): 1125–1142.
C. Jaffrelot. *Pakistan at the Crossroads: Domestic Dynamics and External Pressures*. Columbia University Press, 2016.
Bennett O. Jones, *Pakistan: Eye of the Storm*. Yale University Press, 2002.
John Keegan, *A History of Warfare*. New York: Random House, 1993
Ayub Khan, *Friends not Masters: A Political Autobiography*. New York: Oxford University Press, 1967.
Sumita Kumar, "Pakistan's Strategic Thinking," *Strategic Analysis* 35, no. 3 (2011): 479–492.
Shuja Nawaz, *Crossed Swords: Pakistan, its Army, and the Wars Within*. Oxford University Press, 2009.
Smruti S. Pattanaik, "Civil-Military Coordination and Defence Decision-Making in Pakistan." *Strategic Analysis* 24, no. 5 (2000): 939–968.
Hasan A. Rizvi, "Pakistan's Defence Policy." *Pakistan Horizon* 36, no. 1 (1983): 32–56.
Hasan A. Rizvi, *Military State and Society in Pakistan*. Palgrave Macmillan, 2000.
Kaushik Roy and Scott Gates, *Conventional Warfare in South Asia, 1947 to the Present*. Routledge, 2011.
Yunas Samad, "The Army and Democracy: Military Politics in Pakistan." *Commonwealth & Comparative Politics* 54, no. 1 (2016): 144–146.
Aqil Shah, *The Army and Democracy: Military Politics in Pakistan*. Harvard University Press, 2014.
Ayesha Siddiqa, *Military Inc. Inside Pakistan's Military Economy*. Pluto Press, 2007.
Staniland, P. *et al.*, "Politics and Threat Perception: Explaining Pakistani Military Strategy on the North West Frontier." *Security Studies* (2018): 535–574.
Paul Staniland *et al.*, "Pakistan's Military Elite." *Journal of Strategic Studies* 43, no. 1 (2020): 74–103.

24

AN IRANIAN WORLDVIEW

The Strategic Culture of the Islamic Republic[1]

Ali Parchami

Introduction

For the past four decades, the main exponent of the Iranian worldview has been the governing regime of the Islamic Republic. It has weathered significant external and domestic challenges to this point but has also been the architect of its own most critical weaknesses. Dominated by a clerical hierarchy that draws on Shia Islamic conventions and Iranian national identity, it is a hybrid subculture with its own distinctive outlook, values, and norms. Built around its founding father's *velayat-e-faqih* doctrine, domestically it espouses conservatism through the Islamification of every facet of life.[2] Internationally, the Iranian governing regime is fervently revisionist, with an ingrained hostility toward the West and the United States in particular. Bent upon exporting its revolutionary ideals, it straddles the lines between a modern nation-state and a transnational Islamist polity. This duality is reflected in its structures and a fractious body politic that is characterized by intense institutional and factional rivalries.

The regime has sought to capitalize on Iran's Shia identity by presenting itself as its sole custodian and chief proponent. Adopted in the sixteenth century as the state religion of a reconstituted Persian Empire, Shiism has ever since been used by the country's rulers to mold a unified national identity across Iran's vast territory with its ethno-culturally diverse population. Shia identity imbues Iranians with a sense of exceptionalism: after all, it was the influence of Iranian civilization that transformed an Arab tribal creed into a religion that could be endorsed by other nationalities and cultures. Iranians regard Shiism as the embodiment of the Islam intended by the Prophet and regard themselves as the faith's true torchbearers.[3]

Central to Shiism is *mazloumiat* – the principle of confronting injustice, even against great odds and at the cost of self-sacrifice. It has strong appeal for a nation whose history is punctuated by invasions that laid waste to its cities and saw its population repeatedly massacred. Subjugated by a succession of Arabs, Turks, and Mongols, Iranians never succumbed to their conquerors by surrendering their language and traditions. National identity is, therefore, another incontrovertible driver that shapes the Iranian worldview. At its core is a dichotomy between perpetual belief in victimization and optimism in inevitable resurgence. Pride in the richness of Iran's ancient civilization encourages long-term thinking but can also infuse Iranians with misplaced overconfidence.

DOI: 10.4324/9781003010302-27

The Islamic Republic's Strategic Culture

The Iranian worldview is informed by the perceptions, values, and norms that define its strategic culture.[4] Because of its millennia-spanning history and contemporary strategic significance, Iran's strategic culture has received significant attention from scholars and practitioners. Prominent contributors include but are not limited to Michael Eisenstadt,[5] Nima Gerami,[6] Gregory Giles,[7] W.A. Rivera,[8] and Ilan Berman.[9]

Since 1979, the main conduit of Iran's strategic culture has been the regime that was established by Ayatollah Khomeini. His unique interpretation of the *velayat-e-faqih* doctrine envisaged a polity governed by an Islamic jurist, or "Guardian," with extensive political and theological authority over the state and its people. Khomeini's thesis was controversial, even among Shia clerics, for the powers it granted to the Guardian. Equally contentious was its transnational implications for bestowing on the Guardian a mandate to intervene in the affairs of the wider Islamic community.[10] Railing against the influence of foreign cultures, Khomeini singled out Western imperialism – and pro-Western governments in Muslim-majority states – as the source of moral corruption and Islamic decline. His *velayat-e-faqih* doctrine demanded the total expurgation of non-Islamic influence from Islamic societies, the expulsion of Westerners, and the overthrow of regimes that did their bidding.[11]

The worldview of the Islamic Republic was, therefore, revisionist from its inception. The regime's identity revolves around an in-built hostility toward Western culture, especially its liberal democratic values. It is an antipathy generated partly out of historical grievances – specifically the national and Islamic experience of humiliation by Western imperialists. It is also informed by the Shia principle of *mazloumiat*: the imperative that the oppressed must rise up against oppressors. The regime views the existing international order as intrinsically unjust: a hegemonic construct created and maintained for the benefit of what Khomeini – and his adherents – describe as "world arrogance."[12] Ideally, the Islamic Republic would like this order to be overturned. For its part, it has adopted a resistance culture domestically and, on the world stage, a culture of defiance.

If the regime's rhetoric is to be believed, its overarching objective is to restore the dignity and status of the Islamic world by bringing about conditions that are conducive to a "Muslim awakening."[13] To achieve this, it claims the Islamic Republic will free the "oppressed masses" from the shackles of Western hegemonic influence by exposing the corruption and subservience of governments that serve as Western lackeys. In line with this affectation, the Iranian state media routinely describes the country's Supreme Leader as *Vali Amr Muslemin* – the "leader of all Muslims." It is a self-righteous ideology that may have carried some substance when Khomeini was alive, but the rhetoric quickly betrays its hollowness when weighed against the core value most cherished by the regime.

Core Value(s) of the Regime

Regime security outweighs all other considerations for the Islamic Republic's leadership. Under its *Maslahat-e-nezaam* or expediency diktat, electoral outcomes, the Iranian constitution, Shia traditions, and even Islamic law may be set aside should they contravene the regime's interests.[14] When expedient, the regime has been willing to discard ideology and suspend its antipathy toward even the bitterest of foes. A history of secret negotiations and cooperation with the United States, the provision of safe passage for Al-Qaeda fighters to enter Iraq, and the arming of Taliban insurgents in Afghanistan are examples of singular pragmatism when the regime has felt imperiled.[15] Even as it vehemently condemns the suffering of Muslims elsewhere, in the name of expediency the self-professed "leader of the Islamic world" can turn a blind eye to the internment of Uyghur Muslims in China and oversee its proxies engage in sectarian cleansing in the urban battlefields of Syria.

The nexus between self-preservation and expediency are apparent in the regime's external and internal behavior. Ideologically, the Islamic Republic remains committed to exporting its revolution. But some three decades after the death of Khomeini, the security of the regime is the key variable in all its strategic calculations. Convinced that Washington – and the Saudis and Israelis – are intent on engineering regime change, Tehran is determined to take the war to its enemies in the place and time of its choosing. It is a strategy designed to keep opponents off-balance and provide the regime with leverage in negotiations. Surrounded by hostile states and the might of the US military, Tehran has been using instability as a tactical tool. Knowing that stability is a core US objective in the Middle East – and vital to the security of Israel and the conservative Arab states – the Islamic Republic promotes and aids perpetual radicalization at a sub-state level.

In creating a network of transnational proxies, Tehran has compensated for its conventional military weakness by enhancing its asymmetrical capabilities. Its sub-state affiliates – comprising ideological surrogates, groups with overlapping interests, and mercenaries – are an instrument of deterrence that can be deployed to coerce opponents and project Iranian influence. With thousands of armed militias scattered across the region, Tehran aims to reduce the risk of a direct confrontation with its enemies by making war with Iran a high-cost enterprise.[16] The regime is naturally opportunistic: whenever an opening has presented itself, it has exacerbated and deepened its opponents' difficulties, as with the insurgencies in Iraq and Afghanistan. In other instances, it has ensnared adversaries into protracted conflicts, such as the Saudi-led military campaign in Yemen.

By aiding so-called revolutionary forces, Tehran hopes to further inflame anti-Western sentiment. In the short term, growing hostility may distract and preoccupy the United States and regional opponents, reducing their capital and influence. In the intermediate to long term, Tehran hopes to wear down the United States and its European allies by facilitating their voluntary extrication from the Middle East. Left on their own, Washington's regional partners may either be soft targets or are likely to seek terms and accommodation. In reality, although the Islamic Republic has been a consequential player in major regional events, especially in terms of influencing conflicts, this has entailed considerable political and economic costs without producing the geopolitical outcomes Tehran desires. Its steadfast persistence with a costly policy of limited gains suggests that the expulsion of the West, and the geopolitical reconfiguration of the Middle East, are strategic ideals but regime security remains the core objective.

Regional interventionism is an imperative that is also influenced by domestic developments. With a faltering economy, a rise in social-political discontent, and declining popular legitimacy, the regime has been seeking validation and supporters outside its borders. By pouring money and amenities into mainly Shia Arab communities, the clerical hierarchy has always sought a constituency outside Iran. This goes beyond a desire to project influence: an overseas constituency yields some legitimacy to the regime's pretension that it is a transnational polity dedicated to all Muslims. The resettlement of some of this constituency within Iran also embeds a loyal power base that, if and when necessary, can be deployed against the Iranian population.[17] For these reasons, as its domestic troubles mount the Islamic Republic is likely to entrench itself further across the Middle East.

An extraterritorial constituency is not without challenges and risks. The majority of Shia communities in the Arab world view Iranian intervention with disdain. Angry protestors – from the shores of Lebanon to Iraqi cities and towns – have taken to the streets from time to time to condemn this interference, which they label "Iranian Islam."[18] Ironically, rather than be embraced as the savior it claims to be, the Islamic Republic is often viewed as an imperialist power by the very communities it regards as its natural constituency abroad. Ethno-cultural differences notwithstanding, Iranian machinations in the Arab world have a history of raising the suspicions of Arab governments, leading to an intensification of sectarian violence and the persecution of Shia Arabs.

Financial investments overseas also stir up domestic resentment. With living standards falling, angry Iranian protestors have periodically denounced the regime's regional largesse. Popular chants of "no Gaza, no Lebanon, I give my life for Iran" betray a growing nationalist undertone.[19] When protests arise, including the 2022 swell in response to the death of Mahsa Amini while in the custody of the regime's morality police, the regime has acted quickly and ruthlessly by unleashing violence against unarmed protestors. In the expediency of protecting the regime, the distinction between criminality and the act of sinning has been blurred to dissuade outward expressions of discontent. Even criticism of the clerical leadership can entail prosecution and capital punishment under the crime of "waging war against God."[20]

Identity, Inner Dynamics, and Norms

Even prior to the establishment of the Islamic Republic, Iran's clerics had their own corporate identity: a distinct ethos and outlook shaped by seminary education and manifested by a distinctive dress code and way of speaking. Before 1979, *Ruhaniyat*, the clerical class, was esteemed in Iranian society. With a few notable exceptions, such as the Constitutional Revolution (1905–1911) and Iran's short-lived experimentation with popular democracy (1951–1953), the *Ulema* (senior clergy) viewed interfering in politics as undignified and beneath them.[21] Instead, the religious establishment gave its full support to the reigning monarch and, in return, received royal patronage and the privilege of consultation in decisions pertaining to social and religious norms.

Khomeini changed this long-standing convention by advocating the politicization of the clergy. In doing so, he contributed to a widening rift between the regime's clerical hierarchy and Shia traditionalists in seminaries. Moreover, political empowerment ensured that his clerical followers would be exposed to the same divisions, temptations, and vested interests that tend to plague most governing classes. In the early postrevolutionary years, internal differences were encapsulated by two factions. The first group, radical ideologues, viewed Iran largely as a launching pad for a transnational Islamist struggle. Dismissing Iranian identity and contemptuous of international protocols and borders, ideologues advocated an all-out regional war against the enemies of Khomeinism. The opposing camp, composed of pragmatist conservatives, preferred to consolidate the foundations of the regime before embarking on transnational operations. This group was mindful of the sensitivities of the Iranian public for fear that, if their needs were ignored and national interests not upheld, the regime could be imperiled.[22] Both blocs fully subscribed to Khomeini's worldview, but differed on strategy and timing.

In the post-Khomeini era, factional differences gradually became subsumed in institutional turf wars and interpersonal enmities.[23] From the 1990s onward, the factions coalesced into two broad camps, representing a conservative coalition and a bloc consisting of technocrats and reformers. The former has broadly supported the vast powers entrusted to the Office of the Supreme Leader and has close ties with Shia seminaries, religious foundations, the security services, and the Islamic Revolutionary Guard Corps (IRGC). It encourages securitization and social-cultural Islamification and retains a deep-seated ideological enmity toward the United States and other liberal democracies. The more progressive elements within the reform movement seek to divest the Supreme Leader of some of his powers by granting elected officials the authority to effect real change. To this end, reformers claim to promote a culture of accountability, the liberalization of society and the economy, and the easing of tensions with the West.

Splinter groups within each bloc ensure that factional allegiances are fluid: for instance, a party identified as conservative may align on social issues with reformers but support a hardline position in the United States. Oddly, the reform movement encompasses individuals who in the 1980s would have been recognized as ideologues, while among hardliners are pragmatists from the same era. In contrast to the 1980s, factional tensions are no longer confined to strategy but now reflect a divergence in ideology and norms, including differences over each bloc's ambitions for the character of the regime. Increasingly, these disagreements are underlined by the desire of hardliners to make the final transition to a full-blown revolutionary theocracy by abandoning any residue of republicanism. In contrast, their opponents would like to establish a genuine republic that merely has an Islamist orientation.[24] Following the 2020 victory of conservatives and hardliners in parliament, and the 2021 election of Ebrahim Raisi as president, the reform movement has effectively been sidelined from all key state positions in favor of devotees of the Supreme Leader. This, however, has not ended factionalism as there are significant differences within and between the conservative and hardliners coalition.[25]

The factional rift is exacerbated by a system that allows for limited electoral representation to provide a veneer of popular legitimacy. The conflict between the elected and nonelected organs of the state is often played out in public, though it is not always clear to what extent the drama is real or staged for domestic and international consumption. It is a duality that makes it difficult to understand the processes for decisionmaking and the forces that influence policymaking. For instance, it is recognized that a number of organizations with ties to the Office of Supreme Leader operate outside the remit of the elected government.[26] The executive branch – headed by the president – can neither scrutinize these groups' activities nor control their budget.[27] Among them is the IRGC, whose commanders are only answerable to the Supreme Leader and whose vested interests and priorities, including extraterritorial operations, can contravene the policies of the government of the day.[28]

Further incongruity is caused by parallel institutions – often, though not always, representing the elected and nonelected elements of the regime. The Islamic Republic has numerous departments and agencies with overlapping responsibilities that foster jurisdictional wrangling. This is exemplified by the 2021 struggle between the Ministry of the Interior and the Guardian Council over the required criteria for presidential candidates.[29] Rivalries between parallel organizations that undercut one another have reached such dangerous proportions among the intelligence community that a former minister has publicly warned the authorities about its consequences. In a 2021 interview, Ali Younesi, a former intelligence chief (2000–2005), publicly denounced the culture of "infighting" between "parallel organizations" that attack one another and in so doing allow "foreign infiltrators" to operate with impunity across the country.[30]

These structural shortcomings are accentuated by the regime's penchant for inundating the public with contradictory statements on issues ranging from the rate of unemployment and inflation to who gave the order for the security forces to open fire on unarmed protestors.[31] Secrecy, misdirection, and misrepresentation have become norms for the Islamic Republic's authorities. *Tagiyya* – the practice of resorting to dissimulation – is a well-known principle in Islamic tradition. If Javad Zarif is to be believed, the culture of dissimulation is now embedded within the regime's inner core. As revealed in a leaked interview with Iran's foreign minister at the time, after the Ukrainian passenger flight was shot down by an IRGC missile in January 2020, Revolutionary Guard commanders adamantly denied any responsibility for the incident, and in a closed meeting urged Zarif and the Foreign Office to strenuously deny the regime's culpability.[32]

Officials so routinely make statements that are later exposed as untrue, and habitually attack and contradict one another, that public confidence has been steadily eroded in the regime's ability to govern. National morale was made worse by the mismanagement of the COVID-19 pandemic. Despite the prevailing culture of securitization, the policy of politicizing the pandemic resulted not only in widely suspected distortion of the death rates but also in endemic failures by the authorities to deliver on their vaccination promises – with no single minister or state body assuming responsibility for the fiasco.[33] The bitter blame game across repeated episodes of government failings has fostered a culture of apathy among a disillusioned electorate whose sporadic acts of passive resistance bloomed in 2022–2023 into unprecedented levels of country-wide protests.[34]

Nuclear Controversy

Iran's nuclear program encapsulates many of the characteristics that define the regime's strategic culture. The Islamic Republic vehemently objects to international efforts that seek to monitor and curb its nuclear activities. Its indignation stems in part from a worldview that portrays the West as duplicitous and hypocritical. The regime likes to point out that Iran's nuclear program dates back to the prerevolutionary era when the Pahlavi monarchy was openly expressing a desire to turn Iran into a world power.[35] Yet, despite such blatant ambitions, the country's nascent nuclear program received direct Western assistance from both the United States and the Europeans.[36]

Discarded in the immediate postrevolutionary period, renewed interest in the nuclear program arose in the mid-1980s. Although the Islamic Republic maintains that its program is strictly civilian – with the Supreme Leader issuing a religious fatwa on the matter – it has met with a wall of international condemnation. This plays into the regime's ideological narrative that portrays Iran as a victim of selective application of international law, and the West as hypocritical for ignoring the nuclear activities of other countries, such as Israel and India. The West, according to the regime, is determined to protect its vested interests in a hegemonic order by keeping the Islamic Republic boxed in and preventing its rise from challenging the existing system.[37]

Publicly, the Iranian regime maintains that its nuclear program is necessary for generating power for a large country with a growing population. Its infrastructure, it insists, has been devastated by four decades of sanctions and a lack of foreign investment. When confronted with the fact that Iran has vast reserves of oil and gas, the regime offers further justification that civilian nuclear power is a right under international law; and conforms to the Islamic Republic's mantra of forging an independent path free from foreign influence. While these explanations should not be dismissed out of hand, the controversy surrounding Iran's program often obscures a simple truth: the preservation of the regime is the key to understanding Tehran's nuclear outlook.

Tellingly, it was around 2002 that the scale of Iran's nuclear program was first publicly exposed by an Iranian opposition group that revealed covert facilities and extensive investment in dual-use technology. Iran's nuclear ambitions make sense in light of the geopolitical developments that accompanied the Global War on Terrorism.[38] The US military buildup along the Islamic Republic's periphery, President George W. Bush's "Axis of Evil" speech, which lumped Iran with North Korea and Iraq as potential targets, and Vice President Dick Cheney's ominous threats about "real men" marching on Tehran were all indicative of an existential threat.[39] The fate of Saddam Hussein in Iraq – in sharp contrast to the Kim dynasty in nuclear-armed North Korea – would have been a lesson not lost on the Iranian leadership.

It may seem counterintuitive to propose a link between regime security and nuclear policy when the latter precipitated a raft of secondary US and European punitive measures on top of international sanctions imposed on Iran by the United Nations Security Council. Undoubtedly, it has proved to be a high-cost strategy that yielded little in tangible dividends and instead put the Islamic Republic in an ever-greater bind economically while confirming its status as an international pariah. Yet the fact that Tehran has been willing to accept these costs – even at the risk of further alienating a disgruntled Iranian population – shows the imperative of the nuclear program for the regime.

Alongside Iran's growing ballistic missile capability, nuclear technology provides the regime with leverage in negotiations with the world's leading powers. Under the cover of nuclear talks, the Islamic Republic's representatives can engage in direct discussions on a wide range of issues with their Western counterparts, including the Americans. Drawing on support from Russia and China, albeit intermittently, Iran's nuclear diplomacy looks to secure short-term concessions from the West in search of longer-term political and security settlements. It uses the threat of uranium enrichment, and the installation of ever more advanced centrifuges as bargaining chips to sustain the dialogue. Negotiations may also occasionally provide the regime with an opportunity to play off the European powers against one another and Washington.[40]

Tehran's ultimate objective of securing guarantees against regime change has not so far materialized, but the 2015 Joint Comprehensive Plan of Action (JCPOA) shows that its nuclear policy has not been entirely in vain. Leading world powers know that, short of a major military undertaking, Iran's nuclear program can only be slowed down – mainly through sanctions, sabotage, and targeted assassinations – but cannot be entirely halted nor Tehran's know-how be quickly reversed. Conversely, high-profile nuclear negotiations have given the regime some international prestige, a platform, and for a time a much-needed dose of domestic collateral.

In the West, there have been contrasting opinions over the efficacy of negotiating with the ayatollahs and disagreements over whether Iran has a covert nuclear weapons program or merely dual-use installations that can be converted into something more sinister. According to reports produced over the past two decades by the US Intelligence Community, the aim of the regime may not be to produce weapons per se but to have the capability to do so.[41] This has raised questions about Tehran's intermediate- to long-term objectives, especially as it continues to master the art of ballistic missile technology. Nor are international concerns alleviated when Iran's nuclear program is managed by the IRGC – the organization entrusted with the regime's expeditionary operations, including illicit activities such as money laundering, traffic in small arms, and training of a host of regional sub-state entities.

The ambivalence that surrounds Iran's nuclear intentions is characteristic of the regime. So are Tehran's repeated attempts at dissimilation. Misrepresentations of the program as entirely civilian, and camouflaging of key facilities, are consistent with the norms of the Islamic Republic. They go hand in hand with measures conceived to mislead international inspectors such as the geographical dispersion of nuclear installations, which have the dual purpose of discouraging aerial bombing. Also characteristic of the regime is the bifurcation of nuclear functions between the Office of the Supreme Leader and elected officials. Jurisdiction over nuclear policy lies exclusively with Ayatollah Khamenei. Yet delegations dispatched to nuclear negotiations are diplomats who tend to be nominated by the elected government, even if they operate under strict instructions of the Supreme Leader.

In practice, the executive and legislative branches of the Iranian government have no say in nuclear policy. But this has not stopped rival factions from attacking one another – or the elected government – for perceived shortcomings. The Hassan Rouhani administration, for example, was regularly targeted by parliamentary hardliners for making too many concessions to the West.[42] Surprisingly, the regime has allowed for a lively discourse to play out in the media that is imbued with contrasting points of view regarding Iran's handling of nuclear diplomacy. Naturally, there is never any acknowledgment of the limitations imposed on negotiators by the Supreme Leader or his culpability as the country's ultimate nuclear arbiter. As with other spheres of life in the Islamic Republic, there is palpable dissonance between who controls and shapes policy and the assigning of responsibility and blame.

Ever since President Donald J. Trump unilaterally pulled the United States out of the JCPOA, voices that consider the agreement as "capitulation" to the West have become conspicuously louder in Iran. Instead of the boom in trade and foreign investment that was promised by the JCPOA, the Islamic Republic has been subjected to what the Iranian leadership calls economic warfare. A perpetual sense of victimization and outrage at Western duplicity has led to pressure by hardliners for Iran to abrogate its existing commitments under the JCPOA. In December 2020, a parliamentary bill was passed to stop all international inspections. But such theatrics by Iran's elected institutions do not belie the reality that nuclear policy is made elsewhere. Tehran will remain engaged in nuclear negotiations because it desperately needs economic relief to ease the pressure on the regime and, ideally, a settlement that provides it with security guarantees. The 2021 presidential victory of Ebrahim Raisi – a man with close links to the Supreme Leader and the security services – will not change this imperative. What is different, however, is that Raisi's elevation has been accompanied by a system-wide purge of experienced technocrats. Many seasoned state officials and experts have been replaced by a younger cadre of ideologues whose sole qualification appears to be personal allegiance to the current Supreme Leader.[43] Nuclear discussions with the Islamic Republic have always been difficult, but the new Iranian team may lack even the most basic diplomatic skills necessary to engage in realistic and constructive international negotiations.[44]

Growing Tensions

When considering the Iranian worldview, it is important to acknowledge that modern Iran is a nation-state of around 85 million people. The majority may identify as Persians, but the country is also home to a growing Azeri Turkic population as well as Kurds, Balochis, Lurs, Arabs, and other ethnicities. While predominantly Shia Muslim, Iran has a Christian Armenian and Assyrian community dating back to antiquity, and the largest Jewish population in the Middle East outside of Israel. The Iranian-born diaspora around the world, estimated at around four million, has a diverse worldview shaped partly by its environment. Additionally, observers must be aware of the "Persianate": the term scholars use for societies along Iran's periphery whose history, language, and culture are extensively influenced by Iranian traditions.[45]

Yet since 1979, the Islamic Republic has used the extensive tools at its disposal to submerge Shia and Iranian national identity in subordination to the regime's Khomeinist ideology. In projecting itself as the embodiment of Shia Islam, it wears its international pariah status as a badge of honor by portraying Iran's isolation as the virtue of *mazloumiat* – the fate that befalls the righteous in the struggle against the injustices of tyrannical oppression. As a result, it has formalized a sense of perpetual victimization by portraying the country as being consistently under siege – attacked by domestic and international enemies for no other fault than defying a grievously unjust international order and exposing the malevolence of its hegemonic benefactors.[46]

In response to stringent international sanctions, the regime continuously urges Iranians to accept the sacrifices necessary for economic and cultural resistance.[47] While a convenient rationalization for the regime's domestic failures and its decades of economic mismanagement, ongoing domestic upheaval in Iran makes clear that it is a narrative that is wearing thin. By inextricably identifying itself with Shiism, and by hijacking its rituals and traditions to its service, the clerical hierarchy has contributed to a widening dissonance between public and private religion. Although opinion polls are not readily available in a country that is autocratically governed, reports suggest a growing rift between people's private beliefs and what they disapprovingly regard as *Islam-e-Akhondi* – the "Islam of the clerics."[48] It is not clear whether behind closed doors Iranians are abandoning Islam, Shiism, or merely the regime's representations of it. But 40 years of Islamism is changing popular perceptions of religion.

A sharp decline in mosque attendance and negative attitudes toward the clerical class have been accompanied by a rise in secularism and covert conversions to other religions, including Iran's pre-Islamic Zoroastrian faith and, in particular, Christianity.[49] The regime is also finding itself at the wrong end of the very resistance culture it has been promoting. Not coincidentally, the eternal struggle against tyranny is just as much a central theme in Iranian national mythology as it is in Shiism. Well before the 2022 protests emerged, precursors were evident in the population. From chants of "death to the dictator" by protestors condemning the Supreme Leader to women pushing back against the mandatory hijab, a sizeable segment of the Iranian public exhibited its disenchantment with social restrictions and "clerical Islam" in everyday life.[50]

Not unlike the structures of the regime, duality is a characteristic feature of Iranian culture. Visitors are often surprised by the disparity between the average Iranian's comportment in public and how they conduct themselves in private. The experience of two and a half millennia of turbulence and change has instilled in Iranians a predilection for playing the long game: perseverance in adversity and not revealing too much of one's intentions under duress or in negotiations. These traits are identifiable in the way the Islamic Republic has operated internationally. Domestically, however, the regime is not immune from its dangers. Twentieth-century Iranian history shows that popular expressions of support for a regime can be fleeting, with demonstrators quickly changing sides when expedient.[51] The mass rallies that officials claimed as passionate enthusiasm for the Islamic Republic in the past may have been a product, in part, of handouts from authorities and intimidation of public sector workers to bolster numbers.[52]

Just as problematic for the regime is a steady resurgence in Iranian nationalism. As heirs of a succession of great empires, and as proprietors of a language that was once lingua franca from the Indian subcontinent to Anatolia, Iranians have an exaggerated view of their rightful place in the world.[53] For a time, this played into the regime's interventionist proclivities. While frowning upon manifestations of pride in Iran's pre-Islamic history, the clerical leadership has periodically – and selectively – used nationalism by redefining it within Islamist ideology and incorporating it into its narrative. For example, by framing the country's nuclear policy as a matter of national pride, and by depicting international curbs as efforts to prevent Iran from reclaiming its rightful place as a regional power, the regime has successfully co-opted nationalism in support of its controversial program. Similarly, in drawing attention to the substitution of "Arabian" for the Persian Gulf, the regime insists that its invocation by Western commentators shows that enmity is not confined to the Islamic Republic but is directed more perniciously at Iran's heritage.

By casting itself as the champion of Iranian national unity, the regime has sought to link itself to the survival of the nation-state: without the Islamic Republic, it warns, the West will dismember the country along ethnic lines to prevent Iranian resurgence.[54] There are, however, tensions between the regime's selective application of nationalism and Iranian collective identity and attitudes. For instance, the clerical hierarchy celebrates the Arab-Muslim conquest of Persia as a splendid turning point in history whereas, for a vast majority of Iranians, it was nothing short of a tragedy.[55] The regime's diminishing popularity has seen a commensurate upsurge in public fascination with Iran's pre-Islamic past. Annual gatherings around the tomb of Cyrus the Great – the founder of the Persian Empire – have been accompanied by public defiance in commemoration of Iranianism sans Islam. Notable among them is the Zoroastrian ritual of jumping over fires before the Iranian New Year – a tradition that the authorities have been unable to stop despite concerted efforts.

The struggle to keep the "un-Islamic" behavior of the public at bay has been echoed in political circles by warnings about the return of the "Iranian School" – the fear that the prevailing Islamist ideology might be subsumed into Iranianism.[56] Keeping Iranian nationalism submerged is likely to be a challenge. Iran's population is young and has no memory of the heyday of the revolution. It looks to the West for inspiration, and it is influenced by a large and strident expatriate community that is predominantly secular, nationalistic, and intensely hostile to the Islamic Republic. Since the fall of the Soviet Union, Iranian literature, language, and ancient festivities have also seen a mini-revival in the Persianate – notably the countries in the Caucasus, Central Asia, and Afghanistan. Even as the regime continues to promulgate Islamism in favor of Iranianism, Iranian identity remains a powerful driver in the strategic culture of a people who are more likely to turn their back on Khomeinism than abandon their history and traditions.

Conclusion

Since 1979, the Islamic Republic regime has been the dominant voice in expressing the Iranian worldview. In the service of promoting Khomeinism, its leadership has often demonstrated uncanny pragmatism in drawing selectively from Iran's Shia and national identity to build consensus around itself. But rising discontent over draconian social controls, mismanagement of the economy, and falling living standards have been chipping away at the regime's popular legitimacy, resulting in public outbursts of increasing intensity and duration. The intensification of factional politics, interdepartmental rivalries, and the culture of accusations and recriminations among officials are paralyzing the regime from within. With the old consensus crumbling, recent developments suggest that the Supreme Leader – and his orbit – is intent on bringing an end to structural and factional divisions by replacing plurality with an absolutist theocracy.[57] This may create a more unitary state but at the cost of stripping away any remaining vestiges that connect the regime to the popular will. History shows that no regime that has subordinated Iranian national identity to its ideology has survived for long.

Notes

1 This chapter is adapted from Ali Parchami, "An Iranian Worldview: The Strategic Culture of the Islamic Republic," *Journal of Advanced Military Studies: Special Issue on Strategic Culture* (January 2022): 9–23, https://doi.org/10.21140/mcuj.2022SIstratcul001. Reprinted by permission.

2 *Velayat-e-faqih*, or the "rule of the jurisprudence," is an Islamic doctrine that promotes the idea of government by the learned cleric. By interpreting it in absolute terms, Ayatollah Khomeini's regime endowed Iran's Supreme Leader with guardianship powers over every facet of life.

3 Richard N. Frye, *The Golden Age of Persia: The Arabs in the East* (London: Phoenix, 1987), 150–85.

4 Colin S. Gray, "National Styles in Strategy: The American Example," *International Security* 6, no. 2 (Fall 1981): 22.

5 Michael Eisenstadt, *The Strategic Culture of the Islamic Republic of Iran: Religion, Expediency, and Soft Power in an Era of Disruptive Change.* MES Monographs. No. 7 November 2015. Available at: https://www.washingtoninstitute.org/uploads/Documents/pubs/MESM_7_Eisenstadt.pdf.

6 Nima Gerami, "Iran's Strategic Culture: Implications for Nuclear Policy," in Jeannie L. Johnson, Kerry M. Kartchner, and Marilyn Maines, eds. *Crossing Nuclear Thresholds: Leveraging Sociocultural Insights into Nuclear Decisionmaking* (Cham, Switzerland: Palgrave Macmillan, 2018), 61–108.

7 Gregory F. Giles, "Strategic Personality" *Country Case Study: Iran* (McLean, Virginia: Science Applications International Corporation, 24 January 1996).

8 W.A. Rivera, *Iranian Strategic Influence: Information and the Culture of Resistance* (Lanham, MD: Rowman & Littlefield, 2022).

9 Ilan Berman, *Iran's Deadly Ambition: The Islamic Republic's Quest for Global Power* (New York: Encounter Books, 2015); Ilan Berman, *Tehran Rising: Iran's Challenge to the United States* (Lanham, MD: Rowman & Littlefield Publishers, 2005).

10 Ruhollah Khomeini, *Governance of the Jurist: Islamic Government* (Tehran: TICPIKW, 2002), 7–94; and Mehdi Khalaj, "Iran's Regime of Religion," *Journal of International Affairs* 65, no. 1 (2011): 135.

11 Khomeini, *Governance of the Jurist*, 10–16, 24–28, 85–94.

12 "What Does Arrogance Mean in Imam Khomeini's Statements," Khamenei, Supreme Leader of the Islamic Republic, 22 December 2020: https://english.khamenei.ir/news/8190/What-does-Arrogance-mean-in-Imam-Khamenei-s-statements

13 Khomeini, *Governance of the Jurist*, 6, 78, 89, 94.
14 Said Amir Arjomand, *After Khomeini: Iran Under His Successors* (Oxford, UK: Oxford University Press, 2009), 35.
15 David Crist, *The Twilight War: The Secret History of America's Thirty-year Conflict with Iran* (New York: Penguin, 2013), 440–537; Michael Weiss and Hassan Hassan, *ISIS: Inside the Army of Terror* (New York: Regan Arts, 2016), 20–21, 56; and Trita Parsi, *Treacherous Alliance: The Secret Dealings of Israel, Iran, and the United States* (New Haven, CT: Yale University Press, 2007).
16 *Iran's Network of Influence in the Middle East* (London: International Institute for Strategic Studies, 2019), 11–38.
17 Mehdi Jeddinia and Sirwan Kajjo, "Iran Hints at Using Foreign Militia in Domestic Crackdown," *VOA*, 18 March 2019.
18 Kasra Aarabi, "Iran's Regional Influence Campaign Is Starting to Flop," *Foreign Policy*, 11 December 2019.
19 Joshua Davidovich, "Iranians Target Tehran's Support for Palestinians Amid Massive Protests," *Times of Israel*, 31 December 2017; and "Iranian Protest Chants Lambast Regime's Foreign Interference," *Al Arabiya*, 17 November 2019.
20 Abbas Djavadi, "Who Is Waging War Against God in Iran?," *Radio Free Europe*, 11 March 2010.
21 Majid Yazdi, "Patterns of Clerical Political Behavior in Postwar Iran, 1941–53," *Middle Eastern Studies* 26, no. 3 (July 1990): 286.
22 RK Ramazani, "Iran's Foreign Policy: Contending Orientations," *Middle East Journal* 43, no. 2 (1989): 202–217; and Scott R. Anderson, "International Law and the Iranian Revolution," Brookings, 2 April 2019.
23 The vicious attacks of hardline politicians and newspapers on Hasan Rouhani and his despised foreign minister encapsulate these interpersonal enmities. Hossein S. Seifzadeh, "The Landscape of Factional Politics and Its Future in Iran," *Middle East Journal* 57, no. 1 (Winter 2003): 57–75.
24 Evident in the lead-up to the 2021 presidential elections. Lazar Berman, "Was Khamenei behind the Barring of Moderate Presidential Candidates in Iran?," *Times of Israel*, 27 May 2021.
25 Himdad Mustafa, "The Struggle Among the Political Elite of the Islamic Republic of Iran," *MEMRI Daily Brief 40* (16 November 2022): https://www.memri.org/reports/struggle-among-political-elite-islamic-republic-iran
26 David E. Thaler et al., *Mullahs, Guards, and Bonyads: An Exploration of Iranian Leadership Dynamics* (Santa Monica, CA: Rand, 2010), 60.
27 Ali A. Saeidi, "The Accountability of Para-governmental Organizations (*Bonyads*): The Case of Iranian Foundations," *Iranian Studies* 37, no. 3 (2004): 479–498, https://doi.org/10.1080/0021086042000287541.
28 Frederic Wehrey et al., *The Rise of the Pasdaran: Assessing the Domestic Roles of Iran's Revolutionary Guard Corps* (Santa Monica, CA: Rand, 2009), 55–88.
29 Patrick Wintour, "Iran President Criticises New Rules for Candidates as Election Begins," *The Guardian*, 11 May 2021.
30 "Ex-Intelligence Minister Says Iran Officials Should Fear Mossad," *Iran International*, 29 June 2021.
31 "Vaezee Rejects Ashna's Statement: Report of the Dead Was Given to Rouhani," *Iran International*, 19 August 2020.
32 Mohammad Javad Zarif, audio interview (leaked on 25 April 2021), YouTube, 3:10:41.
33 Golnaz Esfandiari, "Khamenei's Ban on Western Vaccines Blasted as a 'Politicization' of Iranians' Well-being," *Radio Free Europe*, 12 January 2021; and Al-Mashareq, "Decoding Iran's Vaccine Strategy: Denial, Deception, Dishonesty," Diyaruna, 21 June 2021.
34 Christian Oliver, "Iran's Crisis of Incompetence," *Politico*, 14 January 2020; and "Bitter Blame Game in Iran Amid Coronavirus 'Explosion'," *Al-Monitor*, 8 April 2021; *Economist*, "Could Iran's Regime Fall?" (27 October 2022) accessed at https://www.economist.com/middle-east-and-africa/2022/10/27/could-irans-regime-fall.
35 Mustafa Kibaroglu, "Iran's Nuclear Ambitions from a Historical Perspective and the Attitude of the West," *Middle Eastern Studies* 43, no. 2 (March 2007): 223–245.

36 Ray Takeyh, "The Shah, the Mullahs and Iran's Longstanding Nuclear Ambitions," *Wall Street Journal*, 10 December 2020.
37 Homeira Moshirzadeh, "Discursive Foundations of Iran's Nuclear Policy," *Security Dialogue* 38, no. 4 (December 2007): 537; and Parisa Hafezi, "Khamenei Says Iran May Enrich Uranium to 60% Purity If Needed," *Reuters*, 22 February 2021.
38 David Patrikarakos, *Nuclear Iran: The Birth of an Atomic State* (London: I.B. Tauris, 2012), 131–175; and David Hastings Dunn, "'Real Men Want to Go to Tehran': Bush, Pre-emption and the Iranian Nuclear Challenge," *International Affairs* 83, no. 1 (January 2007): 19–38, https://doi.org/10.1111/j.1468-2346.2007.00601.x.
39 David Hastings Dunn, "'Real Men Want to Go to Tehran': Bush, Pre-emption and the Iranian Nuclear Challenge," *International Affairs* 83, no. 1 (January 2007): 19–38.
40 Ali Parchami, "American Culpability: The Bush Administration and the Iranian Nuclear Impasse," *Contemporary Politics* 20, no. 3 (2014): 315–330, https://doi.org/10.1080/13569775.2014.911501.
41 Ken Dilanian, "U.S. Does Not Believe Iran Is Trying to Build Nuclear Bomb," *Los Angeles Times*, 23 February 2012; and Lara Seligman, "How Close Is Iran to a Nuclear Bomb, Really?," *Foreign Policy*, 1 July 2019.
42 "Iran Nuclear Talks: Hardliners Criticise Nuclear Deal," BBC, 3 April 2015; and Kareem Fahim, "Iran's Hard-liners Step Up Attacks on Rouhani Government Sowing Suspicion Over Nuclear Talks," *Washington Post*, 4 May 2021.
43 Saeid Golkar and Kasra Aarabi, *Raisi's Rising Elite: The Imam Sadeghis, Iran's Indoctrinated Technocrats* (Tony Blair Institute for Global Change, November 2021), 4–42, https://institute.global/sites/default/files/inline-files/Tony%20Blair%20Institute%2 C%20Raisi%27s%20Rising%20Elite%2C%20November%202021.pdf
44 Koroush Ziabari, "Raisi's Inept Negotiators Are Sinking Iran Deal Talks," *Foreign Policy* (2 August 2022): https://foreignpolicy.com/2022/08/02/iran-nuclear-deal-talks-jcpoa-revive-raisi-negotiations/.
45 See, for example, Abbas Amanat and Assef Ashraf, eds., *The Persianate World: Rethinking a Shared Sphere* (Leiden, Netherlands: Brill, 2018).
46 "Ayatollah Khamenei Hails Iranians for Frustrating Enemy Plot," U News, 24 February 2020.
47 Ray Takeyh, "Iran's 'Resistance Economy' Debate," Council on Foreign Relations, 7 April 2016; and Kasra Aarabi, "The Fundamentals of Iran's Islamic Revolution," Institute for Global Change, 11 February 2019.
48 Ladan Boroumand, "Iranians Turn Away from the Islamic Republic," *Journal of Democracy*, 31, no. 1 (January 2020): 169–181, https://doi.org/10.1353/jod.2020.0014; and Pooyan Tamimi Arab and Ammar Maleki, "Iran's Secular Shift: New Survey Reveals Huge Changes in Religious Beliefs," *The Conversation*, 10 September 2020.
49 Tamimi Arab and Maleki, "Iran's Secular Shift."
50 Jason Lemon, "Iranians Chant 'Our Enemy Is Right Here' on the Second Day of Anti-Government Protests After Ukraine Plane Shot Down," *Newsweek*, 12 January 2020.
51 Nazilla Fathi, "In Tehran, Thousands Rally to Back Government," *New York Times*, 30 December 2009.
52 Notably during the Constitutional Revolution and the premiership of Mohammad Mossadegh.
53 Thaler et al., *Mullahs, Guards, and Bonyads*, 9.
54 Thaler et al., *Mullahs, Guards, and Bonyads*, 11.
55 Thaler et al., *Mullahs, Guards, and Bonyads*, 8.
56 Robert Tait, "Iranian President's New 'Religious-Nationalism' Alienates Hard-line Constituency," *Radio Free Europe*, 19 August 2010.
57 Kyra Rauschenbach, "The Second Step of Iran's Islamic Revolution: Exploring the Supreme Leader's Worldview," *Critical Threats*, 10 May 2021.

Selected Bibliography

Amanat, Abbas and Assef Ashraf, eds., *The Persianate World: Rethinking a Shared Sphere* (Leiden, Netherlands: Brill, 2018).

Arjomand, Said Amir. *After Khomeini: Iran Under His Successors* (Oxford, UK: Oxford University Press, 2009).
Berman, Ilan. *Iran's Deadly Ambition: The Islamic Republic's Quest for Global Power* (New York: Encounter Books, 2015); *Tehran Rising: Iran's Challenge to the United States*. Lanham: Rowman & Littlefield Publishers, 2005.
Boroumand, Ladan. "Iranians Turn Away from the Islamic Republic," *Journal of Democracy* 31, no. 1 (January 2020): 169–181, 10.1353/jod.2020.0014.
Crist, David. *The Twilight War: The Secret History of America's Thirty-year Conflict with Iran* (New York: Penguin, 2013).
Dunn, David Hastings. "'Real Men Want to Go to Tehran': Bush, Pre-emption and the Iranian Nuclear Challenge," *International Affairs* 83, no. 1 (January 2007): 19–38, 10.1111/j.1468-2346.2007.00601.x.
Eisenstadt, Michael. *The Strategic Culture of the Islamic Republic of Iran: Religion, Expediency, and Soft Power in an Era of Disruptive Change*. MES Monographs. No. 7 November 2015. Available at: https://www.washingtoninstitute.org/uploads/Documents/pubs/MESM_7_Eisenstadt.pdf.
Frye, Richard N. *The Golden Age of Persia: The Arabs in the East* (London: Phoenix, 1987).
Gerami, Nima. "Iran's Strategic Culture: Implications for Nuclear Policy," in Jeannie L. Johnson, Kerry M. Kartchner, and Marilyn Maines eds. *Crossing Nuclear Thresholds: Leveraging Sociocultural Insights into Nuclear Decisionmaking* (Cham, Switzerland: Palgrave Macmillan, 2018).
Giles, Gregory F. "Strategic Personality" *Country Case Study: Iran* (McLean, Virginia: Science Applications International Corporation 24 January 1996).
Iran's Network of Influence in the Middle East (London: International Institute for Strategic Studies, 2019).
Khalaj, Mehdi. "Iran's Regime of Religion," *Journal of International Affairs* 65, no. 1 (2011).
Khomeini, Ruhollah. *Governance of the Jurist: Islamic Government* (Tehran: TICPIKW, 2002).
Kibaroglu, Mustafa. "Iran's Nuclear Ambitions from a Historical Perspective and the Attitude of the West," *Middle Eastern Studies* 43, no. 2 (March 2007): 223–245.
Parchami, Ali. "American Culpability: The Bush Administration and the Iranian Nuclear Impasse," *Contemporary Politics* 20, no. 3 (2014): 315–330, 10.1080/13569775.2014.911501.
Patrikarakos, David. *Nuclear Iran: The Birth of an Atomic State* (London: I.B. Tauris, 2012).
Rivera, W. A. *Iranian Strategic Influence: Information and the Culture of Resistance* (Lanham, MD: Rowman & Littlefield, 2022).
Saeidi, Ali A. "The Accountability of Para-governmental Organizations (*Bonyads*): The Case of Iranian Foundations," *Iranian Studies* 37, no. 3 (2004): 479–498, 10.1080/0021086042000287541.
Thaler, David E., et al., *Mullahs, Guards, and Bonyads: An Exploration of Iranian Leadership Dynamics* (Santa Monica, CA: Rand, 2010).
Wehrey, Frederic, et al., *The Rise of the Pasdaran: Assessing the Domestic Roles of Iran's Revolutionary Guard Corps* (Santa Monica, CA: Rand, 2009).

25

ISRAELI STRATEGIC CULTURE

State of the Scholarship

Itai Shapira

Introduction

Research of Israeli strategic culture uses cultural lenses to study the unique Israeli style of strategy and way of war. It thus manifests the spirit of Hofstede's work on the influence of national cultures over behaviors and institutions,[1] as well as the "cultural turn" in social sciences[2] and strategic studies.[3] However, while strategic culture is still a debated concept,[4] most literature about the Israeli strategic culture does not participate in the theoretical scholarly discussion but applies different interpretations of the concept to specific cases.

This literature is divided into three types. The first studies Israeli strategic culture per se and is limited in volume. Salient scholars are Adamsky,[5] Libel,[6] Giles,[7] Kopeć,[8] and Petrelli.[9] The second type is extensive and studies Israeli strategic culture's origins. Ben-Ephraim[10] and Rickover[11] directly discuss such origins, while many other scholars study the origins of Israeli culture at large. The third type is also comprehensive and focuses on Israeli military culture as a specific facet of strategic culture.

This chapter opens by describing the different theoretical approaches applied in this extant literature, then provides a literature review regarding Israeli strategic culture's origins, traits, and manifestations. The chapter then illustrates the literature's main strengths and weaknesses and concludes by portraying promising avenues for future research.

Theoretical Approaches

In his seminal comparative study on cultures of military innovation, which intensively assesses Israeli strategic culture, Adamsky, drawing on Johnson et al., describes strategic culture as "a set of shared formal and informal beliefs, assumptions, and models of behavior, derived from common experiences and accepted narratives (both oral and written), that shape collective identity and relationships to other groups, and which influence and sometimes determine appropriate ends and means for achieving security objectives."[12] This view manifests the "first generation" school of thought about strategic culture, namely "strategic culture as a context" for a national style of strategy and a way of war.[13] Adamsky's focus on military innovation through cultural lenses also manifests the "third generation" school of thought, studying military matters and not just broad strategic ones.[14]

DOI: 10.4324/9781003010302-28

Adamsky's definition of strategic culture is adopted by Giles, who describes different Israeli strategic subcultures and specifically discusses the Israeli approach toward weapons of mass destruction.[15] Kopeć, in his research of Israeli strategic culture, relied on Johnston's "third generation" definition of strategic culture as establishing "pervasive and long-lasting strategic preferences by formulating concepts of the role and efficacy of military force."[16] Moreover, Kopeć emphasizes the role of elites in shaping strategic culture, the different subcultures that can compose a national strategic culture, and the difference between a strategic culture and a military one.

Libel explicitly engages the theory of strategic culture when studying the Israeli one.[17] He identifies with the "fourth generation" school of thought, emphasizing the function of strategic culture as an explanatory parameter for change in strategic behavior. Libel views strategic culture not as homogenous on the national level nor fixed in time, but rather as "a system of competing subcultures," shaped by elites that create "epistemic communities."

Israeli Strategic Culture – Literature Review

This section portrays the major origins, traits, and manifestations of Israeli strategic culture as described in the extant literature. It also illustrates changes and ambivalences in this culture through a discussion of Israeli national security doctrine and military culture.

Origins of Israeli Strategic Culture

The Jewish experiences of weakness and dependence during 2,000 years of exile from the land of Israel, alongside the horrors of the Holocaust in World War II,[18] brought on the Zionist movement – the driving power behind Israel's independence in 1948 – to aspire not only for an independent Jewish state but also for a new Israeli society and culture. Israel was to be composed of "new Israelis" or *sabras*,[19] never again facing an existential threat.[20] These *sabras* were expected to be productive, practical, pragmatic, authentic, exercising a straightforward approach (*dugri*), getting straight to the point (*tachlis*),[21] and focusing on "doingness" (*bitsuism*).

Society-military interaction also provided an important origin for Israeli strategic culture, as military issues are substantially integrated into Israeli society, politics, and culture.[22] Israel is often considered a "nation in arms"[23] or a "garrison state."[24] For example, the IDF (Israeli Defense Forces) has a broad influence on Israeli politics, economy, and society.[25] Israelis perceive it not just as an armed force but also as a public institution.[26]

Another origin of Israeli strategic culture is the geopolitical context.[27] In its early years, Israel mainly prepared for a surprise military attack by Arab state militaries. Israel had no strategic depth and suffered from quantitative inferiority, seeing itself as the "few against the many"[28] while facing existential threats.[29] In the 1950s, Israel formulated a commensurate national security doctrine, whose main pillars are deterrence, battlefield decision, early warning, and defense (added in the 2000s).[30] Additionally, Israel has developed a qualitative military edge concept, acknowledging that technological superiority should compensate for quantitative inferiority.[31] A strategic partnership with the United States has also become a core pillar of the national security doctrine.[32]

Major Traits in Israeli Strategic Culture

The phenomena described above have created a unique "Israeliness,"[33] characterized by informality, inattention to hierarchy, focus on content rather than style, innovation, simplification, improvisation, and "anti-intellectualism."[34] Some of these traits are typical of Israel's political[35] and management[36] cultures on top of its strategic culture.

In national security matters, Israelis experience a "siege mentality," yet constantly seek international legitimacy.[37] Israel tends to see the world through an "existential threat lens,"[38] perhaps thus reflecting a Jewish tradition: for instance, in the *Hagadah*, read by Jews during Passover, a poignant phrase is: "In every generation, they rise up to destroy us."[39] Israeli national security concepts and deeds are therefore underpinned by a sense of exceptionalism. Israel perceives itself as in need of unique measures to confront unique challenges.[40] Prevention and preemption of existential threats, therefore, are salient traits of Israeli strategic culture.[41]

Israeli strategic culture is also characterized by a mindset of "securitization" as defined by Barak and Sheffer,[42] attributing increased importance to security matters over other areas such as social cohesion, economic stability, and public health. Israeli security establishments have some national-level responsibilities that exceed defense and military duties.[43] Foreign policy is marginalized compared to security matters,[44] and Israeli security establishments rather than the Foreign Ministry are the ones that often conduct clandestine diplomacy.[45]

Israeli strategic culture also prefers informality and oral texts rather than formal and written documentation.[46] A vivid illustration of this is the Israeli national security doctrine discussed earlier: although never published as a formal document, it illustrates a coherent concept.[47]

Manifestations: Operationalization of Israeli Strategic Culture Concepts

Extant literature operationalizes the concept of strategic culture to discuss several Israeli manifestations. The first is the Israeli style of national decisionmaking,[48] sometimes described as "tacticization of strategy."[49] Israeli defense networks, for instance, have a dominant influence on national security and statecraft, effectively acting as "elite epistemic authorities"[50] with a monopoly over national security discourse and practice.[51] Hence, this style of decisionmaking reflects cultural traits such as siege mentality and securitization.

Another manifestation is the Israeli approach toward nuclear weapons, with Israel claiming that it would not be the first to introduce nuclear weapons in the Middle East.[52] Additionally, the "Begin Doctrine," named after Israel's prime minister from 1977 to 1983,[53] guides Israel to proactively prevent adversary countries from acquiring military nuclear capabilities. Israel acted accordingly when destroying the Iraqi nuclear reactor in 1981 and the Syrian nuclear reactor in 2007.[54] This doctrine might also guide the current strategy toward Iran.[55] The Israeli approach therefore reflects traits such as preventive mindset, realistic idealism, siege mentality, securitization, and exceptionalism in the face of existential threats.[56]

Military innovation is another manifestation of Israeli strategic culture. On the one hand, the IDF learns quickly and innovates through adaptation, improvisation, and combat experience.[57] Accordingly, Adamsky showed that when Israel conducted its revolution in military affairs in the 1980s and 1990s, tactical lessons and successful exploitation of advanced technologies were not transformed into an overarching concept of operations. This reflects a preference for practice over theory and doing over understanding. On the other hand, Adamsky himself claimed that a culture of innovation through anticipation, guided by theory and not just relying on adaptation, has also become dominant in the IDF.[58]

Adamsky also discussed Israeli nonnuclear deterrence through the lens of strategic culture.[59] Adamsky claims that Israel has developed a unique interpretation to deterrence through its "campaign between the wars" conducted mainly since 2013 as a continuous campaign under the threshold of war,[60] and through its "deterrence-based operations" against Hamas in the Gaza Strip. Adamsky sees this mostly as a product of operational improvisations rather than a coherent doctrine – thus reflecting an aversion to theory and "anti-intellectualism," a lack of structured military doctrine, a dominance of the military over the political echelon, and exceptionalism.

Another manifestation regards counterinsurgency warfare and low-intensity conflicts (LICs). Petrelli claims that the Israeli aversion toward military theory prevented the IDF from properly adapting its concepts of operations between 1987 and 2005.[61] Additionally, the inclination toward tactical excellence, exploitation of advanced technologies on the tactical level, *bitsuism* ("doingness"), the preference of offensive over defensive operations, and the preference for short conflicts on enemy territory prevented the IDF from practicing a population-centric counterinsurgency and integrating non-military aspects with military ones.

Israeli National Security Doctrine and Military Culture: Ambivalences and Changes in Strategic Culture

Israeli national security doctrine as manifested in the twenty-first century, as well as Israel's military culture, reflect both continuity and change.[62] This reflects an evolving Israeli strategic culture, underpinned by the changing strategic and operational environment.

A scenario entailing a conventional military attack from Syria, the cornerstone of Israel's threat perception for decades, has become obsolete; and since 1979, Israel has maintained a peace agreement with Egypt. However, terrorism challenges have endured over the years, although evolving in nature. Moreover, a major security challenge for Israel in recent decades stems from Iran's ambitions for regional hegemony and military nuclear capabilities. Iran has become "the Archimedean point from which Israel's politics and actual external and domestic policies can be viewed."[63] Iran is perceived by Israel as "the sum of all fears"[64] and is seen to pose an existential threat.[65] This does not only relate to Iran's military nuclear ambitions but also to its precision-guided missiles and unmanned aerial systems projects, whether developed and employed by Iran itself, or by its proxies such as Hizballah in Lebanon.[66]

As of 2023, therefore, Israel is preparing for an unintended escalation[67] that might erupt in multiple theaters.[68] It faces concern emanating from weapons of mass destruction, new challenges in the domains of cyberspace and information, and traditional military and terrorism threats.[69] Israeli national security doctrine's major pillars as described earlier – deterrence, early warning, battlefield decision, and defense – have therefore received new contexts.

Deterrence has become increasingly salient, not just regarding Israel's state adversaries[70] but also as a major goal of the Israeli conflicts with Hamas and Hizballah, in which

Israel utilizes disproportionate power employment.[71] Israel has therefore gradually developed a doctrine for limited deterrence-based conflicts on top of its traditional focus on battlefield decision in large-scale military conflicts.[72]

Battlefield decision is also practiced in a new context. Traditionally, Israeli military doctrine was inclined toward defense on the strategic level but offense and initiative on the operational and tactical levels. It also tended toward short and decisive wars carried out on enemy territory. However, in recent years, Israel has effectively conducted wars of attrition[73] or "mowing the grass" operations,[74] rather than decisive military campaigns. Israel refrains from employing large-scale ground maneuvers,[75] attributing growing importance to airpower for achieving strategic goals.[76]

Defense is also practiced in a new context,[77] no longer merely carried out to prevent a territorial gain from adversary militaries. Israel has constructed fences and walls on all its borders, aiming to counter terrorism and insurgency infiltration.[78] On top of traditional land, air, and maritime threats, Israeli defense must therefore engage the cyber domain, rockets and missiles, and underground tunnels.[79]

Early warning, one of the major roles of intelligence organizations, has also received a new context. It is no longer only about surprise military attacks but also about identifying shifts in the strategic and operational environment. The trauma of the intelligence failure in the Yom Kippur War (1973), therefore, might be fading.[80] For instance, the IDF is recently discussing a "revolution in intelligence affairs," reflecting a change in the integration of intelligence in Israel's military doctrine.[81]

Warfare against hybrid organizations such as Hamas and Hizballah has gradually become the cornerstone of the IDF's force design and "theory of victory."[82] Israel seeks to overcome "the other side's revolution in military affairs"[83] and thus counter adversary firepower.[84] It no longer aims in every military conflict for battlefield decision but often for restoring and improving deterrence,[85] while its military campaigns manifest "post-heroic warfare."[86] Additionally, Israel initiated circa 2013 the "campaign between the wars" already described earlier, mainly to limit Iranian influence. This campaign reflects a preventive mindset and a focus on deterrence, yet in a different context than conventional military conflicts.

This leads to a discussion of Israeli military culture, the most heavily studied topic composing Israeli strategic culture research. Israeli military culture is often described as anti-intellectual, improvisational, aiming at tactical excellence, not developing thought leadership and theoretical frameworks, and lacking robust military education and academic traditions.[87] Cohen, Eisenstadt, and Bacevich frame it as "conservative innovation."[88] Vardi describes it as reflecting a "small wars" improvisational mentality on the one hand and a "big wars" disciplined structure and doctrine on the other.[89] Adamsky described the Israeli approach toward military strategy as a "conceptual salad."[90]

However, Israeli military culture entails several ambivalences. A salient ambivalence can be found, for instance, in the alleged Israeli military "anti-intellectualism," since the IDF does not seem to be purely averted to theory. This can be illustrated by the professional writings of senior acting officers[91] or publications of "IDF Strategy" and "IDF Operational Concept" documents.[92] One might even claim that the IDF has reached an "over-intellectualism" (or "phony intellectualism"[93]), as reflected in the criticism of the Operational Theory Research Institute[94]: the latter's overly philosophical approach was considered as one of the causes for the IDF failures in the Second Lebanon War (2006).[95]

Another ambivalence stands out when discussing organizational and managerial military cultures, since the IDF mitigates tensions between a scientific-engineering approach and an artistic-creative one. This might stem from two different origins of IDF culture: the British military on the one hand; and the *Palmach*, a special unit of the Israeli resistance organization active prior to Israeli independence, on the other hand.[96] While the former underlines hierarchy and structured processes, the latter underlines improvisation and initiative. For instance, "mission command" was an accepted paradigm in the IDF, focusing on an initiative in the face of battlefield uncertainty.[97] However, it has been challenged by the changing nature of conflicts, as well as by advanced technologies that enable centralized control.[98] As another example, a former commander of the Israeli Military Colleges criticized "air-force-influenced managerial patterns" as superseding IDF standards for excellence,[99] while underlining adaptive excellence[100] and creativeness as required skills.[101]

The IDF approach toward force design also reflects an ambivalence between conservatism and innovation[102] or between a scientific and a creative method.[103] Senior officers have recommended moving toward a more innovative and flexible force design concept,[104] and some IDF officers have even recommended a reformulation of the IDF at large.[105] The Israeli approach toward technology is another manifestation of ambivalence. This approach is described by Adamsky as "vigorous technological innovations ... with a conservative military paradigm."[106] On the one hand, for example, the IDF still relies on human superiority.[107] On the other hand, Israel prioritizes technological superiority,[108] as is illustrated by the central role of defense industries in Israeli national security and the private sector.[109]

Another ambivalence can be found in Israeli civil-military interaction, which is more complex than merely reflecting militarism.[110] Israel has a dominant defense establishment on the one hand, but a solid civilian superiority in decisionmaking on the other.[111] Moreover, although Israel has militaristic characteristics, it exercises restraints on military force employment due to fear of casualties.[112]

Main Strengths and Weaknesses of Extant Literature

Undeniably, the extant literature applies different interpretations of the concept of strategic culture to Israeli cases and creates a foundation for an Israeli strategic culture theory. The origins of Israeli strategic culture are extensively researched, even if not always explicitly; this culture's traits are portrayed in many studies; the research of Israeli military culture is rich; and Israeli strategic culture has been operationalized in several fields.

However, the literature also suffers from several weaknesses. It lacks updated primary sources and interviews with practitioners as well as extensive comparative research. It sometimes describes Israeli strategic culture in an oversimplistic way – i.e., homogenous in style and fixed in time – overlooking different subcultures. It fails to sufficiently explain changes in behavior over time and falls short of an updated operationalization.

The gap in operationalization is reflected, for instance, by the lack of sufficient research regarding the Iran-focused Israeli policy, sometimes described as "unusually cohesive" yet manifesting an "aversion to strategy and long-term planning."[113] A cultural perspective has yet to be employed to discuss whether this policy might actually manifest a consistent and long-term vision; the Israeli decision not to implement the "Begin Doctrine," although a military strike against the Iranian nuclear project was considered in the early 2010s[114];

or whether the Israeli threat perception manifests a bias toward capabilities and threats while overlooking Iranian intentions.[115]

The operationalization gap is also pertinent to Israeli policy regarding the turmoil in the Middle East since 2010. On the one hand, Israel adopted a passive approach regarding intervention in the stability of regimes and violation of human rights.[116] On the other hand, Israel conducted a humanitarian operation in the southern part of Syria,[117] and adopted a proactive approach to minimizing security risks[118] and improving its ties with Arab Sunni countries.[119] Strategic culture has not been employed to study this complexity. It has also not been operationalized to understand Israel's policy toward China: Israel engages in various partnerships with Beijing,[120] although the latter is viewed by the United States, Israel's most strategic partner, as a "great-power competitor."[121]

The operationalization gap is also evident regarding Israel's evolving security and military doctrines as described earlier. On the one hand, Israel's conflicts with Hizballah and Hamas manifest an inclination toward deterrence and defense[122] with negativist war objectives,[123] and Israel's approach to the West Bank also manifests a status quo ambition.[124] On the other hand, Israel's "campaign between the wars" and some of its reported operations against Iran reflect an offensive, proactive, and "shaping" approach,[125] aimed at changing the status quo. Strategic culture, therefore, has not been operationalized to understand the way Israel balances defensive and proactive ways of war.

Another issue that has not received sufficient attention might be framed as an IDF tendency to "over-adapt." For instance, Israel was not prepared for the first Palestinian intifada (popular uprising) in 1987.[126] At that time, the IDF was focused on force design for a conventional conflict with Arab militaries and adapting to the insurgency challenge created in Lebanon.[127] In 2000, Israel needed to adapt to the LIC patterns of the second Palestinian intifada, after many years of focus on counterinsurgency and conventional military conflict.[128] However, this "over-adaptation" might have been one of the causes for the IDF deficiencies in the Second Lebanon War in 2006.[129] After this war, the IDF "returned to basics:" design thinking and effects-based operations (EBO) concepts were put aside after many years of use, and a spirit of anti-intellectualism emerged.[130]

Yet in the face of turmoil in the Middle East circa 2010, the IDF acknowledged that it once again needed to adapt, and thus institutionalized the "campaign between the wars."[131] However, this too might have been an "over-adaptation," as IDF senior officers have recently suggested while advocating for restoring capabilities of battlefield decision in large-scale conflicts with Hamas and Hizballah.[132]

Israeli strategic culture research also suffers from securitization, i.e., a predominant focus on security and military topics. Only a few studies have been written, for instance, about Israeli national security agencies besides the IDF, such as *Mossad, Shabak,*[133] the Foreign Ministry, or the National Security Staff. There is also no rigorous academic research of Israeli national intelligence culture.[134]

Finally, as mentioned, two gaps regarding data and methods stand out. First, there is a relative dearth of interviews with acting practitioners, excluding ones conducted by Adamsky and Petrelli. Since Israelis prefer practice over theory and doing over reflecting, the practitioners' point of view is highly relevant. Second, excluding Adamsky's work, comparative research is relatively scarce. Although several studies have compared Israel to other countries regarding specific issues, such as officers' thinking about warfare's future[135] or civil-military relations,[136] this does not amount to a robust use of the comparative method.

Conclusion

Israeli strategic culture research still seems to be in its early stages. Strategic culture has only begun to create "thick descriptions"[137] of Israeli strategic behaviors, let alone to understand changes in Israeli strategy. Israeli strategic culture is in transition, balancing innovative and conservative approaches. It is adaptive and improvisational, but it is also forward-looking and anticipatory, with its practitioners constantly searching for potential disruptions and "relevancy gaps."[138] Moreover, Israel might have a unique perception of the concept of strategy: Israelis seem to focus on the product of strategy rather than on its process, and on strategic practice rather than theory.[139]

Future research has many potential areas for exploration. The most prominent ones – preferably using comparative methods and interviews with practitioners – are the Israeli approach toward the interaction between a theory and practice of strategy, the balance between military and nonmilitary issues, the integration of artificial intelligence in decisionmaking and military doctrine, the evolving security doctrine and its approach toward cyber offense, and the Israeli strategy toward Iran. More broadly, there is a need for research on different Israeli strategic subcultures, such as the national intelligence culture and the subcultures of specific services and directorates in the IDF.

Notes

1 Geert Hofstede, *Culture's Consequences: Comparing Values, Behaviors, Institutions and Organizations Across Nations*, 2nd ed. (Thousand Oaks, CA: Sage Publications, 2000).
2 Victoria E. Bonnell, Lynn Avery Hunt, and Richard Biernacki, eds., *Beyond the Cultural Turn: New Directions in the Study of Society and Culture* (Berkley and Los Angeles, CA: University of California Press, 1999).
3 Theo Farrell, "World Culture and Military Power," *Security Studies* 14, no. 3 (01 April 2005): 448–488. doi:10.1080/09636410500323187.
4 Alan Bloomfield, "Time to Move On: Reconceptualizing the Strategic Culture Debate," *Contemporary Security Policy* 33, no. 3 (2012): 437–461. doi:10.1080/13523260.2012.727679.
5 Dima Adamsky, *The Culture of Military Innovation: The Impact of Cultural Factors on the Revolution in Military Affairs in Russia, the US, and Israel* (Redwood City: Stanford University Press, 2010).
6 Tamir Libel, "Explaining the Security Paradigm Shift: Strategic Culture, Epistemic Communities, and Israel's Changing National Security Policy," *Defense Studies* 16, no. 2 (2016): 137–156. doi:10.1080/14702436.2016.1165595.
7 Gregory F. Giles, "Continuity and Change in Israel's Strategic Culture," in *Strategic Culture and Weapons of Mass Destruction*, eds. Jeannie Johnson, Kerry Kartchner, and Jeffrey Larsen (New York: Palgrave Macmillan, 2009), 97–116. doi:10.1057/9780230618305_7.
8 Rafał Kopeć, "The Determinants of the Israeli Strategic Culture," *Przegląd Narodowościowy* 6, no. 1 (01 December 2016): 135–146. doi:10.1515/pn-2016-0008.
9 Niccolò Petrelli, *Israel, Strategic Culture and the Conflict with Hamas* (London, UK: Routledge, 2018).
10 Shaiel Ben-Ephraim, "From Strategic Narrative to Strategic Culture: Labor Zionism and the Roots of Israeli Strategic Culture," *Comparative Strategy* 39, no. 2 (2020): 145–161. doi:10.1080/01495933.2020.1718988.
11 Itamar Rickover, "The Roots of Israeli Strategic Culture," in *Cultural Crossroads in the Middle East*, eds. Vladimir Sazonov, Holger Molder, and Peeter Espak (Tartu, Estonia: University of Tartu Press, 2019), 246–277.
12 Adamsky, *The Culture of Military Innovation: The Impact of Cultural Factors on the Revolution in Military Affairs in Russia, the US, and Israel*, 7–8.

13 For instance see: Colin S. Gray, "Strategic Culture as Context: The First Generation of Theory Strikes Back," *Review of International Studies* 25, no. 1 (1999): 49–69. doi:10.1017/S0260210599000492. Jack Snyder L., *The Soviet Strategic Culture. Implications for Limited Nuclear Operations* (Santa Monica, CA RAND Corporation,[1977]). https://www.rand.org/content/dam/rand/pubs/reports/2005/R2154.pdf
14 For instance see: Elizabeth Kier, "Culture and Military Doctrine: France between the Wars," *International Security* 19, no. 4 (1995): 65–93. doi:10.2307/2539120. http://www.jstor.org/stable/2539120; Alastair I. Johnston, "Thinking about Strategic Culture," *International Security* 19, no. 4 (1995): 32–64. doi:10.2307/2539119.
15 Giles, "Continuity and Change in Israel's Strategic Culture," 97.
16 Kopeć, "The Determinants of the Israeli Strategic Culture," 136.
17 Tamir Libel, "Rethinking Strategic Culture: A Computational (Social Science) Discursive-Institutionalist Approach," *Journal of Strategic Studies* 43, no. 5 (2020a): 686–709. doi:10.1080/01402390.2018.1545645. Tamir Libel, "Strategic Culture as a (Discursive) Institution: A Proposal for Falsifiable Theoretical Model with Computational Operationalization," *Defence Studies* 20, no. 4 (01 October 2020b): 353–372. doi:10.1080/14702436.2020.1814152.
18 Shlomo Aronson, "Israel's Security and the Holocaust: Lessons Learned, but Existential Fears Continue," *Israel Studies* 14, no. 1 (2009): 65–93. doi:10.2979/ISR.2009.14.1.65.
19 Hebrew for cactus – symbolizing resilience and prickliness; see: Oz Almog, *The Sabra the Creation of the New Jew* (Berkeley, CA: University of California Press, 2000).
20 Yechiel Klar, Noa Schori-Eyal, and Yonat Klar, "The 'Never again' State of Israel: The Emergence of the Holocaust as a Core Feature of Israeli Identity and Its Four Incongruent Voices," *Journal of Social Issues* 69, no. 1 (2013): 125–143. doi:10.1111/josi.12007.
21 Amos Morris-Reich, "End on Surface: Teleology and Ground in Israeli Culture," *Representations* 97, no. 1 (2007): 123–150. doi:10.1525/rep.2007.97.1.123.
22 Yehuda Ben-Meir, *Civil-Military Relations in Israel* (New York: Columbia University Press, 1995).
23 Uri Ben-Eliezer, "A Nation-in-Arms: State, Nation, and Militarism in Israel's First Years," *Comparative Studies in Society and History* 37, no. 2 (1995): 264–285. doi:10.1017/S0010417500019666.
24 Giora Goldberg, "The Growing Militarization of the Israeli Political System," *Israel Affairs* 12, no. 3 (2006): 377–394. doi:10.1080/13537120600744594.
25 Ze'ev Drory, *The Israeli Defence Forces and the Foundation of Israel: Utopia in Uniform* (London: Routledge, 2005). doi:10.4324/9780203646212. Ori Swed and John Sibley Butler, "Military Capital in the Israeli Hi-Tech Industry," *Armed Forces and Society* 41, no. 1 (2015): 123–141. doi:10.1177/0095327x13499562.
26 Meytal Eran-Jona, *Israelis' Perception of the IDF* (Maryland: University of Maryland, 2015). https://cpb-us-e1.wpmucdn.com/blog.umd.edu/dist/b/504/files/2017/08/Meytal-Eran-Jona-paper-2-12cvl4q.pdf
27 Yehezkel Dror, *Israeli Statecraft: National Security Challenges and Responses* (Florence: Routledge, 2011). doi:10.4324/9780203814321.
28 Yigal Elam, "On the Myth of the 'Few against the Many' in 1948," *Palestine-Israel Journal of Politics, Economics, and Culture* 9, no. 4 (2002): 50–57.
29 Steven R. David, "Existential Threats to Israel: Learning from the Ancient Past," *Israel Affairs* 18, no. 4 (2012): 503–525. doi:10.1080/13537121.2012.717386.
30 Yiśra'el Ṭal, *National Security the Israeli Experience* (Westport, CT: Praeger, 2000). Gershon Hacohen, *What Is National in National Security? (in Hebrew)* (Tel Aviv, Israel: Israeli Ministry of Defense, 2014b).
31 William Wunderle and Andre Briere, "Augmenting Israel's Qualitative Military Edge," *Middle East Quarterly* 15, no. 1 (2008): 49–58. https://www.meforum.org/1824/augmenting-israels-qualitative-military-edge
32 Axel Shafer and John Dumbrell, "US–Israel Relations: A Special Friendship," in *America's 'Special Relationships': Foreign and Domestic Aspects of the Politics of Alliance* (New York: Routledge, 2009), 201–217. doi:10.4324/9780203872703-17.
33 Baruch Kimmerling, *The Invention and Decline of Israeliness: State, Society, and the Military* (Berkley, CA: University of California Press, 2001a).

34 Asaf Siniver, "Anti-Intellectualism and Israeli Politics," *British Journal of Middle Eastern Studies* 43, no. 4 (2016): 630–643. doi:10.1080/13530194.2016.1166936.
35 Nissim Cohen, "Solving Problems Informally," *The Innovation Journal* 18, no. 1 (2013): 1–16. doi:10.1007/978-1-4939-1553-8_13.
36 Shlomo Mizrahi and Yizhaq Minchuk, "Performance Management, Gaming and Monitoring in Israeli Organisations," *Israel Affairs* 25, no. 3 (2019): 452–466. doi:10.1080/13537121.2019.1593662.
37 Emanuel Adler, *Israel in the World: Legitimacy and Exceptionalism* (New York: Routledge, 2013).
38 Daniel Bar-Tal and Dikla Antebi, "Siege Mentality in Israel," *International Journal of Intercultural Relations* 16, no. 3 (1992): 251–275. doi:10.1016/0147-1767(92)90052-V. Kobi Michael, "Who Really Dictates What an Existential Threat Is? The Israeli Experience," *Journal of Strategic Studies* 32, no. 5 (01 October 2009): 687–713. doi:10.1080/01402390903189360.
39 Abigail Pogrebin, "Why Is This Year Different From All Other Years?" *The Atlantic,* April 2022. https://www.theatlantic.com/ideas/archive/2022/04/passover-2022-america-anti-semitism/629535/
40 Gil Merom, "Israel's National Security and the Myth of Exceptionalism," *Political Science Quarterly* 114, no. 3 (1999): 409–434. doi:10.2307/2658204.
41 Robert C. Parmenter, *The Evolution of Preemptive Strikes in Israeli Operational Planning and Future Implications for the Cyber Domain* (Fort Leavenworth, KS: United States Army Command and Staff College, 2013). http://apps.dtic.mil/sti/pdfs/ADA589616.pdf; Itai Shapira, "The Dominance of Prevention in Israeli Strategic Culture," *Small Wars Journal*, January 2022. https://smallwarsjournal.com/jrnl/art/dominance-prevention-israeli-strategic-culture
42 Oren Barak and Gabriel Sheffer, "Israel's 'Security Network' and Its Impact: An Exploration of a New Approach," *International Journal of Middle East Studies* 38, no. 2 (May 2006): 235–261. doi:10.1017/S0020743806412332.
43 Uri Bar-Joseph, "Military Intelligence as the National Intelligence Estimator: The Case of Israel," *Armed Forces & Society* 36, no. 3 (2010): 505–525. doi:10.1177/0095327×08330934.
44 Noam Kochavi, "The Waning Influence of Foreign Ministries," *Israel Journal of Foreign Affairs* 12, no. 2 (2018): 159–162. doi:10.1080/23739770.2018.1502977.
45 Clive Jones and Tore T. Petersen, *Israel's Clandestine Diplomacies* (Cary: Oxford University Press, 2013). doi:10.1093/acprof:oso/9780199330669.001.0001.
46 This might even have origins in the Jewish tradition of *Tushba*, namely oral learning of the Torah; see: Talya Fishman, *Becoming the People of the Talmud Oral Torah as Written Tradition in Medieval Jewish Cultures*, 1st ed. (Philadelphia, PA: University of Pennsylvania Press, 2011).
47 Ya'acov Amidror, "Israel's National Security Concept (in Hebrew)," *Between the Poles* 28 (October 2020): 19–34. https://bit.ly/3h6uvTE
48 Charles D. Freilich, *Zion's Dilemmas How Israel Makes National Security Policy* (Ithaca: Cornell University Press, 2013b). Charles D. Freilich, "National Security Decision-Making in Israel: Improving the Process," *The Middle East Journal* 67, no. 2 (01 April 2013a): 257–266. doi:10.3751/67.2.16.
49 Kopeć, "The Determinants of the Israeli Strategic Culture," 144.
50 Kobi Michael, "The Israel Defense Forces as an Epistemic Authority: An Intellectual Challenge in the Reality of the Israeli – Palestinian Conflict," *Journal of Strategic Studies* 30, no. 3 (01 June 2007): 421–446. doi:10.1080/01402390701343417.
51 Baruch Kimmerling, "The Social Construction of Israel's National Security (in Hebrew)," *Democratic Culture* 4/5 (2001b): 267–301. Itamar Rickover, "The Centrality of the Military Echelon in Intelligence Assessments and Political-Strategic Planning: Materialist and Cultural Explanations (in Hebrew)" (Bar-Ilan University, 2011).
52 Giles, "Continuity and Change in Israel's Strategic Culture," 97–116.
53 Avi Shilon, "The Begin Doctrine," in *Menachem Begin: A Life*, ed. Avi Shilon (New Haven, MA: Yale University Press, 2012), 335–347. doi:10.12987/yale/9780300162356.003.0016.
54 Amos Yadlin, *The Begin Doctrine: The Lessons of Osirak and Deir Ez-Zor* (Tel Aviv, Israel: INSS, 2018). https://www.inss.org.il/publication/the-begin-doctrine-the-lessons-of-osirak-and-deir-ez-zor/

55 Whitney Raas and Austin Long, "Osirak Redux?: Assessing Israeli Capabilities to Destroy Iranian Nuclear Facilities," *International Security* 31, no. 4 (2007): 7–33. doi:10.1162/isec.2007.31.4.7.
56 Ariel E. Levite, "Global Zero: An Israeli Vision of Realistic Idealism," *The Washington Quarterly* 33, no. 2 (Apr 01, 2010): 157–168. doi:10.1080/01636601003674038.
57 Raphael D. Marcus, "Learning 'Under Fire': Israel's Improvised Military Adaptation to Hamas Tunnel Warfare," *Journal of Strategic Studies* 42, no. 3–4 (2019): 344–370. doi:10.1080/01402390.2017.1307744. Raphael D. Marcus, "Military Innovation and Tactical Adaptation in the Israel–Hizballah Conflict: The Institutionalization of Lesson-Learning in the IDF," *Journal of Strategic Studies* 38, no. 4 (2015): 500–528. https://doi.org/10.1080/01402390.2014.923767; Lazar Berman, "Capturing Contemporary Innovation: Studying IDF Innovation against Hamas and Hizballah," *Journal of Strategic Studies* 35, no. 1 (01 February 2012): 121–147. doi:10.1080/01402390.2011.608933.
58 Dima Adamsky, "Between the Poles: Israeli Military Innovation between Anticipation and Adaptation (in Hebrew)," *Between the Poles* 21 (July 2019): 167–180. https://bit.ly/3jwESBI; For the theoretical frameworks see: Dima Adamsky and Kjell Inge Bjerga, *Contemporary Military Innovation: Between Anticipation and Adaption* (London: Routledge, 2012). doi:10.4324/9780203112540.
59 Dmitry (Dima) Adamsky, "From Israel with Deterrence: Strategic Culture, Intra-War Coercion and Brute Force," *Security Studies* 26, no. 1 (2017): 157–184. doi:10.1080/09636412.2017.1243923.
60 Gadi Eisenkot and Gabi Siboni, *The Campaign between Wars: How Israel Rethought Its Strategy to Counter Iran's Malign Regional Influence* (Washington, DC: Washington Institute, 2019a). https://www.washingtoninstitute.org/policy-analysis/campaign-between-wars-how-israel-rethought-its-strategy-counter-irans-malign
61 Niccolò Petrelli, "Deterring Insurgents: Culture, Adaptation and the Evolution of Israeli Counterinsurgency, 1987–2005," *Journal of Strategic Studies* 36, no. 5 (2013): 666–691. doi:10.1080/01402390.2012.755923.
62 Dan Meridor and Ron Eldadi, *Israel's National Security Doctrine: The Report of the Committee on the Formulation of the National Security Doctrine (Meridor Committee), Ten Years Later* (Tel Aviv, Israel [2019]). https://www.inss.org.il/wp-content/uploads/2019/02/Memo187_11.pdf; David Rodman, "Review Essay: Israel's National Security Doctrine: An Appraisal of the Past and a Vision of the Future," *Israel Affairs* 9, no. 4 (01 June 2003): 115–140. doi:10.1080/13537120412331321563. Uri Bar-Joseph, "Towards a Paradigm Shift in Israel's National Security Conception," *Israel Affairs* 6, no. 3–4 (01 Mar 2000): 99–114. doi:10.1080/13537120008719574.
63 Gabriel Sheffer, *Israel's Security Networks: A Theoretical and Comparative Perspective* (Cambridge: Cambridge University Press, 2013), 143.
64 Ehud Eiran and Martin B. Malin, "The Sum of all Fears: Israel's Perception of a Nuclear-Armed Iran," *The Washington Quarterly* 36, no. 3 (2013): 77–89. doi:10.1080/0163660X.2013.825551.
65 Chuck Freilich, "The United States, Israel, and Iran: Defusing an 'Existential' Threat," *Arms Control Today* 38, no. 9 (2008): 6–11. https://www.armscontrol.org/act/2008-11/iran-nuclear-briefs/united-states-israel-iran-defusing-%E2%80%9Cexistential%E2%80%9D-threat
66 Udi Dekel, "A Multi-Arena Missile Attack that Disrupts Israel's Defense and Resilience Pillars," in *Existential Threats to the State of Israel in the 21st Century*, ed. Ofir Winter (Tel Aviv, Israel: INSS, 2020), 69–82.
67 Udi Golan, *War without Intention? Unintended Escalation as the Major Driver for War in the Middle East (in Hebrew)* (Tel Aviv, Israel: IDF, 2020). https://bit.ly/3ykIhrF
68 Daniel Byman, "Israel's Four Fronts," *Survival (London)* 61, no. 2 (2019): 167–188. doi:10.1080/00396338.2019.1589094.
69 Gadi Eisenkot and Gabi Siboni, *Guidelines for Israel's National Security Strategy* (Washington, DC, 2019b). https://www.washingtoninstitute.org/uploads/Documents/pubs/PolicyFocus160-EisenkotSiboni.pdf
70 Uri Bar-Joseph, "Variations on a Theme: The Conceptualization of Deterrence in Israeli Strategic Thinking," *Security Studies* 7, no. 3 (1998): 145–151. doi: 10.1080/09636419808429353

71 Sami Cohen, *Israel's Asymmetric Wars*, 1st. ed. (New York: Palgrave Macmillan, 2010), 151–152.
72 Yossi Baidatz and Dima Adamsky, *The Development of the Israeli Approach towards Deterrence (in Hebrew)*, Vol. 8 (Ramat Hasharon, Israel: Israeli National Defense College, 2014).
73 Avi Kober, "From Blitzkrieg to Attrition: Israel's Attrition Strategy and Staying Power," *Small Wars & Insurgencies* 16, no. 2 (June 2005): 216–240. doi:10.1080/09592310500080005.
74 Efraim Inbar and Eitan Shamir, "'Mowing the Grass': Israel's Strategy for Protracted Intractable Conflict," *Journal of Strategic Studies* 37, no. 1 (02 January 2014): 65–90. doi:10.1080/01402390.2013.830972.
75 Itai Brun, "Where Has the Ground Maneuver Disappeared To? (in Hebrew)," *Ma'Arachot* 420–421 (September 2008): 4–15. https://bit.ly/3j2z9Sw
76 Itai Brun, *From Air Superiority to Multi-Dimensional Strike* (in Hebrew) (Tel Aviv, Israel: INSS, 2022). https://www.inss.org.il/he/publication/air-force/; David Rodman, *Sword and Shield of Zion: The Israeli Air Force in the Arab-Israeli Conflict, 1948–2012* (Brighton: Sussex Academic Press, 2012).
77 Meir Finkel, Yaniv Friedman and Dana Freisler-Swiri, "Active Defense as a Fourth Pillar of Israeli Security Doctrine (in Hebrew)," *Between the Poles* 4 (July 2015): 121–136. http://bit.ly/2US0heE
78 Amir Avstein and Meidar Avidar, "Defense Forces: Between Identity and Ethos (in Hebrew)," *Between the Poles* 22–23 (October 2019): 149–163. https://bit.ly/3jfWMam
79 Arnon Gutfeld, "From 'Star Wars' to 'Iron Dome': US Support of Israel's Missile Defense Systems," *Middle Eastern Studies* 53, no. 6 (2017): 934–948. doi:10.1080/00263206.2017.1350844. Lior Tabansky, "Israel Defense Forces and National Cyber Defense," *Connections: The Quarterly Journal (English Ed.)* 19, no. 1 (2020): 45–62. doi:10.11610/Connections.19.1.05. Eran Ortal and Dvir Peleg, "The Misrecognizing of Developments in the General Staff - the Tunnel Threat (in Hebrew)," *Between the Poles* 22–23 (October 2019): 33–61. https://bit.ly/3kc6Fbc
80 Uri Bar-Joseph, "Lessons Not Learned: Israel in the Post Yom-Kippur Era," *Israel Affairs* 14, no. 1 (2008): 70–83. doi: 10.1080/13537120701706005
81 D., A. and R., "A Unique Potential for the Beginning of a "Revolution in Intelligence Affairs" as a Part of "Momentum" (in Hebrew)," *Between the Poles* 28–30 (October 2020): 195–210. https://bit.ly/3qMBswg; Aviv Kochavi and Eran Ortal, "'Ma'Asei Aman': Permanent Change in a Changing Reality," *Between the Poles* 2 (July 2014): 11–85. https://www.idf.il/media/18519/masei-aman-article.pdf
82 Eran Ortal, "Momentum Multi-Year Plan: A Theoretical Framework," *Between the Poles* 28–30 (October 2020): 35–50. https://www.idf.il/en/mini-sites/dado-center/vol-28-30-military-superiority-and-the-momentum-multi-year-plan/going-on-the-attack-the-theoretical-foundation-of-the-israel-defense-forces-momentum-plan-1/
83 Itai Brun, "While You Are Busy Making Other Plans – The Other RMA," *Journal of Strategic Studies* 33, no. 4 (01 August 2010): 535–565. doi:10.1080/01402390.2010.489708.
84 Eran Ortal, "Turn on the Lights, Turn Off the Fire (in Hebrew)," *Between the Poles* 31–32 (May 2021b): 53–69. https://bit.ly/3jeVMTY
85 Avi Kober, "Has Battlefield Decision Become Obsolete? The Commitment to the Achievement of Battlefield Decision Revisited," *Contemporary Security Policy* 22, no. 2 (2001a): 96–120. doi:10.1080/135232605123313911158.
86 Avi Kober, "From Heroic to Post-Heroic Warfare," *Armed Forces and Society* 41, no. 1 (01 January 2015a): 96–122. doi:10.1177/0095327x13498224.
87 Avi Kober, *Practical Soldiers: Israel's Military Thought and Its Formative Factors* (Leiden, Netherlands: Brill, 2015b). Avi Kober, "What Happened to Israeli Military Thought?" *Journal of Strategic Studies* 34, no. 5 (01 October 2011): 707–732. doi:10.1080/01402390.2011.561109.
88 Eliot A. Cohen, Michael A. Eisenstadt and Andrew J. Bacevich, *Knives, Tanks and Missiles: Israel's Security Revolution* (Washington DC, 1998). https://www.washingtoninstitute.org/uploads/Documents/pubs/KnivesTanksandMissiles.pdf.pdf

89 Gil-li Vardi, "'Pounding Their Feet': Israeli Military Culture as Reflected in Early IDF Combat History," *Journal of Strategic Studies* 31, no. 2 (01 April 2008): 295–324. doi:10.1080/01402390801940476.
90 Adamsky, *The Culture of Military Innovation: The Impact of Cultural Factors on the Revolution in Military Affairs in Russia, the US, and Israel,* 111.
91 Aviv Kochavi, "Chief of Staff Introduction (in Hebrew)," *Between the Poles* 28–30 (October 2020): 7–10. https://bit.ly/3jWqX8B; Aharon Haliva, "Denying the Rockets: Initial Thoughts After "Guardian of the Walls" (in Hebrew)," *Between the Poles* 33 (June 2021). https://bit.ly/3hhZRqx; Tamir Hayman, "Chief of the IDI Introduction (in Hebrew)," *Intelligence in Theory and in Practice* 6, no. 6 (December 2020): 4–5. https://bit.ly/3ywRa1i
92 Meir Finkel, "IDF Strategy Documents, 2002–2018," *Strategic Assessment* 23, no. 4 (October 2020b): 3–17. https://www.inss.org.il/publication/idf-strategy-documents-2002-2018-on-processes-chiefs-of-staff-and-the-idf/
93 Kober, *Practical Soldiers: Israel's Military Thought and Its Formative Factors,* 73.
94 Ya'acov Amidror, "The Strike as a Cognitive Paradigm of Effects (in Hebrew)," *Ma'Arachot* 403–404 (December 2005): 54–57. https://bit.ly/3dEPtXU
95 Avi Kober, "The Israel Defense Forces in the Second Lebanon War: Why the Poor Performance?" *Journal of Strategic Studies* 31, no. 1 (01 February 2008): 3–40. doi:10.1080/01402390701785211.
96 Hacohen, *What Is National in National Security? (in Hebrew),* 90–91.
97 Eitan Shamir, "The Rise and Decline of 'Optional Control' in the IDF," *Israel Affairs* 23, no. 2 (04 March 2017): 205–230. doi:10.1080/13537121.2016.1274515.
98 Uzi Ben-Shalom and Eitan Shamir, "Mission Command between Theory and Practice: The Case of the IDF," *Defense & Security Analysis* 27, no. 2 (2011): 101–117. doi:10.1080/14751798.2011.578715.
99 Gershon Hacohen, "IDF Strategy: Force Design (in Hebrew)," *Between the Poles* 7 (April 2016): 119–129. https://bit.ly/3rrj1hi
100 Gershon Hacohen, "From Education for Excellence to Leadership Disability (in Hebrew)," *Ma'Arachot* 457 (October 2014a): 4–9. https://bit.ly/3wOIqCz
101 Gershon Hacohen, "The Leader and the Systemic Planning: Between Engineering and Architecture (in Hebrew)," *Ma'Arachot* 376 (April 2001): 10–15. https://bit.ly/3rjZbnR
102 Meir Finkel, "Flexible Force Structure: A Flexibility Oriented Force Design and Development Process for Israel," *Israel Affairs* 12, no. 4 (01 October 2006): 789–800. doi:10.1080/13533310600890117.
103 Haim Asa, "Force Design: Between Scientific Thinking and Lateral Thinking and Imagination (in Hebrew)," *Between the Poles* 7 (April 2016): 109–118. https://bit.ly/3xVI92c
104 Aharon Haliva, "More of the Same - the Need for a Conceptual Leap in Force Design (in Hebrew)," *Between the Poles* 9 (December 2016): 9–23. https://bit.ly/3jdkCFh
105 Alon Paz, "The IDF is Thus Formulated: The Efforts of Reformulating the IDF (in Hebrew)," *Between the Poles* 7 (April 2016): 9–40. https://bit.ly/3hQd1Ly
106 Adamsky, *The Culture of Military Innovation: The Impact of Cultural Factors on the Revolution in Military Affairs in Russia, the US, and Israel,* 114.
107 Cohen, *Knives, Tanks and Missiles: Israel's Security Revolution*, Yitzhaki Chen, "Technology Is Not Everything (in Hebrew)," *Ma'Arachot* 332 (October 1993). https://bit.ly/3dK1qvf
108 Yoram Evron, "4IR Technologies in the Israel Defence Forces: Blurring Traditional Boundaries," *Journal of Strategic Studies* (2020), 1–22. doi:10.1080/01402390.2020.1852936.
109 Uzi Rubin, "Israel's Defence Industries – An Overview," *Defence Studies* 17, no. 3 (2017): 228–241. doi:10.1080/14702436.2017.1350823.
110 Amichai Cohen and Stuart Alan Cohen, "Beyond the Conventional Civil–Military 'Gap': Cleavages and Convergences in Israel," *Armed Forces & Society* (2020): 1–21.https://doi.org/10.1177/0095327x20903072
111 Yoram Peri, *Generals in the Cabinet Rooms: How the Military Shapes Israeli Policy* (Washington, DC: United States Institute for Peace, 2006). https://www.tau.ac.il/institutes/herzog/generals.pdf
112 Kobi Michael, Limor Regev, and Dudi Kimchi, "Containment Over Decision," *Strategic Assessment* 23, no. 4 (October 2020). https://www.inss.org.il/publication/containment-over-

decision-internalizing-the-limits-of-political-militarization-in-israel/?offset=1&posts=24&type=401

113 Dalia Dassa Kaye and Shira Efron, "Israel's Evolving Iran Policy," *Survival (London)* 62, no. 4 (2020): 9. doi:10.1080/00396338.2020.1792095.

114 Gil Merom, "The Logic and Illogic of an Israeli Unilateral Preventive Strike on Iran," *The Middle East Journal* 71, no. 1 (2017): 87–110. doi:10.3751/71.1.15.

115 This bias was described in: Ken Booth, *Strategy and Ethnocentrism* (London: Croom Helm, 1979).

116 Udi Dekel and Carmit Valensi, *After a Decade of War in Syria, Israel Should Change Its Policy* (Tel Aviv, Israel: INSS, 2021). https://www.inss.org.il/publication/israel-assad/

117 Eyal Zisser, "'Operation Good Neighbor'—Israel and the Rise and Fall of the 'Southern Syria Region' (SSR)," *Israel Studies (Bloomington, Ind.); Israel Studies* 26, no. 1 (2021): 1–23. doi:10.2979/israelstudies.26.1.01.

118 Nitzan Alon and Dana Freisler-Swiri, "The Campaign between the Wars in the IDF (in Hebrew)," *Between the Poles* 22–23 (October 2019): 13–31. https://bit.ly/36hw62I

119 Jeffrey Goldberg, "Iran and the Palestinians Lose Out in the Abraham Accords," *The Atlantic* (September 2020). https://www.theatlantic.com/ideas/archive/2020/09/winners-losers/616364/

120 Assaf Orion and Galia Lavi, *Israel-China Relations: Opportunities and Challenges* (Tel Aviv, Israel: INSS, 2019). https://www.inss.org.il/publication/israel-china-relations-opportunities-and-challenges/

121 Jonathan Hoffman, "The Return of Great-Power Competition to the Middle East: A Two-Level Game," *Middle East Policy* (2021). doi:10.1111/mepo.12539.

122 Oren Barak, Amit Sheniak, and Assaf Shapira, "The Shift to Defence in Israel's Hybrid Military Strategy," *Journal of Strategic Studies* (2020): 1–33. doi:10.1080/01402390.2020.1770090. https://doi.org/10.1080/01402390.2020.1770090

123 Avi Kober, "Israeli War Objectives into an Era of Negativism," *Journal of Strategic Studies* 24, no. 2 (01 June 2001b): 176–201. doi:10.1080/01402390108565557.

124 Udi Dekel and Noa Shusterman, *The Palestinian Arena: Preserving the Status Quo Or Seeking Change?* (Tel Aviv, Israel: INSS, 2021). https://www.inss.org.il/publication/strategic-survey-palestinian-arena/

125 IDF General Staff, *IDF Strategy (in Hebrew)* (Tel Aviv, Israel: IDF, 2018). https://bit.ly/3eIMU7k

126 Avner Barnea, "Strategic Intelligence: A Concentrated and Diffused Intelligence Model," *Intelligence and National Security* 35, no. 5 (2020): 701–716. doi:10.1080/02684527.2020.1747004.

127 Cohen, *Israel's Asymmetric Wars,* 43–56.

128 Shaul Mofaz, "The IDF in the 2000s (in Hebrew)," *Ma'Arachot* 363 (March 1999): 2–9. https://bit.ly/3xOZqJO

129 Dana Freisler-Swiri, "What Can Be Learned from the Development of the Operational Concept? (2006) (in Hebrew)," *Between the Poles* 10 (March 2017): 9–50. https://bit.ly/35TmTNT

130 Assaf Hazani, "Who Is Afraid of French Philosophers? (in Hebrew)," *Ma'Arachot* 437 (June 2011): 80–82. https://bit.ly/3kJU0MY

131 Graham Allison, *Deterring Terror How Israel Confronts the Next Generation of Threats* (Cambridge, MA: Harvard Kennedy School, 2016). https://www.belfercenter.org/sites/default/files/legacy/files/IDF%20doctrine%20translation%20-%20web%20final2.pdf

132 Eran Ortal, "The Fly on the Elephant's Back," *Strategic Assessment* 24, no. 2 (April 2021a): 108–115. https://www.inss.org.il/publication/the-fly-on-the-elephants-back-the-campaign-between-wars-in-israels-security-doctrine/

133 For several exceptions see: Tamir Libel, "Looking for Meaning: Lessons from Mossad's Failed Adaptation to the Post-Cold War Era, 1991–2013," *The Journal of Intelligence History* 14, no. 2 (2015): 83–95. doi:10.1080/16161262.2015.1033238.; Clila Magen and Eytan Gilboa, "Communicating from within the Shadows: The Israel Security Agency and the Media," *International Journal of Intelligence and Counterintelligence* 27, no. 3 (2014): 485–508. doi:10.1080/08850607.2014.900293.

134 For an initial discussion of this culture see: Itai Shapira, "Israeli National Intelligence Culture and the Response to COVID-19," *War on the Rocks* (November 2020). https://warontherocks.com/2020/11/israeli-national-intelligence-culture-and-the-response-to-covid-19/
135 Batia Ben-Hador, Meytal Eran-Jona, and Christopher Dandeker, "Perceptions of the Future Battlefield in Israel Vs. Western Countries," *Israel Affairs* 24, no. 4 (2018): 707–731. doi:10.1080/13537121.2018.1478782.
136 Stuart A. Cohen, *The New Citizen Armies: Israel's Armed Forces in Comparative Perspective* (London: Routledge, 2010). doi:10.4324/9780203861714.
137 This term appeared in Geertz's work on anthropology and cultures. Clifford Geertz, *The Interpretation of Cultures: Selected Essays* (New York: Basic Books, 1973).
138 Meir Finkel, "AI Singularity and the Growing Risk of Surprise – Lessons from the IDFs Strategic and Operational Learning Processes 2014–2019," *Prism* 8, no. 4 (June, 2020a): 31–39. https://ndupress.ndu.edu/Media/News/News-Article-View/Article/2217671/ai-singularity-and-the-growing-risk-of-surprise-lessons-from-the-idfs-strategic/
139 For a broad discussion of strategic theory and practice, see: Hew Strachan, "Strategy in Theory; Strategy in Practice," *The Journal of Strategic Studies* 42, no. 2 (2019): 171–190.

Selected Bibliography

Adamsky, Dima. "Between the Poles: Israeli Military Innovation between Anticipation and Adaptation (in Hebrew)." *Between the Poles* 21 (July 2019): 167–180. https://bit.ly/3jwESBI

Adamsky, Dima. *The Culture of Military Innovation: The Impact of Cultural Factors on the Revolution in Military Affairs in Russia, the US, and Israel.* Redwood City: Stanford University Press, 2010.

Adamsky, Dmitry (Dima). "From Israel with Deterrence: Strategic Culture, Intra-War Coercion, and Brute Force." *Security Studies* 26, no. 1 (2017): 157–184. doi:10.1080/09636412.2017.1243923.

Adler, Emanuel. *Israel in the World: Legitimacy and Exceptionalism.* New York: Routledge, 2013.

Allison, Graham. *Deterring Terror How Israel Confronts the Next Generation of Threats.* Cambridge, MA: Harvard Kennedy School, 2016 https://www.belfercenter.org/sites/default/files/legacy/files/IDF%20doctrine%20translation%20-%20web%20final2.pdf

Amidror, Ya'acov. "Israel's National Security Concept (in Hebrew)." *Between the Poles* 28 (October 2020): 19–34. https://bit.ly/3h6uvTE

Baidatz, Yossi and Dima Adamsky. *The Development of the Israeli Approach towards Deterrence (in Hebrew).* Eshtonot., edited by Aloni-Gorssman, Navah. Vol. 8. Ramat Hasharon, Israel: Israeli National Defense College, 2014.

Barak, Oren and Gabriel Sheffer. "Israel's "Security Network" and Its Impact: An Exploration of a New Approach." *International Journal of Middle East Studies* 38, no. 2 (May 2006): 235–261. doi:10.1017/S0020743806412332.

Barak, Oren, Amit Sheniak, and Assaf Shapira. "The Shift to Defence in Israel's Hybrid Military Strategy." *Journal of Strategic Studies* (2020): 1–33. 10.1080/01402390.2020.1770090.

Bar-Joseph, Uri. "Towards a Paradigm Shift in Israel's National Security Conception." *Israel Affairs* 6, no. 3–4 (01 March 2000): 99–114. doi:10.1080/13537120008719574.

Bar-Tal, Daniel and Dikla Antebi. "Siege Mentality in Israel." *International Journal of Intercultural Relations* 16, no. 3 (1992): 251–275. doi:10.1016/0147-1767(92)90052-V.

Ben-Ephraim, Shaiel. "From Strategic Narrative to Strategic Culture: Labor Zionism and the Roots of Israeli Strategic Culture." *Comparative Strategy* 39, no. 2 (2020): 145–161. doi:10.1080/01495933.2020.1718988.

Ben-Hador, Batia, Meytal Eran-Jona, and Christopher Dandeker. "Perceptions of the Future Battlefield in Israel Vs. Western Countries." *Israel Affairs* 24, no. 4 (2018): 707–731. doi:10.1080/13537121.2018.1478782.

Berman, Lazar. "Capturing Contemporary Innovation: Studying IDF Innovation against Hamas and Hizballah." *Journal of Strategic Studies* 35, no. 1 (01 February 2012): 121–147. doi:10.1080/01402390.2011.608933.

Brun, Itai. "Where Has the Ground Maneuver Disappeared to? (in Hebrew)." *Ma'Arachot* 420–421 (September, 2008): 4–15. https://bit.ly/3j2z9Sw

Cohen, Eliot A., Michael A. Eisenstadt, and Andrew J. Bacevich. *Knives, Tanks and Missiles: Israel's Security Revolution*. Washington DC, 1998.

Cohen, Stuart A. *The New Citizen Armies: Israel's Armed Forces in Comparative Perspective*. London: London: Routledge, 2010. doi:10.4324/9780203861714.

Eisenkot, Gadi and Gabi Siboni. *The Campaign between Wars: How Israel Rethought its Strategy to Counter Iran's Malign Regional Influence*. Washington, DC: Washington Institute, 2019a https://www.washingtoninstitute.org/policy-analysis/campaign-between-wars-how-israel-rethought-its-strategy-counter-irans-malign

Finkel, Meir. "Flexible Force Structure: A Flexibility Oriented Force Design and Development Process for Israel." *Israel Affairs* 12, no. 4 (01 October 2006): 789–800. doi:10.1080/13533310600890117.

Finkel, Meir. "IDF Strategy Documents, 2002–2018." *Strategic Assessment* 23, no. 4 (October 2020b): 3–17. https://www.inss.org.il/publication/idf-strategy-documents-2002-2018-on-processes-chiefs-of-staff-and-the-idf/

Freilich, Charles D. *Zion's Dilemmas How Israel Makes National Security Policy*. Ithaca: Cornell University Press, 2013b.

Giles, Gregory F. "Continuity and Change in Israel's Strategic Culture." In *Strategic Culture and Weapons of Mass Destruction*, edited by Jeannie L. Johnson, Kerry Kartchner and Jeffrey Larsen, 97–116. New York: Palgrave Macmillan, 2009.

Hacohen, Gershon. *What is National in National Security? (in Hebrew)*. Broadcast University. Tel Aviv, Israel: Israeli Ministry of Defense, 2014b.

Kimmerling, Baruch. *The Invention and Decline of Israeliness: State, Society, and the Military*. Berkley, CA: University of California Press, 2001a.

Kober, Avi. *Practical Soldiers: Israel's Military Thought and its Formative Factors*. History of Warfare. Leiden, Netherlands: Brill, 2015b.

Kopeć, Rafał. "The Determinants of the Israeli Strategic Culture." *Przegląd Narodowościowy* 6, no. 1 (01 December 2016): 135–146. doi:10.1515/pn-2016-0008.

Levite, Ariel E. "Global Zero: An Israeli Vision of Realistic Idealism." *The Washington Quarterly* 33, no. 2 (01 April 2010): 157–168. doi:10.1080/01636601003674038.

Libel, Tamir. "Explaining the Security Paradigm Shift: Strategic Culture, Epistemic Communities, and Israel's Changing National Security Policy." *Defense Studies* 16, no. 2 (2016): 137–156. doi:10.1080/14702436.2016.1165595.

Marcus, Raphael D. "Military Innovation and Tactical Adaptation in the Israel–Hizballah Conflict: The Institutionalization of Lesson-Learning in the IDF." *Journal of Strategic Studies* 38, no. 4 (2015): 500–528. 10.1080/01402390.2014.923767.

Meridor, Dan and Ron Eldadi. *Israel's National Security Doctrine: The Report of the Committee on the Formulation of the National Security Doctrine (Meridor Committee), Ten Years Later*. Tel Aviv, Israel, 2019. https://www.inss.org.il/wp-content/uploads/2019/02/Memo187_11.pdf

Michael, Kobi. "Who Really Dictates What an Existential Threat Is? The Israeli Experience." *Journal of Strategic Studies* 32, no. 5 (01 October 2009): 687–713. doi:10.1080/01402390903189360.

Ortal, Eran. "Momentum Multi-Year Plan: A Theoretical Framework." *Between the Poles* 28–30 (October 2020): 35–50. https://www.idf.il/en/mini-sites/dado-center/vol-28-30-military-superiority-and-the-momentum-multi-year-plan/going-on-the-attack-the-theoretical-foundation-of-the-israel-defense-forces-momentum-plan-1/

Peri, Yoram. *Generals in the Cabinet Rooms: How the Military Shapes Israeli Policy*. Washington, DC: United States Institute for Peace, 2006 https://www.tau.ac.il/institutes/herzog/generals.pdf

Petrelli, Niccolò. *Israel, Strategic Culture and the Conflict with Hamas*. London, UK: Routledge, 2018.

Rickover, Itamar. "The Roots of Israeli Strategic Culture." In *Cultural Crossroads in the Middle East*, edited by Sazonov, Vladimir, Holger Molder and Peeter Espak, 246–277. Tartu, Estonia: University of Tartu Press, 2019.

Rodman, David. "Review Essay: Israel's National Security Doctrine: An Appraisal of the Past and a Vision of the Future." *Israel Affairs* 9, no. 4 (01 June 2003): 115–140. doi:10.1080/13537120412331321563.

Shamir, Eitan. "The Rise and Decline of 'Optional Control' in the IDF." *Israel Affairs* 23, no. 2 (04 March 2017): 205–230. doi:10.1080/13537121.2016.1274515.
Shilon, Avi. "The Begin Doctrine." In *Menachem Begin: A Life*, edited by Shilon, Avi, 335–347. New Haven, MA: Yale University Press, 2012.
Ṭal, Yiśra'el. *National Security the Israeli Experience*. Westport, CT: Praeger, 2000.
Vardi, Gil-li. "'Pounding Their Feet': Israeli Military Culture as Reflected in Early IDF Combat History." *Journal of Strategic Studies* 31, no. 2 (01 April 2008): 295–324. doi:10.1080/014023 90801940476.stylefix

26
TURKEY'S STRATEGIC CULTURE
State of the Literature

George E. Bogden

Introduction

Perhaps the most peculiar monument in Ankara is the pagoda in the open park across from the city's main train station. Etched in the stone structure are the names of the Turkish soldiers who perished in the Korean War. Alluding to their valor during the 1950 Chinese breakthrough, when they fought bloody, rearguard bayonet battles to protect retreating American troops, General Douglas MacArthur described each as a "hero of heroes." They were, in his mind, men for whom there was "no impossibility."[1] In Western popular culture, these once celebrated fighters have largely been reduced to "commie-killing" caricatures by *M*A*S*H* reruns. Historian John Vander Lippe fittingly refers to them as "the forgotten brigade."[2]

For students of Turkey's strategic culture, the memorial embodies the uneven contemporary resonance of events that defined the country's position in the twentieth century. The identities, values, and norms among Turkish elites in the 1950s that spurred them to lend military support for the United Nations' "police action" in Northeast Asia have eroded. That gambit, which facilitated Turkey's accession to the North Atlantic Treaty Organization (NATO), demonstrated the core tenets of Turkey's early "European vocation," an essential element of its nascent strategic culture. The state founded and led by Mustafa Kemal Atatürk developed a rigidly disciplined approach to strategic questions, adhered to by many of his successors.[3] Since 2002, however, a parliamentary government has slowly succeeded in altering the Republic founding secular ideals and its previously immovable pro-Western posture.

When the ruling Justice and Development Party (AKP) gained power, a once stable strategic culture began to disintegrate. A tumultuous two decades followed, marked by grand visions and sober reassessments of Turkey's relationship with surrounding regions and the wider world. Along the way, Turkey sought and failed to capitalize on the so-called Arab Spring, radically shifting its relationship to the Middle East. Its government persuaded its electorate to amend Turkey's constitution, fundamentally reorganizing prerogatives distributed between the Republic's prime minister and president. Recep Tayyip Erdoğan, who has occupied these roles sequentially to retain power, withstood a failed attempted coup, prompting a monumental purge of security institutions across Turkey's government. These are only a few of the destabilizing events that have shaken the foundations of the country's strategic culture in the twenty-first century.

 DOI: 10.4324/9781003010302-29

Aims of the Chapter

Unsurprisingly, these transitions have produced scholarly literature as fragmented and contradictory as the events and ideas it seeks to explain. Commentators describing Turkey's strategic culture have tended to reproduce existing narratives about the past. Others, attempting to project future events, given recent changes, have vacillated in their prescriptive, normative, and predictive arguments, often from year to year. Increasingly harsh criticism of President Erdoğan's government has resulted in a consensus in the press about his supposedly "transactional" approach to national security.[4] Observers of the *longue durée* in commentary on Turkey might note the irony of this most recent turn, often invoked in the discussion of its strategic culture.[5] Since the end of the Cold War, Turkish governments are typically cast as fitting the mold of one or another tired trope: its leaders are either eager to serve as a "bridge" or "barrier" between Europe and the Middle East.[6] When Turkey's leaders choose to engage more deeply with a hereditary ally or rival, they instantaneously transform, for many in the West, into puppets of another power, or a fiercely autonomous, inward-looking regime, jealously guarding its sovereignty to avoid falling neatly into any sphere of influence. These assessments betray the difficulties faced by many commentators seeking to account for a complex recent past.

Far from resolving these tensions, this chapter seeks to provide an entry point for students and scholars interested in research on Turkish strategic culture. The first section provides a broad historical foreground; summaries of the strategic approach of the late Ottoman Empire and the Kemalist tradition before 2002 provide a critical backdrop. This prefatory section serves in part to familiarize readers with classic scholarship on Turkish strategic culture, which should be read against the literature described in later sections. Recent scholarly accounts are then reviewed, with attention to their strengths and weaknesses. The chapter also identifies areas – specific subjects and geographic regions – that may be of interest to those seeking to evaluate undercelebrated aspects of contemporary Turkish strategic culture.

Importantly, this chapter does not enter the fray or seek to update the tendentious commentary regarding different ontological approaches within this volume's field. In part, this stems from the deluge of academic works exploring these debates at length.[7] But more importantly, this chapter seeks to integrate and reflect an awareness of how Turkish scholars view the strategic culture of their state.[8] Their scholarship is often in dialogue with American and British scholars who seek to produce academic research for the benefit of audiences outside of Turkey. It is a mistake to focus solely on the latter group, though their work tends to follow preferred disciplinary conventions scrupulously. The major research universities of Istanbul and Ankara that have produced experts on strategic culture tend not to fall prey to the "wars of paradigms" or "wars on paradigms," which occasionally serve as focal points for research elsewhere. Instead, for these scholars, strategic culture largely remains an intuitive concept understood in the way it was originally interpreted by Jack Snyder: as a "set of general beliefs, attitudes, and behavioral patterns … [that have] achieved a state of semipermanence."[9]

Historical Diagnostics

From a Western perspective, Turkey remains a complicated case study in the development of strategic culture. Its landscape and major cities portray a rich mixture of civilizations, bearing the vestiges of multiple empires – Hittite, Phrygian, Byzantine, and Ottoman. To an untrained eye, the modern Republic can appear as an enigma, springing from the seat of the last Caliphate to struggle with a strident tradition of secularism and highly charged questions of identity. Current political change within its borders has highlighted these peculiar dynamics, as well as the historical legacies of Turkey's rapid political transformation in the twentieth century.

Beginning to untangle these influences on strategic culture requires a basic chronology that covers adequate ground. This is provided below, in brief. Throughout these periods, general categories of "republican" and "imperial" paradigms of strategic culture developed within the Turkish state.[10] They have animated decisionmaking for decades. In part, these attributes are the result of interrelated aspirations and strategic assets. As Malik Mufti notes in his magisterial text *Daring and Caution in Turkish Strategic Culture*, "Turkey is a prominent actor in three arenas that are likely to remain centers of great power interest – the Caucasus, Balkans and Middle East."[11] No matter how significantly the country's borders and military strength have fluctuated, its strategists and planners have often faced the same questions about how to deploy confined power in multiple theaters in which they feel obliged to operate.

An Imperial Heritage

Foundational texts describing Turkey's strategic culture have often framed the decades before the modern state as equally formative as the period that directly followed its founding. The last half-century of the Ottoman Empire saw the rise of many ideas which continue to influence current conceptions of international security: "Ottomanism," "secularism," "Islamism," and "Turkism," to name a few.[12] Rather than all influencing strategic culture in one way, these notions have tended to be wielded by one or another opposing camp in Turkey's military establishment to support the case for restraint or adventurism in adjoining regions.[13] For example, arguments tend to characterize the calamitous intrigue that doomed the last imperial court as indicative of the perils of overreach or the need to double down on efforts to influence neighboring societies. Like its many grand accomplishments, the causes and consequences of the Empire's dissembling have been interpreted and reinterpreted, lingering as an inexhaustible source of insights.

From the perspective of later, ambitious Turkish governments, the tragedy of the Ottoman state's decay lay in the realization that it could not be revived in any meaningful sense. Territorially, economically, and politically, the Empire had turned irreversibly unsustainable. Its prestige dwindled as it could not defend its possessions. Ottoman rulers at the end of the nineteenth and beginning of the twentieth centuries essentially sought only to limit rather than to halt the dismemberment of their lands, which they viewed as the unalterable agenda of meddling outside powers.[14] As a result, viziers, ministers, and generals could not effectively retrofit the structures of their state with modern policies because they could never, in the first instance, manage to reestablish external stability. Neither could the court make good on its pretenses of reform. Behind the baroque gates of Dolmabahçe Palace, ostentatious excesses continued unabated, inspiring the slur that the once mighty Sublime Porte had become "the sick man of Europe." The seemingly inexorable decline of the Empire gained a momentum that thwarted movements to save it. Its military fell prey to the same sense of impending collapse, failing to modernize at the rate of its peers.[15] Istanbul's loosening grip on the court's anemic system of taxation, which relied on tributes from faraway provinces, exacerbated the state's cash-strapped finances.

For their part, late Ottoman military elites resented the hollow refrain that echoed from London, Berlin, and Paris: that their state's sovereignty would be respected, despite these circumstances. This assurance often followed aggressive steps by outside powers to support minority religious or ethnic groups under Istanbul's rule. After a failed revolt, a community would appeal to a European government that would in turn insist on measures granting the rebels autonomy. In the eyes of Ottoman rulers, these footholds would lead, eventually, to full independence as soon as they were preoccupied with parallel crises elsewhere.[16] In Serbia, Greece, Romania, Bulgaria, and Albania, this cycle more or less recurred. With each humiliating capitulation, a belligerent resistance to outside interference in the state's domestic affairs grew. This outlook intensified as the Empire's power contracted.[17]

Although the Ottoman court failed in its efforts to balance European powers against each other, this policy was not discredited as a legitimate approach to maintaining security. Common Turkish readings of late Ottoman history describe the fateful maneuver of 1914 as the catastrophe that sealed the Empire's fate. Minister of War Enver Pasha abandoned long-standing policies aimed at exploiting cleavages among European powers and brashly joined those who promised the Ottomans the most for their allegiance.[18] The perennial allure of remaining neutral, outside of transient struggles involving greater powers, has been part of the Turkish security establishment's counterfactual imagination of World War I for many decades.[19]

From the collapse of the Ottoman regime came a restrained approach to defining Turkey's territory and its defense. The government in Ankara would assert control only over those areas that it could reasonably expect to defend alone, without reliance on the goodwill, largesse, or the balancing of interests harbored by outside powers.[20] William Hale notes "for the most part, [this meant seeking to defend] ethnic Turks, or other Muslims who were willing to integrate ... or cooperate." Regarding this lesson, Hale adds "ethnic Turks outside these boundaries could not be protected, except ... where Turkish military force could be brought to bear," citing the Cypriot counterexample.[21]

The last decades of the Ottoman regime saw competing movements for reform aimed at imitating Western technological, economic, and military strength. This emulation, however, did not prevent anger about perceived patronizing acts by outside governments.[22] A deep and abiding fear of separatist movements developed during this period, culminating in atrocities like the Armenian Genocide. The toxic denialism and justification surrounding these episodes relate back to the historical experience about the perceived instrumental role played by minority communities at the hands of imperial competitors. As refugees poured into Anatolia after World War I, non-Muslim and non-Turkish minorities on Turkey's new territory faced the brunt of this suspicion.

The Kemalist Tradition

The leader who most profoundly influenced Turkey's strategic culture was the Republic's founder, Mustafa Kemal Atatürk. A field marshal and statesman so deeply admired that he was given the honorific title "father" of his people ("Atatürk"), he served as Turkey's first president from 1923 until his death in 1938. His broad impact on Turkish society is visible in sharp relief in its security institutions.[23] The so-called Six Arrows of his ideological project – republicanism, populism, secularism, reformism, nationalism, and statism – are too broad and varied to discuss at length here. Yet these ideological projects encapsulated the most important differences between the period preceding his rule and the advent of the modern state. Atatürk succeeded in rapidly creating institutions distinct from the antiquated *ancien régime* atop the long decrepit empire from which Turkey was born.[24]

Central to Kemal's state-building project was his adamant insistence on military self-reliance. This strategic goal has sometimes been mocked with the moniker "the Sèvres Syndrome." This caricature alludes to the stillborn treaty that would have implemented aggressive Entente designs for Ottoman lands. The precept – pathological or not – captures the mentality, ever-present in Turkish military planning, that threats remain poised against the country's contemporary heartland.[25] European governments that prevailed in World War I envisioned a future Turkey composed of only Istanbul and highly circumscribed provinces in central and northern Anatolia. This rump state would have been born bereft, deprived of resources and means of economic advancement.[26] Greece would have gained Eastern Thrace; Armenia would have been formed from the prized provinces of Erzurum, Trabzon, Van, and Bitlis; Kurds would have gained "local autonomy"; and separate bargains among Britain, France, and Italy would have recognized "special interests" of the latter two in portions of southern Anatolia.[27] Illegitimate as the agreement may have been, it signified the potential costs of weakness and lapses in vigilance.

A bloody war of independence ultimately paved the way for a new settlement, the Treaty of Lausanne. This document more or less formalized the borders of contemporary Turkey, granting the definitive international recognition and security that the Ottoman state could not achieve during the final decades of its decline. The new government's failure to attain exclusive control over the Bosphorus – what Russian diplomats perennially call the "Black Sea straits" – demonstrated yet again, in the minds of Turkish leaders, the stubborn insistence of outside powers to intercede in Turkish prerogatives.[28] Lausanne's tragic provisions sanctioning the expulsion of Greeks from Turkey (approximately 900,000 people), and vice versa (approximately 400,000 Turks were forced out of Greece), illuminated the visceral human dimensions of the transformation from Empire to the nation-state. Far from accepting a multiethnic assemblage of many territories, the Republic's government associated the control of lands with the ethnic majorities inhabiting them. The notorious mass movement of populations marked the end of generally graceful coexistence among Greeks and Turks, providing a bitter prelude to disputes between these states since.[29]

The first quarter-century of Turkey's existence saw the profound strategic transition from its cautious neutrality through the outbreak of World War II to its stalwart alignment with the US-led alliance during the Cold War. From 1918 to 1923, its most consequential hereditary adversary, Russia, remained paralyzed by internal tensions. Diplomatic fashions of the 1920s augured against bilateral security pacts in favor of collective security agreements such as the League of Nations. Although Kemal's critics analogized the cult surrounding his personality to those nurtured by ascendant European dictators in the 1930s,[30] Turkey assiduously avoided aligning with them. Self-preservation could be achieved only through neutrality, they presumed. Amid this ferment the phrase that became the Republic's motto – "peace at home, peace abroad" – lodged itself firmly as a key premise of Turkey's strategic culture.[31] Kemal and his first Prime Minister, İsmet İnönü, focused during this period on fashioning secular, nationalist, and modern institutions.

Ultimately, the Nazi-Soviet pact of August 1939 drove Ankara to a tripartite treaty with France and Britain in October 1939. The speed with which Germany overtook France, as well as its fateful decision to invade Russia rather than the Middle East, permitted İnönü's government to sustain a de facto neutral position. He wisely eschewed entering the war. Once the war had concluded, however, Stalin's victorious regime began a long period of pressure, calling for Ankara to capitulate on security questions that would have likely brought it under Soviet dominion. This motivated Turkey's drive to become an integral constituent of NATO. Its government sent the third-largest brigade to fight in Korea, spearheaded the Balkan Pact, and energetically advocated for the installation of Intermediate Range Ballistic Missiles on its territory in the late 1950s.

Until the Cyprus crisis in 1964, Turkey enjoyed a relatively steadfast relationship with NATO. Its military ensured that the state remained a valuable ally during the Cold War, despite its occasional attempts to aggrandize its power through efforts like the failed Baghdad Pact. A government led by Bülent Ecevit from 1978 to 1979 demanded the right to take positions independent from NATO's official line, an approach that recurred in the Middle East and Cyprus until the end of the Cold War.[32] Turgut Özal's sustained attempt to create a "neo-Ottoman" foreign policy represented a later, important break from generally deferential relations with the Western bloc. As prime minister and president of the Republic, he implemented an economic and political program aimed at launching Turkey to a more prominent international position. What some have deemed a "counter-republican paradigm" set the stage for later strategic thinkers who sought a global rather than regional role for their country.[33]

The recurrence of military-led coup d'états in Turkey ensured the priorities of the armed forces routinely overrode competing agendas proffered by politicians. When the military deemed an elected government incompetent, parliamentarians and bureaucrats were removed and replaced, often through a system of tribunals. Leadership reverted to hastily organized councils of state manned by the upper echelons of the armed forces. A common Turkish expression – "fear the colonels, not the generals" – highlights that coups often shook up ranks across the government, not just within civilian institutions. Each military intervention occurred at a unique moment of political and economic unrest, resulting in new political arrangements and, in some cases, new legal frameworks for governance. The specter of coups ossified Turkey's strategic culture, as generations of politicians sought to avoid the catastrophic costs of antagonizing the military.[34]

Tensions within Turkey's strategic culture have often tracked its struggle to deal with political strife among ethnic minorities throughout its territory. In an attempt to forge a "Turkish" identity from the confederation of ethnicities surviving the Ottoman Empire in Anatolia, the government established policies following Kemal's "National Pact" that aimed to incorporate groups of self-ascribed unique ethnic origin. With regard to the Kurds, these policies intermittently entailed violent action by the state to suppress their distinct ethnic identity.[35] The 1983 Language Act sparked a period of greater suppression. It criminalized many aspects of Kurdish cultural life, renewing and enhancing bans on Kurdish language instruction.[36]

Throughout the mid- to late 1980s, militant Kurdish movements responded to these measures by escalating their attacks. Understanding the Turkish government's outlook toward the Kurds through the 1990s is complicated by the presidency of Turgut Özal, who was of partial Kurdish descent. Özal embodied the seeming duality of the relationship between the Turkish government and ethnic Kurds. This relationship is not characterized by an apartheid-like scheme of segregation. Instead, the relationship during Özal's presidency represented the continuance of the Turkish state project to construct a unitary Turkish ethnic identity by denying separate political rights to Muslim non-Arab ethnic minorities in Anatolia. The military has bolstered this dogmatic position, often by inculcating key ideas during the period of national service for young men.[37]

A Horizon-Oriented Literature

The rise of the incumbent AKP, achieved by Erdoğan's parliamentary victory in 2002, represented a fundamental rupture between the will of an elected government and the standard-bearers of Turkey's entrenched strategic culture. The era that has followed has had, for obvious reasons, the greatest influence on contemporary academic literature in the subfield. Much discussion among experts on Turkey's strategic culture has centered on a concerted effort to make sense of the AKP's drive to change the country's relations with the Middle East. However, the overwhelming sense that "history is being made" has focused scholarly inquiry on unfolding events.

Assessing the impact of the AKP's decades-long deviation from past periods of strategic culture should not overcome the imperative of consulting broader historical trends. Indeed, if there might be one lesson from this overarching intellectual project, it may be that the AKP has permanently floundered in its attempt to establish a coherent, alternative framework for strategic questions. Since the failed coup in 2016, paralysis, rather than progress toward coordinated goals, may have become the norm. President Erdoğan's mass expulsion of ambassadors in October 2021 serves as yet another vivid example of a pattern. A Turkish state that had long nurtured allegiances repeatedly has chosen isolation, punctuated by short-sighted, abrasive retribution for perceived slights. Developments of this kind do not and perhaps cannot be fit snugly into narratives about strategic culture.

The AKP's New Vision

Many commentators have noted that the AKP's strategic agenda has remained premised upon an urgent need to contain, control, and ultimately co-opt the institutions it inherited, often from the inside out. Its leaders harbored an overweening suspicion toward the military, driven by the historical experience of the party's leadership. Many faced sanctions for violating laws prohibiting the intermingling of politics and religious faith.[38] Considerable debate has centered on the validity of the AKP's pretense of championing civilian authority in the domain of national security. For decades, Kemalist precepts – especially the military's professed role as a bastion of secularism and a shield from internal threats to the Republic – sat uncomfortably with Turkey's alignment with the democracies of the transatlantic world.[39] The old guard within the general staff, however, did not depart from power through steady processes of reform. Assertions of civilian control often occurred amid abrupt policy changes that upturned existing dynamics. The 2008 removal of all military personnel from Turkey's National Security Council, ongoing show trials for purported seditious plots by the "deep state," and wholesale purges after the 2016 coup attempt – events like these ultimately fulfilled the AKP's agenda. Only after its leadership began to face significant international criticism, following uprisings over Gezi Park, did many scholars begin to second-guess their salutary neglect of the previous decade's struggles within the military establishment.

Few have noted that the early AKP's national security policy did not immediately take on the attributes for which it later became known. On a substantive level, the AKP's initial approach in the Middle East constituted a restrained, evenhanded emphasis on economic advancement through regional economic integration.[40] Overseen by Yaşar Yakış, a founding leader of the AKP who served as its first Foreign Minister, the AKP undertook conciliatory overtures toward former rivals in the Middle East, like Egypt.[41] The most experienced Turkish diplomat in the affairs of that region, Yakış often directly declined to invoke the specter of a "neo-Ottoman" foreign policy.[42] In the wake of the September 11 terrorist attacks, Turkey stood shoulder to shoulder with the United States in its focus on thwarting international terrorism. Turkey supported the invocation of Article V of the North Atlantic Treaty, taking part in successive operations in Afghanistan. The US invasion of Iraq, however, constituted an about-face in Turkey's previous support for US intervention in the region. Famously, the Turkish National Assembly rejected the opportunity to join the US-led coalition forces in Iraq. The AKP demonstrated its willingness to follow public opinion, which overwhelmingly opposed Turkey's involvement, instead of remaining attached to what its leaders saw as a nostalgic and occasionally overbearing US-Turkish alliance.[43]

The rise of one key figure, Ahmet Davutoğlu, has been synonymous with the intellectual revolution perceived to have taken place within Turkey's strategic culture. Before his fall from grace, Davutoğlu served as both Foreign Minister (2009–2014) and Prime Minister of Turkey (2014–2016), taking on that mantle once Erdoğan assumed the presidency. A celebrated academic, he joined the National Assembly and slowly gained clout as a voice in favor of more robust Turkish engagement in surrounding regions.[44] On the one hand, Davutoğlu repeatedly stated that Turkey's primary goal in foreign policy was to join the European Union. On the other hand, the balance of his efforts in office was directed toward the establishment of something akin to a commonwealth of former Ottoman vassals.[45]

An Evolving Role in the Middle East

For a time, Davutoğlu's concept of "strategic depth" came to represent the AKP's signature vaguely imperialist and pan-Islamist notions about the Arab world. Though not always consistent, commentators ascribed considerable credibility to his belief that through hard and soft power, grafted on traditional ties, Turkey could become the arbiter of questions in the Middle East.[46] Turkey's acerbic rift with Israel, its renewed parochialism in the domestic affairs of majority-Sunni countries, and its sudden abandonment of its "zero problems with neighbors" policy – demanding Bashar Al Assad's overthrow – all demonstrated the impact of this new paradigm. The extent to which these policies succeeded or failed remains a highly partisan question. The debate has centered on whether the Arab Spring was the true pivot for Turkey in the Middle East. Some argue that, after civil strife engulfed several important neighbors, the AKP adopted a more aggressive strategic approach, backed by a willingness to intervene militarily.[47] Yet these accounts tend to ignore revealing preludes, like Erdoğan's 2009 string of insults directed at Israel's President Shimon Perez at Davos. Although it was considered uncharacteristic at the time, this incident revealed a more ideological position on behalf of Arab peoples. The Turkish government's bitter anger over the 2010 Mavi Marmara incident occurred long before inter- and intra-state conflict intensified in the region.[48]

Davutoğlu's disappearance from the national scene has not prevented a continued scholarly focus on the change he caused. His resignation as prime minister arrived on the heels of his attempts at autonomy from Erdoğan. The latter, who had elevated his former adviser, deemed Davutoğlu's creeping tendency to act independently to be an unforgivable betrayal. In reality, the episode represented the redistribution of ministerial prerogatives to a newly empowered chief executive. This was, of course, an outgrowth of Erdoğan's plans to rejigger Turkey's constitutional structure in his favor.[49] At the height of Davutoğlu's influence, scholars like Cenk Saraçoğlu described his policy as conceiving of the Middle East in classically Ottoman terms: "nations" ("millets") embedded in broader "civilizations" ("medeniyets") belong to a hierarchy.[50] More recently, revisionist scholarship has attempted to describe Turkey's role as an order-producer (a "*düzen kurucu aktör*") in the region,[51] willing to use considerable military force. Yet it is hard to ascribe legitimacy to this narrative when the Turkish government has, in recent years, armed the Salafi-Jihad organization Ahrar al-Sham and, more dubiously, occasionally assisted the failed Islamic State.[52] These events have generated a resurgence of interest in the genealogy of the AKP's Islamist principles.[53]

The Enigma of Middle-Power Status

Turkey's failure to achieve the high expectations set by the AKP has produced a literature about the real limits of its influence. Skeptics have argued "Ankara's bark is worse than its bite,"[54] whereas others claim that Turkey has adopted a rational response to the return of *realpolitik* in the Middle East.[55] This apology for Turkey's recent ruthlessness has led yet again to the conceptualization of the country as a security "bridge" between East and West, "an interface state or transitory space between the European and Middle Eastern security complexes."[56] Still others have sought to argue that Turkey's government had attempted something of a rapprochement with traditional security partners through decisions to cooperate on the deployment of NATO Patriot missiles near Syria in 2012.[57] Rarely, if ever, have these scholarly contributions looked to other regions. Turkey's sometimes consequential role in Central and Eastern Europe as well as Central Asia is ripe for further inquiry. The recent imbroglio in Afghanistan and Ankara's retreat from its historical advocacy on behalf of Uyghurs in China could also be fruitful areas of research.

Warming Turkish-Russian relations set the scholarly discussion in this area on a different path. In July 2017, Presidents Vladimir Putin and Erdoğan signed a significant military deal. The two leaders announced a long-negotiated transaction to coproduce the S-400 surface-to-air missile system. Accepting Turkish demands for a technology transfer, the agreement made Russia complicit in Ankara's design to create a self-sufficient arms industry, which furnished it with an independent armory of cutting-edge weaponry. The mobility of the Russian system also allows Turkey to overcome constraints imposed by NATO, which has provided similar stationary systems with the proviso that they be deployed neither on the Aegean coast nor at the country's Greek and Armenian borders.

These events suggested Turkey's strategic culture had permanently changed course. A trenchant imperial rivalry, coupled with hostility during the Cold War, had long set Russian and Turkish governments against each other. Their disputes have an elaborate provenance.[58] For a time, the AKP's upper echelons have slowly adopted a position of preferring cooperation with Putin, rather than rancor. Importantly, Turkey was willing to overlook nefarious Russian activity in 2008 and 2014, when Moscow invaded sovereign neighbors militarily. In response, some scholars have adopted arguments about the changing nature of Turkey's "identity" – its desire to take advantage of its middle-power status has led it to deal with Russia in more co-optive terms.[59] Others have argued that Turkey's strategic culture, as it relates to Russia, has taken shape to counter the latter's increasingly assertive foreign policy.[60]

Before the war in Ukraine broke out in early 2022, the overriding conclusion of this area of the literature remained that Turkey's warming relations toward Russia reflected a "tactical" set of decisions. Its middle-power status has led it to recapitulate its old rivalries into working relationships that manage, rather than exacerbate, causes for distrust. However, Turkey's role in the 2020 Nagorno-Karabakh War should give pause to the advocates of these narratives, which ascribe consistency and coherence to the AKP's policy. Despite its professed willingness to work with Russia in certain domains, Erdoğan's government cunningly supplied Azerbaijan with crucial advanced weaponry, vindicating its conquest over disputed territory during the COVID-19 pandemic.[61] Turkey ultimately pushed back Russian influence in the conflict's aftermath by matching its provision of "peacekeeping" troops.[62] The ongoing Turkish efforts to undermine Russia's invasion of Ukraine since February 2022 similarly indicate cooling, skeptical, and even hostile intentions. Altogether, these steps suggest there is more than a bit of daylight between the two governments when it comes to matters of international security.

Conclusion

Turkey's strategic culture remains a complicated subfield in which to find key explanatory variables or overarching themes. Erdoğan's vindication in the 2023 election will likely sustain these circumstances. In this regard, it is difficult to overstate the sense of cynical resentment, bordering on paranoia, that has pervaded the Turkish government since the 2016 failed coup. Rather than ushering in a renewal of Turkey's past, the jets that soared over Ankara on that balmy July night unknowingly sounded the end of an era. Once Erdoğan regained the upper hand, he became determined to root out all enemies, real and imagined. Too many details remain obscured by conspiracy theories and subsequent rhetorical battles over the scheme's significance. Nonetheless, its impact had far-reaching effects. Beyond justifying bloodthirsty reprisals en masse, the foiled plot led the government to suppress the environments in which scholars and practitioners articulate strategic culture. Countless institutions were shut down by government decree. That sneaking, omnipresent fear has stayed a whole generation of Turkish policymakers, academics, and journalists from posing important questions about the country's strategic outlook.

This chilling state of affairs should not discourage those hoping to contribute to this subfield. Instead, new methods and goals should be adopted. In lieu of panacea or totalized historical accounts, the moment has arrived for investigations of specific topics and a broadening of research into geographic regions and subject matter typically not in the news. Developments of the last two decades have stalled innovative assessments of Turkey's military institutions and their interface with the powerful incumbent civilian government. Understanding how Turkey's strategic culture will continue to evolve rests on a wider awareness of how it conceives of threats in an era that has become defined by great power competition. Deriving insights about this international political climate requires an appreciation of Turkey's calculus of self-interest. These circumstances provoke opportunism. Rather than focusing on long-standing, contested stalemates with other states, particularly in the Middle East, scholars and students should look for new avenues – historical, conceptual, and regional – to evaluate ongoing trends.

Notes

1 As late as 2000, these quotations were repeated in the US House of Representatives by Congressman John Murtha. "Remarks on Turkey in the Korean War," 111th Cong., 2nd Sess., *Congressional Record* 156, no. 91 (27 June 2000): E1130, https://www.congress.gov/crec/2000/06/28/CREC-2000-06-28-pt1-PgE1130-4.pdf.

2 John M. Vander Lippe, "Forgotten Brigade of the Forgotten War: Turkey's Participation in the Korean War," *Middle Eastern Studies* 36, no. 1 (2000): 92–102.

3 Except when explaining his honorific name, this chapter refers to the leader by his surname, "Kemal," throughout.

4 This term has been repeated by critics on the left and right. See, for example, Soner Cagaptay and Oya Rose Aktas, "Transactional or Transcendent? Turkey's Ties to the EU," *Henrich Boll Stiftung*, 8 May 2017, https://eu.boell.org/en/2017/05/08/transactional-or-transcendent-turkeys-ties-european-union; Laura Kelly, "Biden Aims to Bolster Troubled Turkey Ties in First Erdoğan Meeting," The Hill, 13 June 2021, https://thehill.com/policy/international/558060-biden-aims-to-bolster-troubled-turkey-ties-in-first-erdogan. See also Samuel Huntington's classic analysis of Turkey as a "prototypical" example of a "torn state," in "The Clash of Civilizations?," *Foreign Affairs* 72, no. 3 (1993): 22–49.

5 For authors aware of the irony of reimagining Turkey's role as a global player, see E. Faut Keyman, "Modernization, Globalization and Democratization in Turkey: The AKP Experience and Its Limits," *Constellations* 17, no. 2 (2010): 312–327; Kemal Kirişci, "The Rise and Fall of Turkey as a Model for the Arab World," Brookings Institute, 15 August 2013, https://www.brookings.edu/opinions/the-rise-and-fall-of-turkey-as-a-model-for-the-arab-world/.

6 In international relations, these ideas emerge in "Regional Security Complex Theory." The use of the label "insulator" country is telling. Barry Buzan and Ole Wæver, *Regions and Powers. The Structure of International Security* (Cambridge: Cambridge University Press, 2003). The term "liminal state" has come to define the opposite conception. See Behar Rumelili, "Liminal Identities and Processes of Domestication and Subversion in International Relations," *Review of International Studies* 38, no. 2 (2012): 495–508.

7 Colin S. Gray, "Strategic Culture as Context: The First Generation of Theory Strikes Back," *Review of International Studies* 25, no. 1 (1999): 49–69; Alastair Iain Johnston, "Strategic Culture Revisited: A Reply to Colin Gray," *Review of International Studies* 25, no. 3 (1999): 519–523.

8 See Patrick Porter, *Military Orientalism: Eastern War through Western Eyes* (New York: Columbia University Press, 2009).

9 Jack Snyder, *The Soviet Strategic Culture: Implications for Limited Nuclear Operations* (Santa Monica, CA: Rand, 1977), 8.

10 These ideas have a storied provenance. Kemal H. Karpat, *Türk Demokrasi Tarihi. Sosyal, Kültürel, Ekonomik Temeller* (Istanbul: Timas Yayınları, 2004), 19.

11 Malik Mufti, *Daring and Caution in Turkish Strategic Culture* (London: Palgrave Macmillan, 2009), 1.

12 M. Hakan Yavuz, "Social and Intellectual Origins of Neo-Ottomanism: Searching for a Post-National Vision," *Die Welt Des Islams* 56 (2016): 438–465.

13 Jeffrey Mankoff, *Empires of Eurasia* (New Haven, CT: Yale University Press, 2022).

14 For a classic Turkish assessment, see F.A.K. Yasamee, *Ottoman Diplomacy: Abdülhamid II and the Great Powers, 1878–1888* (Istanbul: Isis, 1996). Western authors often frame it differently: M. S. Anderson, *The Eastern Question, 1774–1923: A Study in International Relations* (London: Macmillan, 1966).

15 Feroze Yasamee, "Abdülhamid II and the Ottoman Defence Problem," *Diplomacy and Statecraft* 4 (1993): 22–23.

16 Edward Meade Earle, *Turkey, the Great Powers and the Baghdad Railway* (New York: Macmillan, 1923).
17 Ulrich Trumpener, "Turkey's Entry into World War I: An Assessment of Responsibilities," *Journal of Modern History* 34 (1962): 369–380.
18 Frank Weber, *Eagles on the Crescent: Germany, Austria and the Diplomacy of the Turkish Alliance, 1914–1918* (Ithaca, NY: Cornell University Press, 1970), 135–136, 153–154.
19 Mustafa Aksakal, *The Ottoman Road to War in 1914* (Cambridge: Cambridge University Press, 2008), 77.
20 J.C. Hurewitz, ed., *Diplomacy in the Near and Middle East*, vol. 2 (Princeton, NJ: Van Nostrand, 1956), 81, 89.
21 William Hale, *Turkish Foreign Policy since 1774*, 3rd ed. (London: Routledge, 2013), 29.
22 Feroz Ahmad, *The Young Turks: The Committee of Union and Progress in Turkish Politics, 1908–1914* (Oxford: Oxford University Press, 1969), 1–13.
23 For a discussion of this phenomenon, see William Hale, *Turkish Politics and the Military* (London: Routledge, 1994).
24 S.N. Eisenstadt, "The Kemalist Regime and Modernization: Some Comparative and Analytical Remarks," in *Atatürk and the Modernization of Turkey*, ed. Jacob M. Landau (Boulder, CO: Westview Press, 1984), 3–16.
25 Jacob M. Landau, ed., *Atatürk and the Modernization of Turkey* (Boulder, CO: Westview Press, 1984).
26 "Sèvres Baris Antla Çmasi," in *Türk Dis Politikasi,* vol. 1, ed. Baskin Oran (Istanbul: iletişim Yayınları, 2001), 124–138.
27 Harry N. Howard, *The Partition of Turkey: A Diplomatic History, 1913–1923* (New York: Fertig, 1966), 217–241.
28 J.C. Hurewitz, "Russia and the Turkish Straits: A Revaluation of the Origins of the Problem," *World Politics* 14, no. 4 (1962): 605–632.
29 Zürcher, *Turkey: A Modern History*, 171–172.
30 Özay Mehmet, "Turkey in Crisis: Some Contradictions in the Kemalist Development Strategy," *International Journal of Middle East Studies* 15, no. 1 (1983): 47–66.
31 Although attributed to Kemal, this phrase's provenance is unclear. Mehmet Gönlübol and Cem Sar, *Atatürk ve Türkiye'nin Dış Politikasi 1919–1939* (Ankara: Milli Egitim Basimevi, 1963), 90 n86.
32 Suat Bilge, "The Cyprus Conflict and Turkey," in *Turkey's Foreign Policy in Transition, 1950–1974,* ed. Kemal H. Karpat (Leiden: Brill, 1975), 137–143.
33 Mufti demonstrates intellectual consonances with the counter-narrative that initially emerged in the 1950s and was elaborated by Turgut Özal in the 1980s. See Mufti, *Daring and Caution in Turkish Strategic Culture.*
34 Frank Tachau and Metin Heper, "The State, Politics, and the Military in Turkey," *Comparative Politics* 16, no. 1 (1983): 17–33.
35 Kemal Kirişci and Gareth M. Winrow, *The Kurdish Question and Turkey: An Example of a Trans-State Ethnic Conflict* (London: Frank Cass, 1997), 92.
36 Michael M. Gunter, *The Kurds and the Future of Turkey* (New York: St. Martin's Press, 1997), 10.
37 Ioannis N. Grigoriadis, *Trials of Europeanization Turkish Political Culture and the European Union* (New York: Palgrave Macmillan, 2008), 136.
38 M. Hakan Yavuz and Ahmet Erdi Öztürk, "Turkish Secularism and Islam under the Reign of Erdoğan," *Southeast European and Black Sea Studies* 19, no. 1 (2019): 1–9.
39 American officials mocked Turkish counterparts in the 1990s, calling a general "little Atatürk," referring to the Kemalist establishment as "the last Stalinist regime in Europe." Philip Robins, *Suits and Uniforms: Turkish Foreign Policy since the Cold War* (London: Hurst, 2003).
40 Erol Kurubaş, "Arap Baharında Türk Dış Politikası Ve Türkiye'nin Orta Doğu'daki Konumuna Etkis i," *Uluslararası* 1 (2014): 468.
41 Murat Yeşiltaş and Ali Balcı, "AK Parti Dönemi Türk Dış Politikası Sözlüğü: Kavramsal Bir Harita," *Bilgi Sosyal Bilimler Dergisi*, no. 2 (2011): 17–18; Ertan Efegil, "AK Parti Hükümetinin Orta Doğu Politikasıve ABD Yönetimi ile Batılı Uzmanların Eleştirileri," *Akademik Bak ış* 9, no. 19 (2014): 48.

42 Responding to a question calling his foreign policy "neo-Ottomanism," Yakis said "on the contrary, we never use this terminology. We try to avoid it, actually, because the Ottoman legacy is remembered more for its negative side in the Middle Eastern countries...." "Interview: Understanding Turkey's Foreign Policy," Radio Free Europe, 19 September 2011, https://www.rferl.org/a/turkey_foreign_policy_yasar_yakis_interview/24333237.html.

43 Füsun Türkmen, "Anti-Americanism as a Default Ideology of Opposition: Turkey as a Case Study," *Turkish Studies* 11, no. 3 (2010): 329–345.

44 Gökhan Telatar, "AK Parti'nin Düzen Kurucu Dış Politika Söylemi ve Orta Doğu," *Alternatif Politika* 7, no. 3 (2015): 494.

45 Erdoğan conveyed that Turkey could become the world's leading Muslim nation: "Those who want to make Turkey another Andalusia, ... will never abandon their aspirations. The word 'Turk' has an ethnic meaning only in our country, whereas westerners have always used it [to mean] Muslims. This is not to claim any superiority; it is just an historical state of affairs." Türkiye Cumhuriyeti Cumhurba şkanlığı, "Devlet Övünç Madalyası Tevcih Töreni'nde Yaptıkları Konuşma, 16 March 2015, https://www.tccb.gov.tr/konusmalar/353/29802/devlet-ovunc-madalyasi-tevcih-toreninde-yaptiklari-konusma.

46 The scholar-diplomat's dissertation, "a comparative analysis between Western and Islamic political theories and images," argued "the conflicts and contrasts between [them] originate mainly from their philosophical, methodological and theoretical background rather than from only institutional and historical differences." From this idea sprang "strategic depth" – as the most advanced *Muslim* country, Turkey enjoyed a peerless strategic advantage for obtaining influence among its Islamic neighbors and former subjects. Ahmet Davutoğlu, "The Impacts of Alternative Weltanschauungs on Political Theories" (PhD diss. Boğaziçi University, 1990), 3.

47 Jonathan Schanzer and Merve Tahiroglu, "Ankara's Failure: How Turkey Lost the Arab Spring," *Foreign Affairs*, 25 January 2016; Aaron Stein, "Turkey's Failed Foreign Policy," *New York Times*, 22 August 2014.

48 Stephen M. Walt, "The Real Significance of Erdogan's Davos Outburst," Foreign Policy, 2 February 2009, https://foreignpolicy.com/2009/02/02/the-real-significance-of-erdogans-davos-outburst/.

49 Felix Petersen and Zeynep Yanaşmayan, eds., *The Failure of Popular Constitution Making in Turkey: Regressing towards Constitutional Autocracy* (Cambridge: Cambridge University Press, 2020).

50 Cenk Saraçoğlu, "AKP, Milliyet çilik ve Dış Politika: Bir Milliyet çilik Doktrini Olarak Stratejik Derinlik," *Alternatif Politika* 5, no. 1 (2013): 62–63.

51 Bülent Araş, "Düzen Kurucu Aktör," *Sabah*, 30 September 2009.

52 Emrullah Uslu, "*Jihadist* Highway to *Jihadist* Haven: Turkey's Jihadi Policies and Western Security," *Studies in Conflict and Terrorism* 39, no. 9 (2016): 781–802.

53 Ruşen Çakır, "Mill î Görüş Hareketi," in *Modern Türkiye'de Siyasî Düşünce, Cilt 6: İslâmcılık*, ed. Murat Gultekingil and Tanil Bora (Istanbul: İletişim, 2006), 544–575; Bulut Gurpinar, "Turkey and the Muslim Brotherhood: Crossing Roads in Syria," *Eurasian Journal of Social Sciences* 3, no. 4 (2015): 22–36.

54 Alistair Lyon, "Analysis: Syria Crisis Shows Limits of Rising Turkish Power," Reuters, 9 July 2012, https://www.reuters.com/article/us-syria-crisis-turkey/analysis-syria-crisis-shows-limits-of-rising-turkish-power-idUSBRE86808O20120709.

55 Ian O. Lesser, "Turkey's Third Wave—And the Coming Quest for Strategic Reassurance," German Marshall Fund of the United States, 2011.

56 E. Parlar Dal, "The Transformation of Turkey's Relations with the Middle East: Illusion or Awakening," *Turkish Studies* 13, no. 2 (2012): 259.

57 Karabekir Akkoyunlu, Kalypso Nicolaidis, and Kerem Öktem, *The Western Condition: Turkey, the US and the EU in the New Middle East* (Oxford: South East European Studies at Oxford, 2013), 7.

58 Jeffrey Mankoff, "Why Russia and Turkey Fight. A History of Antagonism," *Foreign Affairs*, 24 February 2016, https://www.foreignaffairs.com/articles/turkey/2016-02-24/why-russia-and-turkey-fight.

59 See chapters in Katalin Miklóssy and Hanna Smith, eds., *Strategic Culture in Russia's Neighborhood: Change and Continuity in an In-Between Space* (New York: Lexington Books, 2019).

60 Tarasiuk Yuliia, "Strategic Prospects and the Development of the Bilateral Relations between Ukraine and Turkey," *European Political and Law Discourse* 6, no. 1 (2019): 37–43.
61 Ece Toksabay, "Turkish Arms Sales to Azerbaijan Surged before Nagorno-Karabakh Fighting," *Reuters*, 14 October 2020, https://www.reuters.com/article/us-armenia-azerbaijan-turkey-arms/turkish-arms-sales-to-azerbaijan-surged-before-nagorno-karabakh-fighting-idUSKBN26Z237.
62 "Turkish Parliament Approves Troop Deployment to Nagorno-Karabakh," *Al Jazeera*, 18 November 2020, https://www.aljazeera.com/news/2020/11/18/turkish-parliament-approves-troop-deployment-to-nagorno-karabakh.

Selected Bibliography

Ahmad, Feroz. *The Young Turks: The Committee of Union and Progress in Turkish Politics, 1908–1914*. Oxford: Oxford University Press, 1969.

Akkoyunlu, Karabekir, Kalypso Nicolaidis, and Kerem Öktem. *The Western Condition: Turkey, the US and the EU in the New Middle East*. Oxford: South East European Studies at Oxford, 2013.

Aksakal, Mustafa. *The Ottoman Road to War in 1914: The Ottoman Empire and the First World War*. Cambridge: Cambridge University Press, 2008.

Çakır, Rusen. "Mill î Görüş Hareketi." In *Modern Türkiye'de Siyasî Düşünce, Cilt 6: İslâmcılık*, edited by Murat Gultekingil and Tanil Bora, 544–575. Istanbul: İletişim, 2006.

Cook, Steven A., Madeleine K. Albright, and Stephen J. Hadley. *US–Turkey Relations: A New Partnership*, Independent Task Force Report No. 69. Council on Foreign Relations. May 2012. https://www.cfr.org/report/us-turkey-relations.

Criss, Nur Bilge. *Istanbul under Allied Occupation, 1918–1923*. Leiden: Brill, 1999.

Criss, Nur Bilge. "Parameters of Turkish Foreign Policy under the AKP Governments." *Revista UNISCI* 23 (2010): 9–22.

Criss, Nur Bilge. "A Short History of Anti-Americanism and Terrorism: The Turkish Case." *Journal of American History* 89, no. 2 (2002): 472–484.

Dal, E. Parlar. "The Transformation of Turkey's Relations with the Middle East: Illusion or Awakening." *Turkish Studies* 13, no. 2 (2012): 245–267.

Earle, Edward Meade. *Turkey, the Great Powers and the Baghdad Railway*. New York: Macmillan, 1923.

Efegil, Ertan. "AK Parti Hükümetinin Orta Doğu Politikasıve ABD Yönetimi ile Batılı Uzmanların Eleştirileri." *Akademik Bak ış* 9, no. 19 (2014): 45–58.

Grigoriadis, Ioannis N. *Trials of Europeanization Turkish Political Culture and the European Union*. New York: Palgrave Macmillan, 2008.

Gunter, Michael M. *The Kurds and the Future of Turkey*. New York: St. Martin's Press, 1997.

Hale, William. *Turkish Foreign Policy Since 1774*. 3rd ed. London: Routledge, 2013.

Hale, William. *Turkish Politics and the Military*. London: Routledge, 1994.

Hanioğlu, Şükrü M. *Atatürk: An Intellectual Biography*. Princeton, NJ: Princeton University Press, 2010.

Hurewitz, J. C., ed. *Diplomacy in the Near and Middle East: A Documentary Record, 1914–1956*. Vol. 2. Princeton, NJ: Van Nostrand, 1956.

Isyar, Omer Goksel. "An Analysis of Turkish-American Relations from 1945 to 2004: Initiatives and Reactions in Turkish Foreign Policy." *Alternatives: Turkish Journal of International Relations* 4, no. 3 (2005): 21–52.

Karpat, Kemal H. *Türk Demokrasi Tarihi. Sosyal, Kültürel, Ekonomik Temeller*. Istanbul: Timas Yayınları, 2004.

Keyman, E. Faut. "Modernization, Globalization and Democratization in Turkey: The AKP Experience and Its Limits." *Constellations* 17, no. 2 (2010): 312–327.

Kirişci, Kemal, and Gareth M. Winrow. *The Kurdish Question and Turkey: An Example of a Trans-State Ethnic Conflict*. London: Frank Cass, 1997.

Mehmet, Özay. "Turkey in Crisis: Some Contradictions in the Kemalist Development Strategy." *International Journal of Middle East Studies* 15, no. 1 (1983): 47–66.

Miklóssy, Katalin, and Hanna Smith, eds. *Strategic Culture in Russia's Neighborhood: Change and Continuity in an In-Between Space*. New York: Lexington Books, 2019.

Mufti, Malik. *Daring and Caution in Turkish Strategic Culture: Republic at Sea*. London: Palgrave Macmillan, 2009.
Robins, Philip. *Suits and Uniforms: Turkish Foreign Policy since the Cold War*. London: Hurst, 2003.
Pelt, Mogens. "The Colonels' Coup of 1967 and the Military Takeovers in Turkey in 1960 and 1971." In *The Greek Junta and the International System*, edited by Antonis Klapsis, Constantine Arvanitopoulos, Evanthis Haatzivassiliou, and Effie G. H. Pedaliu, 167–178. London: Routledge, 2020.
"Sèvres Baris Antla Çmasi." In *Türk Dis Politikasi*, vol. 1, edited by Baskin Oran, 124–138. Istanbul: iletişim Yayınları, 2001.
Shaw Stanford J., and Ezel Kural Shaw. *History of the Ottoman Empire and Modern Turkey*. Vol. 2. Cambridge: Cambridge University Press, 1976.
Tachau, Frank, and Metin Heper. "The State, Politics, and the Military in Turkey." *Comparative Politics* 16, no. 1 (1983): 17–33.
Türkmen, Füsun. "Anti-Americanism as a Default Ideology of Opposition: Turkey as a Case Study." *Turkish Studies* 11, no. 3 (2010): 329–345.
Uslu, Emrullah. "*Jihadist* Highway to *Jihadist* Haven: Turkey's Jihadi Policies and Western Security." *Studies in Conflict and Terrorism* 39, no. 9 (2016): 781–802.
Weber, Frank. *Eagles on the Crescent: Germany, Austria and the Diplomacy of the Turkish Alliance, 1914–1918*. Ithaca, NY: Cornell University Press, 1970.
Yasamee, Feroze. *Ottoman Diplomacy: Abdülhamid II and the Great Powers, 1878–1888*. Istanbul: Isis, 1996.
Yavuz, M. Hakan. "Social and Intellectual Origins of Neo-Ottomanism." *Die Welt Des Islams* 56 (2016): 438–465.
Yuliia, Tarasiuk. "Strategic Prospects and the Development of the Bilateral Relations between Ukraine and Turkey." *European Political and Law Discourse* 6, no. 1 (2019): 37–43.
Zürcher, Erik. *Turkey: A Modern History*. London: I. B. Tauris, 1993.

27
UNDERSTANDING NIGERIA'S STRATEGIC CULTURE

Dodeye Uduak Williams

Introduction

Every sovereign country has a unique way of responding to security threats, whether internal or external. Security issues are usually perceived, resolved, and tackled through strategic calculations made within the cultural norms of the people. The calculations, decisions, and actions of sovereign states are shaped by historical experiences, perceptions, and responses of leaders to both internal and external threats. There are three basic aspects of a nation's strategic culture: the people, the government, and the security sector. This chapter explores the extant literature on Nigeria's strategic culture to provide a perspective on how it is defined, shaped, and how it manifests. It further identifies the determinants of the strategic culture of the Nigerian state, the sources of its security threats (as perceived), and locates the responses of Nigerian decisionmakers within their understanding and interpretation of these threats.

The concept of strategic culture, in Ken Booth's terms, refers to "a nation's traditions, values, attitudes, patterns of behavior, habits, symbols, achievements and particular ways of adapting to the environment and solving problems with respect to the threat or use of force."[1] Booth's definition captures the characteristics that create the context of behavior for an entity, or in anthropologist Clifford Geertz's terms, "thick description" – a researcher's meticulous account of field experiences that includes drawing out and contextualizing key patterns of social and cultural relationships.[2] Strategic culture hones in on those aspects of culture that influence the context in which leaders interpret national security issues and make decisions regarding the use of force or the threat of the use of force.[3] It often represents the values of the leadership, because the values of those in leadership – what they say or do not say, do or do not do, reward or punish – are frequently the backbone of that culture.[4] Strategic culture also subsumes the historical experiences, beliefs, and norms of dominant elites and "keepers" of culture in a society, influencing the interpretation and understanding of security issues and functioning as a lens through which policymakers perceive the external security environment. The logic of strategic culture lies in the contention that these shared conceptions in a strategic community play a key role in shaping policy options, military actions, and security decisions – both to enable and constrain the range of motion and the palette of acceptable options available to policymakers. No security decision is taken in a cultural vacuum; rather, all are

DOI: 10.4324/9781003010302-30

made in the inescapable context of the cultural facets of identity, traditions, biases, and beliefs that collectively compose a group or state's strategic culture.

Sometimes missing from the strategic culture literature is a discussion of what "type" of nation is intended to be studied using this paradigm. To illustrate, Kevin Frank argues that "strategic culture is a concept accepted by scholars and practitioners but with problematic applicability to states newly independent or emerging from conflict [given that] the elements that comprise strategic culture in the developed world are not always present in emerging states."[5] Jack Snyder rightly describes strategic culture as existing in the context of members of a security apparatus, that is, the key players and institutions with responsibility over strategic planning and the use of state force.[6] But those security apparatuses and strategic cultures at large differ across nations, sometimes greatly, and are influenced in unique ways by institutions, actors, and forces beyond the government, including popular opinion and influential groups in society. Most of the existing strategic culture scholarship evaluates societies and states that have centuries of some kind of collective identity – presenting challenges for studying those states who only just gained independence decades ago or are emerging from conflict.[7]

Brief Overview of the Nigerian State

Nigeria is an African country – a colonial creation – that gained independence from almost six decades of British colonial rule in 1960. It is located on the Gulf of Guinea and occupies a land area of 923,763 sq. km, out of which 37% is arable land.[8] It is bounded by Cameroon in the south, Niger in the north, Chad in the northeast, and Benin in the west. Nigeria is made up of 36 states, including the Federal Capital Territory, Abuja. It has 774 local government areas and more than 250 ethnic groups with distinct dialects, customs, traditions, and beliefs. The dominant ethnic groups are the Igbo in the southeast, Yoruba in the west, and Hausa/Fulani in the north. Christianity, Islam, and African Traditional Religion (ATR) are the dominant religions.

The Nigerian state was governed by the military for 29 of its first 40 years of existence, until 1999, when the country was returned to civilian rule. About six years into its independence, the country experienced a civil war (1967–1970) and has continued to be challenged by insurrections and threats of secession, terrorism, and violent extremism.[9] Nigeria's experience with military rule is significant because the "dynamics of Nigeria's emergence into nationhood with the attendant long years of military dictatorship have given rise to the emergence and sustenance of several security challenges which the country is bedeviled with today."[10]

Nigeria is heavily endowed with vast natural resources such as petroleum, limestone, tin, columbite, coal, lead, marble, and granite to mention a few, although it is still classified as a low-income economy. According to the World Bank's 2021 data, Nigeria has a national gross domestic product (GDP) of $440.83 billion, GDP per capita of $2,065, and an annual GDP growth rate of 3.6%.[11] Huge oil revenue for the state has not translated into economic development. While the data shows economic growth, that growth is not inclusive and is not driving human development, employment, production, and progress. Weak economic performance means that the majority of Nigerians continue to live below the poverty line.[12]

Nigeria's foreign policy since its independence in 1960 has been governed by an "African consciousness far stronger than any Nigerian consciousness."[13] Nigeria is committed to the economic and political development of Africa as a whole and sub-Saharan Africa in particular, demonstrated by its involvement in the Economic Community of West African States (ECOWAS), the African Union (AU) [formerly the Organization for African Unity (OAU)], other regional and international bodies, and multilateral and bilateral state relations. On a global level, Nigeria is nonaligned. Its geostrategic environment is characterized by the persistent and growing influence of international terrorist networks like ISIS, Al-Qaeda in the Islamic Maghreb, Al Shabaab, and others; the realities of climate change, which are impacting sub-Saharan Africa particularly hard; and globalization and the modern information revolution – all of which present multidimensional challenges for Nigerian national security.[14]

Nigeria's Strategic Culture: Determinants, Threat Perceptions, and Security Policy

The strategic cultures of most African countries are said to be emerging ones, given that many of these states are categorized as developing nations. Contrary to what some argue, the strategic cultures of these countries are not similar just because they are in the African continent. Even among geographically neighboring societies that may have elements of a common strategic culture, different historical, political, strategic, and geographic experiences converge to produce unique aspects of strategic culture between nations. What commonalities do exist are likely to come from sharing common facets of identity that may transcend other differences in states' geography, history, politics, and so forth.[15] While the beliefs, values, and norms that compose a strategic culture shape policy preferences and inclinations among a state's decisionmakers, they do not serve as rigid predictors of behavior, because external factors, obstacles, or intervening variables sometimes interfere with preferred courses of action. Nevertheless, examining the historic, geographic, demographic, political, and strategic factors that serve as determinants of a state's strategic culture can provide key insights for understanding the influences that are likely to bear sway on preferred modes of strategic action and security policy.[16]

Determinants of Nigeria's Strategic Culture

Civil-Military Relations

The idea of civil-military relations comprises the total range of relationships between the military and civilian society at every level.[17] It connotes the existence of a demarcation between the military and civil-political relations. However, in Nigeria, the prevailing and underdeveloped political, social, economic, and material relations in society make this demarcation almost impossible.[18] As Huntington compellingly argues, the military institutions of a state and their functions or relations are determined by the security threats to those states or societies and the internal power structures, authoritative ideologies, and forces that dominate the state or society.[19] The struggle for supremacy and control of politics between the civilian society and the military in Nigeria, manifesting in coups, counter-coups, internal conflicts, civilianization of military presidents, appointments of retired military generals into political offices, and the like is a dominant feature of Nigeria's political life. Furthermore, the increased involvement of the military in the internal security challenges of the country threatens civil-military relations and affects the perception of threats and decisionmaking as well.

Centrality of Ethnicity

Nigeria is rightly described as a plural society with over 250 ethnic groups, more than 900 dialects, and a wide diversity of identities.[20] The Igbo, Yoruba, and Hausa/Fulani are the dominant ethnic groups, and a recurrent feature of Nigeria's political landscape is the agitation by minority ethnic groups and other marginalized groups for more inclusion, or in some cases, secession. As a result, interethnic and intra-ethnic conflicts, communal clashes fueled by political manipulations, and the struggle for control of resources abound, which strain the country's fragile unity. The incessant struggle for positions of political influence among the different ethnic groups in order to wield decisionmaking authority over economic wealth is a threat to the sustainability of democratic peace, stability, and development in the country,[21] with severe implications for regional stability and national security decisionmaking.

Threats of Secession

Closely related to the centrality of ethnicity is the threat of secession. As Oyewo effectively conveys, "there is a considerable popular feeling of exclusion and perceived sense of injustice among various units of the Nigerian federation – a situation that has led to alienation, suspicion, and apprehension among various groups in the country."[22] Unresolved ethnic tensions and divisions rooted in suspicions of domination led to the civil war of 1967–1970. In recent times, separatist ambitions have manifested in the Ogoni nationalism, the Boko Haram insurgency, and the Indigenous People of Biafra crisis, with significant implications for national, regional, and global security.

Nigeria's "Big Brother" Role in Africa

Nigeria places the primary focus of its external relations on Africa, which has always been the centerpiece of Nigeria's foreign policy and remains the focus of its national security and diplomatic interests.[23] Its commitment to African states' political independence from colonialism, unity and peaceful coexistence among African states, and political stability, progress, and socioeconomic development across the continent[24] have earned Nigeria the name "big brother." The implication of this identity and associated efforts has been that Nigeria commits huge amounts of its limited human and natural resources to pursuing this continental agenda. While Nigeria's large population, economic potential, large crude oil deposits, and sizeable army position it to achieve this agenda, its role also has implications for its security policy. Its membership in ECOWAS, AU, the United Nations (UN), and bilateral relations with Britain, the United States, China, and others reflect its interests and its intent to engage at the continental and global level.

Centrality of Religion

Christianity, Islam, and ATR are the dominant religious traditions in the country. Religious values are deeply rooted in the society and are central to the day-to-day lives of the Nigerian people. Religion permeates the social, economic, and political spheres – and as such has also been a potent tool of oppression, deceit, and manipulation in the hands of Nigeria's political elite for selfish gain. As Femi Falana posits, while the general role that religion plays in politics remains debatable, the connection has long been established in Nigeria.[25] Religion and religious conflicts have been a common feature of Nigeria's social and political landscape. Religion can be a mechanism for development, but in Nigeria it has also been divisive with significant implications for national security and cohesiveness. Religious conflicts have claimed thousands of lives and contributed to heightened insecurity through violent extremism, intolerance, and terrorist activities. When conflicts are defined along religious lines it makes them difficult to resolve because they become "value" conflicts marked by suspicion and an unwillingness to compromise.[26] The Boko Haram insurgency, farmers' and herders' conflict, secessionist agitations, banditry, kidnapping, and other vices have religious undertones and are significant factors contributing to insecurity in Nigeria.[27]

Endemic Corruption

Culture comprises norms and behaviors that constitute a way of life. Corruption, defined here as "the abuse of public office through the instrumentality of private agents, who actively offer bribes to circumvent public policies and process for competitive advantage and profit,"[28] is an undeniable feature of socioeconomic and political life in Nigeria. Rose-Ackerman and Palifka rightly argue that transactions that are deemed "corrupt" in robust democracies and highly developed economies may be seen as acceptable and even normatively required in some other societies.[29] Corruption is a systemic factor that has become institutionalized in all government bodies, lying at the root of insecurity in Nigeria[30] and threatening the state's development, peace, security, and stability.[31] The existence of corruption in the security sector has been identified as a major contributor to national insecurity, and the inability of the Nigerian state to decisively tackle insurgency and extremism has been clearly attributed to the scourge of corruption.[32] This habitual practice that pervades all levels of social and political life hinders the adoption of best practices and triggers strategic calculations based on interests other than the national interest.

Nigeria's Colonial History

The historical forces that have shaped and continue to shape Nigeria's socioeconomic and political experience are rooted in its colonial structures.[33] In 1914, a document consolidating the southern and northern protectorate of Nigeria was signed by Sir Frederick Lord Lugard. While colonialism introduced a modern form of government in Nigeria and infrastructural development, including rails, roads, bridges, hospitals, and Western education, the forceful unification of the various ethnic units under a combined administrative system has remained a sore point in Nigeria's political development. It has resulted in constant wars and conflicts, and the forceful integration of the country into the global capitalist economy has also resulted in economic dependence and resource exploitation that continues to contribute to deepening poverty. The imposition of Western culture, beliefs, behavior, language, and forms of governance has threatened indigenous culture, and these still meet with resistance today.

Nigeria's Military Institutions

The history of Nigeria's military can be traced to the colonial period.[34] The Nigerian Army, Navy, and Air Force constitute the military institutions administered under the Federal Ministry of Defence. Three critical issues are of relevance here: political control, funding, and unhealthy rivalry among the military service branches. The Nigerian state has often used the military against the will of the people. The state is the referent object, as the people are not seen as primary beneficiaries of security.[35] While the duties and responsibilities of the military are clearly stated, sometimes they have been deployed to perform duties that should be handled by the police, just to protect political interests. Military budgets and defense spending in Nigeria are fraught with corruption, fraud, and mismanagement. Huge budgetary allocations and spending over the years have not translated into more security and stability, because the funds are often diverted for wrongful or personal use.[36] Furthermore, unhealthy rivalry continues to hinder cooperation and intelligence sharing, with implications for strategic decisionmaking.

Nigeria's Threat Perceptions

As noted previously, strategic culture offers a method of capturing and examining the national security perceptions that are specific to a state's unique history and experience, and that intrinsically vary across states. It is a concept that reminds strategists that each competitor and allied nation views and responds differently to the strategic environment.[37] Threat perception and misperception can have significant implications for security policy as these interpretations of reality form, change, and influence behavior.[38] Threat perception is core to theories of war, deterrence, compellence, alliances, conflict resolution,[39] and strategic culture. Threat perception involves the anticipation by decisionmakers that there is looming harm to a society, often in the form of military, strategic, or economic peril.[40] Jervis argues effectively that the explanation of crucial decisions is not possible without considering the beliefs of decisionmakers about the world and their images of others, and notes that actors with similar perceptions will often respond in the same way to threats.[41] The assessment of a rival actor's capabilities and intentions, internal or external, define threat perceptions and influence the course of action that a state decides to pursue.[42] Given the major determinants of Nigeria's strategic culture discussed above, observers can deduce several key points about what Nigeria considers as threatening behavior.

Externally, Nigeria's "big brother" role conditions it to perceive as threatening any actions that affect the sovereign independence and territorial integrity of African states, or that seek to jeopardize African unity and independence.[43] Any action that threatens regional cooperation and development or impedes security cooperation is met with resistance across the continent, often led by Nigeria.[44] Through the instruments of regional and continental organizations like the AU and ECOWAS, and international organizations like the UN, Nigeria has defended the interest of Africa as a whole, sometimes to the detriment of its national interests. This self-sacrificing involvement of the Nigerian state in the West African subregion and Africa as a whole most of the time bears no fruit with regard to national interest articulations.[45]

Nigeria's foreign policy is therefore very Afrocentric and marked more, as earlier noted, by an African consciousness than a Nigerian consciousness. This Afrocentrism has been described as an Africa-centered diplomacy: "A political construct in which a country perceives the interests and welfare of the African region as critical to its interests and concerns as a nation."[46] In recent times, the inability of the Nigerian military to rout the challenges presented by the Boko Haram insurgency, the farmers' and herders' conflict, banditry, and kidnapping within its own borders has been judged harshly by critics who opine that it has been more effective in dealing with regional and subregional conflicts than with its own internal conflicts. The reason may lie in how Nigeria perceives the threat. Nigeria's involvement in regional peace operations demonstrates the strength and superiority of its military over its regional enemies.[47] Nigeria's commitment to play a dominant role in West Africa contributed to the intervention of the peacekeeping force called the Economic Community Cease-Fire Monitoring Group in 1990 to end the civil war in Liberia and Sierra Leone.[48] Nigeria also actively worked as a "frontline" state against the apartheid regime in South Africa. Outside of boundary disputes with Cameroon and Chad, Nigeria has not really had conflicts with its neighbors, and its armed forces have been more involved than any other state in playing an active role to counter threats, manage order, and maintain stability across the African community of nations.[49] The military doctrine of the Nigerian Army, based on "Active Defence, Flexible Offensive" and building upon a "Responsive Offensive Doctrine," is grounded in the need for the Nigerian military to maintain a defensive posture that prioritizes nonaggression and good neighborliness, consistent with the Nigerian state's foreign policy orientation.[50] Nigeria, as Francis Adeboye posits, does not have any sub-imperial or hegemonic ambition toward its neighbors.[51]

Internally, threatening behavior includes all actions that threaten the indivisible unity of the Nigerian state. Secessionist agitations are not new in Nigeria. As noted above, Nigeria experienced a civil war just a few years after independence. These agitations have continued in the intervening years since 1970 and have been resisted by the state each time. Notable recent agitation has been from the Indigenous People of Biafra; conflicts along religious lines and resource control have been common as well. In addition, the Boko Haram insurgency in the north and the Movement for the Emancipation of the Niger Delta in the south have destabilized whole communities and created insecurity in many parts of the country.[52] The farmers' and herders' conflict also threatens the peace and stability of the state, and as discussed previously, banditry, kidnapping, and ritual killings further add to the threats to internal security.[53] In addition, Nigeria has porous borders and ungoverned spaces that have proven to be a haven for criminal elements, drug trafficking, and illegal trade in small arms and light weapons.[54] Many of these challenges are fueled by poor leadership, extreme poverty, and unemployment among the teeming youth population. It is worth noting that although some of these problems could be more effectively dealt with without recourse to force, the Nigerian state often deploys force because it considers these challenges a threat to its existence – not just political or socioeconomic problems.

Nigeria's National Security Policy

At its core, national security is about the preservation of sovereignty and independence in line with national interests, focused on freedom from threats to the core values of the members of the society.[55] In the words of Martínez and Fernández, "National security is built around three elements: the threat and the hostile agent that causes it, the assets (material and intangible) that are to be protected and finally, the security strategy to be put in place."[56] National security guidelines, responses, policy, finance, border policing, and diplomatic relations within and outside the region and subregion are shaped, as discussed above, by the perception of what constitutes a threat, and the pattern of response to threats is what determines security policy.

As captured by the United Nations Security Sector Reform Task Force, national security policy "is a framework describing how a country provides security for the state and its citizens [and] is a formal description of a country's understanding of its guiding principles, values, interests, goals, strategic environment, threats, risks, and challenges in view of protecting and promoting national security."[57] National security doctrine crystalizes the duties of various state institutions in maintaining the rule of law and carrying out the responsibilities of public protection, security, and law enforcement.[58] Security policy is unique to each country but tends to address three fundamental themes: "The state's role in the international system, perceived international challenges, and the responsibilities of implementing actors in addressing these challenges and opportunities."[59]

Nigeria's national security policy since 1960 has been tied to the challenges presented by the problem of internal upheavals or insurrections.[60] The country spends heavily on defense, the war against terrorism, and other internal security threats. Acute insecurity is a result of all these challenges, and despite its foreign policy orientation toward broader continental affairs, Nigeria appears to have more internal enemies than external. The response of various administrations in Nigeria to security threats reveals an overdependence on the use of force and a preoccupation with a state-centric view of these challenges. While there are undeniable reasons for the use of military force, sometimes the use of force is not suitable given the domestic context within which these security challenges emerge.[61]

Nigeria's 2019 National Security Strategy enumerates the major security concerns of the country and the policy approaches outlined by the government to deal with them. In Chapter 3 of the document, the range and reach of perceived threats are vividly captured:

> [T]errorism and violent extremism, armed banditry, kidnapping, militancy and separatist agitations, pastoralists [and] farmers conflicts, transnational organised crime, piracy and sea robbery, porous borders, cybercrimes and technology challenges ... socio-political threats, fake news and hate speeches, environmental threats, public health challenges, economic challenges ... energy deficit, crude oil related crimes, unemployment and poverty, global economic challenges, [and] regional and global security challenges.[62]

The 2019 National Security Strategy also states clearly its mission "to apply all elements of national power to ensure physical and human security, a just society, peaceful co-existence, national unity, prosperity and sustainable development while promoting Nigeria's influence in regional, continental and global affairs."[63] On paper, the strategy thus indicates an aim to shift away from a state-centric security focus and toward a more holistic and human-centric security orientation. However, in practice, this is far from reality as reluctance, apathy, and the lack of political will from the government to defend the lives and property of Nigerian citizens when the need arises tends to force citizens to take up arms to protect themselves – evidenced in the proliferation of armed non-state actors serving as vigilantes in their own communities.[64]

To summarize, strategic culture in Frank's words is "operationalized through path dependence, in which the accumulation of decisions over time create constraints and restraints upon decisionmakers [and] modes of behavior by the national security apparatus become too difficult or expensive to change." The result, as experienced in Nigeria, is a set of preferred "action templates" and even unhelpful "ruts" for strategic policy and activity that reflect the unique strategic culture of the state.[65]

Conclusion

A wide range of factors contribute to a country's strategic culture, spanning historical experience, beliefs and values, geographic and resource wealth or deficits, norms and customs of behavior, structures of politics and power sharing, civil-military relations, and more – all of which impact its behavior and interactions with other states.[66] The critical significance of history is a particularly constant element drawn out through the scholarship on strategic culture, especially related to those formative past chapters defined by crisis or conflict.[67] Although cultural change is possible, the historical record shows that countries "maintain some persistent and recurrent visions of their security and geopolitical role" that endure over the long term alongside, and sometimes in friction with, near-term considerations of political convenience or expediency.[68] Thus the strategic culture of a state has enduring force but is not static, and is characterized by a dynamic and evolving conception of a country's history and its role in the world – a conception that is sometimes negotiated and reassessed across generations.

Nigeria's strategic culture is a product of several key factors in its history and contemporary experience – a long history of military rule, religious conflict, terrorism, insurgency, ethnic rivalry, a "big brother" role in Africa, an open geographic landscape, and an elite that is more interested in personal gain and accumulation of wealth than nation-building. The state's threat perceptions, rooted in its strategic culture, reflect a degree of continuity between its past and present, especially as its strategic position in Africa may be deteriorating.

Strategic culture is not a monolith, and in some states it is characterized by fragmentation. In the case of Nigeria, there are overlapping subcultures that create a dominant culture marked by a dependence on military force for dealing with internal security challenges. This tendency to gravitate toward military problem-solving for national security challenges – steeped in the values, beliefs, norms, traditions, and identity of the nation – tends not to give room for a wide range of decision options. Yet many hold out hope that Nigeria's strategic culture can evolve to be more responsive to human security challenges in practice. Further research utilizing the strategic culture paradigm can yield valuable insights into the factors that will continue to shape Nigeria's national security policy, strategy, and military action in the consequential years ahead.

Notes

1 Ken Booth, "The Concept of Strategic Culture Affirmed," in Carl G. Jacobsen, ed. *Strategic Power, USA/USSR* (New York: St. Martin's Press, 1990), 121.

2 Clifford Geertz, *The Interpretation of Cultures* (New York: Basic Books, 1993), 5–6.

3 Colin S. Gray, "Out of the Wilderness: Primetime for Strategic Culture," *Comparative Strategy* 26, no. 1 (2007): 1–20.

4 Brandon Smith, "Intentional CSR Culture – What to Consider in CSR Program Strategy," 2016.
5 Kevin Frank, "Strategic Culture in Sub-Saharan Africa: The Divergent Paths of Uganda and Tanzania" (PhD diss., University of Southern Mississippi, 2017), 1.
6 Jack Snyder, *The Soviet Strategic Culture: Implication for Limited Nuclear Operations* (Santa Monica, CA: RAND, 1977), 8.
7 Frank, "Strategic Culture," 5.
8 *National Security Strategy of the Federal Republic of Nigeria*, Federal Republic of Nigeria, December 2019, 4. https://ctc.gov.ng/wp-content/uploads/2020/03/ONSA-UPDATED.pdf
9 John Etebom, "The Long Years of Military Rule in Nigeria: A Blessing or a Curse," *Journal of Public Administration and Governance* 11, no. 2 (2021): 71.
10 Oryina Orkar, Tersoo Shaminja, Nev Solomon, and Timothy Terwase, "Armed Forces, Security Organizations and Nigeria's National Security Challenges: A Case Study," *International Journal of Trend in Scientific Research and Development* 3, no. 3 (2019): 737.
11 World Bank, "Data: Nigeria (Overview)," 2021, https://data.worldbank.org/country/NG.
12 Samson Edo and Augustine Eklegbe, *The Nigerian Economy: Reforms, Emerging Threats and Prospects* (Benin City: Centre for Population and Environmental Development, 2014), vii.
13 Michael Sinclair, *An Analysis of Nigerian Foreign Policy: The Evolution of Political Paranoia* (Braamfontein: The South African Institute of International Affairs, 1983), 1.
14 *National Security Strategy of the Federal Republic of Nigeria*, 5–6.
15 Alastair Iain Johnston, "Thinking About Strategic Culture," *International Security* 19, no. 4 (1995): 56.
16 Frank, "Strategic Culture in Sub-Saharan Africa," ii–iii.
17 Isah Mohammed, "Civil-Military Relations in a Democracy: Challenges and Prospects for a Nigerian Armed Forces" (Bauchi: Nigerian Army Armour School, 2019).
18 Oni Oni, "The Significance and Prospects of Stable Civil-Military Relations in Nigeria," *India Quarterly: Journal of International Affairs* 45, no. 2 (1989): 193–194.
19 Samuel P. Huntington, *The Soldier and the State: The Theory and Politics of Civil–Military Relations* (Harvard University Press, 1957).
20 Adeleke Adegbami and Charles Uche, "Ethnicity and Ethnic Politics: An Impediment to Political Development in Nigeria," *Public Administration Research* 4, no. 1 (2015): 59.
21 Adegbami and Uche, "Ethnicity and Ethnic Politics," 61.
22 Hussain Oyewo, "Threats of Secession: The Biafran Story," *Conflict Trends* 2019/3 ACCORD 2019, 1.
23 *National Security Strategy of the Federal Republic of Nigeria*, 6.
24 Dave Ugwu, "The Nigerian State and the Problematic of Nigeria's Big Brother Role in Nigeria's Foreign Intervention Missions," *International Journal of Innovative Research in Education Technology and Social Strategies* 6, no. 1 (2019): 155.
25 Femi Falana, public lecture delivered at the 51st birthday anniversary of Rt. Rev. Alfred Adewale Martins, 2010.
26 Dodeye Williams, "The Role of Conflict Resolution in Counterterrorism in Nigeria: A Case Analysis of the Movement for the Emancipation of the Niger Delta (MEND) and Boko Haram (BH)," *Peace Research: The Canadian Journal of Peace and Conflict Studies* 48, no. 1–2 (2016): 173.
27 Oluwaseun Afolabi, "The Role of Religion in Nigerian Politics and its Sustainability for Political Development," *Net Journal of Social Sciences* 3, no. 2 (2015): 42.
28 Moses Adagbiri and Ugo Okolie, "Corruption and the Challenges of Insecurity in Nigeria's Fourth Republic," *Journal of Political Science and Leadership Research* 4, no. 3 (2018): 43.
29 Susan Rose-Ackerman and Bonnie Palifka. *Corruption and Government: Causes, Consequences and Reform* (Cambridge: Cambridge University Press, 2016).
30 Adagbiri and Okolie, "Corruption and the Challenges of Insecurity," 53.
31 Eric Adishi, Bianca Torru, and Nduka Oluka, "Strategic Culture and Insurgencies in Northern Nigeria: Challenges and Prospects," *Journal of Political Science and Leadership Research* 4, no. 4 (2018): 69.
32 Adishi, Torru, and Aluka, "Strategic Culture and Insurgencies in Northern Nigeria," 69.
33 Agnes Osita-Njoku, "The Political Economy of Development in Nigeria: From the Colonial to Post-Colonial Eras," *IOSR Journal of Humanities and Social Sciences* 21, no. 9 (2016): 9–10.

34 Jeff McKaughan, "Nigerian Defence Budget – A Critical Review," *African Defence* (January 2014).
35 Temitope Abiodun, Adepoju Asaolu, and Anthony Ndubuisi, "Defence Budget and Military Spending on War Against Terror and Insecurity in Nigeria: Implications for State Politics, Economy and National Security," *International Journal of Advanced Research* 6, no. 7 (2020): 12.
36 Abiodun, Asaolu, and Ndubuisi, "Defence Budget and Military Spending," 13.
37 Frank, "Strategic Culture in Sub-Saharan Africa," 1.
38 Christoph Meyer and Alister Miskimmon, "Perceptions and Responses to Threats: Introduction," *Cambridge Review of International Affairs* 22, no. 4 (2009): 635.
39 Janice Stein, "Threat Perception in International Relations," in L. Huddy, D. O. Sears and J. S. Levy, eds. *The Oxford Handbook of Political Psychology,* 2nd ed. (Oxford University Press, 2013), 364–394.
40 Raymond Cohen, *Threat Perception in International Crisis* (Madison University of Wisconsin Press, 1979), 4.
41 Robert Jervis, *Perception and Misperception in International Politics* (Princeton, NJ: Princeton University Press, 1976), 28.
42 Luis Perez, "Threat Perceptions, Non-State Actors, and U.S. Military Intervention after 9/11" (M.SC Thesis) Virginia Polytechnic Institute and State University Blacksburg, VA, 2016.
43 Dodeye Williams, "Strategic Culture and Nigeria's National Security Policy in the Fourth Republic: A Descriptive Overview," *Journal of Political Science and Leadership Research* 4, no. 2 (2018): 3.
44 Williams, "Strategic Culture and Nigeria's National Security Policy," 4.
45 Joseph Ebegbulem, "Nigeria's National Interest and Foreign Policy: A Critical Evaluation," *Journal of Research in Humanities and Social Sciences* 6, no. 10 (2019b): 49.
46 George Mbara and Nirmala Gopal, "Afrocentrism, National Interest and Citizen Welfare in Nigeria and Foreign Policy Maneuvers," *F1000Research* 9, no. 997 (2020): 1–3, citing M.C. King, *Basic Currents of Nigerian Foreign Policy* (Washington DC: Howard University Press, 1996) as referenced in S. Folarin, "Nigeria's New 'Citizen-Centred Diplomacy': Any Lessons From the United States?" Fulbright American Studies Institute Fellow. *Walker Institute of International and Area Studies WIIAS,* University of South Carolina, Columbia, 2013.
47 I.S. Mohammed, J.A. Abdulmajid, and Yunusu Umar, "Nigeria in African Affairs: The Leading and Dominant Roles," Paper Presented at the Nigerian Army Armor School Tactics – Wing SSCOE CADRE 04/20 Training Programme, Bauchi - Bauchi State (10 July 2020): 1–33.
48 Joseph Ebegbulem, "Nigeria's Leadership Role and Conflict Resolution in West Africa," *International Journal of Research in Humanities and Social Studies* 6, no. 10 (2019a): 22.
49 David Bodunde and Noah Balogun, "An Overview of the Nigerian Defense Policy and the Challenges Against Terrorism," *Journal of Political Science* 9, no. 3 (2019): 525; Francis Adeboye, "The Liberian Conflict and the ECOMOG Operation: A Review of Nigeria's Key Contributions," *Global Journal of Political Science and Administration* 8, no. 3 (2020): 14.
50 Patrick Ogah, "Military Strategy: A Critical Instrument of National Defence," *The Nigerian Army Quarterly Journal* 5, no. 3 (2009): 263–264.
51 Adeboye, "The Liberian Conflict and the ECOMOG Operation," 14.
52 Williams, "The Role of Conflict Resolution in Counterterrorism in Nigeria," 173.
53 Oshita Oshita, Ikenna Alumona and Freedom Onuoha, *Internal Security Management in Nigeria: Perspectives, Challenges and Lessons* (Palgrave Macmillan, 2019).
54 Williams, "Strategic Culture and Nigeria's National Security Policy," 1.
55 Orkar, Shaminja, Solomon, and Terwase, "Armed Forces, Security Organizations and Nigeria's National Security Challenges," 737.
56 Rafa Martínez and Antonio Díaz-Fernández, "Threat Perception: New Risks, New Threats and New Missions," *Contributions to Conflict Management, Peace Economics and Development* (2007): 131.
57 Geneva Centre for Security Sector Governance (DCAF), "National Security Policy," 2021, https://securitysectorintegrity.com/defence-management/policy/.
58 DCAF, "National Security Policy."
59 DCAF, "National Security Policy."

60 Adishi, Torru, and Aluka, "Strategic Culture and Insurgencies in Northern Nigeria," 67.
61 Williams, "The Role of Conflict Resolution in Counterterrorism in Nigeria."
62 *National Security Strategy of the Federal Republic of Nigeria*, 8, 14.
63 *National Security Strategy of the Federal Republic of Nigeria*, xvii.
64 Nkasi Woda, "Nigeria's Internal Security Problem," *Council on Foreign Relations* (CFR), 26 February 2021.
65 Frank, "Strategic Culture in Sub-Saharan Africa," ii.
66 Nayef Al-Rodhan, "Strategic Culture and Pragmatic National Interest," *Global Policy Journal*, 22 July 2015, https://globalpolicyjournal.com/blog/22/07/2015/strategic-culture-and-pragmatic-national-interest?cv=1.
67 Frank, "Strategic Culture in Sub-Saharan Africa," 3.
68 Al-Rodhan, "Strategic Culture and Pragmatic National Interest."

Selected Bibliography

Abiodun, Temitope Francis, Adepoju Adeoba Asaolu, and Anthony Ifeanyichukwu Ndubuisi. "Defence Budget and Military Spending on War Against Terror and Insecurity in Nigeria: Implications for State Politics, Economy and National Security." *International Journal of Advanced Research*. 6 no. 7 (2020): 12–34.

Adagbabiri, Moses M. and Ugo Chuks Okolie. "Corruption and the Challenges of Insecurity in Nigeria's Fourth Republic." *Journal of Political Science and Leadership Research* 4 no. 3 (2018): 41–56.

Adagbiri, Moses and Ugo Okolie. "Corruption and the Challenges of Insecurity in Nigeria's Fourth Republic." *Journal of Political Science and Leadership Research*. 4 no. 3 (2018): 41–56.

Adeboye, Francis Idowu. "The Liberian Conflict and the ECOMOG Operation: A Review of Nigeria's Key Contributions." *Global Journal of Political Science and Administration*. 8 no. 3 (2020): 14–31.

Adegbami, Adeleke and Charles Uche. "Ethnicity and Ethnic Politics: An Impediment to Political Development in Nigeria." *Public Administration Research* 4 no. 1 (2015): 59–67

Adisihi, Eric, Torru Bianca and Nduka Oluka. "Strategic Culture and Insurgencies in Northern Nigeria: Challenges and Prospects." *Journal of Political Science and Leadership Research*. 4 no. 4 (2018): 64–72.

Afolabi, Oluwaseun. "The Role of Religion in Nigerian Politics and Its Sustainability for Political Development." *Net Journal of Social Sciences* 3 no. 2 (2015): 42–49.

Bodunde, David and Balogun, Noah. "An Overview of the Nigerian Defense Policy and the Challenges against Terrorism." *Journal of Political Science* 9 no. 3 (2019): 525–537.

Ebegbulem, Joseph. "Nigeria's National Interest and Foreign Policy: A Critical Evaluation." *Journal of Research in Humanities and Social Sciences*. 6 no. 10 (2019b): 49–60.

Edo, Samson and Augustine Iklegbe. *The Nigerian Economy: Reforms, Emerging Threats and Prospects*. Benin City: Centre for Population and Environmental Development (CPED), 2014.

Etebom, John Monday. "The Long Years of Military Rule in Nigeria: A Blessing or a Curse?" *Journal of Public Administration and Governance*. 11 no. 2 (2021): 71–85.

Kevin, Frank. "Strategic Culture in Sub-Saharan Africa: The Divergent Paths of Uganda and Tanzania Dissertation" (Ph.D. Thesis) University of Southern Mississippi, https://aquila.usm.edu/dissertations/1459.

Mbara, George Chimdi and Nirmala Gopal. "Afrocentrism, National Interest and Citizen Welfare in Nigeria and Foreign Policy Maneuvers." *Flooo Research*. 9 no. 997 (2020): 1–15.

Mohammed, Isah, Abdulmajid and Yunusu, Umar. "Nigeria in African Affairs: The Leading and Dominant Roles." Paper Presented at the Nigerian Army Armor School Tactics – Wing SSCOE CADRE 04/20 Training Programme Bauchi- Bauchi State 10 July 2020.

"National Security Strategy of the Federal Republic of Nigeria." Federal Republic of Nigeria, December 2019. https://ctc.gov.ng/wp-content/uploads/2020/03/ONSA-UPDATED.pdf

Orkar, Oryina, Tersoo Shaminja, Nev Solomon, and Timothy Terwase. "Armed Forces, Security Organizations and Nigeria's National Security Challenges: A Case Study." *International Journal of Trend in Scientific Research and Development* 3 no. 3 (2019): 737–741.

Oni, Oni. "The Significance and Prospects of Stable Civil-Military Relations in Nigeria." *India Quarterly: Journal of International Affairs* 45 no. 2 (1989): 193–213.

Oshita, Oshita O., Ikenna M. Alumona, and Freedom C. Onuoha. *Internal Security Management in Nigeria: Perspectives, Challenges and Lessons.* Palgrave Macmillan, 2019.

Osita-Njoku, Agnes. "The Political Economy of Development in Nigeria: From the Colonial to Post-Colonial Eras." *IOSR Journal of Humanities and Social Sciences* 21 no. 9 (2016): 9–15.

Oyewo, Hussain Taofik. "Threats of Secession: The Biafran Story." Conflict Trends 2019/3 ACCORD 27 November 2019 https://www.accord.org.za/conflict-trends/threat-of-secession/

Sinclair, Michael. *An Analysis of Nigerian Foreign Policy: The Evolution of Political Paranoia.* Braamfontein: The South African Institute of International Affairs, 1993.

Ugwu, Dave. "The Nigerian State and the Problematic of Nigeria's Big Brother Role in Nigeria's Foreign Intervention Missions." *International Journal of Innovative Research in Education Technology and Social Strategies* 6, no. 1 (2019): 155–165.

Williams, Dodeye Uduak. "Strategic Culture and Nigeria's National Security Policy in the Fourth Republic: A Descriptive Overview." *Journal of Political Science and Leadership Research* 4, no. 2 (2018): 1–10.

Williams, Dodeye Uduak. "The Role of Conflict Resolution in Counterterrorism in Nigeria: A Case Analysis of the Movement for the Emancipation of the Niger Delta (MEND) and Boko Haram (BH)." *Peace Research: The Canadian Journal of Peace and Conflict Studies* 48 no. 1–2 (2016): 173–202.

Woda, Nkasi. "Nigeria's Internal Security Problem Council on Foreign Relation (CFR)." Published 26 February 2021 https://www.google.com/amp/s/www.cfr.org/blog/nigerias-internal-security-problem%3Famp

28

VENEZUELAN STRATEGIC CULTURE

Balancing History with Twenty-First-Century Socialism

Megan F. Moore

Introduction

The definition of strategic culture adopted for this chapter is the "[s]hared beliefs, assumptions, and modes of behavior, derived from common experiences and accepted narratives (both oral and written), that shape collective identity and relationships to other groups, and which determine appropriate ends and means for achieving security objectives."[1] Venezuela has not engaged in an armed conflict since the nation's founding in 1830, and yet, strategic culture analysis enriches understandings of its past and present security postures. Strategic culture illuminates factors that helped create and exacerbate the economic crisis that began in Venezuela in 2015. Years of mismanagement and corruption have created mounting tensions in the country that may lead the general population to break from long-held positions and norms, and may even cause new "keepers" of strategic culture to emerge. To understand likely responses to the security challenges presently facing Venezuela, one must first examine the core components of Venezuelan strategic culture.

This chapter explores the sources and main tenets of strategic culture: identity, values, norms, and perceptual lens, and how each has embedded itself in the political system, military, and economy of Venezuela. It identifies the present keepers of Venezuelan strategic culture and how their beliefs have shaped important security matters. A discussion on how different groups' principles have clashed with one another or harmonized around central themes is also provided. The chapter then examines how strategic culture has expressed itself in Venezuelan domestic and international forums, concluding with reflections on areas ripe for further research in Venezuelan strategic culture.

Keepers of Venezuela's Strategic Culture

The work of Harold Trinkunas, political scientist and expert in Latin American politics, provides the foundation for the assessment of Venezuelan strategic culture captured in this

DOI: 10.4324/9781003010302-31

chapter.[2] Trinkunas identifies contemporary Venezuelan strategic culture as being defined by four groups: elites, the general public, Chavistas, and the military. While each group has its own identity, values, norms, and perceptual lens, many features intersect at important nodes. Understanding the various tensions among these keepers of strategic culture and where the nodes meet is important to building a robust analysis of Venezuela's strategic culture.

The Venezuelan elite class, in contrast to the general public, is cosmopolitan and worldly, influenced by international viewpoints and events. Historically, the elite class – wealthy former military and political leaders who served before Hugo Chavez took power – maintained significant strategic culture norms of international cooperation, use of economic tools to increase Venezuela's position in the international community, and partnership with Western nations. When Chavez began his presidency in 1999, he began to erode the elites' position and influence within Venezuela.[3] This action did not fully stymie the elite class, however, and instead sparked efforts to attempt to change Venezuelan politics from the outside, lobbying other governments – namely the United States – to pressure the *Partido Socialista Unido de Venezuela* (PSUV), the political party of Chavez and Maduro, into adopting reforms.

The general public is the traditional keeper of Venezuelan strategic culture, and with comparatively limited exposure to the outside world tends to maintain a more traditional approach to Venezuelan policy postures. Chavista ideology has penetrated some aspects of popular viewpoints, but it has not drastically changed the public's strategic culture attitudes. The changes Chavez and his successor Nicolas Maduro made were possible due to their consolidations of power and the subsequent lack of accountability. The continued absence of checks against the government has begun to normalize Chavista policies.[4] If the ideology's strength continues to progress unabated, there is a distinct possibility that Chavismo will alter the mass population's strategic culture values.[5]

Chavez and his Chavista followers advanced new strategic culture values: ones that were anti-capitalist and anti-world order. Many policies sought by the PSUV conflicted with the strategic culture values of the general population and especially the elite class.[6] Chavista foreign policy is more activist than the position advocated by the general population and seeks to challenge an international status quo that it perceives as favoring the United States. The Chavez and Maduro administrations have joined international alliances to act as a counterweight to US influence, including relationships with China, Russia, Iran, and Cuba.[7]

The Venezuelan military sees itself as the protector of the Venezuelan state, an identity inspired by Simon Bolivar, the man who liberated Venezuela from Spanish rule.[8] The military was an important instrument during the formation of the country, and military rulers exerted their power in the early years of Venezuela's independence to seize their place in Venezuela's history. From 1899 to 1945, Venezuela was ruled by four military dictators in succession. The period from 1945 to the present has been mired by poor governance, in part because of complex civil-military relations,[9] and periods of political unrest have resulted in a belief that Venezuela's main threats are domestic.[10] Venezuela's long history of political instability has affected all keepers of strategic culture.

All four groups serving as keepers of Venezuelan strategic culture source attitudes from Bolivar, who believed in an independent and self-governed nation able to direct its own destiny. Groups across Venezuelan society exhibit an independent spirit and pride in their heritage. Further, they see themselves as uniquely positioned to lead the region, and possibly the world, in important social and economic matters.[11] Chavez's quest to establish a twenty-first-century socialist system within Venezuela and act as a model for current and future socialist states was a manifestation of this self-identified leadership role.[12] Many elements of the same agenda continue under Maduro.

Two other facets of Venezuelan life deserve mention. Venezuelan society at large is significantly influenced by religious faith: a large majority of Venezuelans – 90% – identify with the Christian faith, and base actions and decisions on their belief systems.[13] In addition, the country's vast oil reserves have been a notable factor shaping Venezuelan strategic culture. Chavistas and elites have competed for access to the country's oil resources as they provide one of the only domestic paths to wealth.[14] Thus, Venezuela's vast oil reserves and mineral resources have often led to policies that are inward-looking as a means to secure Venezuela's position in the regional and world stages.[15]

Venezuela's Social and Political Landscape

Shifts in the Venezuelan social and political landscape over the last 20 years have been significant and enduring and may merit consideration as emergent aspects of cultural identity, norms, and perceptions of self and others. For the general Venezuelan public and elites, identity often falls along political party lines. Those who support the ruling party, PSUV, tend to have a positive view of the government and the military. Citizens who identify with the opposition party are more likely to be critical of governmental institutions and policies.[16] Trust, however, emanates from deeper connections and interactions than partisanship. In the case of Venezuela, partisanship is an indicator of personally held beliefs. According to institutional trust theory, trust is often tied to an individual's personal experience with an organization. Perceptions of trust in the PSUV and the Venezuelan armed forces, *Fuerza Armada Nacional Bolivariana* (FANB), are therefore expected to be closely related to interactions a person has had with the institutions. If a person feels their immediate circumstances have benefitted from the policies of the party, they are more apt to support it and place their trust in it.[17]

Social and economic classes also manifest in partisan politics. Political lines tend to be drawn along economic ones with the elite class supporting the opposition party and the working class supporting PSUV. These divisions were present in Venezuela before Chavez rose to power in 1999, and they have been highlighted over the last 20 years because Chavez weaponized the class cleavages and leveraged them to his advantage. In this way, Chavez built a platform on social grievances that existed among Venezuelans and hyperbolized them into an us-versus-them campaign.

The involvement of FANB in the execution of government policies occurred early in Chavez's presidency. In 1999, Chavez created programs called social missions to aid underserved populations, delivering aid, food, medicine, and necessary supplies.[18] Communities that directly benefitted from social missions tend to view FANB in a positive light. However, as unrest and state-sponsored violence have unraveled in the country as a response to the economic crisis, more people have become disillusioned by the party.[19] Social missions were also the beginning of Chavez's vision to fuse civil and military relations. The Venezuelan Secretary of Defense eventually led the national food distribution programs and Chavez appointed hundreds of military personnel, both active duty and retired, to fill national and local government positions, as well as prominent roles in state-owned enterprises. Under Chavez, the armed forces became a politicized body organized to protect the party and its socialist views over nationalist values. He transformed the military from a protector of national security to a domestic security force. In 2008, the armed forces' name was changed from *Fuerzas Armadas Nacionales* to *Fuerza Armada Nacional Bolivariana*, evoking an allegiance to the Bolivarian movement and party.[20]

The biased nature of FANB has created shifts in public sentiment toward the armed forces. In early 2000, the majority of Venezuelans voiced favorable opinions of the military, however by 2014, public opinion flipped with the majority expressing disapproval.[21] Perceptions of partisanship in FANB are difficult to shed, and the organization will continue to struggle with public trust even if measures are taken to depoliticize. Indeed, the public may be willing to accept some level of insecurity in order to have FANB dismantled and reformed, a prospect that could lead to chaos and continued instability. This action may yet be preferred over maintaining the status quo, which the public and elites view as corrupt.[22]

In 2005, Chavez intensified his rhetoric of communism and its superiority to democracy and began to equate socialism with Christianity and characterize capitalism as anti-Christian. He compared his Bolivarian movement with the Christian movement and "emphasized the connections between his administration and the Christian faith."[23] The anti-Christian label expanded not just to capitalist, democratic societies, but anyone who opposed Chavez. In this sense, Chavez created a sentiment of good versus evil in attempts to demonize his perceived enemies.[24] Weaving religious tones into the political debate was a calculated maneuver by Chavez. As noted earlier, Venezuela is a religious nation: 70% of the population identify as Catholic and 20% identify as evangelical Christian.[25] The Catholic Church holds the highest rates of confidence among Venezuelans. The church is favored over the National Assembly, the political opposition to PSUV, and support comes from individuals from all social classes and political leanings.[26]

Key Institutions Influencing Venezuelan Values

Values are closely held beliefs that are often immutable across time and place. As a Catholic-majority country, some tenets of Christian values are embedded in Venezuelan culture. Community is highly prized, and people seek relationships within their identity groups. Supporting members in one's group, including in one's church community, is important to Venezuelans and these ties deepened during the overlapping crises of the late 2010s and early 2020s, as many turned to faith and their local community to seek aid and comfort.[27]

Solidarity is also important to Venezuelan society at large. The idea that Venezuelans are one people, however, is not shared by all. Afro-Venezuelans, for example, have felt marginalized and ignored by the government, particularly after petitioning the government for legal status as a distinct ethno-racial group with its own culture and history and being denied because recognition "could constitute a threat to unity of the nation."[28] The Venezuelan state has ignored racial differences for many years. From 1854 until 2011, the national census did not ask respondents to specify their race.[29] In 2011, the census incorporated questions about racial identity for the first time. The outcome was approximately 52% of respondents self-identified as *morenos*, 43% White, 3% Black, 2.7% Indigenous, and 0.8% as Afro-descendants.[30]

To inspire societal cohesion, military academies teach ideologically driven curricula that are in line with Bolivarianism.[31] Part of the curriculum introduced under Chavez includes teachings about US imperialism, US capabilities, and Venezuela's need to fuse civil and military efforts in order to defeat a superior enemy.[32] The term "forces" to refer to the various military arms was modified to "force" to promote unity.[33] Beginning in 1958, the National Defense Advanced Studies Institute began to open its doors to more economically diverse populations.[34] FANB is seen by middle- and low-income populations as a way to earn a decent and respectable living. Many military members are grateful for the opportunity to progress through the ranks and improve social outcomes for themselves and their children, and are, therefore, willing to support the PSUV and the institutions that maintain the system.[35]

The Venezuelan elite class also favors social cohesion and internal stability because these characteristics directly benefit them. Elites, however, preferred the pre-Chavez status quo when they had more power and control over resources and political decisions. When the Chavez-Maduro administration consolidated power, they simultaneously weakened the other political voices within Venezuela.[36] For instance, two political parties that were once influential in the country, *Acción Democrática* and *Comité de Organización Política Electoral Independiente*, are no longer significant players in Venezuelan politics.[37] The executives of PDVSA, Venezuela's oil company, were an influential force within the elite class until 2004 when Chavez fired almost 20,000 employees and replaced them with loyalists, further concentrating Chavista influence at the top levels of society.[38]

Conflict and Crime: Historical Contexts Shaping Venezuelan Norms

Norms describe what behaviors a group believes to be within the parameters of acceptable, and therefore, legitimate. The general Venezuelan population tends to accept pacifism and reject the PSUV's position that the United States is a foe of Venezuela, with 75% of Venezuelans rebuffing the notion of fighting a war with the United States.[39] Venezuela has not fought a war with another country since its independence from Spain in 1830.[40] Elites are also anti-war and prefer resolving disputes through established channels.[41] An even larger portion of the population has signaled in polls that they do not support deploying the military to support the Bolivarian government.[42] This presents a challenge to PSUV's more hostile posture to the United States.[43]

The military has had a tumultuous history in Venezuela. Since the nation's founding in 1830 through 1958, Venezuela was under dictatorial rule and for over 100 years the military was used as an extension of the national leader. In 1958, the sitting president was ousted by public demand, leaving an opening for new governance of the country. Three of the major political parties began organizing democratic elections and signed a pact to respect the results of the election and initiate a custom of peaceful transitions of power. The pact, known as *Punto Fijo*, also outlined new customs for the military to maintain political neutrality and honor the outcome of elections. The neutrality of the military was so critical that it was codified into the nation's constitution in 1961.[44]

The democratic period from 1958 to 1999 was not without its problems. The signatories of *Punto Fijo* were criticized for using the governance void to consolidate power, and there was rampant corruption among government officials that reached a tipping point in 1989. In 1986, world oil prices plunged to record low levels, cutting off Venezuela's steady stream of revenue. The country responded with sharp cuts to government spending and programs, and people took to the streets to protest the austere measures. The government sent the military to quell the protests and riots, which fueled violence that escalated into military brutality. This was a pivotal moment in Venezuelan history: it not only marked the end of *Punto Fijo*, but it also divided military officials on proper engagement in domestic affairs. Some of the disillusioned military officers joined then Army Lieutenant Colonel Hugo Chavez in a failed coup d'état attempt in 1992.[45]

While oil has been a source of strength for the Venezuelan economy, it has also brought economic insecurity, political infighting, and corruption, generating a cycle of domestic issues that have dominated the political agenda.[46] The lucrative reliance on oil gave elites little reason to diversify and invest in other industries. Thus, Venezuela became accustomed to importing technologies, industrial machinery, and other products. The central government had an early role in regulating the market beginning in the early twentieth century when Venezuela began to commercialize its oil, and the result has been the constant struggle to control the industry, as whoever controls it is guaranteed a path to wealth and political status. The lack of investment in other industries makes oil one of the only paths to wealth in the Venezuelan economy, further compounding the tensions and elite competition to control it. This system also breeds a society where the elites retain power and influence, and the divide between them and the general population grows.[47]

According to data collected by the United Nations Office on Drugs and Crime, Venezuela had the second-highest murder rate in the world in 2019. The state leads Latin America in the most civilian killings by state officials. These figures signal the wider reality that the country struggles to maintain the rule of law.[48] Since 2015, Venezuela has faced an economic meltdown that fuels unrest and uncertainty. The diminished rule of law, however, predates the country's current crisis. After taking office, Chavez began placing military officials in civilian government roles. The militarization of the civil service influenced the militarization of the local police forces, and the normalization of violence coupled with a lack of institutional accountability created a volatile situation. The military has also been tasked with domestic security, blurring the lines between policing and military operations.[49] Gan Galavis describes Venezuela as an anomic state:

> Anomic states are those in which there is no rule of law. Constitutions are permanently infringed; laws are technically deficient and arbitrarily applied by officials; judges and courts are not accessible to the people. An anomic state lacks legitimacy, as it fails in producing a social contract that binds state and social behaviour: not only do ordinary citizens transgress the laws, but also state officials when they use force abusively – especially against poorer populations – and are rarely sanctioned for it. An anomic state becomes evident in the inefficiency of criminal law, the poor performance of police and judicial officials, and the breach of human rights.[50]

Political corruption is very much a norm in Venezuela: it is estimated that $300 billion was diverted and stolen from state coffers over the 2010s.[51] Average citizens tend to respect the rule of law, but the political and economic crises of the late 2010s and early 2020s have thrown the country into turmoil, eroding confidence in government institutions and increasing the likelihood the population will view them as illegitimate.[52] In attempts to regain domestic stability, the police force deployed Operation Liberation and Protection of the People into communities in 2015 with the stated goal of cleansing neighborhoods of criminal gangs and Colombian militia members. The result, however, included persecution of civilians. The pressure to perform from the government was so high that officials were expected to satisfy quotas, leading to arrests of innocent people. The tension created a counter-resistance from the populace, which escalated to police-involved deaths of civilians.[53]

Venezuelan Perceptions and Manifestations of Strategic Culture

Perceptual lens, in strategic culture analysis, refers to the ways in which groups view their role in securing the nation, who they see as friend and foe, and other aspects of worldview and cosmology. The Simon Bolivar spirit is harnessed by the majority of Venezuelans who perceive the country to have a special place in the region. This sense of exceptionalism coupled with Venezuela's abundance of natural resources has historically led many Venezuelans to see the country as a greater force in international politics than reality dictates.[54]

Chavez's brand of Venezuelan exceptionalism centrally cast the United States as the main adversary. The United States attempted to overthrow Chavez in 2002, and he responded by forming alliances with anti-Western nations. Over the years, Chavez worked closely with Robert Mugabe of Zimbabwe, Christina Kirchner of Argentina, Rafael Correa of Ecuador, and Evo Morales of Bolivia.[55] Chavez tried to draw parallels between Spanish imperialism and US imperialism to shape Venezuelan public opinion against the West,[56] negotiated a number of agreements with Iran and Russia, and kept regular dialogues with Syria, North Korea, Libya, Sudan, and Cuba.[57] In death, Chavez is revered as a saint by his party members.[58]

The general Venezuelan public, however, and especially the elites tend to view the United States as a positive force in the international order and would like to see Venezuela aligned more closely with it and European democracies. Despite this openness, the public and elites have largely felt aggrieved by the United States and the larger international community's lack of attention to Venezuela's recent economic and political crises. It is estimated that since 2015 more than 4.5 million people have left Venezuela seeking to flee the food shortages, currency inflation, and overall instability of the country.[59] Venezuelan diaspora communities have grown in number and in strength, especially in the United States. Organized expatriate groups pleaded with the Trump administration to use its power to exert pressure on Maduro to step down. Venezuelan elites have also mobilized by pouring money into political campaigns and lobbying groups to restore the pre-Chavez status quo.[60]

On the Hofstede scale, which offers a framework for cross-cultural comparison, several manifestations of Venezuelan strategic culture can be identified. Venezuela scores 100, the highest score possible, in the Indulgence dimension,[61] reflecting a population that acts on impulse and lives for the moment, likely driven by the significant instability and political turmoil that has made any kind of long-term planning impossible for years. Decisions by the Maduro government have also shown how this characteristic manifests on a regime level, with an economic crisis that has produced hyperinflation and severe food and medicine shortages. The Maduro government's paranoia about stories depicting unfavorable domestic conditions seeping into international newsrooms drove the regime to order the National Guard and local police units to investigate and detain journalists who were reporting on the issue in an effort to preserve the state's image.[62] The Hofstede scale also identifies a high Power Distance score in Venezuela. Communities with high Power

Distance dimensions believe societal inequalities are natural, and this belief leads to acceptance of and an unwillingness to challenge uneven balances of power, opportunities, and access.[63] The persistence of social inequities and divisions may be explained in part by a tolerance of class distinctions, thus somewhat normalizing inequities between the general public and elites.

Venezuela scores high on the Masculinity dimension of the Hofstede scale, characterizing its society as results-driven and competitive, yet it receives one of the lowest scores in Individualism, indicating that Venezuelans are highly collectivist.[64] This combination of cultural orientations manifests itself in many ways. Venezuelans are protective of their group identity and view others in a competitive light. Chavez was known to play to these characteristics by cloaking autocratic policies as measures to further national sovereignty. Anti-US policies were coded as protections of Venezuelan identity, as there were moves to block nonprofit and nongovernmental organizations from operating in the country.[65] Competition also manifests itself in the armed forces, with rivalries arising among the five service branches of the FANB.[66] The Army is the largest military branch, followed by the Navy, Marines, National Guard, and National Militia. The Army views itself as an elite force, and many from within the Army have been appointed into high positions in the government – in part a legacy of Chavez having served in the Army and showing it preferential treatment when selecting individuals to lead agencies. Despite this, the Navy and Air Force view themselves as more skilled than the Army, and the National Guard holds itself in high regard because it is mostly staffed by officers and enlisted men, whereas the Army relies on many conscripts to fill its ranks. The National Guard is perceived by the other branches, however, to be the most corrupt of the branches. Under the Bolivarian system, military promotions are based on party loyalty. This system breeds divisions within the military branches where mid-level officers possess a diversity of ideological beliefs, while high-ranking military officials are devoted Chavistas.[67]

Corrupt military officials also compete with each other in establishing drug trafficking routes and markets.[68] The Cartel of Suns has members in all five branches of FANB, and Venezuela is an active transshipment area for Colombian cocaine traveling to the United States and other markets.[69] The Army and National Guard are believed to have the most members involved in organized crime. Members use their knowledge and authority of roads, air, and seaports to facilitate the movement of drugs from Colombia and the Caribbean and do so with relative impunity. Control of the transshipment routes also means the military can carry out other illicit activities like smuggling in scarce goods, including food and medicine, and charge exorbitant prices for personal profit. Unfortunately, this is a pressing reality in Venezuela.[70]

Areas for Future Strategic Culture Research

The four groups driving current Venezuelan strategic culture have major divergences in critical areas, but when analyzing macro-level issues, the groups align on significant issues as well. Many of the topics on which they align emanate from the origins of Venezuelan strategic culture: the spirit of Bolivar, religion, leveraging oil resources, concern for domestic security, and the acceptance of social divides and hierarchy. The divergences can be explained by the groups' adoption of policies and practices that benefit their own growth and success, sometimes at the expense of other keepers of strategic culture. These similarities and divergences are summarized in the Venn diagram in Figure 28.1.

Figure 28.1 "Keepers" of Venezuelan Strategic Culture.

Source: Author's Creation.

While these enduring features of Venezuela's strategic culture are well researched and documented, assessments of the impact of the present political and economic struggle on Venezuelan identity, values, norms, and perceptions are lacking. These effects deserve further study as they offer potential long-term changes to Venezuelan norms relating to security issues. The ongoing crisis could constitute what Jeffrey Lantis describes as an external shock great enough to cause a "strategic culture dilemma."[71] As of the time of this writing, elements of shifting attitudes have already presented themselves in the general public's disillusionment and distrust of Maduro and the PSUV. Portions of the Venezuelan population who previously supported Chavismo may coalesce with the elite class and favor a more open domestic economic system. Despite its preference for internal cohesion, the Venezuelan population has expressed willingness to endure continued instability if it results in economic, political, and military reforms. Given the severe hardship and sharp decline in gross domestic product, future research could examine whether the shared Venezuelan independent spirit might dim, and whether the current political and economic struggle may cause changes in Venezuelans' perceptions of themselves and others.

The crisis could also produce a new keeper of strategic culture. With a large segment of the population now living outside of Venezuela, research is needed to evaluate whether the Venezuelan diaspora might become a new keeper of the state's strategic culture, and how the diaspora's influence may alter current keepers' identity, values, norms, or perceptual lens. Furthermore, additional work is needed to assess whether the failed policies of Chavismo will challenge twenty-first-century socialism as an enduring force on Venezuelan strategic culture.

Strategic culture analysis provides context to Venezuela's security challenges by identifying the relevant human dimensions influencing decisions and perceptions. It is also a powerful tool for detecting possible future changes to a population's identity, values, norms, and perceptual lens. In the case of Venezuela, signals of new political alliances, shifts in perceptions of who adversaries are, changes in views of the country's role in the region and on the international stage, and even a nascent keeper of strategic culture emerge through a strategic culture assessment. The level and permanence of these shifts merit deeper examination.

Notes

1 Jeannie L. Johnson, Kerry M. Kartchner, and Jeffrey A. Larsen, eds. *Strategic Culture and Weapons of Mass Destruction: Culturally Based Insights into Comparative National Security Policymaking* (Palgrave Macmillan, 2009), p. 9.
2 Harold Trinkunas, "Venezuelan Strategic Culture," Miami: Florida International University, 2009. Retrieved from https://calhoun.nps.edu/handle/10945/43095, 19.
3 Trinkunas, "Venezuelan Strategic Culture," p. 5.
4 Trinkunas, "Venezuelan Strategic Culture," p. 29.
5 Trinkunas, "Venezuelan Strategic Culture," p. 9.
6 Trinkunas, "Venezuelan Strategic Culture," p. 4.
7 Harold Trinkunas, "Venezuela: Between Tradition and Ideology" in Brian Fonseca and Eduardo A. Gamarra, eds., *Culture and National Security in the Americas*, United States: Lexington Books, 2017, pp. 25–26.
8 B. Fonseca, J. Polga-Hecimovich, and H. Trinkunas, "Venezuelan Military Culture," Miami: Florida International University, 2016. http://gordoninstitute.fiu.edu/policy-innovation/military-culture-series/brian-fonseca-johnpolga-hecimovich-and-harold-trinkunas-2016-venezuelan-military-culture.pdf, 9–10.
9 Fonseca, Polga-Hecimovich, and Trinkunas, "Venezuelan Military Culture," p. 6.
10 Trinkunas, "Venezuela: Between Tradition and Ideology," p. 19.
11 Trinkunas, "Venezuela: Between Tradition and Ideology," p. 4.
12 Trinkunas, "Venezuela: Between Tradition and Ideology."
13 F. Freier, "Maduro's Immorality and the Role of the Church in Venezuela," 15 June 2018, Retrieved from Berkley Center for Religion, Peace & World Affairs: https://berkleycenter.georgetown.edu/responses/maduro-s-immorality-and-the-role-of-thechurch-in-venezuela.
14 Harold Trinkunas, "Venezuelan Strategic Culture," Findings Report, Applied Research Center, Latin American and Caribbean Center, Florida International University (2009) 4.
15 Trinkunas, "Venezuelan Strategic Culture," pp. 9–10.
16 John Polga-Hecimovich, "Bureaucratic Politicization, Partisan Attachments, and the Limits of Public Agency Legitimacy: The Venezuelan Armed Forces Under Chavismo," *Latin American Research Review* 54, no. 2 (2019): 476 doi:10.25222/larr.142. http://search.ebscohost.com/login.aspx?direct=true&AuthType=ip,shib&db=f5h&AN=137256823&site=ehost-live&scope=site&authtype=ip,shib&custid=s3555202
17 Polga-Hecimovich, "Bureaucratic Politicization," pp. 479–480.
18 Polga-Hecimovich, "Bureaucratic Politicization," p. 481.
19 Fonseca, Polga-Hecimovich, and Trinkunas, "Venezuelan Military Culture," p. 5.
20 Polga-Hecimovich, "Bureaucratic Politicization," p. 481.
21 Polga-Hecimovich, "Bureaucratic Politicization," p. 483.
22 Polga-Hecimovich, "Bureaucratic Politicization," p. 477.
23 "Constructing the Socialism of the 21st Century on the Airwaves: A Rhetorical Analysis of President Hugo Chávez's Characterization of Venezuela's Socioeconomic Shift on Aló Presidente," *Conference Papers – International Communication Association* (2017): 1–33, 22–25. http://search.ebscohost.com/login.aspx?direct=true&AuthType=ip,shib&db=ufh&AN=135750119&site=ehost-live&scope=site&authtype=ip,shib&custid=s3555202
24 "Constructing the Socialism of the 21st Century on the Airwaves," 25.
25 Freier, "Maduro's Immorality and the Role of the Church in Venezuela."
26 David Smilde and Daniel Hellinger, *Venezuela's Bolivarian Democracy: Participation, Politics, and Culture Under Chávez* (Durham, NC: Duke University Press, 2011).
27 Arelis R. Hernández and Mariana Zuñiga, "Short of Electricity, Food and Water, Venezuelans Return to Religion," *The Washington Post*, 13 April 2019.

28 Krisna Ruette-Orihuela and Hortensia Caballero-Arias, "'Cimarronaje Institucional:' Ethno-Racial Legal Status and the Subversive Institutionalization of Afrodescendant Organizations in Bolivarian Venezuela," *Journal of Latin American & Caribbean Anthropology* 22, no. 2 (2017): 322, 325–326.
29 Cristobal Valencia, "Ethnographies of Energy: Race, Space, and State Formation on the Paria Peninsula," *Journal of Latin American & Caribbean Anthropology* 24, no. 1 (2019): 90–91.
30 Ruette-Orihuela and Caballero-Arias, "'Cimarronaje Institucional:' Ethno-Racial Legal Status and the Subversive Institutionalization of Afrodescendant Organizations in Bolivarian Venezuela," p. 330.
31 Fonseca, Polga-Hecimovich, and Trinkunas, "Venezuelan Military Culture," p. 4.
32 Fonseca, Polga-Hecimovich, and Trinkunas, "Venezuelan Military Culture," p. 11.
33 Fonseca, Polga-Hecimovich, and Trinkunas, "Venezuelan Military Culture," p. 4.
34 Fonseca, Polga-Hecimovich, and Trinkunas, "Venezuelan Military Culture," p. 10.
35 Fonseca, Polga-Hecimovich, and Trinkunas, "Venezuelan Military Culture," p. 17.
36 Trinkunas, "Venezuelan Strategic Culture," p. 5.
37 Trinkunas, "Venezuelan Strategic Culture," pp. 16–17.
38 Trinkunas, "Venezuela: Between Tradition and Ideology," 24, 30.
39 Trinkunas, "Venezuelan Strategic Culture," 5.
40 Trinkunas, "Venezuelan Strategic Culture," p. 4.
41 Trinkunas, "Venezuelan Strategic Culture," pp. 8–9.
42 Trinkunas, "Venezuelan Strategic Culture," p. 23.
43 Trinkunas, "Venezuelan Strategic Culture," pp. 6–7.
44 Polga-Hecimovich, "Bureaucratic Politicization," p. 481.
45 Polga-Hecimovich, "Bureaucratic Politicization," p. 482.
46 Trinkunas, "Venezuelan Strategic Culture," p. 11.
47 Trinkunas, "Venezuelan Strategic Culture," pp. 14–16.
48 Natalia Gan Galavís, "Rule of Law Crisis, Militarization of Citizen Security, and Effects on Human Rights in Venezuela,"*European Review of Latin American & Caribbean Studies* no. 109 (January 2020): 67–86. https://erlacs.org/articles/10.32992/erlacs.10577
49 Gan Galavís, "Rule of Law Crisis, Militarization of Citizen Security, and Effects on Human Rights in Venezuela," 68–69, 71.
50 Gan Galavís, "Rule of Law Crisis," p. 70.
51 R.E. Ellis, "The Collapse of Venezuela and Its Impact on the Region," *Military Review* 97, no. 4 (July 2017): 23.
52 Polga-Hecimovich, "Bureaucratic Politicization," p. 489.
53 Gan Galavís, "Rule of Law Crisis," p. 74.
54 Trinkunas, "Venezuelan Strategic Culture," p. 12.
55 Joseph Kirschke, "Hugo Chavez, Socialism for the 21st Century and Venezuela's Narcissistic Resource Nationalism," *Engineering & Mining Journal* (00958948) 214, no. 7 (2013): 60.
56 Fonseca, Polga-Hecimovich, and Trinkunas, "Venezuelan Military Culture," p. 11.
57 Daniel F. Wajner and Luis Roniger, "Transnational Identity Politics in the Americas: Reshaping 'Nuestramérica' as Chavismo's Regional Legitimation Strategy," *Latin American Research Review* 54, no. 2 (2019): 66.
58 Freier, "Maduro's Immorality and the Role of the Church in Venezuela."
59 Anne Applebaum, "Venezuela Is the Eerie Endgame of Modern Politics," *The Atlantic,* 27 February 2020, https://www.theatlantic.com/ideas/archive/2020/02/venezuelas-suffering-shows-where-illiberalism-leads/606988/.
60 Nicholas Confessore, Anatoly Kurmanaev, and Kenneth Vogel, "Trump, Venezuela and the Tug-of-War Over a Strongman," *The New York Times,* 1 November 2020, https://www.nytimes.com/2020/11/01/us/trump-venezuela-maduro.html.
61 Hofstede Insights, https://www.hofstede-insights.com/product/compare-countries/.
62 William Finnegan, "A Failing State," *New Yorker* 92, no. 37 (2016). https://www.newyorker.com/magazine/2016/11/14/venezuela-a-failing-state
63 Hofstede Insights, https://www.hofstede-insights.com/product/compare-countries/.
64 Hofstede Insights.

65 Timothy M. Gill, "Unpacking the World Cultural Toolkit in Socialist Venezuela: National Sovereignty, Human Rights and Anti-NGO Legislation," *Third World Quarterly* 38, no. 3 (2017): 626–630. doi:10.1080/01436597.2016.1199259. http://search.ebscohost.com/login.aspx?direct=true&AuthType=ip,shib&db=f5h&AN=121362271&site=ehost-live&scope=site&authtype=ip,shib&custid=s3555202

66 Fonseca, Polga-Hecimovich, and Trinkunas, "Venezuelan Military Culture," p. 5.

67 Fonseca, Polga-Hecimovich, and Trinkunas, "Venezuelan Military Culture," pp. 13–14.

68 Fonseca, Polga-Hecimovich, and Trinkunas, "Venezuelan Military Culture," p. 5.

69 Finnegan, "A Failing State."

70 Fonseca, Polga-Hecimovich, and Trinkunas, "Venezuelan Military Culture," pp. 16, 21.

71 Jeffrey Lantis, "Strategic Culture: From Clausewitz to Constructivism," in Johnson, Kartchner, Larsen, eds. *Strategic Culture and Weapons of Mass Destruction* (2009): pp. 33–52, Palgrave Macmillan, New York, https://doi.org/10.1057/9780230618305_3.

Selected Bibliography

Alarcon-Garcia, Gloria, Jose Daniel Buendia Azorin, and Maria Del Mar Sanchez De La Vega. "Shadow Economy and National Culture: A Spatial Approach." *Hacienda Publica Espanola* 232, no. 1 (2020): 53–74.

"Constructing the Socialism of the 21st Century on the Airwaves: A Rhetorical Analysis of President Hugo Chávez's Characterization of Venezuela's Socioeconomic Shift on Aló Presidente." *Conference Papers– International Communication Association* (2017): 1–33.

Ellis, R. E. "The Collapse of Venezuela and Its Impact on the Region." *Military Review* 97, no. 4 (July 2017): 22.

Fonseca, B., J. Polga-Hecimovich, and H. Trinkunas. "Venezuelan Military Culture." Miami: Florida International University, 2016. http://gordoninstitute.fiu.edu/policy-innovation/military-culture-series/brian-fonseca-johnpolga-hecimovich-and-harold-trinkunas-2016-venezuelan-military-culture.pdf.

Freier, F. "Maduro's Immorality and the Role of the Church in Venezuela." 15 June 2018. Retrieved from Berkley Center for Religion, Peace & World Affairs: https://berkleycenter.georgetown.edu/responses/maduro-s-immorality-and-the-role-of-thechurch-in-venezuela.

Gan Galavís, Natalia. "Rule of Law Crisis, Militarization of Citizen Security, and Effects on Human Rights in Venezuela." *European Review of Latin American & Caribbean Studies* no. 109 (January 2020): 67–86.

Gill, Timothy M. "Unpacking the World Cultural Toolkit in Socialist Venezuela: National Sovereignty, Human Rights and Anti-NGO Legislation." *Third World Quarterly* 38, no. 3 (2017): 621.

Kirschke, Joseph. "Hugo Chavez, Socialism for the 21st Century and Venezuela's Narcissistic Resource Nationalism." *Engineering & Mining Journal (00958948)* 214, no. 7 (2013): 60.

Lupien, Pascal. "Ignorant Mobs Or Rational Actors? Understanding Support for Venezuela's 'Bolivarian Revolution'." *Political Science Quarterly (Wiley-Blackwell)* 130, no. 2 (2015): 319.

Polga-Hecimovich, John. "Bureaucratic Politicization, Partisan Attachments, and the Limits of Public Agency Legitimacy: The Venezuelan Armed Forces Under Chavismo." *Latin American Research Review* 54, no. 2 (2019): 476.

Rosales, Antulio. "Pursuing Foreign Investment for Nationalist Goals: Venezuela's Hybrid Resource Nationalism." *Business & Politics* 20, no. 3 (2018): 438–464.

Ruette-Orihuela, Krisna, and Hortensia Caballero-Arias. "'Cimarronaje Institucional:' Ethno-Racial Legal Status and the Subversive Institutionalization of Afrodescendant Organizations in Bolivarian Venezuela." *Journal of Latin American & Caribbean Anthropology* 22, no. 2 (2017): 320–338.

Smilde, David, and Daniel Hellinger. *Venezuela's Bolivarian Democracy: Participation, Politics, and Culture Under Chávez*. Durham, NC: Duke University Press, 2011.

Smith, Robert S. "Cultural Marxism: Imaginary Conspiracy or Revolutionary Reality?" *Themelios* 44, no. 3 (2019): 436–465.

Strinen, Iselin. *Grassroots Politics and Oil Culture in Venezuela*. 2017th ed. Cham: Springer International Publishing, 2017.

Trinkunas, Harold. "Venezuelan Strategic Culture." Miami: Florida International University, 2009.

Trinkunas, Harold. "Venezuela: Between Tradition and Ideology" in Brian Fonseca and Eduardo A. Gamarra, eds., *Culture and National Security in the Americas*. United States: Lexington Books, 2017.

Valencia, Cristobal. "Ethnographies of Energy: Race, Space, and State Formation on the Paria Peninsula." *Journal of Latin American & Caribbean Anthropology* 24, no. 1 (2019): 88–109.

Wajner, Daniel F., and Luis Roniger. "Transnational Identity Politics in the Americas: Reshaping 'Nuestramérica' as Chavismo's Regional Legitimation Strategy." *Latin American Research Review* 54, no. 2 (2019): 458.

PART IV

The Utility of Strategic Culture for Practitioners and Scholars

29

STRATEGIC CULTURE SCHOLARSHIP

A User's Guide to the Cultural Topography Methodology

Jeannie L. Johnson

Introduction

The Cultural Topography (CTops) methodology has been introduced and described in a number of different works: first in an article in *Studies in Intelligence* in 2011, and then as a research approach in two 2018 volumes examining state-based nuclear decisionmaking and military culture, respectively.[1] The purpose of this chapter is to provide a clear step-by-step guide for executing a Cultural Mapping Exercise (CME) – the basic building block of CTops research – as a method within the field of strategic culture and to point scholars and practitioners to existing literature that supplies additional depth on particular aspects of the research process.

The CME was designed as a systematic and deliberate honing method for intelligence analysts seeking to isolate and assess those cultural factors weighing most heavily in the decisionmaking of a particular group, on a particular issue, at a particular point in time. Within this process, analysts employ lines of inquiry regarding identity, norms, values, and perceptual lens with the purpose of better understanding both the motivating factors behind likely behavior for members of the selected group and the cultural constraints that may eliminate options and shape what is perceived as possible or effective. CMEs completed across multiple stakeholder groups may be overlayed to form a Cultural Topography: combined research findings of multiple mapping exercises reveal "peaks" – areas where cultural norms, values, and perceptions are in alignment across key stakeholders – and "valleys," where aspects of an issue cause alignment between groups to break down, revealing divisions in beliefs about appropriate thinking and action on the issue. Most work, to date, has been in the form of single CMEs completed by individual researchers.[2] Their work has delivered timely new insights in intelligence, policy, and academic arenas and offers a valuable resource for any wishing to pursue the analytic insights available through a CTops approach.

The existing descriptions of the CME method have catered to the needs and working realities of intelligence analysts and policy professionals. This chapter will fill a gap by

DOI: 10.4324/9781003010302-33

addressing the methodological requirements of students and academic scholars. In addition to the merit for academic scholarship that CME research can offer, the workforce pipeline for policy and analytic institutions is largely drawn from universities. It therefore benefits the full community of thinkers, researchers, and practitioners for this chapter to provide a guide that will allow students and scholars to pursue the CME as a means of conducting social science research in keeping with the rigor and requirements of their fields. With this objective in mind, this chapter will engage in the academic discourse necessary to appropriately locate the CME within the landscape of social science methods, while holding fast to the pragmatic aim of delivering clear and user-friendly instructions regarding the method's research and analytic process.

Any scholar employing the CME as a primary framework for research will be required to situate this method within the theory and existing methods of an academic field. The various approaches to strategic culture scholarship, effectively cataloged by Brigitte Hugh in this volume,[3] reside most comfortably within the field of strategic studies but have also been employed by scholars of international relations. Depending on the approach a researcher prefers to employ, strategic culture scholarship can serve as a challenge to the theoretical constructs of realism and neorealism, or as a companion theory that agrees with most assumptions of these models but insists that rationality be regarded as culturally encoded. At the heart of strategic culture scholarship is the assumption that encultured decisionmakers weigh costs and benefits in culturally distinct and thus sometimes surprising ways. Colin Gray's seminal work as a theoretical realist and first-generation strategic culture scholar falls within this camp. His scholarship, alongside that of others, will help researchers locate the strategic culture paradigm within the major theories of the social sciences.[4] Once the researcher has positioned the paradigm of strategic culture as either a companion or challenge to an existing theory, the researcher can employ the CME as a specific research method.

In order to understand and correctly employ the social science practices that undergird the CME method, it is important to delineate what it is designed to achieve and what it is not. The CME is designed to deliver critical insights regarding the decisionmaking and behavior of a particular group (transnational, national, subnational, or non-state), at a particular moment in time, facing a specific issue. This stands in contrast to most strategic culture scholarship, which has been conducted on nation-states and has focused rather strictly on states' approaches to defense and security. Researchers employing the CME, conversely, may be focused on the actions of a particular national subgroup or a transnational group and may be tracking issues as wide ranging as corrupt financial practices, disease transmission, responses to internal protests, climate challenges, or cybercrime. The CME is designed to be applied to any group with a degree of shared culture vis-à-vis any issue of policy interest. Even analysis of culturally fragmented groups provides its own sort of utility. In the case of malicious actors, unveiling cultural rifts inside a loosely affiliated or sharply fractured group can identify pressure points that might be leveraged when seeking to disrupt its plans for malevolent action.

The cultural mapping method will be explained here in seven steps. These steps are not immutable and have been collapsed or expanded in previous work based on the specific needs of the researcher.[5] The CME is inductive[6] and interpretivist[7] in nature and draws heavily from the core practices of Grounded Theory, a mode of research and analysis drawn from the field of anthropology.[8] Social science literature supplying further guidance on these aspects of the method is provided in the endnotes of this chapter.

Box 29.1 Cultural Mapping Exercise

Step 1 Identify a research question scoped around a specific issue.
Step 2 Isolate a relevant transnational, national, subnational, or non-state group for focused study.
Step 3 Amass an initial range of cultural domains from which members of this group draw cultural cues.
Step 4 Conduct research cataloging data across categories of Identity, Norms, Values, and Perceptual Lens.
Step 5 Distill findings to Critical Cultural Factors (CCFs).
Step 6 Map CCFs back to cultural domains.
Step 7 Place CCFs within the wider context of operational realities.

Step 1: Identify a Research Question Scoped around a Specific Issue

Purpose

Directing research around a specific problem set or "issue" is critical to identifying the aspects of identity, habits of practice, and key values that are informing group behavior when responding to the issue selected. Human behavior is shaped by cultural codes that stem from multiple cultural domains (defined and discussed in Step 3).[9] Individuals may "code switch" between religious, ethnic, national, workplace, and familial cultural domains in the course of a single day. Often it is the issue at hand that determines which domain's narratives and cultural codes most heavily dominate group thinking, discourse, decisionmaking, and behavior.

Using a specific issue to direct cultural research is an approach that diverges from traditional strategic culture work that seeks to identify dominant narratives and practices of a strategic community found with consistency across time and place. Jack Snyder, the first to use the term "strategic culture," assessed that traits needed to be of near semipermanent quality in order to merit the label "culture."[10] Colin Gray agreed, but caveated that national strategic cultures often house contrasting narratives that emanate from diverse domains and are held in differing levels of esteem by subnational groups and organizations.[11] It is perhaps unsurprising, therefore, that insights produced by CME research have revealed that groups draw on narratives of certain domains (perhaps religious or class-based) in responding to some issues but may draw from entirely different domains (perhaps ideological or generational) to inform thinking and behavior on other issues. Those cultural narratives and practices that may appear most prominent across time in a profile of a national strategic culture, therefore, may not be the most relevant vis-à-vis the particular issue of concern selected by the researcher. For this reason, the issue itself is used across the research and analytic process as the key honing device for narrowing the wide range of possible cultural influences down to those CCFs (Step 5) that will weigh most heavily in the decisionmaking of members of the group.

Practice

Research questions must be scoped to an appropriately specific level in order to deliver findings that contribute usefully to forecasting, foresight,[12] and policy solutions.[13]

For instance, rather than look at the US defense posture as a whole, one might examine the circumstances under which US Cyber Command believes offensive cyber action is justified and which sort of cyber action is perceived as both permissible and effective.

Researchers may pursue various aims when examining group behavior vis-à-vis the issue selected. Scholars may choose to unpack historical case studies by casting a cultural lens on behavior that appears perplexing or that they believe has been misinterpreted within previous literature. Intelligence and policy analysts tend to be concerned with the "now" and seek a degree of predictive rather than explanatory power from their findings. Both are acceptable research paths.

The purpose of the CME is not to test a hypothesis (deductive approach) but rather to use the issue at hand as an entry point into the cultural research process (inductive approach). Researchers are encouraged to revisit the original research question and refine it toward greater utility as the research process delivers new insights.

Step 2: Isolate a Relevant Transnational, National, Subnational, or Non-State Group for Focused Study

Purpose

Most issues of strategic import have the potential to be impacted by a diverse range of actors. The CME method is designed to deliver focused insights on *one* of these. As noted in the introduction, individual CMEs completed on multiple actors surrounding the same issue can yield the overlapping data necessary to construct a full Cultural Topography.

Focusing the analysis on one specific stakeholder group is another way in which the CME departs from some of the traditional approaches within the strategic culture discipline. Most strategic culture work has focused on cultural congruence across elites operating within a national strategic community. This scholarship often notes that national strategic cultures tend to house competing subcultures, and sometimes highlights a few of the most prominent, but pursues the primary objective of identifying cultural trend lines that have held steady across the state's national history. By contrast, researchers undertaking a CME assume that the diverging cultural preferences between critical stakeholders surrounding an issue can be stark enough to translate into differences in policy depending on which national subcomponent takes the lead in determining the national course of action. Because the research emphasis for a CME is often on near-term forecasting around a specific issue, these differences matter.

In some cases, it is not national action that is of concern for the researcher. Research issues that deal with transnational groups or local dissidents require the scalability of the CME method to assess the impact of cultural factors and likely behavioral outcomes at the level appropriate for analysis.

Practice

Choosing a group for a focused study starts with amassing the range of relevant transnational, national, subnational, and non-state groups who may influence outcomes on the selected issue in a significant way. Several of these stakeholder groups may merit a CME study. Which one is selected depends on the needs and interests of the researcher. For instance, a researcher may select the stakeholder group most likely to dominate initial decisionmaking on this issue; the one that will take the lead in the way decisions will be executed; a relatively new, little-known group that is understudied; a key detractor vis-à-vis this issue who might make a useful ally or dangerous enemy; or any number of other considerations.

It is inappropriate to select a single individual (such as a head of state) as the target of a focused study. Cultures are learned, shared behaviors cultivated in groups. If the initial research question revolves around the likelihood of a single actor pursuing a course of action (e.g., the Supreme Leader of Iran crushing internal protests), the researcher is advised to select and analyze a group positioned to act as either a key enabler or potential obstructor to the leader's agenda.

Step 3: Amass an Initial Range of Cultural Domains from which Members of This Group May Draw Cultural Cues

Purpose

The use of the label "domains" is intended here to capture *reasonably cohesive collections of distinctive cultural codes* existing in a wide, interconnected cultural landscape. It is tempting to use the word "subculture" to describe domains. Indeed, many of them are accurately described that way. The reason the term "domain" is preferable for the purposes of this methodology is that many relevant domains are not "sub" in the sense of small or niche. For instance, if we use the domain of national culture as our starting point, many additional domains will be "sub" – smaller, more distinctive clusters of cultural codes – like the regional cultures of the American West or South, or the organizational cultures of each of the institutions in the UK security community. Other cultural domains, however, may not be "sub" but are better understood as "supra" – transcending national culture and shared by a larger group. A religious cultural domain, like Islam or Catholicism, transcends national lines. Ethnic culture shared by a diaspora or the shared ethos of scientists from diverse backgrounds might represent other types of cultural domains that reach beyond national boundaries. The label "domain" is not itself important; it is simply a device to remind analysts that there is a potentially wide range of cultural influences humans experience, ranging from subcultures through supracultures, that may be informing the thinking and behavior of members of the selected group.

Cultural domains inform our identity, provide default settings for behavior, prioritize some values over others, and provide us with stories and experiences that color our perceptual lens. The codes of behavior and belief systems cultivated within these domains are learned (not inborn) and are shared and policed by others who share the respective domain spaces. Taken individually, each domain represents a separate "culture." Common domains may include religious, national, ethnic, social class, geographic (e.g., urban versus rural lifestyles and perspectives), organizational (e.g., formed around a workplace or military service), and peer group domains, including generational cultures or particular subsets like students in an engineering club. Adherence to a particular domain tells us who we are and how we should behave. Shared domains form the cultural connections that help us "get along" with other members of the in-group by establishing understood modes of behavior and lowering transaction costs for cooperation. As Kevin Avruch observes, "for any individual, culture always comes in the plural"[14] – meaning that individuals traverse between different, even contradictory cultural domains and tend to develop patterns of adhering to one set of cultural codes over another depending on situational context and the specific issue being addressed.

Thus, the amassing of potential cultural domains within Step 3 has two core purposes. First, before CME research can narrow cultural influences to those that are relevant to the issue at hand, one must examine the plausible range of influences. Second, once CCFs are identified (Step 5), the researcher is asked to map these back (Step 6) to the cultural domains identified here. This mapping process provides insights into the extent to which cultural influences impacting group decisionmaking are likely to be shared and supported by wider populations, or whether the key cultural drivers identified are largely confined to group membership and run counter to cultural codes emanating from other domains.

Practice

If the researcher is completing scholarly work, theories regarding group behavior, including existing strategic culture scholarship, will have been addressed in a literature review and are a useful starting point in amassing a range of cultural influences. Some analysts have found it helpful to craft a visual representation of various domains in concentric circles around the group, or in a Venn diagram depicting likely overlap between domains that share significant cultural similarities. As research progresses and insight into relevant cultural domains surfaces and becomes more sophisticated, new domains may be added and original domains may be disaggregated into appropriate subcomponents.

A few litmus tests may be helpful when attempting to tease out the various subcultures and supracultures that represent domains of potential influence on the group. Humans cultivate cultural constructs in order to impose meaning or "make sense" of their experiences. In doing so, the resultant culture helps define what is right and wrong, important and unimportant. Therefore, an ingredient like "history" is not itself a cultural domain; rather, it supplies the shared events that cultural domains (like religion or military service) have selected from, interpreted, and formed into internal narratives that help define identity and prescribe certain norms as effective or honorable. If a cultural domain does not supply a rather clear sense of "us/them" boundaries and enforce some taboos, it may not meet the thresholds required to merit inclusion in the researcher's assembly of domains.

Step 4: Conduct Research Cataloging Data across the Categories of Identity, Values, Norms, and Perceptual Lens

Purpose

The core purpose of Step 4 is to engage with multiple distinct cultural data sources in a way that reveals patterns in thinking and likely behavior for members of the group regarding the selected issue. As researchers wade into diverse sources of cultural data, their findings are cataloged within the four macro categories of Identity, Values, Norms, and Perceptual Lens (IVNPL). These categories inspire usefully distinct lines of research and also serve as the primary sorting silos for the resultant data. The definitions provided here are a refined version of those offered in the original 2011 article outlining the CME method:

- **Identity:** The character traits the group assigns to itself, the reputation and role it pursues, and the "identity goods" – individual roles and statuses – designated to members.

- **Values:** Material goods or personal characteristics that result in increased status for members, including deeply held beliefs about what is desirable, proper, and good that serve as broad guidelines for social life.
- **Norms:** Accepted, expected, and preferred modes of behavior, including shared understandings concerning taboos.
- **Perceptual Lens:** The cognitive filter through which this group views the world, including the default assumptions that inform its opinions and ideas about self and others.[15]

Prior work detailing the CME method has provided research prompts and a discussion regarding potential data sources for each of these four categories.[16] This chapter acts as a companion to that work rather than duplicating it. It is worth providing a brief word, however, on the research purpose behind each definitional category.

Repeated practice with the CME method has revealed that *identity* often acts as the "deep anchor" or guiding justification for group norms, prioritized values, or worldviews. It is listed as the first research category by design. Humans often join groups for the cultural "identity goods" membership offers them. As noted by Theo Farrell, identity goods are powerful and precious. Threatening them can mean war.[17]

Values are defined within the CME in a slightly unusual way. The definition is driven by a need to separate out what is authentically valued from what is purportedly valued by the group. Groups often designate a set of core values that are celebrated in slogans, mission statements, or religious texts. These may or may not be the values actually influencing behavior. A sounder way to track what is "valued" in a group or society is to observe and catalog the personal attributes, life experiences, social connections, or material goods (among other possibilities) that increase the status of group members. Collectively conferred status is a group's way of assigning value. For those conducting research aimed toward designing policy solutions, understanding which ideational and material goods are authentically valued within the group will help identify salient incentives and will point interlocutors toward rhetoric and engagement strategies that enhance dignity.

The concept of *norms* is defined in the social science literature in a wide variety of ways. The purpose of its definition here is to provide a focused look at behavior. Habitual behaviors practiced by a group may or may not be rooted in a principled belief system. Thus, the ubiquity of behavior may prove to be a false friend when testing for robustness. Some norms are simply a product of habit, disconnected from the original logic or belief system that gave rise to them. Because the purpose of CME research is to narrow the range of likely cultural influences on the decisionmaking and *action* to be taken by a group, it is important to place a specific focus on accepted, expected, preferred, and taboo behaviors. In a later stage, these norms will be assessed for robustness by tracking connections (or the lack thereof) to identity, values, and the belief systems which inform the group's perceptual lens.

Perceptual Lens is the term of art employed here for a category that many might call "worldview" or "mindset." The aim is to capture outward thinking – group notions about outsiders and how the world works – but also inward thinking: beliefs about one's own experience and history. Perceptual lens includes an examination of cognitive filters that assign strong meaning to some aspects of observed data and ignore others.[18]

During the research process, analysts should expect to see a cultural trait surface in slightly different ways across categories. For US Marines, the *identity* of being "disciplined" is primarily associated with the *norm* of coolness under fire. It surfaces as a *value* when it is used to compliment or confer status on another Marine. As will be discussed in Step 5, a cultural trait signals robustness when it surfaces in multiple ways across categories.

Practice

The active process of research within Step 4 is a close companion to the core tenets of Grounded Theory: to seek clarification regarding the ways in which group member behavior is shaped by context.[19] Researchers should, to the extent possible, suspend prior biases and assumptions about the group under study and the respective cultural domains being investigated. The research process unfolds through immersion in cultural data from multiple sources. Termed "thick description" by Clifford Geertz,[20] the idea is to throw the research net widely in order to capture material and discursive cultural cues produced within the overlapping cultural contexts experienced by members of this group. It is more fun than it sounds. Researchers may catalog themes within popular novels or folklore, study national narratives within educational texts, read field notes from recently returned doctoral students completing ethnographies, watch local sitcoms in order to recognize norm violations, conduct interviews, and analyze political speeches. Other potential sources of cultural data are discussed at some length in prior publications.[21]

Given its comprehensive nature, the early stages of CME research may seem somewhat directionless and overwhelming. This is a positive sign. It means the researcher has done the hard work of attempting to set aside prior assumptions and hypotheses and is approaching data with an open mind. The four categories of IVNPL are supplied as a basic framework for guiding lines of research and cataloging the results. As researchers come across distinctive or pronounced cultural references within a political speech, or repeatedly represented in artwork in personal homes, these shared cultural references (or "memes"[22]) are recorded within one or more of the IVNPL categories as appropriate. As data accumulates within each of the four categories, patterns and themes will begin to emerge. Researchers are encouraged to use the linguistic terms employed by members of the group under study when labeling these patterns and themes. For instance, as noted above, consistent reference to the terms "professionalism" and "discipline" within US Marine Corps discourse signals that these are important concepts within Marine Corps identity and value structure. Using these terms as labels within both the identity and value categories allows the researcher to begin to assemble the specific cultural content associated with those labels and better discern their constructed meaning within the group. Grounded Theory scholars call this process "memoing and constant comparison" and note that data gathering and analysis are happening simultaneously.[23]

It is important to allow cultural narratives, themes, and patterns to unfold naturally from the data. Researchers discovering the ubiquity of a certain behavior might be tempted, for instance, to press artificially hard to designate a belief system in which it is rooted. Some norms, however, are the equivalent of what Avruch calls "cultural clichés" – artifacts floating on the surface of cultures but largely untethered to meaningful belief systems or aspects of identity.[24] "Knocking on wood" is a case in point. Americans invoke the term (and the action) with some frequency as a superstitious practice to ward off the "jinxing" that may occur when a favorable prospect is said out loud. While the behavior of knocking on wood is common, very few Americans can explain its origins or any belief system that may accompany it. Further, most Americans would recoil from being identified as "superstitious" despite the fact that this is the most accurate identity descriptor of the behavior. Americans may be more superstitious than they would like to admit, but it is not the sort of identity trait they will be keen to defend.

Most intelligence and policy analysts utilizing the CME have employed simple and straightforward sorting and labeling techniques in order to identify emergent patterns. These have been sufficient to deliver prescient and accurate insights into finished intelligence. For those wishing to pursue an advanced degree of rigor, Grounded Theory scholars offer a range of coding techniques.[25] Grounded Theorists note with pragmatism that research may be bounded by the feasibility of collecting data (on closed societies or vulnerable populations, for instance) and the limited time frames that may be available to researchers. With these realities in mind, Grounded Theorists offer the goal of "saturation" as a descriptor of when the research process is complete. Saturation is achieved when engagement with new data sources validates and reinforces cataloged themes rather than delivering new insights or surprises.[26]

Step 5: Distill Findings to Critical Cultural Factors (CFFs)

Purpose

The purpose of Step 5 is to separate the *relevant* from the *merely interesting* across research findings. While this may sound rather obvious and is certainly what the researcher intended to achieve at the outset of the CME process, it can be surprisingly painful to do. Sifting out data means setting aside insights discovered at rather steep personal cost in time and effort. Beyond these investments, researchers simply become attached to the data. Belief systems and behavior that are strikingly different from one's own are intrinsically fascinating. Not sharing it with prospective readers of the research seems like a crime.

In order to fulfill the CME research mandate of delivering a degree of actionable insight regarding the CCFs weighing in on group thinking and behavior on the specified issue at the designated moment in time, researchers must submit to a honing process that examines the accumulated cultural data for *robustness*, *relevance*, and likelihood to provoke a *response* (cooperative or conflictual) when this group engages on this issue. Without such a culling process, researchers employing the CME are likely to reproduce the complexities found in national strategic culture profiles – competing internal narratives, potentially contradictory cues concerning identity, and overlapping cultural domains which suggest divergent courses of action – and will consequently fail to deliver sufficiently specific research findings or contributions to the formation of effective policy.

Practice

Relevance

Researchers might consider a number of factors when seeking signposts of *relevance* for particular cultural codes in relation to the selected issue. In some cases, the issue itself is preloaded with cultural default settings stemming from the domain – national, religious, ethnic, organizational – that is the traditionally understood reference point for informing appropriate thinking and action on this issue. In other cases, recent events may have impacted which one of a possible set of cultural codes and narratives have become "relevant" for a nation or a subgroup. After the events of 9/11, the United States pursued a significantly more nationalist and interventionist set of cultural codes, accompanied by narratives celebrating American democracy and the prospect of its transportability. These fueled early decisionmaking and behavior as the United States pursued its invasions and nation-building projects in both Afghanistan and Iraq.[27]

Particularly charismatic opinion leaders often play a role in moving a latent cultural narrative into a more dominant position of relevance within group discourse on a particular issue. The leader may leverage a recent event, demographic shift, or long-standing frustration experienced by the target group in order to rally support for a narrative that serves as both an explanation and a call to action for the group. It is important to note, however, that the leader who leverages a preexisting narrative to serve their own ends is bound by the group's understanding of the acceptable courses of action captured in the narrative. Publicly held narratives are powerful tools for concerted action precisely because they are widely held and understood. A leader who invokes one as a call to action is imposing soft boundaries around what legitimate action within that narrative space is understood to be. Cultural narratives are a shared group commodity. They cannot be endlessly manipulated for personal ambitions.

In all cases, the evaluation of relevance for each cultural factor is determined by a clearly evidenced tie to its influence on the group's thinking and process of decisionmaking. This may include the specific ways in which the decision is likely to be carried out and the group's perceptions concerning what they are likely to achieve. No matter how intrinsically interesting, if a trait or combination of traits cannot be defensibly traced to group beliefs, discourse, or likely behavior on the selected issue, it does not qualify for inclusion as a CCF.

Robustness

The cultural factors designated as relevant by the researcher may not be in perfect alignment and may in fact seem to encourage various alternative courses of action. When cultural cues seem in conflict with one another, an examination of *robustness* may aid the researcher in previewing the prioritization that members of the group will place on particular factors. Robustness cannot be determined by any one metric, but the following signals are helpful:

- How frequently did this cultural factor appear across multiple and diverse data sources?
- How enduring is this factor? Is it relatively new or an embedded aspect of historic belief or practice?
- How widely shared is this factor across cultural domains experienced by members of the group?
- To what extent is opinion or behavior that is in line with this factor rewarded and that is out of line punished?
- Is this cultural factor prominent within the discourse of key decisionmakers within the group?
- Is this cultural factor manifest in multiple ways across the IVNPL research categories?

Likelihood to Provoke a Response

Built into most CME research questions is a specific focus on the behavior the researcher seeks to predict or explain. Thus the final honing stage of Step 5 focuses heavily on cultural factors connected to plausible group *responses*. The rough ranking of cultural factors according to relevance and robustness in the previous two stages offers the researcher a distilled set of factors to work with in determining important responses: red lines, powerful incentives, and likely calls to action for the group. Leveraging the question sets provided in prior publications for each research category (IVNPL) will help the researcher through the final phase of narrowing accumulated data to a refined set of CCFs for this group, on this issue, at this moment in time.[28] A sampling of questions for evaluating likely responses includes:

Identity

- What identity goods for members of this group are likely to be disrupted or displaced by proposed action on this issue?
- Is group cohesion strong along identity lines in response to this issue? What would cause the group to fracture or to unite behind a common front?
- What sorts of problems does this group see itself as responsible for solving?
- To what extent would specific trajectories on this issue threaten the future role this group perceives for itself?

Norms

- What is the order of activities when solving a problem (called *action chains* or *strategies for action*) for this group?[29] Does face-to-face confrontation happen first or last? Is violence used as a signal or is it an endgame?
- What are legitimate pathways to power for this group?
- What modes of punishment for detractors are this group willing to pursue?
- What are the group's taboos surrounding this issue?

Values

- What is considered honorable behavior or heroic action for this group?
- Which are the group's "sacred" (nonnegotiable) values that if transgressed are likely to cause revolt?
- In which circumstances does this group reward action? Restraint?
- Which values are emphasized when leadership is encouraging cooperative behavior on this issue?

Perceptual Lens

- What defines "victory" for this group on this issue?[30]
- What does this group believe are its red lines?
- How does this group assign intentions? What motives make the most sense to them?[31]
- What does this group's history tell it about "dangerous" behaviors and circumstances for a society?

Step 6: Map CCFs Back to Cultural Domains

Purpose

Step 6 is designed for those researchers who seek to assess the degree of support from surrounding communities that the selected group may experience in their decisionmaking and action. Tracking CCFs back to relevant domains (where possible) sheds some light on how widely shared the group's interpretation of the issue may be and the likelihood of support or rejection from adjacent communities.

As noted in the introduction to this chapter, a refined understanding of the cultural positioning of multiple relevant stakeholders around the selected issue would require completing individual CMEs on several stakeholder communities and assembling the overlayed data as a Cultural Topography. Within a single CME, Step 6 provides a preview of what the advanced research of a full Topography might reveal by tracking CCFs back to relevant domains and noting the areas where group CCFs are either in line or strikingly out of sync with surrounding communities. Matthew Brummer and Eitan Oren's work on Japan provides a useful case in point. Japan's Self-Defense Forces (SDFs) draw heavily from the domain of their own organizational culture when considering the issue of expanding Japan's role in offensive and defensive military action. Their perceptions of the benefits to Japan of a marginally more aggressive role for the military stand in contrast to those traditionally held by the general public. A researcher forecasting the potential for a policy shift in Japan toward a more engaged role for its military forces (issue selected in Step 1) might focus on the cultural codes and preferences being cultivated within the organizational culture of Japan's SDFs (the actor selected for focused study in Step 2) and the ways in which these preferred courses of action are likely to be constrained by the traditional preferences of the voting public (Step 6).[32]

Practice

The comprehensive research pursued in Step 4 typically provides the data necessary to locate CCFs within their respective cultural domains. This may be the CCF domain of origin or a domain that has since adopted the same cultural value or aspect of identity. A single trait may be found across multiple domains. The aim of mapping CCFs to domains is not to establish the genesis of the trait but rather to mark how widely it is shared. For the purposes of the CME, it is not important to delineate whether the norm of hospitality found in many countries of the Middle East originated with Arab culture (ethnic domain) or with Islam (religious domain). It is sufficient to note that it is a robust cultural factor in both.

It is unlikely that the range of cultural factors distilled to CCFs in Step 5 will collectively track back to a single domain. Rather, group members may be drawing from multiple cultural domains when responding to the issue in question. CME research investigating the form of counterinsurgency practiced by the US Marine Corps revealed that some CCFs strongly influencing decisionmaking and behavior were informed by internal Marine Corps culture; others could be tracked to US military culture writ large; and some of the most consequential were values and beliefs shared across American national culture.[33]

In addition to evaluating potential support for group action, mapping CCFs back to domains may produce insights across a number of other research objectives; for example, assessing prospects for group narratives to succeed in recruiting efforts from adjacent communities. An outside group messaging through the narratives of one domain (e.g., national) may be speaking past group members who are interpreting the issue through an alternative domain (e.g., tribal). Researchers may also identify groups who are likely to become stridently defensive of their own perceived cultural territory in response to action on this issue, and can deliver more sophisticated insights regarding productive pathways in messaging or negotiation with this group.

Step 7: Place CCFs within the Wider Context of a Group's Operational Realities

Purpose

As has been noted by several strategic culture scholars across this volume, cultural predispositions alone do not determine behavior. Isolating relevant CCFs helps the researcher see the decisionmaking context more clearly and understand "rationality" – the cost/benefit analysis of key decisionmakers – from the group's point of view. It is not, however, determinative of behavior. The group's ability to act on cultural preferences is advanced or restrained by a number of operational realities that enable or disrupt the preferred course of action. These may include access to material resources, geographic limitations, the availability of technological expertise, and the promise of aid – or threat of opposition – by outside groups, among other factors.

Practice

Capturing the existence of the wider context of operational realities is an essential ingredient informing the researcher's development of plausible scenarios for the group's actions on the selected issue. It may be beyond the scope of work for a CME researcher to address the full range of operational realities faced by the group in question, but in order to protect the validity and integrity of cultural research, it is incumbent on the researcher to note that such constraints exist and to highlight areas where they surfaced during the CME process. Exploring the contours of specific constraints on group behavior may be left to future researchers, but it is important that CME researchers supply appropriate warnings and caveats about the group's ability to act on its culturally preferred path.

Conclusion

Because the CME was designed by practitioner-scholars, it responds to the call for demonstrated policy relevance in social science work.[34] At the heart of the CME research method is a firm commitment to engaging with the most critical problem sets of the day and producing insights that illuminate potential solutions. For the practitioner, it provides actionable data on a contemporary issue to help planners more deftly navigate the complex landscape of interstate (and non-state) relations. For the scholar, the research process outlined above has the potential to advance our understanding of underexamined groups, provide structure for the analysis of competing cultural narratives, and leverage the predictive power of empirical findings. Within the field of strategic culture, the CME provides a scalable method that usefully addresses the complexities and competing influences that produce a degree of flux in the character of security policy emanating from a national strategic community.

Notes

1 Jeannie L. Johnson and Matthew T. Berrett, "Cultural Topography: A New Research Tool for Intelligence Analysis," *Studies in Intelligence* 55, no. 2 (2011): 1–22; Jeannie L. Johnson and Marilyn J. Maines, "The Cultural Topography Analytic Framework," in Jeannie L. Johnson, Kerry M. Kartchner, and Marilyn J. Maines, eds., *Crossing Nuclear Thresholds: Leveraging Sociocultural Insights into Nuclear Decisionmaking* (Palgrave, 2018), 29–60; Jeannie L. Johnson, *The Marines, Counterinsurgency, and Strategic Culture: Lessons Learned and Lost in America's Wars* (Georgetown University Press, 2018).

2 For examples available outside of classified settings, see the *Journal of Advanced Military Studies Special Issue on Strategic Culture* (2022), Katie C. Finlinson, "The United Arab Emirates as a Case Study in Assessing Over-the-Horizon Nuclear Proliferation," 112–129; Benjamin Potter, "Unrecognized Republic, Recognizable Consequences: Russian Troops in 'Frozen' Transnistria," 168–188; Jeffrey Taylor, "Deterring Russian Nuclear Threats with Low-Yield Nukes May Encourage Limited Nuclear War," 207–229; Emilee Matheson, "Lord's Resistance Army Culture Provides Opening to Prevent Attacks and Advance Humanitarian Efforts," 189–206; and Matthew Brummer and Eitan Oren, "'We Must Protect This Peace with Our Hands': Strategic Culture and Japan's Use of Force in International Disputes as Depicted in Ministry of Defense Manga Promotional Materials," 88–111.

3 Brigitte E. Hugh, "From First Generation to Fourth: Growth and Refinement Across the Field of Strategic Culture," in Kerry M. Kartchner, Briana D. Bowen, and Jeannie L. Johnson, eds., *Routledge Handbook of Strategic Culture* (Routledge, 2023).

4 Gray advances the case for *realpolitik* within his assessments of strategic culture in "Comparative Strategic Culture," *Parameters,* 14, no. 1 (Winter 1984): 27. For additional points of view see John Glenn, Darryl Howlett, and Stuart Poore, eds. *Neorealism Versus Strategic Culture* (Burlington, VT: Ashgate Publishing Company, 2004).

5 See, for example, Johnson and Maines, "The Cultural Topography Analytic Framework"; Johnson, *The Marines, Counterinsurgency, and Strategic Culture*; and Finlinson, "The United Arab Emirates as a Case Study in Assessing Over-the-Horizon Nuclear Proliferation."

6 For an introduction to inductive approaches, see Georg Henrik von Wright, *Explanation and Understanding* (Ithaca: Cornell University Press, 1971) and J. Donald Moon, "The Logic of Political Inquiry: A Synthesis of Opposed Perspectives," in Fred I. Greenstein and Nelson W. Polsby, eds., *Political Science: Scope and Theory, Handbook of Political Science Vol. 1* (Reading, Massachusetts: Addison-Wesley, 1975), 132–133.

7 Peter Wilson explains the value of an interpretivist approach in better understanding international institutions. The logic applied to institutions also applies to cultural groups: Peter Wilson, "The English School Meets the Chicago School: The Case for a Grounded Theory of International Institutions," *International Studies Review* 14, no. 4 (December 2012): 567–590.

8 For researchers new to Grounded Theory, Timonen et al. provide an excellent starting point. See Virpi Timonen, Geraldine Foley, and Catherine Conlon, "Challenges When Using Grounded Theory: A Pragmatic Introduction to Doing GT Research," *International Journal of Qualitative Methods* 17 (2018): 1–10. For a review of the original theory, see Barney G. Glaser and Anselm L. Strauss, *The Discovery of Grounded Theory: Strategies for Qualitative Research* (Chicago: Aldine Publishing Company, 1967).

9 Kevin Avruch, *Culture & Conflict Resolution* (US Institute of Peace Press, 1998), 15, 18–19.

10 Jack L. Snyder, "The Soviet Strategic Culture: Implications for Limited Nuclear Operations," R-2154-AF (Santa Monica, CA: RAND Sept 1977), v.

11 Colin S. Gray, *Perspectives on Strategy* (Oxford: Oxford University Press, 2013), 98–99.

12 While "forecasting" and "foresight" are sometimes treated as interchangeable terms, the field of anticipatory intelligence makes an important distinction between the two in definition and practice. Strategic culture research can offer utility for both. For further discussion see Josh Kerbel, "Coming to Terms with Anticipatory Intelligence," *War on the Rocks,* 13 August 2019, https://warontherocks.com/2019/08/coming-to-terms-with-anticipatory-intelligence/; and Briana D. Bowen, "Strategic Culture and Anticipatory Intelligence," in Kerry M. Kartchner, Briana D. Bowen, and Jeannie L. Johnson, eds., *Routledge Handbook of Strategic Culture* (Routledge, 2023).

13 Questions that focus on the near term and examine tightly scoped scenarios are more likely to yield predictive data as discussed in Edwin Bacon, "Comparing Political Futures: The Rise and Use of Scenarios in Future-Oriented Area Studies," *Contemporary Politics* 18, no. 3 (September 2012): 270–285.
14 Avruch, *Culture and Conflict Resolution*, 15.
15 Johnson and Berrett, "Cultural Topography: A New Research Tool for Intelligence Analysis."
16 Johnson and Berrett, "Cultural Topography: A New Research Tool for Intelligence Analysis," includes an Appendix that provides research prompts for each of the four research categories and provides a discussion of cultural data sources; and Johnson, *The Marines, Counterinsurgency, and Strategic Culture* offers a thorough discussion of each of the four research categories and documentation of their utility in research practice.
17 Theo Farrell, *Norms of War: Cultural Beliefs and Modern Conflict* (Boulder, CO: Lynne Rienner Publishers, 2005), 1.
18 Dima Adamsky provides strategic culture scholarship which does an excellent job of exploring the cognitive domain in *The Culture of Military Innovation: The Impact of Cultural Factors on the Revolution in Military Affairs in Russia, the US, and Israel* (Stanford, CA: Stanford University Press, 2010).
19 Timonen, et al., "Challenges When Using Grounded Theory," 6.
20 Clifford Geertz, *Interpretation of Cultures* (New York: Basic Books, 1973), 5–10.
21 Johnson and Berrett, "Cultural Topography: A New Research Tool for Intelligence Analysis"; and Jeannie L. Johnson, "Conclusion: Toward a Standard Methodological Approach," in Jeannie L. Johnson, Kerry M. Kartchner, and Jeffrey A. Larsen, eds. *Strategic Culture and Weapons of Mass Destruction: Culturally Based Insights into Comparative National Security Policymaking* (New York: Palgrave Macmillan, 2009), 243–257.
22 Susan Blackmore, "Imitation and the Definition of a Meme," *Journal of Memetics – Evolutionary Models of Information Transmission* 2, no. 2 (1998): 1–13.
23 Timonen, et al., "Challenges When Using Grounded Theory," 7.
24 Avruch, *Culture and Conflict Resolution*, 19.
25 Timonen, et al., "Challenges When Using Grounded Theory," 5.
26 Timonen, et al., "Challenges When Using Grounded Theory," 8.
27 Johnson, *The Marines, Counterinsurgency, and Strategic Culture*.
28 See Appendix in Johnson and Berrett, "Cultural Topography: A New Research Tool for Intelligence Analysis."
29 Ann Swidler, "Culture in Action: Symbols and Strategies," *American Sociological Review* 51, no. 2 (April 1986): 273–286; Edward T. Hall, *Beyond Culture* (New York: Anchor Books, 1989), 157–159.
30 Dominic D.P. Johnson, and Dominic Tierney, *Failing to Win: Perceptions of Victory and Defeat in International Politics* (Cambridge, MA: Harvard University Press, 2006).
31 Glen Fisher, *Mindset: The Role of Culture and Perception in International Relations*, 2nd ed. (Yarmouth, ME: Intercultural Press, 1997), 52.
32 Brummer and Oren, "'We Must Protect This Peace with Our Hands.'"
33 Johnson, *The Marines, Counterinsurgency, and Strategic Culture*.
34 Michael C. Desch, "How Political Science Became Irrelevant: The Field Turned Its Back on the Beltway," *The Chronicle of Higher Education* (27 February 2019), https://www.chronicle.com/article/how-political-science-became-irrelevant/.

30
STRATEGIC CULTURE AND TAILORED DETERRENCE

Nicholas Taylor[1]

Introduction

To most people, including a large proportion of both military and civilian defense professionals, deterrence was a strategy of the Cold War. The bipolar security architecture of that era endured for nearly half a century due to what appears to many, in retrospect, to have been a largely stable deterrence relationship between the United States and the Soviet Union. Since the Cold War ended, the understanding of deterrence in many parts of the world has atrophied. In the United Kingdom, for example, deterrence was rarely the primary objective of defense activities after 1990; indeed, the term was largely used to refer to nuclear deterrence. The UK 2010 Strategic Defence and Security Review had an entire section titled "The Deterrent," reinforcing the perception that deterrence was a nuclear mission.[2] However, deterrence is, and has always been, a broadly defined and frequently used tool of foreign policy, employed when states wish to convince an actor not to behave in a particular way. Examples of "non-nuclear" deterrence include the strong messages conveyed to Saddam Hussein by the United States prior to the first Gulf War, highlighting the grave repercussions that would result if Iraq employed chemical weapons[3]; or efforts undertaken by the international community to convince states such as North Korea not to develop nuclear weapons. While usually viewed as counter-proliferation activities, these cases have clear deterrence elements.

In the last decade – and given greater focus since the Russian annexation of Crimea in 2014 – NATO nations and others in the West have paid much greater attention to deterrence. Many states have attempted to reinvigorate and refresh their understanding of and approach to the subject. This chapter proposes that strategic culture is a field of study that can make a significant contribution to the development of successful deterrence strategies. The chapter begins by providing a high-level summary of classical deterrence theory, followed by a brief overview of some of the recent developments in deterrence thinking. A potential approach that one might take to the development of a deterrence campaign or strategy is then described, identifying how and where strategic culture might contribute. While not arguing that deterrence is ever easy, it is proposed that strategic culture can make deterrence planning less difficult; in particular, in understanding the adversary's perceptions of the situation at hand and, crucially, whether they hold any sacred or protected values that might make deterrence and influence strategies inherently challenging.

DOI: 10.4324/9781003010302-34

Deterrence Background

As might be expected, it is important to start with a definition of deterrence. Many can be found in the numerous publications available on the subject, either academic, government, or military. Your author proposes the following: deterrence is the persuasion of an actor not to conduct a course of action that they otherwise intend to undertake. Much has been, and continues to be, written about deterrence as the concept was a central feature of the post-World War II security environment. The standoff between NATO and the Warsaw Pact, led by the United States and the USSR, respectively, was based on a recognition of the Mutually Assured Destruction (MAD) that the nuclear weapons of each side were capable of causing. Thomas Schelling's seminal work *Arms and Influence* remains a key reference to this day.[4] Given the dominance of nuclear weapons in strategic thinking at the time, research and public debate resulted in "deterrence" becoming synonymous with nuclear deterrence. The deterrence research of the Cold War was characterized by Robert Jervis as consisting of three waves[5]; however, the significant changes in the defense and security environment since the Cold War have necessitated the development of "classical" approaches to deterrence, which Knopf labeled as "the fourth wave."[6] This section summarizes some of the main concepts and approaches to deterrence theory.

Classical Deterrence Thinking

Deterrence by Punishment

Deterrence by punishment is the best-known form of deterrence and the one most often described. Here, an actor threatens to inflict costs on an aggressor through retaliation after an attack (or other unwanted action) has taken place. This was the basis of MAD during the Cold War; neither of the superpowers could defend themselves from a sizeable nuclear strike by the other, but it was the near-guarantee of massive retaliation – and certain destruction on both sides – that deterred both parties. Deterrence by punishment still seems to be the default form of deterrence endorsed by many academics and governments, but there are other approaches.

Deterrence by Denial

The other well-known approach to deterrence is that of deterrence by denial.[7] This form of deterrence aims to dissuade a potential aggressor by convincing them that their efforts will not succeed, and they will not achieve their objective. This is illustrated by, for example, a ballistic missile defense (BMD) system: rather than promising swift retaliation in response to a ballistic missile attack, BMD in theory protects its user from the effects of a missile attack but inflicts no harm on the aggressor. The attacker is hopefully deterred by the realization that its missiles will not get through and their attack will fail. The BMD operator may choose to take punitive action at a later stage, but this is not part of the deterrence strategy. Deterrence by denial has been described as more reliable than deterrence by punishment because, as Lawrence Freedman puts it, "with punishment, the [adversary] is left to decide how much more to take. With denial, the choice is removed."[8]

Recent Developments in Deterrence Thinking

In the last two decades, the rational actor perspective that dominated Cold War deterrence thinking has been complemented with a broader set of approaches from fields such as psychology, anthropology, and the study of decisionmaking. The key themes are summarized here.

The Fourth Wave

As Knopf describes, new deterrence research emerged after the 11 September 2001 terrorist attacks.[9] In contrast to the Cold War threat environment, this focused on asymmetric threats principally from non-state actors, some of whom were (and continue to be) willing to die for their cause. It was during this period that the term "tailored deterrence" began to enter the lexicon of military and policy analysts. A number of other terms have also been used increasingly since the 2010s, which could broadly be considered to expand the categories of deterrence by punishment and deterrence by denial.[10]

First, *deterrence through counternarrative and delegitimization* challenges terrorists' justification for violence. This often relies on challenging the religious and sociopolitical rationales that participants in violence may use to justify their actions. This approach focuses on communications and messaging that counter the adversary's narrative.[11] It would perhaps be unreasonable to expect this strategy to deter on its own, but it can make an important contribution. As Knopf highlights, support from a proportion of the population is often crucial to terrorist groups; if such groups expect key audiences will respond negatively to their activities, this may restrain their behavior.[12]

Cumulative deterrence builds on repeated victories, either physical or full spectrum, over the short, medium, and long terms that gradually wear down the enemy.[13] This is a deterrence strategy often ascribed to Israel in its conflict with Arab states and non-state groups. In these conflicts, successive Israeli victories eventually persuade the other side that they cannot defeat Israel, and thus are deterred from further attacks.

Self-deterrence[14] is an unwillingness to use military capabilities (or impose other costs) on an adversary, despite a declaratory threat to do so. This is not due to any external constraints but is voluntary. This could be the result of many factors, including proportionality, respective values, and legal constraints. Self-deterrence is often used with reference to state actors and the potential employment of nuclear weapons, given the significant effects this would have.

Finally, *deterrence by entanglement*[15] relies on the development of relationships and dependencies between actors, which increase the potential costs – and reduce the potential benefits – of aggressive action, as the aggressor would incur costs too. Given the size of China's economy, deterrence by entanglement is often described when discussing China's external relationships, but it can apply to other relationships.

Broadening of Deterrence Concepts

These developments in the conceptualization of deterrence have resulted in many authors and academics proposing that the classical concepts of deterrence be broadened to include a much wider scope of activity.[16] No real consensus has been achieved on this, but it is argued that rather than simply relying on threats of punishment to deter, deterrence strategies should also include inducements and reassurances intended to achieve compliance with the rules-based international system. Additionally, the use of all levers of national power, not just the military, and the role of soft power are also now considered to play a role in deterrence. Consequently, there has been a general blurring of "deterrence" with the terms "dissuasion" and "influence," but all seek to impact the activities of another actor. It bears remembering that deterrence has always had a reassurance element. As Schelling highlighted, an adversary has to know that, just as they should anticipate costs being imposed if they do carry out the action a state seeks to forbid, they should also be reassured that those costs will *not* be imposed if they do *not* carry out the action.[17]

The Practicalities: How to "Do" Deterrence

The changes in the international security landscape since the Cold War resulted in a slow evolution of Western deterrence thinking. Keith Payne's seminal 2001 book and accompanying article, both titled *The Fallacies of Cold War Deterrence and a New Direction*, continue to offer an invaluable starting point for thinking about deterrence planning.[18] While not directly using the term "strategic culture," Payne's publications challenge the assumption of "Cold War-style analyses of deterrence" that assumed all leaders to be rational actors that would engage in the same cost-benefit calculations.[19] Payne argued that this fails to take into account other factors, including "idiosyncratic leadership beliefs;" he also describes, in quite some detail, an approach for "tailoring deterrence policies," proposing a set of "areas of enquiry" that would provide a much improved "deterrence framework."[20]

In the mid-2000s, the US Department of Defense published a new Joint Operating Concept for Deterrence Operations (colloquially referred to as the DO JOC).[21] While not doctrine, which reflects "fundamental principles by which military forces guide their actions,"[22] the DO JOC "propose[s] solutions for innovative ways to conduct operations."[23] Further developing some of the key themes in Payne's publications, the DO JOC identified "a new concept for 'waging' deterrence" that would integrate all levers of national power – not just the military – and "develop strategies, plans and operations that are tailored to the perceptions, values, and interests of specific adversaries."[24] Another important contribution made by the DO JOC is the description of the adversary's decisionmaking calculus: a way to estimate the adversary's perceptions of a particular course of action.[25] For the small group of academics, researchers, and analysts (both civilian and military) that was still thinking about deterrence, developing tailored deterrence strategies for specific actors in specific contexts was gaining traction.

Tailored Deterrence

The term "tailored deterrence" appeared in an increasing number of papers and articles throughout the 2000s. This included the term being outlined by the Bush administration in the 2006 *Quadrennial Defense Review*, which described how the United States intended to "shift from 'one size fits all' deterrence toward more tailorable capabilities."[26] Elaine Bunn's 2007 paper on the topic provides an excellent summary that includes a list of what should be known about the group that one is trying to deter.[27]

This tailored approach recognizes that different adversaries, and adversary decision-makers, have different motives, rationales, and drivers, and thus any deterrent actions need to be tailored to specific actors in specific situations in order to maximize their chance of succeeding. Some authors have challenged the utility and highlighted the limitations of tailored deterrence, notably Larkin's 2011 paper "Cracks in the New Jar: the Limits of Tailored Deterrence."[28] One of Larkin's key criticisms of tailored deterrence is the "assumption that the US can reliably assess adversaries' decision calculus."[29] What must be acknowledged is that such assessments never claim to be 100% accurate. What tools such as decision calculus assessment do provide is a systematic way to record everything that is known about an adversary in order to estimate the drivers of their decisions. What is most important is that the improved understanding of adversaries that tailored deterrence planning provides decreases the risk of "mirror imaging:" of assuming that other decisionmakers think and act like us.[30]

Maximizing Understanding and Developing Deterrence Activities

The chances of deterrence succeeding are increased if a state maximizes its understanding of the adversary. It is difficult to predict with accuracy what any actor – even one a state believes it knows well – will do, but an increasing understanding of adversaries makes it possible to better identify what they seek to gain, what they fear they may lose, and what may be acceptable and unacceptable outcomes from their perspective.

This understanding can be broken down into two categories: first, gaining a broader understanding of the group's aims, objectives, general patterns of behavior, capabilities, and so on; in sum, overarching elements of their strategic culture.[31] These can be captured in an adversary profile that summarizes the key information. Second, context-specific detail regarding the immediate deterrence situation at hand can help operationalize this information. What is the adversary seeking to achieve by carrying out the action in question? This can be captured in a decision calculus, as described in the DO JOC.[32] What are the adversary's perceived costs and benefits of taking this action and, importantly, of not taking it?

This understanding of the adversary informs the development of a deterrence campaign or strategy. Planners will develop a range of candidate deterrence activities that are intended to influence the adversary's perceptions in the desired direction. This should include activities that will impose costs on the adversary (deterrence by punishment), usually involving holding the adversary's interests at risk; decreasing the adversary's perceived benefits (deterrence by denial); incentivizing restraint; and proposing alternative acceptable outcomes or solutions.

The Relevance of Strategic Culture

In the words of Johnson and Maines, "conceptualizing and operationalizing the study of strategic culture continues to be a complex undertaking."[33] A common criticism has been that, by their very definition, cultural models are so broad and ill-defined that they are unable to provide suitable granularity and specificity to aid foreign policy planners. Indeed, critics of strategic culture regularly argue that there is not even an agreed definition of the term. With this in mind, your author's preferred definition is that offered by Johnson, Kartchner, and Larsen:

> [Strategic culture is] that set of shared beliefs, assumptions and modes of behavior, derived from common experiences and accepted narratives (both oral and written),

> that shape collective identity and relationships to other groups, and which determine appropriate ends and means for achieving security objectives.[34]

Noting these criticisms, it is apparent that research has highlighted the valuable contributions that strategic culture can provide to understanding the "other." In the context of tailored deterrence, strategic culture must therefore be of some utility to the deterrence planner. In particular, work by Johnson, Kartchner, and Maines has provided an approach to identifying the dominant factors of a group's strategic culture, which can be used to inform foreign policy decisions – including deterrence strategy development.[35] This approach proposes four factors of group strategic culture, namely *identity*: "the character traits that a group assigns itself and the reputation it pursues"; *norms*: "accepted, expected, or customary behaviors within a society or group"; *values*: "deeply held beliefs about what is desirable, proper, and good [including] both secular and sacred values"; and *perceptual lens*: "the 'filters' through which individuals or group members determine 'facts' about themselves and others."[36]

Using this approach provides a useful insight into how the structured analysis of strategic culture might provide a coherent, standardized way to improve understanding of adversaries – while acknowledging that strategic culture is but one of many variables that inform group decisions. When this approach is fused with existing deterrence planning methods, significant improvements can be made to the ensuing deterrence strategies. The next section describes how an improved understanding of each of these four factors of strategic culture might contribute to some of the critical stages in deterrence strategy development, namely adversary profile development, decision calculus, and deterrence activity development.

The Adversary Profile

It has proven useful at the outset of deterrence strategy development to create a suitable profile of the adversary. This is an assessment tailored specifically to the requirements of deterrence planning and contains an array of material on the actor in question that is usually not found in one single source.[37] Typical content might include what is known of the adversary's overarching aims and objectives; their military (and any other significant) capabilities; the decisionmaking structure(s), including military command and control; and if possible, individual profiles of senior decisionmakers in the group. The "Deterrence Framework" proposed by Payne goes some way toward organizing this material, but a structured analysis of the adversary's strategic culture provides insights that may not automatically be apparent to the outside observer at first look.[38] This should include how factors such as geography, history (particularly of conflict and/or occupation), religious beliefs or other belief systems, and numerous others shape the adversary's views on defense and security, and the habits of practice the adversary has developed across its security institutions.

The Decision Calculus

As previous sections have described, one of the significant changes in deterrence thinking brought about by the adoption of tailored deterrence is the recognition that not all actors see the world the same way. Consequently, the aim of a deterrence campaign should be to influence the adversary's perceived costs and benefits of acting in the direction that a state desires. There is a clear overlap here with the understanding of the adversary perceptual lens that is proposed by Johnson and Maines in the Cultural Topography Analytic Framework.[39] The perceptual lens describes the "filter" through which members of the other group – the adversary – see the world, including their opponents and themselves.

Making the best estimate of the adversary's perceptual lens will begin with the development of the adversary profile, which should include general statements about the adversary's worldview and perceptions. However, these will be developed further, within the specific context of the deterrence scenario at hand, in the decision calculus. Analysts must make every effort, using all sources of information, to estimate what the adversary decisionmaker perceives as being the costs and benefits of acting – and of not acting – in the light of the understanding provided by the adversary profile.

Deterrence Activity Development

Possibly the most crucial benefit that improved understanding of strategic culture can provide to deterrence (and wider influence) planning is in the development and implementation of deterrence activities or courses of action. All the research and analysis captured in the decision calculus is purely to enable the development of candidate deterrence activities that have been assessed to be likely to influence the adversary's decision in the desired direction. These activities should use all levers of power, not just military (in other words, diplomatic, informational, and economic as well), and be effects-based: they are intended to achieve a cognitive effect.

In particular, having an understanding of the adversary's values – which are often strongly linked to identity – will increase the chances of developing effective deterrence activities. This is due to the role played by sacred values,[40] also referred to as protected values, which many groups hold.

The Importance of Sacred Values

Decisionmaking is generally assumed to involve cost-benefit calculations that help the decisionmaker achieve their goals. In general, if the costs outweigh the benefits, most people will adjust or even abandon their goals. This is known as "instrumental" decisionmaking.[41] Sacred values differ from instrumental values by incorporating moral, often religious, beliefs. These beliefs drive decisions and actions in ways that are not associated with the likelihood of the decisionmaker achieving their goals.[42]

Unlike instrumental values, sacred values include what anthropologist Scott Atran calls "inscrutable propositions that are immune to logic or empirical evidence."[43] Consequently, one of the defining properties of sacred values is the reluctance or even the refusal of their holder to negotiate or trade them.[44] Examples of sacred values include the status of cows in Hindu culture or Jerusalem in Judaism, Christianity, and Islam. As these examples show, sacred values often have their basis in religion, but secular values such as belief in fairness, reciprocity, or collective identity might also be considered sacred values.

How to Identify Sacred Values

Baron and Spranca propose that the defining property of sacred values is *absoluteness*, or the characteristic that people will resist trading-off sacred values for any form of compensation; in this sense, their sacred values are absolute. People are likely to refuse to discuss or enter into negotiations involving sacred values. In addition to absoluteness, Baron and Spranca propose a number of other properties that are present in most cases.

First, sacred values often reflect *quantity insensitivity*: the quantity of any consequence of a decision or action is irrelevant to holders of sacred values; the sacred value applies to the act itself, not the result. This can also apply to the likelihood of a consequence occurring: holders of sacred values may not accept any likelihood of a particular event occurring, no matter how slim. Second, sacred values are *agent-relative* (rather than agent-general), meaning that the participation of the decisionmaker involved is important, rather than what the actual consequences of the decision may be. Third, actions required by sacred values are seen as *moral obligations*; they are not personal preferences or conventions but are seen as compulsory behaviors. Fourth, the *denial of trade-offs* means that holders of sacred values generally tend to deny even the existence of real and sometimes significant policy trade-offs with their sacred values. Finally, the role of *anger* is pronounced as, perhaps crucially, people may become angry at the very thought of any violation of their sacred values. This is because, from their perspective, it would be a moral violation.[45]

Although sacred values are often derived from religious beliefs, research suggests that they can emerge around issues with relatively little historical background and significance; in particular, this can occur when they become linked to conflicts over group identity (another of the four variables Johnson and Maines recommend in the study of strategic culture).[46] The research published in 2009 into Iran's nuclear program agreed with this premise.[47] It concluded that, despite the program being a relatively recent development, some Iranians treat it as sacred. From this, it might be concluded that relatively recent sovereignty-related issues can become bound up with issues of collective identity and, in turn, become sacred or protected values.

Implications of Sacred Values

If material incentives are offered to someone in return for them compromising their sacred values, the likelihood of violent opposition often increases.[48] People will not only reject any type of compensation for dropping their commitment to the sacred values, but they will defend such values regardless of the costs.[49] Indeed, some groups are willing to endure very high costs and for long periods of time.[50] It is this absolutist view that runs directly counter to theories of rational choice and rational play in negotiations. Consequently, in conflicts where one of the parties involved holds sacred values, standard approaches to negotiations are highly likely to backfire. Material offers including money and other incentives will be interpreted as morally taboo and insulting.[51] For holders of sacred values, even contemplating proposals to exchange those values for secular ones is unacceptable. That very idea is something to be condemned if one is the observer and something to be concealed if one is the decisionmaker.[52]

Implications for Deterrence Planning

While the academic research into sacred values does not, on the whole, directly address deterrence and influence issues, there are clear implications for this policy domain. As has been discussed, deterrence often relies on the threat of imposing costs on an actor; in other words, deterrence by punishment. In the context of armed conflict, this frequently refers to the destruction or other form of attack on specific items of property or territory.

A wide variety of things may be considered sacred; they may be abstract or ideational, for example, the concepts of equality or freedom; or physical, such as the status of Jerusalem to Jews or Mecca to Muslims. The status of any of these things is, to the holder of such a sacred value, nonnegotiable. As Atran highlights, "standard business-style negotiations in such ... conflicts will only backfire."[53] Therefore, in a deterrence context, it can be deduced that any threat to such a sacred value will not achieve the desired effect; in fact, multiple pieces of research conclude that such threats can result in anger and violence.[54] Unintended escalation is likely to result.

Additionally, many deterrence and coercion strategies seek to achieve the desired objective by imposing unacceptable costs on an adversary; however, some people and groups are willing to endure very high costs and for long periods to defend their sacred values.[55] In such a situation, it may be very difficult, if not impossible, to impose sufficient costs on an adversary to make them change their behavior. The consequences of deterrence and influence planning are clear. If the deterrence objective concerns an adversary's sacred values, or if a deterrence strategy involves threatening any beliefs or physical objects that the adversary holds as sacred, then the strategy is likely to fail and, as noted, may result in unintended escalation.

This emphasizes the value of understanding an adversary's strategic culture and, crucially, any sacred or protected values the group may hold. If it is established that an adversary holds such sacred values, then alternative approaches and engagement strategies may produce outcomes that are more favorable. Such strategies may include a different emphasis to achieve deterrence, focusing more on deterrence by denial, delegitimization and counternarrative, or any other suitable means. Furthermore, consideration of other strategies such as containment and/or reassurance, if applicable, may be merited. Finally, alternative approaches to conflict resolution might be pursued, which should involve attempts to identify suitable "off-ramps," or outcomes that are acceptable to all parties.

Ways to Engage with Holders of Sacred Values

Improving understanding of other groups' sacred values may help to resolve, and even avoid, conflict. There are a number of ways in which sacred values might be understood and, subsequently, how deterrence strategies may be adjusted accordingly.[56] Although they may not be applicable to some deterrence scenarios, two bear highlighting for their potential to de-escalate tensions and build relationships.

The first is what Atran refers to as "symbolic concessions."[57] This involves identifying something that is sacred to the other group but that means little to one's own side. Atran uses the example of "ping-pong diplomacy," in which the United States lost match after match of table tennis to China in 1971. This provided something of great symbolic value to China, where table tennis is a sport of national prestige, at little cost to the United States.[58]

Another important act highlighted by Atran is the simple apology.[59] He accepts that, in international politics, apologies by themselves may not be dealmakers, but are more of a means of facilitating compromise. A symbolic gesture such as an apology can redefine the scope and limits of subsequent engagement and possible material transactions. There are a number of pitfalls to avoid – notably that qualified apologies can be worse than no apology, and without the acceptance of responsibility, apologies do not work – but an apology can be an important starting point to the development (or rebuilding) of trusted relationships.

Conclusion

Strategic culture has long been criticized for its lack of definition and the difficulty in applying it meaningfully to foreign and defense policy challenges. Similarly, one of the challenges put to deterrence in the twenty-first century, and specifically to tailored deterrence, is the inherent difficulty ascribed to "knowing" the other. It seems long overdue to bring these two disciplines together, notwithstanding previous important contributions from scholars, including Adamsky and Lantis, among others.[60]

As Colin Gray did so well, it is essential to acknowledge the limitations of both practices. Strategic culture is by no means a silver bullet and will be just one of a number of variables that informs a group's defense and security decisionmaking. Deterrence, as Gray put it, is "inherently unreliable" and thus one must be able to tolerate any failures of deterrence that may occur.[61] But what strategic culture can offer to the deterrence planner – and a wide array of defense professionals – is a rich and potentially powerful means of improving understanding of other groups, which should be maximized to inform our engagement strategies. This chapter has identified a number of ways in which this can support tailored deterrence strategy development and influence, but it seems clear that strategic culture can also support a wide range of foreign policy activities.

There is still some way to go to bring strategic culture into mainstream foreign policy and defense activities; notwithstanding, the structured approach developed by Johnson and Maines appears to be a significant step in operationalizing the discipline and making it more "user friendly." More research and analysis is required to fully test and explore the utility of strategic culture in deterrence and wider influence strategy development; for example, in improving our ability to understand which elements of strategic culture drive specific policy and strategy decisions, and identifying group sacred values. Nevertheless, your author believes that continuing to develop the concept of strategic culture into an applicable tool – albeit one of many for foreign policy specialists – has the potential to improve diplomatic, foreign affairs, and military engagements at a relatively low cost.

Notes

1 This chapter represents the views and analysis of the author, and should not necessarily be construed as representing the UK government, or any of its agencies.

2 UK Government, *Securing Britain in an Age of Uncertainty: The Strategic Defence and Security Review*, Cm 7948 (October 2010), pp. 37–40.

3 Freedman and Karsh, *The Gulf Conflict* (London, UK: Faber and Faber, 1993), p. 257

4 Schelling, T., *Arms and Influence* (Yale University Press, 1 Jan 1966).

5 Robert Jervis, "Deterrence Theory Revisited," *World Politics* Vol. 31, No. 2 (January 1979).

6 Published over a decade ago, Jeffrey Knopf's article provides an excellent review of developments in deterrence thinking since the end of the Cold War; Jeffrey Knopf, "The Fourth Wave in Deterrence Research," *Contemporary Security Policy* Vol. 31, No. 1 (April 2010), p. 1.
7 See, for example, Glenn Snyder, *Deterrence by Denial and Punishment*, Research Monograph No. 1 (1959) and *Deterrence and Defense: Toward a Theory of National Security* (1961).
8 Lawrence Freedman, *Deterrence* (Cambridge, UK: Policy Press, 2004), quoted in M. Gerson, "Conventional Deterrence in the Second Nuclear Age," *Parameters* (Autumn 2009).
9 Knopf, p. 1.
10 As argued by Tim Sweijs and Samuel Zilincik, "The Essence of Cross-Domain Deterrence," in Osinga & Sweijs (eds.), *NL ARMS Netherlands Annual Review of Military Studies 2020: Deterrence in the 21st Century-Insights from Theory and Practice* (Springer, 2020), p. 147.
11 Knopf, p. 10.
12 Ibid., p. 18.
13 Doron Almog, "Cumulative Deterrence and the War on Terrorism," *Parameters*, Vol. 34, No. 4 (Winter 2004/2005), p. 6.
14 See, for example, T.V. Paul, "Self-deterrence: Nuclear Weapons and the Enduring Credibility Challenge," *International Journal: Canada's Journal of Global Policy Analysis*, Vol. 71, No. 1 (first published online 30 November 2015), pp. 20–40.
15 For example, see Joseph S. Nye Jr., "Deterrence and Dissuasion in Cyberspace," *International Security*, Vol. 41, No. 3 (Winter 2016/2017), pp. 58–60.
16 Useful examples include De Spiegeleire, Holynska, Batoh, and Sweijs, *Reimagining Deterrence: Towards Strategic (Dis)Suasion Design*, The Hague Centre for Strategic Studies (March 2020); and Davis, O'Mahony, Curriden, and Lamb, *Influencing Adversary States: Quelling Perfect Storms*, RAND Corporation (2021). We should recall that, writing in 1974, George and Smoke emphasized that "deterrence should be viewed not as a self-contained strategy, but as an integral part of a broader, multifaceted influence process." George and Smoke, *Deterrence in American Foreign Policy: Theory and Practice* (Columbia University Press, 1974), p. 591.
17 Schelling, *Arms and Influence* (1966).
18 Keith Payne, *The Fallacies of Cold War Deterrence and a New Direction* (University Press of Kentucky, 2001) and "The Fallacies of Cold War Deterrence and a New Direction," *Comparative Strategy* Vol. 22, No. 5 (2003), pp. 411–428.
19 Payne, "The Fallacies of Cold War Deterrence and a New Direction," *Comparative Strategy*, p. 411.
20 Ibid., pp. 424–425.
21 US Department of Defense, *Deterrence Operations Joint Operating Concept*, Version 2.0 (December 2006).
22 Taken from UK Ministry of Defence, *Developing Joint Doctrine Handbook*, Fourth Edition (November 2013); this publication is now withdrawn by UK MOD but the definition is still valid.
23 US DoD, *DO JOC*, p. iii.
24 Ibid., p. 3.
25 Ibid., p. 5.
26 US Department of Defense *Quadrennial Defense Review Report* (6 February 2006), pp. 4 and 49.
27 Elaine Bunn, *Can Deterrence Be Tailored?*, National Defense University Institute for National Strategic Studies Strategic Forum No. 225 (Washington, DC, 2007), 3, https://www.files.ethz.ch/isn/31364/SF225%20new.pdf
28 Larkin, S.P., "Cracks in the New Jar: The Limits of Tailored Deterrence," *Joint Force Quarterly* 63 (October 2011), pp. 47–57.
29 Ibid., p. 7.
30 Mirror imaging can be defined as "filling gaps in … knowledge by assuming that the other side is likely to act in a certain way because that is how (we) would act under similar circumstances." CIA, *Psychology of Intelligence Analysis*, accessed online 5 November 2019.
31 One of the challenges leveled at strategic culture is the lack of an agreed definition for the term; this author's preferred definition is provided in the next section.
32 US DoD, *DO JOC*, pp. 5, 19–20, 47.

33 Johnson and Maines, "The Cultural Topography Analytic Framework," in Jeannie L. Johnson, Kerry M. Kartchner, and Marilyn J. Maines (eds.), *Crossing Nuclear Thresholds: Leveraging Sociocultural Insights into Nuclear Decisionmaking* (Palgrave Macmillan, 2018), p 9.
34 Jeannie L. Johnson, Kerry M Kartchner and Jeffrey A. Larsen, *Strategic Culture and Weapons of Mass Destruction: Culturally Based Insights into Comparative National Security Policymaking* (Palgrave Macmillan, 2009), p. 9.
35 Jeannie L. Johnson, "Conclusion: Toward a Standard Methodological Approach," in Johnson, Kartchner, and Larsen (eds.), pp. 243–255; and Johnson and Maines, pp. 40–42.
36 Johnson, Kartchner, and Larsen, op cit; Johnson and Maines, op cit. This approach is based and expands upon the Cultural Topography methodology originally proposed by Johnson and Berrett: Jeannie L. Johnson and Matthew T. Berrett, "Cultural Topography: A New Research Tool for Intelligence Analysis," *Studies in Intelligence,* Vol. 55, No. 2 (Extracts, June 2011), pp. 1–22.
37 US DoD, *DO JOC*, p. 29.
38 Payne, "The Fallacies of Cold War Deterrence and a New Direction" (2003), pp. 424–425.
39 Johnson and Maines, op cit.
40 A concept introduced to the author by Kerry Kartchner, who has amassed much of the research on this angle of deterrence theory.
41 Ginges, Atran, Medin, and Shikaki, "Sacred Bounds on Rational Resolution of Violent Political Conflict," *Proceedings of the National Academy of Sciences* 104 (2007), pp. 7357–7360.
42 Atran, Axelrod, and Davis, "Sacred Barriers to Conflict Resolution," *Science* 317 (2007), pp. 1039–1040.
43 Scott Atran, *Talking to the Enemy*, Penguin (2010), p. 38.
44 Baron & Spranca, "Protected Values," *Organizational Behavioral and Human Decision Processes*, Vol. 70 (1997), pp. 1–16.
45 Ibid., pp. 4–6.
46 Atran, p. 381.
47 Dehghani, Iliev, Sachdeva, Atran, Ginges and Medin, "Emerging Sacred Values: Iran's Nuclear Program', *Judgement and Decision Making*, Vol. 4, No. 7 (December 2009), pp. 930–933.
48 Ginges, Atran, Medin, and Shikaki, op cit.
49 Atran, p. 375.
50 Varshney, "Nationalism, Ethnic Conflict, and Rationality," *Perspectives on Politics* Vol. 1, No. 1 (March 2003).
51 Atran, p. 377.
52 Tetlock, Lerner, and Peterson, "Revising the Value Pluralism Model: Incorporating Social Content and Context Postulates," in Seligman, Olson, and Zanna (eds.), *The Psychology of Values: The Ontario Symposium, Volume 8* (Erlbaum: Hillsdale, NJ, 1996), pp. 25–51.
53 Atran, p. 377.
54 Baron and Spranca, op cit; Atran, op cit; Ginges, Atran, Medin, and Shikaki, op cit.
55 Varshney, op cit.
56 Ibid., pp. 388–389.
57 Atran, p. 379.
58 Ibid., p. 388.
59 Ibid., pp. 389–392.
60 Dmitry Adamsky, "From Moscow with Coercion: Russian Deterrence Theory and Strategic Culture," *Journal of Strategic Studies*, Vol. 41, No. 1 (July 2017), pp. 1–28 and Jeffrey S. Lantis, "Strategic Culture and Tailored Deterrence: Bridging the Gap between Theory and Practice," *Contemporary Security Policy,* Vol. 30 No. 3 (December 2009), pp. 467–485.
61 Colin Gray, "Deterrence and the Nature of Strategy," *Small Wars and Insurgencies* (2000).

Selected Bibliography

Adamsky, Dmitry, "From Moscow with Coercion: Russian Deterrence Theory and Strategic Culture," *Journal of Strategic Studies* 41(1) (July 2017).

Almog, Doron, "Cumulative Deterrence and the War on Terrorism," *Parameters* 34(4) (Winter 2004/2005).
Atran, Axelrod, and Davis, "Sacred Barriers to Conflict Resolution," *Science* 317 (2007): 1039–1040.
Atran, Scott, *Talking to the Enemy*, Penguin (2010).
Baron and Spranca, "Protected Values," *Organizational Behavioral and Human Decision Processes*, 70 (1997): 1–16.
Bunn, Elaine *Can Deterrence Be Tailored?*, National Defense University Institute for National Strategic Studies Strategic Forum No. 225 (Washington, DC, 2007), https://www.files.ethz.ch/isn/31364/SF225%20new.pdf.
Davis, O'Mahony, Curriden, and Lamb, *Influencing Adversary States: Quelling Perfect Storms*, RAND Corporation (2021).
De Spiegeleire, Holynska, Batoh and Sweijs, *Reimagining Deterrence: Towards Strategic (Dis) Suasion Design*, The Hague Centre for Strategic Studies (March 2020).
Dehghani, Iliev, Sachdeva, Atran, Ginges and Medin, "Emerging Sacred Values: Iran's Nuclear Program," *Judgement and Decision Making* 4(7) (December 2009).
Freedman and Karsh, *The Gulf Conflict*, London, UK: Faber and Faber (1993).
George and Smoke, *Deterrence in American Foreign Policy: Theory and Practice*, Columbia University Press (1974).
Ginges, Atran, Medin, and Shikaki, "Sacred Bounds on Rational Resolution of Violent Political Conflict,", *Proceedings of the National Academy of Sciences* 104 (2007): 7357–7360.
Gray, Colin, "Deterrence and the Nature of Strategy," *Small Wars and Insurgencies* (2000).
Jervis, Robert, "Deterrence Theory Revisited," *World Politics* 31(2) (January 1979).
Johnson and Maines, "The Cultural Topography Analytic Framework," in *Crossing Nuclear Thresholds: Leveraging Sociocultural Insights into Nuclear Decisionmaking*, Jeannie L. Johnson, Kerry M. Kartchner, and Marilyn J. Maines eds., Palgrave Macmillan (2018).
Johnson, Jeannie L., Kerry M. Kartchner, and Jeffrey A. Larsen, *Strategic Culture and Weapons of Mass Destruction: Culturally Based Insights into Comparative National Security Policymaking*, Palgrave Macmillan (2009).
Knopf, Jeffrey, "The Fourth Wave in Deterrence Research," *Contemporary Security Policy* 31 (1) (April 2010).
Lantis, Jeffrey S. "Strategic Culture and Tailored Deterrence: Bridging the Gap between Theory and Practice," *Contemporary Security Policy* 30(3) (December 2009): 467–485.
Larkin, S.P., "Cracks in the New Jar: The Limits of Tailored Deterrence," *Joint Force Quarterly* 63 (October 2011).
Nye, Jr., Joseph S., "Deterrence and Dissuasion in Cyberspace," *International Security* 41(3) (Winter 2016/17): 58–60.
Paul, T.V., "Self-deterrence: Nuclear Weapons and the Enduring Credibility Challenge," *International Journal: Canada's Journal of Global Policy Analysis* 71(1): 20–40 (first published online 30 November 2015).
Payne, Keith, "The Fallacies of Cold War Deterrence and a New Direction," *Comparative Strategy* 22(5) (2003): 411–428.
Payne, Keith, *The Fallacies of Cold War Deterrence and a New Direction*, University Press of Kentucky (2001).
Schelling, Thomas C., *Arms and Influence*, Yale University Press (1966).
Snyder, Glenn, *Deterrence by Denial and Punishment*, Research Monograph No. 1, Princeton: Woodrow Wilson School of Public and International Affairs (1959).
Snyder, Glenn, *Deterrence and Defense: Toward a Theory of National Security*, Princeton University Press (1961).
Sweijs, Tim and Samuel Zilincik, "The Essence of Cross-Domain Deterrence" in *NL ARMS Netherlands Annual Review of Military Studies 2020: Deterrence in the 21st Century-Insights from Theory and Practice* Osinga & Sweijs eds., Springer (2020).
Tetlock, "Thinking the Unthinkable: Sacred Values and Taboo Cognitions," *Trends in Cognitive Science* 7 (2003): 320–324.
Tetlock, Lerner, and Peterson, "Revising the Value Pluralism Model: Incorporating Social Content and Context Postulates," in *The Psychology of Values: The Ontario Symposium, Volume 8* Seligman, Olson & Zanna eds., Hillsdale, NJ: Erlbaum (1996).

UK Government, *Securing Britain in an Age of Uncertainty: The Strategic Defence and Security Review*, Cm 7948 (October 2010).
UK Ministry of Defence, *Developing Joint Doctrine Handbook*, Fourth Edition (November 2013).
US Department of Defense *Quadrennial Defense Review Report* (6 February 2006).
US Department of Defense, *Deterrence Operations Joint Operating Concept*, Version 2.0 (December 2006).
Varshney, "Nationalism, Ethnic Conflict, and Rationality," *Perspectives on Politics* 1(1) (March 2003).

31

STRATEGIC CULTURE AND THE US INTELLIGENCE COMMUNITY

Roger Z. George and James J. Wirtz[1]

At a conference on analysis sponsored by the Central Intelligence Agency (CIA) in the mid-2000s, an invited senior investment manager made an interesting assertion about her firm's prowess at predicting markets: "We are so good at foreign market forecasts because our financial analysts carry 60 different passports." At that moment, it was clear how culture limited the ability of not only the CIA but also the entire US Intelligence Community (IC) to understand foreign governments' behaviors. Every officer in the IC carries an American passport – one of the prerequisites for obtaining a security clearance is a willingness to give up any pretense of dual citizenship, especially holding onto a foreign passport. The fact that the IC does not allow its employees to conduct this common international business practice demonstrates how strategic and organizational culture shapes the IC's ability to understand developments in foreign lands. On the one hand, the ban on holding several passports represents a commitment to protect and support US national interests; on the other hand, it can hamper the IC's ability to understand the policies and goals of other states. Strategic culture comes with preferences and cognitive biases that affect how the IC collects and analyzes information. It bounds what is perceived as permissible, desirable, and expected out of the full panoply of what is possible.

This chapter describes how US intelligence is built to operate within an American way of thinking about foreign affairs. It will also delve into how the intelligence enterprise – comprising 18 separate agencies spread across half a dozen federal departments – reflects and reinforces a collection of individual organizational cultures that also shape the operation and effectiveness of intelligence. Put somewhat differently, the relationship between an American strategic culture and the IC might be pictured as a series of matryoshka dolls nestled inside each other. The outer, largest doll represents the broad American strategic culture. Inside that doll rests another representing the American style of intelligence that is shared across the IC, which collectively differentiates US intelligence organizations from British or French intelligence agencies that conduct similar functions. Then come smaller dolls that might be unique to the CIA – or any of the other 17 national intelligence agencies like the Department of Defense (DOD), Defense Intelligence Agency (DIA) or the State Department's Bureau of Intelligence and Research (INR). There are even tinier dolls within large organizations like the CIA reflecting its separate directorates with distinct missions of

DOI: 10.4324/9781003010302-35

collection or analysis – coexisting and competing subcultures within the broader strategic culture that take on their own organizational personalities. Strategic culture and the multiple organizational cultures that exist within it become apparent only in a relative sense. Observers can only identify their own cultural milieus by noticing the differences contained in the strategic and organizational culture of others.

A complete explanation of how strategic and organizational culture affects the performance of intelligence agencies would require a book, not a chapter. Nevertheless, the IC is clearly shaped by American strategic culture, and its own competing cultures influence and magnify that strategic culture in important ways. To simplify matters and to cope with the issue of "cultural relativity," the CIA will serve as the touchstone not only to illustrate how the IC is nestled within American strategic culture but also to highlight differences across the organizational cultures of other "three letter" agencies.

Getting Past Definitions

Few intelligence practitioners are well versed in the growing intelligence studies literature regarding strategic culture. If asked what a strategic culture is, intelligence officers would probably look puzzled and could not recite any specific definition. Instead, they might say that strategic culture is perhaps the national character of a state or the common narrative that foreign government officials hold in their collective minds. Or, they might respond that it is how an opponent – like Russia or China – thinks about its core interests and preferred national security strategies. For intelligence professionals something close to Jack Snyder's early definition of strategic culture likely would suffice:

> Strategic culture is a set of general beliefs, attitudes, and behavior patterns regarding strategy that has achieved a state of semi-permanence that places them on the level of "cultural" rather than just policy.[2]

Most practitioners – not being dogmatic neorealists who believe that states consistently act to maximize power – also would be prepared to accept that each state behaves somewhat differently depending on its unique history, geography, political system, and many other factors. Indeed, when intelligence analysts are assessing the possible courses of action that a foreign government or leader might take, they typically have to integrate into their own cognitive processes a wide variety of factors (see Figure 31.1).

Most analysts would agree, for example, that Russia's national narrative of its past empire, its Great Patriotic War against Nazism, and the Soviet Union's ignominious collapse have been important to President Putin and his advisers. In addition, however, intelligence officers acknowledge that other inputs matter as well, such as the influence and organizational interests of key institutions like the Russian military, its intelligence services, and the foreign ministry. Recognizing that these strategic and organizational cultures are semipermanent also implies that both can change over time, often spurred by major events like won or lost wars and other major geopolitical changes or shocks. Indeed, in the American context, there is a long-standing debate about the exact nature, stability, and influence of an "American way of war" on US battlefield performance and strategic preferences.[3] One also thinks immediately of post-1945 West Germany, which eschewed its history of militarism, subordinated its citizen soldiers within the North Atlantic Treaty Organization (NATO) military framework, and allowed its Foreign

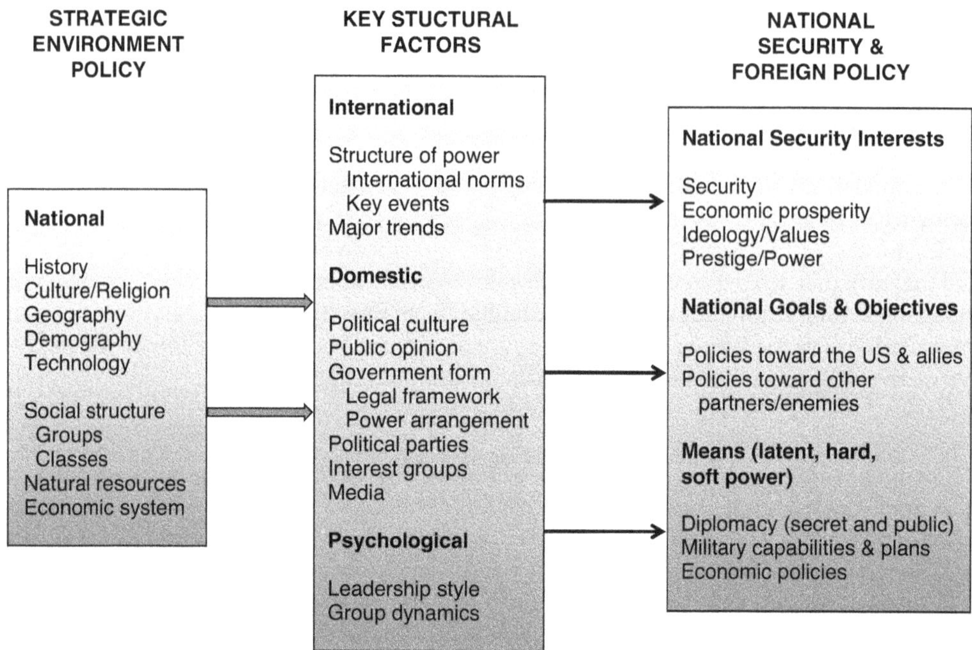

Figure 31.1 Determinants of Foreign Actors' National Security Policy.

Source: Author's Creation.

Minister – vice the weaker Defense Minister – to be the principal national security decisionmaker. What is tremendously hard for American intelligence officers to determine – having not lived in those foreign cultures – is how strong or salient these presumed strategic cultures are and exactly how they might limit the perspectives of foreign elites and leaders. No amount of linguistic skill, travel, or overseas assignments can fully replicate the experience of living and thinking within those foreign cultures. While analysts may be aware of these different strategic cultures, they are also surrounded and constrained by their own American style of foreign policymaking.

Strategic Culture's Impact on US Intelligence

Other scholars have described American strategic culture more completely than is possible in this brief chapter.[4] Key elements include its once revolutionary democratic form of governance, the century of expansion westward that expelled overseas foreign presence on the North American continent, and the achievement of great power status in the twentieth century. In addition to that early history, winning two world wars as well as creating a postwar liberal international order, which replaced an earlier "isolationist" sentiment that became untenable in the face of fascism and later international communism, have become part of American strategic culture.[5] American intelligence was born alongside the strategy of US containment of the Soviet Union, which became an almost Manichean battle of good versus evil – hence President Ronald Reagan's moniker of the "evil empire" to capture the essence of the Soviet Union. The CIA thus became a key institution for combatting communism, extending American influence, and protecting the nation's vital interests.

American "exceptionalism" – the notion that American democracy is unique and justifies Washington's efforts to project its so-called universal values – also drove the recruitment of America's elite to its national security agencies, including intelligence. One is reminded that the nation's first post-1945 strategists and intelligence officers were mostly drawn from the same Ivy League schools and held similar views of the challenges that America faced. The phrase "white, male, and Yale" has long been associated with American intelligence, especially the CIA.[6] For example, Sherman Kent, the Yale historian who served in the Office of Strategic Services and later in the CIA, wrote the first major book on US intelligence and is considered a father of the CIA's analytic culture. He was also a carpool-mate and National War College faculty colleague of George Kennan, the influential writer of America's containment strategy.[7] The roster of the CIA Hall of Fame – termed Trailblazers[8] – lists many Harvard, Yale, and Columbia graduate and law school alumni who first built and then ran the CIA throughout the early Cold War years. Suffice it to say, intelligence officials were as imbued with the notion of America's unique international role as were the likes of George Marshall, Dean Acheson, and Paul Nitze.

One way to explain how the US intelligence culture emerged from and was shaped by American strategic culture is to acknowledge that intelligence by law is designed to serve the president and his advisers who sit on the National Security Council (NSC). The 1947 National Security Act – periodically amended to reflect changes in American policies and governmental structure – dictates that the senior intelligence leadership will advise the president and the NSC on all intelligence matters. Since the 1980s, every president from Reagan to the present also has issued a National Security Strategy. Much of the ethos of an American strategic culture is embedded in those documents. Virtually every National Security Strategy defines US national interests in a similar fashion: physical security, economic prosperity, preservation of national values, and a stable international order. They also embody a distinctly Mahanian view of a world in which American interests and power are global in reach and appeal. Critics of various twists and turns in American foreign policy have often complained that these strategy documents contain too much moralism or legalism. Some presidents and administrations are more explicit about the "exceptional" nature of America's role. At the end of the Cold War, President Clinton defined America's role as the "enlargement" of democracy. Later, President George W. Bush's 2006 National Security Strategy was equally expansive: "It is the policy of the United States to seek and support democratic movements and institutions in every nation and culture, with the ultimate goal of ending tyranny in the world."[9] This messianic strain in presidential national security documents waxes and wanes depending on the state of US domestic and foreign policies. This crusading impulse can be tempered following defeats like Vietnam or Afghanistan, but the urge to reengage with the world is almost always there, as is the impulse to take a leadership role, especially when it comes to "fixing" other nations' foreign policies, politics, and even traditional social relationships.

This strategic culture has a profound impact on American intelligence. First, given America's prominent role in global affairs, US intelligence must maintain what it views as global coverage, a luxury few foreign intelligence services can afford. Especially in the CIA, there is a view that at any moment a crisis or opportunity in any part of the world may excite a president or his advisers who will be expected to be briefed on those happenings. Accordingly, the US IC is expected to warn about potential threats emerging from parts of the world that the White House has defined as critically important, be they in Europe, Asia, the Middle East, or other regions while responding to the "crisis du jour." The Director of National Intelligence (DNI) as head of the IC presents an unclassified Annual Worldwide Threat Assessment brief to the intelligence oversight committees of Congress each February that recites a long list of potential problems.[10] This unclassified 40–50-page presentation, along with the testimonies of other intelligence chiefs from the State and Defense Department as well as the Federal Bureau of Investigation (FBI), informs the public of the wide range of issues that intelligence is following to alert the congressional and executive branches of any danger facing the nation. A citizen listening to or reading the testimonies would be impressed with the scope of the problems facing the policy and intelligence communities. This public display is somewhat unique both in its candor and its scope. Few other foreign intelligence services can presume to follow so many issues in such detail as the US IC, and by contrast, intelligence and defense communities operating within other strategic cultures are usually tight-lipped about threat assessments, probably out of a shared belief that it is not wise to "tempt fate" by talking about troublesome neighbors or defense vulnerabilities. In any event, this global intelligence focus reinforces the notion that the United States is a superpower with worldwide interests and reach – all part of the American strategic culture that has persisted over time.

Second, this belief that the United States must have the best and most comprehensive intelligence of any nation has driven up the size, complexity, and cost of US intelligence. What began as a modest intelligence function in 1945 – mostly limited to the military services – evolved into a sprawling "enterprise" that now includes half a dozen large agencies inside the Pentagon, a large and independent CIA, along with more modest intelligence units within the departments of State, Justice, Treasury, Commerce, and Homeland Security. Commensurate with the expanding size of the IC, the national intelligence budget is one of the largest, behind that of the DOD. The sheer size, multiplicity of agencies, and the inevitable competition for resources also explain the clashes among the many organizational cultures found inside the IC.

Third, the post-1945 US IC only emerged reluctantly within a strategic culture that saw democracy and intelligence as almost antithetical to each other. While fighting a revolutionary war to be free from the corrupt power politics of Europe, General Washington was also the first US "spymaster" dispensing funds for monitoring British troops and undermining their plans. Nevertheless, secret diplomacy was condemned in the United States and spying was dismissed as late as the early twentieth century, despite the fact that most great powers had been conducting espionage for centuries. "Gentlemen do not read each other's mail," was Secretary of State Henry L. Stimson's comment about his decision in 1929 to end the US government's practice of intercepting, deciphering, and reading foreign cable traffic.[11] Thus, the creation of the CIA in 1947 was undertaken with trepidation, and the US intelligence culture reflects deep suspicion within the body politic about its misuse and danger to American freedoms. No comparable caution is seen in many foreign intelligence services, where intelligence can be directed toward foreign adversaries as well as domestic opponents – for example, in Russia and China and even in a few American allies. The impact of this suspicion is that the US intelligence culture contains a high degree of accountability and oversight within all three branches of government. For example, there are inspectors general in all intelligence agencies, a presidentially directed Intelligence

Oversight Board, two congressional oversight committees, as well as a special Foreign Intelligence Surveillance Court that must approve domestically collected intelligence on US citizens under very specific conditions and for limited time frames. While far from perfect, such oversight and adherence to the law has been engrained and enculturated into the agencies and officers serving in the IC.

While the broadly defined US strategic culture has shaped the US intelligence culture in some unique ways, the IC does share at least one commonality with virtually every other foreign intelligence service: intelligence culture begins with secrets. The ability to create and selectively share secret information is often treated as the raison d'être of intelligence agencies. As Michael Herman wrote decades ago: "Within the intelligence system itself secrecy shapes the organizational culture at all stages; handling of secret material determines procedures and sets attitudes even when this analysis is based largely on open data."[12] US intelligence is no different with respect to its efforts to protect "sources and methods," especially when agents' lives are at stake. Moreover, like other services, CIA case officers operate undercover to preserve their anonymity and hide their true purposes from hostile counterintelligence efforts. Within the intelligence literature, there is also a lively debate over whether all data used to inform policymakers is considered "intelligence" or whether the definition should be limited to only classified information.[13] In an era of ubiquitous information, it would seem that secrets are becoming a smaller proportion of the universe of intelligence sources. Nevertheless, the operational security demands of the classified world still shape the organizational cultures of most intelligence services with a few exceptions.

Like other strategic cultures, America's tends to shape the perceptual lens that intelligence officers use to observe and assess the outside world. These perceptual lenses, more than reality itself, color American intelligence assessments. Historical narratives, experiences, and beliefs can lead anyone – whether as decisionmakers or intelligence analysts – to filter out information that is discordant with those notions and put added weight on those that fit their own preconceived ideas about rational behavior on the part of foreign leaders and governments.[14] Hence, the CIA dismissed the "crazy" idea that Moscow would place missiles in Cuba in 1962, because it was inconsistent with past behavior and would obviously provoke such a strong US reaction that Moscow should never have considered such a risky gambit.[15] In a similar vein, Bush advisers and most US intelligence analysts could not conceive why Saddam Hussein would not try to hide weapons of mass destruction (WMD) in 2002 when he did it so cleverly up to the 1990 Gulf War.[16] Such strong mindsets are difficult to dislodge, partly because they are unconscious and so deeply embedded that they are hard to identify, much less root out. These mindsets also have a way of reinforcing themselves in the intelligence process. Analysts and their assessments are warmly embraced by policymakers wanting confirming information about their suspicions of an adversary. Disconfirming information is seldom explicitly requested. Likewise, intelligence collectors are rewarded for providing reports that can be used as proof of an adversary's perfidy. Collectors themselves also can be biased in the way they view a target. Some former analytic officers have commented to the authors that senior Directorate of Operations (DO) managers who have spent a lifetime wrestling against the Soviet Committee for State Security (KGB), whose successor agencies include the modern Russian Foreign Intelligence Service (SVR), tend to have a far darker view of the once Soviet and now Russian capabilities and intentions than their analytic counterparts.[17]

Unique Elements of US Intelligence Culture

Just as there is a unique American strategic culture, US intelligence since 1945 has developed several distinguishing characteristics. First, the DNI serves as the president's senior intelligence advisor and participates in deliberations of the NSC; however, the DNI is not coequal to the Secretaries of Defense and State or the National Security Advisor. In the American system, the DNI or any other IC leader *is not* a decisionmaker or policymaker. Intelligence officials, unlike other NSC members, are not formally asked their opinions about which policies or courses of action a president should select. A few presidents have even asked intelligence officials to brief the NSC and then leave the room when the policy was discussed. It is more typical for a DNI or CIA Director to remain at the table as policymakers discuss a potential course of action; however, it is also understood that there is a bright line between giving intelligence advice and venturing into policy recommendations, which is against the ethos of the US IC.

This is a two-way street. Policymakers have their own norms to follow and generally remain aloof from the analytical process, avoiding participating in or influencing intelligence assessments. This arms-length relationship is intended to prevent politicization of intelligence, whereby policymakers overtly or inadvertently shape analyses to meet their preferences.[18] It is insightful to contrast this clear separation with the British system of having foreign office and defense ministry officials join the Joint Intelligence Committee proceedings to hammer out assessments that reflect the combined views of both policy and intelligence officers. In the American system, the effect of separating intelligence from policy so clearly is designed to protect the integrity and objectivity of intelligence. Yet, the reality of close working relations between producers and consumers of intelligence often blurs this line, and charges of politicized intelligence are not infrequent – when intelligence has an impact and someone's ox is gored, it often leads to accusations of all sorts of wrongdoing.[19]

A second feature that is somewhat novel is how the US intelligence enterprise distributes its functions across different parts of the IC. Collection efforts that produce HUMINT (human intelligence) are found in the CIA and the DIA, the NSA produces SIGINT (signals intelligence), the National Reconnaissance Office (NRO) produces IMINT (imagery intelligence) while operating SIGINT satellites, and OSINT (open-source intelligence) is housed inside CIA's Open-Source Enterprise, which operates as a community-wide service. In addition, the DIA oversees the smaller and exotic MASINT (measurement and signatures intelligence) program. The analysis also is distributed across a number of agencies. All-source analysts are found at the CIA, DIA, and INR, while single-source SIGINT analysts operate in the NSA, and GEOINT (geospatial intelligence) analysts are located in the National Geospatial-Intelligence Agency (NGA).[20] This plethora of separate collection programs creates, in its own way, multiple organizational cultures based on different collection technologies, priorities, and procedures. Likewise, the proliferation of analysts across the community creates a mosaic of analysis and makes it hard at times to know who is working on what topics. This can lead to duplication of effort and a cacophony of views, which can be confusing to policymakers wishing for a clear answer to a policy problem. At other times, the IC argues that these different perspectives produce a type of "competitive analysis" that sharpens the thinking of all analysts by challenging their analytical assumptions, thus guarding against groupthink. Nevertheless, analytical talent is so dispersed around the IC that it is sometimes hard to know where all the expertise resides. Early in its existence, the Office of the DNI (ODNI) attempted to identify individual analysts within the IC to establish a register that could identify each analyst's target accounts and expertise.[21] Yet, differing definitions of what constitutes "analysis" frustrated this effort.

A possible third feature of the American intelligence culture is the central role played by technology in the collection, analysis, and dissemination of intelligence.[22] There is often a tendency to favor technical intelligence over human sources among some analysts as well as consumers. For example, military intelligence analysts might tend toward greater reliance on the Defense Department's satellite and airborne collectors than the CIA's human sources.[23] For their tasks, they may judge imagery to be the most reliable information for military operations that require details regarding an adversary's capabilities, while downplaying human intelligence reporting that seems less reliable.[24] Military preparations can be seen from the air or space, while an adversary's intentions are inherently opaque, and what HUMINT contributions exist on such matters are more limited and less verifiable. In the Jimmy Carter administration, for example, CIA Director Stansfield Turner – himself a former commander of NATO – favored the development of more technical capabilities at the expense of CIA HUMINT, which he proceeded to cut down.[25] At the other end of the spectrum, civilian analysts at the CIA generally are in a better position to exploit, as well as vet, HUMINT sources precisely because they can more readily task case officers in the CIA's DO to develop human sources.

The emphasis placed on analysis as opposed to just collection of intelligence is also much greater in the United States than in other foreign intelligence cultures. For example, the analytic function was extremely limited within the KGB, which largely relied on espionage reports as its primary deliverable to the officials it served.[26] From the beginning, the CIA's mission was to gather, collate, and evaluate intelligence that was drawn from across the other departments and military services. Analysis was seen as a distinct function. Rather than presenting a president or Secretary of State with a raw intelligence report, information is first reviewed for its validity and accuracy and then combined with other reports to produce a "finished intelligence product." In some intelligence cultures, raw reports from the field and individual reports are "fed up the chain" – leaving analysis to the prime minister, general staff officers, or other decisionmakers. Indeed, illustrative of the extent to which the CIA has defined its analytic service, it created a Sherman Kent School for Intelligence Analysis. This is a separate training institution from the DO's own training program. The discrete educational programs speak to the distinct efforts to inculcate different values as well as practices even within the same agency. Intelligence scholars will also be aware of the continuing debate over whether intelligence analysis has achieved the status of a "profession," but in the minds of CIA practitioners, it is certainly treated as such.[27]

The Impact of Organizational Cultures on Intelligence and Strategy

The preceding discussion has underlined various aspects of the American intelligence culture, many of which are shaped by the overarching American strategic culture. The complexity of the IC also gives rise to many internal factors that shape the American way of intelligence, particularly its multiplicity of collection and production organizations, which contain their own separate cultures. Members of both the NSA and the CIA, for instance, each like to refer to themselves as working for "the Agency." There is a large literature on the nature and impact of organizational culture on policy output along with another considerable literature on how organizational cultures influence national security decisionmaking.[28] There is far less written about intelligence agencies' specific cultures, most likely because of the classified nature of their work and the unwillingness of former officials to say much about the internal workings of their former employers. Prepublication review procedures also limit insights gleaned from memoirs and articles written about intelligence practices. Nevertheless, the subcultures that inhabit the IC – which some former officials have likened to "tribes" – still help to shape the products produced for intelligence consumers.[29]

Given its centrality to the US IC, the CIA offers a good touchstone not only to compare how intelligence cultures can vary across the IC but also to illustrate how the dominant culture can vary even within units in a single organization. The CIA grew out of the earlier Office of Strategic Services, which had a focus on wartime intelligence gathering and special covert operations. A part of the CIA's "DNA," so to speak, thus has a paramilitary operational code. The Directorate of Operations' Special Activities Division, which is responsible for covert operations, continues in the tradition of the Office of Strategic Services, drawing many of its officers from the Special Forces. The collection of foreign intelligence by using case officers to recruit secret agents is another part of the DO's mandate. CIA stations maintain worldwide coverage and the DO is responsible for managing liaison relations with the security services of foreign governments. A distinct part of the CIA's ethos is centered on producing all-source intelligence and supplying it to senior government officials. Since the presidency of Harry Truman, the CIA has been preparing daily intelligence summaries for presidential review, which eventually evolved into the President's Daily Brief of today.[30] Until the creation of the DNI in 2004, the CIA was solely responsible for this publication, and the CIA's analytic directorate took pride in producing it along with other less sensitive daily publications. These three core missions – covert operations, espionage, and intelligence analysis – helped to form the organizational culture and subcultures within the agency. They provided reasons for why the CIA was "central," in the minds of agency officers – they developed all-source finished intelligence and they directly served the president. They were a "one-stop shop" for intelligence matters within the US government.

Imagine, then, how this organization might enculturate its employees through its recruitment, training, and promotion of individuals who excelled at achieving results in special operations, or HUMINT recruitments, or presenting sophisticated analyses to the highest levels of government. As one former senior operations officer noted: "the DO culture is one of risk, salesmanship, foreign environments, energy, and secrecy."[31] Recruitments, not substance, mattered to case officers. They have a license to break the law in other countries to recruit foreign spies, sometimes without the knowledge of US ambassadors or senior State Department officials who might worry about the blowback from disclosure or counterintelligence scandals. Covert operations officers also sometimes prefer not to discuss the risks involved and restrict the planning of such activities on the basis of a "need to know." In the past, covert operators have been overly optimistic about their plans' efficacy and have stopped CIA analysts from evaluating them. Moreover, secrecy also pervades the agency, especially inside the DO that steals secrets and conducts covert operations. As a cultural norm, it helped to shield the review of such activities, which in some cases eventually led to huge problems. At other times, secrecy hampered the development of a learning culture because many critical postmortems of failed operations, poor tradecraft, or misleading forecasts were also hidden from most agency employees.

CIA analysts, in contrast, are all about the information and how to fit it together to solve intelligence puzzles and mysteries. They are also prone to believing that their work constitutes the "gold standard" when it comes to intelligence assessments. Nevertheless, we know that there have been numerous intelligence failures, in part due to analysts' overconfidence in their knowledge of a target, reliance on dated or deceptive information, or flawed mindsets about an adversary's preferences, intentions, and behavior. More than a few analysts at other IC agencies have bridled at the seeming arrogance of CIA analysts as well as the CIA's reluctance to assign their junior analysts to community-wide training opportunities because of the belief that CIA training was superior.

Periodically, however, the CIA's tribes have faced the need to adapt their cultures and standard operating procedures in response to changes in the primary intelligence challenge of the day. New actors, technologies, and issues can alter the strategic landscape, undermine collection and analysis ecosystems (i.e., relationships among other government agencies, target prioritization, and customer preferences), and can render bureaucratic norms and organizational procedures obsolete. Under these circumstances, specific types of expertise are sometimes no longer required. At the end of the Cold War, for example, CIA cultures became out of date. The persistent focus on the Soviet hard target highlighted how the DO had to reform and prioritize other targets. As one senior CIA official noted in 1995, the DO was slow to adapt:

> The clandestine service of the Cold War does not serve present needs ... Much of its "culture" – independence and flexibility – must be maintained.... But ... it needs an overhaul... The targets of the clandestine service have changed.... The clandestine service has not valued assignments outside its core activities.[32]

At the same time, CIA analytic managers began to realize that their analysis was missing the mark. Then Deputy Director Robert Gates had concluded that analysts were out of touch with policymakers, did not understand how the policy process worked, and needed to get out of their offices and "downtown" to better serve the president and his principal advisers. Very soon a cultural shift occurred in which the President's Daily Brief became the most important metric for success, making it most analysts' goal to write for this publication. The entire analytic business model began to revolve around this one daily publication. Depending on analysts' targets and discipline, their promotions could be tied to superior production performance as well as substantive expertise and innovative analytical practices; however, "feeding the beast" as the continuous President's Daily Brief production cycle was known became an analytic directorate priority. This was a huge cultural shift away from the lengthy research papers of the Cold War to daily "actionable" intelligence geared toward helping a twenty-first-century president make a policy decision.

These examples are important because the 9/11 attacks and the flawed 2002 Iraq WMD estimate brought further challenges to those operational and analytical cultures. The creation of the Director of National Intelligence produced a shock wave through the CIA.[33] It demoted the role of the CIA director, who was no longer the president's intelligence advisor to the NSC. The CIA was pushed further into the background when the DNI took control over the President's Daily Brief process, opening it up to the work of other analysts at the DIA and the State Department's INR Bureau. The Office of the DNI also took over the National Intelligence Council and the National Intelligence Estimate process that the CIA had previously dominated. In addition, the DNI assigned the National Intelligence Council to prepare the intelligence briefing books for NSC principals and deputies meetings, formerly done by CIA's analytic offices. Intelligence reforms also included an expanded defense HUMINT program, which initially challenged the unique role played by the DO in recruiting spies overseas.

In the bureaucratic tussle over how these changes were operationalized, the CIA won back some of its prestige and control. In fact, CIA analysts continue to prepare the bulk of the President's Daily Brief articles, although they currently must go through the DNI for review. CIA analysts also continue to prepare the briefing books for the National Intelligence Council, and both the DNI and the CIA Director are usually included in key NSC principals committee meetings. The DO – renamed the National Clandestine Service for a time – was able to retain its key role of training and setting the priorities of both Defense Department and CIA case officers. But in each case, the CIA tribes saw their missions and raison d'être challenged and had to adjust their expectations and cultures to suit the times. A former senior analytic manager has said that many CIA analysts had approached him after the DNI's appointment to ask, "[I]s the CIA still central?" Likewise, DO officers fretted when a later DNI, former Admiral Dennis Blair, battled to control the assignment of chiefs of station overseas – traditionally an assignment left to the CIA Director and the DO to determine. Although then CIA Director Leon Panetta won this bureaucratic battle, it was another case of how the CIA has been forced to defend its bureaucratic turf and prerogatives.

These examples of organizational cultures within the CIA illustrate how each of the IC agencies approaches their missions with unique procedures and characteristics. The Pentagon-focused agencies have battled to retain control over technical collection priorities that are most useful to defense planners and battlefield commanders, while the CIA and other civilian agencies want to protect some of those collection priorities for the more political and economic targets that non-DOD users value. In this sense, the organizational struggle over the collection of information will inevitably shape the quality and scope of intelligence provided to civilian and military decisionmakers. In the field of counterterrorism, the FBI and the CIA have had to learn to understand each other's cultures and adapt. A senior CIA operator and later counterterrorism coordinator at the State Department explained that when initially assigned to lead joint counterterrorism efforts at the FBI, he was shocked when attending his first meeting: "I sat there like a tribal outsider, observing a new culture as participants slung around facts, figures, guesstimates, and other details around a tiny room." Having come from a tribe that valued the written report, he had stepped into a tribe that instead valued oral history.[34]

Another comparison demonstrates how the secrecy mantra at the CIA is not shared to the same extent by the State Department's INR Bureau, where analysts can more easily meet and communicate with outside experts and even foreign nationals in true name and affiliation, without the same level of security-consciousness that permeates the CIA tribes.[35] The State Department's mission is to represent and meet with foreign officials; hence, INR culture is part of that tribe. One can find many more examples in other agencies, but the important point is that all of these cultures and subcultures operate with different priorities and perceptual lenses, which contribute to the information environment in which US decisionmakers operate.

Conclusion

The prevailing American strategic culture and the American style of intelligence are inherently connected and have reciprocal effects on each other. As the American strategic culture shifts, so too will the American style of information gathering and analysis. In turn, as the US IC evolves, the information environment, which the US strategic culture inhabits, will also shape evolving worldviews. The conclusion that policymakers and intelligence analysts are educated and enculturated in the same worldview and share a set of common cultural values, preferences, and expectations is not that surprising and facilitates the symbiotic relationship that exists between the policy and intelligence communities. The fundamental raison d'être of the IC is to provide policymakers with the information and analysis needed to protect the nation and achieve policy success, while policymakers rely on intelligence professionals to supply them with the knowledge and foresight needed to succeed in office by advancing effective policies. Without this symbiosis, there would be far more disputes between the two than have already occurred. The danger created by this common cultural framework lies in the difficulty of the IC recognizing and getting outside its own strategic and organizational cultures so it can guard against too much ethnocentric thinking, mirror-imaging, obsession with protecting organizational missions and prerogatives, or obsolete mental models of foreign behavior to which an agency has become wedded.

The IC has adopted several reforms that attempt to counter these cognitive and organizational biases. First, there are a growing number of analytic tools that can expose and challenge an analyst's preconceived notions. Termed "structured analytic techniques," they are regularly used to examine persistent lines of analysis and subject them to "devil's advocacy" or other forms of assessment to ensure that analytic assumptions have not become obsolete.[36] Second, the ODNI continues to support more use of outside consultants and regards analysts' mission to include developing an understanding of how non-government experts might see the same set of intelligence problems in different ways. Third, the IC is trying to overcome some of the bad habits of its own internal tribes that can make it less efficient and prescient. For example, the 2015 "modernization" plan of then CIA Director John Brennan brought together the operations and analytic offices for each specific region or functional topic – for instance, Asia or Weapons of Mass Destruction – under a single Mission Center. This modernization plan was intended to maximize the combined expertise of both collectors and analysts, remove unnecessary barriers to information sharing, and create what was termed a "One Agency, One Mission" mentality among analysts and managers. Such reforms, however, must balance removing the negative aspects of competing subcultures, while preserving the positive features that each contains. As the ODNI similarly strives to integrate the diverse IC agencies into a single enterprise, it too must work to overcome the limitations of operating inside an American strategic culture and a bevy of organizational cultures without weakening the effectiveness of the IC's essential missions.

Notes

1 The opinions expressed are those of the authors alone and do not reflect the position of any government, government agency, or commercial enterprise.
2 See Jeffrey S. Lantis and Darryl Howlett, "Strategic Culture," in John Baylis, James Wirtz, and Colin Gray eds., *Strategy in the Contemporary World*, 6th ed. (Oxford: Oxford University Press, 2019), 92.

3 Russell F. Weigley, *The American Way of War: A History of United States Military Strategy and Policy* (Bloomington, IN: Indiana University Press, 1977); Brian M. Linn, *The Echo of Battle: The Army's Way of War* (Cambridge, MA: Harvard University Press, 2007); and Colin Gray, "Irregular Enemies and the Essence of Strategy: Can the American Way of War Adapt" (Carlisle, PA: Strategic Studies Institute, US Army War College, 2006).

4 Thomas G. Mahnken, "United States Strategic Culture," Prepared for the Defense Threat Reduction Agency, 13 November 2006, 25pp., at: https://irp.fas.org/agency/dod/dtra/uspdf.

5 To be sure, there are discordant strains within the American strategic culture and one form is described by Walter Russell Mead's characterization of four schools of thought: the more isolationist being Jeffersonian and Jacksonian and the more internationalist being the Hamiltonian and Wilsonian principles for conducting American security policy. See Walter Russell Mead, *Special Providence: American Foreign Policy and How It Changed the World* (New York: Routledge, 2002).

6 Robert Callum, "The Case for Cultural Diversity in the Intelligence Community," *International Journal of Intelligence and Counterintelligence* Vol. 14, No. 1 (2001): 25–48.

7 George F. Kennan, *Memoirs 1925–1950* (New York: Pantheon Books, 1967), 306.

8 Created in 1997 at the 50th anniversary of the CIA, The Trailblazer Award recognized 50 individuals whose actions, example, innovations or initiatives have taken the CIA in important new directions and helped to shape the Agency's history. As of its 70th anniversary in 2017, there were 83 individuals inducted into the Trailblazer group. See Tammy Waitt, "CIA Honors 'Trailblazer and Trailblazer Museum for 70th Anniversary," *American Security Today*, 25 September 2017, at: https://americansecuritytoday.com/cia-honors-trailblazer-museum-70th-anniversary/.

9 *The National Security Strategy of the United States* 2006, Overview, 1, at: https://wwwcomw.org/qdr/fulltext.nass2006.pdf.

10 For example, see Office of the Director of National Intelligence, *Annual Threat Assessment of the US Intelligence Community*, 9 April 2021. https://www.dni.gov/files/ODNI/documents/assessments"ATA-2021-Unclassified-Report.pdf.

11 Henry L. Stimson and McGeorge Bundy, *On Active Service in Peace and War* (New York: Harper & Brothers, 1948), 199.

12 Michael Herman, *Intelligence Power in Peace and War* (Cambridge: Cambridge University Press, 1996), 92.

13 See Michael Warner, "Wanted: A Definition of Intelligence," *Studies in Intelligence*, Vol. 46, No. 3 (2020): 21, at: https://apps.dtic.mil/sti/pdfs/ADA525816.pdf. Warner notes that Abram Shulsky has put forward the notion that a commonality of intelligence organizations is that they conduct their actions secretly: "Secrecy is essential because intelligence is part of the ongoing struggle between nations."

14 Robert Jervis, *Perception and Misperception in International Politics* (Princeton, NJ: Princeton University Press. 1976); and Richards J. Heuer, Jr., *The Psychology of Intelligence Analysis* (Washington, DC Center for the Study of Intelligence, 1999).

15 Special National Intelligence Estimate 85-3-62, "The Military Buildup in Cuba" (September 1962): 1–2, 8–9; and James J. Wirtz, "Organizing for Crisis Intelligence: Lessons from the Cuban Missile Crisis," in James G. Blight and David Welch eds., *Intelligence and the Cuban Missile Crisis* (London: Frank Cass, 1998), 120–149.

16 Robert Jervis, "Reports, Politics, and Intelligence Failures: The Case of Iraq," *Journal of Strategic Studies* Vol. 29, No.1 (2006): 23.

17 Personal communication with a former deputy chief of the CIA Mission Center on Russia-Eurasia, 27 January 2022.

18 James J. Wirtz, "The Intelligence-Policy Nexus," in Loch Johnson ed., *Strategic Intelligence Understanding the Hidden Side of Government*, Vol. I (Westport, CT: Praeger Security International, 2007), 139–150.

19 Richard Betts, *Enemies of Intelligence: Knowledge & Power in American National Security* (New York: Columbia University Press, 2007), 66–103; and James J. Wirtz, "How Could Getting It Right Go So Wrong? The 2007 Iran NIE Revisited," *Intelligence and National Security* Vol. 36, No. 2 (2021): 157–159.

20 For example, one can observe that the NSA is a major employer of linguists, mathematicians, and computer specialists, while the National Reconnaissance Office hires engineers, astrophysicists, and other space science specialties.

21 See Nancy Tucker, "The Cultural Revolution in Intelligence: An Interim Report," *The Washington Quarterly*, Vol. 31, No. 2 (January 2010): 50, at: https://www-tandfonline-com.oxy.idm.ocic.org/doi/abs/10.1162/wash2008.31.2.47. The Analyst Resource Catalogue (ARC) lists some 17,000 names, allowing analysts to find each other as well as managers to identify analysts with certain skills.

22 Observers have long noted America's preoccupation with technology in all matters, including national defense and strategy see Mahnken, "United States Strategic Culture," 12; and Colin Gray, *Modern Strategy* (Oxford: Oxford University Press, 1999), 147.

23 Since the late 1990s, there has also been a Defense HUMINT Service, which the CIA reluctantly agreed to, but it insisted that there be joint training by CIA standards and overall control lodged with the clandestine service at Langley.

24 Military commanders are most likely to build their war plans based on what they "know" are an adversary's capabilities, when they have less information and confidence in what an opponent's intentions are. Witness the 2022 statements by senior US defense officials who have highlighted Russia's military buildup bordering Ukraine – highly publicized through released photo intelligence of Russian deployments – in the face of great uncertainty regarding President Putin's intentions vis-à-vis Kyiv.

25 Stansfield Turner, *Secrecy and Democracy: The CIA in Transition* (Boston, MA: Houghton Mifflin, 1985), 195–205.

26 Raymond Garthoff, "Soviet Leaders, Soviet Intelligence, and Changing Views of the United States, 196501991," in Paul Maddrell ed., *The Image of the Enemy: Intelligence Analysis of Adversaries Since 1945* (Georgetown: Georgetown University Press, 2015), 28–67.

27 For a critical review see John Gentry, "The 'Professionalization' of Intelligence Analysis: A Skeptical Perspective," *International Journal of Intelligence and Counterintelligence* Vol. 29, No. 4 (2016): 647–676; on the Kent School see Stephen Marrin, "CIA's Kent School: Improving for New analysts," *International Journal of Intelligence and Counterintelligence* Vol. 16, No. 4 (2003): 609–637.

28 One of the most widely known such studies was Graham Allison's, *Essence of Decision: Explaining the Cuban Missile Crisis* (Boston: Little Brown Publishers, 1972), which evaluated decisionmaking using three different levels of analysis, including one which was explicitly tied to organizational cultures, their key roles/missions and standard operating procedures. For a discussion of national security organizational cultures, also Roger George and Harvey Rishikof, eds., *The National Security Enterprise: Navigating the Labyrinth*, 2nd edition (Georgetown University Press, 2016).

29 Mark M. Lowenthal, "Tribal Tongues: Intelligence Consumers, Intelligence Producers," *Washington Quarterly* Vol. 15, No. 1 (1992): 157–168.

30 Adrian Wolfberg, "The President's Daily Brief: Managing the Relationship between Intelligence and the Policymaker," *Political Science Quarterly* Vol. 132, No. 2 (Summer 2017): 225–258.

31 Personal communication with a former COS and division chief, 26 January 2022.

32 Richard Kerr, Deputy Director of CIA, Testimony Before the House Intelligence Committee, 16 November 1995, at: https://irp.fas.org/congress/1995_hr/kerr1116.htm.

33 See Peter C. Clement, "Impact of Intelligence Integration on CIA Analysis," *Studies in Intelligence* Vol. 65, No. 3 (September 2021): 25–33, at: https://www.cia.gov/static/75321ecc43daacde5345441d3244f3c/Article-Clement-The EarlyYearsofIntelIntegration.pdf.

34 Henry A. Crumpton, *The Art of Intelligence: Lessons from a Life in the CIA's Clandestine Service* (New York: Penguin Books, 2012): 108.

35 See Susan Nelson's discussion of her pursuit to expand analytic outreach from the Bureau of Intelligence and Research, which found resistance from other agencies like the CIA. Susan H. Nelson, "Analytic Outreach: Pathway to Expertise Building and Professionalization," in Roger George and James Bruce, eds., *Analyzing Intelligence: National Security Practitioners' Perspective* (Washington: Georgetown University Press, 2014), 319–336.

36 Randolph H. Pherson and Richards J. Heuer Jr., *Structured Analytic Techniques for Intelligence Analysts* 3rd edition (Washington, DC: CQ Press, 2020).

Selected Bibliography

Callum, Robert. "The Case for Cultural Diversity in the Intelligence Community," *International Journal of Intelligence and Counterintelligence*, vol. 14, no. 1, 2001: 25–48.

Clement, Peter C. "Impact of Intelligence Integration on CIA Analysis," *Studies in Intelligence*, Vol. 65, No. 3, September 2021: 25–33, at: https://www.cia.gov/static/75321ecc43daacde5345441d3244f3c/Article-Clement-TheEarlyYearsofIntelIntegration.pdf.

Garthoff, Raymond. "Soviet Leaders, Soviet Intelligence, and Changing Views of the United States, 196501991," in Paul Maddrell ed., *The Image of the Enemy: Intelligence Analysis of Adversaries Since 1945* (Georgetown: Georgetown University Press, 2015): 28–67.

Gentry, John. "The 'Professionalization' of Intelligence Analysis: A Skeptical Perspective," *International Journal of Intelligence and Counterintelligence*, Vol. 29, No. 4 (2016): 647–676.

George, Roger, and Harvey Rishikof, eds. *The National Security Enterprise: Navigating the Labyrinth*, 2nd ed. (Georgetown University Press, 2016).

Gray, Colin. "Irregular Enemies and the Essence of Strategy: Can the American Way of War Adapt," (Carlisle, PA: Strategic Studies Institute, U.S. Army War College, 2006).

Lantis, Jeffrey S., and Darryl Howlett. "Strategic Culture," in John Baylis, James Wirtz, and Colin Gray eds., *Strategy in the Contemporary World*, 6th ed. (Oxford: Oxford University Press, 2019).

Lowenthal, Mark M. "Tribal Tongues: Intelligence Consumers, Intelligence Producers," *Washington Quarterly*, Vol. 15, No. 1, 1992: 157–168.

Mahnken, Thomas G. "United States Strategic Culture," Prepared for the Defense Threat Reduction Agency, 13 November 2006, 25pp, at: https://irp.fas.org/agency/dod/dtra/uspdf.

Mead, Walter Russell. *Special Providence: American Foreign Policy and How It Changed the World* (New York: Routledge, 2002).

Tucker, Nancy. "The Cultural Revolution in Intelligence: An Interim Report," *The Washington Quarterly*, Vol. 31, no. 2 (January 2010), 50, at: https://www-tandfonline-com.oxy.idm.ocic.org/doi/abs/10.1162/wash2008.31.2.47.

Weigley, Russell F. *The American Way of War: A History of United States Military Strategy and Policy* (Bloomington, IN: Indiana University Press, 1977)

Wirtz, James J. "How Could Getting It Right Go So Wrong? The 2007 Iran NIE Revisited," *Intelligence and National Security*, Vol. 36, No. 2, 2021: 157–159.

Wirtz, James J. "The Intelligence-Policy Nexus," in Loch Johnson ed., *Strategic Intelligence Understanding the Hidden Side of Government*, Vol. I (Westport, Conn: Praeger Security International, 2007): 139–150.

Wolfberg, Adrian. "The President's Daily Brief: Managing the Relationship between Intelligence and the Policymaker," *Political Science Quarterly*, Vol. 132, No. 2, Summer 2017: 225–258.

32
THE IMPACT OF US INTELLIGENCE COMMUNITY STRATEGIC CULTURE ON INTELLIGENCE ANALYSTS

Rob Johnston and Judith Meister Johnston

Strategic Culture and US Intelligence Analysts

This chapter is intended to examine how the strategic culture of the US Intelligence Community impacts the individual intelligence analysts within it, with a particular eye toward the influence of the heavily elite-based membership of the Intelligence Community, and how individual analysts, in turn, impact and carry on the organizational and strategic culture of the community. While some of the details will seem overly generalized, this work is derived from personal fieldwork as an anthropologist in the Intelligence Community and from experience as a former member of that community.[1] This chapter is also based on personal experience working for and with a number of those constituent agencies. We will often talk about organizations as if they are independent of individuals. This is not meant as a sleight of hand to imbue organizations with a specific agency,[2] but rather a recognition that organizations are tribes – that is, like any other tribe, they are an established "thing." They are a collective agreement that lives within and between organizational members but also serves as a compact with the past and the future. The promise of membership includes an implicit agreement to try to ensure that the organization will survive in the future.

An organizational identity becomes part of an individual's sense of self. That is, what might be considered acculturation (the process of learning another culture) in many professional settings is better understood as enculturation (the process of learning one's own culture) in closed or highly selective communities. The same can be said for physicians and nurses, astronauts, special operating forces, and other completely immersive professions that require years of vetting, training, and practice. The ramifications of individual enculturation into the US Intelligence Community, and that community's strategic culture, are profound. Intelligence professionals are confronted by the forces of enculturation within the Intelligence Community. The lifelong imprint that professional identity has on an individual needs to be deconstructed and explored to be understood.

DOI: 10.4324/9781003010302-36

Colin Gray tells us that the master narrative surrounding strategic culture in the Intelligence Community:

> ... is the disarmingly elementary, even commonsensical, idea, that a security community is likely to think and behave in ways that are influenced by what it has taught itself about itself and its relevant contexts. And that education, to repeat, rests primarily upon the interpretation of history and history's geography (or should it be geography's history?) My geopolitical friends favor the aphorism that "geography is destiny." Perhaps the dedicated culturalist will offer the rival dictum that "culture is destiny."[3]

Additionally, Jack Snyder, in his seminal 1977 RAND report, defines strategic culture in the national security community as:

> ... the sum total of ideas, conditioned emotional responses, and patterns of habitual behavior that members of a national strategic community have acquired through instruction or imitation and share with each other with regard to nuclear strategy. In the area of strategy, habitual behaviour is largely cognitive behavior.[4]

Separating intelligence officers from the strategic culture of the Intelligence Community or its specific member organizations is an impossible and rather fruitless task. Intelligence officers are the product of the organizations they serve. Starting with the individual selection of new officers, training those officers, and bringing those officers into the fold of a closed and insulated, or even isolated, profession all contributes to the process of enculturation of individuals into a new and often dominant identity. The identity of any organization leaves an imprint on a person; often one is what one does. In some cases, organization can take the place of an individual's identity, or so influence it as to make it inseparable from other concepts of self. The mantra "a United States Marine is a Marine until death," for instance, is a well-understood and recognizable phenomenon. Ask any Marine, and they are likely to tell you that there are no "former" Marines.

The enculturation of Marines is purposeful and successful by the standards of that organization. Identity formation in this case is through basic training and is designed explicitly to be the best and shortest path between an individual's identity as a civilian and then their modified identity of being a Marine. The organization looks on at this phenomenon as a point of pride. The organization has created a lifelong family-like construct of individuals with shared beliefs and values. The organization seeks to build commitment to serve the organization's goals and, by extension, the nation and explicitly the Constitution through oaths and ceremonies. The organization's heritage and history become critical knowledge that everyone is responsible to carry with them throughout their lives and to pass onto others who are deemed worthy of carrying on the organizational culture.

In the case of the Intelligence Community, this process includes an oath to protect and defend the Constitution from all enemies, both foreign and domestic. It is critical to note that the oath does take place in a physical space that is exclusive. Outsiders are not allowed to attend or participate in the ritual; the oath is reserved for those individuals who are becoming part of the community. The organization works in a classified environment; meaning, by definition, that the work is secret and the import of the ceremony of taking the oath begins to take on additional meaning. The oath, the secrecy, the exclusivity of the environment, and the act itself are only the beginning for any new individual organizational member. The enculturation starts there and continues throughout an individual's career.

Origins of Strategic Culture in the US Intelligence Community

Intelligence organizations, with roots in existential national security crises like war, will imagine that their postwar survival is itself a sufficient explanation to demonstrate that the organization must carry on as before. In addition to the notion that tomorrow will undoubtedly be like today, recreating organizational culture feels safe, and unchallenged, and occurs with minimal internal conflict. This comfort, in many situations, might be good. It is not wholly unreasonable to believe that tomorrow will be pretty much like today. This means that recruiting new members to the organization will be done in known and comfortable environments that are well trodden, and those environments are populated by individuals who have already been screened by society for future elite status, so to speak.

That comfort also means that an individual has guardrails for their behavior within an organization, with guidance for future decisionmaking based on past performance, and a way, however clumsy, to bring order to chaos. All these reasons anchor a culture in a particular place and time with a particular fixed version of itself. The comfort, however, is often overtaken by events. The ideal version of an intelligence officer changes slightly over time with focused effort, but this version is ultimately anchored to the history of previous performance. It is important here to mention that much like financial due diligence, the biggest caveat with organizational performance is that *past performance does not guarantee future results*, a fact that organizations seem to forget with great regularity.

Today is mostly like yesterday until it is not. Consider 7 December 1941, 19 October 1950, 4 November 1956, 20 August 1968, 24 December 1979, and of course 11 September 2001.[5] While this list is not all inclusive, all those dates were most certainly not like the day that proceeded them, illustrating the notion that there are days where past performance did not guarantee future results; a simple proof that the past is not always a faithful guide for organizational behavior. That is where problems begin to seep into organizational culture and specifically the strategic culture of intelligence organizations. Organizations need to adapt and evolve to meet new and different environmental pressures.[6]

The history of the US Intelligence Community is well covered by recognized professional historians and runs from George Washington's spy ring to Lincoln and most importantly to Franklin D. Roosevelt and Harry Truman's era. Without revisiting that entire history, we will discuss the World War II period as a foundational element in today's Intelligence Community and then the immediate aftermath of 11 September 2001 as the Intelligence Community began its own reformation.[7]

Strategic Culture Influencers in the Intelligence Community

The origins of strategic culture in intelligence lead us, naturally, to try to understand the individuals who make up these intelligence organizations,[8] including all the relevant institutions and levers of power in American decisionmaking. Understanding the individuals helps us to acquire a better window into that culture. Influencers in general are discussed in this section and specific influencers are discussed in the following section. C. Wright Mills describes these individuals as:

> The power elite is composed of men whose positions enable them to transcend the ordinary environments of ordinary men and women; they are in positions to make decisions having major consequences. Whether they do or do not make such decisions is less important than the fact that they do occupy such pivotal positions: their failure to act, their failure to make decisions, is itself an act that is often of greater consequence than the decisions they do make. For they are in command of the major hierarchies and organizations of modern society. They rule the big corporations. They run the machinery of the state and claim its prerogatives. They direct the military establishment. They occupy the strategic command posts of the social structure, in which are now centered the effective means of the power and the wealth and the celebrity which they enjoy.[9]

This functionalist definition helps us understand the roles that people assume to carry in the function of governance. There are other historical views of power elites, some competing and some complementary.[10] That said, Mills' work is still as relevant today as it was 70 years ago. What is interesting in this case is that there is an Intelligence Community where someone can work and gain status as an elite. Applicants may or may not know in advance what elite status conveys to them once they are a part of the organization. Arguably, many likely do not think about their elite status at all, or if they do, they think of their status as a transitive function of their job rather than their personhood. It is also equally possible that elite status is what is being sought by applicants and therefore is something for the organization to consider during their selection process. In either case, elites emerge.

How the Strategic Culture Intelligence Elites Evolved

Colin Gray explains:

> Strategic culture is of interest because the concept suggests, perhaps insists, that different security communities think and behave somewhat differently about strategic matters. Those differences stem from communities' distinctive histories and geographies.[11]

After the founding of the Office of Strategic Services (OSS) during World War II, amid the postwar period of bureaucratic instability, analytic elements of the OSS moved from department to department and the organization finally emerged as a key analytic element of the Central Intelligence Agency. What is critical during this period is the focus on and the solidifying of a culture of elites who were white, educated, predominantly male, and upper or upper-middle class. The process, and even the reason, is not overly complicated. During World War II it was understood that there was an existential threat at the United States' front door. In response to this threat, William Donovan, a law school classmate of Franklin D. Roosevelt, was able to persuade the president to establish an organization that operated without the same oversight as the calcified bureaucracy that had formed in the State Department, the Army, and the Navy. It would be an agile and nimble organization without much red tape that comprised society's best and brightest, as conceived of at the time. This meant people from socioeconomic positions of power – white, upper class, educated, patriotic, and adventurous, preferably Christians of one sort or another, and only very occasionally female.

The analytic cadre, where we will focus, was no exception. Friends and friends of friends began to call upon one another to create this analytic workforce. As one would expect, these friends and their friends all personally knew or at least were aware of each other. Often, these individuals were related. These were mostly men, white men, steeped in the lofty traditions of the Ivy League: well-educated professors, and their students in many cases, recruited on a "friends and family" plan, so to speak.

William Langer, an OSS veteran analyst and manager, and later Central Intelligence Agency (CIA) veteran analyst and manager, as well as a professor of history at Harvard, was the primary individual charged with developing the underlying methodology for intelligence analysis in the Research and Analysis (R&A) Branch of the OSS. He was also the manager, and one of the principal recruiters, of additional manpower and he had a lasting effect on the profession of intelligence analysis by setting the standards to create an intelligence analysis methodology. William Langer had a younger brother, Walter, who was a psychoanalyst at Harvard and was given the task of writing what turned out to be a remarkably prescient profile of Adolf Hitler for the OSS.[12] William Langer's older brother, Rudolf, was a professor and then Chair of Mathematics at the University of Wisconsin during World War II. During the war, two of Rudolf's most prominent graduate students went to work for key institutions – one at the Civil Aeronautic Authority on his way to eventual employment at the Naval Research Lab and then NASA, and the other at the Office of Scientific Research and Development, the precursor to the Manhattan Project. During the Vietnam War Rudolf was selected to run the Army's Mathematics Research Center.

One Lutheran and Moravian family of German immigrants, brothers and first-generation Americans, veterans of World War I, Harvard educated men all, left a collective and significant imprint on the Intelligence Community, NASA, and what would become the Nuclear Regulatory Commission. Not surprisingly, these founding members of the Intelligence Community began hiring friends, family, and other scholars they knew or had worked with in the past. As an interesting aside, the Langers' power pales in comparison to that of the Dulles brothers, John F. and Allen, both of whom were Princeton alumni. The Dulles brothers were grandsons of former Secretary of State John W. Foster and nephews of former Secretary of State Robert Lansing. John F. Dulles would go on to become the Secretary of State himself, and Allen Dulles would go on to become the Director of the Central Intelligence Agency. In other words, here was another highly impactful family of elites, considerably more connected to the American political power system than the Langer brothers.[13] This tendency of hiring family members for sensitive positions was an effective screening method for an emerging community just coming to grips with the reality of counterintelligence and hostile behavior by adversaries. The unintended consequences of these hiring strategies would play out in the future as a specific hiring bias.

More Influencers and Shortcomings in the Evolution of Strategic Culture

Due to the number and individual egos of the Research and Analysis (R&A) Branch of the OSS, which was staffed largely by Ivy League professors and their students – including seven future presidents of the American Historical Association, five of the American Economic Association, and two Nobel Laureates – William Langer had created an organization, as good as it was, made up of America's elite. William Langer had hoped that by hiring the "right" sort of minds (or people, to be more accurate) he could overcome any problem America and its allies might face. This model did not spend time worrying about ethnocentrism or systemic racism in the creation of its own organizational culture nor the lasting power it would have on the composition of the analytic workforce.[14]

Sherman Kent, an OSS veteran analyst and manager, and later Intelligence Community analyst and manager, was a professor of history at Yale University and an intellectual leader and founder of US intelligence analysis focused on the methods and communication of analysis to policy elites outlined in his 1949 book *Strategic Intelligence for American World Policy*.[15] The domestic politics of the day required someone with a recognized voice on intelligence analysis to write a book to inform the American public and policymakers about the usefulness of intelligence and the critical role analysts played in the policymaking process. At that time, he conceived of intelligence analysis as a combination of newspaper reporting, business processes, and university research. Kent's own concept of research was informed by his professional training as a historian at Yale and from serving under William Langer during and after the war. Kent eventually succeeded Langer to lead analysis through the difficult period of the Bay of Pigs disaster and the Cuban Missile Crisis. Kent himself focused his analytic skills mainly on archival evidence, his own standards for intellectual rigor, stating and trying to avoid his known biases, and avoiding having any personal policy agenda.[16] Those biases, however, were limited to the recognized and acknowledged biases of his time, not the biases we might consider today like gender, race, and sexuality.

Kent's influence on intelligence analysis cannot be overstated. The Sherman Kent School for Intelligence Analysis is an Intelligence Community schoolhouse for training analysts. The Sherman Kent Center is the internal think tank within the schoolhouse. The peer-reviewed journal *Studies in Intelligence* was founded at CIA by Sherman Kent in 1955 to focus on the profession of intelligence, and the idea for an Institute for Advanced Study, based on the Princeton model and focused on research about all aspects of the profession of intelligence, finally came to fruition as the Center for the Study of Intelligence at CIA under the direction of Director of Central Intelligence James Schlesinger in 1973. Sherman Kent has been a legend and a thought leader for intelligence analysts for more than 50 years.

Willmoore Kendall, a contemporary of Sherman Kent, was a State Department Office of Research and Intelligence veteran analyst, an Intelligence Community veteran analyst, and later a professor of political science at Yale University. Interestingly, he graduated high school at 13 and college at 18, was a Rhodes Scholar at Oxford, Trotskyist, and veteran of the Spanish Civil War. Based on his experience in the Spanish Civil War, Kendall renounced communism and obtained his doctorate at the University of Illinois, and worked for the OSS during World War II. While Kendall viewed the work of intelligence analysts as one similar to that of social scientists, he ultimately took the view that intelligence should be focused on policy advocacy rather than merely reporting objective data to policymakers, as Kent argued.[17] Kendall went on to create the *National Review* with one of his students, William F. Buckley, Jr., making clear his intent to influence policymakers and policy decisions rather than just informing them. Much like the history of intelligence analysis, the relationship with policymakers has vacillated between science and art, but in this case, it has vacillated between informing consumers and influencing them. As a result, Kendall was often viewed as a policy advocate rather than an intelligence scholar.

All of this is a very small window into the people and culture of the OSS and its impact and influence on the future of the Intelligence Community. This snapshot does not include a discussion of white elite women working in consultation with the OSS, like Ruth Benedict, a graduate of Vassar and Columbia, and Margaret Mead, a graduate of Barnard and Columbia. Despite their historically perceived lower social status as women, they were still socially elite and white all the same. That status helped them interact with the elites in the Intelligence Community and policymakers, even if through intermediaries. It is important to recognize that the OSS was critical in the foundation and history of much of the Intelligence Community: at the lowest count, 40% of the agencies in the Intelligence Community have roots that go back to the OSS. Reviewing the history of many of the other agencies, it would be just as easy to draw a line from white, elite, Ivy League members to today's organizations. This history of elite status is fundamental to understanding the identity and culture of these organizations.

Elites In Practice – Growing the Organization and Culture in Its Own Image

"Legacies" in the Intelligence Community (families or friends who follow in the footsteps of an Intelligence Community member), a type of elite, are a valued given. Because of the perceived importance of working in the Intelligence Community, where secrets are sacred and national security is the ultimate goal, the encouragement of legacy hires is very popular. Family members are already aware of the challenges of working in the Intelligence Community and usually are easy to get through the clearance process. They are typically not problematic hires and they easily fit into the organizational and strategic culture, as it has already been part of their family life. There is little pushback among legacy hires regarding usual hiring practices, workforce makeup, and job requirements. The strategic culture of the community thrives.

There is not necessarily a problem with employing elites per se. Right or wrong, organizations aspire to hire elites all the time. The morality of doing so is a debate outside of this chapter. The real problem is being stuck in a *weltanschauung* (or a person's "look out upon the world" perspective) wherein the utter lack of diversity and all that this means goes too often unnoticed, as does the potential impact that this has on all elements of analysis. This relates not just to diversity of gender, race, or ethnicity, but also the diversity of experience and thought. Hiring and hoarding a bunch of like-minded Ivy League brains means hiring a particular view of the world and how that world works.

American elites may well understand some key facets of how elites operate in foreign countries, and that has value. Unfortunately, the understanding of elites seldom extends to understanding how normal non-elites operate in and navigate the world. How non-elites may organize and how they might perceive any given set of circumstances in their own country is often an elusive set of facts: one that may have a serious negative impact on analytic findings and policy recommendations. Those sets of facts need to be understood through the lens of "the woman on the street" and how she may perceive the world around her. Understanding how she may make decisions about her family and the value of the family in her current circumstances may well be outside of any elite's capacity to comprehend. It is possible, likely even, to imagine a world where the non-elites are making decisions about the facts on the ground that stand in stark opposition to what elites may believe to be true.

Elites' Impact on Intelligence Community Strategic Culture

New hires in the Intelligence Community may not consider themselves elite, although these organizations do impress upon recruits the fact that they are elite as they share statistics about the large number of applicants versus the actual number hired. Stories are shared frequently about employees – "just like you" – doing amazing, heroic, important things to protect the nation's security. New hires are introduced to the history, hero stories, and myth-building that happens throughout the Intelligence Community.

New hires may not think about social status at all. They may consider themselves simple representatives of America, strategic culture and all. What they experience once inside the organization is the pronouncement that they were selected because they were the best and brightest. They were chosen to join the elite organization because they were perceived to be people who could adjust to and become elite themselves. Even if not spelled out explicitly, they certainly understand how someone can join an organization and vault from the student or corporate or academic status to being someone "in the know." Someone who engages policy and decisionmakers. Someone who matters on the national stage.

The strategic culture in which intelligence functions occur has a significant impact on who produces it, how it is produced, and how it is consumed. Just as importantly, the interaction between producers and consumers is immersed in and surrounded by strategic culture. There is no way to carve strategic culture out of the equation. Rather than treating culture or strategic culture as confounding variables to be pried away, isolated, and controlled for, strategic culture should be thought of as a logical extension of *weltanschauung*, and specifically that "look into" the world that is shared within a tribe. Cultural analysis for intelligence analysts is both an approach to analytics and also the universe in which intelligence analysis and policymaking take place.

Strategic Culture in the Intelligence Community: To Shape or Not to Shape

The cost of any organization choosing to be elite is reflected in a lost opportunity. The opportunities to find new and creative solutions, to view the world through the lens of others, and to seek and find strategic choices and advantages may well be lost given the lack of cognitive diversity. When we think about strategic culture and ethnocentrism the risks are apparent. Yet, organizations continue to exist in an elite status. That choice impedes the product that the Intelligence Community exists to produce – insight into adversarial behavior – and it prevents those products from reaching the fulsome results that will help customers and consumers alike.

Over the years, and particularly since the pivotal intelligence failure of 9/11, recommendations have been delivered to improve the Intelligence Community consisting of specific techniques within tradecraft or efforts that will help the analysis be more objective, or more thorough, or more logically consistent.[18] These approaches clearly assume that changes should be made and that other past efforts have failed to address what could be seen as root causes of intelligence challenges – those that stem from the acculturation of the Intelligence Community's strategic culture. The challenge for members of the Intelligence Community and those who rely on them for information, and those who direct them, is to evaluate the conditions of strategic culture, determine what is valuable and what is detrimental, and determine how to keep and grow the valuable while diminishing the detrimental.

Individuals in any type of organization are inherently trying to recreate their own experience with that organization by recreating themselves over time and through their explicit and implicit values. These are reflected as organizational culture and, while not necessarily instilled through the arduous, intentional processes the military and intelligence organizations employ, are seeking much the same ends through the euphemistic (or odious) title of "on-boarding." This jargon signifies a purposeful way an organization tries to bring people within the cultural fold of that organization. As such, organizations take on a life of their own. They purposely recreate their culture and values through indoctrination and enculturation. They recreate themselves, in part, because there is an organizational sense that this specific clustering of people and skills, in this specific way, with these specific features, seems to have been an approach that was successful in the past.

Unfortunately, this notion can lead to failure over time. Organizations that fail to evolve and change to meet new threats and challenges tend to fail. Often the solutions seem to be to hand the problem over to Human Resources. Too often, one-dimensional approaches to helping strategic culture evolve, such as job rotations, joint duty assignments, and diversity hires, are employed by human resources and management scenarios. While important in the organizational culture at large, these practices do not necessarily lead to improving and evolving strategic culture in the Intelligence Community. The Intelligence Community certainly recognizes this problem and has numerous reviews and data regarding diversity.[19] Naturally, these organizations, or rather the people inhabiting them, think that they will be successful in the future, based on today's success. That thought is reassuring and comfortable. The process, however, has yet proved to be effective if judged by the outcomes of the 9/11, Iraq Weapons of Mass Destruction, and Arab Spring fiascos.

Research tells us a great deal about how to overcome biases in thinking and writing. However, is there value to be found in embracing an agency's strategic culture and incorporating it into the analytic process, findings, and reporting? If so, can analysts treat strategic culture as qualitative researchers treat personal biases – state them up front and let the reader assess the impact of these biases, or cultural influences, for themselves? The shortcoming here is that, unlike in some academic settings, consumers of intelligence are not always adept at evaluating the impact of cultural influences on their own work. Much like citing sources in finished intelligence products, perhaps it would be useful to identify the origin story of each analytic theme.

Conclusion

The challenges related to the strategic culture of the US Intelligence Community are many and profound. On the grand scale, the Intelligence Community could be looking at challenges like: do elite organizations change their own culture and values? If so, is there mission value in making that change, and how would that be accomplished? At a more tactical level, the Intelligence Community faces questions that include: how does the Intelligence Community move into the future with a workforce of the past? What are holistic approaches that would ensure success in such an endeavor? Piecemeal programs like diversity recruitment are not working as well as was hoped. Diversity retention is an ongoing problem and seems to suggest a ceiling for diverse hires. Some Intelligence Community organizations have made remarkable strides in gender equity, but there is considerable work left to be done.

There are additional problems, particularly related to the criteria for security clearances. For example, getting young new hires means adjusting recruitment norms to account for such things as a nation where 16 states have legalized recreational marijuana and 37 states have legalized medical marijuana, while federal law has yet to legalize marijuana use. This means the Intelligence Community is going to have to face the fact that this issue has ceased to have any meaning when it comes to potential security risks. As difficult as it was to abandon, the Intelligence Community did get around to changing its view on Communist sympathizers after the collapse of the Soviet Union. No longer concerned with failed ideology, the community focused on espionage and violent behavior in its screening process rather than political theory. A similar case can be made for sexual orientation or gender identity.

In a related vein, as anger and frustration have roiled the country in the twenty-first century, the idea of arrests for things like protests and dissent no longer matter. People who have the temerity to give voice to their opinions in the open have been subject to arrest and detainment. Even the Ivy League has struggled to manage the last decade of American life. The traditional indicators of an individual embedded in the fabric of the power elite do not reflect the American people's experience. Americans are no longer – if they ever were – the Platonic ideal of that once cloistered group of OSS officers. Ethnic and racial equality is an ongoing challenge, as is diversity writ large. The Intelligence Community is making efforts at reaching out to Historically Black Colleges and Universities and trying to recruit from all over the nation. Those efforts are impacted by the choice people have over their own lives and destiny. The incentives of stability and mission may not be enough to convince someone to agree to conformity. Individuals may no longer want a lifetime affiliation with the Intelligence Community, and in the end, that may be a good thing. The Intelligence Community may want to impact the broader population by encouraging people to enter and leave and still be in touch with the organization, much like corporate America has embraced "company alumni."

These problems also strike at the heart of incentives. If people of difference (as opposed to the elite model) are not given the opportunity to succeed and prosper in an organization, they are going to leave. The difficult truth of diversity in the face of modernity needs to be addressed now, rather than later. The Intelligence Community is going to need to find ways to behave that reflect its actual interests in diversity – not as a form of sloganeering, but rather as a way of sincerely representing America and perhaps even becoming better at the task the community is charged with performing. The strategic culture of the United States is increasingly less tightly coupled with elite "friends and family" of the Intelligence Community. Now is a time for the Intelligence Community to grapple with and reflect that change.

Notes

1 Rob Johnston, *Analytic Culture in the US Intelligence Community: An Ethnographic Study* (Washington, DC: Center for the Study of Intelligence, 2005).

2 The term agency as used here is meant in the anthropological or sociological sense, that is, the capacity of individuals to act independently and to make their own free choices. Thomas Dietz and Tom R. Burns, "Human Agency and the Evolutionary Dynamics of Culture," *Acta Sociologica* 35, no. 3 (1992): 187–200. https://journals.sagepub.com/doi/10.1177/000169939203500302

3 Colin S. Gray, *Out of The Wilderness: Prime Time for Strategic Culture,* Report prepared for Defense Threat Reduction Agency Advanced Systems and Concepts Office, Comparative Strategic Cultures Curriculum, 2006, p. 7.
4 Jack L. Snyder, *The Soviet Strategic Culture: Implications for Limited Nuclear Operations*, R-2154-AF (Santa Monica, CA: Rand, 1977), p. 8.
5 In chronological order, the events include the Japanese attack on Pearl Harbor attack, Chinese troops advancing beyond the Yalu River, the Soviet invasion of Hungary, the Soviet invasion of Czechoslovakia, the Soviet invasion of Afghanistan, and terrorist attacks on the United States. The literature is full of examples of intelligence failure: see Abraham Ben-Zvi, "The Study of Surprise Attacks," *Review of International Studies* 5, no. 2 (1979): 129–149.; Avi Shlaim, "Failures in National Intelligence Estimates: The Case of the Yom Kippur War," *World Politics* 28, no. 3 (1976): 348–380; Roberta Wohlstetter, *Pearl Harbor: Warning and Decision* (Stanford University Press, 1962); Frederick L. Shiels, *Preventable Disasters: Why Governments Fail* (Rowman & Littlefield, 1991); James J. Wirtz, *The Tet Offensive: Intelligence Failure in War* (Cornell University Press, 1994). The most useful text from our experience is: Charles Perrow, *Normal Accidents* (Princeton University Press, 2011).
6 Personal communications with former Director of Analysis at CIA Doug MacEachin between 2004 and 2010.
7 The quality and diversity of these histories range from misinformed and poorly researched books like Weiner's to thoughtful and well-researched texts by Immerman and Robarge. See Tim Weiner, *Legacy of Ashes: The History of the CIA* (Anchor, 2008); Richard H. Immerman, *The Hidden Hand: A Brief History of the CIA* (John Wiley & Sons, 2014); and David Robarge, "Leadership in an Intelligence Organization: The Directors of Central Intelligence and the CIA," in *The Oxford Handbook of National Security Intelligence* (Oxford University Press, 2010).
8 Bruce D. Berkowitz, and Allan E. Goodman, *Strategic Intelligence for American National Security* (Princeton University Press, 2021).
9 C. Wright Mills, *The Power Elite* (Oxford: Oxford University, 1956), pp. 3–4.
10 Tom B. Bottomore, *Elites and Society*, 2nd ed. (London: Routledge, 1993); Pierre Bourdieu, *Distinction: A Social Critique of the Judgement of Taste,* trans. Richard Nice (Cambridge, MA: Harvard University Press, 1984); William G. Domhoff, *Who Rules America?* (Englewood Cliffs, NJ: Prentice-Hall, 1967); Gaetano Mosca, *The Ruling Class* (London: McGraw-Hill, 1939); Vilfredo Pareto, *The Mind and Society: A Treatise on General Sociology* (New York: Harcourt, Brace, 1935).
11 Colin S. Gray, *Out of The Wilderness: Prime Time for Strategic Culture.* Report prepared for Defense Threat Reduction Agency Advanced Systems and Concepts Office, Comparative Strategic Cultures Curriculum, 2006, 8.
12 W.C. Langer, H.A. Murray, E. Kris, and B.D., Lewin, *A Psychological Analysis of Adolph Hitler: His Life and Legend* (MO Branch, Office of Strategic Services, 1943). https://ia801304.us.archive.org/33/items/APsychologicalAnalysisofAdolfHitler/A%20Psychological%20Analysis%20of%20Adolf%20Hitler.pdf
13 Robin W. Winks, *Cloak & Gown: Scholars in the Secret War, 1939–1961* (Yale University Press, 1996).
14 Ken Booth, *Strategy and Ethnocentrism* (London: Routledge, 2014).
15 Sherman Kent, *Strategic Intelligence for American World Policy* (Princeton University Press, 2015).
16 Kent, *Strategic Intelligence for American World Policy.*
17 Jack Davis, "The Kent-Kendall Debate of 1949," *Studies in Intelligence* V36:5-91-103 (1992); Jack Davis, "Sherman Kent and the Profession of Intelligence Analysis," *Kent Center Occasional Paper*, 1:5-1-16 (November 2002).
18 Thomas Kean and Lee Hamilton, *The 9/11 Commission Report: Final Report of the National Commission on Terrorist Attacks upon the United States.* Vol. 3 (Government Printing Office, 2004); Charles Duelfer, *Comprehensive Report of the Special Advisor to the DCI on Iraq's WMD, with Addendums* (Central Intelligence Agency, 2005).
19 Alan Ott, Intelligence Community Diversity and Equal Opportunity Intelligence Community Diversity and Equal Opportunity, Congressional Research Service, Washington DC, 2020.

33

ARTIFICIAL INTELLIGENCE AND STRATEGIC CULTURE

Theo Farrell and Kenneth Payne

Introduction

How does strategic culture shape the development and use of military technologies and vice versa? In this chapter, we consider these questions in the context of artificial intelligence (AI), a technology that is widely anticipated to transform warfare as well as societies and civilian economies more broadly.

As we enter the 2020s, AI is already capable of remarkable feats, with new accomplishments arriving almost daily. The strength of AI lies in pattern recognition, searching through huge volumes of data for useful connections. This has allowed a narrow, modular intelligence that, in certain domains, outperforms humans and some of those domains have utility in war. Militaries are eager to sponsor AI research and to instrumentalize it in weapons and other systems. Much hype attends AI, perhaps fueled by Hollywood visions of the Terminator. Yet today's systems are rather limited; their narrow intelligence is very different from our human ability to adapt from one task to another. An algorithm might excel at classifying an image but need extensive retooling before it can fly a helicopter. A more general AI capable of human-like intelligence remains a distant prospect, although that is still the goal for many investors in this technology. AI research is an international scientific endeavor, but it is also emerging from different national and organizational cultures, which will shape how AI is developed and used in different military contexts around the world.

Strategic culture comprises beliefs that are shared by security policy and military communities about their self-identity, the world in which they operate, and given these things what is possible in war. Strategic culture has been used to explain state choices in national security, such as the United States versus Soviet approaches to unlimited nuclear war.[1] Scholars have also examined the cultures of military organizations to explain, for example, why the Royal Navy was slower to pursue submarine warfare than the German Navy in World War II.[2] The normative landscape of war also includes beliefs shared by states and codified in international law about the moral boundaries of acceptable military action, and beliefs shared transnationally by military professionals about what a modern military should look like and how it should act.[3]

 DOI: 10.4324/9781003010302-37

Strategic culture therefore exists and operates at many different levels – organizational, national, and international – and this results in similarities and differences in security policy and military behavior.[4] In turn, it may be expected that culture will operate at all these levels to shape the development and use of AI for military purposes. As discussed below, there are some early indicators of the differences between the United States, Russia, and China. Over time, we may also see common understandings develop across groups of states and scientific communities, and eventually internationally, with regard to what is appropriate and effective with military AI.

In the next section, we explore the relationship between culture and technology in military affairs. The third section then examines how strategic culture has and will continue to shape the development and application of AI in war, and the fourth section considers how strategic cultures may, in turn, be shaped by AI. In the final section, we draw out some conclusions.

Technology and Strategic Culture

From earliest times technology has been central to the conduct of warfare. Indeed, on one common reading, the history of war has been a succession of revolutions in military organization, war finance, fortification, logistics, communications, and systems and tactics for land, air, and sea warfare.[5] The pace of technological change in war accelerated following the industrial revolution of the mid-nineteenth century and again a century later with the electronics revolution. The major Western states excelled at harnessing new technologies for the production and organization of military power, and over the centuries these technologies have been emulated by extra-European states seeking prestige as well as protection from unwanted European encroachment.[6] In recent decades, the United States, with its technology-intensive force structure has become the dominant model for militaries around the world requiring very large capital investment.[7]

It is clear that technology is a major driver of change in war. However, the problem with focusing on successive military revolutions and the rise of the West is that it encourages a technological-deterministic view of the history of war, which underappreciates the importance of culture (including non-Western cultures) in shaping how the tools of war are forged and used.[8] Indeed, strategic culture shapes all aspects of the development and employment of armed forces; below we outline three ways that it does so.

First, culture operates at the microlevel, through overlapping social networks of those who design, build, fund, and use technology, to shape what is developed and how it is developed.[9] Here, it is necessary to recognize that technology includes both material things – such as the M16 rifle – and the social system that makes possible the manufacture of material things.[10] From this perspective, technological development does not follow a natural trajectory toward ever-greater performance improvement. Rather, it is shaped by social forces – including scientific communities, corporate interests, and consumer habits – that determine what constitutes better performance, what is worthy of funding, and what is worth purchasing.[11] In his classic sociological study of nuclear missile guidance, Donald MacKenzie shows how this extreme feat of engineering emerged from the complex and conflictual interaction over decades of scientific, military, and policy communities. The point is that increasing nuclear missile accuracy required huge investment and was only one of many ways that the United States could have addressed the strategic challenges it faced during the Cold War.[12]

Second, culture operates at the meso-level to shape how military technology is developed and used, through the beliefs and preferences of military communities. Thus, the officer culture of the British Army, which emphasized social connection and character over technical skill, impeded the army's willingness and ability to embrace mechanized warfare in the interwar period.[13] In contrast, social class was less significant to the culture of the young Royal Air Force, and far more value was placed on technical know-how and pragmatism.[14] Where new technologies threaten core organizational identities, they are likely to take longer to be adopted by militaries. This was the case with cruise missile technology in the 1970s, which threatened the role of pilots in air forces.[15] However, air forces and navies were in general more techno-centric than armies for most of the twentieth century, with the increasing integration of electronics in land platforms from the early 1980s onward eroding this difference.[16]

Third, culture operates at a macro-level through national identity and popular imagination to align military power with national purpose, and shape when and how countries fight wars. Here, we share John Duffield's criticism of most scholarship on strategic culture, namely that it focuses on the shared beliefs of elites in policy and military communities and neglects to take account of the wider domestic sources of strategic culture.[17] For example, the rise of US naval power in the twentieth century was driven by more than bureaucratic politics or the onset of World War II. Rather it can be traced back to broader commercial and political forces in late nineteenth-century America, including the rise of a Hamiltonian vision of a muscular republic, pressure from business groups to expand into Asian markets, the growing popularity of the "science" of geopolitics, and the need for a new outward-looking national identity for industrializing post-Civil War America.[18] Similarly, the rise of the US Air Force in the twentieth century, and specifically the development of technologies for strategic bombing, was rooted in popular visions of future war invoked by those in the military and industry seeking to promote the airplane.[19]

The latest revolution in military affairs (RMA) driven over the past 50 years by new digital technologies and led by the United States demonstrates the impact of strategic culture on military technology. The strategic cultures of the United States, Russia, and Israel resulted in each focusing efforts on different aspects of this RMA, with the US military leading in capital investment in digital technology, the Russian military investing more effort in developing the conceptual understanding of this RMA, and the Israeli military concentrating on the practical use of digital systems in operations.[20] Within the North Atlantic Treaty Organization (NATO) there has been broad acceptance of the thrust of this US-led military transformation with some national variance, reflecting a general isomorphic tendency of strategic cultures within NATO reinforced by decades of institutional integration and military collaboration, as well as the robustness of the strategic cultures of the major powers such as Britain, France, and Germany leading to a determination to do things somewhat differently.[21]

AI technologies are sweeping the world, fueled by private investment and promising to revolutionize business processes across the civilian economy. Similarly, an AI revolution in military affairs is much anticipated, and to this end states and companies are investing in the development of AI military applications. Against this backdrop of sweeping worldwide change, strategic culture is likely to produce variance between states, and between military communities within states, regarding how AI is imagined, developed, and integrated into armed forces.

How Strategic Culture Is Shaping AI

For a long time, the strategic culture that shaped AI was overwhelmingly American. AI was a product of what Eisenhower famously termed its "military industrial complex" and of the various subcultures and epistemic communities the United States spawned. Research happened elsewhere too – notably in the United Kingdom, France, and Japan. The United Kingdom especially has produced some landmark thinkers on AI, and more broadly on computer science and philosophy of mind. But it had less success in translating basic research into applied technologies. And, compared to the United States, it lacked scale, both in defense and industry.

Not all American research was driven by defense money and goals. The private sector plainly had an interest in AI with, for example, early research by Bell Laboratories for use in communications. But the US federal government, through specialized research offices of the Department of Defense (DOD) and Central Intelligence Agency poured money into civilian university research projects during the Cold War.[22] They also acted as a major customer for the development of computer technologies that would be architecture for AI. Key military tasks were image recognition and natural language processing, both useful for intelligence and surveillance. But even with the scale of funding, progress was slow as the computers simply were not up to the task.

Popular culture combined with business interest to influence AI research in the United States. A rich ecosystem emerged during the 1980s with many ingredients; the technology counterculture that gave rise to cyberpunk was one part of this. So too the entrepreneurial vibrancy of Silicon Valley, with small companies bubbling up, sometimes spinning out from university research, sometimes funded by the Pentagon.[23] The rise of the gaming console and the home computer in the 1980s created a vibrant computer games industry blending creativity and increasingly large budgets; a similar story to Hollywood special effects, including in big-budget films made with DOD cooperation. By the 1990s, the scene was set for the internet era: exploding computer power, increasingly fast network connections, and an economic boom as capital piled into online ventures.

At the start of the new century, the Pentagon's advanced research agency remained a leading player in AI research – its Grand Challenge competition produced a breakthrough in automated cars, and it funded a deep learning helicopter that could swoop acrobatically with inhuman dexterity. But a big change was coming. In AI, the dominant historical paradigm has been dramatically overturned since 2010 by the rise of machine learning and "connectionist" approaches loosely modeled on the neurons of the human brain.[24] The rise of internet giants in the 2010s – Google, Facebook, and Amazon – has altered the AI landscape profoundly through massive investment in machine learning. Their deep pockets rival many governments and even the Pentagon's spending on AI. This has allowed them to hoover up talent from the university sector, competing internationally for the best minds. They have acquired smaller companies too – perhaps most notably the United Kingdom's DeepMind, home of many recent breakthroughs – with the explicit goal of producing truly flexible, general intelligence.[25] AI research is costly: both the human capital and the computer power required in deep learning are exorbitantly expensive. The result has been a concentration in a handful of established, mostly American-owned, corporations that can bear the sobering costs of basic research – approaching $2 billion in losses for DeepMind so far.

It is these features of US strategic culture – the level of integration between military and civilian sectors combined and the sheer size and vibrancy of the US high-technology sector – that has enabled the United States to dominate the development of military technologies such as AI. The United States also dominates through large multinational collaborations on the development of next-generation combat platforms that will increasingly integrate AI capabilities, such as the extravagantly expensive F-35 Lightning II multirole combat aircraft, jointly developed by eight partner countries, and sold on to a handful more.

Western militaries have also begun to develop concepts and programs for wholly new capabilities incorporating AI applications. This has included the emergence of shared beliefs about how AI should be used to support and conduct operations.[26] A key issue concerns the extent to which AI will augment or replace humans in military tasks. In the civilian economy, AI has already enabled the automation of business processes, and this indicates considerable potential for many military administrative tasks similarly to be automated, offering cost savings. There is also scope to enhance training and exercises through AI. Where these tasks concern military operations the matter becomes more contested, both because of the extreme challenges that combat environments present for digital systems, and because of the unique ethical challenges of conducting operations that involve lethal use of force.[27] There are also concerns that while AI can process huge volumes of data with incredible speed, it will be less able to use this information to make appropriate judgments in war.[28]

AI support for operations includes applications that will provide decision-support for tactics, targeting, and planning, and AI-enabled autonomous noncombat systems for logistics, engineering, and medical support in the field. Militaries must equip and prepare to operate in environments that are challenging by virtue of geography, climate, and enemy action. For example, extreme heat and dust caused significant problems for military electronic systems in Iraq and Afghanistan. The current RMA has involved significant Western investment in digital communication and sensor systems. However, here too, the wars in Iraq and Afghanistan have revealed limitations, especially in achieving the necessary bandwidth to move data around the battlespace. A particular challenge concerns the electronic signature emitted by human-operated digital systems, which exposes those operating such systems to attack. By enabling autonomy in military systems, AI could alleviate bandwidth and signal issues as the volume of battlespace communication would be greatly reduced and its efficiency would be greatly increased.[29]

Much of the public focus has been on how AI will be combined with advanced robotics to produce wholly Autonomous Weapons Systems (AWS). From a military perspective, this offers numerous advantages. Expensive and vulnerable human-operated platforms may be replaced by smaller, cheaper, and expendable AWS. Additive manufacturing, or 3D printing, may permit self-repairing and even self-replicating AWS. AI will be able to take and execute fire decisions on a far faster cycle than humans; this is commonly called the "OODA loop" (referring to "observe-orientate-decide-act"). Thus AI-enabled AWS would be able to engage in an ultrafast tactical battle. However, there is considerable concern among Western militaries and defense policymakers with the ethical implications of taking humans "out of the loop." Some existing military systems are already fully automated, such as active defense systems on military platforms, but humans monitor these and can override fire decisions and therefore in principle remain "on the loop."[30]

The dominant view among Western militaries is that war will remain fundamentally a human activity.[31] Given concerns about the technical limitations of operating AI in the battlespace, especially the bandwidth issue and vulnerability to enemy action, and the moral implications of giving AI free rein to kill, Western militaries are expecting AI to support and augment humans in combat in the coming decade. Consensus is emerging around the concept of the "human-machine team," which would include AI tactical decision-support and AI-enabled robots supporting human combatants.[32]

The strategic cultures of Russia and China, heavily influenced by Marxist ideology, incline both countries to view the conduct of war as rooted in the dominant economic forces of the time. Chinese military leaders expect that AI will lead to the transformation of the current digital RMA into a new form of what they call "intelligentized" warfare. In 2017, China's leader, Xi Jinping, instructed the People's Liberation Army to "accelerate the development of military intelligentization."[33] According to the Russian President, Vladimir Putin, "Whoever becomes the leader in this sphere will become ruler of the world."[34] Russia and China both see AI as presenting a key opportunity to close the military technology gap with the United States. However, each state is taking a different approach to developing AI, reflecting their respective strategic cultures.

Russian strategic culture privileges the army which brought victory in World War II, and strategic nuclear forces that underpin Russia's status as a world power. Russia's Military Industrial Committee has established the target for 30% of Russian military equipment to be robotic by 2025. Russia is developing AWS for all domains but is world-leading in robots for ground combat; it has built the world's first robot tank equipped with a 30 mm cannon and anti-tank guided missile. Russia is also developing a robotic submersible able to launch nuclear weapons. Russia is very interested in AI decision-support for military commanders, which reflects a bias in its strategic culture toward scientific approaches to the management and conduct of war. Russia will certainly seek to exploit AI to conduct gray-zone warfare (i.e., aggressive acts against another state below the threshold of war using military and cyber assets) in which it already excels.[35] Techniques useful here include AI-assisted offensive cyber warfare and AI-enabled information warfare, for example, via the creation of realistic social media bots and video "deepfakes."

However, at present, Russia is overclaiming success in military AI.[36] Russian combat robots are uninhabited vehicles rather than true AWS. Russian state investment in AI research has been modest, at only $360 million from 2007 to 2017; in comparison, the US government invested $200 million in AI research in 2017 alone.[37] This reflects an underlying problem for Russian strategic culture, which is that its self-identity as a great power is not matched by the requisite economic power: its defense budget is relatively small and stretched by sustained deployments overseas and, since 2022, its full-scale invasion of Ukraine; its economy is no bigger than Italy's; and it has a poor track record in computer innovation.[38]

In contrast to Russia, China's AI ambitions are matched by investment.[39] China spent $12 billion on AI systems in 2017 and this was expected to rise to $20 billion in 2020. China is developing a range of ground, aerial, and maritime AWS. It arguably leads the US military in the deployment of automated active defense systems on its main battle tanks and automated patrol boats for fleet and harbor security. China is also world-leading in the development of swarm technologies and tactics and is investing in AI battlespace decision-support and automation of logistics. Chinese strategic culture is shaping its approach to military AI in two respects. First is a skeptical approach to the introduction of full autonomy. Given the emphasis on central control and soldier discipline in Chinese strategic culture, there is an ongoing debate within the People's Liberation Army as to whether AI can be trusted. Second is a focus on developing AI for surveillance. This reflects the emphasis on internal security in Chinese strategic culture and huge investment in infrastructure for social control, which includes over 500 million security cameras.[40] AI technologies for population monitoring are very similar to those for battlespace targeting. This focus on population control is enabling China to acquire the largest volume of data in the world, providing the resource to be world-leading in big data and potentially therefore in machine learning.[41]

China has some signal advantages, including its economic scale and ability to invest in AI. However, it also has many drawbacks, including widespread corruption and limited soft power to attract and retain scientific talent.[42] The centralization of the Chinese economy, even after post-Deng reforms, ensures that the Chinese Communist Party retains a strong influence in the direction of AI research, which enables the concentration of investment but is likely to stifle scientific initiative. There are technology startups and internet giants in China, and it would be imprudent to discount the capacity for genuine innovation. At the same time, for all the surge in Chinese papers and patents, so far the big breakthroughs in basic AI research have come from the West.

How AI Could Shape Strategic Culture

By 2050, it is likely that AI will have transformed the conduct of war. Present Western concerns with keeping humans in or on the loop will clash with larger concerns about how less scrupulous opponents are seeking to harness AI to gain military and strategic advantage. The speed, precision, physical control, and recall of AI systems and the scalability of machine learning offer almost boundless possibilities for future military operations. Collectively these attributes suggest a shift toward quantity, disposability, saturation, and distribution of military assets.[43] The quality of AI, rather than other platform attributes, will be increasingly important and absorb development costs and effort.

Different forms of combined arms warfare are likely to emerge, and armed forces are already experimenting with the possibilities.[44] Swarming is one standout possibility: it permits distributed assets to rapidly concentrate for attack and then disperse to avoid counterattack.[45] AI swarms exploit novel tactics to mass against enemy weaknesses, creating the possibility of a first-mover advantage. With no crew to sustain, AI should enhance platform endurance and range which will increase the depth of the battlefield. The increased quantity of cheap assets will include proliferating surveillance capabilities, making it harder to hide and survive on the battlefield and further accentuating the shift toward wider dispersal and disposability. Concealment will be challenging and so may be sacrificed for some roles. Stealth demands performance trade-offs, for example, in payload and maneuverability. But with no crew to protect and lower unit cost, stealth may be less useful than speed or firepower.

The overall effect, especially given the shrinking size of armed forces and risk aversion of many leading AI states, may be to depopulate the battlefield of humans – first on the air and sea, and later on land as combat robotics improves. Large, expensive, prestige platforms, like aircraft carriers, strategic bombers, and main battle tanks, will in time give way to smaller, cheaper AWS. In the meantime, the shelf life of existing large naval, sea, and land platforms will likely be extended by assigning clouds of "semi-slaved" AWS to each, to extend their reach and provide defensive screens.[46]

One important aspect of military AI, both in combat and more broadly in strategic confrontation, will be the struggle in cyberspace for information security. This too will be automated. The balance between attack and defense will be a feature of combat here as much as in the physical world.[47] On that important point, however, much is uncertain: encryption and air gapping currently favor the defense, but AI hacking seems to favor the offense, or at least anticipatory self-defense.

AI for intelligence analysis and strategic level decision-aid are long-standing goals of researchers. Today's AI excels at pattern recognition and correlation, but it lacks human-like understanding and is susceptible to biases in training data, often reflecting the human biases it samples.[48] On prediction in complex environments, AI has yet to demonstrate human-level intuition in forecasting, and even hybrid human-machine teams do not currently outperform the best forecasters when predicting geopolitical events. Hence while AI will increasingly dominate tactical action and even operational planning and execution in the coming decades, its strategic level role will be constrained, perhaps acting as machines do in "centaur chess" to complement the human creativity of strategists with its capacity to simulate and explore options.

Nonetheless, the changes described above will have a profound impact on the culture of the armed forces. Existing service proclivities – for aviators, submariners, infantry, and so on – will be challenged by the shift away from human combatants. Modern armies are heavily reliant on technology but still venerate the combatant, via a "warrior ethos." Recruitment and training are geared toward physical fitness, strength, and aggression. Leaders are selected for moral and physical courage. Air forces extol the dexterity and multitasking required of pilots for aerial dogfights. AI, by contrast, will promote technologists, able to develop, maintain, and monitor autonomous systems.

Those championing AI are likely to encounter resistance from traditional, hierarchical organizations, especially from vested interests in the status quo. Leadership from the top will be key to realizing transformative change.[49] Here it should be noted that the digital transformation of Western militaries has been driven forward since the early 2000s by civilian leaders and military chiefs.[50] However, the AI transformation will require deeper cultural change because it will reorder the balance between traditional warfighting and support branches of the military. In this, it will be much like earlier revolutions in naval warfare – with the introduction of steel hulls, coal and oil-fired turbines, submersible boats, and eventually nuclear reactors – which elevated the demand for and prestige of engineers.

Security dilemma dynamics between states will also drive forward transformative change. The prospects for arms control in this area are poor, because states cannot agree on terms, and cheating on agreements would be easy: monitoring would be immensely challenging, and a decisive capability upgrade could be one upload away. For similar reasons, states are most unlikely to agree a ban on AWS.[51]

As battlefields are depopulated, and the pace of automated action accelerates, AI will demand new approaches to leadership and command. States that extol "mission command" with highly delegated authority are better placed to adapt than those with centralized and hierarchical structures. This favors Western military cultures over Russian and Chinese military cultures. Increasingly, the tactical initiatives will be autonomous. Physical courage may be a less useful leadership attribute when leaders are further from the action, and moral courage in military leaders may be less salient when ethical decisions are taken *ex ante* at the strategic level, rather than *in bello* by operational leaders. Managerialism to keep the automated system functioning will be a prized skill; operational artistry is less so when autonomous systems are knitting together tactical actions for operational effect. Logistical acumen becomes the center of gravity, rather than élan and coup d'œil.

Fighting power will increasingly rest on the ability to innovate and instrumentalize the latest technology. In the United States and elsewhere, military visionaries are keen to adopt the perceived flexibility and responsiveness of Silicon Valley. Here the United States can leverage its techno-centric strategic culture, in terms of the strengths of and ties to civilian scientific communities.[52] But there are dangers – not only that "bureaucracy does its thing" in impeding change but also that many if not most tech startups fail. Speeding up the acquisition cycle will be critical, as AI capability will only remain cutting edge for months, not decades. Equally important will be a more holistic approach to procurement that will deliver integrated AI architecture, as opposed to digital systems that are not able to communicate with one another.

One important cultural change wrought by AI will be in what Clausewitz called the "grammar" of war; that is, the principles and norms governing the conduct of war. The pace of novelty, as in the cybersecurity struggle, will require new thinking about the meaning of strategic behaviors. Established norms shape responses to adversary actions. As norms are unsettled, there is ambiguity as to the meaning and implications of activities involving AI. For example, is it permissible to capture and exploit AWS? These questions have been asked already with respect to uninhabited aerial and submersible vehicles. There is also the risk that flaws in algorithms or machine learning will lead to military activity that carries excessive strategic or ethical risk.[53]

Ultimately, AI may alter the social fabric that underpins the strategic cultures of all states.[54] One obvious impact is the social compact between governed and government – AI permits snooping on the lives of citizens, all of whom leave an indelible digital trace of their behaviors and attitudes. Individual privacy is challenged as never before.[55] As the case of China shows, AI conjures the possibility of a totalitarian surveillance state that can compel acceptable behaviors and, more insidiously, shape imaginations. States have long sought to influence attitudes, and now AI permits tailored propaganda. The broad parameters are already apparent in deepfake videos and Amazon recommendations. And what happens to war when the "passionate hatred" of the people can be unleashed at a distance, by uninhabited systems?

A more fundamental change in our social structure is emerging from the marriage of AI and biotechnology. This change is already underway, as scientists employ AI to understand and edit the human genome. The possibilities edge into science fiction. One result will be augmented intelligence – first to eliminate debilitating genetic conditions; later to select eugenically for superior cognitive attributes. The latter offers military applications that some states may well pursue. Popular imagination and business interests may align to result in more widespread AI augmentation of human intelligence. Ethical considerations will constrain this research in some states more than others, perhaps to the advantage of authoritarian governments like China and Russia.

Clausewitz famously drew out the distinction between character and the nature of war. Revolutionary military technologies such as mechanized vehicles, airplanes, submarines, and nuclear weapons have the capacity to change the character of war; that is, the ways it is fought. However, the nature of war is ineluctable, timeless, and biological – rooted in our distinctive humanity. In essence, war is violent, uncertain, social, and psychological. Humans make and enact decisions to achieve their shared goals against enemies. As the above discussion suggests, if AI is used to create new systems of political control and military strategy, it will alter aspects of the nature of war – specifically, the extent to which human passion and cognition shape warmaking.[56] At the same time, fundamental elements of the nature of war will remain unchanged, namely, its political purpose, the centrality of violence, and the role of friction and chance.[57]

Popular imagination already foresees the possibility of agency for artificial minds, with their own goals, their own community, and their own culture.[58] Thoroughgoing materialists, which include almost all scientists, hold that the mind is a "meat machine" that is a product of materials and structure. It follows that other such machines are possible, in terms of entirely artificial emulations of human brains or hybrid human-machine minds.[59] These innovations seem far distant, though not entirely far-fetched. Meanwhile, modern AI is entirely mindless in that it lacks emotion and self-awareness. But already it engages in collaborative and adversarial activities; already it "evolves" competitively; already it designs its own AI. The endpoint, theoretically, of AI is artificial strategic culture.

Conclusion

Strategy, politics, and culture all shape how states create military power.[60] Strategic competition is unquestionably driving forward the development of military AI. Reflecting a view common to major militaries worldwide, the British Ministry of Defence considers that "we are in a race with our adversaries to unlock this advantage. The clock is ticking."[61] Equally, as noted, Russia and China see AI as offering the potential to help them catch up militarily with the West.

Domestic politics and institutional arrangements, including civil-military and bureaucratic politics and state-defense industry relations, are also significant in how states are investing in and integrating AI into their respective armed forces. Nondemocratic states are more able than democratic states to drive investment in particular military technologies, in essence by diktat to defense industries. However, the top-down character of such relations between state and industry hinders the failure-tolerant creative processes that are required for scientific discovery. Hence, democratic states tend to be better at developing cutting-edge military technologies.[62] State military capacities are also very relevant where the acquisition and integration of AI involves emulating the major powers, in terms of the ability to absorb new technologies and master new tactics.[63]

This chapter shows how strategic culture matters as much as these material factors when it comes to the development and use of military AI. AI offers a vision of future warfare that builds on the current US-led digital RMA, which in turn reflects the techno-centrism of US strategic culture. Transnationally, the emerging concept of human-machine teaming reflects a consensus among Western militaries that AI will support rather than replace humans in combat. Military AI plans and developments in Russia and China reflect the strategic cultures of each country, with a focus on ground and strategic forces in Russia, and on surveillance and big data in China. In addition to military and policy elite beliefs, strategic culture also includes wider societal views and especially popular imagination and future visions of war.

In time, it is most likely that AI will displace humans from combat functions and in so doing will necessarily alter military cultures that currently privilege the warfighter. Similarly, AI insertion in strategic policymaking also holds the potential to alter national strategic cultures. For some time now, science fiction has presented a future where machine minds use lethal force independent of human control, sometimes for human benefit and other times not. This would require advances in AI technology that are not expected in the coming decades, but if realized could change the very nature of war.

Some experts worry that in the future, AI may increase international instability and cause inadvertent wars.[64] Automated use of force could cause conflicts to escalate rapidly and slip out of human control. Dependency on AI decision-support could reduce the capacities of human leadership in crises.[65] The nightmare scenario of a nuclear war caused by AI, once the stuff of science fiction, is now the subject of serious analysis.[66] In the past, most attempts to predict the future of war have got it wrong.[67] One can take little comfort from this: it remains to be seen if our current visions have exaggerated or underestimated the danger AI presents to humanity.

Notes

1 Jack L. Snyder, *The Soviet Strategic Culture. Implications for Limited Nuclear Operations* (Santa Monica, CA: Rand Corporation, 1977); Colin Gray, *Nuclear Strategy and National Style* (Lanham, MD: Hamilton, 1986).
2 Jeffrey W. Legro, *Cooperation Under Fire* (Ithaca, NY: Cornell University Press, 1995).
3 Theo Farrell, *The Norms of War* (Boulder, CO: Lynne Rienner, 2005).
4 Peter J. Katzenstein (ed.), *The Culture of National Security* (New York: Columbia University Press, 1996). Farrell, *The Norms of War.*
5 William H. McNeill, *The Pursuit of Power* (Oxford: Blackwell, 1983). Archer Jones, *The Art of War in the Western World* (Oxford: Oxford University Press, 1987). Martin Van Creveld, *Technology and War: From 2000 B.C. to the Present* (London: Collier Macmillan, 1989).
6 David B. Ralston, *Importing the European Army* (Chicago: Chicago University Press, 1990). Emily Goldman, "The spread of Western military models to Ottoman Turkey and Meiji Japan," in Theo Farrell and Terry Terriff (eds.), *The Sources of Military Change: Culture, Politics, Technology* (Boulder, CO: Lynne Rienner Publishers, 2002), pp. 41–68.
7 Chris Demchak, "Complexity and Theory of Networked Militaries," in Theo Farrell and Terry Terriff (eds.), *The Sources of Military Change* (Boulder, CO: Lynne Rienner, 2002), pp. 221–264. Ann Hironaka, *Tokens of Power* (Cambridge: Cambridge University Press, 2017).
8 Jeremy Black, *War and the World* (New Haven, CT: Yale University Press, 2000). John A. Lynn, *Battle* (Cambridge, MA: Westview Press, 2003).
9 Jon R. Lindsay, "War Upon the Map: User Innovation in American Military Software," *Technology and Culture,* 51, no. 3 (2010), pp. 619–651.
10 Matthew Ford, *Weapon of Choice* (London: Hurst, 2017).
11 Merrit Roe Smith and Leo Marx, *Does Technology Drive History?* (Cambridge, MA: The MIT Press, 1994).
12 Donald MacKenzie, *Inventing Accuracy* (Cambridge, MA: The MIT Press, 1990).
13 Elizabeth Kier, *Imagining War* (Princeton, NJ: Princeton University Press, 1997).
14 David Stubbs, "The Culture of the Royal Air Force, 1917–1945," in Peter R. Mansour and Williamson Murray (eds.), *The Culture of Military Organizations* (Cambridge: Cambridge University Press, 2019), pp. 403–425.
15 Theo Farrell, *Weapons without a Cause* (London: Macmillan, 1997), pp. 122–123.
16 Carl H. Builder, *The Masks of War* (Washington, DC: John Hopkins University Press, 1989). Chris Demchak, *Military Organizations, Complex Machines* (Ithaca, NY: Cornell University Press, 1991).
17 John S. Duffield, *World Power Forsaken* (Stanford, CA: Stanford University Press, 1998).
18 Peter Trubowitz, Emily O. Goldman, and Edward Rhodes (eds.), *The Politics of Strategic Adjustment* (New York: Columbia University Press, 1999).
19 Michael S. Sherry, *The Rise of American Air Power* (New Haven, CT: Yale University Press, 1987).

20 Dima Adamsky, *The Culture of Military Innovation* (Stanford, CA: Stanford University Press, 2010).
21 Terry Terriff, Frans Osinga, and Theo Farrell (eds.) *A Transformation Gap?* (Stanford, CA: Stanford University Press, 2010).
22 Nils Nilsson, *The Quest for Artificial Intelligence: A History of Ideas and Achievements* (Cambridge: Cambridge University Press, 2010).
23 Margaret O'Mara, *The Code: Silicon Valley and the Remaking of America* (London: Penguin Books, 2020). Thomas Rid, *The Rise of the Machines* (New York: W.W. Norton, 2016).
24 Martin Ford, *Architects of Intelligence* (Birmingham: Packt Publishing, 2018). Michael Wooldridge, *The Road to Conscious Machines: The Story of AI* (London: Pelican Books, 2020).
25 Greg Williams, "Inside DeepMind's Epic Mission to Solve Science's Trickiest Problem," *WIRED*, 6 August 2019.
26 US Army, *The U.S. Army Robotic and Autonomous Systems Strategy* (Fort Eustis: Training and Doctrine Command, 2017). UK Ministry of Defence (MoD), *Joint Concept Note 1/18: Human-Machine Teaming* (Shrivenham: Development, Concepts and Doctrine Centre, 2018). Australian Army, *Robotic & Autonomous Systems Strategy* (Canberra: Future Land Warfare Branch, 2018).
27 Michael C. Horowitz, "The Ethics and Morality of Robotic Warfare: Assessing the Debate over Autonomous Weapons," *Daedalus* 145, no. 4 (2016), pp. 25–36.
28 Avi Goldfarb and Jon R. Lindsay, *Artificial Intelligence in War* (Washington DC: Brookings Institution, 2020).
29 Sydney J. Freedberg, "A Twentieth Century Commander Will Not Survive: Why the Military Needs AI," *Breaking Defense*, 12 January 2021, https://breakingdefense.com/2021/01/a-20th-century-commander-will-not-survive-why-the-military-needs-ai/
30 Paul Scharre, "How Swarming Will Change Warfare," *Bulletin of the Atomic Scientists*, 74, no. 6 (2018), pp. 385–389.
31 Brad Dewees, Chris Umphres, and Maddy Tuny, "Machine Learning and Life-and-Death Decisions on the Battlefield," *War on the Rocks*, 11 January 2021, https://warontherocks.com/2021/01/machine-learning-and-life-and-death-decisions-on-the-battlefield/. Alex Neads, Theo Farrell, and David Galbreath, *From Tools to Teammates* (Canberra, ACT: Australian Army Research Centre, 2021).
32 UK Ministry of Defence (MoD), *Joint Concept Note 1/18: Human-Machine Teaming*. Mick Ryan, *Human Machine Teaming for Future Ground Forces* (Washington, DC: Centre for Strategic and Budgetary Assessments, 2018). Paul Scharre, *Autonomous Weapons and Operational Risk* (Washington, DC: Centre for a New American Security, 2016).
33 Elsa B. Kania, "Chinese Military Innovation in the AI Revolution," *RUSI Journal*, 164, nos. 5–6 (2019), pp. 26–34, at p. 27.
34 James Vincent, "Putin Says the Nation that Leads in AI 'Will Be Ruler of the World'," *The Verge*, 4 September 2017. https://www.theverge.com/2017/9/4/16251226/russia-ai-putin-rule-the-world
35 Forrest E. Morgan, Benjamin Boudreaux, Andrew J. Lohn, Mark Ashby, Christian Curriden, Kelly Klima, and Derek Grossman, *Military Applications of Artificial Intelligence: Ethical Concerns in an Uncertain World* (Santa Monica, CA: RAND Corporation, 2020), pp. 83–99
36 Jeffrey Edmonds, Samuel Bendett, Anya Fink, et al., *Artificial Intelligence and Autonomy in Russia* (Arlington, VA: Centre for Naval Analyses, May 2021).
37 Morgan et al., op. cit., p. 91.
38 Samuel Bendett, "Russia's New 'AI Supercomputer' Runs on Western Technology," *DefenseOne*, 4 March 2019. https://www.defenseone.com/technology/2019/03/russias-new-ai-supercomputer-runs-western-technology/155292/
39 Gregory C. Allen, "Understanding China's AI strategy," *Centre for New American Security*, 6 February 2019. Morgan et al., op. cit., pp. 60–71. Huw Roberts, Josh Cowls, Jessica Morley, *et al.* "The Chinese Approach to Artificial Intelligence: An Analysis of Policy, Ethics, and Regulation," *AI & Society* 36 (2021), pp. 59–77.
40 Morgan et al., op. cit., pp. 60–71.
41 Rand Waltzman et al., *Maintaining the Competitive Advantage in Artificial Intelligence and Machine Learning* (Santa Monica, CA: RAND Corporation, 2020).

42 Stephen G. Brooks and William C. Wohlforth, "The Rise and Fall of Great Powers in the Twenty-First Century," *International Security* 40, no. 3 (2015/2016), pp. 7–53.
43 Kenneth Payne, *I Warbot: The Dawn of Artificially Intelligent Conflict* (London: Hurst & Co, 2021), pp. 81–136.
44 James Johnson, "Artificial Intelligence and Future Warfare: Implications for International Security," *Defense & Security Analysis* 35, no. 2 (2019), pp. 147–169. Kenneth Payne, *Strategy, Evolution and War: From Apes to Artificial Intelligence* (Georgetown: Georgetown University Press, 2018).
45 Scharre, "How Swarming Will Change Warfare."
46 Neads et al., *From Tools to Teammates.*
47 Ben Garfinkel and Allan Dafoe, "How Does the Offense-Defense Balance Scale?," *Journal of Strategic Studies*, 42, no. 6 (2019), pp. 736–763.
48 Gary Marcus and Ernest Davis, *Rebooting AI: Building Artificial Intelligence we can Trust* (New York: Pantheon Books, 2019).
49 Stephen Peter Rosen, *Winning the Next War* (Ithaca, NY: Cornell University Press, 1991).
50 Theo Farrell, Sten Rynning, and Terry Terriff, *Transforming Military Power* (Cambridge: Cambridge University Press, 2013).
51 Ingvild Bode and Hendrik Huelss, "Autonomous Weapons Systems and Changing Norms of International Relations," *Review of International Studies* 44, no. 3 (2018), pp. 393–413. Matthijs M. Maas, "How Viable Is International Arms Control for Military Artificial Intelligence? Three Lessons from Nuclear Weapons," *Contemporary Security Policy* 40, no.3 (2019), pp. 285–311.
52 Christian Brose, *The Kill Chain: Defending America in the Future of High-Tech Warfare* (New York: Hachette, 2020).
53 Benjamin Jensen, Christopher Whyte, and Scott Cuomo, "Algorithms at War: The Promise, Peril and Limits of Artificial Intelligence," *International Studies Review* 22, no. 3 (2019), pp. 526–550. James Johnson, "Delegating Strategic Decision-Making to Machines: Dr Strangelove Redux?," *Journal of Strategic Studies; 45, no. 3* (2022), pp. 439–477.
54 Kai-Fu Lee, *AI Superpowers: China, Silicon Valley and the New World Order* (New York, Houghton Mifflin, 2018). Brad Smith and Carol Ann Browne, *Tools and Weapons: The Promise and the Peril of the Digital Age* (London: Hodder & Stoughton, 2019).
55 Arthur Holland Michel, *Eyes in the Sky: The Secret Rise of Gorgon's Stare and How It Will Watch Us All* (New York: Houghton Mifflin, 2019).
56 Kenneth Payne, "Artificial Intelligence: A Revolution in Strategic Affairs?" *Survival*, 60, no. 5 (2018), pp. 7–32.
57 Frank G. Hoffman, "Will War's Nature Change in the Seventh Military Revolution?" *Parameters* 47, no. 4 (2017–18), pp. 19–31.
58 Iain M. Banks, *Consider Phlebas* (London: Macmillan, 1987).
59 Mark O'Connell, Mark. *To Be a Machine: Adventures Among Cyborgs, Utopians, Hackers, and the Futurists Solving the Modest Problem of Death* (London: Granta, 2017). Max Tegmark, *Life 3.0: Being Human in the Age of Artificial Intelligence* (London: Allen Lane, 2017).
60 Risa A. Brooks and Elizabeth A. Stanley (eds.) *Creating Military Power* (Stanford, CA: Stanford University Press, 2007).
61 UK Ministry of Defence (MoD), *Joint Concept Note 1/18: Human-Machine Teaming*, at iii.
62 Matthew Evangelista, *Innovation and the Arms Race* (Ithaca, NY: Cornell University Press, 1988).
63 Michael C. Horowitz, *The Diffusion of Military Power* (Princeton, NJ: Princeton University Press, 2010).
64 Jurgen Altmann and Frank Sauer, "Autonomous Weapons Systems and Strategic Stability," *Survival*, 59, no. 5 (2017), pp. 117–142.
65 Michael C. Horowitz and Paul Scharre, *AI and International Security* (Washington, DC: Centre for a New American Security, 2021).

66 James Johnson, "'Catalytic Nuclear War' in the Age of Artificial Intelligence & Autonomy: Emerging Military Technology and Escalation Risk between Nuclear-Armed States," *Journal of Strategic Studies*, 2021.
67 Lawrence Freedman, *The Future of War: A History* (London: Allen Lane, 2017).

Selected Bibliography

Adamsky, Dima. *The Culture of Military Innovation* (Stanford, CA: Stanford University Press, 2010).
Allen, Gregory C. "Understanding China's AI Strategy," *Centre for New American Security*, 6 February 2019.
Altmann, Jurgen and Frank Sauer. "Autonomous Weapons Systems and Strategic Stability," *Survival*, 59, no. 5 (2017), pp. 117–142.
Bode, Ingvild and Hendrik Huelss. "Autonomous Weapons Systems and Changing Norms of International Relations," *Review of International Studies*, 44, no. 3 (2018), pp. 393–413.
Brose, Christian. *The Kill Chain: Defending America in the Future of High-Tech Warfare* (New York: Hachette, 2020).
Edmonds, Jeffrey, Samuel Bendett, Anya Fink, Jeffrey Edmonds, Samuel Bendett, Anya Fink, Mary Chesnut, Dmitry Gorenburg, Michael Kofman, Kasey Stricklin, and Julian Waller. *Artificial Intelligence and Autonomy in Russia* (Arlington, VA: Centre for Naval Analyses, May 2021).
Farrell, Theo. *The Norms of War* (Boulder, CO: Lynne Rienner, 2005).
Farrell, Theo, Sten Rynning, and Terry Terriff. *Transforming Military Power* (Cambridge: Cambridge University Press, 2013).
Goldfarb, Avi and Jon R. Lindsay. *Artificial Intelligence in War* (Washington DC: Brookings Institution, 2020).
Hoffman, Frank G. "Will War's Nature Change in the Seventh Military Revolution?" *Parameters*, 47, no. 4 (2017–2018), pp. 19–31.
Horowitz, Michael C. "The Ethics and Morality of Robotic Warfare: Assessing the Debate over Autonomous Weapons," *Daedalus*, 145, no. 4 (2016), pp. 25–36.
Horowitz, Michael C. and Paul Scharre. *AI and International Security* (Washington, DC: Centre for a New American Security, 2021).
Jensen, Benjamin, Christopher Whyte, and Scott Cuomo. "Algorithms at War: The Promise, Peril and Limits of Artificial Intelligence," *International Studies Review*, 22, no. 3 (2019), pp. 526–550.
Johnson, James. "Artificial Intelligence and Future Warfare: Implications for International Security," *Defense & Security Analysis*, 35, no. 2 (2019), pp. 147–169.
Kania, Elsa B. "Chinese Military Innovation in the AI Revolution," *RUSI Journal*, 164, no. 5–6 (2019), pp. 26–34.
Lee, Kai-Fu. *AI Superpowers: China, Silicon Valley and the New World Order*. (New York, Houghton Mifflin, 2018).
Maas, Matthijs M. "How Viable Is International Arms Control for Military Artificial Intelligence? Three Lessons from Nuclear Weapons," *Contemporary Security Policy*, 40, no.3 (2019), pp. 285–311.
Michel, Arthur Holland. *Eyes in the Sky: The Secret Rise of Gorgon's Stare and How It Will Watch Us All* (New York: Houghton Mifflin, 2019).
Morgan, Forrest E., Benjamin Boudreaux, Andrew J. Lohn, Mark Ashby, Christian Curriden, Kelly Klima, and Derek Grossman. *Military Applications of Artificial Intelligence: Ethical Concerns in an Uncertain World* (Santa Monica, CA: RAND Corporation, 2020), pp. 83–99.
Nilsson, Nils. *The Quest for Artificial Intelligence: A History of Ideas and Achievements* (Cambridge: Cambridge University Press, 2010).
Payne, Kenneth. *Strategy, Evolution and War: From Apes to Artificial Intelligence* (Georgetown: Georgetown University Press, 2018).

Payne, Kenneth. "Artificial Intelligence: A Revolution in Strategic Affairs?" *Survival*, 60, no. 5 (2018), pp. 7–32.
Payne, Kenneth. *I Warbot: The Dawn of Artificially Intelligent Conflict* (London: Hurst & Co, 2021).
Roberts, Huw, Josh Cowls, Jessica Morley, *et al.* "The Chinese Approach to Artificial Intelligence: An Analysis of Policy, Ethics, and Regulation," *AI & Society*, 36 (2021), pp. 59–77.
Ryan, Mick. *Human Machine Teaming for Future Ground Forces* (Washington, DC: Centre for Strategic and Budgetary Assessments, 2018).
Scharre, Paul. *Autonomous Weapons and Operational Risk* (Washington, DC: Centre for a New American Security, 2016).
Waltzman, Rand, Lillian Ablon, Christian Curriden, Gavin S. Hartnett, Maynard A. Holliday, Logan Ma, Brian Nichiporuk, Andrew Scobell, and Danielle C. Tarraf. *Maintaining the Competitive Advantage in Artificial Intelligence and Machine Learning* (Santa Monica, CA: RAND Corporation, 2020).

34

STRATEGIC CULTURE AND ANTICIPATORY INTELLIGENCE

Briana D. Bowen

The distinct fields of strategic culture and anticipatory intelligence share a common problem in that both struggle with significant divergence over their foundational definitions and conceptual boundaries, but both offer highly useful approaches for making better sense of complex systems and problem sets. Strategic culture and anticipatory intelligence are interdisciplinary fields born in the nexus of policy and academia, oriented toward delivering actionable insights for decisionmakers, and drawing on a cheerfully multifarious range of theories and methods to try to better grapple with complex realities in the security domain. Understanding in one field enhances insight and effectiveness in the other, and each field offers useful tools and approaches to complement the other.

This chapter first provides a brief overview of anticipatory intelligence, then explores three key domains of useful intersection with the field of strategic culture.[1] The first of these is an examination of the value of utilizing strategic culture to assess the approaches of traditional strategic actors – nation-states – to the complex security challenges addressed by anticipatory intelligence. The chapter argues that an understanding of the strategic cultural conceptions of identity, values, norms, and perceptions is essential for making sense of states' approaches and responses to twenty-first-century problem sets. It further evaluates the utility of a strategic culture lens in assessing how different states make sense of and navigate complexity and the phenomenon of emergence, and how divergent perceptions of these developments can lead to dangerous misreading and miscalculation in the international security environment, particularly in the era of resurgent great power competition.

Second, this chapter argues for the application of strategic culture tools and approaches to an increasingly broad set of actors not traditionally considered to be part of the "strategic" set. The forces driving unprecedented complexity and the democratization of technology have empowered informal networks, loosely defined groups, and even individuals to have a historically unmatched impact on regional and global affairs. This argument further considers the broadening "securitization" of problem sets across society, which is locating issues such as food and water security, supply chain resilience, and informational struggle in the global communication commons under the label of national (or other) security, requiring the consideration of a much wider group of actors relevant at the strategic level.

DOI: 10.4324/9781003010302-38

Finally, this chapter suggests areas for future research in the nexus between strategic culture and anticipatory intelligence, including advancing the research focus on strategic cultures as complex systems and leveraging tools from complexity science for cultural analysis.

Defining Anticipatory Intelligence

Anticipatory intelligence is a relatively young discipline within national security intelligence practice[2] and an even newer domain within academia.[3] Captured in the 2019 US National Intelligence Strategy as a top mission priority, if a nebulous one, the raison d'être of anticipatory intelligence is to:

> Identify and assess new, emerging trends, changing conditions, and underappreciated developments to challenge long-standing assumptions, encourage new perspectives, identify new opportunities, and provide warning of threats to U.S. interests.... Anticipatory intelligence usually leverages a cross-disciplinary approach, and often utilizes specialized tradecraft to identify emerging issues from "weak signals," cope with high degrees of uncertainty, and consider alternative futures. Anticipatory intelligence looks to the future as foresight (identifying emerging issues), forecasting (developing potential scenarios), or warning. Anticipatory intelligence explores the potential for cascading events or activities to reinforce, amplify, or accelerate conflict.[4]

Despite a degree of buzzword enthusiasm for the concept over the past decade, anticipatory intelligence has struggled with poorly defined parameters and functions, particularly in distinguishing itself from the well-established domain of strategic intelligence, defined in the same 2019 National Intelligence Strategy as the charge to "[i]dentify and assess the capabilities, activities, and intentions of states and non-state entities to develop a deep understanding of the strategic environment, warn of future developments on issues of enduring interest, and support US national security policy and strategy decisions."[5] Some practitioners have argued that all intelligence is meant to be forward-looking, that the "warning" function in strategic intelligence essentially is anticipatory intelligence, and therefore that little significant distinction is merited between the two.

Yet others, foremost Josh Kerbel, the "godfather" of anticipatory intelligence, have made the case that anticipatory intelligence is and must be a purposely distinct approach. Kerbel's case is based on an assessment of the twenty-first-century global security environment as a meaningfully different creature from the global dynamics of the Cold War, the era in which the US Intelligence Community was born and experienced formative adolescence, including the development of its approach to strategic intelligence. Offering a definition of anticipatory intelligence as "[t]he intelligence process or practice whereby potentially emergent developments stemming from the increasingly complex security environment are foreseen via the cultivation of holistic perspectives," Kerbel's approach focuses on the distinction between "complicated" and "complex" security environments, drawing on parameters based in the field of complexity science.[6]

Complicated security environments and problem sets, typified by the bipolar contest between the United States and the Soviet Union during the Cold War, are more linear, are assumed to have a degree of proportionality in action and reaction, often reflect intentional moves from state actors, and can underpin an international dynamic akin to a two-body problem (e.g., the stable orbitational relationship between the earth and moon). Complex environments and problem sets, by contrast, are less linear and monocausal, may experience exponential escalation due to unintended or unanticipated intervening variables, and are more likely to reflect disproportionate action and reaction between diverse, connected, and interdependent actors at the state, non-state, and supra-state levels. Illustrated by the Arab Spring uprisings across the early 2010s, this type of environment may more closely resemble the dynamics of a three-body problem (e.g., the chaotic and unpredictable relationship between three proximate celestial bodies). Furthermore, complex systems are predisposed to spawn the phenomenon of emergence, wherein a macro-level outcome that no willful actor intended emerges from the collective dynamics of a system.[7]

Human societies have long faced messy security challenges, and elements of both complicated and complex security environments as characterized above appear across the long span of human history. In addition, "simply" complicated problem sets do persist into the twenty-first century. But a compelling case can and should be made that over the past century and particularly since the end of the Cold War, the evolution of global political power distribution; the rising economic power of select individuals amid intensifying wealth inequality; the interconnection of states and the global population through manufacturing, supply chains, and the internet; and the democratization of technology, particularly ubiquitous communications platforms, have created a world defined by meaningfully unprecedented complexity, however familiar some of the enduring challenges faced by humanity.[8] In the face of this era's security environments and problem sets, inertia-anchored attachments to the assumptions, methods, and approaches of a less complex past – one not many decades removed – can prove a dangerous blind spot leading to one unpleasant strategic surprise after another.[9] Traditional strategic intelligence retains an important, even essential, role in tracking and delivering actionable insights on the security challenges of this era, but the value of anticipatory intelligence, developed and practiced as a distinct domain with tools oriented toward tackling modern elements of unprecedented complexity, should not be overlooked.

Here the conversation returns to strategic culture. For their apparent differences in focus and approach, anticipatory intelligence and strategic culture share important common genetic traits. Both may be thought of, and have been variously argued, as either a field or a method.[10] Both pragmatically harvest insights, approaches, and tools from a wide range of disciplines – in the case of strategic culture: anthropology, sociology, history, international relations, strategic studies, and more; in the case of anticipatory intelligence: complexity science, futures studies, security studies, resilience modeling, predictive computing, and so forth. Further, both fields accept rather elastic definitional bounds and motley applied methods for the sake of establishing practical approaches to better tackle the consequential security questions of the era. Jack Snyder's foundational 1977 piece on Soviet strategic culture sought a new approach to glean much-needed insights on the question of Soviet nuclear behavior[11]; anticipatory intelligence was born into practice within the US Intelligence Community in response to the complex emergence of the Arab Spring and other turbulent developments of the early twenty-first century.[12] Recognizing the resemblance in structure and purpose between these fields illuminates the value of jointly leveraging approaches across the two.

Anticipatory Intelligence: Traditional Strategic Actors

Because anticipatory intelligence relates to emergent and often as-yet poorly conceptualized security challenges and disruptions, understanding the cultural forces shaping, enabling, and constraining state decisionmaking in this context can offer significant value in the pursuit of foresight, or "imagining how a broad set of possible conditions ... might interact and generate emergent outcomes."[13] Of the four strategic cultural aspects identified in Johnson and Berrett's Cultural Topography methodology – identity, values, norms, and perceptual lens[14] – insights relating to the perceptual lens are especially valuable in assessing how members of strategic communities across different nations will perceive, conceptualize, and respond to complex future security challenges.

Anticipating States' Responses to Complex Problem Sets

Many complex emerging issues, such as climate change and linked issues of food and water security, present states with distinct threats. Yet the way different states' strategic communities perceive a complex problem set may vary significantly in terms of the scope, severity, salience, and options for response to the problem, and at times will reflect even more fundamental divergences in cosmology and paradigm. For example, American strategic culture tends to see the world as changeable, and malleable to the imposition of human will. It assumes that most extant problems have a solution if only it can be found and implemented, probably through the clever invention and use of technology.[15] Despite the intensive politicization of climate change in the United States which has undeniably fragmented policy responses over recent decades, these culturally informed worldviews have been visible in US policy approaches to climate change during the Obama and Biden administrations. When action has been politically possible, it has prioritized investment in technological interventions like green energy, carbon capture technologies, net-zero infrastructure, and resilience-oriented retrofitting of critical infrastructure like defense installations.[16]

By comparison, Russian strategic culture has inclined Moscow's elites to even more limited engagement with the climate crisis across recent years. This is in part because of politicized skepticism of largely Western-led climate science and international collaborations, but also in part because of a differing orientation to the relationship between nature and human society, including a lesser inclination to believe that humanity can successfully impose its will on the environment or even perhaps, normatively, that it should. One analyst tracking Russian discourse notes that Vladimir Putin "has actually argued that climate change is a global phenomenon that cannot be countered and that it will help Russian economic development in the long run."[17] Although this view is not universal, the Russian action that has taken place on climate mitigation efforts has been very lackluster.[18] Russians' experience with the vast forces of nature across eleven time zones and latitudes ranging from Arctic to subtropical are likely to contribute to this perspective, as do the essential supporting roles of "General January and General February"[19] in the storied defeats of Napoleon and Hitler during their respective invasions of the Russian homeland in the so-called Patriotic War (1812) and Great Patriotic War (1941–1945). Both victim to and protected by the harsh forces of nature across its history, modern Russian strategic culture seems to reflect a degree of popular ambivalence as to whether climate change is even something Russians should attempt to counter. This combined with the pragmatic recognition of the distinct advantages climate change may bring Russia – much more arable cropland, more temperate winters, access to rich Arctic oil and gas fields, and a thawing and highly lucrative Northern Sea Route through the Arctic Ocean – means that Russian leaders are both materially and ideationally disposed toward minimal action on climate crisis mitigation. Importantly, this may hold true even as Russia suffers from the very real negative effects of rising global temperatures that in much of its Arctic region are rising twice to four times the global rate of increase.[20]

In addition to problem sets that present some type of evident threat, anticipatory intelligence also dedicates focus to the unintended consequences or unforeseen malicious uses of emerging technologies and the broader potential impacts of these on societies. In many of these cases, technological advances themselves may not be classified as a threat and may be viewed by many as positive and even altruistic innovations. For example, rapid advances in genetic engineering hold significant promise for medical advancements in the treatment of cancer and genetic diseases, but also present mounting concern about the (as yet over-the-horizon) capabilities of malicious actors to genetically manipulate existing pathogens to enhance virulence or synthetically create new or once-eradicated pathogens like smallpox.[21] The rapid democratization of many biotechnological advances fuels these concerns, despite the practical limits still barring many of the worst envisioned abuses of biotechnology.[22] In this technological domain and many others, utilizing tools of strategic culture analysis can offer an important vehicle for assessing the cultural bounds that will influence the approach of different states to the development, adoption, and use of novel and potentially disruptive technologies.

This is in part because rational calculus about risk-benefit tradeoffs for the adoption of new technologies varies across cultures. Consider the development and deployment of artificial intelligence-boosted surveillance and coercion infrastructure across portions of China's population under (as yet limited) "social credit" pilot schemes,[23] and the accelerating global trend of "dataveillance" infrastructure of various provenances being acquired and adopted not only by authoritarian states but democracies struggling to keep law and order in increasingly fragmented societies.[24] Consider too the frontier of biotechnological enhancement of the human form, especially in the context of military forces, which is fueling a nervous and periodically halting scientific race between at least Russia, China, and the United States[25] – no state wishing to be caught behind in a potentially critical area of future competition and each grappling with their own encultured ethical codes that alternately constrain or loosen the bounds of what lines may permissibly be crossed in the advancement of science and strategic advantage.

In addition, there is particular historical significance associated with the "norms-setting" done by the first state to achieve a major technological breakthrough. For example, the dawn of the nuclear era in 1945 placed the United States – and by implication American strategic culture – as the sole possessor of the Allied bomb. The tumultuous fluctuation in US nuclear policy that ensued ultimately (though not inevitably) gave birth to the US "nuclear taboo," a normative inhibition against the use of nuclear weapons after the bombings of Hiroshima and Nagasaki.[26] While acknowledging that the nuclear taboo is neither a global norm nor reverenced by all eight of the other states that have since developed nuclear weapons, it is probably true that US actions – and inactions, in the case of Harry Truman's choice not to use nuclear weapons in the Korean War – during the early, precarious years of the nuclear era fundamentally shaped the world's nuclear history in subsequent years. With an eye toward this critical norms-setting role of technological breakthrough states, it is of great interest to scholars and practitioners of anticipatory intelligence to understand the prevailing elements of identity, values, norms, and perceptions that define American, Russian, and Chinese strategic communities as they relate to the development, application, and idiosyncratic ethics surrounding artificial intelligence, mass dataveillance, biotechnological advances, geoengineering, quantum computing, and over-the-horizon technologies.

Understanding How Other States Make Sense of Complexity

In the progressively complex global environment anticipatory intelligence seeks to address, where intentionality and proportionality are unlikely to follow predictable linear trends and the phenomenon of emergence can yield profound impacts on the international system, utilizing strategic culture to understand the way traditional strategic actors are perceiving and interpreting complexity holds significant value. States naturally react differently to actions or disruptions believed to be willful rather than accidental or natural developments, and wrongly sizing up the intentionality and proportionality of other states' actions can fuel dangerously cascading events.

This dynamic is well illustrated by the divergent American and Russian interpretations of the "colour revolutions" of the early 2000s, a series of popular uprisings within former Soviet and Middle Eastern states, and the Arab Spring revolutions across the Middle East and North Africa in the early 2010s. These events were interpreted by American elites as the boil-over of long-standing popular grievances in mostly authoritarian states – driven by grassroots activism from large underemployed youth populations, poor economic opportunity, political repression, and the birth of social media platforms that allowed local organizers to bring together huge crowds in sites like Egypt's Tahrir Square.[27] This sensemaking of a series of political events spread over more than a decade reflected a basic American perception of these events as movements indigenous to each state, driven by sincere local economic and political grievances, inspired by resonance with populations across neighboring states, and positively enabled by global advances in technology. Where the colour revolutions and Arab Spring movements fell short in effecting lasting democratic change – which they did in most cases – or devolved into far worse outcomes like the Syrian Civil War, an air of wishful regret for these "failed movements" tends to characterize American perceptions of the period.[28]

By contrast, Russian elites through the 2000s and early 2010s perceived the colour revolutions and Arab Spring through a radically different perceptual lens, interpreting these movements not as locally inspired but externally engineered through the artifice and covert machinations of the West, in particular the United States. Western nongovernmental organizations supporting the development of civil society in post-authoritarian societies were labeled by Russian elites as pernicious Western agents, and narratives and social media platforms (from 2009 on) utilized by protestors were seen as weapons of informational struggle with which the United States was arming local pretenders and would-be revolutionaries. All of this was done, in the Russian strategic elites' view, with the purpose of overthrowing stable authoritarian regimes and either installing Western-leaning governments or willfully allowing terrorist groups and political chaos that would suit US interests to flourish in Russia's backyard, presenting multiple levels of threat to the stability and security of the Russian state.[29] While most American elites scoff at this interpretation of events, it is widely and sincerely held among most Russian elites, reflected not only in seminal speeches and publications of the 2010s, including the founding article of the so-called Gerasimov Doctrine,[30] but actively and explicitly leveraged in the justification for Russia's 2014 and 2022 invasions of Ukraine.[31]

American perceptions of the colour revolutions and Arab Spring are tinted by the United States's own experience, narratives, and culture, and this is certainly true of Russia's interpretation.[32] Russia's well-established siege mentality and defensive paranoia – enduring features of its strategic culture[33] – have driven its interpretation of and response to a case of complex emergence that US elites would largely interpret as innocent grassroots activism, and in so doing have contributed to an escalation in Russian aggression that at the time of this writing is driving Europe's largest land war since World War II. Understanding the culturally encoded perceptions of states' strategic communities regarding complexity and emergence will become yet more important in coming years as an acceleratingly complex global system, further vexed by the resurgence of great power competition, spawns more unanticipated, unintended, or disproportionate effects that states will struggle to interpret and respond to. Critically, better insights here can help decisionmakers identify the lurking tripwires and flashpoints that may tip other states into startling reactions based on divergent interpretations of complex phenomena.

Recognizing States' (Mis)perceptions of Non-state Actors

The democratization of technology has served as one of the foundational forces driving rising complexity: in the same way that "the real use of gunpowder is to make all men tall,"[34] the internet, smartphones, and social media have given every phone user a potentially global microphone and audience – a revolution unparalleled in any previous generation of telecommunications technology. But the familiar example of viral online posts that have changed the world is only one dimension of the extraordinary forces of technological democratization, which are also pushing cyberweapons, gene editing tools, alternative currencies, supercomputing capacities, and potent artificial intelligence applications into the hands of ordinary people. The power for good of this movement can and should be recognized. Equally should be its enormous potential to further complexify the world by knighting ordinary individuals and once-unremarkable groups as actors that have sudden relevance on a strategic level. A key area for anticipatory intelligence and strategic culture to jointly tackle in this domain is a better understanding of the way nation-states' security elites – with the weight and inertia of 400 years of the Westphalian system (and centuries more before of state and imperial power) bearing on them – will variously interpret and misinterpret the unprecedented power of individuals in a world of rapidly democratizing technology. Understanding these perceptions takes on special importance in a returned era of great power competition, where solo or non-state actors with little attentiveness to escalating tensions in the international arena may lob disruptions into that domain that catalyze significant and ill-understood consequences.

One example of this dynamic is the American hacktivist self-dubbed P4x, who took it upon himself to launch a slew of blackout-inducing cyberattacks against North Korea's internet infrastructure after having been targeted by North Korean hackers as part of a broader campaign to steal cyber tools and information from US security researchers. Feeling that he had received no meaningful assistance from US government entities, P4x decided to individually impose a form of retributive justice on the Kim regime with his one-man cyber offensive.[35] In the United States, the technological access and education (or cyber street smarts) necessary for this individual actor to take on such a campaign all exist and are, in fact, remarkable only in their application against a hostile foreign government. Yet in a political context like North Korea, where less than 1% of the population has access to the internet[36] and the state plays an exceptionally heavy role in setting the agenda and bounds of individuals' actions, it may be inconceivable for North Korean elites to perceive P4x's and other cyber vigilantes' actions as those of a plucky lone-wolf hacktivist rather than the willful and coordinated campaign of the US national security apparatus and its agents. This may be especially true in authoritarian contexts with strategic communities that are predisposed to mirror-image their own organizational structures, authorization processes, and norms onto other states.

Another example might be drawn from the phenomenon of the nontrivial number of foreign fighters – including US and West European nationals, many of them veterans of their respective states' armed services – that poured in to support Ukrainian defenders in the early, wildly uncertain weeks of Russia's 2022 invasion.[37] North Atlantic Treaty Organization (NATO) governments variously urged their citizens not to participate in the conflict or at the least make explicit that they had no affiliation with their national governments, based on the anxiety that if Russia perceived NATO to have de facto "boots on the ground" in Ukraine it might respond to this – or use it as a pretext – for escalation against NATO member states.[38] While this case was notable in its scale, foreign fighters joining a conflict outside of the mandate of a national military is far from unprecedented. Rather, what presented a more exceptional development in early 2022 was the parallel "hacktivist" war being fought over Ukraine between a vast and very loosely defined conglomerate of hackers from around the world – some in support of Ukraine and others in support of Russia. Quasi "wartime declarations" from groups like Anonymous accounted for a portion of this activity, but cyber actors even more disparate and disconnected from larger groups joined the fray.[39] This phenomenon was notable in that there was, in a sense, a "world war" being contested between individuals hailing from dozens of nations wielding weaponized keyboards from afar, despite there being only two formal participants in the war and an anxious, explicit effort by most onlooking states to avoid formally entering the fray.

The question of how Russian security elites perceive the actions of this loose global conglomerate of hacktivists – still unfolding at the time of this writing – is consequential because of the role cyber and information warfare play in Russian strategic culture, which is defined by a vastly greater sense of holism and integration between warfighting tools and domains than American strategic culture.[40] Per Pallin, "Russia sees information warfare as an integrated entity, where propaganda, electronic warfare and IT operations are all used simultaneously."[41] Elements of the informational, psychological, mass-media, energy-based, and even facets of the biological and chemical domains are included under the label "information-psychological weapons."[42] Russian strategic culture views informational warfare as a particularly dangerous domain that presents a unique threat to societal and state security,[43] predisposing Russian elites to see outsized peril associated with foreign action in this domain – however coordinated and nefarious, or scattered and improvisational. While US security elites have their own anxieties about amorphous and unaccountable cyber vigilante movements,[44] the Russian perceptual lens may elevate the actions of a loose and leaderless pack of hacktivists – present or future – to a more pernicious or even existential threat to Russian security. Novel approaches to strategic culture study can help illuminate the anticipatory intelligence "threatscape" by offering a better understanding of how states' strategic communities are predisposed to perceive events and actors – including the messier variety progressively endemic to the twenty-first century – and therefore better anticipate the scale and nature of reactions.

Anticipatory Intelligence: Nontraditional Strategic Actors

As the systematic assessment of "new, emerging trends, changing conditions, and underappreciated developments [that] challenge long-standing assumptions ... and provide warning of threats,"[45] anticipatory intelligence both reflects and perhaps to an extent contributes to a growing trend in twenty-first-century societies to expand the definition of security and the range of problem sets that fall within its ambit. Securitization theory serves as a useful device for understanding how an issue is being conceptualized in a society and bears particular relevance for understanding the range of security issues and nontraditional strategic actors that are increasingly populating the anticipatory intelligence horizon.

Conceptualizing the Effect of Expanding Securitization

The concept of securitization, developed in the 1990s by the Copenhagen School of security studies, argued that "security" should be thought of as a function of rhetorical framing: when a problem set is successfully framed as a (potentially existential) security issue, it warrants less debate and more extraordinary action than typical political issues on which normal debate, divergence of opinion, and incremental action might be taken.[46] The securitization of an issue is itself neither intrinsically "good" nor "bad" and may reflect a range of motivations from the figure advocating for an issue's securitization that spans pragmatic resilience planning to populist fearmongering. Although variation exists, a common trend being seen in many areas of the world as of the 2020s is to respond to the rising complexity of the international arena by moving more substantive issue areas into the securitized space – in part because some areas are seen to be facing new or resurgent threats; in part due to recognition of the vulnerabilities that are being increasingly created by profound interconnection on the global stage. For example, the rise of interstate cyber competition in the 2000s and the resurgence of information warfare in the 2010s expanded many Western national security strategy documents to include discussions on the national security threat of "active measures" that fall below the threshold of kinetic conflict from Russia and, in more recent years, Iran and China.[47] The COVID-19 pandemic forced a focus on public health as a national and human security issue, fueled realizations about the degree to which national stability and economic viability in the twenty-first century are tied to convoluted transnational supply chains, and presented uncomfortable evidence of the degree of societal harm that can be done by fragmenting trust in government, science, and local communities. And the Russian invasion of Ukraine that began in 2022 has threatened not only the devastating consequences of direct conflict but has driven multidimensional supply chain crises around the globe and placed energy and food security at the top of national agendas.[48]

If "everything is security," the utility of security as a construct itself is diluted. Not all problem sets in the modern world belong under the umbrella of security, and there are risks to society of over-securitization that fall beyond the scope of this discussion.[49]

But in a world that is becoming increasingly interconnected and interdependent, with a diverse range of willful and adaptive traditional and nontraditional actors, and with disproportionate consequences and entirely unintended cases of emergence sometimes manifesting in the global system,[50] there is a defensible case to be made for assessing a wider swath of issues through the lens of national and societal security. Doing so incorporates a significantly greater number of actors within the domain of strategic significance, and by implication, signals that these actors merit strategic culture analysis as a means of providing a better understanding on the enabling and constraining elements of culture that are impacting them.

Assessing a Broader Range of Figures as "Strategic" Actors

Strategic culture as a field originated with a focus on members of a strictly limited "strategic community" – a state's military and defense establishment, or even more narrowly its nuclear establishment, as in Jack Snyder's foundational 1977 piece on Soviet strategic culture.[51] Across four generations of strategic culture scholarship, scholars and practitioners have expanded and contracted the focus of strategic culture study, some defending a purist focus on states' traditional strategic communities, others advocating for broader national profiles of strategic culture, yet others promoting the assessment of sub-state groups that may indirectly impact a state's strategic decisionmaking.[52] After the 9/11 terrorist attacks instigated the Global War on Terror, some scholars turned to applying a strategic culture lens to non-state actors that hold obvious significance for security affairs, including terrorist organizations like Al-Qaeda.[53] Others within the fourth generation of strategic culture scholarship have employed methodologies like the Cultural Topography framework to examine a wider range of sub-state and non-state actors, groups, and organizations with an eye toward gleaning an understanding of how specific enabler, obstructor, or influencer groups will act on a particular intelligence issue at a specific point in time.[54] This trend aligns with key priorities in the field of anticipatory intelligence, which calls for analysis of a yet broader spectrum of groups that fall outside the traditional definition of strategic communities but have emerged as potential critical actors.

Neither anticipatory intelligence nor strategic culture as fields can afford to accept the assumptions embedded in realism and neorealism that states are uniform and undifferentiated actors and the only ones that matter in the international arena. To illustrate, the rapid elevation of new actors to the level of strategic influence includes major multinational corporations like Microsoft and Meta, which have found it necessary to adopt elements of "corporate foreign policy" and through the 2010s and 2020s have routinely discovered and asserted themselves as decisionmakers at the crux of issues like freedom of speech and press – as well as detection and response to crime and terrorism – that would once have been the sole domain of states.[55] In an era where corporations are absorbing some of the functions of government and individual tech billionaires have a net worth exceeding the gross domestic product of midsized countries, the power of corporations and private-sector decisionmakers starts to rival that of some traditional state actors.

It is in keeping with the spirit of strategic culture to examine the identities, values, norms, and perceptual lens of unconventional entities that are newly significant in their impact on national and global society, and the argument can be made that it is even more important to understand elements of a group's enabling and constraining culture when these actors do not face the typical constraints, deterrent realities, or public responsibilities of a nation-state. The case for studying corporate strategic culture in this light requires little daring. If one accepts that strategic culture is "that set of shared beliefs, assumptions, and modes of behavior, derived from common experiences and accepted narratives ... that shape collective identity and relationships to other groups, and which determine appropriate ends and means for achieving security objectives,"[56] many major corporations carry a potent enough corporate culture and some interpretation of "security objectives" – or the company's de facto impact on national or international security affairs, whether pondered or not – to qualify for consideration.

But far less structured groups are also wielding increasing influence at the strategic level, illustrated by the example offered previously of the scattered conglomerate of hacktivists in the shrouded cyber conflict surrounding the Russia-Ukraine War. Even highly amorphous groups that span national borders and share few common aspects of identity, like subscribers to the QAnon conspiracy theory popularized in the United States in the early 2020s or strident advocates of the myriad COVID-19 conspiracy theories, might be considered a form of emerging strategic actor group. As the mounting political fragmentation of societies and the weakening of geographic communities[57] gives way to the rise of narrative-based online communities that in turn have significant impacts on political discourse, societal cohesion, and governing institutions, societies may have to confront the rise of surprisingly powerful and frustratingly nebulous actor groups.

Some of these nontraditional strategic actors may fail conventional benchmarks for "having" a strategic culture. Climate activist supporters of Extinction Rebellion, for example, or dedicated subscribers to a conspiracy theory might better be considered followers of a social movement than an identity-based collective. A fair criticism can be lodged that it is a challenging and uncertain undertaking to study loosely defined groups that may not share a strong sense of identity, mutual values, or common experiences by virtue of being scattered across nations, hailing from vastly dissimilar backgrounds, and perhaps having never met in person. Yet the reality may be that loose, narrative-based communities that are primarily bound together by a virulent shared perception and narrative – powerful enough to drive real-world action as a consequential collective group, for good or ill – may represent one of the holistic societal trends of the 2020s and beyond,[58] and that the novel technological and other forces empowering these individuals and groups require that they be treated as legitimate strategic actors. If unfolding trends continue to support this premise, neither practitioners of anticipatory intelligence nor strategic culture can ignore this domain.

Despite the challenges presented above, studying the nascent strategic cultures of nontraditional strategic actors can offer key insights on how group culture "bounds the possible"[59] for group members and can enhance understanding of what elements do exist in groups' identity, collective norms or values, and perceptions and views of the world, including intent, motivation, drivers, and limits of virtual and physical action. An increasingly complex global playing field requires the willingness to pursue questions of great consequence even when the object of study is not cleanly bounded. In the field of anticipatory intelligence, this is a necessary approach, if not a comfortable one.

Areas for Future Research

This chapter has presented a case for the valuable fusion of tools and insights from the fields of anticipatory intelligence and strategic culture. Building on this overarching point, the academic and professional literature in this domain would benefit from future work in several additional directions. First, building on previous academic research exploring culture and international relations through the lens of complexity science,[60] further work examining strategic cultures as complex systems and utilizing structured cultural analysis techniques like the Cultural Topography methodology to "map" these systems may offer valuable insights and refinements of tools both within complexity science and strategic culture.[61] Thinking about strategic cultures and complex systems in tandem may help scholars and practitioners understand both better, especially within the context of anticipatory intelligence.

Second, important further work could be done to expand the role of narrative research in strategic culture, refining the field's toolkit for capturing and contextualizing the constructed "stories" relating to security matters (and securitized issues) that resonate across a strategically significant group. Approaches to narrative research are well established in anthropology, sociology, communication studies, and other fields,[62] but remain underdeveloped – and the extant literature underutilized – in the field of strategic culture.[63] This gap bears particular significance for the future of anticipatory intelligence studies as many of the nontraditional strategic actors noted in the section above may be united less by traditional forces of proximity, identity, and norms, and instead may be more fueled by a narrow set of common perceptions that are being driven by a limited constellation of particularly compelling and virulent narratives. Better integrating the toolkits in the strategic culture and narrative research domains would deliver powerful new mechanisms for identifying and assessing the narratives that are being constructed by, and in turn driving, important nontraditional strategic actors in the anticipatory intelligence domain.

Finally, valuable work could be done to enhance the utility of turning strategic culture and anticipatory intelligence assessment tools inward, in the spirit of Sun Tzu's mandate to "know thyself,"[64] by bridging anticipatory intelligence's focus on resilience-building – the principal policy response available in answer to complex emergent security challenges[65] – and strategic culture's assessments of state, non-state, or supra-state actors. These assessments may include unpacking one's own or an allied organizational culture to understand its components that contribute to or detract from the organization's resilience, and can also yield important insights on the roles likely played by various actor groups in securing (or threatening) the resilience of critical political, economic, and social systems. Successful resilience-building requires stakeholders across multiple boundaries and sometimes with widely diverging interests to collaborate in the name of constructing systemic resilience. Strategic culture assessments of diverse stakeholders may illuminate areas of cultural alignment that could serve as promising starting points for consensus-building, and can also reveal areas where cultural accord among stakeholders is weak and likely to fracture under pressure. Both are key insights in assessing the resilience potential of a system and crafting productive paths of action. Across these and yet other domains, partnering the better innovations of strategic culture and anticipatory intelligence is likely to strengthen the insights delivered by both fields for the multidimensional security challenges of the twenty-first century.

Notes

1 The intersection of the fields of strategic culture and strategic intelligence is excellently covered in two earlier contributions to this *Routledge Handbook of Strategic Culture*: see Roger Z. George and James J. Wirtz's Chapter 31, and Rob Johnston and Judith Meister Johnston's Chapter 32.
2 Office of the Director of National Intelligence (ODNI), *National Intelligence Strategy of the United States of America,* 2014, https://www.dni.gov/files/documents/2014_NIS_Publication.pdf.
3 Briana D. Bowen, "Lessons from Anticipatory Intelligence: Resilient Pedagogy in the Face of Future Disruptions," in Travis Thurston, Kacy Lundstrom, and Christopher González, eds., *Resilient Pedagogy: Practical Teaching Strategies to Overcome Distance, Disruption, and Distraction*, Empower Teaching Open Access Book Series, 2021.
4 Office of the Director of National Intelligence (ODNI), *National Intelligence Strategy of the United States of America,* 2019, 9. https://www.dni.gov/files/ODNI/documents/National_Intelligence_Strategy_2019.pdf.
5 ODNI, *National Intelligence Strategy,* 2019, 8.
6 Josh Kerbel, "Coming to Terms with Anticipatory Intelligence," *War on the Rocks,* 13 August 2019, https://warontherocks.com/2019/08/coming-to-terms-with-anticipatory-intelligence/.
7 Scott E. Page, *Understanding Complexity,* The Great Courses/The Teaching Company, 2009 [Audiobook]; Kerbel, "Coming to Terms with Anticipatory Intelligence."
8 Josh Kerbel, "It's True, the World Always Has Been Complex — But Not Like This," *The Hill,* 9 May 2022, https://thehill.com/opinion/national-security/3478119-its-true-the-world-always-has-been-complex-but-not-like-this/.
9 Josh Kerbel, "The Cognitive Crucible: #85 Josh Kerbel on Anticipatory Intelligence and Complexity," *Information Professional Association,* 2022, https://information-professionals.org/episode/cognitive-crucible-episode-85/.
10 Jeannie L. Johnson, "Bounding Strategic Culture: Method, Not Theory," commissioned by the USAF Institute for National Security Studies, October 2009.
11 Jack L. Snyder, *The Soviet Strategic Culture: Implications for Limited Nuclear Operations*, RAND Corporation, Santa Monica, CA, 1977, https://www.rand.org/pubs/reports/R2154.html.
12 Kerbel, "Coming to Terms with Anticipatory Intelligence."
13 Kerbel, "Coming to Terms with Anticipatory Intelligence."
14 Jeannie L. Johnson and Matthew T. Berrett, "Cultural Topography: A New Research Tool for Intelligence Analysis," *Studies in Intelligence,* Vol. 55, No. 2 (Extracts, June 2011), https://www.cia.gov/static/811a5292007dbd5e9a73844c113f257b/Cultural-Topography.pdf.
15 Colin S. Gray, "British and American Strategic Cultures," (unpublished paper) prepared for the Jamestown Symposium, "Democracies in Partnership: 400 Years of Transatlantic Engagement," 18–19 April 2007; see also Jeannie L. Johnson's Chapter 11 in this volume.
16 One example among many: Energy.gov, "U.S. Department of Energy Announces Up to $96 Million to Advance Carbon Capture Technologies," 10 February 2022, https://www.energy.gov/fecm/articles/us-department-energy-announces-96-million-advance-carbon-capture-technologies-natural.
17 Stacy Closson, "The Impacts of Climate Change on Russian Arctic Security," 304, in Roger E. Kanet, ed., *Routledge Handbook of Russian Security,* Routledge, 2019, citing Alžběta Jurčová, "The Consequences of Climate Change: Will Russia Emerge as an Unlikely Winner from Lack of Action?" BlogActive EU, 10 July 2017, http://europeum.blogactive.eu/2017/07/10/the-consequences-of-climate-change-will-russia-emerge-as-an-unlikely-winner-from-lack-of-action/.
18 "Russian Federation: Policies & Action," *Climate Action Tracker,* accessed 15 February 2023, https://climateactiontracker.org/countries/russian-federation/policies-action/.
19 A reference to the quip attributed to Tsar Nicholas I: "Russia has two generals in whom she can confide—Generals January and February," As quoted in Punch, 10 March 1855.
20 Paul Voosen, "The Arctic Is Warming Four Times Faster Than the Rest of the World," *Science*, 14 December 2021, https://www.science.org/content/article/arctic-warming-four-times-faster-rest-world; Closson, "The Impacts of Climate Change on Russian Arctic Security."
21 National Academies of Sciences, Engineering, and Medicine, *Biodefense in the Age of Synthetic Biology*, Washington DC: National Academies Press, 19 June 2018.

22 Barry Pavel and Vikram Venkatram, "Facing the Future of Bioterrorism," *Atlantic Council,* 7 September 2021, https://www.atlanticcouncil.org/commentary/article/facing-the-future-of-bioterrorism/.
23 Kai Strittmatter, *We Have Been Harmonized: Life in China's Surveillance State,* Custom House, 1 September 2020.
24 Steven Feldstein, "The Global Expansion of AI Surveillance," *Carnegie Endowment for International Peace,* 17 September 2019, https://carnegieendowment.org/2019/09/17/global-expansion-of-ai-surveillance-pub-79847.
25 See Ronald O'Rourke, "Great Power Competition: Implications for Defense—Issues for Congress," *Congressional Research Service* Report R43838, updated 8 November 2022.
26 Nina Tannenwald, "The Nuclear Taboo: The United States and the Normative Basis of Nuclear Non-Use," *International Organization* Vol. 53, No. 3, Summer 1999: 433–468, http://www.jstor.org/stable/2601286.
27 See Michael McFaul, "Transitions from Postcommunism," *Journal of Democracy* Vol. 16, No. 3, July 2005; Stephanie Schwartz, "Youth and the 'Arab Spring,'" *United States Institute of Peace,* 28 April 2011, https://www.usip.org/publications/2011/04/youth-and-arab-spring.
28 For example, Ben Hubbard and David D. Kirkpatrick, "A Decade After the Arab Spring, Autocrats Still Rule the Mideast," *New York Times,* 14 February 2021, https://www.nytimes.com/2021/02/14/world/middleeast/arab-spring-mideast-autocrats.html.
29 Charles K. Bartles, "Getting Gerasimov Right," *Military Review* Vol. 96, No. 1, 2016: 30–38.
30 Valery Gerasimov, "The Value of Science Is in the Foresight" [in Russian], *Military-Industrial Courier (VPK),* 26 February 2013, https://www.vpk-news.ru/articles/14632.
31 Lincoln Mitchell, "Putin's Orange Obsession: How a Twenty-Year Fixation with Color Revolutions Drove a Disastrous War," *Foreign Affairs,* 6 May 2022, https://www.foreignaffairs.com/articles/russia-fsu/2022-05-06/putins-orange-obsession.
32 For an excellent discussion on the paradigmatic differences between "free speech" vs. "information counter-struggle" in the respective US and Russian worldviews that partly account for these divergent perspectives, see Katri Pynnöniemi, "Information-Psychological Warfare in Russian Security Strategy," in Roger E. Kanet, ed., *Routledge Handbook of Russian Security,* Routledge, 2019.
33 Dima Adamsky, "Cultural Underpinnings of Current Russian Nuclear and Security Strategy," in Jeannie L. Johnson, Kerry M. Kartchner, and Marilyn J. Maines, eds., *Crossing Nuclear Thresholds: Leveraging Sociocultural Insights into Nuclear Decisionmaking,* Palgrave Macmillan, 2018, 173–198.
34 A quip attributed to Scottish essayist Thomas Carlyle.
35 Andy Greenberg, "North Korea Hacked Him. So He Took Down Its Internet," *WIRED,* 2 February 2022, https://www.wired.com/story/north-korea-hacker-internet-outage/.
36 "Countries with the Lowest Internet Penetration Rate as of April 2022," *Statista,* https://bit.ly/3M5siEc.
37 Lisa Abend, "Meet the Foreign Volunteers Risking Their Lives to Defend Ukraine—and Europe," *TIME,* 7 March 2022, https://time.com/6155670/foreign-fighters-ukraine-europe/.
38 Dan Lamothe, et al., "Despite Risks and Official Warnings, U.S. Veterans Join Ukrainian War Effort," *The Washington Post,* 11 March 2022, https://www.washingtonpost.com/national-security/2022/03/11/americans-veterans-ukraine-russia/.
39 DarkOwl, "Impact of Russia's Invasion of Ukraine Across the Internet and Darknet," [Presentation] 2022; Joseph Menn, "Hacking Russia Was Off-Limits. The Ukraine War Made It a Free-for-All," *The Washington Post,* 1 May 2022, https://www.washingtonpost.com/technology/2022/05/01/russia-cyber-attacks-hacking/.
40 Dima Adamsky, *The Culture of Military Innovation: The Impact of Cultural Factors on the Revolution in Military Affairs in Russia, the US, and Israel,* Stanford University Press, 2010, 39–42.
41 Carolina Vendil Pallin, "Russian Information Security and Warfare," 211, in Roger E. Kanet, ed., *Routledge Handbook of Russian Security,* Routledge, 2019.
42 Pynnöniemi, "Information-Psychological Warfare," 220, citing V.K. Novikov, *Informatsionnoe Oruzhie – oruzhie sovremennyh i budushchii voin.* Moskva: Goryachaya liniya – Telekom [in Russian]. Third Edition, First published 2011.

43 Pynnöniemi, "Information-Psychological Warfare," 214–223.
44 Claudia Grover, "Ukraine Hacktivism 'Problematic' for Security Teams Says NSA Cyber Chief," TechMonitor, 11 May 2022, https://techmonitor.ai/technology/cybersecurity/ukraine-hacktivism-problematic-nsa-ncsc.
45 ODNI, *National Intelligence Strategy*, 2019.
46 Barry Buzan, Ole Wæver, and Jaap De Wilde, *Security: A New Framework for Analysis*, Lynne Rienner Publishers, 1998.
47 See, for example, the unclassified Annual Worldwide Threat Assessment Reports released by the US ODNI: https://www.dni.gov/index.php/newsroom/reports-publications.
48 Jeff Tollefson, "What the War in Ukraine Means for Energy, Climate and Food," Nature Vol. 604, No. 7905, 5 April 2022: 232–233.
49 For a thought-provoking discussion on this subject, see Rosa Brooks, *How Everything Became War and the Military Became Everything,* Simon and Schuster, 2016.
50 Page, *Understanding Complexity.*
51 Snyder, *The Soviet Strategic Culture.*
52 See Brigitte E. Hugh's Chapter 3 in this volume for a useful survey of this literature.
53 See Edward D. Last's Chapter 10 in this volume for a discussion of these studies.
54 See Jeannie L. Johnson's Chapter 29 in this volume for a "user's guide" to the Cultural Topography method.
55 Rodrigo Fernandez, et al., "How Big Tech Is Becoming the Government," *Centre for Research on Multinational Corporations (SOMO),* 5 February 2021, https://www.somo.nl/how-big-tech-is-becoming-the-government/.
56 Jeannie L. Johnson, Kerry M. Kartchner, and Jeffrey A. Larsen, eds., *Strategic Culture and Weapons of Mass Destruction: Culturally Based Insights into Comparative National Security Policymaking,* New York: Palgrave Macmillan, 2009, 9.
57 Office of the Director of National Intelligence (ODNI), *Global Trends 2040: A More Contested World,* March 2021, https://www.dni.gov/index.php/gt2040-home.
58 See ODNI, *Global Trends 2040: A More Contested World.*
59 See Jeannie L. Johnson, *The Marines, Counterinsurgency, and Strategic Culture,* Georgetown University Press, 2018, 37–52.
60 See, for example, Steen Bergendorff, *Simple Lives, Cultural Complexity: Rethinking Culture in Terms of Complexity Theory,* Rowman & Littlefield, 2009; Walter C. Clemens, Jr., *Complexity Science and World Affairs,* State University of New York Press, 2013; Melanie Mitchell, *Complexity: A Guided Tour*, Oxford University Press, 2009.
61 Thanks and credit for this suggested area of future research belong to my colleague Dr. Bradley Crookston.
62 Neil J. Salkind, "Narrative Research," in *Encyclopedia of Research Design*, Thousand Oaks, CA: SAGE Publications, Inc., 2010. DOI: 10.4135/9781412961288.
63 An area of focus also advocated by my colleague Dr. Kerry Kartchner, whom I thank for thought-provoking discussions on the subject.
64 Jeannie L. Johnson, "The 'Know Thyself' Conundrum: Extracting Strategic Utility from a Study of American Strategic Culture," in *Strategy Matters: Essays in Honors of Colin S. Gray,* ed. Donovan C. Chau, Air University Press, August 2022.
65 Bowen, "Lessons from Anticipatory Intelligence."

Selected Bibliography

Adamsky, Dima. *The Culture of Military Innovation: The Impact of Cultural Factors on the Revolution in Military Affairs in Russia, the US, and Israel*, Stanford University Press, 2010.

Bergendorff, Steen. *Simple Lives, Cultural Complexity: Rethinking Culture in Terms of Complexity Theory*, Rowman & Littlefield, 2009.

Bowen, Briana D. "Lessons from Anticipatory Intelligence: Resilient Pedagogy in the Face of Future Disruptions," in Travis Thurston, Kacy Lundstrom, and Christopher González, eds., *Resilient Pedagogy: Practical Teaching Strategies to Overcome Distance, Disruption, and Distraction*, Empower Teaching Open Access Book Series, 2021.

Buzan, Barry, Ole Wæver, and Jaap De Wilde. *Security: A New Framework for Analysis*, Lynne Rienner Publishers, 1998.

Clemens Jr., Walter C. *Complexity Science and World Affairs*, State University of New York Press, 2013.

Johnson, Jeannie L. "Bounding Strategic Culture: Method, Not Theory," Commissioned by the USAF Institute for National Security Studies, October 2009.

Johnson, Jeannie L., and Matthew T. Berrett, "Cultural Topography: A New Research Tool for Intelligence Analysis," *Studies in Intelligence*, Vol. 55, No. 2 (Extracts, June 2011), https://www.cia.gov/static/811a5292007dbd5e9a73844c113f257b/Cultural-Topography.pdf.

Kerbel, Josh. "Coming to Terms with Anticipatory Intelligence," *War on the Rocks*, 13 August 2019, https://warontherocks.com/2019/08/coming-to-terms-with-anticipatory-intelligence/.

Kerbel, Josh. "It's True, the World Always Has Been Complex—But Not Like This," *The Hill*, 09 May 2022, https://thehill.com/opinion/national-security/3478119-its-true-the-world-always-has-been-complex-but-not-like-this/.

Kerbel, Josh. "National Security Language is Stuck in the Cold War," *Slate*, 5 October 2021, https://slate.com/technology/2021/10/national-security-language-cold-war-sloppy-thinking.html.

Mitchell, Melanie. *Complexity: A Guided Tour*, Oxford University Press, 2009.

Office of the Director of National Intelligence (ODNI), *Annual Worldwide Threat Assessment* Reports (see repository): https://www.dni.gov/index.php/newsroom/reports-publications.

Office of the Director of National Intelligence (ODNI), *Global Trends 2040: A More Contested World*, March 2021, https://www.dni.gov/index.php/gt2040-home.

Office of the Director of National Intelligence (ODNI), *National Intelligence Strategy of the United States of America*, 2019, https://www.dni.gov/files/ODNI/documents/National_Intelligence_Strategy_2019.pdf.

Page, Scott E. *Understanding Complexity*, The Great Courses/The Teaching Company, 2009.

PART V

Conclusion

35

CONSOLIDATING AND ENRICHING THE FIELD OF STRATEGIC CULTURE

Key Themes and Recommendations

Kerry M. Kartchner, Briana D. Bowen, and Jeannie L. Johnson

In this concluding chapter of the *Routledge Handbook of Strategic Culture*, we extract and engage with many of the key themes from the preceding chapters with the aim of providing a blueprint for consolidating and enriching the field of strategic culture. For maximum utility, we have organized these major themes into eight sections, which might prove helpful as an organizing device for those teaching strategic culture in universities and other settings. The eight sections include: discussions on strategic culture as causation, context, cognition, and culprit; defining and scoping strategic culture; analytic frameworks and approaches; levels of analysis including supra- and subcultures; sources of strategic culture; manifestations of strategic culture; modalities of transformation in strategic culture; and areas for further strategic culture research. Finally, we offer the editors' reflections on the future of strategic culture in scholarship and applied practice. In consolidating these insights, we hope to capture a valuable snapshot of the state of play in the field, highlight areas of productive consensus and enduring debate, and offer a research agenda for the future of strategic culture as a powerful analytic tool for students, scholars, and practitioners.

Strategic Culture as Causation, Context, Cognition, and Culprit

This volume has engaged strategic culture from a number of angles, levels, and perspectives while drawing on insights across four generations of strategic culture scholarship, variously considering strategic culture as causation, as context, as cognition, and as culprit.

Strategic Culture as Causation

The case for strategic culture as a linear determinant of behavioral outcomes remains elusive. Existing scholarship, including the contributions to this volume, has identified

DOI: 10.4324/9781003010302-40

strong correlations between strategic culture and national and organizational behavior, but cannot definitively establish causal relationships (which traditionally would require empirical trials and retrials with quantitatively established parameters and finely tuned inputs compared to resultant outputs). Further research and continued development of analytical methodologies may yet illuminate pathways and modalities of causation and of relationships between strategic cultural cause and behavioral, policy, doctrinal, or grand strategy effect. Conversely, some strategic culture scholars may determine that linear causality models are simply a poor fit given the comprehensive nature of the strategic culture subject matter and the strong likelihood of intervening variables (e.g., material realities, pressure from external actors, martial readiness) bearing sway on the actors under study. Fourth-generation scholars and beyond may assert that the significant utility found in narrowing the range of probable action is the most that can be asked of strategic culture scholarship.

Strategic Culture as Context

The case for strategic culture as constitutive of context, rather than a single variable that might be extracted out and tested within a positivist model, has come to dominate the field. Enculturation is pervasive, ubiquitous, and enduring, as Colin Gray (Chapter 4) reminds us.[1] As context, strategic culture is, of all social science fields, best able to shed light on intentions. For example, strategic cultural context may clarify and help resolve the well-known conundrum in intelligence assessment that is the dual-use dilemma, where the intentions of an actor need to be determined vis-à-vis a technology that may have a plethora of peaceful uses while at the same time serving as a precursor to weapons capability. Strategic culture can help explain why a certain policy was adopted, or why a course of action was pursued. As importantly, strategic culture as context can help explain why certain policy options, doctrinal postures, or courses of action were not taken – or were not considered, conceived, or even imagined, because the strategic culture in question was locked into certain historical narratives or constrained by a set of embedded norms that determined what was expected or accepted as feasible actions by that culture. Strategic culture as context informs us that it is ethnocentrically hard to anticipate actions by others that we ourselves would not even consider, or to assess intentions by others to do something that we would not contemplate.

Strategic Culture as Cognition

Different cultures think and process information differently. Social psychologists have demonstrated that people from different cultures actually see and perceive reality in divergent ways. People from certain cultural backgrounds will not share the same perceptual distortions or biases as those from other cultures. Some cultures think and process information holistically, while other cultures have more particularistic ways of thought. Some communicate in high context styles, others in low context styles. Some think polychronically, others are more monochronic. Strategic culture as cognition is reflected in the perceptual lens through which a culture views the world and its place in that world, whether in a hierarchical sense or a more equitable relational sense. Reasonable consensus holds that each of these culturally encoded cognitive presets and filters bear into thinking, decisionmaking, and action in consequential matters of security and strategy.[2]

Strategic Culture as Culprit

Kerry Kartchner's introductory contribution to this volume (Chapter 1) suggests the thesis that strategic culture blinds as well as binds and that in doing so it can contribute to social and organizational catastrophes just as much as it contributes to strengthening and enhancing any collective. Such catastrophes are likely to occur when the bounded perceptions endemic to a strategic culture contribute to misperception, miscalculation, and surprise. When confronted by geologic or geostrategic calamities for which a culture lacks predetermined response scripts, or for which the preferred courses of action do not match the circumstances, mistakes happen. When this occurs, culture is often tagged as the culprit. Less drastically, strategic culture can be blamed for suboptimal outcomes. As Jeannie Johnson (Chapter 11) notes in her discussion of US strategic culture: "security practices and tactics that are a clear match with organizational or national identity and practice are privileged over potentially more effective policies that fall outside the strategic culture mainstream."

Defining and Scoping Strategic Culture

As noted in the early chapters of this volume, the definitional diaspora around strategic culture itself is an enduring conundrum for building consensus in the field. Amid the plethora of definitions circulating in the broader literature,[3] we note that among our contributors to this handbook, at least 14 cite or refer to the original definition proposed by Jack Snyder in 1977, and virtually all contributors anchor their conceptualization of strategic culture in some way on this definition: "the sum total of ideas, conditioned emotional responses, and patterns of habitual behavior that members of a national strategic community have acquired through instruction or imitation and share with each other with regard to [national security] strategy ... a set of general beliefs, attitudes, and behavioral patterns with regard to [national security] strategy [that] has achieved a state of semipermanence that places them on the level of 'culture' rather than mere 'policy.'"[4] Several other contributors cite the definition from the 2006 Defense Threat Reduction Agency's project to develop a strategic culture curriculum: "strategic culture is that set of shared beliefs, assumptions, and modes of behavior, derived from common experiences and accepted narratives (both oral and written), that shape collective identity and relationships to other groups, and which determine appropriate ends and means for achieving security objectives."[5] A number of our authors also suggest distinctions between strategic culture and other related concepts such as political culture, national culture, military culture, collective mentality, or national character.

There is utility for the future of the field in promoting consensual validation regarding the definition and scope of strategic culture, but it bears recognizing that both of the commonly used definitions above and most other variants through this volume center around several core themes, including the ideational character of strategic culture, its collective nature, its emphasis on military and national security, and its conveyance through myths, narratives, and instruction. Although the definition and scoping of strategic culture chosen for any given study will be determined by its chosen topic, level of analysis, or paradigmatic approach, by utilizing existing definitions or at least explicitly addressing the deficiencies of existing definitions, scholars and practitioners can help strengthen cohesion and robustness in the field of strategic culture.

A further enduring debate in the field relates to describing the scope of strategic culture: what it encompasses and what it does not, what can it be used to explain and what it cannot, what its dimensions and manifestations are, and what constitute its modalities of change, assimilation, adaptation, or evolution. Building off a survey of fourth-generation literature, Brigitte Hugh (Chapter 3) refers to strategic culture as "a paradigm and method employed to help scholars and practitioners understand interlocutors on the international stage more completely and better inform a spectrum of effective policy options." With an eye toward practitioner utility, Nicholas Taylor (Chapter 30) explores the value strategic culture can offer in providing insights into an adversary's strategic culture that may not be readily apparent to outside observers – a function that holds particular efficacy for designing successful deterrence strategies.

Several authors scope strategic culture by demarcating its limits, both conceptually and operationally. In Chapter 11, Jeannie Johnson observes that strategic cultural influences do not necessarily provide a clear-cut script for action "but do tend to bound [actors'] beliefs about the range of effective and appropriate options available in a given situation." Christoph Meyer (Chapter 8) conceptualizes strategic culture as "a contributing but rarely sufficient causal factor in explaining foreign, security, and defense policy," but he finds the concept "particularly useful in situations of high uncertainty and crisis when political actors tend to rely more on their conventional analytical prisms and the instruments they trust, rather than having time to question the adequacy of their preexisting worldviews and beliefs about what works." In a similar vein, Jeffrey Larsen's case study on NATO (Chapter 7) characterizes strategic culture as "a significant unseen binding force for the members of the Alliance, often only recognized following a crisis, or found in regularly published official statements from Brussels." These examples and myriad more across this volume underscore that, while the scope of strategic culture may be cast more broadly by some scholars than others, its application to the dynamics of threat assessment and choices in security policy are apparent.

A final compelling area for thought here is the concept of *equifinality*, or "the property of allowing or having the same effect or result from different events."[6] For our purposes, equifinality refers to achieving similar analytical outcomes using two or more different analytical methodologies. Consider, for example, that some of our authors have noted that policy outcomes within their respective countries or groups under study may seem to be ultimately dictated by realist considerations. Sometimes a strategic cultural analysis indeed will reach the same conclusion or prescription regarding a given policy outcome that would be predicted or indicated by a realist or neorealist analysis. Yet this does not diminish the value of a strategic cultural approach as a complement to more traditional theoretical avenues of research. Even in those instances where a particular policy or action appears to have been driven more by power politics than ideational factors, the path leading to that outcome – the choices that were considered and discarded, the options that most resonated with existing and historical narratives – can often be best illuminated by a strategic cultural analysis. The courses of action a given polity considers may have been constrained by those that could be legitimized by a nation or group's past experiences. The terms used to convey and propagate policy decisions, even those in the vein of *realpolitik,* will be culture-specific and their meaning only fully comprehended by the members of that culture. In other words, even if realist impulses compelled a country or organization to pursue a certain path, strategic culture may best explain how that path was chosen, communicated, and carried out. And, strategic culture can offer crucial insights for better anticipating and making sense of that consequential range of state and non-state actions that seem to defy "rational" calculus or practical considerations.

Analytic Frameworks and Approaches

Robust discussion, comparison, and experimentation with analytical and methodological frameworks represent the ongoing development of the field of strategic culture. No single methodological approach was imposed on the contributors to this volume; the editors instead emphasized producing a handbook that broadly represents current scholarship and approaches in the field. Consequently, the various analytical frameworks engaged by authors in this volume reflect to a large degree the diversity of methodologies and approaches in the field. As a high-level survey, five groupings of methodological approach can be observed among our contributors. Some of these approaches are more taxonomical – that is, they create categories for distinguishing between groups or concepts – while others constitute more multifactorial frameworks or hypotheses. None offer themselves up as a fully formed theory of strategic culture, which the field may yet concede as unachievable but perhaps unnecessary for its practical utility.

The National Level Profile Approach

The origins of the field of strategic culture are firmly grounded in national level analysis, sometimes called "national character profiling." While some fourth-generation strategic culture scholars may consider this level of analysis archaic and overtaken by the finer granularity of group and organizational level analysis, there remains a valid and important role for national level strategic cultural profiles in contemporary strategic culture scholarship. The national level approach is ably defended in this volume by Jeffrey Lantis (Chapter 9), who states that "[a]t its core is the principle that historical experiences and culture help define national strategic orientations that favor certain military-security policies over others." Lantis notes that many of the most critical sources of strategic culture, such as climate, national geography, and natural resources, all impact the nation as a whole and are not necessarily limited to subgroups. The idea that a state has a "way of war" is a manifestation of engaging strategic culture at the national level. Threat perceptions, strategic doctrine, and matters of grand strategy are often most suitably addressed at the national level. Dependent on the research question being pursued, then, the strategic culture scholar may find that a national level perspective offers the analytical results they need.

A version of the national level approach is reflected in the chapters within this volume that capture and describe key elements or instrumental features, normative principles, and main characteristics of a state. Many of our contributors to Part III, writing on the state of the scholarship with regard to national level strategic culture resources, utilize this approach. Writing on Turkey, George Bogden (Chapter 26) examines how the state's Ottoman imperial heritage and twentieth-century Kemalist tradition have profoundly influenced Turkey's present strategic culture and conduct. Neil Munro (Chapter 13) takes a long view of Chinese history to identify idiosyncratic aspects of modern China's sense of identity, examining how "tensions have emerged between feelings of superiority and inferiority, between the needs for development and equality, between demands for freedom and order, and between China's territorial ambitions and geopolitical reality."

The Case for a "Holistic Approach"

Mette Skak (Chapter 12) is among those contributors to this volume who contend that strategic cultural analysis should incorporate a holistic approach to the subject, utilizing strategic culture to provide context or assess "complex causality." Invoking precepts first enunciated by Colin Gray, Skak also asserts that societal perceptions cannot be separated from national behavior and that addressing both ends and means is essential when analyzing strategic culture. Beatrice Heuser (Chapter 2) offers another take on a more holistic approach by utilizing the concept of collective mentality, advancing that while "'strategic culture' in Snyder's original conception applies to strategic communities ... 'collective mentality' applies to entire nations or subcultures within them. ... While many scholars tend to blend the two under the strategic culture label, collective mentality is the larger whole encompassing strategic communities with their particular 'strategic cultures,' but also many other subcultures." In common with the national level profile school of thought, scholars who advocate for a holistic approach may see a primary purpose of strategic cultural analysis as identifying the grand strategy of a given actor.

The "Issue-Based Approach"

Brigitte Hugh's (Chapter 3) review of fourth-generation strategic culture scholarship examines approaches utilized by those seeking more actionable data than a national level profile may offer, noting that "some more recently developed methodologies choose to assess focused aspects of strategic culture by directing research with a specific issue in mind. This type of analysis is also referred to as issue-based strategic culture." Harrison Menke (Chapter 16) provides an example of this type of study, taking an issue-based approach to assessing "the impact of Swedish strategic culture on two key case studies: Sweden's decision to pursue and abandon nuclear weapons and its decision not to join the Treaty on the Prohibition of Nuclear Weapons." One notable issue-based methodology is the Cultural Topography framework, first introduced within the US Intelligence Community by Jeannie Johnson and Matt Berrett.[7] In this volume, Jeannie Johnson (Chapter 29) provides a user's guide to the Cultural Topography method for scholars seeking to systematically assess the impact of cultural factors on a specific actor, regarding a specific issue, at a specific point in time.

Approaches Derived from the Cultural Topography Methodology

Approaches utilizing four particular cultural dimensions – identity, values, norms, and perceptual lens – are adopted by several authors in this volume to present broader strategic culture profiles of states or actors. These four dimensions were first introduced by Jeannie Johnson in early work seeking a standard methodological approach[8] and later formed the core of the Cultural Topography framework. *Identity* relates to how groups designate themselves and the distinctions they make between in-groups and out-groups. *Values* are beliefs about what is important or sacred to the group, or what enhances status within the group. *Norms* are the accepted, expected, or customary rules of behavior within a group, some of which are intended to "enact" values. *Perceptual lenses* are filters through which members of a group view the world and their role in it and affect how the group cognitively processes information about the world. Chapters incorporating this framework include

Harrison Menke's on Sweden (Chapter 16), Thomas Biggs' on Germany (Chapter 17), Briana Stephan's on South Korea (Chapter 19), Michael Barron's on Japan (Chapter 20), Nathan Fedorchak's on Singapore (Chapter 21), and Megan Moore's on Venezuela (Chapter 28).[9] Kerry Kartchner's (Chapter 5) discussion on religion as a source of strategic culture is also organized primarily around how religion is constitutive of these four dimensions.

Other Approaches Mentioned, But Not Implemented in This Volume

Mette Skak (Chapter 12) recalls the 1950s approach developed by Russian-born sociologist Nathan Leites termed operational code analysis,[10] defining it as "the art of distilling rules and norms for political action from the pattern of actual action, strategies, and key explicit statements." Operational code analysis is not explicitly employed by any author in this volume but has seen use in other contemporary strategic culture studies.[11] Some authors call for the integration of approaches from other fields and domains into strategic culture analysis. Briana Bowen (Chapter 34), for example, advocates drawing in approaches from the field of anticipatory intelligence to glean added insights by modeling strategic cultures as complex systems. Other authors urge expanding approaches of narrative and discourse analysis to illuminate features of strategic culture. Maria Hellman (Chapter 15) suggests that incorporating narrative theory into strategic culture analysis could "show how strategic cultures come into being, how they evolve or remain unchanged, and how they become powerful tools in exercising international influence and attaining domestic legitimacy ... [pointing] to the salience of collective identity and identity formation and how these link to power positions of actors in a system." Given that political and military leaders use narratives to legitimatize interventions and other strategic actions, narrative research could be a valuable tool for exploring intentions, national self-images, and actor perspectives.

Levels of Analysis: Supra- and Subcultures

Several points emerge from surveying our authors' discussions about strategic culture levels of analysis and the presence of supracultures and subcultures. First, as Michael Eisenstadt avers in Chapter 6, "Cultures are neither monolithic nor deterministic or immutable." For the purposes of "strategic" culture, with a focus on national security ends and means, the attention of the chapters in this handbook has naturally gravitated toward larger groups, often at the national level, with nominal stakes in security policy outcomes as the given level of analysis. Notwithstanding that predominant focus, various chapters have touched upon or emphasized manifestations of strategic culture at the civilizational (e.g., East, West), regional (e.g., Europe, South Asia), supranational (e.g., NATO, the European Union [EU]), national (e.g., China, Russia), non-state actor (e.g., ISIS, AQIM), and organizational levels (e.g., military service branches, intelligence agencies). Two of our authors explicitly focus on strategic culture at the supranational or international organizational level: Jeffrey Larsen describes NATO strategic culture in Chapter 7 and Christoph Meyer discusses EU strategic culture in Chapter 8. At the other end of the spectrum, Edward Last (Chapter 10) demonstrates that even non-state actors have distinctive and legitimate strategic cultures, evincing ideational and behavioral elements equivalent to those of state or supranational entities. In Last's view, "Any collective able to threaten and use violence can have strategic culture."

Second, many authors explicitly identified subcultures within their respective areas of research, including different military branches of countries under study. Writing on the United Kingdom, Alastair Finlan (Chapter 14) states that "No highly developed state has a homogenous military culture. Instead, they possess several distinctive military cultures represented variously by the air force, army, navy, marines, special forces and, more recently, space force, all in competition with each other for resources and primacy." Itai Shapira (Chapter 25) describes Israeli strategic culture "not as homogenous on the national level nor fixed in time, but rather as 'a system of competing subcultures,' shaped by elites that create 'epistemic communities'" (networks of policy-relevant subject matter experts).[12] Shapira also cites the description of four subcultures in Israel developed for the 2006 Defense Threat Reduction Agency's strategic culture project by Gregory Giles.[13] Examining the US Intelligence Community, Roger George and James Wirtz (Chapter 31) observe the proliferation of organizational subcultures across the vast enterprise, noting that the "complexity of the IC also gives rise to many internal factors that shape the American way of intelligence, particularly its multiplicity of collection and production organizations, which contain their own separate cultures."

Third, policy differences alone would not normally be sufficient to delineate or "speciate" the existence of subcultures in a national polity. But when those subcultures engender stark differences in identity, values, norms, or perceptual lenses, strategic cultural analysis may require that new "species" of subcultures be designated. In those cases, differences over policy and grand strategy preferences can be the basis for identifying and assessing subcultures within a national polity.[14] At least one chapter in this volume engaged such a policy metric for distinguishing between subcultures. Harrison Menke (Chapter 16) describes two subcultures in Sweden that exhibit distinctive patterns of decisionmaking: "traditional neutralists emphasize altruistic traditions and are reluctant to make any compromises about Sweden's foreign policy orientation. Oppositely, constructive neutralists are more flexible in their interpretations and focus more on issues of national interest.... While each shares a common foundation – that Sweden should be nonaligned and independent – they differ on how best to achieve said objectives."

Fourth, subcultures can be ephemeral and may coalesce or dissolve based on external crises, internal political realignments, or other exigent circumstances. Jeffrey Larsen (Chapter 7), writing on NATO's collective strategic culture, distinguishes between "older and newer members of the Alliance," noting that "this split approximates to a substantial degree the parallel differentiation between those states who emphasize transatlantic relations with North America, and those that prefer an approach to security that is more Eurocentric." Interestingly, Larsen attributes "many of the divisive issues that concern NATO in normal times" to these distinctions, but finds that "all of these concerns seem to fall by the wayside during a crisis when allied cohesion is required." Ultimately, the level of analysis chosen for any study should be determined by the issue of interest, the scope of the study parameters, and the scale of inquiry that will deliver the greatest utility.

Sources of Strategic Culture

The questions of where strategic culture comes from and what features or characteristics produce it are central to any project aiming to systematize the study of strategic culture or to map a structured path toward effectively employing strategic cultural analysis. No single author in this volume offers an exhaustive list of such sources, but their contributions collectively cover those sources that may form the basis of any meaningful strategic culture project. Colin Gray (Chapter 4) emphasizes the "deep roots" that cultures typically have in geography and history. The survey conducted by Brigitte Hugh (Chapter 3) suggests that common further sources of strategic culture include "resources, myths and symbols, political structure and defense organizations, climate, habits and traditions, and rituals and ceremonies." Michael Barron (Chapter 20) asserts that in addition to these, demographics, economic development, and international relationships – including buy-in to powerful transnational values like Japan's strong "subscription to the prevailing liberal international order" – can be formative sources of strategic culture.

Kerry Kartchner (Chapter 5) makes the case for considering religion as an overlooked yet foundational source of strategic culture, pointing to "the influence of religion on the formation of identity, on the creation and shaping of values, on the delineation of norms (especially with respect to issues of war and peace, the nominal purview of strategic culture), and on molding the prism of perception and cognition, with special emphasis on how those dimensions are manifested at the strategic level." Dodeye Williams (Chapter 27) explores the impact of religion in the context of Nigerian strategic culture, noting its duality as a force for good or ill and potential to be a "potent tool of oppression, deceit, and manipulation in the hands of Nigeria's political elite." In his comparison of Arab, Israeli, and Iranian cultures, Michael Eisenstadt (Chapter 6) observes that "tribal and religious values influence the political and strategic cultures of the Middle East" more than many other parts of the world, but asserts that the conduct of diplomacy and war are also influenced by other variables such as the personality of leaders, organizational structures and processes, human and material resource endowments, physical geography, and the operational environment – acknowledging that some of these factors may in turn be shaped by culture.

Manifestations of Strategic Culture

Manifestations of strategic culture are the visible and often tangible expressions of strategic culture. They may be the physical representations in ritual and practice of ideational dimensions of strategic culture like identity, values, norms, and perceptions. They are the externalization of implicitly shared ideals, aspirations, hopes, and intentions. Virtually every chapter in this volume has discussed ways in which strategic culture is manifested within the scope of their subject country, organization, or grouping.

One of the most prominent manifestations of strategic culture addressed by several of our authors is a state or non-state actor's way of war, which in itself reflects the actor's defense and security policies, the totality of its strategic choices, and the product of its doctrine, training, and threat perceptions. Writing on the predominant cultural groupings of the Middle East, Michael Eisenstadt (Chapter 6) argues that "differences in national 'ways of war' are often ascribed to efforts by states to adapt to, counter, or 'design around' an adversary's strategy or way of war," but he further asserts that "ways of war are based, first and foremost, on cultural templates that incorporate assumptions about the character of war, onto which these adaptive responses are grafted or overlaid." Edward Last (Chapter 10) explores the way of war concept in the context of non-state actors, including Islamist extremist groups, considering "the strategic culture of the various iterations of Al-Qaeda in Iraq/ISIS [to be]

... 'a particular and idiosyncratic interpretation of an Islamic way of war as perceived by Salafi-jihadists.'"[15] Examining a state led by a single family's interpretation of war and security matters, Steve Sin (Chapter 18) assesses the impact of regime-created myths and symbols on North Korean strategic culture. Sin advances that "one of the best ways to glean a country's national identity is through the examination of its myths and symbols ... [which are] both spatial and temporal. They are spatial because they form complex multidimensional and multilayered puzzle pieces where each piece represents a specific myth and symbol that applies to a specific group or segment of the society ... [and] temporal because they are born out of (and change based on) the country's history."

Others address approaches to diplomacy, style of decisionmaking, alliance formation, and creating and employing technologies as clear manifestations of distinct strategic cultures. Ali Parchami (Chapter 24) discusses Iran's nuclear program as a manifestation of its strategic culture and particularly its national identity, observing that "by framing the country's nuclear policy as a matter of national pride, and by depicting international curbs as efforts to prevent Iran from reclaiming its rightful place as a regional power, the regime has successfully co-opted nationalism in support of its controversial program." Itai Shapira (Chapter 25) offers Israel's style of national decisionmaking and its distinct approach toward deterrence and nuclear weapons as examples of strategic cultural manifestations. He lists other manifestations of Israeli strategic culture as the state's practice of proactively preempting or preventing adversaries from acquiring nuclear capabilities of their own (the so-called Begin doctrine), its approach to military innovation, and its doctrine and practices for counterinsurgency warfare and low-intensity conflicts. Approaches to military innovation as a manifestation of strategic culture are a theme also explored by Theo Farrell and Kenneth Payne (Chapter 33). They find, first, that American strategic culture has to date overwhelmingly shaped the direction of research and development of artificial intelligence (AI), and that as Western militaries have undertaken new concepts, programs, and capabilities integrating AI, shared beliefs have emerged "about how AI should be used to support and conduct operations." Just as the US-led approach to developing AI has been driven and bounded by American strategic culture proclivities, Farrell and Payne advance that Russia and China each have approaches to developing AI that reflect their own respective strategic cultures, which may suggest that "strategic culture matters as much as ... material factors when it comes to the development and use of military AI."

Modalities of Transformation in Strategic Culture

A defining feature of strategic culture is that it resists change; that it is embedded in deep habits of mind and practice that can endure for ages and across generations. While allowing for reluctant evolution in strategic culture over time, Colin Gray (Chapter 4) is decidedly skeptical on the ability of societies to willfully force cultural change, asserting: "we may change our opinions, but to change our culture would be a task of far greater magnitude ... Strategic and other dimensions of culture are not acquired lightly and casually, and neither can they be casually discarded or changed by a policy decision." Michael Eisenstadt (Chapter 6) observes that "cultural attributes can persist for decades, centuries, or even millennia, and can live on long after the formative events that created them." In noting the staying power of demographic, geographic, religious, and linguistic continuities for Arab, Iranian, and Jewish civilizations across thousands of years, Eisenstadt reinforces that relatively fixed ecological and geographic factors along with persistent cultural elements such as language and religion may combine to create "enduring attitudes toward power, authority, and the use of force." Thus, "a people's self-conception, as embodied in its collective beliefs and national myths, may also cause it to discern echoes of the past in the present and to see in its history a guide to the future, producing continuities in perception and conduct." This attribute of strategic culture scholarship – to make explicit enduring patterns of thought and behavior – is one of the field's most important contributions to understanding persistent patterns of war and peace that may not be explained by other theoretical approaches.

Nevertheless, the contributions to this volume also reflect a new consensus around the idea that strategic cultures do change over space and time, and that there is an adaptability and evolutionary quality to aspects of any given strategic culture. Jeffrey Lantis (Chapter 9) sees this new consensus on strategic culture's mutability as an advancement in the scholarship and demarcates three principal modalities of change in strategic culture: (1) "external shocks [that] fundamentally challenge existing beliefs and undermine past historical narratives"; (2) "systemic changes or new demands [that] create situations where primary tenets of strategic thought in a state come into direct conflict with one another (strategic cultural *dissonance*)"; and, (3) "elites [whose actions] play a special role in strategic cultural continuity and change." Change in strategic culture can originate from exogenous factors as well as from endogenous ones. Outside influences can loom large in the evolution of strategic culture, and strategic cultures can learn from their own history. Christoph Meyer (Chapter 8) attributes some elements of change in strategic culture at both the EU and member state levels to "learning from operational experiences and experiences of strategic surprise in Europe's neighborhood and broader security environment."

In elaborating on this theme of change, some authors have deliberately identified and distinguished between mutable and immutable dimensions of various strategic cultures, as Steve Sin (Chapter 18) does in his examination of the "(relatively) immutable" features of North Korean strategic culture. Others have drawn distinctions between residual "old versus new elements," suggesting that such elements "often coexist and comingle" and implying that an uneven evolutionary process defines certain aspects of a given strategic culture. Illustrating the coexistence of old and new elements, or the endurance of vestigial elements of a previous strategic culture, Mette Skak (Chapter 12) examines the concept of "Soviet atavisms," which include strategic cultural remnants of Leninism and Stalinism, and asserts that "all such Soviet atavisms matter as they make Russian strategic culture different from Western strategic cultures." Other authors, including Michael Barron writing on Japan (Chapter 20) and Briana Stephan writing on South Korea (Chapter 19), draw attention to potentially consequential generational changes or prospects of cultural transformation that may be wrought by recent upheavals in the global rules-based order.

Change and adaptation may be forced upon a given strategic culture by changes in its threat environment, or sometimes even by deliberate choices made by keepers of the strategic culture. North Korea offers the most vivid and exceptional modern instance of a strategic culture virtually constructed by a single leader and propagated by his descendants, reflecting a series of syncretic and deliberate policy choices. One author raises the intriguing notion that such deliberate strategic culture changes can lead to unintended consequences. Alastair Finlan (Chapter 14) mulls whether "the 2016 Brexit referendum that ultimately took the United Kingdom out of the European Union may well be the catalyst for the future break-up of the United Kingdom, which is exactly the opposite of the Vote Leave campaign's dream to return Britain back to its pre-European Union glory days." In the most drastic future scenario, "Brexit's repercussions could herald a return to a situation not seen since the Middle Ages with an independent England, Wales, and Scotland with profound consequences for the idea of a unified UK strategic culture."

A few of our authors have suggested or implied that cultures that have more or less embraced change as a consistent feature of life may have distinct advantages in a world of rapid technological and geostrategic turmoil. Itai Shapira (Chapter 25), for example, characterizes Israel's strategic culture as one that is adaptive and improvisational, in transition, balancing innovative and conservative approaches, and which is also forward-looking and anticipatory, with the keepers of Israeli strategic culture constantly searching for "potential disruptions and 'relevancy gaps.'" Writing on Venezuela, Megan Moore (Chapter 28) suggests that strategic culture can be an important analytic tool for detecting possible future changes to a population's culturally bounded identity, values, norms, and perceptual lens. While the enduring features of strategic culture remain a foundational contribution of the field, the challenge in present and future scholarship is to determine and effectively capture the modalities and scope of these evolutionary tendencies.

Areas for Further Strategic Culture Research

With an eye toward progressively refining and enriching the field of strategic culture, many of our authors offer recommendations and suggestions for the research agenda of future strategic culture scholarship. Five common themes for the next generation of work in the field emerge across the volume: (1) enlarging the scope of factors considered in strategic culture analysis; (2) elucidating modalities of change in strategic culture; (3) exploring and clarifying the influence of strategic culture on specific policy outcomes; (4) evaluating the cascading effects on strategic cultures of emerging global challenges, and (5) enhancing cumulativeness and analytical effectiveness in strategic culture scholarship.

Enlarging the Scope of Factors Considered in Strategic Culture Analysis

Several authors critique the tendency of some strategic culture scholarship to rely on simplistic or overly narrow explanations and call for greater emphasis on more complex analyses and approaches that take a more holistic perspective on the sources, functions, and manifestations of strategic culture. Itai Shapira (Chapter 25) observes that the literature on Israeli strategic culture can be somewhat shallow, exhibiting an approach that assumes Israeli strategic culture is homogenous and fixed in time. Harrison Menke (Chapter 16) and Nathan Fedorchak (Chapter 21) find that the literature on Sweden and Singapore's strategic cultures, respectively, has tended to focus on narrow and discrete elements rather than holistic and interconnected features of each state's strategic culture. Others note the need to address gaps in building conceptual bridges (to borrow Thomas Biggs' metaphor from Chapter 17) between a country or organization's political, national, and security cultures or between the norms, perceptions, and values of civilian versus military leadership, particularly in countries like Pakistan that are defined by a military-centric strategic culture, as Arshad Ali addresses in Chapter 23. Constructing such bridges would contribute, among other things, to more holistic approaches to strategic culture analysis and more accurate and actionable insights into actors.

Elucidating Modalities of Transformation in Strategic Culture

The consensus noted above among most contributors to this volume that strategic culture is neither monolithic, static, or constant – as may have been assumed in first- and second-generation strategic culture scholarship – underpins an equally firm consensus that the exact modalities of change need further exploration and synthesis. Many of our authors call for a better understanding of how specific strategic cultures evolve, adapt, assimilate, transform, or integrate, including across multiple subcultures. In this vein, Maria Hellman (Chapter 15) reflects on the need to expand strategic culture scholarship to include attention on "internal threats and political violence ... [which] can serve to highlight adaptions and changes needed for improved resilience, conflict prevention, and post-conflict management." Evaluating the impact of US Intelligence Community strategic culture on intelligence analysts, Rob Johnston and Judith Johnston (Chapter 32) explore the perennial question of whether organizations can willfully revise their own culture and values – and if such outcomes are possible, what the means for accomplishing them are in the face of the sometimes overwhelming inertia of culture.

Exploring and Clarifying the Influence of Strategic Culture on Specific Policy Outcomes

Several authors advocate for further research refining the understanding of direct or indirect causal relationships between the norms, perceptions, and values permeating dominant strategic cultures (or subcultures) and how and why leaders make specific strategic decisions, with an eye toward strengthening the utility of strategic culture in anticipating likely future behavior – or least identifying tighter constellations of plausible action. For example, writing on Indian strategic culture in Chapter 22, Muhammad Shoaib Pervez calls for further "analysis of Indian strategic thinking by linking the practices of its elites with that of sociocultural norms of Indian society" and examination of the en-cultured "role of Indian political parties in propagating norms of behavior as well as the socialization of elites." More broadly, Nicholas Taylor (Chapter 30) calls for further research and analysis "to fully test and explore the utility of strategic culture in deterrence and wider influence strategy development; for example, in improving our ability to understand which elements of strategic culture drive specific policy and strategy decisions, and identifying group sacred values."

Evaluating the Cascading Effects on Strategic Cultures of Emerging Global Challenges

Russia's brutal and unprovoked invasion of Ukraine beginning in February 2022 sent shock waves across the globe, including the foreign policy and international relations communities. The new trends and reverberations of this catalyst and its ensuing developments have given several authors in this volume pause to consider how these might impact their respective countries and organizations of interest. The timing of preparation and submission for this volume's manuscript precluded allowing each author to offer more fully developed thoughts, but several authors updated their chapters in the early months after the breakout of war with brief considerations about the possible ramifications of these developments and called for future scholarship to account for these new circumstances and their impact on various strategic cultures.

Maria Hellman (Chapter 15), for example, wonders about the impact of external pressures on French strategic culture from the unfolding and changing dynamics between the United States, Russia, and China. Briana Stephan (Chapter 19) notes that South Korea's strategic culture will face the need to adapt to complex strategic challenges related to China's economic and military growth and its ties to Russia, and suggests further study on how these challenges – including the possibility of a Chinese move against Taiwan – will affect South Korea's perceptions of the credibility of its alliance with the United States. Arshad Ali (Chapter 23) poses many of the same questions with regard to Pakistan's strategic cultural norms and perceptions, both in light of the US-Russia dynamic and looming challenges from India and China.

In addition to Russia's war on Ukraine and the rising pitch of tripolar great power competition, the enduring reverberations of the COVID-19 pandemic, global food and water insecurities mounting from climate change, and impacts rippling from the democratization of increasingly powerful technology all hold the weighty potential to have cascading effects on strategic cultures around the world. Against this backdrop, Briana Bowen (Chapter 34) advocates for integrating approaches from the fields of anticipatory intelligence and strategic culture, noting that "strategic culture study can help illuminate the anticipatory intelligence 'threatscape' by offering a better understanding of how states' strategic communities are predisposed to perceive events and actors – including the messier variety progressively endemic to the twenty-first century – and therefore better anticipate the scale and nature of reactions."

Enhancing Cumulativeness and Analytical Effectiveness in Strategic Culture Scholarship

Finally, some authors note that the field of strategic culture has been vulnerable across time to charges of lacking coherence, systemization, and consensus, which is exacerbated by the absence of unity on its definition, scope, and operationalization. A productive debate exists in the community over the value of standardizing a definition and scope of strategic culture to promote the cohesion and consensual validation of the field, encouraging restraint in the proliferation of new definitions without contextual justification; or alternatively, the value of prioritizing exploration and expansion of the definition and application of strategic culture in the pursuit of practical utility, even if challenging for the academic cohesion of the field. Both schools of thought may find benefit in bolstering the analytical effectiveness of strategic culture by more fully drawing from the repertoire of tools employed by the range of disciplines from which strategic culture has been constructed, including anthropology (e.g., Grounded Theory, "thick description"), history (e.g., narrative analysis), psychology (e.g., perceptual cognition, biases, and social reasoning), geography (e.g., modeling societal impacts of natural disasters), and political science and international relations (e.g., theory building).

The Future of Strategic Culture in Scholarship and Applied Practice

The field of strategic culture offers significant utility for scholars and practitioners taking on security issues ranging from enduring challenges of war and peace to emergent challenges of modern technology and climate change. The range of themes and perspectives captured in this volume collectively offer a snapshot of both the remarkable strengths and unresolved debates within the field of strategic culture as a tool for gleaning insights into consequential actors' motivations, intentions, justifications, and upper limits of action regarding security issues. We cherish our late colleague Colin Gray's characteristically epigrammatic point that "most good ideas become bad ideas if too much is asked of them." In that spirit, we offer three suggestions for contemporary and future domains where the scholarship and professional application of strategic culture might be particularly valuable.

First, strategic culture has an important future returning to its roots in nuclear weapons issues. The difference in global dynamics is not a small one between Jack Snyder's world of the Cold War and the contemporary world of tripolar great power competition overlaid on the rising strategic significance of non-state actors ranging from religious and political extremist organizations to massive multinational corporations. This does not diminish the global risk and difficulty faced during the Cold War but rather acknowledges that current decisionmakers in nuclear weapons states around the world face a metastasis of actors, factors, and unpredictable dynamics that are in meaningful ways more convoluted than earlier eras. What has not changed is the inimitable destructive power of nuclear weapons, and the shared effects that humanity faces if misperception, misinterpretation, miscalculation, or misalignment of realities leads to a nuclear exchange at any scale. Strategic culture analyses, particularly leveraging fourth-generation methods and approaches to hone in on the range of enabling and constraining groups, empowering narratives and restraining taboos, and idiosyncratic beliefs about nuclear use, may yield powerful insights to analysts and decisionmakers around the world in better anticipating, identifying, and managing nuclear risk and the potential after-effects of nuclear use.

Second, the same utility bears out for applying a strategic culture approach to the nebulous dynamics of geostrategic and great power competition in contests below the threshold of war. Utilizing a strategic culture approach to hone in on state and non-state actor groups' perceptions of and action templates in response to emerging security challenges, including cyberwarfare, weaponized narrative campaigns, mass proliferation and abuse of big data, the militarization of outer space, airborne and seaborne trespassing, economic brawls, supply chain scrabbling, and exponential leaps forward in dual-use technologies, to name a few, may offer insights on how contemporary members of various strategic communities around the world are culturally predisposed to understand risk, security, and intolerable threats. And, importantly, strategic culture can help analysts, diplomats, negotiators, crisis managers, and policymakers extract a better understanding of which "rung" of the escalation ladder a particular security issue may stand on in the perceptions of various actors, and whether stepping up or down a rung is likely to be seen as the sensible response to that challenge.

Third, strategic culture can offer significant value for scholars and policymakers seeking to better navigate our contemporary challenges that have no willful, or at least intentional, actors driving events but that require navigating complex human landscapes to manage or mitigate. Climate change-driven migration, for example, is likely to displace tens of millions around the globe within decades. The human displacement experiences of the 2010s, and again of the early 2020s, should vividly illustrate that mass migration driven by any cause has far-reaching and sometimes transformative effects on societies, and requires societal-level stakeholder buy-in and cooperation to manage productively. Natural disasters, from the devastating 2023 Turkey/Syria earthquake to increasingly ferocious heatwaves, tornados, and hurricanes, do more damage as they befall a more populated and displaced world. Effective international cooperation reacting to unforeseeable disasters, and attempting to forestall or mitigate the unpleasantly foreseeable ones, requires interlocutors equipped with the cultural insights necessary to stitch together collaborative policies from varied cultural fabrics. A strategic culture approach can empower international aid workers, advocates, policy architects, and domestic decisionmakers to better understand the cultural landscapes they will need to traverse to build international consensus, domestic tolerance, or emergency responses in a more chaotic world.

Strategic culture cannot offer a single-solution concept for those searching to make sense of all human decisionmaking. Instead, strategic culture as context and cognition is ultimately about understanding habits of mind and practice. It is a product of perceptions, cognitive processing, and socialization. It is a manifestation of intent and intentions, values and ideals, norms and rules. Strategic culture scholarship is rightly wary of making hard predictions and of proffering overdetermined outcomes. Contemporary strategic culture scholarship offers value by identifying and contextualizing tendencies and usefully narrowing the range of the probable. We hope this volume serves as an effective and timely handbook in consolidating and advancing scholarship across the field of strategic culture.

Notes

1 Colin S. Gray, "The Nature and Utility of Strategic Culture Scholarship," Chapter 4 in Kerry M. Kartchner, Briana D. Bowen, and Jeannie L. Johnson, eds., *Routledge Handbook of Strategic Culture,* Routledge, 2023. (NB: Chapters referenced here forward are from this volume.)

2 See, for example, Dima Adamsky's excellent work on this subject, including *The Culture of Military Innovation: The Impact of Cultural Factors on the Revolution in Military Affairs in Russia, the US, and Israel* (Stanford University Press, 2010).

3 A list of the most common definitions of strategic culture is provided by Kerry Kartchner in Chapter 1 of this volume.

4 Jack L. Snyder, *The Soviet Strategic Culture: Implications for Limited Nuclear Operations* (Santa Monica, CA: RAND Corporation, September 1977), v and 8.

5 Jeannie L. Johnson and Jeffrey A. Larsen, Comparative Strategic Cultures Syllabus, prepared by SAIC for the Defense Threat Reduction Agency, Fort Belvoir, Virginia, 20 November 2006.

6 Merriam-Webster Dictionary, "Equifinality," https://www.merriam-webster.com/dictionary/equifinality.

7 Jeannie L. Johnson and Matthew T. Berrett, "Cultural Topography: A New Research Tool for Intelligence Analysis," *Studies in Intelligence* 55, no. 2 (June 2011): 1–22.

8 Jeannie L. Johnson, "Conclusion: Toward a Standard Methodological Approach," in Jeannie L. Johnson, Kerry M. Kartchner, and Jeffrey A. Larsen eds., *Strategic Culture and Weapons of Mass Destruction: Culturally Based Insights into Comparative National Security Policymaking* (New York: Palgrave Macmillan, 2009), pp. 243–257.

9 Virtually all of these chapters are based on student case studies originally prepared as part of the requirements for strategic culture courses taught by Kerry Kartchner, which incorporate the Cultural Topography analytical approach as a pedagogical framework and were adapted for this volume.

10 Nathan Leites, *The Operational Code of the Politburo* (New York: McGraw-Hill, 1951).

11 For a contemporary example of operational code analysis, see Huiyun Feng, "A Dragon on Defense: Explaining China's Strategic Culture," in Jeannie L. Johnson, Kerry M. Kartchner, and Jeffrey A. Larsen, eds., *Strategic Culture and Weapons of Mass Destruction: Culturally Based Insights into Comparative National Security Policymaking* (London: New York: Palgrave Macmillan, 2009), pp. 171–187.

12 Itai Shapira, citing Tamir Libel, "Explaining the Security Paradigm Shift: Strategic Culture, Epistemic Communities, and Israel's Changing National Security Policy," *Defense Studies* 16, no. 2 (2016): 137–156.

13 See Gregory F. Giles, "Continuity and Change in Israel's Strategic Culture," in *Strategic Culture and Weapons of Mass Destruction*, edited by Jeannie Johnson, Kerry Kartchner, and Jeffrey Larsen (New York: Palgrave Macmillan US, 2009), pp. 97–116.

14 Scholars and instructors of American foreign policy will be familiar with Walter Russell Mead's four models or schools of US foreign policy in *Special Providence: American Foreign Policy and How It Changed the World* (New York: Alfred A. Knopf, 2001).

15 Last, citing Ahmed Hashim, *The Caliphate at War: The Ideological, Organisational and Military Innovations of Islamic State* (London: Hurst, 2018), pp. 289–291.

INDEX

Note: Page numbers in *italic* refer to figures, and page numbers followed by 'n' refer to notes.

www.ingramcontent.com/pod-product-compliance
Lightning Source LLC
Chambersburg PA
CBHW081215220326
41598CB00037B/6787

* 9 7 8 1 0 3 2 5 6 5 1 4 9 *